Lyman Cast Bullet Handbook

Third Edition

C. Kenneth Ramage, Editor

#9837249

The Cast Bullet Handbook, 3rd Edition

Published by Lyman Publications
Rt. 147
Middlefield, Ct. 06455
Printed in the U.S.A.

EDITOR'S FOREWORD

Today's cast bullet shooter benefits from a number of technological advances unknown to his grandfather. Pistol shooters can accomplish, with cast bullets, most objectives they seek with jacketed bullets. Riflemen, too, can benefit tremendously from the use of cast bullets in a wide variety of applications.

One of the true pleasures a reloader can bestow upon himself is to become involved in bullet casting. For those of experimental and inquiring bent—or those seeking new challenges—the answer lies in this fascinating hobby of casting your own projectiles.

No special firearms are needed; the ones you currently own will be fine. The press and dies you now have are fine, although you will want to add Lyman's "M" die to expand and flare rifle case mouths...and perhaps a new seating screw to perfectly match the cast bullet's nose.

Other than that, less than one hundred dollars is needed to gather the pot, ladle, mould, handles and related paraphernalia needed to produce good bullets; sized, lubed and ready to go.

This third edition of the Lyman CAST BULLET HANDBOOK is the most ambitious of our efforts to present the various facets of cast bullet shooting to both the inexperienced and more advanced cast bullet shooter.

It is our hope that you, the reader, will understand that it is relatively easy to get very good results from the very first if certain broad guidelines are followed. For the advanced reloader and bullet caster we have greatly expanded the scope of past editions in every respect and are confident that even the most experienced cast bullet shooter will be satisfied with his investment in this handbook.

A person's enjoyment of his hobby can often be enhanced by membership in an organization composed of like-minded individuals. The early tangible benefit to these memberships is receipt of the newsletter or magazine, usually filled with articles and interesting tidbits. Later, with increased involvement, friendships are formed that can last a lifetime. One or more of the following organizations will interest most cast bullet shooters. Write for membership information:

Cast Bullet Association, Incorporated
Sidney F. Musselman
5522 Trent Street
Chevy Chase, MD 20015

International Handgun Metallic Silhouette Association
Box 1609
Idaho Falls, ID 83401

National Muzzleloading Rifle Association
Box 67
Friendship, IN 47021

American Single Shot Rifle Association
c/o L. B. Thompson
318 Washington
Salen, OH 44460

Preparation for this book has covered a span of several years and included a thorough review of our bullet designs, dimensions and standards to ensure that bullets cast and processed by Lyman equipment will give good results.

There are more than 5,000 tested cast bullet loads in our data tables. Included are both "fast" and "slow" powders in good working combinations.

As in each Lyman handbook, we endeavor to explain the basic operations involved in casting and sizing bullets, selecting components and reloading rifle and pistol cartridges. Beyond that we have prepared sections of interest to the advanced bullet caster.

Students of exterior ballistics will find 119 pages of trajectory and wind drift tables for cast rifle bullets prepared by Ted Almgren and Dr. Bill McDonald. Space limitations precluded the inclusion of similar tables for cast pistol bullets and the muzzleloading conicals and roundballs. These complete tables can be found in our PISTOL & REVOLVER HANDBOOK and the BLACKPOWDER HANDBOOK, respectively—as can extensive load tables and related information.

One of the recent developments of vital interest to cast bullet shooters is the metallurgical research of lead alloy and the resultant information. We are pleased to offer a substantial article on the properties of molten lead by Dennis Marshall, a cast bullet shooter who happens to be a research scientist with a large metallurgical firm.

This Lyman Handbook is the most substantial reloading reference published, to date, for cast centerfire bullets. We hope you find it's contents useful and that you will write to our Technical Department if you have questions.

Yours truly,

C. Kenneth Ramage

C. Kenneth Ramage, Editor
Vice President,
Publications & Technical Services

TABLE OF CONTENTS

ACKNOWLEDGEMENTS

Lyman would like to thank the following firms whose contributions of materials and/or assistance implemented the production of this handbook:

Brass Extrusion Laboratories, Ltd.
800 West Maple Lane
Bensenville, IL 60106

Cast Bullet Association Inc.
c/o Sidney Musselman
5522 Trent Street
Chevy Chase, MD 20015

E.I. Dupont De Nemours & Co.
Explosives Dept. Chemical Sales Div.
10th and Market St.
Wilmington, Delaware 19898
Attn: Mr. L.W. Werner
(Manufacturer of Dupont Powders)

Federal Cartridge Corporation
Dept. L
2700 Foshay Tower
Minneapolis, Minn. 55402

Hercules, Inc.
910 Market St.
Wilmington, Delaware 19899
Attn: Mr. John C. Delaney
(Manufacturer of Hercules Powders)

Hodgdon Powder Company
7710 W. 50 Highway
Shawnee Mission, Kansas 66202
Attn: Bob Hodgdon

Omark-C.C.I., Inc.
Box 660
Lewiston, Idaho 83501
Attn: Mr. Arlen Chaney
(Manufacturer of C.C.I. Primers & Speer Bullets)

Remington Arms Company, Inc.
939 Barnum Ave.
Bridgeport, Conn. 06602
Attn: Mr. Joe Callahan

Savage Arms Corporation
Westfield, Mass. 01085

Service Armament/Navy Arms Company
689 Bergen Blvd.
Ridgefield, NJ 07657

Shiloh Products, Incorporated
37 Potter Street
Farmingdale, (L.I.) N.Y. 11735

Smith & Wesson
2100 Roosevelt Ave.
Springfield, Mass. 01101

Thompson-Center Arms Company
Rochester, New Hampshire 03867

Triple-S Development Co., Inc.
1450 E. 289th Street
Wickliffe, OH 44092

Winchester-Western
275 Winchester Ave.
New Haven Conn. 06504

Lyman's Handbook Staff

Editor:
C. Kenneth Ramage,
Vice President
Publications & Technical Services

Associates:
Raymond I. Cowles
Manager, Ballistics Laboratory
Walter M. Florczak
Technical Assistant
Thomas G. Roncaioli
Technical Assistant

C. Kenneth Ramage
Editor

Raymond I. Cowles
Manager, Ballistics Laboratory

Walter M. Florczak
Technical Assistant

Thomas G. Roncaioli
Technical Assistant

BULLETS TO 1775

Primitive man was weaker, slower and less well provided with natural weapons than many beasts. Humans, however, destroyed or bent to their will all other animals. The domination of the planet by man was based upon the use of weapons. Man, alone among all animals, had the hands and brain to make and use tools and weapons. Even in earlier times, there were two types of weapons. The club was undoubtedly the first direct-contact arm. The thrown stone was the first missile, the lineal antecedent of the bullet.

With the passing of centuries, man evolved new types of arms; as his manual skills increased, his weapons became more efficient. Direct-contact arms progressed from stone axes, bone knives and fire-hardened wood spears to bronze and iron axes, swords and pikes. Missile arms kept pace. Thrown rocks and stones, although still sometimes used, gave place to stone-pointed and metal-tipped javelins and arrows; a sling could be employed to increase greatly the velocity and range of the thrown stone.

Throughout classical times, both missile weapons and direct-contact arms continued to be employed in war. The winning armies were generally those which employed an effective combination of both types of arms. Macedonian pikemen were supplemented by archers, slingers and men armed with javelins. The Roman Legion had as well-rounded a balance between contact and missile arms as ever existed. The heavy-armed infantry had weapons of both types. Their pilum, or heavy javelin, had tremendous power, but short range. The Roman short sword was deadly. Legions had attached to them light auxiliary slingers and archers. The Roman slinger used a cast-lead pellet very like a bullet; they had tremendous range, but only fair accuracy and little ability to penetrate heavy shields.

During the Middle Ages, armor developed to such an extent that missile weapons were not usually decisive. In the 14th Century, however, the British archer dominated Western Europe. Missile weapons gained an ascendency at this time which they have never lost, for the true bullet was just appearing.

The First Gunpowder Arms

The battle of Crecy in 1346 saw the most efficient use of arrows in all history. Coincidentally, gunpowder weapons were used here for the first time. The small artillery pieces employed were not effective, but they were a start. Projectiles from firearms were to make all other missile weapons obsolete within 100 years.

Gunpowder artillery was first used about 1300 for siege and garrison purposes. The larger wrought-iron cannon were effective. They fired from fixed beds at stationary targets and were both cheaper and more easily transported than the classical war engines, which they replaced and which threw large masses of stone, monster arrows, and beams of wood. Scaling down an artillery piece to an arm which could be carried and used by one man was tried; however, these hand cannon were not really successful. They had to be ignited with a piece of glowing match held free in one hand of the user, while he tried to aim the weapon with the other. Accuracy was impossible, save at the closest range.

Firearms Projectiles

Artifically rounded stone balls were used as projectiles in war engines in order to give more uniform trajectory and accuracy. The first gunpowder artillery probably used similar projectiles, although early illuminated manuscripts sometimes show stubby arrows being fired. Perhaps the early hand cannon fired pebbles and stones. However, cast-metal bullets soon became standard.

Projectiles were cast of lead and iron in the 14th Century. Iron balls were best for the smaller cannon, although stone balls continued to be used in the larger sizes for some time. Lead was ideal for small-arms bullets and did not have a serious rival.

Why was lead so right for small-arms bullets? First, it was cheap. Second, it was soft and didn't hurt the fairly delicate iron barrels of the early small arms. Third, a lead bullet could be produced easily in any required shape and size. Fourth, lead was heavy, much the heaviest of all the common metals. It has a specific gravity of 11.35 based on water as 1.0; iron has a specific gravity of 7.84. This means that a lead bullet, once driven at a given muzzle velocity, will retain its velocity and energy better than a bullet made of any other common metal. Even today, lead is still preeminent for the same reason, although we usually alloy it with hardeners. For military purposes, a jacket is generally put around a lead core.

Lead is one of the commonest metals and has been produced from the mineral galena, or lead sulphide, in almost every country of the world. The metal was well known by the 8th Century B.C. The Romans cast articles of lead, both pure and in combinations with tin. The already-mentioned cast-lead pellets for the slingers sometimes bore the insigne of the individual Legion. The large number of seals and tokens cast by the Romans suggests iron, or bronze, moulds not unlike modern bullet moulds. By the 13th Century, hard-metal moulds were certainly being used extensively for casting lead signs, or tokens, for pilgrims. A gang mould of this type, casting six at once, is preserved at Dundrennan in Britain. Bullets could, of course, be produced by the same general process. Cast-lead bullets were to remain for 500 years the major small-arms projectile.

Other projectiles have been used at various times. Occasionally, cast-iron bullets have been tried in small arms; however, wear has always been excessive. Special projectiles such as arrows have been fired in small arms. Apparently, Elizabethan sea captains though a good deal of arrows for use in muskets aboard ship. It's probably that these arrows were used to set fire to opposing vessels; however, the quantities ordered seem excessive, if this were their only purpose. Grenades were being fired by muskets at least two centuries ago; they are still being projected from infantry rifles today.

The Matchlock Musket

A bullet from a hand cannon was less accurate than an arrow from a longbow or a quarrel from a crossbow, because of the difficulty of aiming and, at the same time, placing the flowing match on the touchhole by hand. Human eyesight cannot really observe two things at once. In order to direct a bullet at a target, one had to look at the target; in order to fire the weapon, one had to look at the touchhole. The invention of the matchlock, however, changed this. The glowing match was held securely in a pivoted arm, or serpentine, and moved precisely to the touch-

Matchlock and Wheel-Lock Muskets. Top: This matchlock, throught to have been made about 1540, is Cal. 790, barrel length 40.3", length over-all 54.4", weight 9¼ pounds. Bottom: Made in The Netherlands, or the South-German Rhineland, this wheel-lock is Cal. 736, barrel length 47.8", length over-all 61", weight 13.4 pounds. Both pieces were used in the British Civil War. *From the Weller Collection.*

hole when the trigger was pressed. The user of the weapon could give his undivided attention to his target, once his weapon was ready. Matchlock muskets began to be used in quantity about 1450.

The Bullet Makes Democracy Possible

A single, relatively small pellet of lead from a matchlock musket could penetrate the finest armor and kill the strongest man inside it. The bullet tended towards an equality of men in combat long before it was achieved politically. A democratic government was probably impossible when wealth men could buy armor for themselves and their retainers that made them almost impervious to the weapons of the poor. So long as offensive weapons depended upon physical strength and skill acquired by long training, the common man was at a distinct disadvantage. However, in a few hours the average man could learn to load, aim, and fire a musket. Bullets, not laws, made men substantially equal.

Flintlock and Wheel-Lock Weapons

The matchlock musket, although in use before 1450 and still employed in 1700, never remained unchallenged as the major weapon in war. At first, there were longbows and crossbows; pikes were used as long as the matchlock musket itself. Later, there were flintlock and wheel-lock firearms.

The matchlock system depended upon a glowing, smelly, sputtering match. It was slow to light, took a lot of adjustment and wouldn't function at all in the rain. Surprise, or defense against surprise, was impossible with such an arm. Yet is was cheap, rugged and effective under the formal conditions of European battles.

Wheel-locks and rudimentary flintlocks came into limited use within a relatively short time after the matchlock was invented. Both the new-type arms could be carried loaded and ready to fire. However, the wheel-lock was delicate, expensive and fouled quickly. Even though the flintlock finally replaced the matchlock for infantry use about 1700, it had initially several disadvantages. The flintlock was more complicated, more expensive and not so reliable for continuous volley firing.

Wheel-lock weapons were widely accepted for use by cavalry and for hunting considerably before 1600. Tactics involving wheel-lock petronels and long pistols by mounted soldiers did much to destroy the power of pikes and establish the infantry musket. The wheel-lock sporting rifle was very effective indeed. It finally displaced both the longbow and the crossbow for shooting at game; the matchlock smoothbore never could have done this.

Early European Target Shooting

The gunmakers of Central Europe and their customers were interested in the accuracy of their firearms. Early in the 15th Century they took to target shooting to determine this and continued it as a popular sport. Shooting matches were common. A match held at Augsburg in 1508 attracted 919 firearms contestants, as well as 533 men with crossbows. Apparently, some events allowed the use of rifles and some did not. Ranges were sometimes as long as 805 feet. Prizes were magnificent in comparison to anything offered today. For example, a fine horse and accouterments was given in Nuremburg in 1433; 210 Imperial gulden was the first prize in Strasburg in 1576.

Sometimes Americans think that accuracy with rifles originated over here. This is definitely not so. We don't know exactly what targets were used, nor the significance of the scores made, in the early European shooting, but we do have some of the weapons. Cleves Howell, Jr., an able ordance engineer, has made five-shot groups of 8 inches or less at 100 yards with a wheel-lock rifle unchanged since it was made in 1671. He sent me one three-shot group which measured 1¾ inches, center to center.

Similar firing with an approximately contemporaneous flintlock rifle was equally accurate. These rifles did not have heavy iron ramrods, which would have been required to load bare lead bullets. Their rammers are of rather light wood, obviously designed to be used with patched lead bullets. Mr. Howell remarks that he can do no better shooting with some modern open-sighted sporting rifles than he can with these two rifles made almost 300 years ago.

American Colonial Firearms

Firearms in Europe were used by large landowners and their gamekeepers, or jaegers; the majority of the population know little of them. In America, from the founding of Jamestown in 1607 and Plymouth Colony in 1620, they were the constant companions of all colonists and were used almost daily. White men depended upon their firearms for protection against the then-formidable Indians and for food. As already suggested, European armies used pikes, swords and protective armor until about 1700. The matchlock musket was their major infantry arm during this period. In America, contact arms were never important; bullets from firearms were paramount.

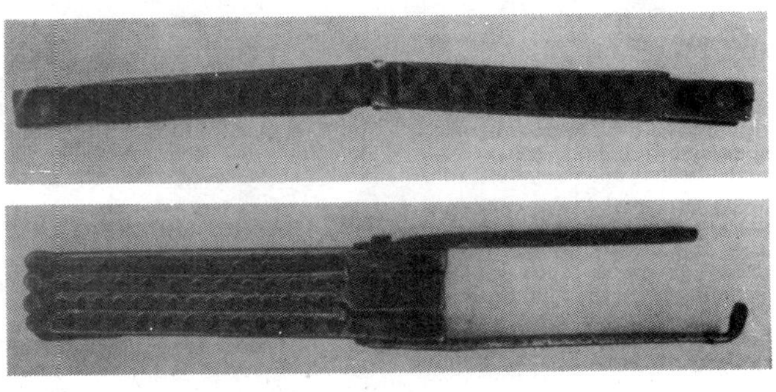

British Four-Fold Mould Circa 1700. This mould, shown open and closed, cast oval slugs for the blunderbusses of the day as well as swan shot. *From the Keith Neal Collection.*

British Three-Fold Mould, Circa 1675. Buck, swan and goose shot were cast by this mould, shown open and with handle folded in traveling position. *From the Keith Neal Collection.*

The wheel-lock was complicated, expensive and easily injured; repair in the wilderness might be impossible. The matchlock with its smell, noise and glow was even less useful. For at least 125 years before the Revolution, flintlock shoulder weapons, then called firelocks, supplanted other firearms almost completely. They provided not only security, but a large part of the fresh meat and much of the sport of the Colonies. Shooting has never been so universally popular anywhere, at any time.

Early American Bullets

Although both lead and gunpowder were produced in America before the Revolution, far more of both was brought in from abroad. These commodities formed a substantial part of total Colonial imports. Both powder and lead came in bulk. Generally, the ultimate consumer cast his own bullets.

Many bullet moulds probably were lost or wore out before the muskets which they originally accompanied. American-made flintlock rifles, which played an important part on the frontiers in the 18th Century, were commonly supplied with moulds made by the same gunmaker that produced the rifle itself.

American bullet moulds were made of iron, brass, stone, and, in a few cases, of wood lined with clay. Since many of the smoothbore weapons for which bullets were made were not particularly accurate, small irregularities in the bullets themselves were of little importance. Spherical lead balls are best produced with two pieces of metal or stone which come together to form the cavity so that each piece contains half of the final bullet. The pincer-type was best for single-cavity moulds, but was not ideal for larger capacity moulds. Two metal leaves hinged at one end were often used for gang moulds. Stone moulds were usually fashioned of two separate halves, not permanently connected together, but located by pins. Sometimes, moulds were made of three or four leaves hinged together. These were efficient for the casting of buckshot-type bullets in large quantities.

Perhaps originally, the cavities in the two halves of bullet moulds were hollowed out with chisels by hand and then lapped together; however, long before the Revolution, these cavities appear to have been made with rotary cutters turned by hand, called cherries. So long as the mould metal was reasonably soft, a blacksmith could make a fair round-ball cherry without much trouble. If a cherry was available, moulds of soapstone and slate were satisfactory since, even though they wore more rapidly than brass or iron, they were easy to replace. Brass moulds could be cast and then lapped to finished size, or cut with cherries.

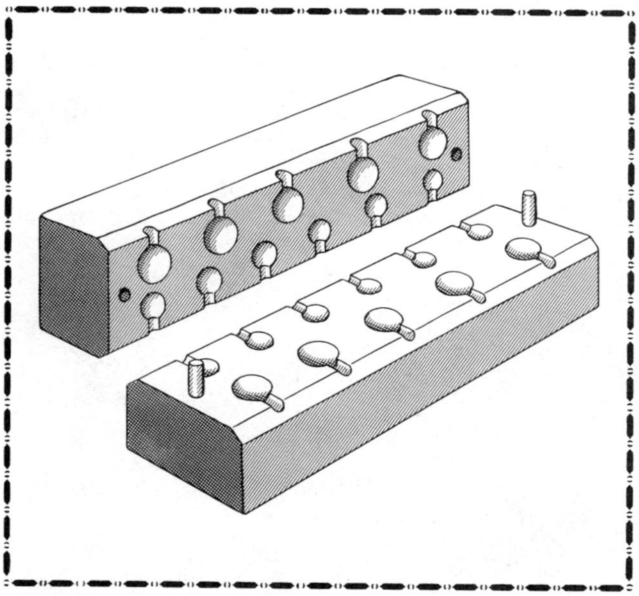

Early Stone Mould. This is typical of the kind of stone mould that might have been made by an early settler for his own use. Mould casts five round bullets ⅜″ in diameter and six smaller bullets about ¼″ in diameter.

Early American Iron Mould. This Cal. 54 forged-iron pincer-type mould was recovered from a Revolutionary War campsite. *From the McMurray Collection.*

Bullets of Two General Types

Colonial firearms shot bullets of two general types. First, there were the full bore-sized balls which were fired one at a time. Second, there were smaller projectiles fired several to a discharge. Many moulds produced both full size bullets and buckshot in the same operation. A single ball was usually fired in rifles. Loads for smoothbores composed in part, or entirely, of buckshot were effective in both war and hunting.

Neither a single ball, nor a small number of buckshot, was efficient for hunting fowl. Cast shot considerably smaller than the buckshot was often employed against swans, geese and turkeys. There are Colonial moulds still in existence which cast 40 or more so-called swan shot.

For smaller shot, lead sheets were cut into cubes. Corners were rounded off by placing the cubes between two hardwood boards and revolving the top board by hand.

Pre-Revolutionary Brass Gang Mould. A most interesting mould that casts 5 round balls of these diameters: .50, .55, .62, .57, and .67, plus a cylindrical slug Cal. 68 measuring ¾″ in length. The latter could hardly have been accurate, but at close range it was certainly deadly. *From the McMurray Collection.*

BULLETS OF THE REVOLUTION

The American Revolution was fought mainly with firearms. The artillery fired cast-iron solid shot, iron shell and iron grape and canister shot. The muskets, rifles and, occasionally, the smaller artillery pieces fired one or more lead bullets. Many hundreds of lead bullets were fired for each iron projectile used. The Revolution was truly a lead-bullet war.

Smoothbore Small Arms

Muskets and other similar smoothbores were the most frequently used weapons in the American Revolution. There were four major types. First, the British Brown Bess was standard for their army throughout; early in the war, it was important on the American side also. Second, French muskets were standard in the American armies after 1778. Third, other European muskets were used to some extent throughout the Revolution. Fourth, American-made smoothbore small arms were always present, at least in small numbers.

A general type of weapon known as the Brown Bess Musket came into use in the British Army in the late 1720's. Actually, there were several types and models. The muskets used in the Revolution were of three barrel lengths: 46, 42, and 39 inches. These represent three chronological models. Other changes were made in fittings, shape, and mechanical details at about the same time that the barrel lengths were changed; these are known as the Models I, II, and III. Model IV, with a 39-inch barrel and reinforced cock, probably saw limited service toward the end of the war. All were nominally Cal. 75, but had bores which were frequently considerably larger. All probably weighed, when the wood was new, about 10 pounds. They were sturdy, reliable arms particularly adapted for use with bayonets.

In addition to these muskets, there were similar British carbines, musketoons and fusils. Carbines and musketoons were the same length as muskets, but lighter. All these arms were similar to the Brown Bess Muskets in appearance. In general, they were carbine bore, or Cal. 65, although some of each variety were made full musket bore, or Cal. 75.

Since virtually all Brown Bess type weapons were of two caliber, only two bullet sizes were necessary. British bullets were greatly undersize to facilitate loading in a barrel already fouled by frequent firing. The musket-bore projectiles for barrels with a minimum inside diameter of .750″ were about .685″ in diameter. The carbine-bore bullets for Cal. 65 barrels were usually about Cal. 60.

The French muskets which came to this country during the Revolution varied so much that it's difficult to classify them into definite models. More than 100,000 were sent over; some of these seem to have been used for considerable periods in the French armies, returned to various arsenals, repaired, and reassembled in a somewhat haphazard fashion. Barrel lengths and fittings are various; however, all are nominally Cal. 69. Apparently, they were intended to fire Cal. 65 bullets.

A few Dutch, German and Spanish smoothbores of various types were present in America at the start of the Revolution. Others were brought in from the West Indies and Europe during the war. Some French and Spanish arms had been taken by Colonial forces in the French and Indian Wars. A Spanish musket believed to have been captured by a Massachusetts Colonial at Havana in 1763 has a bore of .728″. German muskets used by the Hessian soldiers in the pay of Britain were numerous. Apparently, all these muskets could use one or other of the British bullets or the French bullet already referred to, although a sloppy fit meant poor accuracy.

The production of American gunmakers before the Revolution was not inconsiderable. Perhaps about one third of all firearms needed in the Colonies were made here. However, not nearly so high a proportion of the weapons actually used in the war were so produced. Small arms made in America before 1783 were usually patterned after British military arms. Some few were copies of French arms. Bore sizes varied widely, however, even from gun to gun produced by the same maker. Many of these semi-military smoothbores were even smaller than Cal. 60.

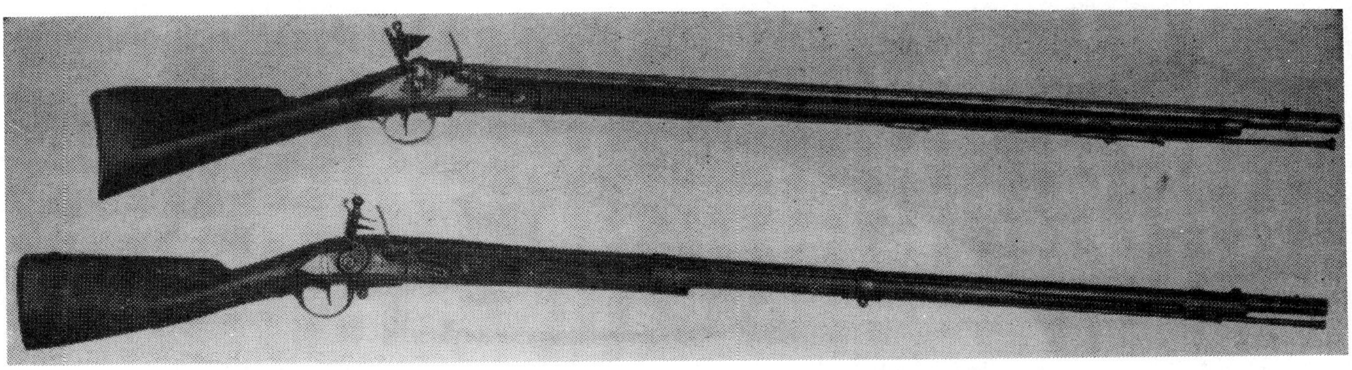

Typical Weapons of the Revolution. Top: Model III Brown Bess Musket, bore .761″, barrel length 39¼″, length over-all 55″, weight 9¼ pounds, walnut stock. Lock marked "D. Egg." Bottom: Committee of Safety Musket, bore .730″, barrel length 42″, length over-all 57¼″, weight 9 pounds, walnut stock. Lock marked "C Isch", breech plug marked "C.I." *From the Weller Collection.*

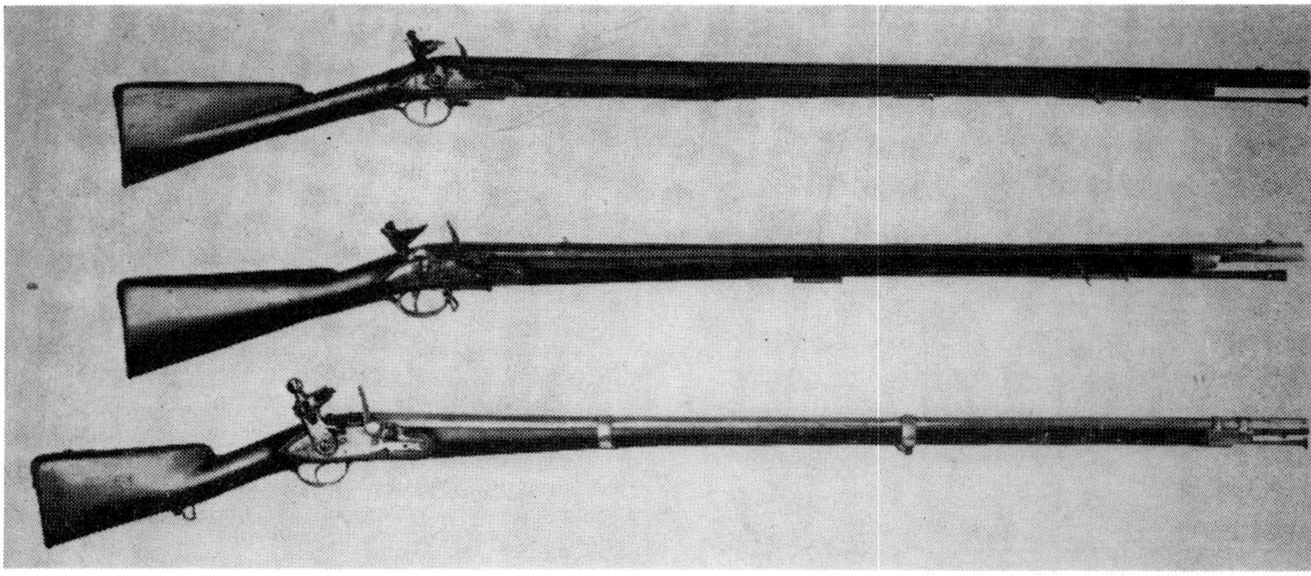

Typical Weapons of the Revolution. Top: British Model III Fusil, bore .775", barrel length 39", length over-all 55", weight 8¼ pounds, walnut stock. Escutcheon plate marked "6/58," lock marked "W. Parker," barrel marked "Holborn," Middle: British Rifled Brown Bess, bore .704, nine narrow, deep grooves, barrel length 39", length over-all 54½", weight 10 pounds, walnut stock. Note iron ramrod and rear sight. Lock marked "D. Egg, Crown GR," Barrel marked "D. Egg London." Bottom: Spanish Light Musket, bore .725, barrel length 43½, length over-all 59", weight 8¼ pounds, walnut stock, band fastened, brass mounted. Note typical Spanish migulet lock. *From the Weller Collection.*

Smoothbore Ammunition

Paper cartridges were used very early in the history of firearms. The Swedish infantry under Gustavus Adolphus were provided with these before 1600. At first, cartridges were just a paper package of powder which was easier and quicker to load than any form of measured loose powder, particularly in a close military formation. A separate ball was then rammed down on top. It wasn't long, however, before the bullet was included in the cartridge. Apparently, very early cartridges of this type had the bullet bare and the end of the paper cylinder tied with string about the sprue of the cast bullet. Later, the paper packet contained both bullet and powder.

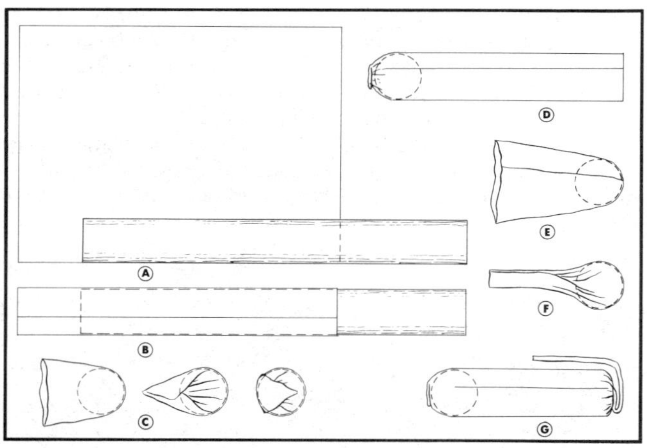

French Paper Cartridges. The basic steps used by the French in making their paper cartridges are illustrated above. The French used a rectangular piece of paper quite different in shape from the paper used in the British method (see "A"). The paper is wrapped around the dowel and the longitudinal seam pasted (see "B"). Then the open end is folded against the butt of the dowel and pasted down (see "C"). Dowel is removed, the ball dropped in, and powder charge placed on top (see "D"). Tube is pinched together above the powder and bent 90 degrees and the base of the cartridge tapped against the table (see "E", "F"). Finally, the pigtail is folded again and left parallel to the body of the cartridge (see "G").

Long before the Revolution, cartridges were standard in both British and French armies. Even Colonial militia was supposed to have cartridges, although a powder horn and loose bullets seem to have been acceptable at militia drills. Cartridges were made in about the same way everywhere, although there was a definite difference in construction between the British and French variety. Both were formed by wrapping a piece of paper around a cylinder of wood slightly larger than the bullet to be used. In the French type, both the bullet end and the straight side seam were pasted. In the British type, the bullet end was tied with string and the closing of the outside seam made unnecessary by using paper cut on a diagonal. If both, a charge of powder was measured into the paper cylinder back of the bullet, and the loose paper folded into a sort of pigtail.

Cartridges were considered necessary for efficient infantry. A musketeer could fire at least twice as fast with cartridges as with a powder horn and loose bullets. A cartridge could be grabbed quickly and was all in one piece. In an emergency, the whole thing could be rammed down the barrel unopened. The standard cartridge paper was sufficiently weak so that it could be burst at the breech with a single blow of the rammer. It was considered better, however, to tear off the end of the paper and pour the powder down the barrel. The lead bullet was then pushed down on top of the powder, either with the paper still wrapped around the bullet, or the paper stuffed in over the top of the bullet as a wad. The paper, of course, could be left out entirely, save if the musket was to be fired downhill.

Tending the Lock

At the time of the Revolution, the cartridge contained not only the propellent charge, but sufficient additional powder for priming. Some vents between the chamber and the pan may have been large enough to admit sufficient powder into the pan for priming when the bullet was seated firmly on top of the powder charge; however, this method wasn't standard. Soldiers were taught to put a bit of powder in the pan from the cartridge and close the pan cover and frizzen before loading. A reliable half-cock position to prevent the hammer falling during the loading process was necessary for safety. However, the expression "going off half-cocked" indicates that some weapons failed in this respect.

The most difficult task for continuous fire was not the load-

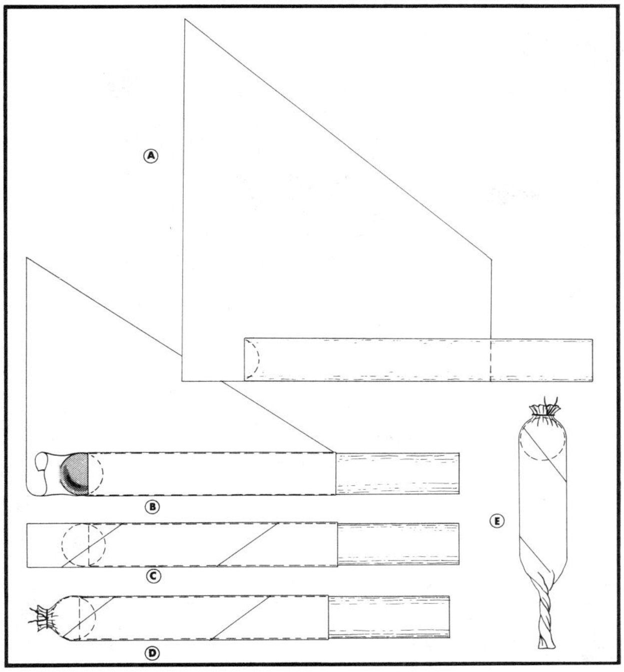

British Paper Cartridges. The British style of paper cartridge had the ball end tied off with twine and the other end twisted to hold the powder. The basic steps are shown in the illustration above. Brown paper of this shape is wrapped about a dowel. The end of the dowel has been hollowed out to fit the ball (see "A"). The ball is inserted and the remainder of the paper wound on the dowel and tightly tied to hold the ball (see "B", "C", "D"). Dowel is removed, powder inserted, the end of paper twisted to hold the gunpowder (see "E").

ing operation, or even keeping the vent open by "picking" when necessary, but the proper adjustment of the flint so that it would produce the necessary shower of hot sparks. Individual judgment was required to know when to reset, or change, a flint. These small pieces of stone were very important. At the time of the Revolution, many flints were good for only three or four discharges.

Loading Drills

Muskets were usually loaded by an individual soldier doing the entire job on his own musket himself. This procedure had been reduced to a ritual followed precisely on a "by-the-numbers" basis in European armies. Von Steuben taught the same techniques to the American Continental soldiers. Loading was not difficult; iron ramrods, introduced into European armies between 1725 and 1760, were almost universally employed at the time of the Revolution. They greatly improved both speed and reliability of musketry fire. However, practice was still important. Loading and firing had to become instinctive before it could be relied upon in action. A good soldier could fire four times in a minute.

There was another system of musketry fire used in the British armies before the Revolution. It may easily have been used in this country to a limited extent. By means of this, a battalion could deliver almost continuous fire, but in reduced volume. A file, generally consisting of a front-rank man, a middle-rank man, and a rear-rank man, would work as a team. The front man would cock and fire all three muskets in sequence. The middle-rank man might clean the vent, place the weapon on half-cock, prime from a separate flask, and close the pan cover. The rear-rank man would load. James Wolfe, the victor at Quebec, taught his regiment this technique and aided greatly in breaking the army of Prince Charles Edward at Calloden in 1746. It had disadvantages, however, and was probably not used often in the Revolution.

The Manufacture of Cartridges

In the British service, small-arms cartridges were usually made up in Britain and shipped in casks and kegs to the places where they were to be used. In an emergency, cartridges could be made up locally by the soldiers themselves. The French followed the same general procedure. Initially, American soldiers received paper, powder, and lead and made up their own cartridges. Later on, civilian contractors loaded a portion of the American cartridges. One of the early ordnance activities at Springfield was the making of small-arms cartridges.

American cartridge manufacture was complicated by the fact that even in small units there would be several different calibers of muskets. Bullets to fit the British Cal. 75 and Cal. 65 weapons and the French Cal. 69 muskets was not enough. Pennsylvania troops needed seven different sizes of smoothbore ammunition, varying apparently from Cal. 52 to Cal. 80.

Multiple Loads

Buckshot was popular in America before the Revolution. Charges of buckshot alone, or three to ten buckshot in addition to a musket ball, were frequently used in the war. Buckshot varied considerably in size. The largest were cast so that a layer of three filled the bore in which they were to be fired. Smaller sizes were also used; sometimes as many as 20 would be fired at once. Buckshot was loaded into cartridges in front of the single ball, although the procedure was not so standardized as it became in the 19th Century.

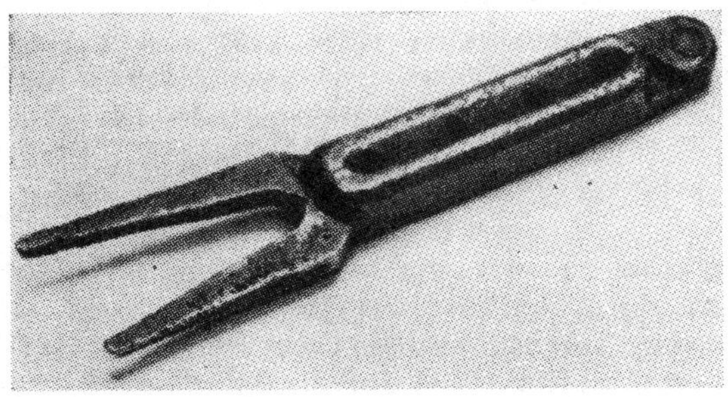

Seth Pomeroy Bullet Mould. Pomeroy was a Brigadier General in the Continental Army and died on active duty in 1776. He was also a distinguished gunmaker, farmer and politician. His mould is one of the most interesting in American history. It originally cast four spherical lead bullets of different sizes, the largest of which was probably for the Brown Bess musket. *From the Peterson Collection.*

Bullet-Casting Procedures

Until long after the Revolution, all bullets were cast in moulds. Gang moulds were common for military use. Contractors supplying muskets to Colonial governments usually supplied a proportion of moulds with the weapons. For instance, Virginia specified one iron gang mould casting 16 full sized bullets at a time for every 40 muskets. Maryland wanted one brass mould casting 12 bullets on one side and as many buckshot as possible on the other for every 80 muskets. It was, of course, possible to fill cavities on one side, turn the mould over, fill the cavities on the other side, and then dump all bullets from the mould. These bullets were cast of lead as pure as happened to be available. There was no thought of hardening the lead by adding tin or other alloy. A team of six men working together with gang moulds under the best conditions could produce 20,000 or more bullets in a day.

The Rifle and the Revolution

The importance of the rifle during the Revolution has been debated at length. At one time, some rather imaginative historians said that the Colonies won because of the superior accuracy and range of the American rifle. More recently, careful research has shown rifles to have played no important part in many actions. However, this view has also now been overemphasized. In the South particularly, the rifle was often quite important. A good rifle was far superior to any musket for long-range accuracy.

Even in the North, rifle companies were of considerable value for special services. Tim Murphy of Dan Morgan's riflemen seems to have shot the British General Frazier at the battle of Freeman's Farm at a range in excess of 200 yards. However, the rifle had many disadvantages and was not used as a basic weapon for even light infantry in either the British or American services by the end of the war. American rifle companies were in many instances rearmed with muskets after 1777.

Rifles were loaded in two ways during the Revolution. The American system called for the patching of a lead bullet slightly smaller than the bore, either with linen or some animal membrane. The so-called European system of loading required that a slightly oversized bare lead bullet be forced into the muzzle of the rifle and then pushed down the bore so that the bullet took the rifling positively. It would appear in the light of surviving evidence that the American system was known and widely used by the German jaegers, as well as certain British soldiers armed with muzzleloading rifles. However, the European system was neither so laborious, nor so imbecilic, as it has been portrayed.

Muskets were loaded intentionally with bullets smaller than their bores in order that, even when the weapon became foul, the bullet would still slide down the barrel easily. Rifles required a fairly tight fit on both the patched ball and the bare lead bullet.

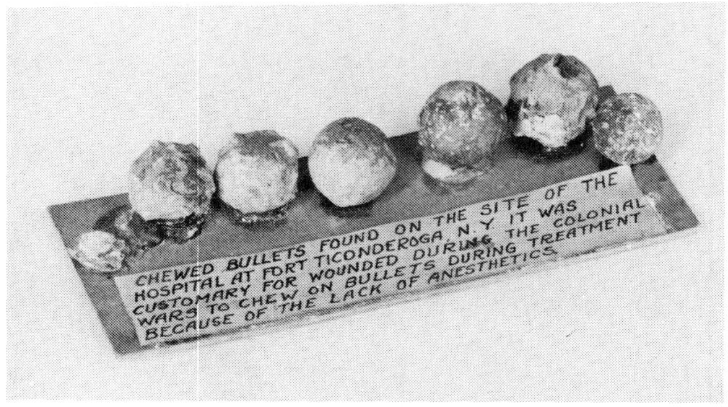

Chewed Bullets From The Revolution. Anesthetics were unknown during the Revolution, but surgeons and their assistants did the best they could with what they had and what they knew. A patient undergoing an operation, or an amputation, was generally given a bullet to be placed between the teeth. The tooth-marks on these bullets are graphic pictures of agony almost to the limit of human endurance. *From the Dowling Collection.*

Loading began to give trouble after the second or third shots. Even though an experienced man could continue to fire for many rounds, this individual skill was seldom available in recruits and undoubtedly led to the abandonment of the rifle by the Continental infantry.

Mention must be made of the British Ferguson rifle which loaded from the breech. These were not present on any battlefield in sufficient quantity to be significant; however, in the hands of skillful soldiers, they could deliver a substantial volume of accurate fire for some time. Since they loaded at the breech, a groove diameter bullet could be used, insuring positive taking of the rifling.

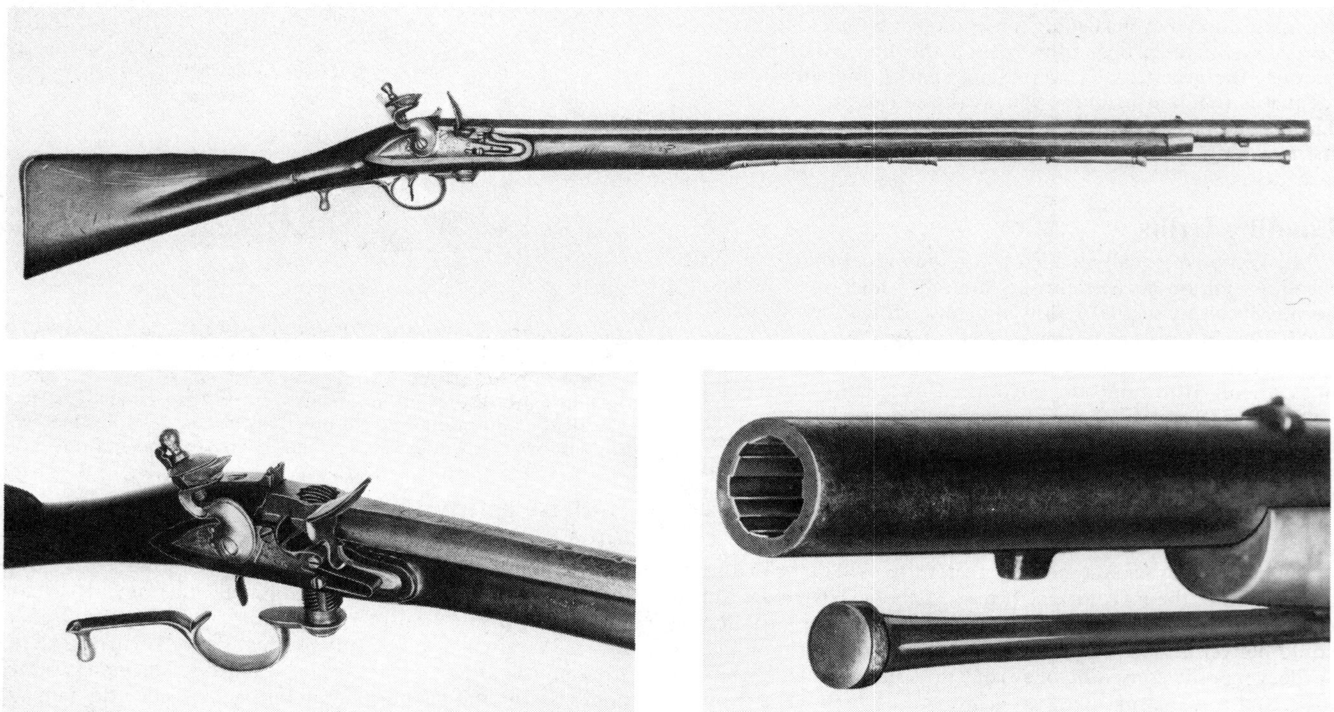

The Ferguson Rifle. Ferguson developed the breech-loading military rifle bearing his name during six years of British Army service in the West Indies about 1770. Ferguson perfected a breech screw that functioned at right angles to the axis of the bore. One revolution of the trigger guard opened the breech and cleared it for loading (see close-up). These rifles had accuracy equal to any of their contemporaries and far surpassed all in speed of reloading. This particular rifle is one of the few contemporary military Fergusons known to be in existence. *From the Weller Collection.*

Rifle Ammunition

American rifles were no-standard as to bore size; each had to be loaded with bullets of the proper size for it. No effort seems to have been made to provide cartridges for riflemen. Each one probably cast his own lead bullets with his own single-cavity mould. He cut his patches beforehand, or at the time of loading, according to personal preference. He carried powder loose in a powder horn. The Hessian jaegers seem to have followed a similar system.

The Ferguson rifles would function with British carbine bore paper cartridges. However, the cartridge itself could not be inserted into the weapon. The bullet had to be squeezed out of the end and put in the circular hole at the top of the breech; the remainder of the powder was then poured into the orifice behind the bullet. It was possible to load this weapon in such a way that the extra powder would be thrown into the pan when the breech plug closed. Ferguson himself seems to have preferred to use, however, balls from a special pouch and a spring-type powder flask which would throw a uniform charge into the breech time after time.

Ammunition Containers

Paper cartridges were carried in the British armies in leather cartridge boxes; usually each cartridge was placed in a separate hole bored in a wooden block which filled the inside of the box. A large leather flap fitted tightly at the top and lapped far over the side. It would protect ammunition even from a downpour. French cartridge boxes were usually similar.

The Continental armies had great trouble in getting enough of these fine leather and wood boxes. Many substitutes were tried. Sometimes an earlier British expedient of a bored wood block with a flap of leather nailed to it would be used. Sheet-metal cannisters for the carrying of paper cartridges stacked together without the wooden block were also made and issued.

Riflemen were not the only American soldiers to carry their powder in horns, or even leather-covered rum bottles. Paper cartridges were not by any means universal in even the Contitental infantry, much less the militia. On occasion, individual charges of powder were poured into muskets from horns by eye, and a ball taken from the bullet pouch, or even a trouser pocket. However, paper cartridges and good cartridge boxes were always preferable.

Light Infantry Cartridge Box. A very satisfactory dragoon-type cartridge box made of leather and believed to have been issued near the end of the 18th Century. Contains row of 24 tin cartridge cases, tin powder cannister. *From the West Point Museum Collection.*

Revolutionary War Cartridge Box. Three tin loops 7/8-inch wide are riveted to sides and bottom. Inside dimensions 2 7/8" x 3 3/4" x 6 1/2". *From the West Point Museum Collection.*

THE ELONGATED BULLET

Spherical lead bullets were used almost exclusively in war and in hunting from the 14th Century until about 1850. The United States Army fought the Mexican War using round lead balls in both the Cal. 69 flintlock muskets and the newer Cal. 54 percussion rifles. The rifle was more accurate than the musket; however, neither was dangerous beyond 300 yards. Round bullets had such a poor aerodynamic shape that they just wouldn't carry much further.

The superior flight characteristics of elongated bullets had been known for a long time; they could be fired with reasonable accuracy at ranges up to 600 yards. Benjamin Robbins writes of their superiority in a series of papers originally published in the 1720's. Before the Revolution, Patrick Ferguson seems to have employed elongated bullets for his own use in his breech-loading rifles. Cylindrical bullets must fly through the air point-foremost, however, if they are to be either accurate or long-ranged. The only practical means of obtaining point-foremost flight was by firing these bullets from a rifle. Loading troubles, bad enough with round balls, were insurmountable with ordinary elongated bullets in muzzle-loaders.

The 19th-Century Inventions

The Industrial Revolution was in full swing. Men were looking for new ways to do old things, even in the field of small arms. The rifle had made a name for itself in the American Revolution, at New Orleans in 1814 and in the hands of the British Rifle Corps against Napoleon. Most European armies set about arming at least part of their force with rifles. The old method of forcing either an oversized bullet, or an oversized package consisting of bullet and patch, down a bore from the muzzle was impractical. Somehow, an arrangement had to be thought of which would give speedy loading, accuracy, and long range.

Three Types of Bullets

Literally scores of bullets were invented during the first half of the 19th Century. These fall into three general classes. First, bullets were cast in such a form that they would fit the grooves of the rifling, both on being loaded down the barrel and on being fired. Second, bullets were cast smaller than bore size, but in some way expanded after being loaded so that they filled the grooves when fired. Third, there were breechloaders which used larger-than-bore-diameter bullets. Both round and elongated types were present in all three types, although all realized the superiority of the latter for long range.

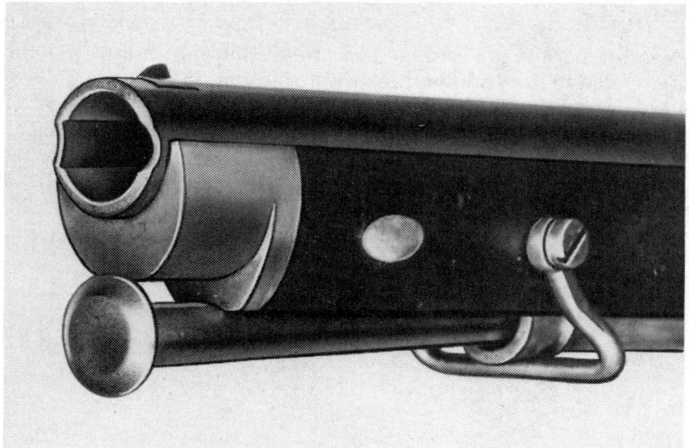

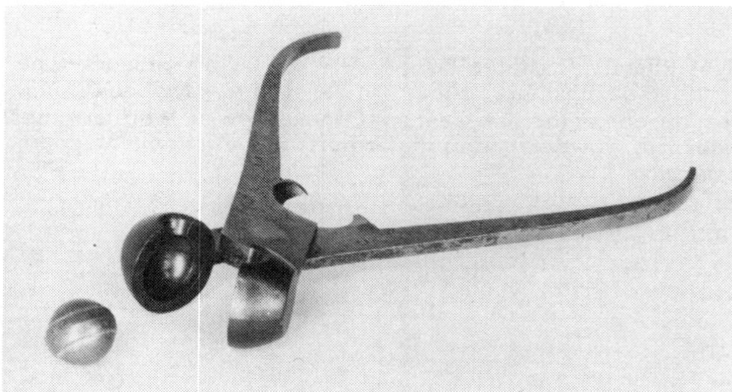

Brunswick Mould and Bullet. The ring on the bullet fits the two rifling grooves vibible in the illustration of the muzzle of the Brunswick Rifle. *From the Weller Collection.*

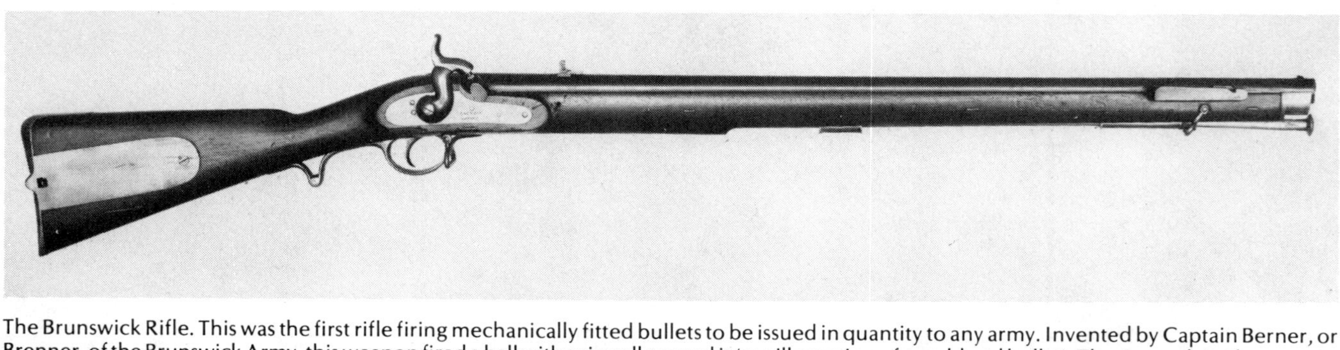

The Brunswick Rifle. This was the first rifle firing mechanically fitted bullets to be issued in quantity to any army. Invented by Captain Berner, or Brenner, of the Brunswick Army, this weapon fired a ball with a ring all around it (see illustration of mould and bullet). The ring rode in the grooves visible in the illustration of the muzzle. *From the Weller Collection.*

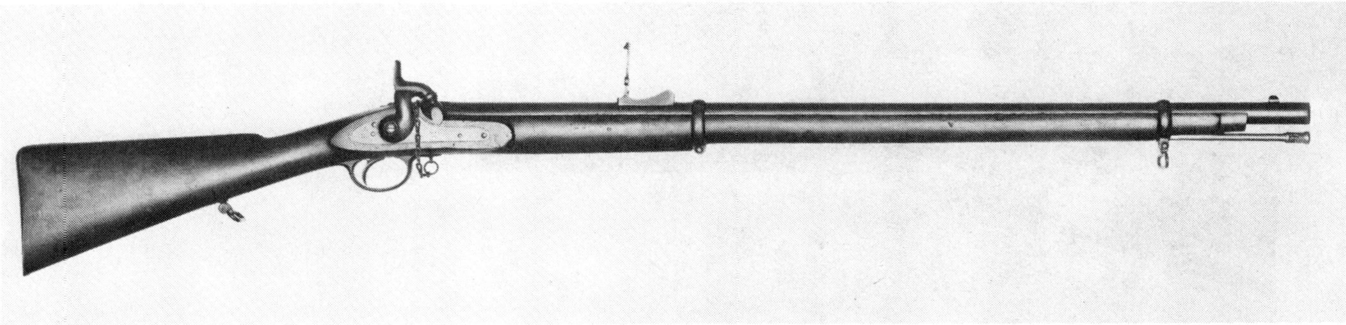

Diamond-Bored British Military Rifle. This Lancaster military rifle of the 1850's was made up with a so-called diamond bore (see close-up). *From the Weller Collection.*

The Shaped Bullets

The first really successful shaped bullet was that for the British Brunswick rifle, the invention of a Captain Berner of the Brunswick Army. Similar rifles were used in other European countries. The barrels of all Brunswick-type rifles have two deep grooves opposite each other. These fired a spherical bullet that had a raised ring around it; the ball fits both the bore and the grooves on the way down and on the way out.

General John Jacob of the Anglo-Indian Army evolved a similar four-groove rifle which at first fired a spherical bullet with two rings around it at 90 degrees to each other. However, Jacob soon changed to elongated, or cylindro-conoidal, bullets with four ribs designed to fit the four grooves of his special rifles made for him by Daw, the London gunmaker. These Jacob's rifles are powerful, accurate, and extremely long-ranged. An ammunition wagon was blown up in action by an explosive bullet from a Jacob's rifle at approximately 2,000 yards; he duplicated this feat several times on his enormous Indian ranges.

A somewhat similar weapon was the so-called Cape rifle, which was highly thought of by British sportsmen. This was a two-groove rifle, with grooves somewhat shallower, but fully as wide as the Brunswick. It took an elongated bullet cast with two projections to fit these grooves. The name "Cape" probably came from the popularity of these weapons in the Cape of Good Hope where Englishmen, even 100 years ago, were hunting the biggest of game. Purdey used this system for some of his best muzzle-loading sporting rifles.

Perhaps the best known of the mechanically fitted bullets were the hexagonal Whitworth projectiles to fit the special bores of his very accurate long-range rifles. These will be dealt with at length in another chapter. Polygonal bore rifles and bullets to fit them have been made in from three to eight sides. Some other fitted shapes have also been tried. The writer has a British rifle from this period with a bore which resembles a lengthwise cross section of a modern American football.

The Expanding Bullets

The second general type of new rifle bullets were those that could be loaded fairly easily down even fouled rifle bores because they were undersize, but were expanded so that they took the rifling positively on the way out. The important research in connection with these bullets was done in the French service, and extended over a period of almost a quarter century beginning about 1825. The names of four French officers are generally associated with this step-by-step development. They are Delvigne, Pontchera, Thouvenin and Minie. An Englishman, William Greener, did some important work along this line and was suitably rewarded by his government. Both the British Lancaster and Wilkinson systems worked fairly well. However, the French research and development seems to have been complete within itself and will be described in some detail.

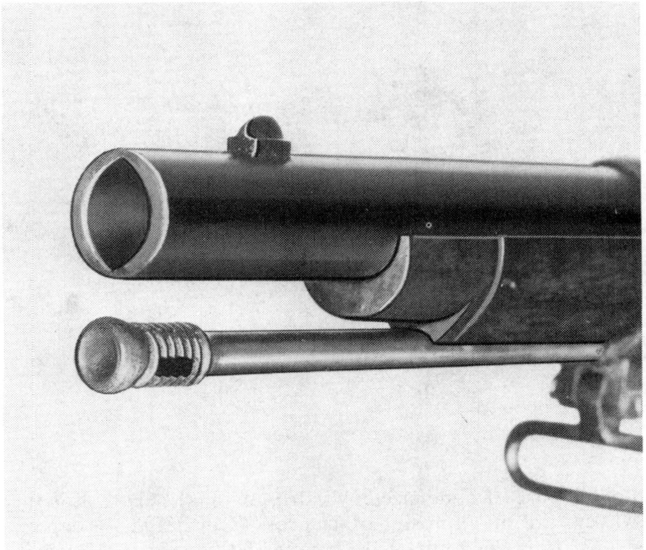

In 1828, Captain Delvigne perfected a means of loading a spherical lead ball, either with or without a greased patch, down a barrel through which it would pass easily, and then expanding it against the top of a chamber smaller than the bore proper. The bullet would swell into the grooves of the rifling under repeated blows of the rammer. It was difficult, however, to expand it the same amount each time. Further, the bullet itself became wider than it was long and lost velocity quickly. Finally, the bullet tended to be pushed down into the chamber, crushing the powder and becoming T-shaped.

To correct the last-mentioned trouble, Colonel Pontchera added a hard-wooden sabot to keep the bullet out of the chamber entirely and cause expansion into the grooves between the rammer and sabot, both hollowed out to spherical shape, at a point above the chamber. This worked well when the rifle was clean. When the bore fouled, however, the sabot tended to stick and other troubles developed.

These chambered rifles were tried with cylindro-conoidal bullets; if the bullets were made small enough to pass down a fouled barrel, they gave erratic results when expanded unevenly into the grooves. However, Colonel Thouvenin substituted a central steel stem at the breech for the chamber and rammed his bullets into the rifling more or less concentrically about this pointed stem by repeated rammer blows. These bullets gave better accuracy than the round bullets expanded by means of a chamber rim, even with the wooden sabot, particularly at long range. They still couldn't beat, however, carefully loaded patched round bullets up to 200 yards.

Captain Minie, as well as apparently the other three, had been experimenting for years with a clean bore and tight-fitting elongated bullets. Sometimes these gave great range and accuracy. They weren't practical, however, because of loading troubles. Minie removed the stem from the breech of some experimental Thouvenin rifles and placed an iron cup on the

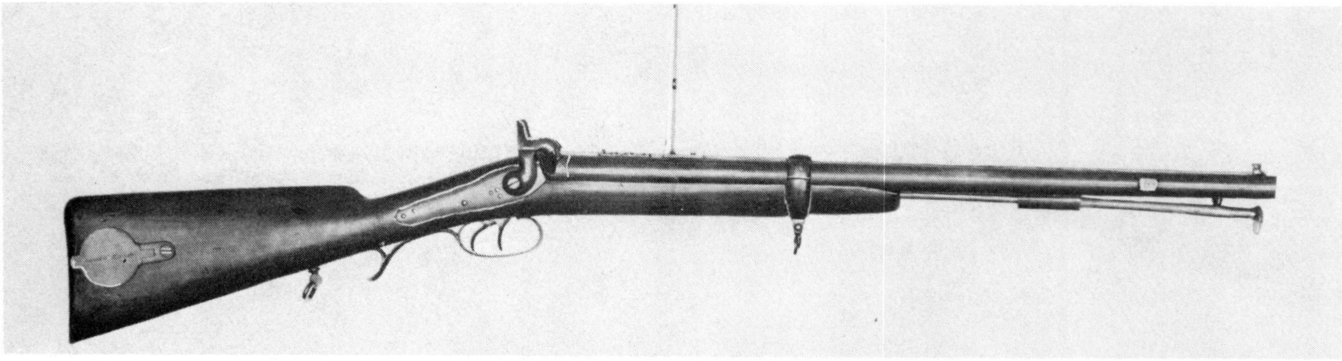

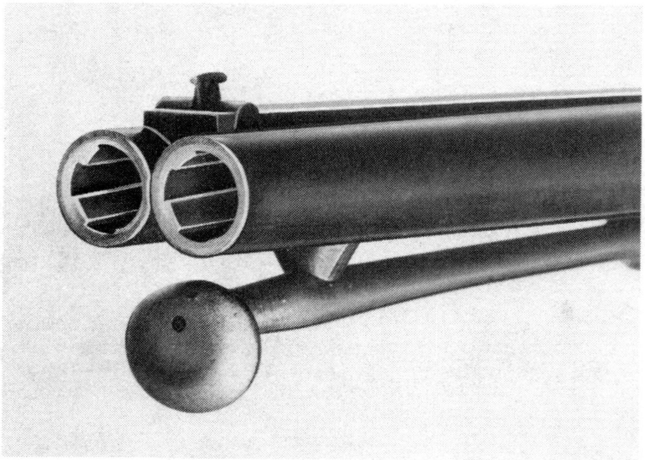

Invented by General John Jacob of the Indian Army, this is a most interesting weapon. The Jacob's rifle fired a four-ribbed elongated bullet, either solid or hollow-point with an explosive charge, with astonishing accuracy. This is the type of rifle that Jacob used in blowing up an ammunition wagon at a range of about 2,000 yards. *From the Weller Collection.*

hollow base of undersized cylindro-conoidal bullets. Loading was easy; at the explosion of the propellent charge, the cup was driven into the bullet base cavity and caused the bullet to expand and fill the grooves. Both accuracy and range were superior to anything yet achieved. Minie bullets were as accurate as the round bullets in any rifle at 100 and 200 yards. Man-sized targets could be hit at 400 yards consistently; inherent accuracy and power were still apparent even at 1,000 yards.

The fundamental idea of the Minie bullet swept the Western world. A metal cup or plug was used in the base of the bullet in some services, as Minie had done himself. Other countries filled the base cavity with a wooden or clay plug, or even nothing at all. The pressure of powder gases alone was sufficient to do the necessary expansion. Most important of all, accuracy and range were not greatly impaired by having the bullets small enough in diameter to be loaded down bores fouled by as many as 100 discharges without cleaning. Loading was easier than it had been with any other muzzle-loading rifle ever invented.

Breech-Loading Rifle Bullets

The third general type of new rifle bullets were those for breech-loading rifles. Since these bullets were often even larger than groove diameter, they filled the bore completely. They gave a few loading troubles, since they were loaded into the breech. The Dreyse needle rifle of Prussia is the best known of these which used self-primed cartridges; it was issued in quantity before 1856. Norway had a similar rifle operating on a totally different system. These rifles were not popular in Britain, France, or America, however, because they were complicated and leaked gas badly at the breech.

Both Britain and the United States had far simpler and more rugged breech-loaders. The Sharps was perhaps the best known; there were others, including the excellent Wesley Richards, Green and Terry; all using linen, or paper, cartridges and separate percussion caps. However, truly efficient breech-loading depended upon a sound, powerful metallic cartridge which was not perfected until after 1865.

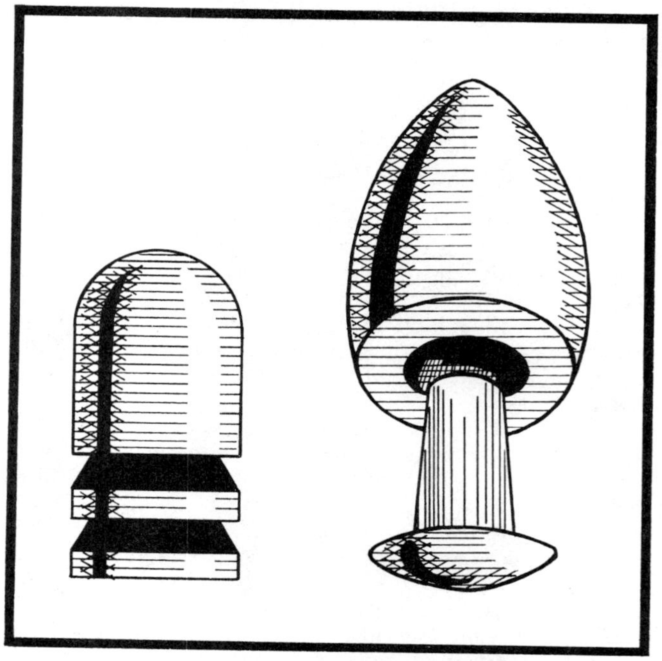

Wilkinson and Greener Systems of Expanding Bullets. Left: The very deep cannelures in this soft lead Wilkinson bullet caused the two rear sections to be crushed and squeezed out into the rifling grooves upon discharge, even though the base of the bullet was entirely flat. Right: The Greener bullet had a tapered hollow in its base into which an expansion plug was driven by the force of the explosion. Both these systems are interesting mainly as variation on the more successful Minie system.

The American Experiments

Minie produced his most successful bullets about 1848; Britain had a Minie-type rifle by 1851. There seems to have been little effort made by either country to keep secret their new small arms. The United States Army was on friendly terms with both the British and French services. The development of the Minie bullet and its adoption in the British and French armies was well known here. The American small-arms experiments, which lasted apparently from 1853 through 1855, were conducted at Harper's Ferry and confirmed the superiority of the Minie bullet to any other.

United States shoulder arms in 1853 consisted of three calibers with bore diameters of .54 inches (the rifles), .58 inches (the cadet muskets) and the .69 inches (the regular muskets and some carbines). Muskets had been rifled experimentally as early as 1842; now many more of both calibers underwent the same change. Many types of cylindro-conoidal bullets were tried in each of the three calibers. The Harper's Ferry Minie-type bullet was finally adopted for all three. It had a medium-sized conical base cavity with no filler and three cannelures around the middle.

In addition to bullets for existing arms rifled where necessary, a whole new family of weapons was produced to fire the Cal. 58 Harper's Ferry Minie-type bullets; these were a rifle, a rifle musket and a pistol carbine, all known as Model 1855's. These weapons were the only new arms to be produced in the future. However, extensive modernization was to be done on older small arms. Flint Cal. 69 muskets were rifled, equipped with new rear sights and changed to percussion. Percussion smoothbore muskets were rifled and equipped with elevating rear sights. New rear sights were installed on Cal. 54 rifles to take care of the new ammunition.

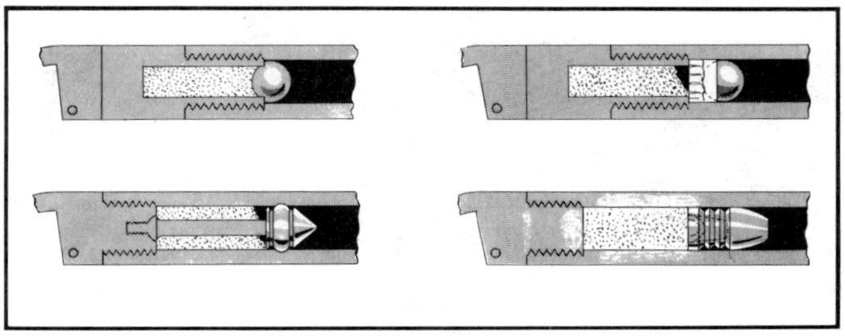

A Complete Change

Within a dozen years, the 500-year-old round bullet was suddenly obsolete. The elongated bullets took over for both military and sporting purposes. The expanding Minie-type projectiles were the most common, but both formed bullets and those loaded from the breech were also used. The bullet change was to have a profound effect upon military tactics; the American Civil War was the first major conflict fought with the new rifles.

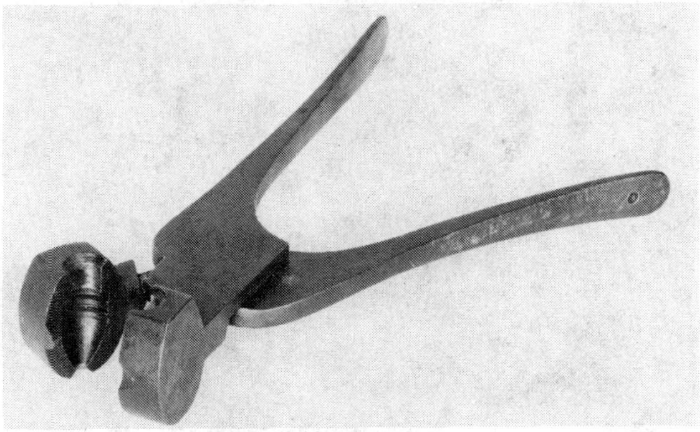

Harper's Ferry Mould. This iron mould casts a Cal. 69 Mini-type bullet. Many Civil War Minie bullets were cast by running the lead into the moulds through hollow base plugs. The Cal. 69 mould was originally of this type but was changed in order to get more uniform bullets. *From the Weller Collection.*

BULLETS OF THE CIVIL WAR

In 1861, the Minie bullet was incomparably superior to all round balls. Its range, accuracy and ease of loading made cavalry charges impractical on battlefields. Artillery in the open was forced back such a distance from its target as to be ineffective. Infantry with Minie rifles was indeed the "Queen of Battles."

Considering the very recent evolution of the Minie bullet, there were a great many rifles of this type available in the United States. Very similar British, Belgian, and Austrian rifles were imported. Both sides set about manufacturing a lot more. All American officers appreciated the new weapons and what they could do.

Obsolete Bullets

Mobilization took place on both sides with great speed and on a very large scale. The supply of modern weapons was insufficient. Early in the war, both the North and the South were forced to use many obsolete small arms. The South was using flintlock smoothbore muskets made almost a half century before at the Virginia Manufactory in Richmond. The North was also in some cases using smoothbore muskets and importing a large reserve of flintlock weapons, apparently to prevent them from being bought by the Confederacy.

Smoothbore percussion muskets, particularly the U.S. Model 1842, were used longer in the Southern armies than in the Federal forces. One of the proofs that Stonewall Jackson's wounds at Chancellorsville on 2 May 1862 were inflicted by his own troops was that a spherical lead bullet fired from a Cal. 69 smoothbore musket was found in his right hand. This weapon, unrifled, had disappeared from the Army of the Potomac, but was still used in some units of A.P. Hill's Confederate division.

A Minie-Bullet War

The Civil War, however, was fought mainly with rifles and the new expanding bullets. Perhaps better than 95% of the bullets fired were of the general Minie type. We are fortunate in having not only old records, surviving unfired ammunition, bullet moulds, and the like, but also thousands of the bullets themselves. The actual battlefield recoveries are the best evidence of what was used.

Since World War II, a new technique of relic recovery has been perfected. The author is indebted to Sydney C. Kerksis, Frank Hatch and Captain William Gavin, who have developed recovery procedures using a modern mine detector to an astonishing degree. (A word of caution: an army surplus mine detector and ordinary GI training to operate it won't work.) These men built their own detectors and have real skill in using them. They recover dozens of bullets per hour, as well as other relics. Fortunately, they have kept the large numbers of bullets taken from the battlefields in Virginia, Maryland and Pennsylvania, as well as lesser quantities from those of the Western armies, separate. Further, they have tabulated their findings. It's relatively easy to tell what was and was not used in battle at any given time during the war.

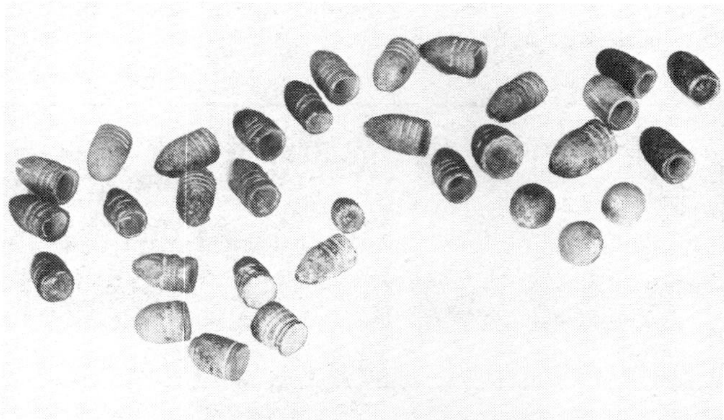

Bullets Recovered From Civil War Battlefields. These bullets are representative of the kinds being recovered, even now, from various battlefields of the Civil War. *From the Weller Collection.*

The Harper's Ferry Bullets

The most common bullet used in the Civil War was the Harper's Ferry adaption of the original Minie idea. This bullet had a hollow base, but no cup, or other filler. The base cavity is sometimes conical and sometimes in the form of a truncated cone. It had three grooves, or cannelures; these were normally not filled with lubricant. However, the whole bullet was dipped in tallow and beeswax. Bullets of this general type were made by both North and South in several calibers. In Cal. 58 (indistinguishable now from Cal. 57), they are by far the most numerous of all bullets fired. There are also many Cal. 69 Harper's Ferry bullets among recoveries from early battlefields, as well as moderate quantities of Cal. 54 Harper's Ferry bullets from Austrian and Mississippi (U.S. Model 1841) rifles. Cal. 70 or Cal. 71 Harper's Ferry bullets are found in small numbers. These were probably for British, Belgian and Austrian rifles of these calibers or larger.

The British Pritchett Bullets

The British Enfield rifle fired in the British service a Minie-type bullet of a distinctive appearance. It had a more rounded nose and no grooves, or cannelures, whatever. The base cavity is usually in the form of a truncated cone and is larger than in the Harper's Ferry type. This leaves quite a thin edge. When these bullets were made in Britain, the hollow base was filled with a boxwood plug, mainly to save the easily bent base edge from being damaged in handling. These bullets are most frequently recovered in Cal. 57, but both Cal. 54 and Cal. 69 are found. Some Pritchett bullets were made in Britain and imported in the form of loaded paper cartridges; however, far more were produced here, perhaps with imported bullet-making equipment.

The Confederate Fowler Bullets

Another common recovery from battlefields is the Confederate Fowler bullet. This projectile actually follows the original

Minie idea and has a separate base cup made of lead. It has two cannelures instead of three, and is said to have had lubricant placed in these. Mr. Kerksis writes that this bullet "very rarely . . . tumbled in flight. This is not true of Harper's Ferry and Pritchett bullets." Fowler bullets are found in Cal. 54, Cal. 57, and rarely in Cal. 69. These bullets were perhaps the most costly to make of all bullets used in the war, but may easily have been the most reliable under battlefield conditions.

The Union Williams Bullets

The North used large quantities of bullets which were designed to scrape out the fouling which adhered to the insides of the barrels after several shots. These were known as Williams bullets, after their inventor, and were of three types. The scraper, a convex zinc washer placed so as to be flattened into the grooves of the barrel by the inflammation of the powder charge, was the same in all three. It was secured to the base of a special Cal. 58 bullet by a pin in Type I and a plug with full bore-size head in Type II. Type III was similar to Type II, but lighter and shorter for use in the pistol-carbine. These bullets were issued about one in ten in 1862; the ratio was increased to about four in ten later on. They are seldom found, however, on the late '64 and '65 battlefields.

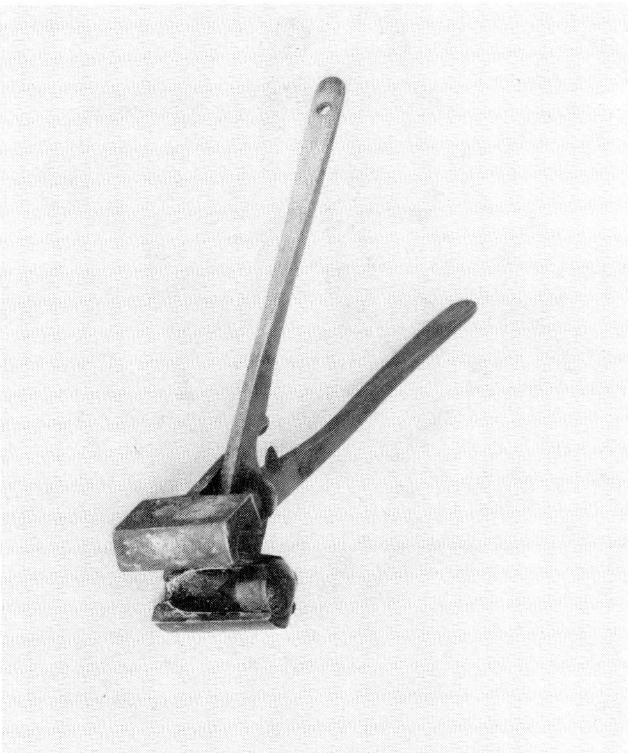

Pritchett Mould. This brass mould casts a Cal. 577 Minie-type bullet. Note the plug that gives this bullet its characteristic hollow base. *From the Weller Collection.*

Miscellaneous Bullets

Sharps breech-loading rifles and carbines usually fired solid-base bullets; other bullets for breechloaders were similar. Quite a variety of this type are recovered and are difficult to identify as to the weapon which fired them. Spencer cartridges are sometimes recovered unfired; both the Civil War Spencer bullets and cartridge cases are different from ammunition loaded later. An unfired bullet from a short copper case recovered from the Five Forks battlefield below Petersburg measures about .545" in diameter. Post-Civil War Spencer bullets are sometimes .515" in diameter. There were many other breech-loading Union carbines; however, recovered bullets from them are rare and difficult to identify.

A number of round bullets are found of a type that could have been used in the Cal. 69 smoothbores; they measure .650" to .660" in diameter. Occasionally, the small buckshot from the buck-and-ball cartridges of this size are also recovered. However, not all of the spherical lead balls that could have been used in these muskets were actually used in small arms. Artillery shells and cannister charges were sometimes loaded with pellets of this type.

Both round and conical revolver bullets are fairly common where cavalry fought. A few individual specimens of a large number of other bullets have been recovered. Whitworth bullets, as well as other Cal. 45 British rifle bullets, are rare, but have been found. So have the long, relatively small-caliber Union sharpshooter's bullets. Sometimes a recovered bullet will defy explanation and not fit into any known classification.

Loading Troubles Then and Now

Perhaps the most surprising facts one learns from examining battlefield recoveries are that unfired bullets were .010" to .030" undersized, and that soldiers still had loading troubles. Recovered bullets sometimes indicate that they had to be literally pounded down the barrel, probably by hitting the ramrod with a rock. The problem was, of course, black-powder fouling building up in bores, particularly on dry days. Some officers during the war tried to have Cal. 54 ammunition supplied to Cal. 58 rifles so as to avoid all loading troubles. The British service Minie bullets at one time measured .550" in diameter for use in .577" bores.

Modern experimental firing with Civil War rifles indicates that a moderately undersized bullet works well, but that a Cal. 54 bullet will almost always tumble from a Cal. 58 rifle. The Lyman Cal. 58 Minie-type bullet measures about .575". This gives fine accuracy, but is too tight for comfortable loading in a Cal. 58 rifle after a number of shots. It would have been difficult to use bullets of this size in action. Lubricating the bullets after the modern fashion will help. For best accuracy, it's desirable to clean the bore fairly frequently, at least every 20 rounds.

Ammunition for Minie Rifles

Paper cartridges for muzzle-loaders varied little from those used in the Revolution. In most cartridges, the Minie bullet merely took the place of the round ball. The tying up was done a bit better; pasting seems to have disappeared. The same type of buck-and-ball cartridges and all-buckshot cartridges were made, although they were not nearly so numerous. Surviving Civil War cartridges differ in individual appearance; some are hard to tell at a glance from those of the Revolution.

Usually, the Minie bullet was placed in the cartridge so that the whole thing could be loaded into the barrel in one package and the paper smashed with a blow of the ramrod. However, some British cartridges had the bullet pointing toward the powder. The reason for this was that the cartridge should be opened and the powder poured down the barrel; the bullet and paper then were pushed down on top of the charge with the torn ends of the cartridge uppermost. The Confederate Fowler bullets were generally exposed outside the paper cartridge; one end of the tube of paper was placed between the base cup and the bullet itself.

Small-arms ammunition was made up in laboratories. In both the North and South, these were efficiently run and produced millions of paper cartridges. Once made, they were suitably packaged, usually ten rounds to the package, with percussion caps often included, and printed labels describing the contents pasted on. A number of these packets would be assembled together in perhaps a metal-lined box for shipping.

These laboratories were usually located on the edges of towns and cities and used largely semi-skilled labor, a majority of which was frequently female. Although more of the loading and packaging was done by hand then, than would be done today, the production was carefully planned and well carried out. The laboratory in Richmond did explode tragically on Friday, 13 March 1863, killing and badly burning a number of workers.

Unexploded Shell Recovered From Civil War Battlefield. This cast-iron shell was recovered in a place which indicates that it could have been fired at First or Second Manassas. The Bormann fuse had been cut but the shell failed to explode. The shell was soaked and then cut apart with a hack saw. It was found to contain a mixture of small-arms bullets and black powder in a most unusual arrangement. One shell half with projectiles has been left as it was. The other has had all projectiles removed and carefully cleaned. Most of the projectiles, consisting of pistol bullets, round balls, solid-base bullets and Harper's Ferry Minie Balls, were obsolete at the time this shell was fired. This shell is believed to be Confederate. *From the Weller Collection.*

Bullet Making

Large numbers of bullets were still cast during the Civil War. A number of single-cavity moulds for Harper's Ferry and Pritchett bullets have survived. Further, there were some patented gang moulds casting hundreds of bullets an hour, as well as the smaller hinged type of gang moulds. Apparently, the majority of Minie bullets, however, were pressed or swaged into final shape. Short slugs of lead were formed in dies under hydraulic pressure to the desired shape.

A careful examination of battlefield recoveries appears to indicate both swaged and cast Minie bullets. However, an examination of a fired bullet is frequently not conclusive. Swaged Pritchett bullets often have "57" on the bottom of the truncated cone base cavity; cast bullets sometimes have a typical hollow left when the metal solidified. Lead erodes slowly over the years. In many instances, one just can't tell; the ridge of lead where the moulds joined or the swaging dies came together isn't conclusive, even when it can be found.

The swaging method did not produce better bullets than could be cast with care; however, it considerably reduced the amount of skill required. Operations which could be performed by semi-skilled labor were, of course, desirable. Swaging with power was easier. The hexagonal Whitworth bullets are very hard to produce in any other manner, even today.

BULLET MOULDS, GUN APPENDAGES

BULLET MACHINE, casting upwards of 100,000 bullets a day. Used by U. S. Army. Has 8 moulds, 10 balls each. Is operated by handle and casts upwards of 300 balls a minute, as quick as a man can pour in the lead and man can turn the handle. Is complete with doubled sprue cutting knife, which removes the surplus lead. Is complete with moulds for casting .69 calbier bullets.

WHITWORTH AND BERDAN BULLETS

A good Minie rifle, when clean and loaded with a well-designed bullet close to bore size, was very accurate. However, under combat conditions these weapons were not nearly so precise. Issue bullets were, as already suggested, considerably undersized and still gave loading troubles. Some tumbled in flight; a tumbling Minie bullet is even less accurate than a round ball. You literally can't hit the broad side of a barn at 200 yards with one.

The spirit of mechanical enquiry that had produced the Minie bullet was soon busy remedying its shortcomings, and inventing other weapons to take its place. The desire for extreme accuracy at long range, first made practical by elongated bullets only a few years before, swept over Britain and America during the 1850's and 1860's. Along with it, there was also an equally strong desire for ease of loading and rapidity of fire. At first, these two aims were not compatible; we will discuss extreme accuracy in this chapter and metallic cartridges which finally made possible really rapid fire in the next.

The British Minie rifles produced after 1851 were fairly accurate. Some new Enfield Cal. 577 rifles were very good compared to any other weapons then known. With continued firing, however, accuracy deteriorated and loading trouble mounted. The reasons for this were fairly obvious. First, the black-powder fouling tended to cake inside the barrel, particularly in the grooves. Second, the soft lead bullets had slivers of metal torn off them in the bore, probably by the hard fouling. Third, soft bullets were deformed in ramming down fouled bores.

Sir Joseph Whitworth

About this time in Britain, Joseph Whitworth had one of the most outstanding reputations ever achieved by an engineer. Whitworth is still remembered for his steel, and his screw threads; he was the first Britisher to measure precisely. Furthur, he seems to have had a flair for personal publicity. He received a commission in 1854 to make extremely detailed and lengthy experiments with small arms entirely at Government expense, although Whitworth refused personal compensation. He built a fully enclosed range, apparently on his own property at Birmingham, where shooting up to 400 yards could be done indoors. Whitworth invented a precise machine rest which greatly decreased the importance of the human element in experimental shooting. Initially, he tested Enfield rifles, as well as many others.

The individual Enfields were found to vary considerably, both in accuracy and in physical dimensions. Poorly made barrels shot poorly. Even in the best Enfields, a fair percentage of bullets did not actually take the rifling in the way that they were supposed to, particularly when the barrels were dirty. Severe leading destroyed accuracy. The short, fat Enfield Minie bullet was not well designed for accurate flight at long range.

Other contemporary rifles were found to have at least as many disadvantages. Apparently, however, Whitworth did not fire the Jacob's rifle. Perhaps none was available at the time. After disposing of all rifles already made. Whitworth produced a number of rifled barrels of different types and fired them in his machine rest. He was particularly fond of polygonal bores, that is, bores that were regular polygons in cross section. He seems to have experimented with polygonal bores of from five to ten sides, but finally chose the hexagon.

The hexagonal-bored rifle did not originate with Whitworth. Actually, a hexagonal bore had been proposed for use in British small arms by a celebrated shot, Sergeant Major Moore of

Semi-Military Whitworth Rifle. This is one of several models of military and semi-military Whitworth rifles of hexagonal bore (see close-up of muzzle) that fired both hexagonal and cylindrical bullets. *From the Weller Collection.*

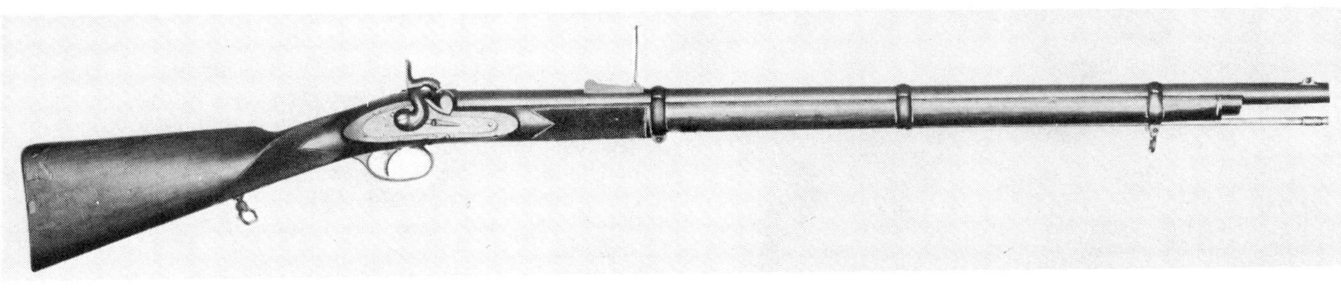

the Royal Artillery, in 1839. He is said to have made up a single rifle of this type himself. Whitworth, however, brought to the gunmaking industry both precision and a scientific approach. He carefully evaluated all his long and costly experiments. His findings improved all rifles, as well as methods of testing them.

Hexagonal-bore barrels were made up with widely varying twists, as fast as one turn in one inch, and as slow as one turn in sixty inches. The Whitworth rifles finally produced in quantity were Cal. 45 with a twist of one turn in twenty inches. The bore measured .451 inches across the flats and had a diameter across the rounded corners of .490 inches. These rifles were designed for hexagonal bullets swaged to exact shape under hydraulic pressure. The original hexagonal bullet weighed 530 grains, which was the same as the Enfield Cal. 577 Pritchett bullet then in use. Comparative shooting of the Whitworth and the Enfield was undertaken; the results were in favor of the Whitworth which had a "Mean Radial Deviation" from the exact center of the target fired at, of only 3.86 inches at 300 yards, and 23.13 inches at 1,000 yards. Enfields, even though clean and carefully loaded, averaged 12.69 inches and 95.01 inches at the same ranges. Modern shooting with these same weapons verifies approximately these figures.

It was found that cylindro-conoidal bullets slightly less than .450 inches in diameter, with hollow bases, shot well in Whitworth rifles. These weighed, according to surviving specimens, 480 grains, 500 grains, and rarely 530 grains. Whitworth himself seems not to have liked cylindrical bullets, even though some of the rifles he produced were furnished with moulds to cast these. He preferred the positive fit of a hexagonal bullet. He formed his bullets of a hard alloy of lead and tin, since they required no deforming to fit the bores. These hard bullets gave greater penetration and less leading. Tallow or grease wads were loaded between the powder and the bullet to lubricate the bores, as well as cut down on the amount of fouling.

The Whitworth rifles were found to foul badly, however, under certain circumstances. Even though the hexagonal bores had no sharp groove corners for fouling to accumulate, ramming of bullets became difficult. Whitworth designed a scraper which was mounted on the end of the ramrod. It was possible to scour out the bore after each shot and then lubricate it with a new wad. This worked well, but was slow. However, long range accuracy, and not rapidity of fire, was most desired.

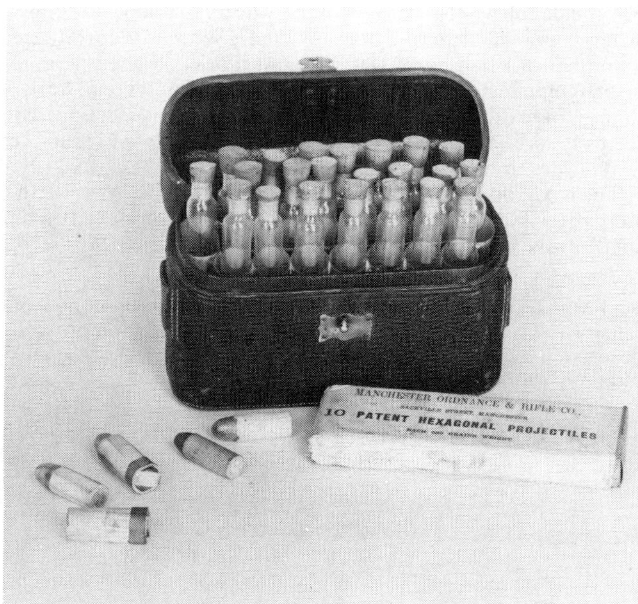

Whitworth Cartridge Box. The leather box contains glass vials, each of which held a measured charge of powder. Shown also are round and hexagonal bullets. *From the Weller Collection.*

Rifles of Berdan's Sharpshooters

Not all Confederate sharpshooters used the imported British Whitworth and other rifles, which were comparable in weight to the Enfield. A number of heavy muzzle-loading rifles were made in the Confederacy and used by Southern marksmen. However, these weapons were far more common in the hands of Union sharpshooters, particularly those under the command of General Hiram Berdan. These weapons sometimes weighed as much as 30 pounds, and fired long, relatively small-caliber lead bullets in front of heavy charges of fine-grain black powder. Apparently, each man had a slightly different technique for loading. The weapons certainly didn't use the same size bullets, or the same weight of powder. The bores required cleaning frequently.

The American Berdan-type muzzle-loading rifles were to continue in use long after the end of the Civil War. The best targets fired with these heavy rifles are not bad, even compared to modern bench rest shooting. Some of these old fellows can still shoot.

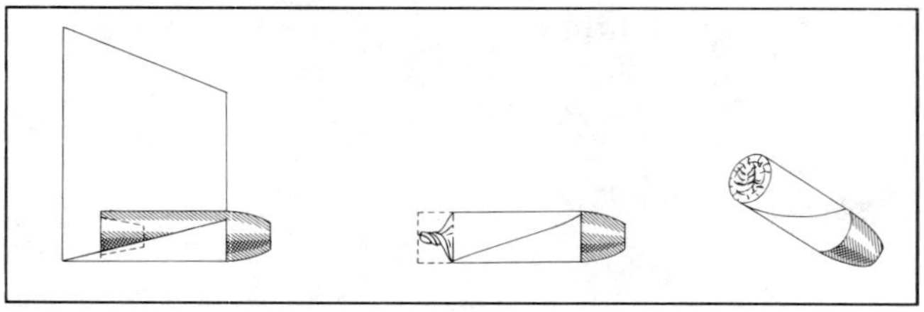

Paper-Patched Bullet. These sketches show a common method of paper-patching bullets. Smooth bullets from .003 – .006 undersize were wrapped with fine, strong paper similar to banknote paper. The patch was cut to a length that circled the bullet twice and almost, but not quite, permitted the ends to butt together. Paper, dampened before being applied, clung closely to the bullet when rolled on. The extra paper projecting at the base, was folded up into a pigtail and pressed into the base cavity.

The Paper-Patch

In both Whitworth and Berdan rifles, bullets patched with paper were superior to either those having smooth lead, or even lubricated cannelures. The rolled paper-patch seems to have been evolved in both countries at about the same time. Generally, precisely two turns of moistened paper were wrapped tightly around a bullet, usually leaving only a point exposed. A bit of extra paper was left over at the base to twist into a tight little spiral. If the bullet had a hollow base, this pigtail could be pushed into it without hurting the functioning of the hollow base in expanding under pressure.

Loading these paper-patch bullets down cleaned Whitworth bores presented no problems; the muzzles were hexagonally chamfered to receive them. However, in America the paper-patched bullets had to be a pressed fit into the rifling for best accuracy. This was done most easily by means of a false muzzle, which was just a short piece cut off the end of the barrel during manufacture that could be replaced in its exact original position by means of locating pins when the weapon was to be loaded. The outer end was counterbored to take the paper-patched bullet and hold it snugly. A short rammer with large palm-piece was then used to seat the bullet a few inches into the rifling; it could then be pushed down with the ramrod after the false muzzle had been removed.

The records established by Whitworth rifles at ranges of 900, 1,000 and 1,100 yards and even beyond, as well as the excellent shooting done in America at about the same time, depended to a large extent on paper-patched bullets, grease wads and bores cleaned after each shot. Sometimes, instead of having the paper wrapped about the circumference of the bullet in the American rifles, two paper ribbons were placed under the false muzzle at 90 degrees from each other, and the bullet pushed down the bore enveloped in the two strips.

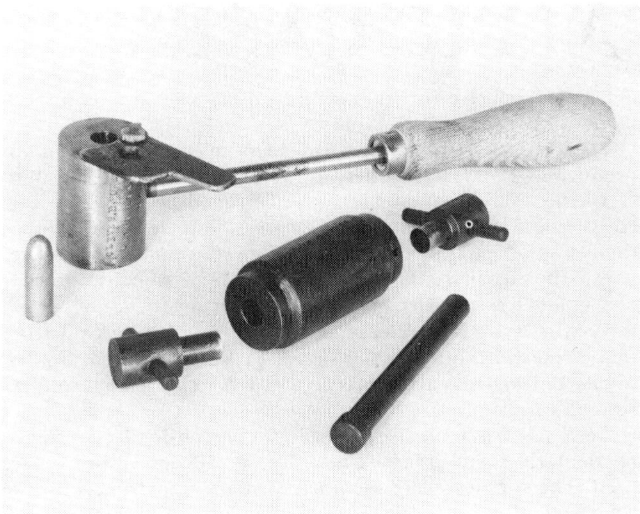

Modern Whitworth Mould and Swaging Equipment. Modern shooting with Whitworth rifles bears out their reputation for accuracy. Mould and swaging equipment illustrated here was made for Cleves Howell.

Muzzle-Loading Cartridge Rifles

In the late 1870's and 1880's, the muzzle-loading of these fine heavy rifles became needlessly laborious, compared to the new breech-loading rifles using metallic cartridges. Besides, a breech-loader was considerably easier to clean. The loading from the muzzle, however, with a carefully positioned paper-patched bullet through a false muzzle had many disadvantages. A compromise was reached giving the advantages of both. Breech-loading cartridge rifles were made with the same heavy barrels and false muzzles. The metallic cartridges used contained only the powder charge and a wad. The paper-patched bullets were still loaded from the muzzle. The same engraving process was used to seat the bullet in front of the powder charge, now in a removable metallic cartridge case. The distance between bullet and case mouth was precisely the same each time. These rifles could be cleaned from the breech after each shot, and could have the bores oily or dry, depending on the desires of the shooter. Only a dry bore can be used with a muzzle-loader.

A variation of this false-muzzle, breech-loading idea was sometimes used in which a paper-patched bullet was introduced from the breech and forced into the throat of the rifling to precisely the same distance each time with a special tool, called a Schutzen loader. It was necessary, of course, that the actions employed for this type of loading allow free access to the breech in order that the tool could be simple and powerful. The same type of cartridge with powder charge and wad only was used.

BULLETS FOR METALLIC CASES

The foremost weapons lesson of the Civil War was the need for a breech-loading rifle firing self-primed cartridges of full infantry power. Both the Spencer and Henry carbines, using rim-fire ammunition of medium power, had been fairly successful; only a little further improvement was required. Manufacturing and industrial techniques could now cope with the problems of making both breech-loaders and metallic cartridges practical. The Spencer carbine could be fired in an emergency as fast as 21 shots a minute; the Henry was even faster. The nation which first equipped its infantry with a rifle of full power and this capacity for rapidity of fire would have a great advantage over those which did not. Even the perfecting of a breech-loader and full-power metallic cartridges, without the repeating features of the Spencer and Henry, would be a great step forward.

The problems were really manufacturing rather than inventing. Both breech-loaders and metallic cartridges are very old indeed. A wheel-lock breech-loading rifle using metallic cartridges, made in the reign of Henry VIII, is now on exhibition at the Woolwich Arsenal Museum outside London. The relatively light iron chamber pieces fit entirely within the barrel; the action is almost exactly like that patented by Snider at the time of the American Civil War. A number of iron cartridges were undoubtedly carried by the soldier using this piece; he could load and fire reasonably fast for a time. However, the weapon undoubtedly leaked gas at the breech, was costly to produce, and gave difficulty in extracting and loading after the action fouled. These three disadvantages kept breech-loaders from being practical for more than 300 years. Some easily manufactured, reliable method had to be developed to keep the powder gases and products of combustion entirely within the cartridge and bore. A metallic cartridge which contained its own priming and expanded on discharge to give positive sealing of the bore, or obturation, was the obvious answer.

The Self-Primed Expanding Cartridge

Cartridges of this type are older than many people realize. A Swiss by the name of Pauli (Pauly) working in Paris, probably with assistance from a Frenchman by the name of Pottet, produced between 1812 and 1829 self-primed centerfire cartridges which seem astonishingly modern, but achieved little popularity. These cartridges were too difficult to make and too expensive for military purposes. The simple muzzle-loaders were superior for sport.

The Prussian army adopted instead an invention of an apprentice of Pauli and Pottet, the Dreyse breech-loading military rifle. This weapon fired a self-primed combustible cartridge containing an elongated groove-diameter bullet. Since there was no cartridge case, obturation depended on features within the breech itself. A long firing pin penetrated the powder charge and exploded a cap held in a papier-mache sabot behind the bullet; the popular name "needle gun" was derived from this. It met with only partial success; it could be fired rapidly, but leaked so badly at the breech with a full charge that soliders refused to aim.

The Civil War gave rise to dozens of breech-loading systems, particularly for carbines. Cartridges of all descriptions were used. The simplest were the linen, or paper, cartridges for the Sharps and Sharps-type weapons which required outside priming similar to muzzle-loaders. The Burnside cartridge was made of copper, but had no self-priming; it required a standard percussion cap to direct a jet of flame into the hole at the base of the cartridge. The Burnside bullet is particularly interesting since it appears to be the first bullet in which cannelures were habitually filled with lubricant. The Smith carbine required rubber cartridges and separate priming. There were others also, some even more unusual. All these weapons used some theoretically effective means of obturation; none was really satisfactory for as many as 50 consecutive rounds. The Sharps was the simplest and worked after a fashion, since the jet of flame from the breech was usually well in front of a careful marksman's face and right hand.

Rim-fire cartridges of medium and low power, on the other hand, were quite successful. Smith and Wesson revolvers were chambered for .22 and .32 rim-fire cartridges from 1858 on. The .44 Henry rim-fire cartridge was well known in the Western theater during the Civil War. Similar Spencer cartridges were used by the millions.

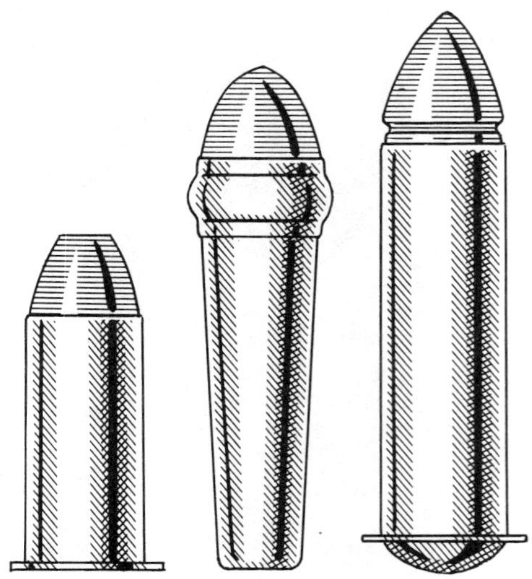

Early Metallic Unprimed Cases. These are typical of the early cases that did not have internal priming. Left: Cal. 50 Maynard, best known of the early Maynard cartridges, was ignited by the flash of a percussion cap. Middle: Cal. 54 Burnside, one of the earliest of American brass cases, was similarly fired. Right: Cal. 56 Billinghurst Requa was used in a volley gun with twenty-four barrels mounted in a row. One of the first of a quick-firing weapons, all barrels of this gun were fired by a single train of priming powder.

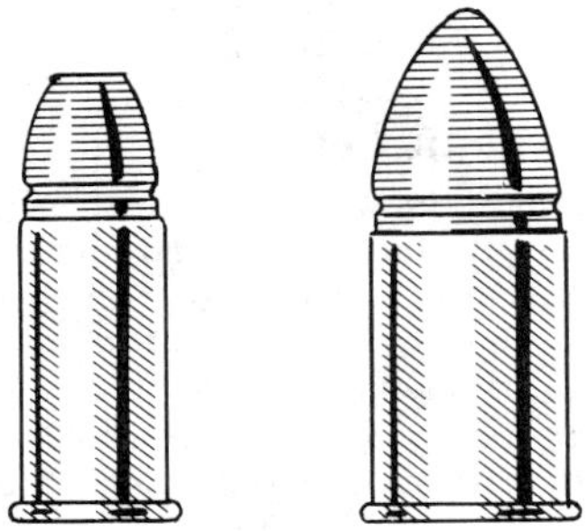

Early Rim-Fire Cartridges. These are noteworthy examples of early rim-fire cartridges. Left: Cal. 44 Henry, used in the famous rifles that led to the development of the Winchesters. Right: Cal. 56 Spencer carbine cartridge. Empty cartridge cases are still found where historic Indian fights and buffalo hunts took place.

The Infantry Breech-Loader

The stage was set. Within a decade, every major army adopted some form of breech-loading rifle and a self-primed cartridge to fit it. In general, some method of converting muzzle-loading weapons into cartridge breech-loaders was used. The United States converted the Cal. 58 rifle musket by the Allin system into single-shot breechloaders similar to the famous "trap door" Springfields. Britain used the Snider system to convert the Enfields.

In most of these early conversions, the muzzle-loading Minie bullet was merely placed in a new metallic cartridge; it still depended upon the powder gases to expand it into the rifling. Bullets for the British Cal. 577 Snider were of this type. They were more accurate when loaded into the .577 Snider (or Boxer) cartridge than they were when loaded from the muzzle. A part of the answer probably lay in the lubricant now placed in the cannelures of bullets for metallic cartridges. Bullets for our Cal. 58 Allin patent single-shot rifles were also of the Minie type, but had a very short life for the caliber was reduced to .50 inch and a new cartridge, the .50-70 Army, introduced.

Groove-Diameter Bullets

The bullet for the .50-70 Army, like almost all other for the dozens of different metallic center-fire cartridges which began to appear, was groove diameter. In fact, bullets were usually slightly larger than the groove diameter. Tolerances in the manufacture of bores and the depth of grooves were not always what they are today. Factory-loaded bullets had to be large enough to fill the grooves in the largest barrel in which they might be fired.

These new metallic cartridge bullets were, of course, elongated; sometimes they were quite long in proportion to their diameter, particularly for the powerful Sharps and similar rifles. They were generally hardened with tin and/or antimony to resist as much as possible a tendency to lead the bores. Bullets were now always lubricated by means of cannelures filled with various greasy substances, save for the paper-patched bullets already discussed. The paper-patched bullets were more accurate, but far less rugged; they were definitely not moistureproof.

Swaging of Bullets

Until about 1840, all bullets were cast. After that time, some military bullets began to be swaged. As we have seen, this process was common during the American Civil War. Although, the very accurate lead bullets of the 1880's and later, including all those for Harry Pope's rifles, were cast, most factory loaded bullets began to be swaged in much the same manner as they are today.

This swaging is an awe-inspiring sight. Lead of the proper alloy is extruded cold in the form of wire. The machinery is unbelievably powerful and expensive, but the production is great. The wire is fed into a machine which cuts it into proper lengths to be "headed up" into bullets in the next operation. The cannelures are pressed into the bullet sides and filled with lubricant when the bullet is sized. All operations, including final loading, are semi-automatic; very little hand labor is required.

The Effect of Smokeless Powder on Bullet Design

During the 1880's, several different types of smokeless powder propellents began to be used to a considerable extent in small arms. These propellents did away with the fouling problems inherent with black powder. Higher velocities could be obtained without exceeding permissible pressures. Higher velocity with the same weight bullet meant unnecessary power and punishing recoil. In the last days of black powder, most armies had reduced calibers to about 11mm or .45. Now a further reduction was possible. Small bore military rifles swept the world. Some nations used bullets as small as 6mm (Cal. 243); none used larger than 8mm (Cal. 323). Because of their higher velocities, these new lighter, smaller bullets were as powerful as their predecessors. They had many advantages.

However, the new smokeless powders and the higher velocities meant very considerably higher temperatures. Under certain circumstances, even tin-alloyed hard lead bullets could not withstand the heat from the powder gases and the friction against the barrel. After more than 500 years, the lead bullet began to have limitations.

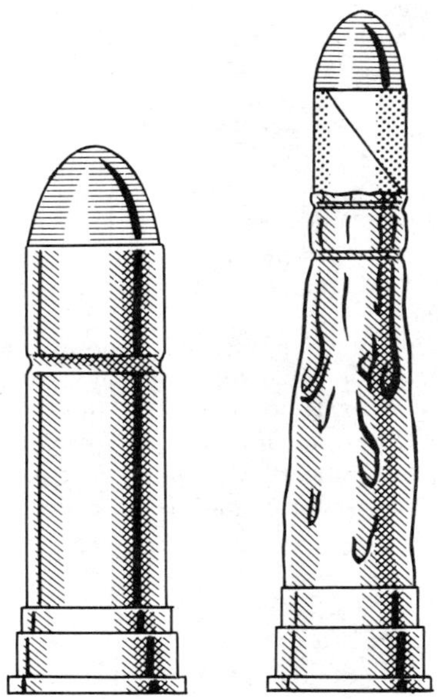

Early Center-Fire Cartridges. Especially successful examples of early center-fire cartridges. Left: Cal. 577 Snider-Boxer was made considerably smaller than the breech diameter for easy loading, expanded upon ignition to seal the breech and prevent the escape of gas. Right: Cal. 450 Martini-Henry cartridge with thin coiled-brass case was easily damaged if roughly handled.

Hard Bullets

The first answer to the problems of smokeless powder, high velocity and small bores was to encase the lead in a copper jacket. Captain Rubin of the Swiss Army is given gredit for this idea. Quite soon thereafter the French appear to have used

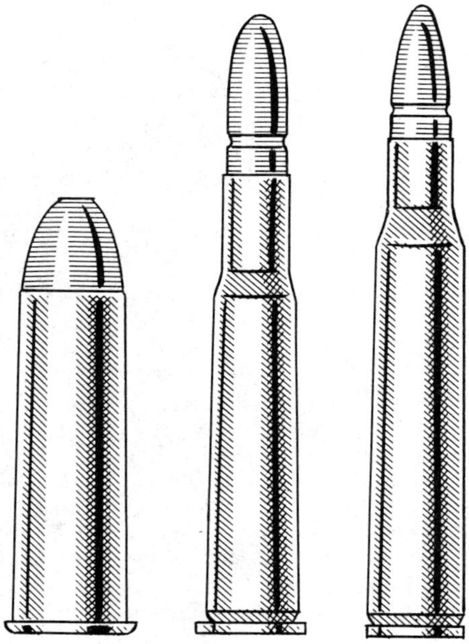

Evolution of the Military Center-Fire Cartridge. Left: Cal. 50-70, issued between 1869 and 1873. Middle: Cal. 30-40 Krag, used in the U.S. Krag-Jorgensen rifle during the Spanish-American War. Right: The famous Cal. 30-'06 is the best known military cartridge and veteran of two World Wars and the Korean War.

homogeneous solid copper or solid bronze, bullets in their new 8mm Lebel cartridge. Other bullets were developed with hard metal envelopes of bronze, cupronickel, gilding metal and steel, as well as some other compositions. The term "steel-jacketed" has been applied to bullets having jackets of other metals besides steel. However, actual steel jackets were used to a considerable extent around 1900, both bare and with some form of soft metal coating. They were again made and used in both World War I and World War II because of the wartime copper shortage. Some very fine match ammunition is presently loaded with mild steel bullet jackets. Tests indicate that these do not wear bores more than other jacket materials; the coatings, when present, are mainly to prevent rust and maintain appearance.

Occasionally, bullets have been made of other substances. Kirksite bullets were used for target practice by U.S. forces during World War II where penetration was not desired. Remington makes some of their metal-piecing bullets for pistols of almost pure zinc. Plastic bullets have been made for target shooting. Sintered iron and plastic bullets which fly apart easily are used in shooting galleries to make ricochets impossible. The highest velocity ever achieved by a bullet, about 15,000 feet per second, involved a small nylon ball, a lot of pressure and an evacuated bore.

Cores Other Than Lead

Lead cores are best for long range, since they give the flattest trajectory for any given bullet size and muzzle velocity. However, in wartime lead is scarce in the United States. Our lead production is enormous, but there are so many other uses for the metal that our rifle bullets in World War II were generally cored with steel. These steel cores can be produced amazingly cheap by modern methods. If extra armor-piercing qualities are needed, the cores are made of high-carbon steel and hardened.

For extreme penetration, the cores of modern bullets re sometimes made of stellite, or tungsten carbide. The relatively soft metal jackets are completely smashed and evaporate upon contact with an armored target; however, the hard core will penetrate even better than heat-treated steel, particularly at medium velocities. At ultra-high velocities, 3,500 feet per second and above at the target, core material seems to make little difference. Penetration is great even with lead cores.

Special Bullets

A great deal of research has been done with various types of points to make bullets more effective on game, particularly middle-sized, thin-skinned game. A full metal-jacketed bullet will pass entirely through such an animal and out the other side without being immediately fatal. Hollow points, copper-tube points, exposed-lead points and other patented points in endless variety have been tried.

Hollow-pointed bullets were found to be very accurate because their center of mass was well to the rear. Britain placed a small cone of aluminum inside the cupronickel jacket and in front of the lead core of their Mark VII service bullets in use until after 1929.

The variety of bullets, both of cast and swaged lead and with metal jackets, which have been used in metallic cartridges is enormous.

The Gas Check

The gas check and its application to hand loading has been dealt with at length. It's a compromise between the jacketed bullet, which can withstand the heat of high velocity firing, and the cast lead bullet which is so satisfactory for many purposes. A lead bullet has a copper cup pressed on its base. Since this cup goes only a short way up the sides of the bullet it gives the advantages of lubricating cannelures and reduces barrel wear, but permits considerably higher velocity than can be achieved with bare lead alone. The gas check has done its job when the bullet is out of the barrel; if it comes off then, no harm is done.

BULLETS IN SHOTGUNS

Shotguns have been used more by Americans than all other sporting firearms combined. Millions of shotshells are fired annually at both animate game and clay birds. Relatively tiny round lead pellets are by far the most numerous bullets made and used.

What is a shotgun? Today, the answer is obvious; anyone can tell a modern shotgun from a rifle at a glance. Historically speaking, however, the answer is not so easy. In the muzzle-loading era, any smoothbore gun could be used to fire small shot reasonably effectively. The difference between a musket and a fowling piece was not clear; even experts sometimes don't agree about a relic which survives. An early 18th Century fowling piece usually had a longer barrel and was lighter in weight than a musket. In most instances, however, the same weapon was used in militia drills and fighting as was used for killing duck, geese and other fowl. The New England "long fowlers" were used with single balls at Bunker Hill to do great damage to the British.

This modern shotgun appearance, that is, thin barrels of relatively large bore, was seldom encountered before 1800. Double-barreled fowling pieces were not common until the first quarter of the 19th Century; fairly suddenly, they became almost the only accepted sporting arm. Manton, Purdey and their London contemporaries started a trend in shotguns which reached individual farmers in America. In a surprisingly short time, these relatively light side-by-side double-barreled guns became standardized in appearance, and haven't changed much since. Double-barreled flint and percussion fowling pieces inscribed "Joseph Manton, London," but made cheaply in Liege, Belgium, were widely sold by crossroads stores throughout America.

There are men who claim that the flint shotguns really made by Manton are the most beautiful and effective scatterguns ever devised. By modern standards, however, these fine old weapons do not shoot very well, even disregarding the flint-ignition handicaps. When carefully loaded, they can give concentrations of shot at reasonably close range that are hard to beat, even with some modern weapons. However, it's necessary to use weak charges of powder which limits effectiveness to about 20 yards.

Spherical Bullets

During the musket era, a single bullet was, of course, merely a bore-size musket ball. With the coming of the true shotgun, a single bore-size ball was still frequently employed. It was most effective for certain purposes, both in muzzle-loaders and breech-loaders, until the advent of choked bores. These single spherical bullet loads, called in this country "Pumpkin Balls," lost whatever popularity they had over here when they had to be made almost as much undersized as Brown Bess bullets for fear someone would shoot them in a full-choke gun. American game doesn't really require them. In the Eastern United States, a charge of buckshot was usually considered superior for deer hunting. The larger Western game, such as buffalo, bear, elk and the like, weren't usually hunted with shotguns.

In British India and Africa, however, smoothbore guns were very often loaded with a single bore-sized ball, both in the muzzle-loading and early breech-loading era. Since these balls were a tight fit in the bore, they gave fair accuracy up to 100 yards. The real reason for their popularity, however, was the tremendous shock effect at close range. Fired at relatively high velocities, they would stop dangerous game dead in its tracks.

These spherical balls from smoothbores, with full charges of powder, probably gave muzzle velocities near 1,800 feet per second. The .577 Snider military rifle didn't produce much more than half that; even fine sporting rifles with heavy charges couldn't equal the smoothbore in velocity. The reason was that both rifles and smoothbores used black powder. The

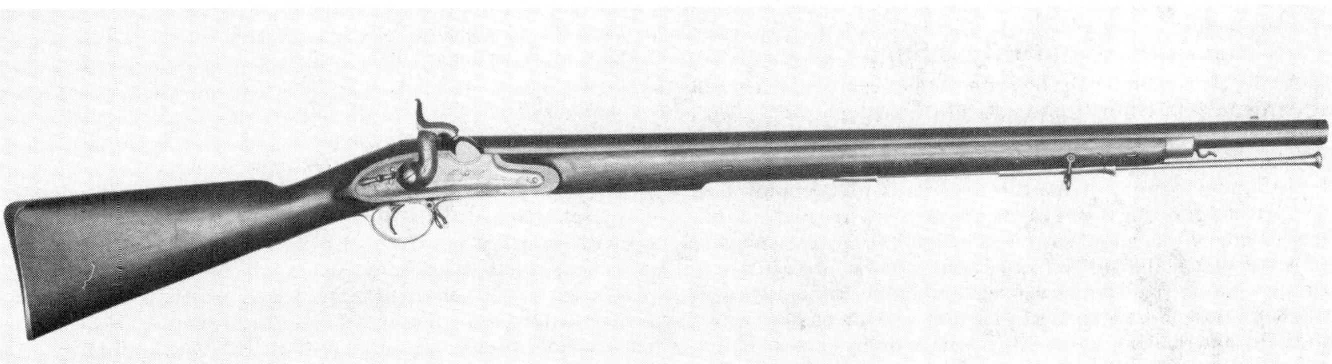

British Tower Musketoon. For an all-round muzzle-loading gun, using the simplest ammunition, this strong, rugged, and extremely well-made weapon is hard to beat. This short musket was produced at the Enfield Arsenal and is one of the first weapons made there. Its quality is considerably above that usually found in military issue weapons. This particular weapon performs creditably with a 12-gauge unrifled slug cast in a Lyman mould and is accurate beyond 100 yards. It also gives a good shot pattern with 1 1/8 ounces of No. 6 shot at 35 yards. *From the Weller Collection.*

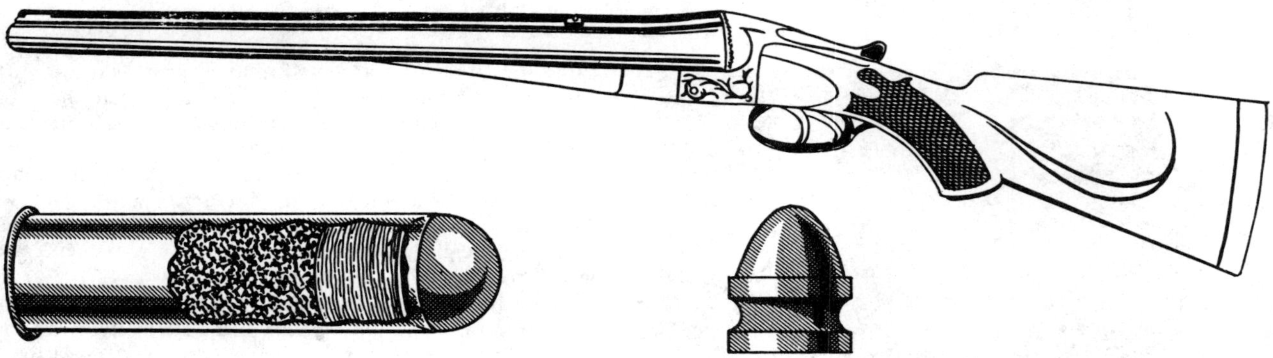

19th Century Ball Guns and Rifles. Illustrated are an 8-bore double Elephant Rifle, cross-section of an 8-bore cartridge with spherical bullet, and an 8-bore conical bullet. These short, powerful guns were popular in India for shooting dangerous game. The 8-bores were usually rifled and fired either spherical or short conical bullets. Sometimes the so-called paradox type of boring was used--smooth for most of the barrel, but with a short section at the muzzle choked and rifled. The 4-bores were generally smooth and fired a 4-ounce spherical ball at velocities up to 1,800 fps with charges of as much as 14 drams of powder. All these guns with full charges of powder were deadly at both ends.

spherical balls were relatively light compared to even short cylindrical bullets for rifles. The shotguns could withstand the pressures of the largest useful charges of black powder, and produced higher velocity. These smoothbores were not light, seven-pound twelve bores, but were likely to weigh ten pounds or more, and were usually ten bore, or larger. An Englishman of moderate means could shoot either ducks or tigers with the same gun without being at a too great disadvantage compared to a brother officer with a whole battery of sporting weapons.

An unusual development along with this line was the so-called "paradox" bore. A heavy smoothbore shotgun, usually a side-by-side double, would have perhaps three inches of the muzzle end of the barrel reduced slightly in size, and rifled. This rifling was neither very deep, nor of a rapid twist. It was enough, however, to seize upon a round ball and cause it to settle down and spin gently about the axis of the bore. This ball was fairly accurate to perhaps 150 yards, and appears to have been just as destructive to tissue as similar bullets from smoothbores, since it had almost as high a velocity. Sometimes short cylindrical bullets were employed.

These heavy paradox guns could be relied upon for any form of close range shooting at large and dangerous game. They were equal to double rifles in accuracy and power, and probably patterned with shot as well as contemporary shotguns. The reduced section of the barrel seems to have acted more or less like a choke. However, the weapons were so heavy that they were not really shotguns at all. Who could swing such a muzzle-heavy weapon at flying game? I suspect that their effective use with small shot was largely confined to sitting targets.

Cup-Like Bullets and Rifled Slugs

A bullet cast in the shape of a cup, with its center of mass well towards the closed or front end will fly through the air with closed end foremost. Such bullets from smoothbores are said to be more accurate than round balls. This cupped-bullet idea has been know for more than a century and is now employed in special shotgun slugs. If fairly heavy grooves are pressed into the outside of this bullet, air resistance will cause it to spin and introduce the stabilizing effect of rifling. These projectiles are supposed to have advantages over single spherical bullets.

They can be bore-size to fit the barrel during passage down the bore, and then, because of their almost hollow construction, be swaged down in passing through the choke, if any. Spherical balls must be no larger than the smallest diameter of any choke in which they might conceivably be fired. Some factory-loaded rifle slugs for 12-gauge shotguns are .725 inch in diameter at the rear skirt. Some full-choke bores measure as little as .685 inch in diameter. This swaging in the choke of as much as .040 inch is certainly drastic, particularly for a really fine gun, yet no warning about their use in light full-choke barrels is given on the box.

Some rifle slugs, particularly those made by Brenneke in Germany before World War II, had hardened points of other metals. It would appear that such refinement is unnecessary. Only an idiot would go intentionally after rhinoceros and other thick-skinned game with a modern 12-gauge shotgun.

Shot

The primary purpose of a shotgun is, of course, to fire charges of bird shot. Intermediate-sized spherical bullets are also sometimes employed. Buckshot has always been popular in America, as already described at length. However, the basic use for shotguns is to fire at birds. Relatively small shot is needed to kill effectively even the largest geese.

Few things in firearms have varied so much as classifications of shot. As many as four different tables of shot sizes were in use in Britain at one time; three were used in America until a few years ago. The present American standard of shot sizes is quite simple for No. 1 shot and smaller. Shot size subtracted from 17 will give diameter in hundredths of an inch. For instance, No. 2 shot: 17 less 2 equals .150 inch, which is the average diameter of these pellets. Similarly, for No. 7½ shot: 17 less 7½ equals .095 inch, which is again their diameter.

At one time, swan shot was quite popular, and ranged in size between buckshot and bird shot. Most swan shot was probably around .200 inch in diameter. The only shot in common use today larger than No. 2's is BB which is .175 inch in diameter. Probably 95 per cent of the shotgun shells sold today are loaded with shot, between No. 4's and No. 9's inclusive.

The Manufacture of Shot

As already discussed, small shot was made at one time by cutting a sheet of lead into cubes and then trying to round off the corners, either by rubbing with a board, or tumbling in a barrel. However, in 1782, an Englishman by the name of William Watts dreamed of dropping molten lead through a sieve, down from a church steeple into cold water at the bottom. The principle was that the individual pellets of molten metal would have sufficient time to assume a spherical shape because of surface tension during their fall. They would solidify when they entered the water. Almost overnight, this relatively simple process revolutionized shotgun shooting. For the first time, really uniform shot could be made. Watts is reported to have sold his patent rights for thousands of pounds of sterling.

Shot larger than about .200 inch in diameter continued to be cast; however, all smaller sizes were dropped down the insides of special shot towers, sometimes as much as 200 feet tall. These were built in fair numbers in both the United States and Europe in the 19th Century. Lead for this purpose was alloyed with arsenic to make it flow through sieves properly. At first, the lead was not otherwise alloyed; this shot was quite soft.

The Balling of Soft Shot

The great disadvantage of soft shot was that the pellets, which were forced together in the barrel, would sometimes adhere to each other in flight; this phenomenon is calling balling. In one instance, in Britain over a century ago, almost the entire shot charge from a ten-bore stuck together and broke a clergyman's jaw at a range of more than 300 yards. Even small balls greatly interfered with evenness of pattern.

If tin or antimony was added in sufficient quantities to the molten lead, the shot would be quite hard. Hard shot was not so susceptible to balling. The term "chilled shot" means hard or alloyed shot; sudden cooling by cold air in the shot towers has no effect on final hardness of lead alloys.

The shot tower made available shot of any size whatever by varying the size sieve and the distance through which the molten lead fell. Shot, once produced, was easily checked for roundness by rolling it down an incline. Spherical shot would gain sufficient velocity to jump a gap at the bottom; those not truly round would fall into the gap and be remelted. Separation into various sizes was done on screens. Uniform, chemically impervious surfaces were produced by tumbling with plumbago, or graphite.

Modern Shot Making

Certain advances in techniques have been made over the years. Sometime shot, to increase the hardness and decrease even further the tendency to ball, was plated with copper. The basic shot tower continued in use for more than a century, even though it had several disadvantages. Its initial cost was high; it was relatively inefficient, if the sizes produced were changed frequently. It was found possible to dispense with the shot tower entirely, and use instead an updraft of air. By varying the velocity of the air, individual molten lead globules falling a distance of perhaps ten feet could be made to think that they fell from 40 feet to 200 feet, depending upon the sizes of shot desired and the sieves used. After World War II, some manufacturers abandoned the dropping of shot entirely and swaged even the smallest sizes. Automatic machines appear to have been in operation to do this in Italy at least as early as 1948.

In Quest of Long Range

Throughout the time that sportsmen have fired at game with charges of small shot, they have endeavored to increase the range and killing power of their weapons. The simplest expedient is to increase the size of the shot and the weight of the charge. For instance, two ounces of No. 2 shot contains approximately the same number of pellets as one ounce of No. 5 shot. A No. 2 shot will maintain a killing potential at least 15 yards further than a No. 5 shot. With a gun appropriately choked, it would be possible to fire as tight a pattern at 70 yards with a ten-gauge Magnum firing two ounces of No. 2 shot as at 55 yards with a standard 16-gauge firing one ounce of No. 5 shot.

The increasing of the size of the shotgun, and the size of the individual pellets of shot, can be used to increase range only to a limited extent. There are several disadvantages inherent in this. Few men shoot enough today to swing a 10½ pound gun. However, in the past certain arrangements have been tried for much greater increases in effective range, even to more than 100 yards. During the muzzle-loading era in Britain, a kind of wire cup made by Eley retained the shot all in a cluster inside the cup for perhaps 50 or 60 yards, and then let them fly free.

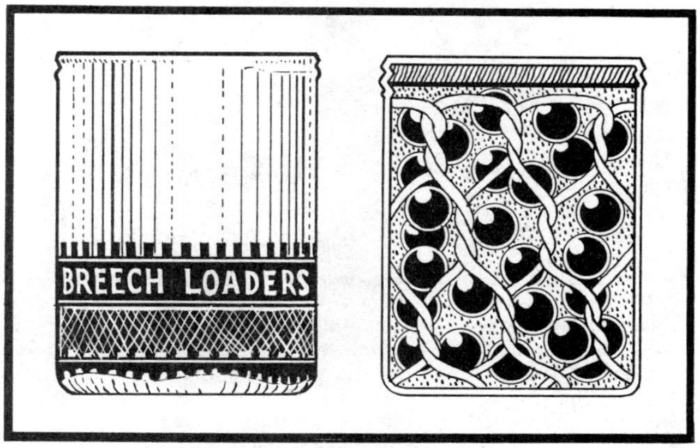

The Eley Basket for Holding Shot Together. This device was designed to hold the shot together in a wire basket for the first 50-60 yards of its flight, then release it. Its purpose was to produce killing patterns at extremely long ranges.

A bit later, a patented shell was used to contain the entire shot charge. The two halves of the shell did not come apart and release the shot inside until a predetermined range had been reached. This range could be varied when the shell was loaded; in theory at least, a sportsman could change the effective range of his weapon at will. The advertisers of this scheme claimed that the entire shot charge could be put in a 30″ circle at 130 yards. It would seem highly unlikely that this device worked in the way that its proponents said that it did, even though William Greener spoke well of the idea. Besides, who can hit any form of shotgun target at 130 yards?

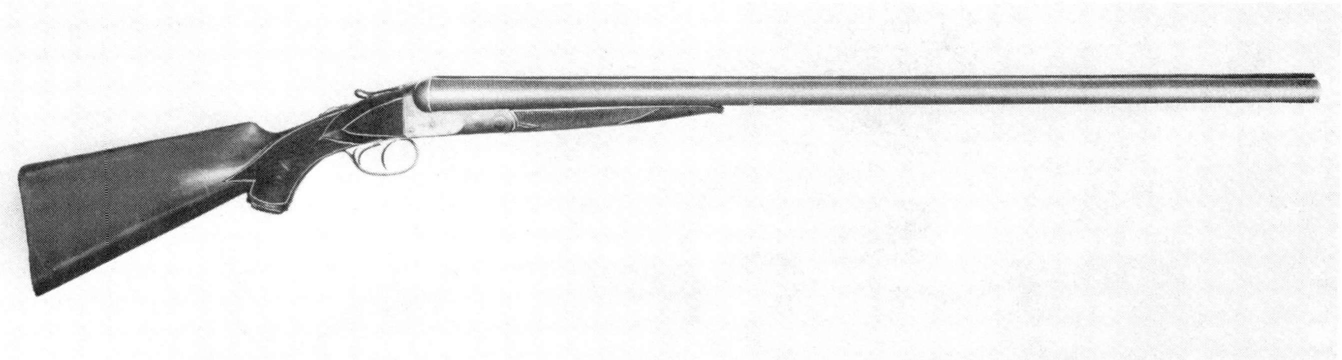

Grover Cleveland's 8-Gauge Shotgun. This superbly finished Colt shotgun was presented to Grover Cleveland on his second inauguaration by the Colt company. Measurements of the gun indicate that it was made for a man with short arms who fired from an erect position. The gun weights 11 3/4 pounds and has 34-inch modified-choke Damascus barrels. *From the Weller Collection.*

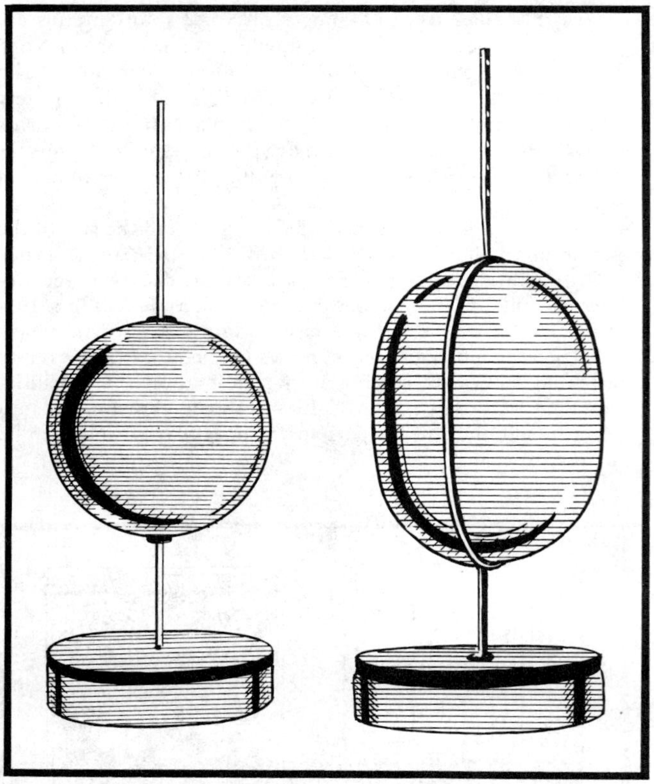

Shrapnel Spherical and Elongated Shot "Shells." Two shell segments filled with shot were held together by a light wire spindle. The entire assembly was loaded into the shotshell with wad end down. When fired, the device traveled intact for a distance and then the wire spindle separated from the shell segments, releasing the shot. An effective range of 95-140 yards was claimed.

Punt Guns and the Market Hunters

In the era of unlimited slaughter of waterfowl before 1900, shotguns the size of small artillery pieces and mounted on boats were used; these increased the effective range and the number of birds killed per discharge. Some of these guns fired more than a pound of shot at a time. It could, of course, be of any desired size. Apparently, the optimum range was generally around 100 yards with shot roughly equivalent to modern BB. Dozens of waterfowl were sometimes killed by a single shot. Many of these weapons, although used in the breech-loading era, appear to have been muzzle-loaders.

These same market gunners frequently fired from the shoulder four-gauge and six-gauge shotguns. These weapons were particularly effective against birds sitting, or for killing cripples. At one time, eight-gauge double-barreled shotguns were extremely popular with gentlemen wild-fowlers and fired usually no more shot than is currently thrown by a ten-gauge Magnum, which is now the largest size that may legally be used against game anywhere in the United States.

The Modern Shotgun

The modern shotgun has reached an advanced state of development. Ammunition for it is extremely reliable. Its efficient use, however, requires agility, coordination and practice. Since the usual targets are objects flying in the air, either real birds or clay disks, most targets must be led. Unlike the rifle, a shotgun cannot be carefully aimed; it must be pointed and swung. A sportsman has to practice these things until the whole process becomes instinctive. Most skillful shotgun shooters seem to feel that uniformity of weight, stocking, charge, and the like are more important than the relatively small advantages obtained by increasing gauge and, consequently, the weight of the gun.

Because of the drastically reduced bag limits, few hunters today can hope to learn to use a heavy ten-bore Magnum efficiently on ducks and still fire well with a seven pound twelve-bore at upland game. The trend is towards smaller bores and, if possible, shorter ranges, using reasonable amounts of medium-sized shot. Individual skill is far more important than extra gun performance. A shotgun usually has a best load for each size of shot. The man who finds out what this is and sticks to it with plenty of practice will be hard to beat.

This history of cast bullets is reprinted from our CAST BULLET HANDBOOK, 1st edition, published in 1958.

THE EARLY YEARS
John Barlow and Ideal: 1884-1925

When Phineas Talcott sold the Ideal Reloading Tool Company to the Lymans in 1925, it had already established its own colorful history over more than 50 years of shooting development.

The Ideal company was founded in 1884 by John H. Barlow, a pioneer in the reloading tool business who may have practiced his art as early as 1875, according to some sources. Early Ideal Handbooks illustrate Barlow's reloading tools with a patent date of March 11, 1884 lettered on the cut. Barlow also refered to 1884 as the year his reloading tool business began in a farewell statement upon his retirement, published in Ideal Handbook No. 22.

In bidding goodbye to his friends in the trade Barlow wrote in part:

"Monday, May 16th, 1910, The Marlin Firearms Co. take (sic) over all the machinery, tools, stock, fixtures and good will of the Ideal Mfg. Co..... I have been manufacturing these tools that have been furnished to shooters through the various channels for the past twenty-six years, and I have naturally made some friends among the jobbers, small dealers and individual shooters."

Barlow, an experienced machinist and toolmaker, acquired his love of the shooting sports during service with the Army. He was a sergeant with Company D, 14th U.S. Infantry and apparently was discharged in 1865. He had fond memories of his service days and dedicated several early Ideal Handbooks to his former Captain, George K. Brady, then a Lieutenant Colonel.

This cartoon appeared in Ideal Handbook No. 4.

In Barlow's day, black powder was the propellant readily available and most rifle makers furnished reloading tools with their rifles for loading centerfire cartridges. It was nearly impossible to load too much black powder in a case and most arms were designed to accommodate maximum charges—when a case was filled to the top and compressed by the seated bullet.

But when smokeless powder came along, everything changed. A shooter could very easily load too much smokeless powder in a case—enough to blow up a rifle and himself along with it if he wasn't careful.

About this time, many of the big gun companies stopped producing reloading tools. What was left of the frontier was fast becoming civilized and shooters began to rely increasingly on factory ammunition since they rarely needed their firearms to provide food or protection on a daily basis.

But Barlow pushed on. He experimented with loads, noted performance, and described the pitfalls along the way. He worked closely with ammunition and gun makers and soon his line of reloading tools began to grow. Later he took his observations and published them in a paperbound book—the first Ideal Handbook—much to the delight of the shooting fraternity which had little to guide it in the pursuit of reloading cartridges. Today's Lyman Handbooks are direct descendants of Barlow's early manuals.

The job Barlow chose for himself was no easy matter. To keep shooters in one piece with the bewildering array of new powders on the market was a formidable challenge.

Barlow issued an appropriately stern warning in Ideal Handbook No. 11 saying in part:

"It is a recognized fact...that nitro powders have come to stay, and at the present time there is a great variety of them and that the number is increasing every day, continually adding to the confusion...Some of them cannot be used without great caution and absolute knowledge of how to handle them. Those who have been accustomed to black powder...must not think for a moment that they can purchase any of the nitro or smokeless powders and proceed to use them the same as they may have been accustomed to use the Hazard or other standard black powders.

"The charge of black powder may be increased to expel a heavier bullet from an ordinary barrel; to do so with some of the nitro powders might cause the whole charge to burst sideways, shattering the arm and preparing the shooter for burial. Over the grave of such a person it would be well to erect a monument with the inscription that 'He died from monkeying with a nitro powder about which he knew nothing.'"

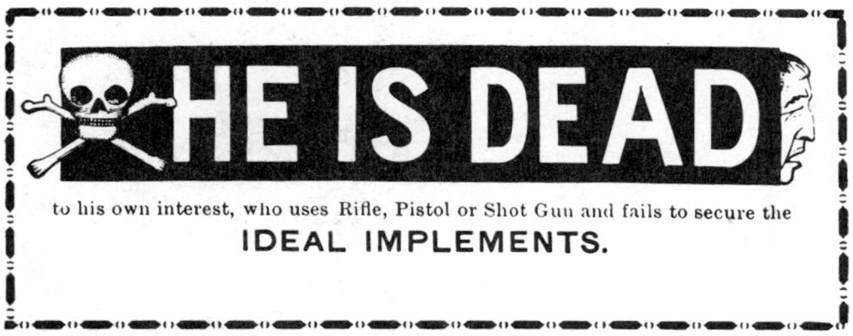

Cartoons and quips frequently illustrated Barlow's Ideal Handbooks and these are two fine examples from Ideal Handbook No. 4.

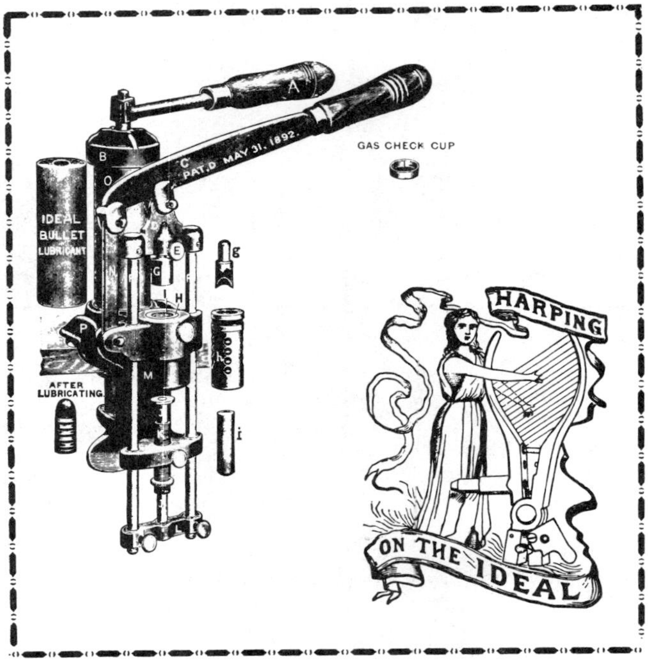

This illustration of the No. 1 Ideal Sizer-Lubricator and the gas check is from Ideal Handbook 17. The cartoon "Harping on the Ideal" accompanied comical quotes about "Ideal" and was found in Handbook No. 4.

In addition to his authoritative reloading manuals, Barlow also made another contribution to the shooting world—he introduced the gas check, a shallow copper cup that fits over the base of a lead-alloy bullet to protect it from the heat of smokeless powder. The gas check was first illustrated in Barlow's No. 17 Ideal Handbook, published in 1906. That same year he introduced the No. 2 lubricator-sizer, "designed for continuous hard work." This was a heavy-duty model of the No. 1 lubricator sizer, patented in 1892, and destined to be the forerunner of today's No. 450. A shell chamfering reamer and the Ideal Bullet Seater for Scheutzen rifles were added the same year.

Old editions of the Barlow-period handbooks are certainly valuable collectors' items and a reprint of his Fourth, originally published in 1890 in Lyman's 39th Ideal Handbook, shows the man as a colorful and informed shooter. His reloading tips are generously interspersed with cartoons, quips and comments from satisfied customers.

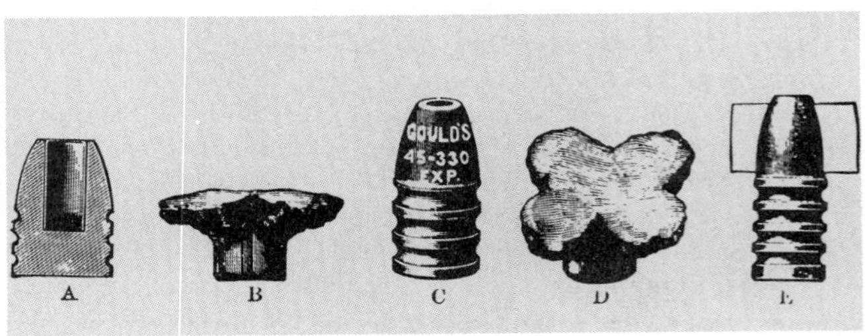

This illustration from Ideal Handbook No. 4 accompanied Barlow's description on how to make Express bullets more effective using hollow points or splits to enhance expansion.

By 1890, Barlow's line of reloading tools included eight tong toolmold combinations; three types of molds; plus a variety of accessories for the shotshell and metallic cartridge reloader.

For pistol clubs and military organizations, Barlow offered the Ideal Cartridge Loading Press and Ideal Armory mold, each with the capacity to produce ammunition on a volume scale.

Within 10 years the reloading products line also included universal powder measures (No.'s 1 and 2), a bullet lubricator-sizer, and the improved Ideal Loading Machine suited to load rifle, pistol and shotgun shells.

In addition to describing Barlow's reloading tools in detail, the Handbook touches on subjects such as how to roll paper-patched bullets, the use of hollow-pointed express bullets, and the practice of making some express bullets explosive by inserting a .22 caliber blank in the bullet's nose. He discusses the proper way to reload "Everlasting" centerfire shells and methods to break in a bullet mold. His "Hot Doughnuts" (do-nots) are as timely today as when he first penned them.

HOT DOUGHTNUTS.

Dough-nut give up a match until the last shot is fired.
Dough-nut stand within range of an inexperienced person when he is handling a loaded weapon.
Dough-nut pull a gun over a fence, out of a boat or off a wagon, with the muzzle pointed toward you.
Dough-nut borrow a rifle, rod or shot-gun to send them home uncleaned and rusty.
Dough-nut you know that practice makes perfect, and that it is within the province of every man to become at least a fair shot?
Dough-nut become disgusted if a poor score is made.
Dough-nut condemn your loading tools, when the fault is in yourself.
Dough-nut leave loaded firearms within reach of children.
Dough-nut handle new firearms without first receiving proper instructions.
Dough-nut overload a gun of any sort to spoil the weapon and injure the shoulder.
Dough-nut blame a gun or ammunition when poor shots are made.
Dough-nut change the sights of a rifle for every bad shot.
Dough-nut buy firearms because they are cheap, for they are expensive in the end.
Dough-nut put a gun away uncleaned. The cleaner the gun the cleaner the score.
Dough-nut imagine the gun is not loaded.
Dough-nut look down or blow into the muzzle of any weapon.

But Barlow's active years were on the decline and, faced with old age, he opted to sell the company and retire. The Marlin Firearms Company was a logical successor to control the Ideal line since that firm had worked closely with Barlow through the years. So, on May 16, 1910, Ideal changed ownership and came under the control of Marlin, which continued its association with the Ideal Handbook until Lyman bought the company in Oct. 1925.

Marlin eventually sold Ideal to Phineas Talcott during World War I and the Lymans bought the company when it was not doing very well. It is said the Lymans bought Ideal at the request of the export manager of Winchester Repeating Arms. His Australian customers were demanding reloading tools with the Winchester rifles they bought, but production at Ideal was not meeting the demand. So the Lymans agreed to buy the firm and increase its production rate.

Along with the tool line came responsibility for the Ideal Handbook. No. 27 was the first edition to carry the Lyman name, but it was not very much different from the previous edition. Beginning with No. 28, however, there was a new editorial emphasis. Col. Townsend Whelen was enlisted to edit this edition with assistance from Major Julian S. Hatcher and Captain Charles Askins. The foreword to that book read: *"An up to date book on reloading of ammunition is essential to the correct, safe and efficient use of reloading tools. This Ideal Handbook supplies such information. The old Ideal Handbook was more or less out of date, and moreover it was based primarily on the use of black powder, which can not be regarded as more or less obsolete."*

Bullet Casting Past to Present

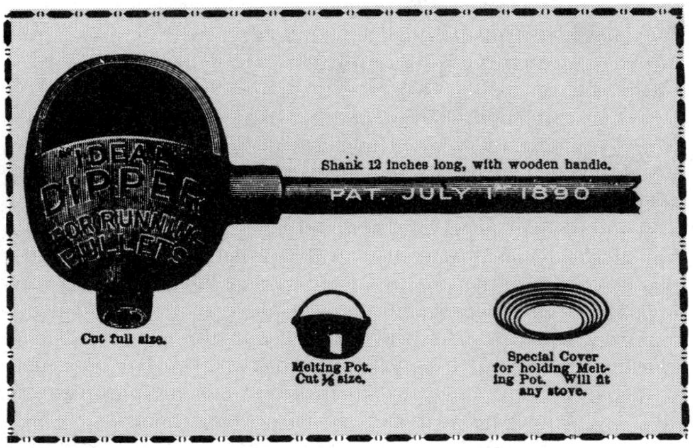

Ideal Dipper. The Ideal Dipper, seen here from Ideal Handbook No. 4, carries the patent date July 1, 1890 on its handle. Today's dipper is a virtual stand-in for the original. Both proved indispensable in casting good lead alloy bullets. It cost $.50 at this time.

Ideal Melting Pot. The Ideal melting pot was intended for use with a wood burning stove and held 10 pounds of bullet metal. It cost $.50 when Ideal Handbook No. 4 was published.

Ideal Melting Pot Holder. The holder was expressly designed to accommodate the melting pot, fitting the kitchen wood burning stove and allowing the pot to reach deeper than without the holder. The cover cost $.50 according to the above illustration (right) from the Ideal Handbook No. 4.

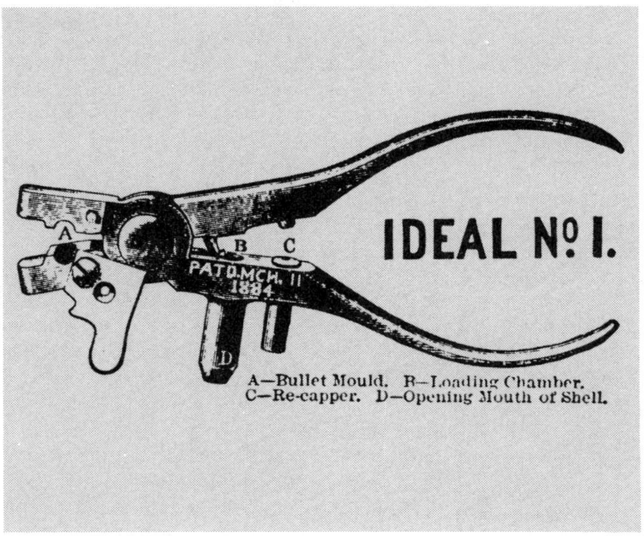

Ideal No. 1 Reloading Tool. The Ideal No. 1 tool, patented March 11, 1884 was a light, compact and complete little tool designed for loading the smaller pistol cartridges of the day. It was capable of performing all the operations required in reloading. It moulded the bullet (A), deprimed the shell, seated a new primer, forced the bullet in place and crimped the shell. It did not have a bullet sizer, however. Illustration above, from the No. 4 Ideal Handbook, lists the tool's features. It weighed 20 ounces and cost $2.25 at that time. It was then available for these cartridges: ".22-10-45, .22-15-45 W.C.F., .32 Short, .32 Long, .32 S&W, .32 Ex. Long, .32 H&R, .32 M&H, .38 Short, .38 Long, .38 Ex. Long, .38 S&W, .38 M&H, .41 Short—Colt's D.A., .41 Long—Colt's D.A."

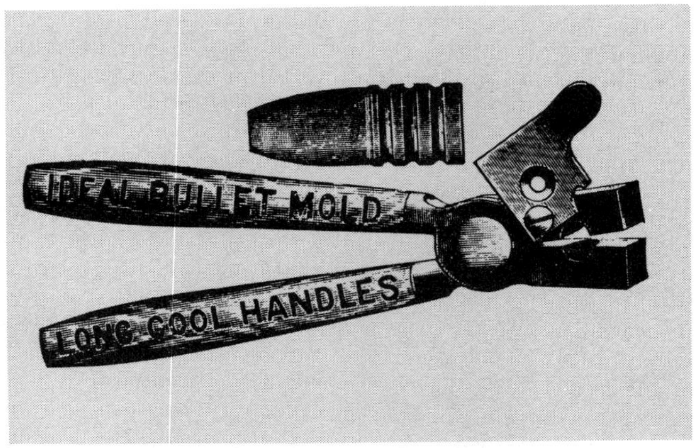

Ideal Bullet Mould. The standard Ideal Bullet mould with its long cool handles was the model by which others were judged. Illustration from the Ideal Handbook No. 4 was accompanied by this description: "These moulds have wood handles, the shanks are securely fastened. The joint face is broad, and is secured by a large pivot pin, which prevents them getting out of place. Moulds are made for all standard sizes, and we have cherries for many special bullets (see Ideal table of bullets). Single moulds will be invariably made to cast bullets as near the standard size as possible. If it is desired to have the mould cast bullets large so as to size them, it must be ordered so specifically." The mould weighed 12 ounce and cost $1.10 for regular grooved bullets, $1.50 for round balls, $1.70 for express balls and $1.10 for a blank ready to be cut.

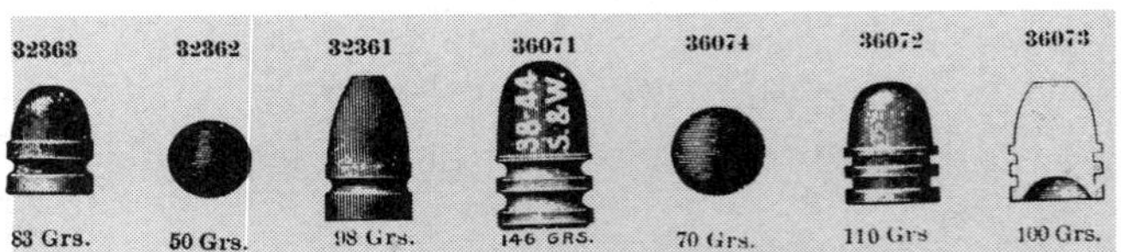

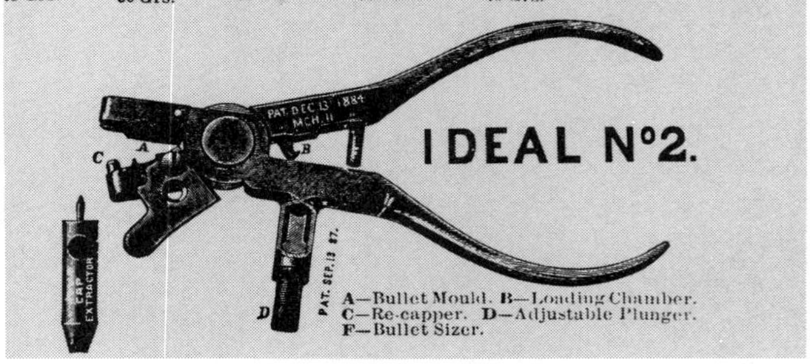

Ideal No. 2 Reloading Tool. The No. 2 tool was a target pistol cartridge tool intended to load cartridges exactly like factory ammunition for the S&W and Colt target pistols on the market. An excerpt from the No. 4 Ideal Handbook says: "With it shells can be loaded with either round or conical bullet; the cuts of bullets show the different ones that can be loaded with this tool. The 100 Grs. .38 Cal. is a special of our own that has met with much favor...When the mould is for round ball the tool is called Gallery, when for the conical bullet it is called Target. The purchaser can have which he desires, and with the extra moulds he can have the variety, as the tool will seat all that are of the same caliber, with the exception of the 98 Grs. (32S. & W. Rifle), the Rifle tool will seat the other 32 Cals. with the addition of an extra adjusting screw and vice versa. The adjustable plunger will seat the ball on any charge of powder."

The No. 2 tool was then made for the .32-44 S&W Target, grooved ball; .32-44 S&W Gallery, round ball; .38-44 S&W Target, grooved ball; .38-44 S&W Gallery, round ball; .32 S&W Pocket Pistol; .32 S&W Rifle; and .38 S&W Pocket Pistol. In its nickel-plated version it cost $3.50. Separate moulds were $1.50, the hollow base 100 grain mould cost $2.00 and the extra seating screw was $.50.

Tips on Bullet Casting. The Ideal Handbooks always carried good tips on the proper way to break in and use a bullet mould. Illustration above is from Ideal Handbook No. 4. Part of the descriptive copy accompanying this illustration follows. "New moulds will not cast good bullets until they have been used long enough to become oxidized. There must not be a particle of oil or grease in them...Have metal and moulds very hot...to be more explicit, we would say that the lead and moulds should be kept at a temperature that will require a few seconds for the lead in the spew hole to solidify after the nozzle has been separated from the mould...A small percentage of tin with pure lead will make the metal flow better and will produce a harder bullet. About one part tin to fifty of lead will make a good mixture.

"If a harder bullet is desired, add more tin...;one part tin to sixteen of lead will be found hard enough for the hardest bullet, but this must be decided by the shooter himself...An old pair of buck-skin gloves is just what you want when you get to business... Never strike the Mould with a hammer or other metallic substance. To cut off the spew use a billet of wood...The Hot Bullets should not be dropped onto a pine board or other hard substance; an old woolen cloth is good...A new mould is a fractious thing and sometimes it taxes the patience of the best of them, but when once got in order will stay so and should not be abused by allowing it to get rusty."

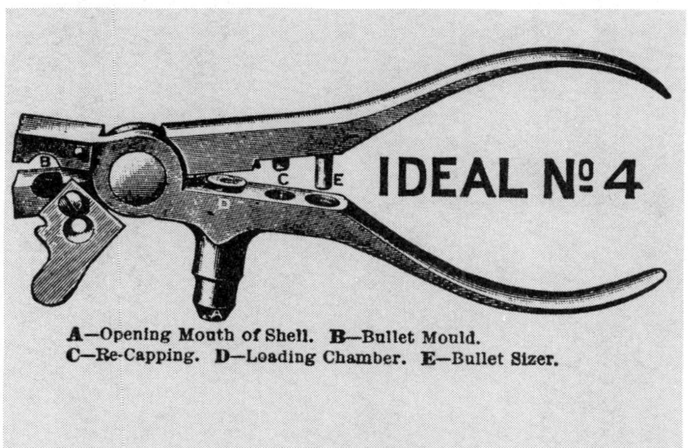

No. 4 Reloading Tool. The Ideal No. 4 tool moulded a bullet slightly above the standard size. After the bullet was lubricated, it was forced through the tool's sizing die to pack the grease firmly in the grooves. The No. 4 Handbook listed the following bullets as being available with the tool: ".25-20-77 Ideal Bullet, .25-20-86, .32 Colt's Lightning, .32 W.C.F., .32-20 Marlin, .32-30 Remington, .38 W.C.F., .38 Colt's Lightning, .38-40 Marlin, .44 Colt's O.M. Heel Bullet, .44 Colt's Lightning, .44 W.C.F., 44-40 Marlin, .44 S&W Russian, .44 S&W American, .44 M&H O.M., .45 Colt's, .45 S&W." The nickel-plated tool cost $2.50 at that time.

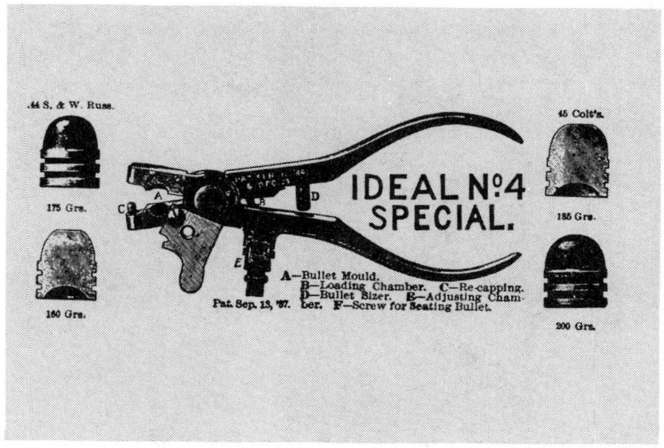

No. 4 Special Tool. The Ideal No. 4 Special was a pistol cartridge tool. Each set was supplied with adjustable screws for round or standard balls. It was designed to load cartridges for the following revolvers: .44 Colt Frontier, .45 Colt, .44 S&W Russian, .44 M&H Army, and the .32-20 W.C.F. Colt and Marlin cartridges. In nickel-plated form it cost $4.00 when Ideal Handbook No. 4 was published.

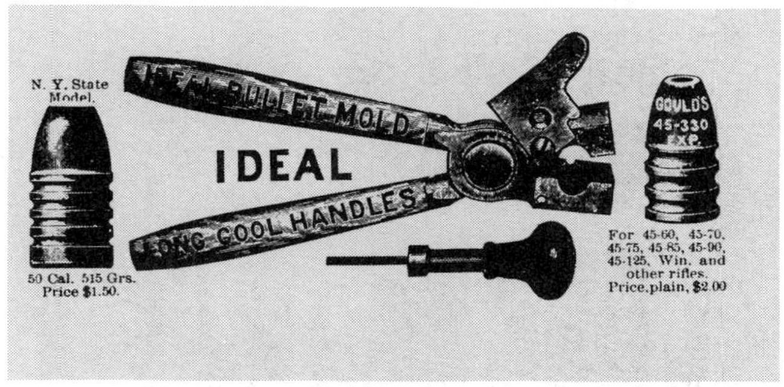

"Special" Bullet Mould. This special Ideal mould, with its long cool handles, is illustrated from Ideal Handbook No. 4. The description from that edition reads: "The lovers of extra fine tools are growing in numbers every day. To such we would say that we make (when ordered) fine, heavy extra finished moulds with polished cocobolo handles." Regular size bullet moulds cost $1.75 apiece while Express bullet moulds cost $2.50 each.

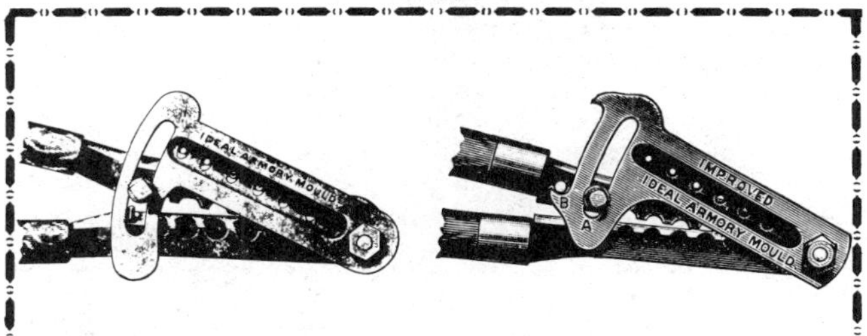

Ideal Armory Mould. Designed to complement the Ideal cartridge-loading press, the Ideal Armory mould was described in Ideal Handbook No. 4 as "a strong, heavy well-made bench mould. It is cut for six of the largest military and sporting bullets. Length of mould complete, about 13 inches; weight, about 4 pounds. Military companies, clubs, or those using large quantities of bullets and ammunition will find the Loading Press and Mould a great time saver and economizer." The mould cost $10. Most often ordered for .45 and .50 caliber bullets (six bullets per mould) the Armory mould could also be ordered in .38 caliber (seven bullets), .32 caliber (eight bullets) and .25 or .22 caliber (nine bullets). Originally the mould featured only a square headed screw in the cutoff slot (left illustration) which often broke. The mould was later redesigned (right illustration from Ideal Handbook No. 11) to feature a separate stop pin "B" which protected the screw from damage.

The Perfection Mould. The Perfection mould was an adjustable mould for producing grooved bullets. The adjustments could be varied one groove at a time, varying the length or weight 5-10 grains at a time. This illustration, from Ideal Handbook No. 4, listed the bullets with hollow or flat bases. It was not made for all calibers—just for the most popular sizes. It was made to cast bullets slightly above size. At this time it was available in the following diameters: .257, .311, .319, .323, .375, and .457 with each costing $3.00.

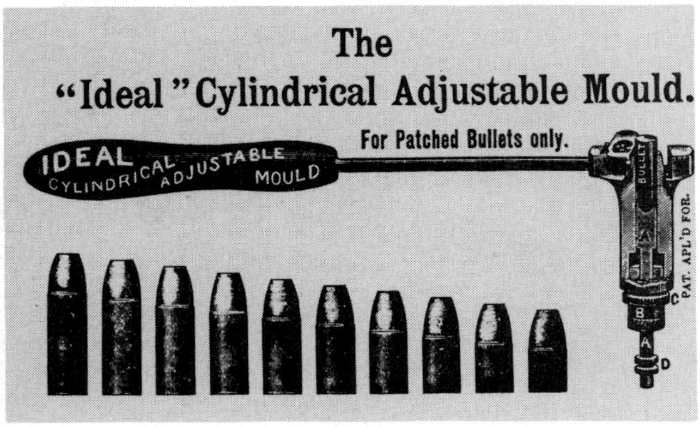

Cylindrical Adjustable Mould. The Ideal Cylindrical mould was ideal for shooters who favored paper-patched bullets. The description which accompanied this illustration from Ideal Handbook No. 4 says: "This NEW MOULD is a perfectly true cylinder...The adjustable former not only enables the user to get different lengths of bullets so he can vary the weight to his desire, but with it the bullets are pushed out of the mould leaving them perfectly true without taper, assuring the whole length of the bullet bearing in the barrel of the rifle, therefore not requiring so much upsetting and obviating any possible chance of it being started in a tipping or crosswise manner, so common with the tapered bullet...These moulds can be furnished for the following calibers; vis., .25, .32, .48, .40 and .45." Price for all calibers was $3.00.

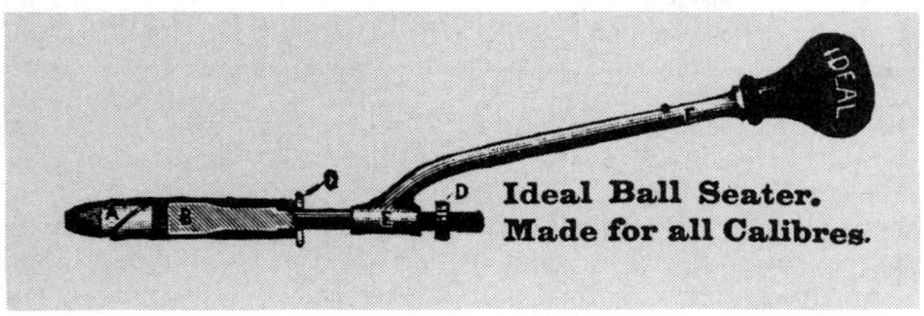

Ideal Bullet Seater. The Ideal Bullet Seater, seen here from illustration in Handbook No. 4, featured this description. "With the Single-Shot Rifle **at the range**, the desire is to get the best possible results that can be obtained with rifle, powder and bullet. The most accurate shooting that has ever been done thus far to our knowledge, was with the bullet seated into the barrel about 1/32 of an inch ahead of the shell. The bullet is thus well in the rifling, before the explosion takes place, obviating any possible chance of it being started in a tipping or crosswise manner. The shell is then filled with powder (a wad on top or not as desired), and inserted in the chamber after the bullet. The **Ideal Bullet Seater** has an adjustable plunger, 'B' that can be set with check nut, 'D' to seat the ball any depth desired; and they will be absolutely the same depth, which **must** be, to insure uniform shooting. The implement weighs but four ounces. It will be made for all calibers from .22 up." Price for the tool was $1.

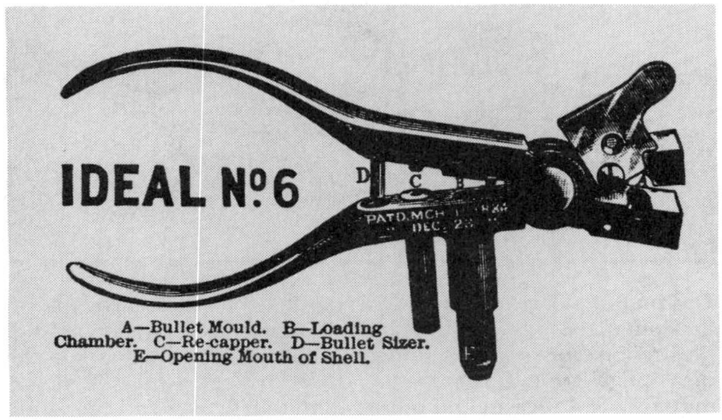

Ideal No. 6 Tool. The No. 6 Ideal Tool was designed to load the larger military and sporting cartridges of the day. Like the No. 4 it performed all the reloading operations required and featured a built-in bullet sizer. When Ideal No. 4 Handbook was printed, this tool cost $3.00, $3.50 with the double adjustable chamber and $.50 extra for the Express bullet mould. It was offered in some 38 popular bullet styles.

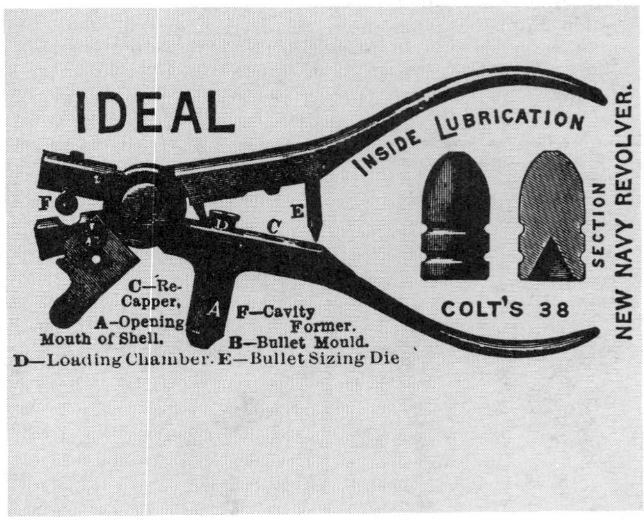

Ideal No. 8 Tool. The No. 8 tool, seen here in an illustration from Ideal Handbook No. 4 was described in that edition as "a special tool for reloading the Union Metallic Cartridge Co.'s new .38 cal. long, inside lubricated ammunition....This tool will not load the old model .38 long shells as they are shorter than the new; the bullet is seated in the new shell deep enough to cover the grooves which hold all the lubrication. The new shells and the hollow based bullets (see cut) as made and loaded with this tool will, however, be all right for any pistol or rifle using the old outside lubricated .38 long, and will be found superior." The tool sold for $3.00 in nickel-plated form. It cost $2.00 for an addition mould for this bullet.

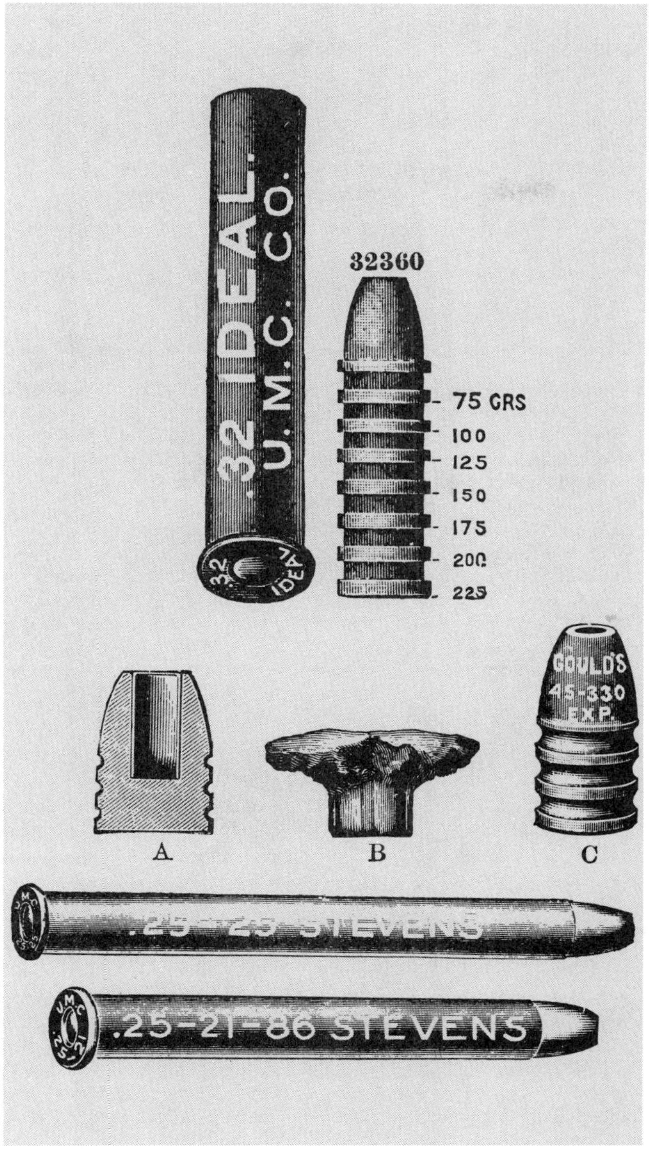

Unique Bullet Designs By Barlow. John Barlow was directly responsible for some of the best-designed bullets of his day as can be seen from the above illustrations from Ideal Handbook No. 11. At left is Barlow's .32 Ideal cartridge and the bullet Barlow made for it. Part of the description reads: "The shells and bullets here shown are new ones designed by our Mr. Barlow and first made for us by the Union Metallic Cartridge Co....The shell is 1¾ inches long, is straight inside and outside, has a solid head and strong pocket equal to the everlasting shells...The standard .32 Ideal cartridge...is this shell with 25 grains of powder and 150 grain bullet seated to cover all grooves, and no crimp. The diameter of the bullet is .323". Barlow also designed the Gould's 45-330 Express bullet and according to the Ideal Handbook No. 11, "This bullet was first made by our Mr. Barlow for A.C. Gould, editor of the New York Shooting and Fishing. It has given universal satisfaction as an accurate flyer and a great killer of game...The cavity is simply filled with wax or tallow."

Barlow also made the first .25-25 shell according to Ideal Handbook No. 11 where it states: "The Crank Rifle Shooter is a tireless fellow, always after something different. Not being satisfied with either of the above .25 calibers (25-20 CF and 25-20 Marlin were pictured) Capt. W.L. Carpenter of the 9th U.S. Infantry called on our Mr. Barlow, who made for him, from a solid brass rod, the first model shell, which has since become famous as the .25-25 Stevens; so called because the Stevens Arms Company made the first rifle for it. There is now another modification of it called the Stevens .25-21, which is the same shell shortened and loaded with 21 grains of powder instead of 25. Both of these cartridges use the same bullets as the .25-20's."

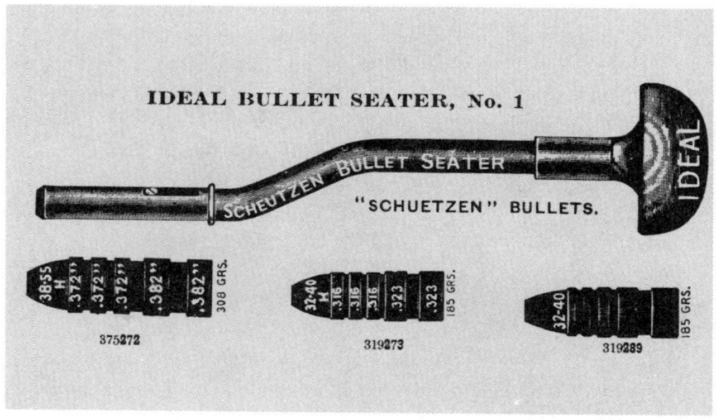

The popularity of Scheutzen rifles led to the introduction of this specialized bullet seater by 1906. Illustration above is from Ideal Handbook 19 and was accompanied by this description: "No. 1 Bullet Seater is made especially for seating the Scheutzen bullets here illustrated, which were designed by Dr. W.G. Hudson. These bullets are made of a hard alloy, and require an extra strong implement to seat them in the barrel, most of which have to be throated at the breech...No. 1 Bullet Seater will be made only for .32-40 and .38-55 single shot rifles." It cost $1.50.

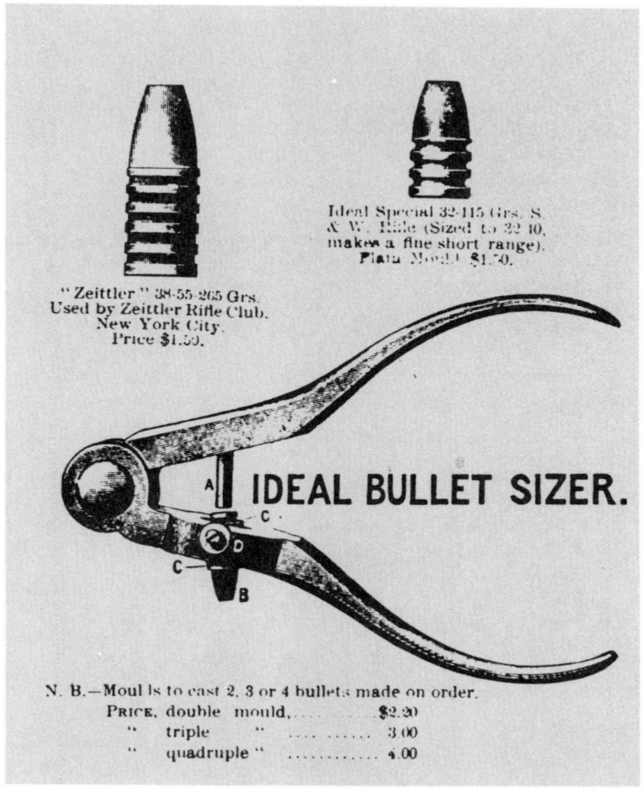

Ideal Bullet Sizing Tool. This Ideal tool was made so that dies of any caliber could be used in it. Description from Ideal Handbook No. 4, from which this illustration is taken, reads: "It will be noticed that the die swings upon centers, which are located near the top of the die, the presser punch is also swung upon a pivot which, while forcing the bullet through the die will keep it in perfect alignment with the pressure. This construction also permits placing the die near the joint, thereby giving the required power with shorter levers, which makes the tool a handy and convenient length. The tool will be the same for all calibers. The die only being made for the caliber desired, they will be interchangeable and for all standard sizes, from .22 to .50 caliber." Barlow also made dies of special dimensions. The tool with one standard die sold for $1.75.

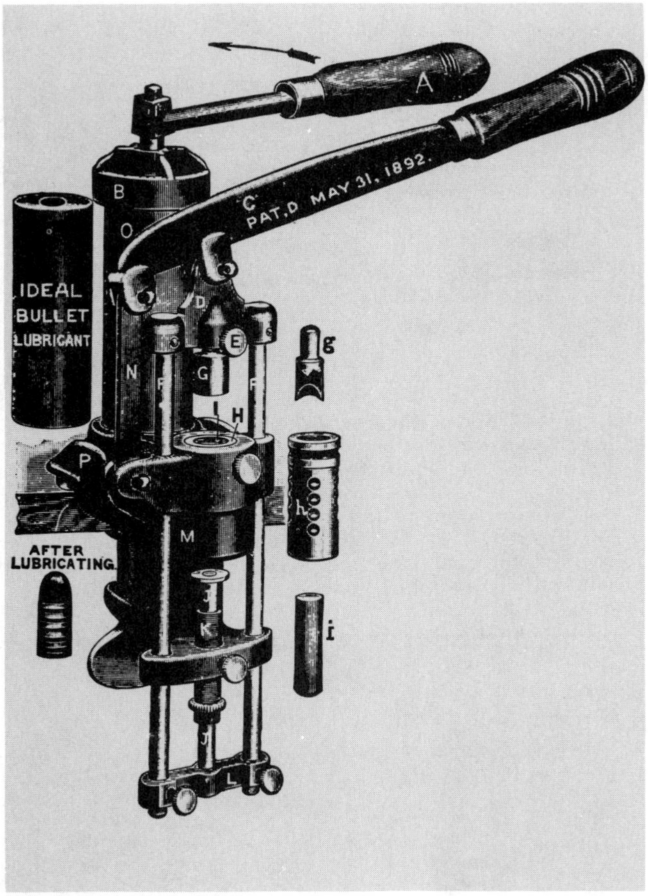

Ideal Bullet Lubricator and Sizer. This unique Ideal tool, patented May 31, 1892, and illustrated from Ideal Handbook No. 14, lubricated and sized bullets in one operation. It was designed to use prefabricated sticks of Ideal Banana Lubricant "prepared especially for us by E.A. Leopold," according to Ideal No. 14. Artist's illustration listed the die, bottom and top punch as **g, h** and **i**—designations that stuck through the years and are used today without a second thought about their possible origin.

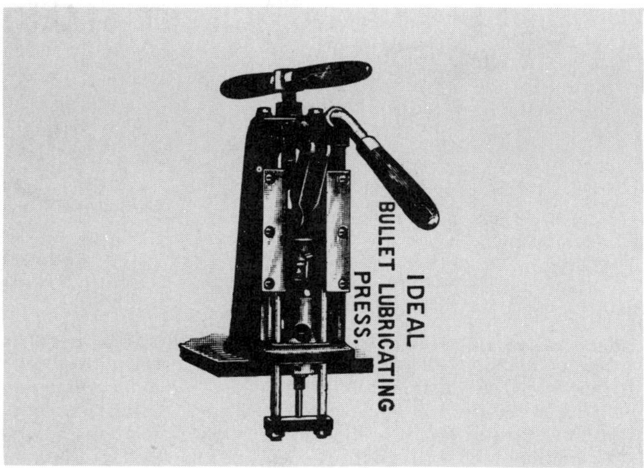

Ideal No. 2 Lubricator and Sizer. The No. 2 model of the Ideal lubricator and sizer was designed for "continuous hard work. It is heavy, strong and powerful. It will stand hard usage and should last a lifetime. Weight is 25 lbs. Tools, dies and punches for the No. 1 Lubricator are interchangeable with those for Lubricating Press No. 2," according to the description accompanying this illustration from Ideal Handbook 19. The press complete with dies and punches cost $20.

Improved Lubricator Pressure Nut. By 1906 the Ideal lubricators and sizers were outfitted with this "improved" pressure nut. The description noted that "When the nut is forced down, the grease presses against the inside of the metal ring, causing the ring to expand against the walls of the grease tube. This prevents the nut from turning thereby permitting it to travel down the grease pressure screw and exert full pressure on the grease. The pressure of the ring packing against the walls also prevents entirely, escapement of grease up past the nut." The unit cost $.75 and could be adapted to old Ideal lubricator-sizers for the additional charge of $.35.

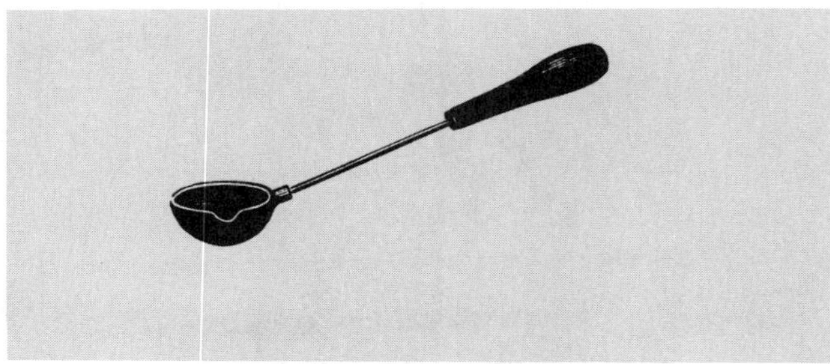

Large Ideal Dipper. The large ideal dipper had been in use since the Armory Outfit was offered to military units and sporting clubs in 1904. The cup of the dipper measured 2½ inches in diameter and the dipper was intended to be used to fill the Ideal Armory Mold.

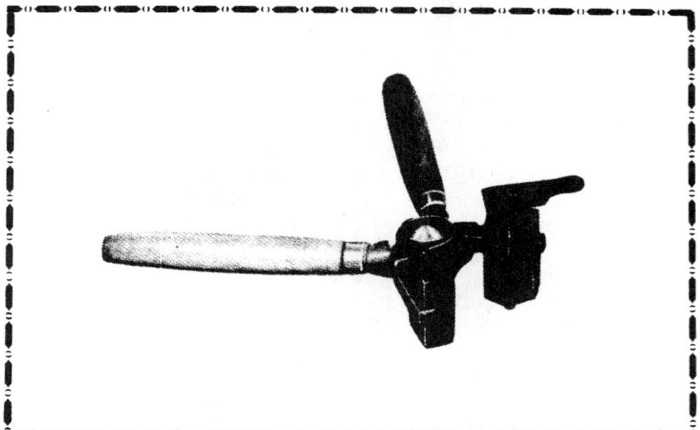

Adjustable Bullet Swaging Core Mould. These single-, double- and four-cavity moulds were made for the shooter who swaged his own bullets. It cast a precision core with perfectly flat ends, accurate to 3/10th of a grain. An adjustable base screw permitted cores of variable length and weight to be cast. It was available in .22 (.186), .30 (.239), .38 (.316), .44 (.392) and .45 (.409) calibers. The single cavity mould and handles cost $14.50 while the double-cavity mould and handles was $18.50 and the four-cavity mould and handles cost $25.00.

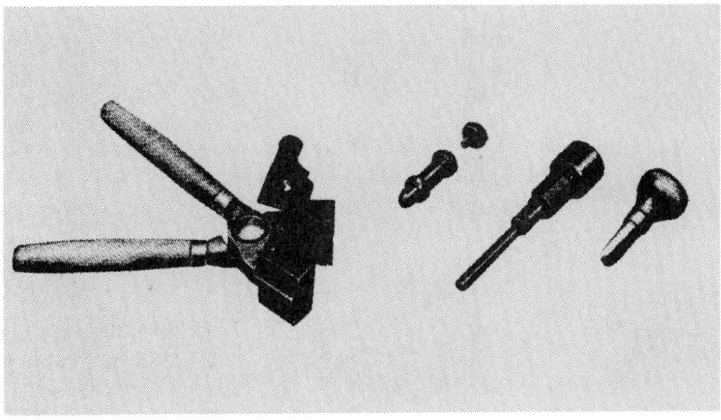

Shotgun Slug-Swaging Die Set. In the late 1950's and early 1960's shotgun shooters believed that rifled slugs were necessary if any accuracy was to be obtained from a slug—a view that later lost credence. Slug shooters could buy a complete set to make rifled slugs from Lyman. It included a hollow-base slug mould (choice of 12, 16 or 20 gauge) and a swaging die set that rifled and swaged the slug in one motion. The dies could be fitted to any turret of Comet press in the line. The complete set cost $21.75 with a conversion kit available for $15.50.

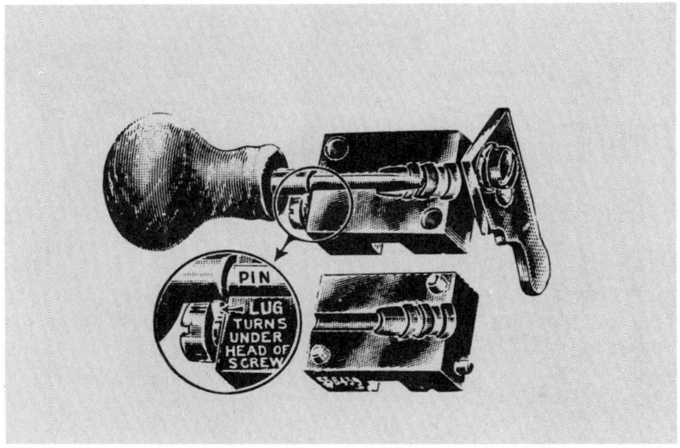

Hollow Point Mould and Pin. Early in the 1940's, Lyman's Ideal bullet moulds were available with a new hollow point pin complete with lug. The lug held the pin securely in place when casting bullets and allowed the pin to be easily removed after the metal had hardened. This illustration is from Ideal Handbook No. 34.

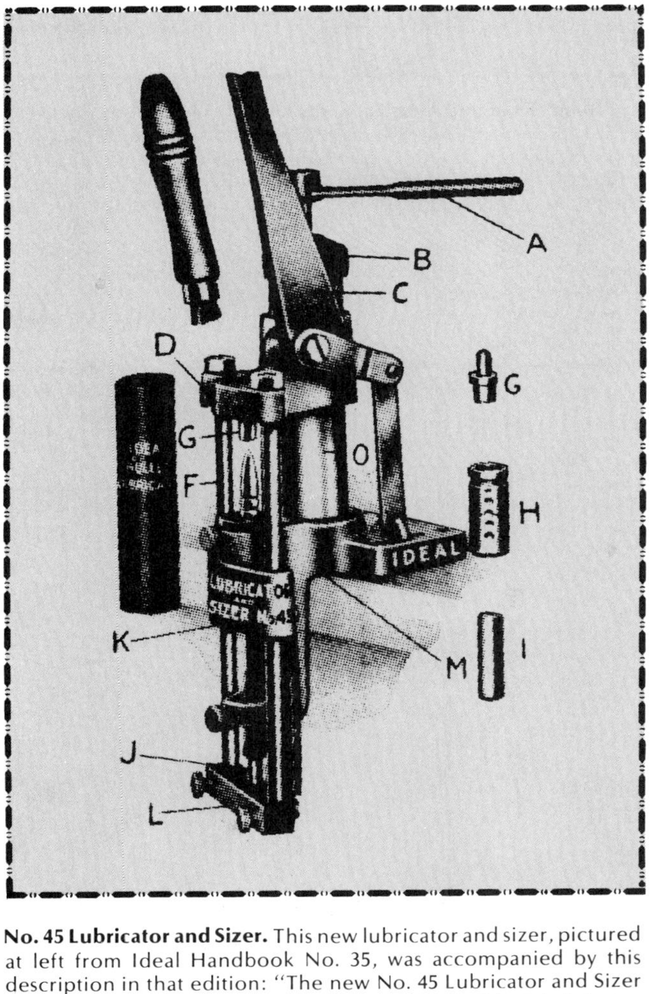

No. 45 Lubricator and Sizer. This new lubricator and sizer, pictured at left from Ideal Handbook No. 35, was accompanied by this description in that edition: "The new No. 45 Lubricator and Sizer employs the same type of sizing dies used in the No. 1 Lubricator and Sizer, enabling owners of the older sets of dies to use them in the No. 45 Tool. Increased leverage makes possible easier operation on large bullets while large guide rods preserve the alignment of the tool. This new tool is fitted with a steel grease tube which is used as an additional guide to help preserve alignment." By this time Lyman had two types of lubricant, replacing the original "Banana" lubricant. There was a regular stick and a special stick featuring graphite "for conditions where barrel leading is a problem.

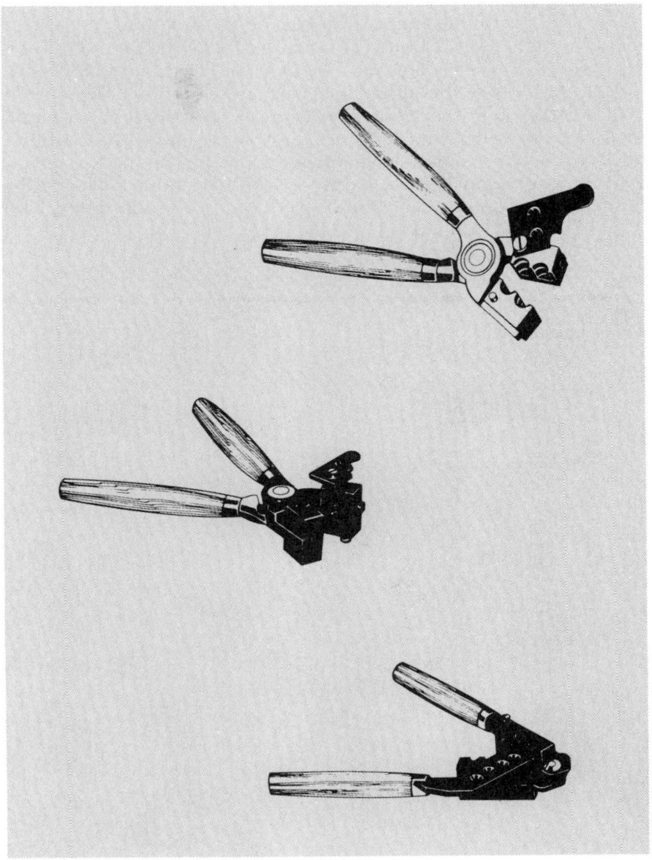

Ideal Bullet Moulds With Interchangeable Blocks. The ideal single cavity mould was improved upon by making the mould blocks interchangeable and thereby requiring the shooter to purchase only one set of wooden handles for his bullet casting needs. Then as now, blocks were attached to the handles by screws using a screw driver and once the blocks were in place they aligned true and securely. The handles were designed to fit the hand comfortably when closed. By 1949 double cavity mould blocks were in the line to be followed by four-cavity mould blocks in the 1950's.

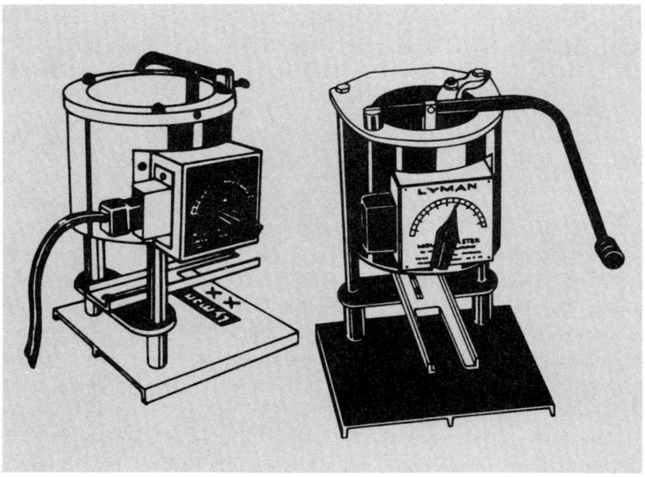

Ideal Electric Lead Melting Furnaces. Electrically heated lead pots were listed in Ideal Handbooks as early as 1927 but many years were to pass before Lyman produced its own version. In Ideal Handbook No. 28, the Trent Electric Melting Pot was listed for shooters "where electricity is available." It had low, medium and high temperature settings, weighed 17 pounds empty and held 20 pounds of lead. Ideal Handbook No. 33, published in 1939, listed the Type "A" Potter Improved Electric Melting Furnace, "a self-contained unit which may be plugged in to any electric outlet...The furnace holds 2½ pounds of bullet metal and may be used with any single or double cavity mold." Potter also made a "B" Type pot with twice the capacity as the "A" furnace. By 1960 the Mould Master furnace was in the Lyman line. It had an 11-pound capacity, bottom pour spout and a calibrated thermostat listing 450-850 degrees farenheit. It was accurate to within 20 degrees of its setting. A 20-pound capacity lead furnace of Lyman manufacture, the Mould Master XX, was introduced in 1976. It featured a calibrated thermostat and a bottom pour spout. Each furnace could also be equipped with Lyman's mould guide and ingot mould.

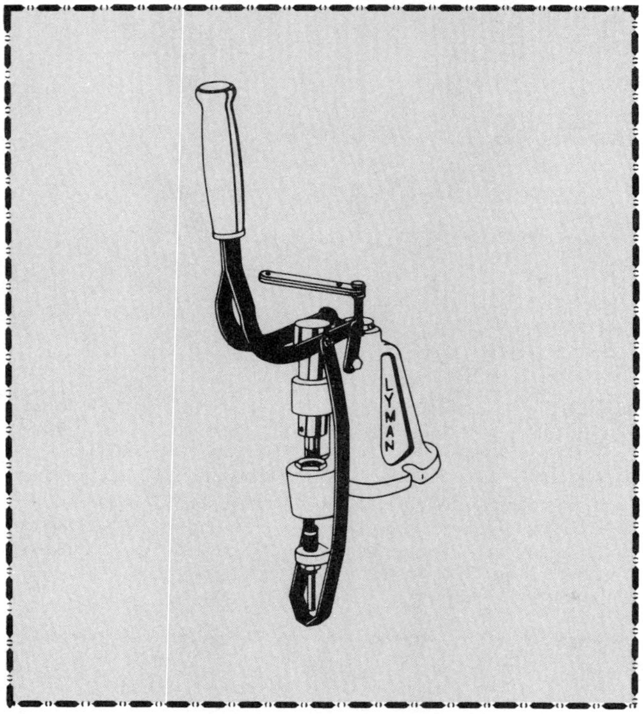

450 Bullet Sizer and Lubricator. By 1968 the 450 Bullet Sizer and Lubricator was in the Lyman line. Its short-stroke, power-link leverage allows the largest cast bullets to be sized and lubricated with ease. A large C-frame iron-steel casting completely encloses the grease reservoir. Precision in-line boring insures absolute die alignment. It is adaptable to all bullets by changing the bullet sizing die sets. These consist of a top punch "G" and a bullet sizing assembly "H" and "I". When introduced the unit cost $29.00 complete with a set of dies.

THE METALLURGY OF MOLTEN LEAD ALLOYS

by Dennis Marshall

Introduction

The art of bullet casting has made significant progress over the years, especially in the area of rifle accuracy. Shooters have discovered or are rediscovering the importance of blending bullet hardness and dimensions, powder type, lubricant and a host of other variables to suit the intended purpose of the ammunition. Attention to such detail has generally raised the performance level of cast bullets to the extent that minute-of-angle accuracy is becoming increasingly common-place.

Unfortunately, the science of bullet casting has not faired so well over the same period. Relatively little information has been published on the metallurgical properties of lead alloys which are important to cast bullets, or ammunition in general. Instead, the firearms literature is laced with numerous errors and old wives' tales which are unable to answer new or recurrent technical problems with cast bullets.

One area in particular where there is a virtual void of technical information is the metallurgy of the liquid state. Since the metals which are put into the pot have a direct bearing on bullet quality and performance, an understanding of melt behavior is essential to all subsequent cast bullet metallurgy. Therefore, what follows here is a description of the metallurgical properties of molten lead alloys. The discussion will include the basics of melt formation, phase diagrams, the benefits of tin, high temperature casting, impurity effects and some comments on fluxing. Much of this information will contradict accepted "theories" on the behavior of lead melts but all of it is traceable to scientific books and periodicals, and adequate references are provided for those who would like a more rigorous account.

Metallic Solutions

Perhaps the single most significant error in all the bullet casting literature is the misconception that lead-tin-antimony melts gravity segregate. Bullet casters have been lead to believe that unless they flux a melt on a regular basis, the less dense tin and antimony will separate from the lead and rise to the surface where they will no longer be available to harden the alloy. This is absolutely wrong, and in fact quite the opposite is true; tin and antimony, either as individual additives or in combination, literally dissolve in molten lead to form true, stable solutions, just as table salt or sugar will dissolve in water. And, with the exception of oxidation or an electrochemical potential, once the solutioning has occurred, there is no force, gravitational or otherwise which can separate the constituents. This situation is not unique to lead and its alloys, but is common to most metals. For example, if zinc were not soluble in molten copper it would be far more difficult and costly to produce cartridge brass or guilding metal. Similarly, carbon, nickel, chromium, molybdenum, tungsten and other elements dissolve in molten iron to give us various grades of steel for our firearms.

The factors which govern the mixing of metals have nothing to do with density. Indeed, if mixing depended on density then why should table salt with a density of 2.165 dissolve in water which has a density of 1.0? A density difference will mechanically separate materials which are *not* soluble in one another, but it cannot separate the constituents of a solution. Instead, mixing is determined by energy, or more specifically, the energy difference between the mixed and unmixed metals (1). Now this may sound rather messy and abstract but the basic principle is easily understood by way of analogy.

All materials contain energy, and any real, physical process which involves materials must be accompanied by a release of energy. For example, blackpowder is a mechanical mixture of materials, potassium nitrate, sulfur and charcoal, which as a whole contain a certain amount of energy. If the powder is ignited, energy, most of which is in the form of heat, is released and we are left with some by-products of combustion such as potassium sulfide(2), which also contain energy. We can now write a very simple energy equation for this reaction as follows:

$$\text{ENERGY OF BLACKPOWDER} - \text{ENERGY OF BY-PRODUCTS} = \text{ENERGY RELEASED}$$

As long as the right hand side of this equation is positive, the reaction can take place, and for this to occur the energy of the products must alwys be less than the energy of the starting materials. That is to say, the energy of unburnt blackpowder is greater than burnt blackpowder.

Similarly, mixing of tin or antimony with molten lead is a real process and must be accompanied by the release of some energy. This means that two pounds of a 50/50 lead-tin alloy mixture at 750° F must have less energy than one pound of lead plus one pound of tin unmixed at the same temperature. Now the energy release is not all that great when lead-tin-antimony alloys are mixed, but a measurable quantity of energy is released just the same(3) and a solution is formed.

This is not to say that all attempts to mix metals will result in the formation of a solution. *Depending on the composition and the temperature*, there are many instances where the metal mixture has a higher energy content than the unmixed constituents. In these instances, a solution cannot be formed and the metals remain separated in the liquid state. Examples of metals which form immiscible melts with lead are aluminum,

Figure 1. Freshly skimmed surface of a lead-9% zinc alloy at 850°F showing melt segregation. An elongated puddle of zinc liquid is floating on the lead melt. If the temperature is raised high enough, the zinc puddle will dissolve in the lead and disappear. This situation is in sharp contrast to the homogeneous melts used to cast bullets.

copper, and zinc. For example, at 850° F, which is within the normal range of bullet casting temperatures, lead will dissolve a little over 2% zinc. If more zinc is added, an immiscible melt is formed and a puddle of the lower density zinc liquid will form on the surface of the molten lead, and no amount of stirring or fluxing will make it disappear. An example of this is shown in **Figure 1**. If the temperature is raised, the solubility of zinc in lead increases and the zinc puddle will gradually dissolve. At 1468° F, zinc and lead are completely soluble in one another but on cooling, any alloy containing more than 2% zinc will again separate to form two immiscible liquids.

The practical consequence of a metallic solution is that bullets of consistant composition and weight can be cast. Discounting losses due to oxidation, the compositions of the first and last bullets from a pot of alloy are the same. If segregation did occur, composition would vary from end to end or from one side of the bullet to the other and this would lead to unbalanced bullets. Indeed, the evidence against segregation is overwhelming.

Phase Diagrams

At this point, it is convenient to introduce the phase diagram. A phase diagram graphically illustrates how metals mix as a function of temperature and composition. For our study of molten alloys, the phase diagram shows the range of temperatures and compositions over which liquid lead-tin-antimony solutions are stable and the temperatures at which the various alloys begin to solidify during casting.

Binary Alloys:

The binary (two component) phase diagrams for lead-tin and lead-antimony alloys are shown in **Figures 2** and **3**(4). The temperature at which any alloy begins to solidify is given by line **ABC** in each figure and is referred to as the liquidus line. The region above the liquidus defines the range of temperatures and compositions over which a homogeneous liquid solution can exist. This broad region for the formation of homogeneous alloys is a real asset to the bullet caster who can add to or dilute

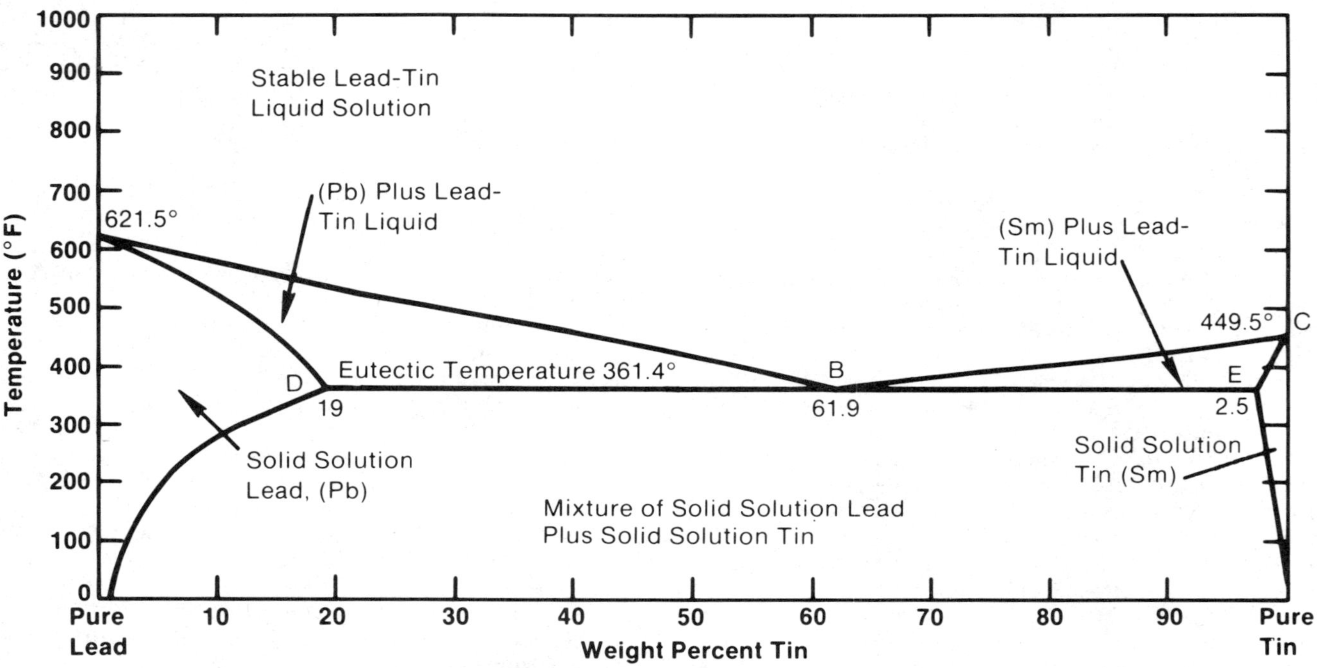

Figure 2. Binary lead-tin phase diagram from the Metals Handbook(4).

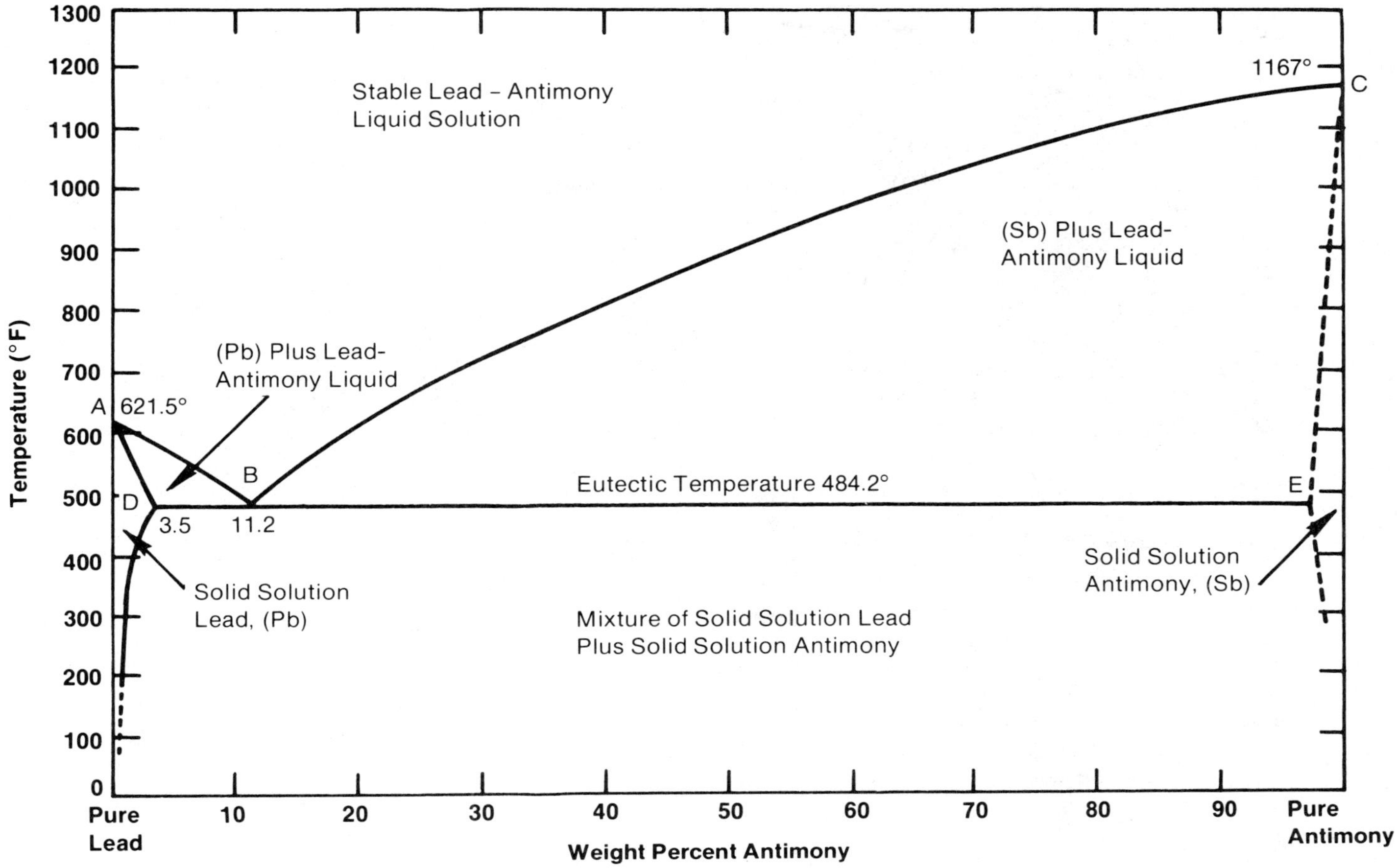

Figure 3. Binary lead-antimony phase diagram from the Metals Handbook(4).

the alloy composition, or vary the casting temperature to optimize mould fill out without sacrificing the solution character of the melt. The liquidus also shows that any practical alloy the caster might desire is within the temperature capability of all normal bullet casting equipment.

To exemplify the advantages of broad range solubility, it is interesting to contrast the molten behavior of lead-zinc alloys with that of lead-tin or lead-antimony. **Figure 4** shows the lead-zinc phase diagram(4). Note that within the range of normal bullet casting temperatures, the solubility of zinc is constrained by a portion of the liquidus indicated by line **ABC**. At temperatures greater than 783.8° F, solubility increases but is limited by the miscibility gap. Any alloy which falls within the shaded area in **Figure 4** will divide into two discrete liquids and would be unsuitable for casting. For example, if we melt an alloy with a composition given by point 1 we will form a true solution. If zinc is added to shift the composition to point 2, the liquid solution will break down to form two separate, immiscible liquids, the compositions of which are given by points **x** and **y**. Since liquid **y** is richer in zinc, it is less dense than liquid **x**. Thus, **y** will float on **x** and we have the exact situation which was photographed in **Figure 1**.

If you trace the liquidus line in **Figures 2** and **3** you will notice that there are three points at which an alloy will freeze at a single, sharp temperature, the pure elements and the eutectic composition indicated by point B. At all other compositions, alloys solidify progressively over a range of temperatures which begin at the liquidus and is complete at the solidus, line **ADEC**.*

The first solid to form when an alloy begins to solidify (i.e. when the temperature falls below the liquidus) is called the primary phase. An important feature of the eutectic is that it divides the liquidus into two separate primary fields. From **A** to **B** the first solid to crystallize during cooling is lead and is symbolized as (Pb). This region of the liquidus is referred to as the primary field of crystallization for lead. The parenthesis around the chemical symbol indicates solid solubility, e.g., (Pb) means that some antimony or tin is literally dissolved in the solid lead which formed from the melt. On the other side of the eutectic, the region from **B** to **C** is the primary field of crystallization for tin (**Figure 2**) and antimony (**Figure 3**).

Since the composition of the lead-tin eutectic is 38.1% lead and 61.9% tin, it is highly unlikely that any bullet will contain primary crystals of tin, (Sn). However, it is entirely reasonable and common to run across bullets which contain more than 11.2% antimony. Some years ago a batch of "linotype" was acquired which, through repeated use or faulty alloying, had a composition of 13.5% Sb and 0.5% Sn. Microscopic examination of bullets cast from the alloy showed primary crystals of antimony in the microstructure. Bullets which contain primary antimony are characterized by extreme brittleness. Some will even fracture if dropped on a hard surface.

The eutectic composition possesses another unique feature. Referring to **Figure 3**, since the primary fields of crystallization for lead and antimony meet at the eutectic, an alloy containing 11.2% Sb will solidify at the eutectic temperature and will simultaneously form (Pb) and (Sb).

Ternary Alloys:

The liquidus for the lead corner of the ternary system is shown in **Figure 5** and covers all the practical lead-tin-antimony compositions which are likely to be encountered in bullet casting. The diagram was taken from reference(4), point **B** was corrected to agree with more recent data(6) and the entire diagram converted from Centigrade to Fahrenheit. The remainder of the ternary phase diagram is unimportant to bullet casting and contains features which are considerably beyond the scope of this article.

The horizontal lines in **Figure 5** represent the antimony

*The relationship between complete solidification and the phase diagram is explained elsewhere(5).

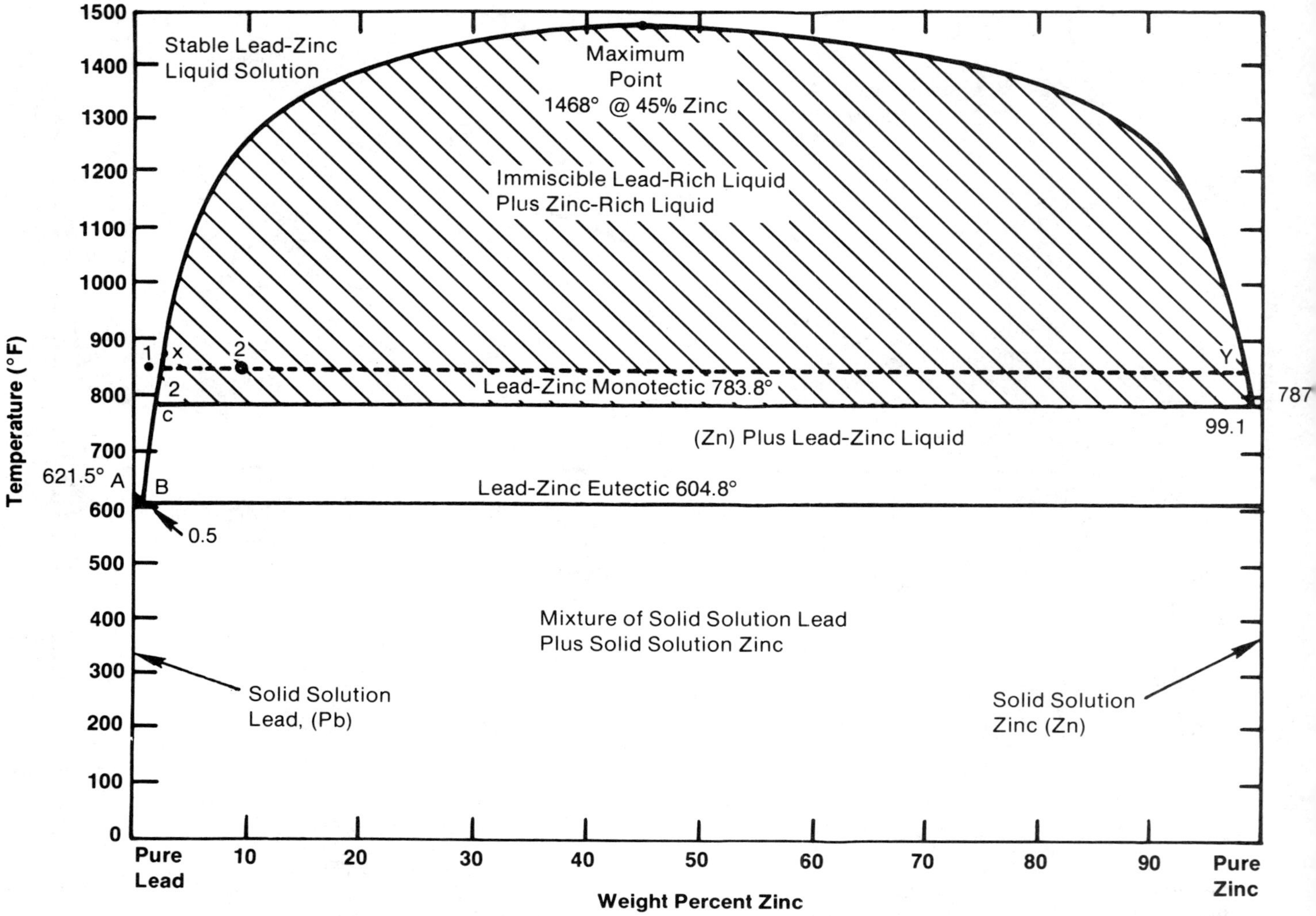

Figure 4. The more complex lead-zinc phase diagram is shown(4). Over the usual range of bullet casting temperatures, the composition of lead-zinc solutions is constrained by the liquidus line, ABC. Point 2 is the location of the alloy photographed in Figure 1.

composition, the vertical lines slanting 30° to the right represent the tin concentration, and the remaining lines slanting 30° to the left represent the concentration of lead.

At the lead corner, indicated by the symbol, Pb, the composition is 100% lead. Moving away from the corner, each lead line represents a 2% decrease in the lead content, e.g. the 92% lead line runs between 8% Sb and 8% Sn. Thus an alloy at point **J** contains 92% **Pb**, 4% **Sb and** 4% **Sn.**

The temperature lines on the phase diagram define the shape of the liquidus surface. The liquidus surface in the ternary has the same meaning as the liquidus line in a binary; it represents the temperatures at which alloys *begin* to solidify. For example, an alloy containing 14% antimony and 8% tin (point **K**) will being to solidify at 500°F. At all temperatures above the liquidus surface, antimony and tin are in liquid solution.

Point **B** in **Figure 5** is the location of the lead-antimony eutectic and corresponds to point **B** in **Figure 3**. The heavy line from **B** to **H** is called a trough in ternary phase diagrams and its purpose is essentially the same as point **B** in **Figure 3**, it separates the primary fields of crystallization for lead solid solution, (Pb), and antimony solid solution, (Sb). Similarly, the trough from **H** to **F** separates the primary fields of crystallization for (Pb) and SbSn, while the trough from **H** to **E** separates the primary fields for (Sb) and SbSn. The chemical symbol, SbSn, represents an intermetallic compound. Just as sodium and chlorine combine to form a compound we call table salt, antimony and tin combine to form an intermetallic compound which has physical properties very similar to pure antimony. Another intermetallic compound which is more familiar to most of us is silicon carbide, symbolized by SiC, which is used in various types of abrasive paper.

The symbols (Pb) and (Sb) and SbSn are shown in the phase diagram and each is located within its primary field of crystallization. (Pb) forms over the region **BHIFGA**, (Sb) forms over the region **BHEDC** and SbSn forms over the region **EHIF**. By inspection it is easy to determine the primary field of crystallization for any alloy. For example, (Pb) is the first solid to form when casting wheel weights or Lyman No. 2 alloy, SbSn will be the first solid to form when casting ordinary monotype which contains 15% antimony and 7% tin (near point K). As with the binary lead-antimony alloys, ternary compositions which have (Sb) or SbSn as the primary phase are characterized by extreme brittleness.

Point **H** is the location of the ternary eutectic composition which solidifies at a single sharp temperature, 464°F. It corresponds to the lowest melting point alloy in this region of the lead-tin-antimony system. Just as a binary eutectic represents the junction of two primary fields of crystallization, so point **H**, a ternary eutectic, is the junction of three primary fields of crystallization. An alloy containing 84% lead, 12% antimony and 4% tin will solidify at 464°F to form (Pb), (Sb) and SbSn simultaneously. In metallurgy, eutectic alloys are generally known for their good casting qualities, and the eutectic at point **H** is no exception. It is the composition of linotype.

The temperature at point **I** is 473°F and is the high point along the trough **HF**. (Note that the arrows on **HF** point away from **I**). It corresponds to a pseudo binary eutectic between (Pb)

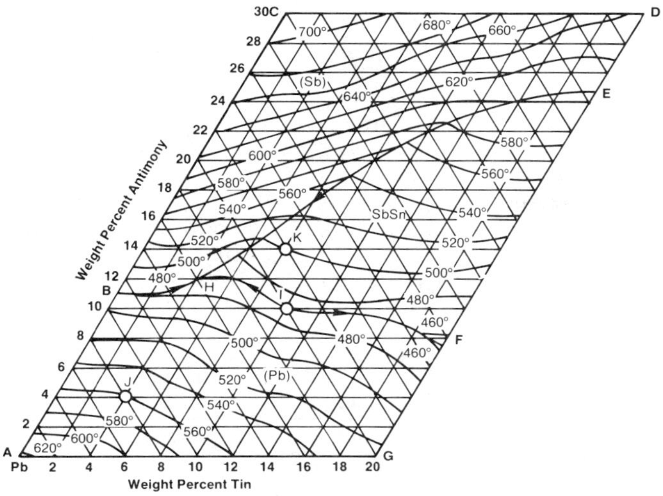

Figure 5. Liquidus surface in the lead corner of the ternary lead-tin-antimony phase diagram adapted from the Metals Handbook(4). The heavy curved lines, called troughs, separate the primary fields of crystallization for (Pb), (Sb) and SbSn. The arrows on the troughs point in the direction of decreasing temperature.

and SbSn, particles of antimony will not be present in an alloy solidified at **I**.

There is additional significance to point **I**. A line drawn between the lead corner and **I** represents equal proportions of Sb and Sn. As with point **I**, all alloys along this line will contain only (Pb) and SbSn, hence SbSn is the material which strengthens Lyman No. 2 alloy, *not antimony or tin*. If the Sn/Sb ratio exceeds one, tin will form as (Sn) after solidification and will not be efficiently utilized. From the standpoint of maximizing cast bullet strength, tin is useful only when it combines with antimony to form SbSn.

A benefit of the phase diagram is that it provides a guide to casting temperatures when alloy composition is known. Since molten metal beings to cool the instant it leaves the pot, it is recognized that the alloy must be heated somewhat above its liquidus to obtain good fill out. A good starting temperature is about 100° F above the liquidus. For bullet cavities which are long in relation to diameter, heat extraction is somewhat faster due to the greater surface to volume ratio of the bullet and casting temperature will generally have to be raised. In any case, increase the temperature only as needed to improve castability, and no more.

Tin Improves Castability

Industry and bullet casters alike have long recognized the beneficial effect tin has on castability. Because of their commercial importance, scientists have studied the effect of tin on antimonial alloys whose compositions are from the primary field of crystallization for lead, see **Figure 5**. Improved castability has been attributed to two effects(7). The first of these relates to a grain refining mechanism, but to understand it, we must first gain an understanding of how lead grains form when an alloy is cast into a bullet mould.

As stated previously, when an alloy cools to its liquidus temperature the primary phase begins to solidify. This initial solidification represents the beginning of grain formation. Perhaps the best analogy to grain formation from the melt is to picture the growth of pine trees where each tree is a grain. Metallurgists refer to the "metal trees" as dendrites(8). A single dendrite is schematically depicted in **Figure 6**. The growth of large "metal trees' is a factor which detracts from castability. While the "tree" is growing, liquid metal must flow between its branches to reach other parts of the bullet cavity. Just as pine trees restrict the flow of wind, so the formation of overly large dendrites restricts metal flow.

Higher tin concentrations reportedly cause the branches of the growing lead dendrites to detach from the main trunk during the solidification process. From our pine tree analogy,

Figure 6. Schematic representation of a dendrite "tree" which grows during the solidification of a lead alloy. Present theory suggests that when these dendrites grow too large inside the mould cavity, they choke off the flow of liquid to other parts of the mould, inhibiting castability. In the final solidified bullet, each dendrite represents a metal grain.

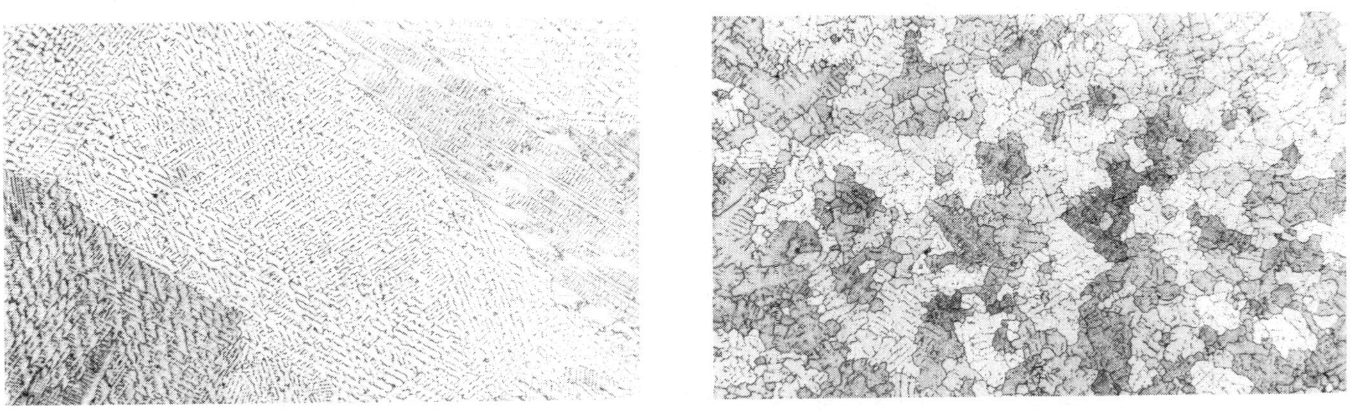

Figure 7. The grain refining effect of tin additions to wheel weight metal is illustrated in these two photomicrographs: 1) microstructure of a bullet cast with straight wheel weights-castability was fair, b) microstructure of a bullet cast from the same pot of wheel weight metal to which 2% tin was added in the form of bar solder-castability was excellent. Original magnification 100X, enlarged 1.5 times.

the ability of a tree to inhibit wind flow is reduced if some of its branches are removed or the tree made smaller. Similarly, the ability of the metal to flow into the fine details of a mould is less restricted if the branches either break or melt off, thus limiting the size of the "tree". One manifestation of limited dendrite growth is illustrated in **Figure 7** which shows a substantial decrease in grain size on adding tin to wheel weight metal.

The second mechanism thought to improve casting relates to the general reduction in dross when tin is added to molten antimonial alloys. We shall refer to it here as the skin effect.

Experiments have shown(9) that as little as .02% tin reduces the rate of dross formation in antimonial alloys. A layer of tin oxide, presumably SnO_2, forms a tight layer over the melt surface which limits contact between the metal and air. The effect is most pronounced up to 750° F and then degrades as the temperature is elevated.

When an alloy fills a mould, the advancing front of liquid metal is constantly breaking through a skin of oxide on its surface which then reforms and again breaks until metal movement stops. If the alloy oxidizes rapidly, the oxide film is relatively thick and it requires more energy to repeatedly break the oxide skin and this detracts from flow. When tin is added to the melt the rate of skin formation drops and flow is improved.

This basic mechanism of skin formation has been demonstrated in industry where moulds were flushed with flue gas prior to pouring the metal. With the oxygen purged from the mould, the advancing metal front was not oxidized and castability improved.

There is, therefore, good reason to continue the practice of adding tin to bullet metal. However, tin has become very costly and these additions should be as conservative as possible. To improve the castability of wheel weights, a highly recommended mix is 1/2 lb. of 50/50 solder to every 10 lbs. of metal(10). The composition of this mix comes out to approximately 3% antimony and 2.4% tin. If you get good results with lesser amounts of tin, so much the better. If any alloy is an unknown, and is troublesome to cast, you can add tin in small increments. *If* the problem is due to a lack of tin, a noticeable improvement should be evident by the time 1% tin has been added. If improvement is noted, continue small additions to refine castability but it is seldom necessary to add more than 2 or 3%, no matter what the alloy. If noticeable improvement is not evident after adding 1% tin, stop. There may be another casting variable you have overlooked and this should be cleared up first so as not to waste a very costly ingredient.

High Temperature Casting

There are numerous instances where the ability to cast bullets at high temperatures would be an advantage. For muzzleloaders, the castability of Minie-design slugs is often impaired by the inability to fill the thin base skirt region. Fill-out of the skirt area is difficult at lower casting temperatures since the ratio of mould surface area to metal volume is large and, and consequently, heat is rapidly extracted from the molten metal. In the study, "Cast Bullets for Hunting", which is described elsewhere in this book, some of the alloys tested had a very low antimony content, about 1%, and were difficult to cast in .30 cal. moulds at ordinary temperatures. In both of these cases, fill-out is enhanced by significantly raising the pot temperature. Unfortunately, this also introduces some nasty problems with dross and composition control.

Most of the alloys used in bullet casting contain some tin. As previously indicated, at temperatures not exceeding about 750° F the oxidation of a small amount of tin in solution leads to the formation of a protective oxide barrier across the surface of the melt which retards further oxidation. While this protective mechanism is operative during all types of casting, it is at its best when the surface is left undisturbed as when using a bottom pour pot.

At higher temperatures, the protective mechanism begins to break down by an as yet undefined mechanism and drossing increases. Laboratory experiments have shown that perceptible drossing (what you can weigh, not just see) of undisturbed linotype melts begins at 842° F(9), while for Pb - 5% Sn melts the majority of the tin is drossed off in as little as 30 minutes at 1050° F(11). Translated into practical terms, this means composition control can be a real problem when casting hot for an hour or more.

The solution to this casting dilemma is to sprinkle a small quantity of boric acid over the surface of the melt. When boric acid is heated, e.g., sitting on the surface of molten lead, it decomposes to water vapor and boron oxide according to the following reaction:

$$2H_3BO_3 = 3H_2O + B_2O_3.$$

The water vapor dissipates harmlessly since it is not trapped below the surface, leaving small beads of B_2O_3 glass on the melt. (WARNING: Do not push the boric acid under the surface of the melt. This accomplishes nothing and may cause a vapor explosion if water is still being evolved.) The beads then spread over the surface of the melt forming a continuous layer of glass which prevents air from contacting the alloy. Incidentally, the glass layer must remain intact to be effective and this precludes the use of a ladle.

The rate at which the glass spreads depends on temperature. The lowest practical temperature is about 850° F, at which temperature B_2O_3 is very viscous. As temperature increases, the glass becomes less viscous, producing a more coherent layer. In fact, the effectiveness of the glass improves with temperature and there really is no practical upper bound since your bullet casting pot will be destroyed long before the B_2O_3 is degraded.

To date, B_2O_3 has been used at temperatures up to 1150° F. While the glass is stable at these temperatures, the viscocity drops sufficiently that there is a tendency for it to contract into a puddle on the surface of the melt rather than remaining as a thin layer. This necessitates frequent additions of boric acid and detracts from the general utility of the layer. The best operating range appears to be between 850° and 1050° F, the latter temperature being considerably above the normal range of casting temperatures.

Figure 8. Boric acid provides excellent oxidation protection of lead alloy melts for high temperature casting. A salt shaker or similar container can be used as an applicator.

The glass layer is quite flexible and may be added to or removed at any time. If you notice any open spots in the layer, simply sprinkle on a little more boric acid; a salt shaker like that in **Figure 8** makes a handy applicator. When adding alloy to the pot, the layer can be removed by collecting the glass on an old spoon. The solidified glass can later be scraped off the spoon. Just be sure to let the glass cool before discarding it as it is hot enough to start a fire. After adding the alloy, sprinkle on a little more boric acid and you're back in business.

Use of boric acid also provides a few safety advantages: 1) it eliminates the need to flux at high temperatures which can be dangerous. Beeswax and other commonly used fluxes ignite rather forcefully when very hot and can cause burns. 2) B_2O_3 will dissolve a portion of the oxides which do form on the melt, and those which are not dissolved are agglomerated. This eliminates the usual powdery residue of toxic oxides and facilitates safer disposal. 3) A layer of glass over the melt surface reduces the bullet caster's exposure to metallic vapors which are ever present during casting, especially at high temperatures. However, it is still wise to cast with generous ventilation.

Impurities

The purpose of this section is to relate the effects certain impurities have on the melt behavior of bullet alloys. With the exception of tin, antimony and arsenic, almost anything else which will dissolve in lead can be considered an impurity. For the most part, the stray elements in bullet alloys are too dilute to be of any consequence, and can be kept that way by paying careful attention to the scrap alloys put into the pot.

Table 1 is a list of elements which may find their way into bullet alloys. The table compares their solubilities in pure lead with typical concentrations in some commercial alloys, although the table is by no means comprehensive. It should be noted that while the solubilities listed provide a general guide to the solution behavior of impurities, these values can either increase or decrease in the presence of tin and antimony. The table also illustrates that many impurities are limited simply because of their sparing solubility in molten lead.

Arsenic:

Arsenic was included in **Table 1** solely for purposes of comparison; IT DEFINITELY IS NOT AN IMPURITY. Arsenic is every bit as important to bullet alloys as antimony and tin, especially where hardness is concerned.

Arsenic by itself is of no great utility to the bullet caster. It does not significantly harden lead, and binary lead-arsenic melts are actually quite dangerous owing to the generation of arsenic fumes. An interesting benefit of high arsenic alloys is that the arsenic oxidizes to form As_2O_3 which dissolves other oxides and continuously fluxes the surface of the melt(9). For lead shot, arsenic is added in concentrations not exceeding 1%

Element	Chemical Symbol	% Solubility in Pure Lead at 752°F(400°C)	Typical Concentration in Common Lead Alloys	Alloy Source
Antimony	Sb	22.8	1 – 6	Cable sheathing, batteries, WW
			Up to 25	Type metals
Tin	Sn	Unlimited	0.1 to 0.5	Batteries, WW
			2.5 to 20	Type metals
			Up to 63	Solder
Arsenic	As	12.6	0.1 to 0.22	WW
			Up to 1.0	Shot
			Up to 3	Arsenical lead babbitts
Chromium	Cr	L.T. 0.01	L.T. 0.01	
Iron	Fe	L.T. 0.0002	L.T. 0.002	Used type metals
Nickel	Ni	0.2		
Bismuth	Bi	Unlimited	0.1	Cable sheathing
Barium	Ba	8.2	2.0	Very old bearing alloys
Aluminum	A1	0.02	L.T. 0.01	
Calcium	Ca	0.35	0.04 to 0.25	Batteries
Cadmium	Cd	Unlimited	Approx. 1.0	Batteries
			12.5	Wood's metal
Copper	Cu	0.13	0.06 to 0.25	Batteries, WW
Zinc	Zn	1.7	L.T. 0.002	
Sodium	Na	Unlimited		
Silver	Ag	6.5	Up to 2.5	Some electronic solders
Cobalt	Co	L.T. 0.01		
Platinum	Pt	13.4		
Gold	Au	48		
Magnesium	Mg	9	L.T. 0.001	
Manganese	Mn	0.02	L.T. 0.005	
Silicon	Si	L.T. 0.005	L.T. 0.005	
Strontium	Sr	0.5	Up to 0.3	Batteries

Table 1. List of elements found in commercial lead alloys which can be regarded as impurities in bullet metal. Antimony, tin and arsenic were included for comparison. **L.T.** means less than, **WW** signifies wheel weight metal.

where the self-fluxing action of As_2O_3 allows falling lead droplets to assume a spherical shape. This is why high arsenic shot is so desirable.

In most antimonial alloys, arsenic ranges from 0.05 to 0.5%; the typical concentration in wheel weights is about 0.17%. At these concentrations, the arsenic is no more a toxic hazard than the antimony and its presence permits wheel weights and other such alloys to be heat treated to 30BHN or more(5).

Calcium and Strontium:

The most likely source of these two contaminants is the new maintenance-free batteries which are on the market(12). While lead-calcium and lead-strontium alloys have properties which are well suited for use in batteries, they have no place in bullet casting. If either alloy is accidentally mixed with regular bullet alloys, a dangerous situation can arise. For example, if lead-calcium and lead-antimony alloys are melted together, calcium and antimony will combine in the melt to form an intermetallic compound by the following reaction:

$$2Sb + 3Ca = Sb_2Ca_3.$$

This compound, and others like it, have low solubility in lead, and since the compound is less dense than lead, it will float to the melt surface where the unsuspecting bullet caster will remove it with the dross, see **Figure 9**. The result is a melt with a reduced antimony content.

The second, and more critical aspect of the problem relates to the subsequent handling of the dross. Either in moist air or in contact with water, the intermetallics contained in the dross can react to form poisonous gases such as stibine (SbH_3) or its arsenic counterpart, arsine (AsH_3). Some sources indicate that as little as 50 parts per million (ppm) of AsH_3 is injurious to the system while the concentrated gas is fatal. Calculations show that if the antimony and arsenic from 3 lbs. of wheel weights are combined with calcium, the intermetallics formed in the dross can generate 0.3 cubic feet of gas. This is enough to contaminate the air space in a small room to dangerous levels. It is also interesting to note that stibine has been used as a fumigating agent where concentrations in the order of 10ppm are lethal to mice.

Figure 9a. An undisturbed, freshly skimmed melt of wheel weight metal remains bright and clear for some time.

For these and other reasons, batteries are no longer regarded as a practical source of bullet alloy.

Barium and magnesium are in the same chemical family with calcium and strontium and might be expected to cause similar problems. However, neither element is presently used to any significant extent in lead alloys. Barium was once used in bearing alloys but this practice has been discontinued. If barium were mixed with an antimonial alloy there would be an immediate increase in dross levels. However, there are no data available to indicate whether a "Sb_2Ba_3" type compound would be generated in the dross. The only conceivable source of magnesium might be some industrial castings which got mixed with lead scrap. Magnesium in a low tin, antimonial alloy would cause a large increase in drossing but existing phase diagram information does not indicate any problem with intermetallics. In lead-tin alloys the presence of magnesium increases dross and would also cause cast bullets to fall apart by intergranular corrosion in a matter of days or weeks.

Figure 9b. A tell-tale indicator of mixing antimonial and calcium alloys is shown above. The lumpy texture of the dross is produced by small clusters of the intermetallic compounds, Sb_2Ca_3 and As_2Ca_3, rising to the surface.

Copper:

Copper is present in almost all antimonial alloys in amounts ranging from .01 to about .25% depending on antimony and tin concentrations. Industry derives a number of benefits from adding copper to lead alloys, particularly in the area of chemical processing, but for the bullet caster the matter is more or less academic. In small concentrations, copper has no significant effect where cast bullets are concerned, but in high concentrations it will interfere with casting.

The only commonly available alloy which would be useful in bullet casting and which also contains a large percentage of copper is tin based babbitt. Tin babbitts typically contain 4 to 8% antimony, 3 to 8% copper and less than 1% of other minor additives. Some also contain nickel which has a function in the babbitt similar to copper. The cost of tin being what it is, if you can acquire tin babbitt for a reasonable price, don't pass it up. The copper content can be reduced with a little effort.

With the pot adjusted to normal casting temperature, about 750° F, add the desired quantity of tin babbitt to your lead alloy, then stir, flux and skim the pot as usual. Now turn the pot temperature down to about the melting point of lead, 621° F. It is preferable to use a thermometer for this, but one if it is not available the appropriate setting can be determined beforehand by adjusting the pot to just keep plumbers lead molten. As the melt slowly cools, the solubility of copper will diminish and small crystals of a copper rich intermetallic compound will rise to the surface giving the melt a lumpy appearance. Do not stir or flux, but periodically skim the surface to remove these lumps. When the temperature of the melt finally stabilizes at the melting point of pure lead, the copper content will be approximately 0.1%; the exact level will depend on alloy composition and how thor-oughly the melt was skimmed. Return the pot to normal casting temperatures and you're all set.

Zinc:

Zinc raises the strength of pure lead a small amount and has seen very limited industrial use in lead alloys, but for the bullet caster, zinc is a headache, pure and simple. The presence of zinc in an antimonial alloy causes an apparent thickening of the dross. When the alloy is cast the skin of dross on the advancing stream of metal prevents fill out of mould details. The most likely source of contamination is zinc die casting metal. For those who sometimes cast zinc bullets, it's best to keep the zinc alloy well labeled and far away from your supply of lead. Once lead has been contaminated by zinc there's nothing that can be done to recover the alloy.

Aluminum:

The phase diagram for lead-aluminum looks very much like the lead-zinc diagram in **Figure 4** except that line **BC** is very close to the lead side of the diagram which restricts solubility. When an aluminum contaminated melt cools, aluminum dendrites (trees) are the first thing to precipitate. These dendrites tend to clog small orifices which operate below the temperature of the melt, e.g., the spout on a bottom pour pot. The problem is sometimes encountered in the printing industry when casting type metal which has been contaminated with aluminum.

Aluminum also increases the skin effect mentioned previously and interferes with castability at normal bullet casting temperatures. However, this skin does not contribute to increased drossing problems. The skin forms rapidly but remains thin and actually provides oxidation protection which improves as temperature increases!! A number of patents have been granted on alloys which utilize aluminum for oxidation protection. Most intriguing is the fact that lead-tin alloys can be operated for days on end at 1200° F without appreciable dross build-up or loss of tin. Thus aluminum has the potential for providing a benefit similar to boric acid. But, aluminum is not easily put into solution with lead; it's not a simple matter of throwing a chunk of aluminum in the pot. The details for proper alloying for bullet casting purposes have not yet been worked out, but if and when they are, aluminum may find limited application for high temperature casting.

Other Impurities:

Of the remaining impurities listed in **Table 1**, cadmium is the only one which is likely to give trouble. Cadmium is used in some battery alloys and in low melting point fusible alloys like Woods metal. Cadmium causes some trouble with increased dross generation but the real hazard is its toxicity. The vapor pressure of cadmium is quite high which means it tends to evaporate from the pot, much more so than lead or antimony. Cadmium containing drosses are also extremely toxic. If you avoid battery scrap, you'll avoid this problem.

Iron, manganese, chromium and other elements typically found in steels are sparingly soluble in lead and usually only effect the drossing to a minor extent. Iron is often found in type metals due to the attack of the pot by tin and antimony; remember that industrial pots are exposed to lead alloys for much longer periods of time than those used in bullet casting.

Fluxing

With the realization that fluxing has nothing to do with keeping bullet alloys mixed, it is fair to ask - why do we flux? For bullet casting, the principle function of the flux is to clean the metal of dross or dirt. In addition to the obvious benefit of reducing casting defects, recent microscopic examinations of cast bullets by this author show that dross inclusions can cause localized recrystallization and softening like that shown in **Figure 10**.

Prior to fluxing, dross exists as a mechanical mixture of metal and metal oxides; the metal content ranges up to 90% by weight. It is clear that fluxing separates the dross from the metal, allowing the metal fraction to return to the melt and facilitating the easy removal of the oxides. However, the mechanism by which this separation occurs is not understood, nevertheless, fluxing provides a valuable function.

Experience indicates that fluxing is most effective at temperatures where a layer of liquid flux can remain in contact with the melt for a minimum of one minute. Simply throwing a pea-size lump of paraffin or beeswax on the surface, sitrring for a moment and then skimming does little to lower the dross content of a melt and may even make matters worse, especially at high temperatures. The reason for this relates to the fact that the cleaning action of the flux only takes place on the surface of the melt, while the dross iteself is distributed both on and below the surface. Since the flux is not soluble in the melt, it cannot reach the dross which is below the surface so the dross must be brought to the flux by generous stirring.

There are three basic locations for dross in a melt. The first and most obvious is that which accumulates at the surface. Depending on temperature, composition and the amount of agitation, surface dross can range from an insignificant scum to a layer of suspended oxide which extends some distance below the melt surface. This sort of dross is most troublesome to the ladle caster. A second type is that which clings to the wall of the pot. Hammer(13) has indicated that melt pressure can pin residue against the wall of the pot such that it is unable to rise to the surface. During bottom pour casting, this material is often swept into the discharging stream of metal and ends up coating the exteriors of bullets. A third and relatively innocuous form of dross remains suspended in the melt. It is known(14) that very small oxide particles, in the order of 0.00004" in diameter, can be trapped indefinitely below the melt surface because they lack the buoyancy to ovecome thermal currents.

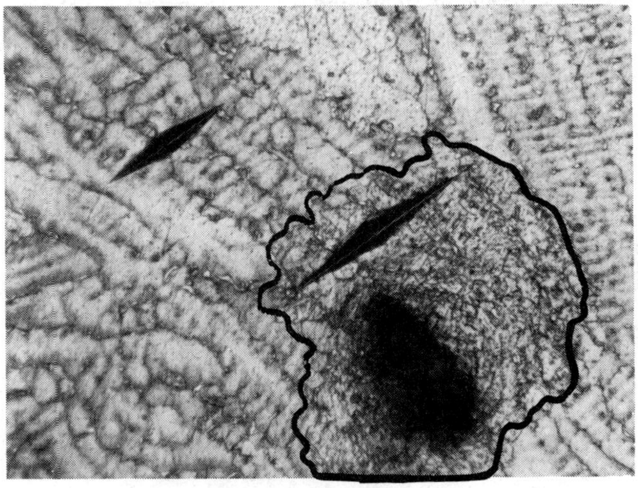

Figure 10. The dark oval in the photomicrograph is a dross inclusion in a nominal 3% Sb alloy which caused the surrounding metal to recrystallize and soften. The recrystallized region is outlined. The diamond-like impressions were made by a Knoop microhardness indentor using a 2 gram load. The hardnes at the larger indent in the recrystallized metal was 5.2; the unaffected metal (smaller indent) had a hardness of 10.9. (500X).

The best time to flux a melt is just after it reaches the liquidus temperature. Add enough paraffin, candle wax or similar substance to cover the surface with a 1/4" layer of liquid. The alloy must now be stirred and the walls and bottom of the pot scraped to release trapped dross. An old tablespoon with a wooden handle affixed makes a handy tool for this. If the metal was excessively dirty to begin with, such as when melting old wheel weights, the process should be repeated until the melt surface is free from any noticeable particles.

As valuable as fluxing is, it can be overdone. Remember, fluxing cleans the metal of dirt and dross; if the metal isn't dirty, why keep cleaning it? For bottom pour casting, fluxing is only necessary when adding alloy to the pot, and then only if the added material is dirty. If the new ingot was cast from a well cleaned batch of metal and its surface is not dirty or corroded, a light skinning is all that is necessary. Even if you're addicted to a ladle, like me, drossing can be controlled to the extent that little or no fluxing is required. When ladle casting, it is especially important to cast at the lowest practicable temperature since the surface is constantly being disturbed. Of the dross which is generated, the fraction which clings to the ladle can be periodically removed by drawing the hot ladle across a wire brush attached to your casting table or fume hood. The ladle should be drawn toward you so the spring in the wire kicks the loosened dross in the opposite direction. A box placed in back of the brush will catch the small spatters. Following this procedure, and periodically skimming the melt, will maintain acceptable dross levels. It is true that skimming without fluxing removes metal as well as oxide, but the amount of metal is small relative to the convenience of not having to flux. The microstructures of bullets cast by the above methods are consistently free of dross.

- - - *Epilogue* - - -

Now when a bullet casting friend comes along and says that unless you flux the melt, tin and antimony will gravity segregate, what are you going to tell him?

REFERENCES

1) L. S. Darken and R. W. Gurry, *Physical Chemistry of Metals*, McGraw-Hill, New York, 1953.

2) *The NRA Handloaders Guide*, National Rifle Association of America, Washington, D.C., 1969.

3) R. Hultgren, et al, *Selected Values of Thermodynamic Properties of Metals and Alloys*, John Wiley and Sons, Inc., New York, 1963.

4) Metals Handbook, 8th Ed., Vol. 8, Metallography, Structures and Phase Diagrams, American Society for Metals, Metals Park, Ohio, 1973.

5) Col. E. H. Harrison, *Cast Bullets*, National Rifle Association of America, Washington, D. C., 1979, p. 118.

6) R.P. Elliott, *Constitution of Binary Alloys*, First Supplement, McGraw-Hill, New York, 1965.

7) U. Heubner, et al, Z. Metallkde., 1976, *67*, 277.

8) *Solidification*, Proceedings, ASM Seminar, Oct. 11 & 12, 1969, American Society for Metals, Metals Park, Ohio, 1971.

9) Wilhelm Hofman, *Lead and Lead Alloys*, English Translation ed. By Dr. Gerwig Vibrans, Springer-Verlag, N. Y., 1970.

10) Frank Marshall, Jr., "Adequate Accuracy is Attained Simply", The Fouling Shot, Published by the Cast Bullet Association, No. 22, Oct., 1979.

11) D. Marshall, unpublished research.

12) D. Marshall, "Battery Plates: Bad New for Bullet Casters", American Rifleman, Jan., 1979, p. 16.

13) A. J. Hammer, "Dross and Fluxes", Precision Shooting, Nov., 1964, p. 4.

14) H. Krug, Z. Metallkunde, 1974, *65*, 337.

BULLET CASTING, SIZING AND LUBRICATION

Understanding the Bullet Mould

Lyman moulds are the most widely known and asked-for bullet moulds in the world. This company is the oldest current manufacturer of bullet moulds in the United States—or anywhere else, for that matter. We like the cast bullet business and make every effort to offer a well-made product.

Through the periodic changes in our mould line we can see how shooting has progressed over the years. This includes the variety and quality of the guns being used, as well as increased knowledge and expertise on the part of the reloading shooter.

Before the brief life of the muzzleloaders requiring specialized, close-tolerance projectiles, there was really very little requirement for diameters to be held within several thousandths of a specification. To do such a thing would require that a mould be custom made to fit each barrel—since the bores varied greatly, one to another, even models of the same nominal caliber! In fact, riflemakers did often make custom moulds to fit just-finished rifles.

Today our rifle and handgun barrels and chambers are much improved, of course. However, we do not have "perfection" by any stretch of the imagination. Many shooters, and even some reloaders, assume that their modern centerfire rifle or pistol is perfectly—and uniformly—bored and chambered. Yet these same shooters are probably aware that it is common for devotees of the modern muzzleloading sport to find it necessary to have a smaller/larger ball diameter or thicker/thinner patch to adjust their projectile to a good fit with their particular bore.

This bore/chamber variance isn't at all unusual in muzzleloaders; certainly not any more than in our U.S. made centerfire rifles and pistols. They share the same basic problem—tolerance.

Modern centerfire rifles, pistols and factory ammunition and components are produced under standardized cartridge, chamber and barrel specifications. Strict compliance with these standards, which include all dimensions and pressure factors involved in a given chambering, is voluntary. However, all U.S.—and most foreign manufacturers—appreciate this centralized standards system and subscribe to its criteria in their manufacturing processes. In each step taken to produce a finished barrel, cylinder or cartridge, there is an allowable dimensional variance for our old friend, tolerance.

Until the modern muzzleloading era began, most shooters, and a lot of reloaders, didn't own a micrometer and wouldn't have known one if it crawled up and sat in their lap. Today, that situation is far from reversed, but there has been encouraging progress. A good micrometer is almost as valuable as a good bird dog. Every bullet caster should own a micrometer.

Today there are increasing numbers of centerfire shooters measuring their chambers and barrels. Handgun silhouette shooters are measuring revolver chamber mouths, as well as barrel dimensions; some discovering that chamber mouths *do vary* within the same cylinder and that the bullet emerging from them just might not fit the bore properly.

Furthermore, cagey centerfire rifle shooters are measuring alignment of chamber with the bore and discovering problems can occur there. The old ritual of "slugging the bore" has been improved by measuring an upset lead slug from *both* the muzzle and breech areas. Surprising variance can occur between the two points—and has, even on new barrels from reputable makers.

It is vital that the cast bullet shooter understand "tolerance" and the various places that dimensional variance can appear. Cast bullets are not nearly as tough and resilient as their jacketed descendants and can suffer from this characteristic to a disproportionate degree. Often the reloader can adjust his handloads to minimize the effects of any variation. Like factory firearms and ammunition, Lyman moulds incorporate a tolerance factor in their various dimensions. They always have and always will.

Periodically, various groups of mould designs are reviewed to evaluate the validity of basic dimensional specifications. In fact, the entire Lyman line has been reviewed recently. Changes were made, where needed, to accomodate dimensional standards found in current production rifles and pistols. Some diameters remained unchanged; some increased while other decreased.

Lyman realized years ago that the diameter of the bullet which dropped from the mould varied with the composition of the bullet metal poured into the mould. Years have passed and this basic law of metallurgy holds just as true today as it did then.

To eliminate confusion—at least internally—Lyman settled on #2 Alloy as it's standard bullet metal for centerfire projectiles; pure lead for the muzzleloading balls and conicals—and shotgun slugs, too.

It is so today. Standards for Lyman centerfire moulds are based on the use of #2 Alloy. Dimensions of the muzzleloading and shotgun slug designs are predicated upon the use of pure lead.

The alloy composition makes a dimensional difference and some bullet casters have been slow to understand the subject. Complicating things has been the fact that, oft-times, bullet metal is not all it is supposed to be.

For example, when we buy a quantity of alloy for lab use and want to be absolutely certain of its composition, we send out a number of ingots (from the same shipment, of course) for independent analysis. Often the results amaze us. Ingots of "scientific quality" bullet metal purchased direct from a reputable foundry can vary tremendously.

It is no wonder that bullet casters are sometimes disappointed with accuracy and confused by as-cast weights and diameters which vary significantly from a published reference. Those variances aside, the truly important thing is that the bullet be sufficiently hard for the desired velocity level and of a diameter to fit *your* gun.

This leads up to "as-cast" diameter and the process of "sizing."

"As Cast" diameter is, literally, the diameter of the bullet as it drops from the mould. Ideally, it would exactly match sized diameter or be only a bit larger so that very little, if any, metal would be displaced during the sizing and lubricating operation.

Lyman plans the "as-cast" diameter of each caliber grouping to permit maximum utility within reasonable sizing ranges. We do not claim our moulds will cast to a single guaranteed diameter or that said cast bullets will be perfectly round. Instead, we state that we manufacture each of our moulds to a specified size in #2 Alloy with a manufacturing tolerance for both diameter variation and out-of-roundness. Each run of blocks is inspected at random by both dial indicator systems and actual casting and measurement of resultant bullet.

The production of moulds to a customer's specification is the domain of the custom mould maker—and even then there are few (if any) who will guarantee a perfectly round cast bullet from their mould. And, of course, the cost of such a custom mould is much greater than those produced for general consumption.

So the bullet which drops from a Lyman mould is intended to

be reduced in diameter and trued up by the sizing process.

Sizing the cast bullet has been done a number of ways over the years. Today all sizing dies feature tapered leads which allow excess metal to be swaged, rather than shaved, into dimensional conformity.

During the recent Lyman review of specifications (which included the sizing dies as well), care was taken to maintain a satisfactory relationship between maximum as-cast diameter and minimum sizing diameter commonly used in a given caliber. Again, the standard metal is #2 Alloy.

Lyman sizing dies, like the moulds, are made within certain size and manufacturing tolerances.

How Lyman Moulds Are Made

1. Cold-rolled steel blocks of a special formula are purchased by Lyman Products for its complete line of moulds.

2. The rough blocks are milled so that the sides and faces are squared and correctly dimensioned.

3. The second cut is the groove which eventually will accommodate one of the pincer-type handles. At the same time the ends of the blocks are squared and milled to dimension.

4. The facing surfaces of the blocks are ground uniformly to a perfect match. They are then paired and numbered, and air-vent lines are milled on the faces.

5. Holes are drilled and tapped for dowel pins and setscrews.

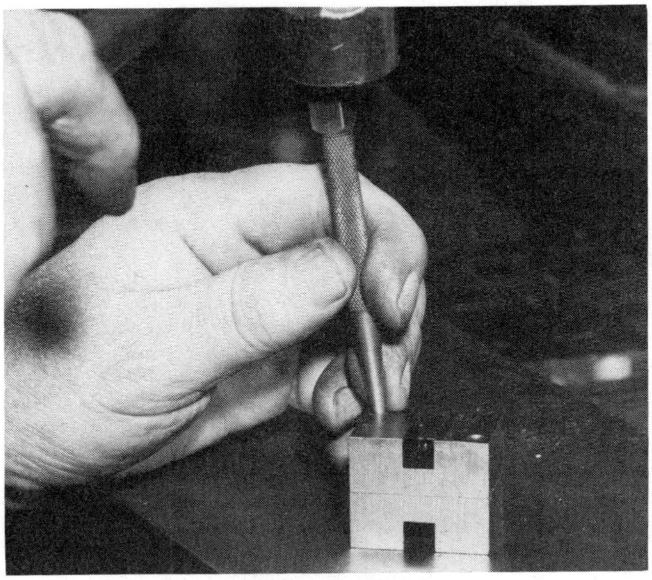

6. At this stage the dowel pins are driven home and the mould blocks checked for alignment.

7. Cherrying drills, prepared by a toolmaker, cut the mould cavities. Hundreds of these drills, or "cherries", are needed to supply the variety of calibers and bullet designs within the Lyman line.

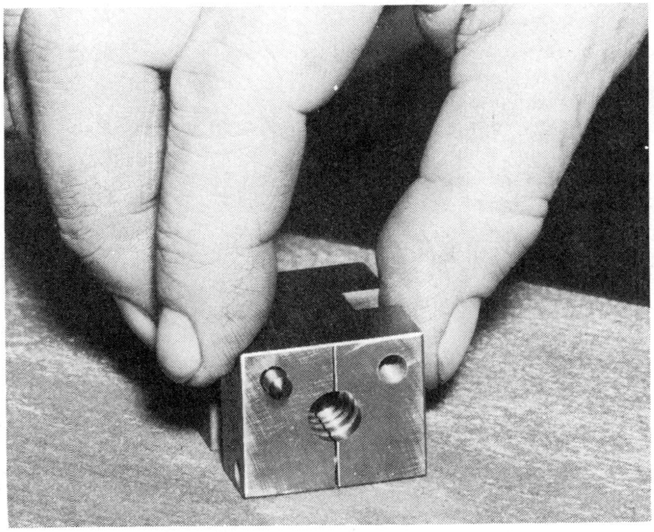

8. At several stages each block is strapped to remove any burrs that may have resulted from earlier handling and sharp edges are slightly broken.

9. The blocks are heat-cured to oxidize the metal. This "blues" the mould and minimizes the time required for the breaking-in process.

10. A screw fastens the sprue cutter in place and the blocks are ready for final inspection, packaging and shipment.

Tuning and Trouble Shooting Your Moulds

From time to time it may be necessary to correct certain conditions which commonly occur on mould blocks either as a result of manufacturing processes or usage by the caster.

There is nothing especially difficult about working on a set of mould blocks; the procedures are simple and logical. All that is required is patience and a delicate hand on the file or stone.

Most of us are aware that stoning the innards of our favorite firearm can improve functioning and trigger pull tremendously. All that is done, usually, is to remove minor burrs and polish sliding surfaces to minimize friction.

That's the same purpose of working on the moulds—tune and smooth to enhance performance and enjoyment.

So, with that in mind, we have prepared a discussion of common conditions which can be remedied by the individual.

1. The new moulds should be taken apart and thoroughly degreased.
2. With blocks only assembled, hold mould up to a light source and check for light between the blocks (looking at the sides where they join as well as into the cavity).
   If the blocks are not in complete contact, determine if it is caused by:
   (a) raised metal on block interfaces
   (b) excessive pin projection
   (c) foreign matter in pin cavities
   Raised metal or burrs should be *carefully* stoned away or taken down with a jeweler's file, taking extreme care not to damage the cavity.
   In the case of excessive pin protrusion, the offending pin should be set back slightly by placing the block on a firm support and lightly striking the pin with a ballpeen hammer. Check for fit until no light shows between blocks. Remove foreign matter in pin cavities with Q-tip, match stick, pipe cleaner, etc.
3. Check mould block top for burrs or raised metal, and the sprue cutter screw hole as well. Any irregularity here will raise, or angle, the sprue cutter plate; causing galling/marring of mould top and/or build-up of metal between sprue cutter plate and mould block top.
4. Inspect the vent lines. These should enter cavity on both sides. If blocked, *very carefully* open up with fine jeweler's triangular file (vent lines are only .003"-.005" deep, so be careful).
5. The sprue cutter plate must lie flush with, and parallel to, the top of the blocks. Check for flatness and, if required, polish with 400 grit wet or dry paper which has been placed on plate glass. Since the sprue cutter plate is a stamping, check for burrs or raised metal around the bottom edges. Any which are present should be carefully removed with a jeweler's file.
6. Assemble blocks only. Holding each block in one hand, gently try to rotate one against the other. Any movement seen or felt should be eliminated by tapping alignment pins slightly deeper into the facing pin cavity.
7. Inspect handle channels for burrs, especially around screw holes. Remove any which are found with a jeweler's file. Handles should pivot freely on these screws to facilitate easy opening and joining of the blocks. Casting can begin once you are satisfied with the blocks and reassemble them.

The following tips cover common casting situations:

8. Care should be used to avoid getting any metal on block interfaces when casting. Presence of this causes separation of interfaces with consequent increased and distorted bullet diameter. Any spots present should be wiped off with some sort of natural fiber rag (like cotton) while mould is hot—don't use synthetic cloth for this.
9. After dropping a bullet, first close blocks then close sprue cutter plate. Reversing this procedure may result in the plate striking the right block, causing burrs or marring.
10. Random hang-up of bullet in right or left block may be considered normal (depending on bullet configuration). Persistent hang-up in either block is usually caused by a tiny burr or other imperfection. Look for this with a magnifier. Often these may be removed by gently rolling a plug of steel wool (#00 or #000) in the cavity.
11. The sprue cutter plate should rest flush on block tops and turn readily; but not loosely. Excessive pressure of cutter against mould may prevent proper venting; a sloppy fit may cause raised sprues on bullet bases, as well as an accumulation of metal between the cutter plate and mould top.
12. Moulds should, of course, never be permitted to become rusty. Should any be present, gently rubbing with fine steel wool (#00 or #000) will remove it. However, should extensive rust occur, all may not be lost. Clean the mould, as above, the best you can. Then, cast with it. You may find it casts easier and the bullets shoot just as well.
13. Since rust is to be prevented, a preferred method to the practice of leaving the last bullet in the cavity is to apply rust preventive oil to the mould. Better still is to wrap the cool mould in VPI paper. VPI crystals with the mould in a closed container may also be used. This latter method obivates the necessity of degreasing the mould before its' next use.
    The new Lyman mould box, which handles all but four-cavity blocks, is an excellent—and virtually airtight—storage container. A strip of tape around the lid/body joining line seals the box completely.

Bullet Casting

Bullet Metal

For years, Lyman has used #2 Alloy as the standard for its centerfire bullet moulds. This alloy is quite hard—but not as hard as Linotype metal. Shotgun slugs and all muzzleloading projectiles are cast in pure lead, a very soft substance.

Over the years, shooters have logically combined the hardest practical alloy, Linotype metal, with the finest high-speed lubricant. Alox/beeswax, to achieve high velocity with little or no sacrifice in accuracy.

Linotype metal is an excellent choice for maximum rifle and pistol velocities. Happily, it is very similar to #2 Alloy as far as shrinkage goes. Chances are, there will be little dimensional variance between the two.

Over the past few years we have noticed an interesting situation unfold concerning accuracy of bullet metal alloys. The Lyman Technical Department buys its bullet metal—pure lead, #2 Alloy and Linotype metal—from a foundry.

Contents of the metals are clearly specified and the metal should be of "scientific" quality.

To shorten a long story, we found, via independent analysis, a significant variation from specification in the #2 Alloy and Linotype metals in lots purchased over a span of several years. More recently, we have tried to define the contents of wheelweights—and found not only a great deal of variation in the formula across the country but a universal decrease in the antimony content. We have adjusted our do-it-yourself #2 Alloy formulas accordingly.

We report the foregoing simply to alert you, the bullet caster, to the fact that all might not be as it is labeled. So if one alloy fails where you think it shouldn't, perhaps it is a bit out of spec.

#2 Alloy

As previously stated, #2 Alloy is a rather hard bullet metal which contains three ingredients: lead, tin and antimony.

While we have recommended that Lyman #2 Alloy be used for the cast bullet loads in this Handbook, this does not necessarily mean the bullet caster has to purchase this specific metal in order to cast good bullets. Lyman used this bullet metal **to furnish a known standard for bullet weights, loads, etc.** For example, a bullet will vary in weight (and performance) in relation to the blend of the alloy. If a standard alloy of a specific blend is not specified and followed, consistent performance will not be possible.

The reloader can, and should, make his own bullet material by blending the required metals to the proper proportions. The metal required can be gathered from junk yards, gas stations, automotive supply houses, plumbing supply houses, hardware stores and printing shops.

Composition & Hardness of Common Bullet Metals

Alloy	Lead	Tin	Antimony	BHN
Linotype	86%	3%	11%	22
#2 Alloy	90%	5%	5%	15
16-1	94%	6%	0%	N/A
10-1	91%	9%	0%	11.5
Wheelweights	95.5%	0.5%	4%	9
Pure Lead	100%	—	—	5

Recipes For Mixing Your Own # 2 Alloy

9 pounds wheelweights 1 pound 50/50 bar solder	Makes 10 pounds # 2 Alloy
4 pounds Linotype 1 pound 50/50 solder 5 pounds pure lead	Makes 10 pounds # 2 Alloy

Some of the materials gathered from junk yards and such will contain small traces of other metals, but for the purpose of making bullets they may be considered as pure. Bar solder containing 50% tin and 50% lead is available from plumbing houses and is an excellent source for tin. Linotype is available from printing shops and while it is a little too hard in its existing form (11% antimony, 3% tin, 86% lead), it can be cut back by blending the proper proportions of tin to the mix. Wheelweights contain about 4% antimony and make an excellent base material to blend with the tin. Pure lead can be purchased either in ingot form or salvaged from a junk yard in the form of pipe, etc. *Caution: Do not salvage metal from auto or marine batteries. There has been a change in the material which can be hazardous to your health. Avoid batteries, period.*

To more clearly illustrate the effect of alloy variation on as-cast diameters and sized-to diameters in given moulds and sizing dies, we have prepared the chart on the next page.

While this data is based on extensive research which took place over the last two years, the table title carries the word "predicted" since minor variances—even in measuring from one person to the next—will always be with us.

Melting and Fluxing Bullet Metal

When working with bullet metal, one of the first requirements is a suitable heat source capable of heating the metal to about 750° or 800° Fahrenheit. A simple cast iron pot to hold the metal and most any heat source (kitchen stove, etc.) will suffice. Improved equipment, such as the Lyman Electric Furnace, is more desirable. The electric furnace is cleaner, safer and more convenient and its adjustable thermostat allows best control of the melt temperature. Whether you are blending the various metals into an alloy or actually casting bullets, the same melting and fluxing procedure is followed.

Heat the metal for about twenty minutes until it becomes liquified and flows freely. It is then ready for fluxing. As the metal melts, a gray scum will rise to the surface, contrasting sharply with the silver brightness of the molten lead. **DO NOT REMOVE THIS SCUM.** This contains tin, the most valuable component of the bullet metal. Fluxing will recombine the tin-lead-antimony mixture. This operation is extremely important and should be done carefully.

To flux the metal merely drop a small bit of tallow, beeswax, or bullet lubricant into the mixture. A smokey gas will rise from the top of the pot and this gas should be immediately ignited with a match. This will eliminte the smoke.

A more modern-and much more pleasant-fluxing procedure is to use a dry substance, such as the product called MARVELUX. Smoke and greasy fumes are eliminated—an important benefit to those casters with wives and/or mothers in residence—and a good flux is obtained. This method is much preferable to the foregoing traditional technique.

Whichever fluxing substance you choose, be sure to stir the mixture with the dipper. As you stir, hold the dipper so the cup side is down and raise it out of the metal with each stirring stroke. This scoops air into the mixture and helps the flux.

Predicted Physical Characteristics of Bullets Cast in Various Lead Alloys

		Predicted As-Cast Characteristics								H&I	Predicted Dimensions - Sized			
	Bullet	Lead		Wheelweights		#2 Alloy		Linotype		Sizing		Wheel-	#2	Lino-
Caliber	Example	Wt. grs.	Dia.	Wt. grs.	Dia.	Wt. grs.	Dia.	Wt. grs.	Dia.	Die	Lead	weights	Alloy	type
Rifle														
.22	#225415	47	.2246″	46	.2246″	45	.2250″	44	.2252″	.224	.2237″	.2237″	.2240″	.2240″
.243, 6mm	#245496	87	.2437″	85	.2441″	83	.2445″	81	.2447″	.243	.2427″	.2427″	.2430″	.2430″
.25	#257464	92	.2576″	91	.2581″	88	.2585″	87	.2588″	.257	.2567″	.2567″	.2570″	.2570″
.270	#280473	129	.2776″	127	.2781″	123	.2785″	121	.2788″	.277	.2765″	.2767″	.2770″	.2773″
.280, 7mm	#287308	170	.2846″	167	.2850″	162	.2855″	158	.2858	.284	.2836″	.2837″	.2840″	.2844″
.30	#311291	176	.3095″	173	.3100″	168	.3105″	163	.3108″	.308	.3076″	.3077″	.3080″	.3081″
8mm	#323470	167	.3234″	165	.3240″	160	.3245″	155	.3248	.323	.3226″	.3227″	.3230″	.3233″
.35	#358315	215	.3578″	212	.3584″	206	.3590″	200	.3594″	.357	.3566″	.3567″	.3570″	.3573″
.375	#375449	276	.3773″	272	.3779″	264	.3785″	256	.3789″	.376	.3756″	.3757″	.3760″	.3763″
.45/70, 458	#457193	439	.4575″	433	.4583″	420	.4590″	407	.4595″	.457	.4566″	.4567″	.4570″	.4574″
Pistol														
.25	#252435	53	.2522″	53	.2526″	51	.2530″	49	.2533″	.251	.2507″	.2507″	.2510″	.2510″
.32	#311252	80	.3125″	79	.3130″	77	.3135″	75	.3138″	.309	.3086″	.3087″	.3090″	.3100″
9mm	#358242	96	.3558″	95	.3564″	92	.3570″	89	.3574″	.356	.3556″	.3557″	.3560″	.3563″
.38/357	#358156	156	.3578″	153	.3580″	149	.3590″	144	.3594″	.357	.3566″	.3567″	.3570″	.3573″
.41	#410610	225	.4101″	221	.4108″	215	.4115″	209	.4119″	.410	.4096″	.4097″	.4100″	.4103″
.44	#429421	256	.4296″	252	.4303″	245	.4310″	238	.4314″	.429	.4286″	.4287″	.4290″	.4294
.45	#452374	235	.4515″	232	.4523″	225	.4530″	218	.4535″	.451	.4506″	.4507″	.4510″	.4514″

Metal that has been properly fluxed will leave the surface almost mirror bright and flecked with small particles of black and brown impurities. Skim off and discard these impurities. Flux the metal whenever, by its appearance, it seems to need it. While the dipper is not in use, it should be left in the molten metal to keep it hot.

Casting Bullets

After the metal has been fluxed and is hot enough to pour easily through the dipper, it is ready for casting. In addition to the mould, you should have on hand a hardwood stick (about 10" long) to be used for opening the mould. Also, pad a small area of your bench with an old piece of cloth material. This will soften the fall of the hot bullets as they drop from the mould and prevent them from being damaged.

While there is no set way to cast good bullets, we do offer the following as a suggested method. Fill the dipper half-full of metal and place the spout of the dipper against the pouring hole as shown. Holding the mould and dipper together, slowly turn them to a vertical position with the dipper on the mould as shown.

The extra metal that runs over the top of the mould is called sprue. When it hardens, which takes only several seconds, pick up the hardwood stick and tap the sprue cutter sharply. This will separate the sprue from the base of the bullet. Drop the sprue into a cardboard box, or other receptacle. Open the mould and let the bullet fall to the pad. If the bullet does not drop out readily, use the stick to rap the hinge pivot sharply, as shown in the illustration. Use only wood for this purpose and never strike the mould block themselves.

It is very important to pour a generous sprue and allow it to harden. As the bullet cools, it draws metal down from the sprue. If it cannot (i.e. the sprue has already hardened or is of insufficient size) then internal voids will form.

Further, cutting a still-molten sprue damages the bullet's base and often creates a buildup of smeared alloy between the cut-off plate and mould blocks.

As the mould will be cool, your first bullets will be imperfect. Casting bullets, one after the other, will bring the mould to the proper temperature. If you wish, the mould can be preheated by placing it on the rim of your furnace, or along side your lead pot on the stove. Never, under any circumstances, dunk the mould in the molten metal!

Wrinkled bullets indicate that the mould, and/or metal, is too cool. Frosted bullets indicate that the mould, and/or metal, is too hot. Good bullets should be clean, sharp and fill the mould. Imperfect bullets should be collected and, along with the sprue, returned to the pot.

Bullets selected for accuracy shooting should be carefully weighed on the reloading scale. This reveals air pockets that may have formed in the bullet, lightening or unbalancing it. The actual weight of your bullets will depend pretty much on the consistency of your alloy material, and it may vary slightly from lot to lot. Weigh about twenty or thirty bullets out of each group and determine an average bullet weight. Reject any bullets that vary more than 1/2 grain (+ or –) of this weight. Rejected bullets may be saved and recast at a future date.

Heat Treatment of Wheelweight Alloy

Over the years there has been periodic discussion of hardening otherwise soft bullet alloy by heat-treating, quenching, etc. Some of the techniques were elaborate; one which comes to mind required 72 hours of continuous oventime! Others were written by non-shooters who couldn't quite tie the technique to the application.

Recently, additional and substantial work has been done by Dennis Marshall centering on plain old wheelweight metal. Only an hour of oventime can triple, at least, the hardness of typical wheelweight alloy!

Wheelweights, which lie at about 9 on the Brinnell scale, can reach a hardness exceeding 30 if the proper degree of heat is applied and subsequent procedures done correctly.

Beyond the nominal alloy contents of lead, tin, and antimony, wheelweights always contain traces of other substances. Key to the success of heat treating is arsenic. The amount of arsenic present doesn't seem to be too critical—just that there be some there.

An old kitchen spoon makes a fine tool to use in cleaning the surface of the molten lead. The spoon's shape and size make its use a bit handier than a lead dipper.

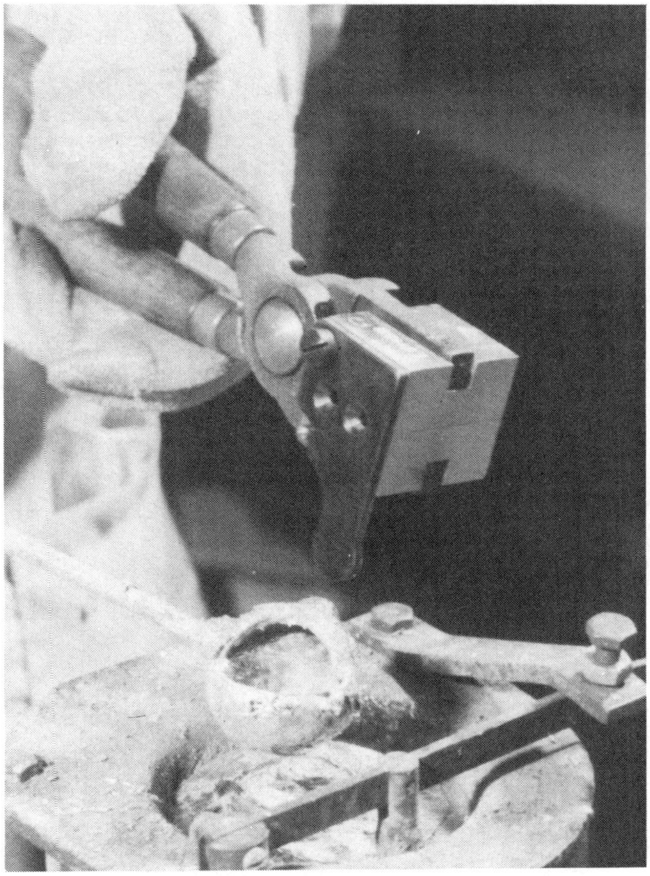

The pouring spout on the ladle and the pouring hole on the sprue cutter are designed to fit closely together.

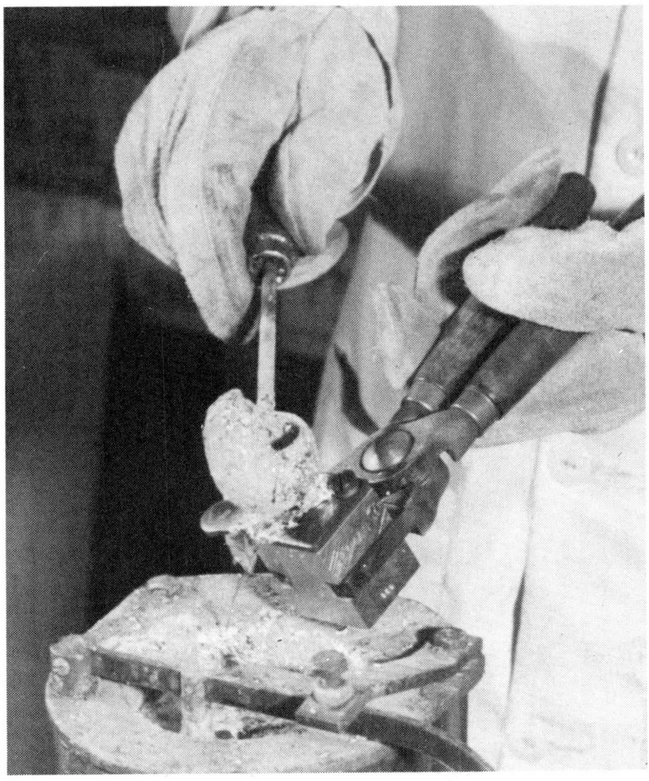

After the spout and mould are snugly together pour the lead into the mould by slowly rotating the block/ladle unit with the ladle on top. Experiment with your rate of pour to obtain the best results.

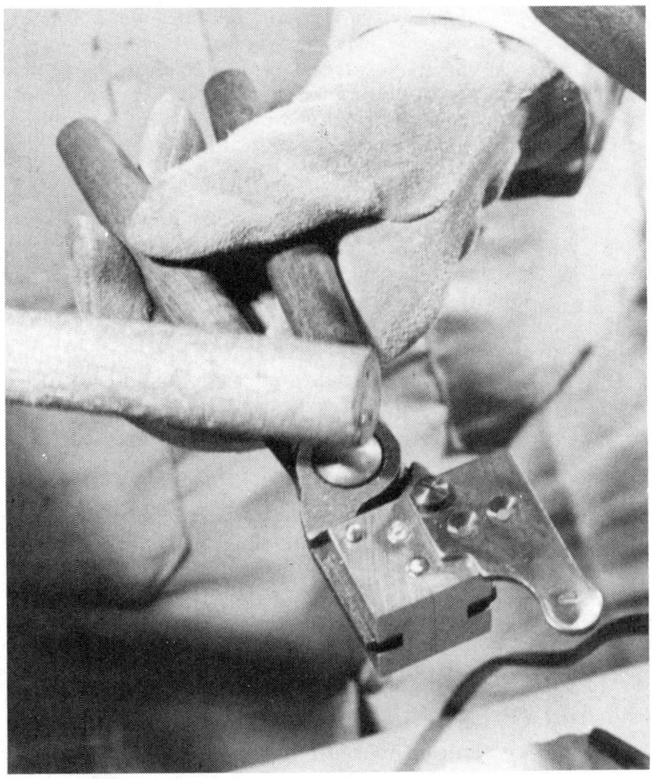

Hitting the hinge pin is the best way to shake loose a sticking bullet. Don't land random blows on the blocks or sprue cutter since this could cause deformation and affect the quality of bullets.

The procedure goes like this:

1. Cast your bullets in the normal manner, saving several scrap bullets.
2. Size bullets. *Do not lubricate.*
3. Place several scrap bullets in your oven and increase the temperature until you notice the bullets starting to melt or slump. Start at 450° F, using an accurate oven thermometer.
4. Once the test bullets start to melt, back off about 5°-10° F and slide in the first batch of good castings. There seems to be no benefit to keeping a batch in the oven more than an hour. In fact, a half-hour does just as well, if not better.
5. Remove the bullets from the oven and plunge them into cool water. Allow them to cool thoroughly.
6. When you are ready to lubricate, install a sizing die .001″ larger than the one used for sizing. This is to prevent the sides of the bullet from work-softening from contact with the sizing die. Apply gas checks and lubricant.
7. Load normally.

Editor's Note: Several questions came to mind concerning various aspects of the foregoing procedure:

1. If the bullet can be damaged by a second pass through the same sizing die, what happens to metal strength when the bullet goes down the barrel?
2. Again, what happens to hardness when an oversize bullet is seated in a case neck expanded with either a standard diameter ball or M-Die?

When asked, Marshall stated that while the questions were logical, the bullet leaves the barrel long before the metal can react and soften. In the second case, there simply appears to be no affect.

This procedure, as reported by Dennis Marshall, seems the best and most thought-out presented to date. We have done some testing at Lyman, raising the Brinnel hardness to three times that of the original metal—from 9 BHN to 28 BHN. Marshall, by using the highest possible temperature, has achieved a hardness of BHN 39.

Considering that Linotype is rated at BHN 22, this heat treatment of wheelweight metal offers a substantial hardness increase at very little additional effort and cost.

Lyman Lead Melting Equipment

Experienced bullet casters recognize the importance of top quality equipment when casting really precise bullets. Although the mould makes the bullet, it must have plenty of alloy held at the correct temperature to perform to its proper specifications. That's why the right furnace is so important, and why the serious caster relies on Lyman. Take a look at our offering of furnaces and then choose the one that best suits your needs.

Improved Mould Master XX

Lyman's improved Mould Master XX is made for the serious bullet caster who needs an electric furnace that will hold up to twenty pounds of molten lead. Featuring a greatly increased capacity over older designs, it still operates on household current and will provide years of reliable service. The thermostat housing has been relocated to one side to allow the caster a better view of the bottom-pour spout. Further, access to the pot for ladle casting has been greatly improved by replacing the old style over-arm stop with a metering thumbscrew in the lever hinge.

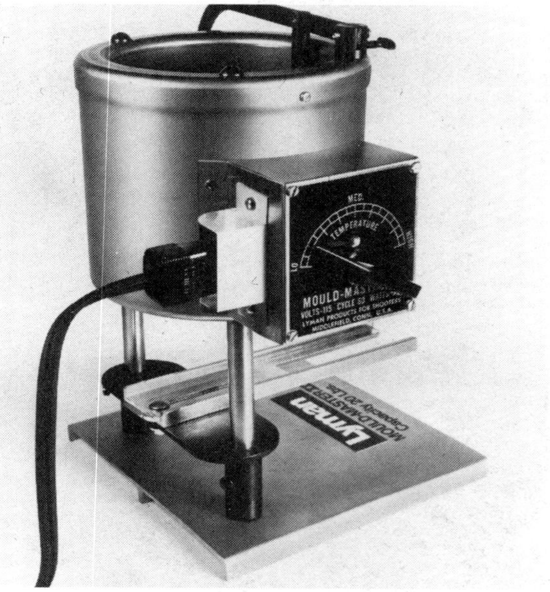

Mould Master XX

Features include:

- 20 pound pot capacity.
- Improved steel crucible.
- Calibrated thermostat to permit controlled heat throughout the casting spectrum.
- Solid, sturdy base.
- Bottom-pour spout to speed casting process.
- Works with Lyman's mould guide, ingot mould, and other Lyman casting products.
- Available in 115V A.C.

Mould Master

A heavy duty, 11 pound capacity electric furnace which comes complete with one ingot mould, The Mould Master is a safe and reliable furnace designed for years of flawless service. It operates on standard household power — 115 Volts, A.C., 1000 watts. Calibrated thermostat regulates heat from 450° to 850° F, and controls temperature to within 20° F. A lever-operated valve controls discharge spout.

Mould Master

Mould Guide

For use with bottom pouring electric furnaces, this accessory cuts casting time in half. It holds and aligns bullet moulds and takes the weight off the operator's hands.

May be used with Lyman and similar moulds and mould handles up to and including four cavities.

Sizing and Lubrication

The sizing process is merely a method of swaging cast bullets to a standard size that corresponds to or slightly above the groove diameter of the gun. Bullet sizing also ensures that the bearing bands of each and every bullet in the group are perfectly round in shape. As no metal is removed from the bullet (bullet is swaged to shape and size), sizing does not alter the bullet's cast weight.

The actual process of sizing and lubricating cast bullets is straight-forward and covered in the furnished instructions; but there are a few things to consider.

1. Set up your lubricator/sizer according to furnished instructions. Lyman recommends that sizers not be used to apply lubricant when the temperature of the work area is below 60° F. The reason is that cold lubricant is much more difficult to force through the sizer and into the bullet's grooves.

   A caster can actually damage his sizer by applying too much pressure on the lubricant in the reservoir.
2. Be sure to use the correct top punch for the bullet design. It's possible to push the bullet into the sizing die with any old punch which isn't grossly oversize. However, only the correct top punch can ensure the bullet's alignment with the sizing die's centerline. Lose that alignment and you've lost a lot.
3. Occasionally, you may find that lube is building up under the bullet base. This condition is usually caused by excessive lube pressure and may be remedied by reducing said pressure for subsequent bullets.

   However, reduced lube pressure may then leave you with only partially-filled lube grooves. If that happens, turn the bullet 90 degrees and push it into the sizing die a second time. That should do it.

   Be sure to wipe the lube from bullet bases as the propellant within the cartridge case might be contaminated, particularly in warm weather or after extended storage.
4. While bullets can be cast and safely stored indefinitely, you should size and lubricate only the number of bullets you need for the planned production of finished cartridges.

   Allowing lubed bullets to sit around invites them to accumulate dust and grit—all the foreign matter guaranteed to scratch your bore.

   Should you find yourself in that situation, the lube can be removed with a solvent and the clean bullets safely stored for later use.

Effect of Bullet Alloy on As-Cast & Final Size Through .308 Die

#311291 Bullet

	Lead	Wheel-weights	#2 Alloy	Lino-type
Avg. dia. as-cast	.3090	.3095	.3100	.3103
Avg. dia. sized through .308 die	.3078	.3079	.3082	.3083

Sizing Bullets

As stated previously, the cast bullet must be properly sized to or above the groove diameter of the gun. To determine the exact groove diameter, the barrel should be slugged. Lyman "H & I" sizing dies corresponding to the groove diameter of the firearm are available. The Lyman 450 Bullet Sizer and Lubricator is designed to hold these dies and to supply the necessary leverage to complete the sizing operation.

Basic groove diameters

RIFLE

Caliber	Groove
All 22 cal. (except 22 Hi-power)	.224, .225†
.243, .244, 6 M/M	.243, .244†
.256 Win. & All 25 cals.	.257, .258†
.264 Win. Mag. & 6.5 mm.	.264†
.270 Win.*	.277, .278†
7 M/M, .280 Rem., 284 Win.	.284, .285†
30 cals.	.308, .309†, .310
7.62 Russian*	.310
32/20 Win.	.311, .312†
7.65 Mauser*	.311
.303 British*, 7.7 M/M Jap.*	.313, .314†
8 M/M Mauser (S Bore)	.323, .325†
32 Win. Spec. 32 S.L. & 32 Rem.	.321, .322†
338 Win. & 33 Win.	.338†
.348 Win. & .351 Win. S.L.	.350, .352†
9 x 56 M/M & 9 x 57 M/M	.354, .355†, .356
35 cal.	.357, .358†, .359
375 H & H Mag. & 375 Win.	.375, .377†, .378
38/55*	.379
38/40	.400, .401†, .403
44/40	.427, .428†
.444 Win. SPL & .44 Mag.	.429, .430†
444 Marlin	.430, .431†
45/70 & 458 Win.	.457, .458†, .459
.50 Cal. Rifle	.509, .512†, .515

PISTOL

Caliber	Groove
22 Jet & .221 Fireball	.224†, .225
25 A.C.P.	.251†
30 Luger	.310†
9 M/M Luger*, 38 A.C.P., 380 Auto	.354, .355†, .356
38 Spec., 357 Mag.*, 38 S & W*	.357, .358†, .359, .360
38/40	.400†, .401, .403
41 S&W Mag.	.410†
44 S&W Spec. & 44 Mag.*	.429†, .430, .431
45 A.C.P., 45 Auto Rim, 45 Colt* (Post WW II)	.450, .451†, .452
.256 Win.	.257†, .258
.32 A.C.P., 32/20, .32 S&W	.311, .312, .313†, .314
.45 Colt (Pre WW II) & .445 Webley	.454†

*Wide variations in Groove Dia. Suggest you slug barrel.
†Indicates suggested useful diameter.
This may vary from gun to gun.

Determining the "Size To" Diameter

Generally speaking, the less a bullet must be sized, the more accurately it will shoot. Some cast bullet shooters feel that .002" is the most a bullet can be sized and retain the ability to perform accurately.

Whether .002" is, indeed, the magic number is only part of the overall picture.

The real question is the true dimensions of your gun. Rifles should be measured both at the muzzle *and* just past the chamber mouth. Handguns should receive the same bore measurements while revolvers should have *each* chamber mouth measured. The results of these tests should show you where to start.

To obtain these measurements you must drive a pure lead slug into the barrel or chamber mouth. Use two slugs on the barrel (revolver owners may want to use just one; maneuvering around the frame to introduce that second slug is tough) and a separate one for each chamber in a revolver's cylinder—and identify them with the cylinder from which they came.

Perhaps the best system is to use the substance Cerrosafe, measuring the resultant cast with a good micrometer.

Normally, you define your initial "size to" diameter as one which will match, or slightly exceed, the groove diameter just ahead of the chamber. Sometimes revolver shooters have the matter confused by having a groove diameter larger than that of their chamber mouths. They should select their first sizing die on the basis of barrel, not chamber, dimensions.

Bullet casters should expect to experiment with several sizing diameters, just as they'll use several types of propellant, in a given load development program. Normally, these other diameters will be greater than the groove diameter as discussed above. To shoot an undersize cast bullet is to invite almost certain leading and inaccuracy.

Gas Checks

These little cups of gilding metal or copper protect the bases (and sides) of a cast rifle or pistol bullet from gas cutting. In turn, the bore is protected from receiving minute globules of molten tin, etc., which are ironed onto the bore walls as the bullet overtakes them. These deposits then grow by the same mechanism and soon begin to scuff off metal as each bullet passes—and thus a leaded bore is born.

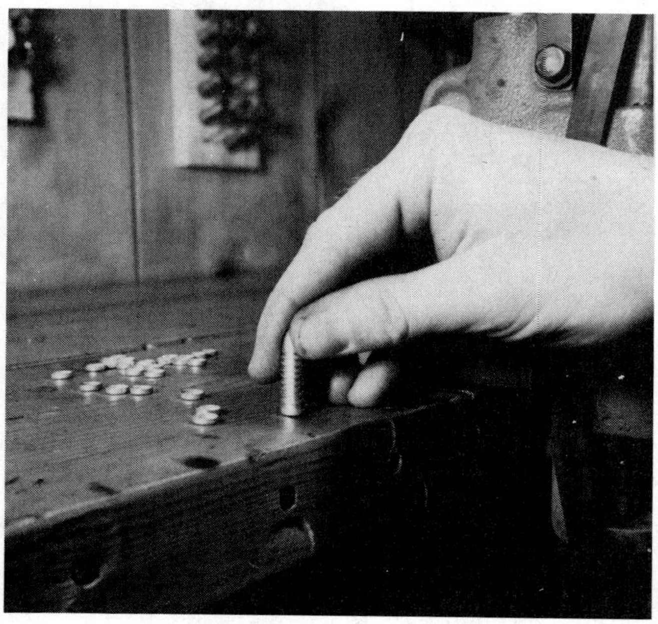

Gas checks, regardless of type, must be square with the bullet base. Sometimes ensuring the squareness requires an extra step or two--don't begrudge it. You will benefit from the enhanced accuracy of your cast bullet reloads.

Align the cast bullet as best you can to give it a true start into the sizing die. Extra care at this stage will pay dividends at the range.

Of course, there are other causes of bore leading—and even combinations of causes, but these little bits of hard metal—used with hard bullets (22BHN or better) and Alox lubricant—permit velocities up to 2,500 fps and beyond without leading.

Gas checks are normally applied as a part of the sizing operation. The important factor with gas checks is that they be squarely seated on the bullet. A crooked gas check means a "flyer".

What's important is to develop a seating technique that *you* can successfully manage. Some casters set the check in the mouth of the sizing die and let the lubricator/sizer handle the entire job. Others carefully press the bullets into the gas checks, by hand, as a separate operation; the result of which goes into the lubricator/sizer.

There are two types of gas checks: those that do not crimp-on and those that do. Both have their advantages. Use the one which works best for you; for most people, that would be the crimp-on style.

The finished product. Sized evenly with the gas check squarely in place, this bullet will shoot accurately.

Diagnosing Your Castings

Sprues can tell tales. This is a good sprue (left), cut smoothly from a square, filled-out base. The other sprue (right) was torn from, rather than cut from, the bullet's base because the cutter plate was struck before the metal had hardened. This kind of tear can also occur when hard alloy (like linotype) sprues are cut. This is not cause for rejection in most cases.

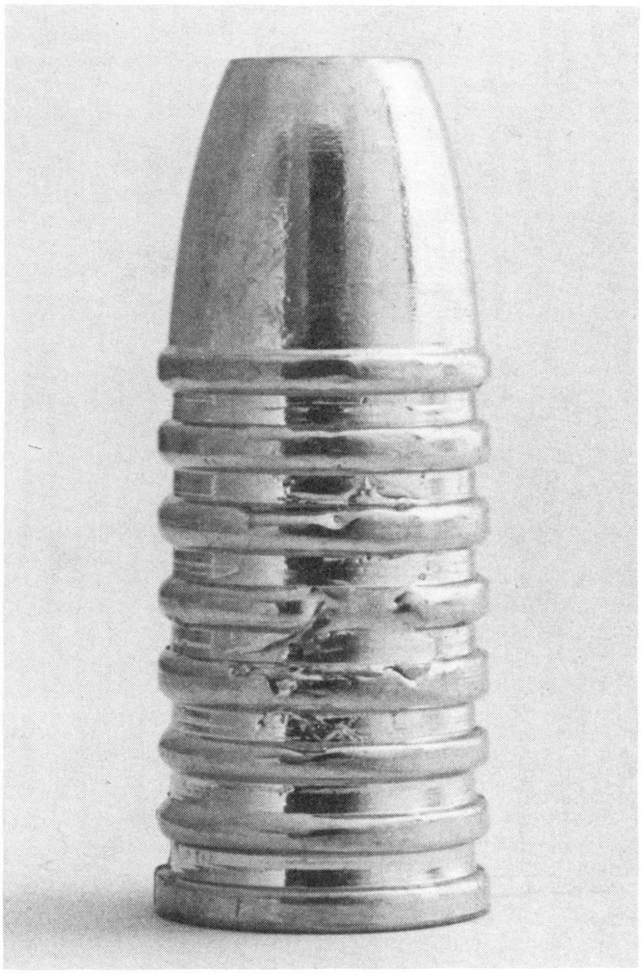

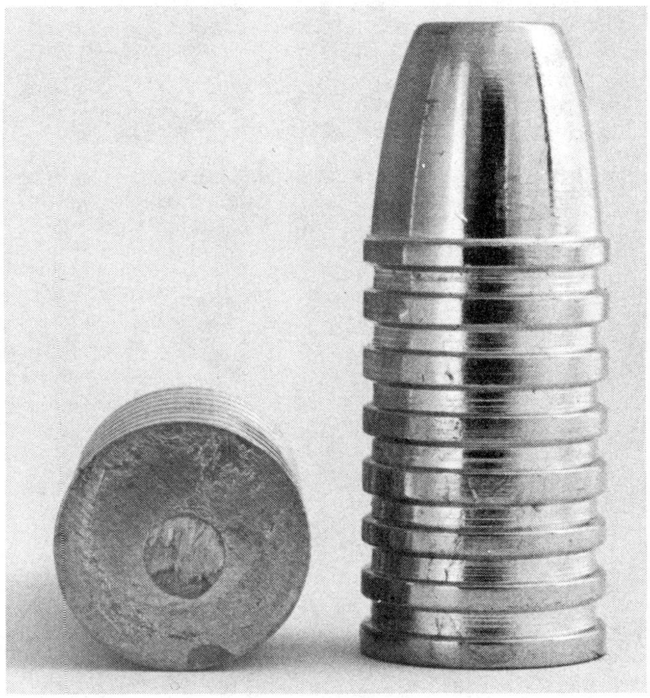

The effects of mould temperature are shown clearly on these specimens of #457121. Rounded bands, incomplete fill-out (left) indicate a cool mould (with possible impaired venting) while (right) despite minor surface blemishes, this is a good casting, suitable for all but the most demanding applications. The base (center) is clean and sharp, with a smoothly-cut sprue.

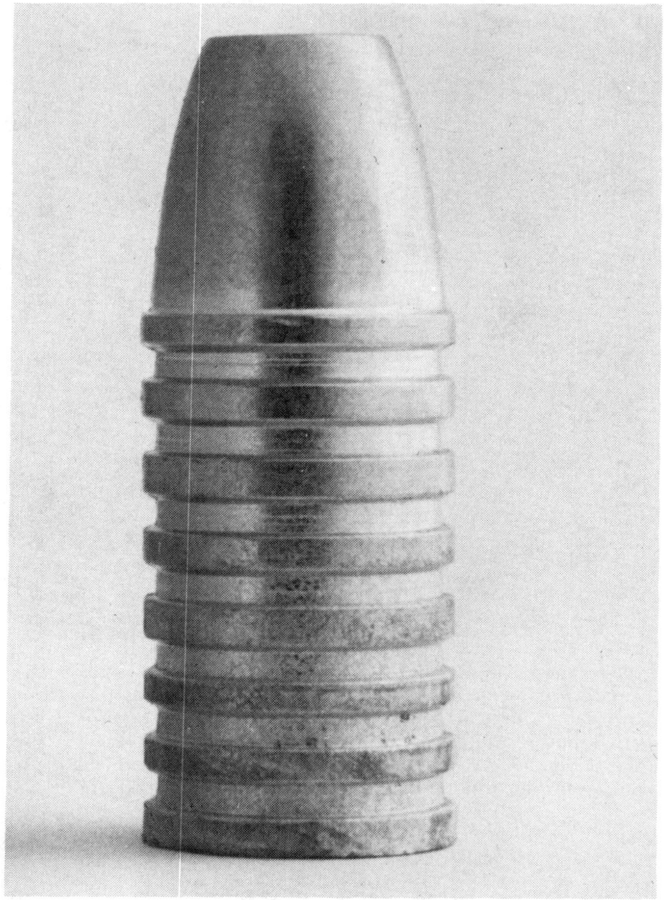

Products of mould temperature extremes. #429251 (left) dropped from a cold mould. #457121 (right), on the other hand, exhibits the typical surface "frosting" of high temperature casting—which is sometimes necessary and usually does not impair accuracy.

The casting of #31141 (left) is good, demonstrating proper filling and crisp edges; the other casting, also #31141, exhibits a rounded base which indicates one or more of the following problems: inadequate venting, inadequate metal supply or cool mould/metal.

"Flashing" is a gross condition indicating incomplete block closure; possibly due to lead on block interfaces, burrs at pin bases, dents at the mouths of (or debris within) pin cavities, loosely held handles or improper pin/cavity alignment. These castings of #429434 should be rejected and the blocks from which they came examined and corrected.

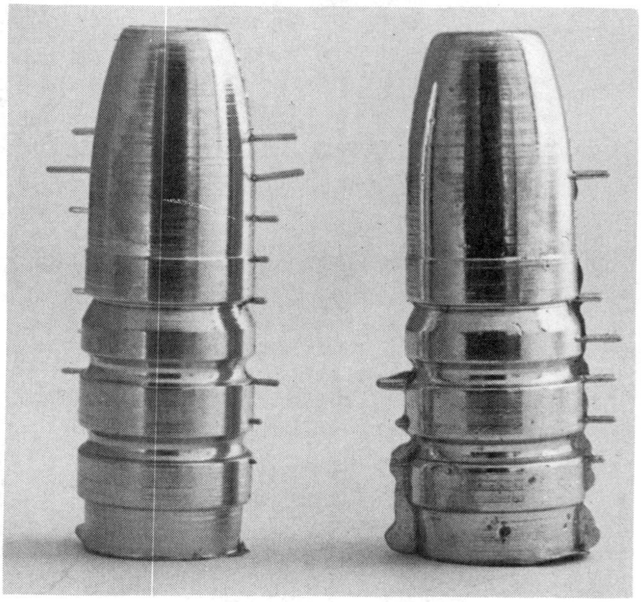

"Finning", as shown here on #31141, can be caused by several factors or combinations thereof. Incomplete block closure is a minor possibility, noting the flashing at the base of the right-hand bullet.

More likely is a high-pressure fill, uncommon with dipper casting, in which the alloy weight in a fully charged bottom-pour pot forces the bullet metal into the vent lines.

Another possible factor is the alloy tin content; the more tin the easier the metal flows, to a point. "Whiskered" bullets are usable after rubbing them clean with your fingers. "Flashed" bullets may not be, though.

RELOADING PREPARATION: EQUIPMENT AND COMPONENTS

Getting Started

Making the move into bullet casting is rather easy and not all that expensive considering that you'll be deriving additional enjoyment of your hobby—plus increased per-shot economy which quickly defrays equipment cost. If you are a muzzleloader, your start-up costs will be greatly reduced by the elimination of sizing and lubricating equipment.

The key to an easy introduction is starting out with the proper equipment. We feel that the beginning caster needs the following basic casting equipment to obtain good initial results: mould & handles, melting pot, pouring ladle, casting mallet, ingot mould and a lubricator/sizer fitted with appropriate top punch and sizing die. Muzzleloaders excused on the last item, as noted.

1. We feel our mould and handles offer the best investment to bullet casters. It is quite common to hear of Ideal/Lyman moulds which have been going strong for years. Some of our competitors offer less expensive moulds, but consideration of their rather short service life makes their purchase a questionable economy.
2. The melting pot is an item on which you can economize if you wish. Lyman offers not only the more expensive electric furnaces of 11 and 20 pound capacities, but also a simple iron pot holding about eight pounds of metal, which can be used on the kitchen stove or with a campstove out back or in the garage. The mobility and versatility of the small pot can be a real blessing should there be a wife or mother in your household.
   The Lyman electric pots offer a great deal more capacity and sophistication. A major feature on both is the option of dipping bullet metal with a ladle or pouring from the bottom of the pot directly into the mould using the bottom-pour valve system. Additionally, the thermostats in these pots are calibrated with the scale on the control plate, permitting temperature settings to be noted for future casting use. A handy option for both the 11 pound Mould Master and the 20 pound Mould Master XX is the "mould guide." This device is attached to the furnace legs and serves to align and support blocks for volume casting from the bottom-pour valves.
3. The pouring ladle, or dipper, is a good, inexpensive buy. These are offered by several firms including Lyman. You should have one of these even if you buy one of the bottom-pour electric furnaces. Put ten bullet casters in a room, return in an hour, and you'll find two distinct groups: one in favor of bottom-pour; the other praising the dipper.
4. A casting mallet of some sort is needed with which to strike the mould's sprue plate to cut off the bullet's sprue. A hammer handle will do, as will a number of other implements which have non-marring faces. You cannot use any object which will mar or dent the mould. Lyman offers a casting mallet which is weighted, by design, towards the striking end for ease of use.
5. The ingot mould, like the dipper and iron pot, should be a basic item in your initial investment. This mould allows you to empty your pot or furnace at the end of a casting session and produce easily-managed one pound ingots which can be marked as to metal type and stored. These small ingots fit into all lead pots and furnaces without any problem.

This is a typical casting set-up. Little space is required and an evening's work can produce hundreds of good bullets.

6. The lubricator/sizer is, we feel, essential to the bullet caster reloading centerfire bullets. The Lyman #450 not only swages the bullet to size, but applies a gas check (if needed) and cleanly lubricates from it's ample reservoir. Ours is a sturdy tool, designed for bench-mounting, which has proved itself over years of use. Thanks to the interchangeability of sizing dies and top punches, our sizer will handle just about any sizing situation, producing clean, concentric cast bullets ready for loading.

Additionally, every caster should include basic safety garb. We recommend that no casting be done unless the caster is protected by sturdy gloves and eye protection. An accident with molten lead can have serious consequences.

One last item for riflemen, while not essential to initial efforts, is the Lyman "M" Die which flares the case mouth just a bit. This makes starting the cast bullet that much easier. Lyman pistol die sets already include this type of expander.

Bullets can be cast and processed in the same area in which you reload. All primers and propellant should be cleaned up, containers sealed and put away. There are no special fixtures

The Lyman 450 Sizer and Lubricator performs three separate operations:

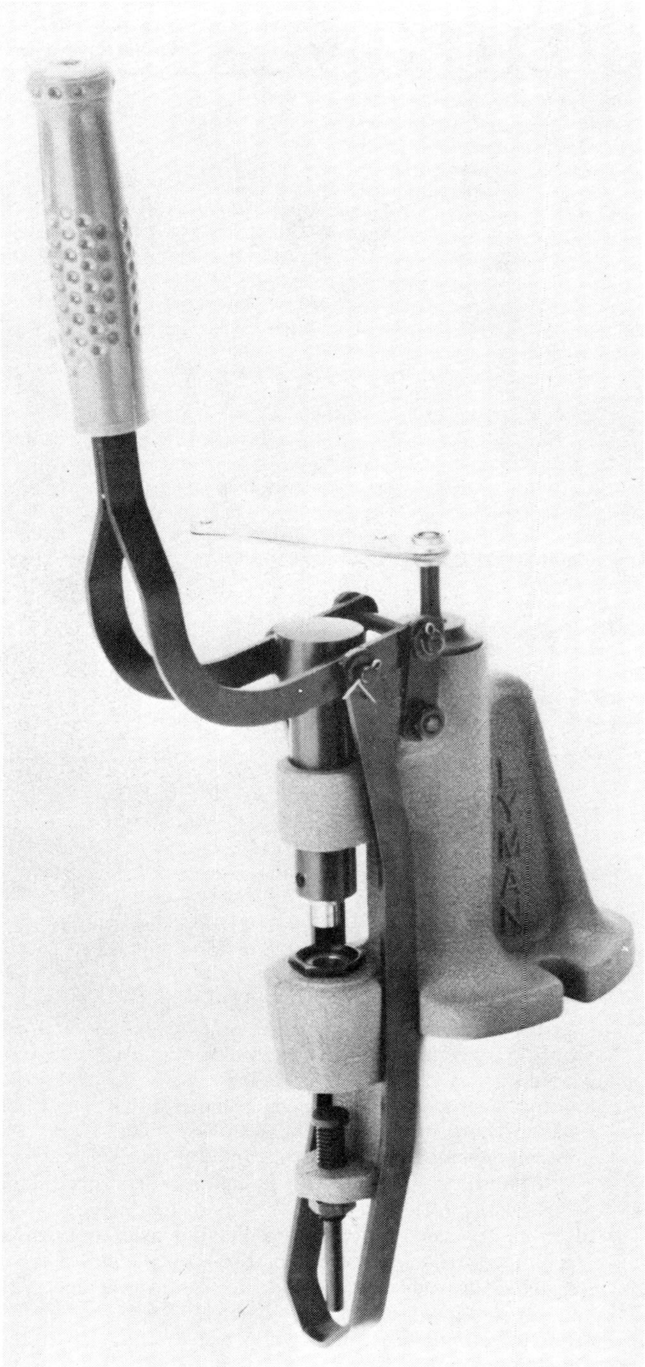

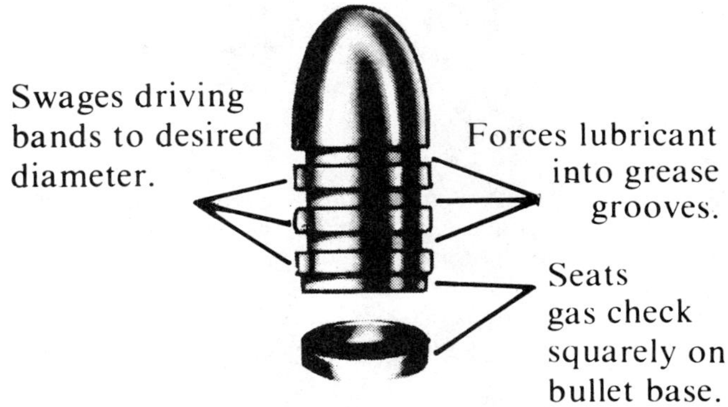

Melting Pot, Ingot Mould, and Casting Dipper

needed...a folded towel will cushion the bullets as they fall from the mould. Surely there's an old box that will serve to catch sprue and rejects.

Venturing into bullet casting is really a rather simple task. Sure, there are plenty of variables with which to wrestle, but excellent results can be had right from the first by choosing good equipment and following the instructions for its use.

Because of the nature of this book we have, so far, assumed you are a reloader already in possession of the equipment needed to reload centerfire cartridges with jacketed bullets. Usually, new casters of centerfire projectiles are already reloaders.

However, should you be new to reloading, please consider Lyman equipment. We manufacture a complete line of reloading presses, reloading dies suitable for cast bullets-plus all needed accessory items.

Reloading

When confronted with the considerable variety of well designed tools available to today's reloader, the decision as to which are best to buy may be somewhat perplexing. Naturally, we hope that your final choice will be Lyman. With Lyman tools, you are assured of guaranteed satisfaction backed by over ninety years of service to shooters. Also, the instructional text in this Handbook is related directly to the use of

Lyman tools. Using the tools pictured will help you in following the text and in understanding the illustrations.

The O-Mag and Spar-T presses pictured in this section are extremely rugged reloading presses. Simplicity of operation and reasonable price make them particularly attractive to both the novice and advanced reloader. For a complete description of the tools shown and current prices, send for a **free** Lyman catalog. Simply fill out and mail the card which is bound inside the back cover of this manual.

The O-Mag Press

The O-Mag Press, introduced by Lyman in 1979, is one of the best press buys on the market. The massive frame is cleanly machined and accurately bored for true concentricity of the cartridge and die in each step of the reloading process. This press will handle all resizing projects and is strong enough for most swaging needs

A particularly notable feature of the O-Mag is the enlarged frame opening which accepts even the largest cartridge without the reloader having to resort to tipping and wiggling in order to slip the case and bullet into the seating die.

Other features include fixturing for an auto-primer feed mechanism, three bench-mounting bolt slots (instead of two) and a new sure-grip handle. The O-Mag, like the Spar-T, can be adjusted for operation on either up-or down-stroke. This press will accept any brand of 7/8″x14 dies.

The O-Mag is available by itself or as a set. Our O-Mag set includes the O-Mag Press, plus the tools you need to turn out top-quality reloads for your rifle or handgun: O-Mag Press, O-Mag Primer Catcher, Universal Primer Arm, Ram, detachable Shell Holder and complete set of standard AA Dies.

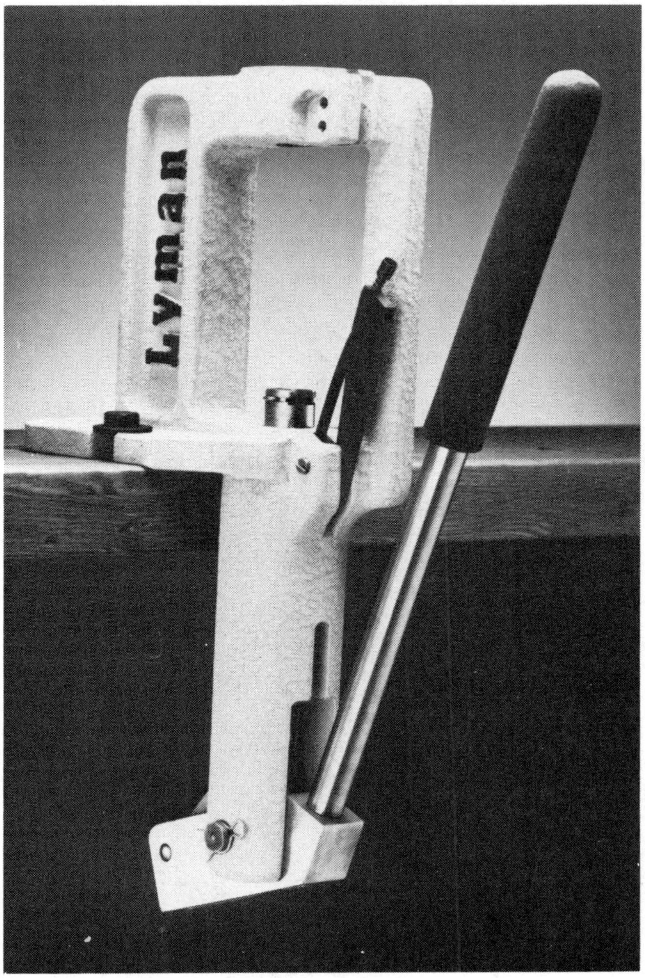

The Spar-T Press

Our Spar-T Press combines the high speed of turret loading with operating ease and strength. The massive cast frame, ruggedly constructed of high-silicon iron-steel, firmly holds the 6 station Verti-Lock Turret in place with a heavy duty 3/4″ steel stud. A positive stop, audible click action ensures foolproof cartridge to die alignment and rapid operation.

The Spar-T Press has found particular favor with pistol shooters because of its turret system which permits mounting two 3-die sets for frequently reloaded cartridges. Once adjusted, these die sets can be left in the turret, ready for immediate use.

Reloaders of rifle cartridges use the press in the same way, often attaching a #55 powder measure in a die station by using a 7/8″x14 adapter.

The Spar-T is available by itself or as a set which includes the most wanted accessories.

The Set provides all the tools you need to turn out top-quality reloads for your rifle or handgun: Spar-T Press, Spar-T Auto-Primer Feed, Universal Primer Arm, Spartan Ram, Spartan Shell Holder and, complete set of All-American Dies.

Reloading operations are demonstrated, later in this book, on our new O-Mag press and the Spar-T turret press.

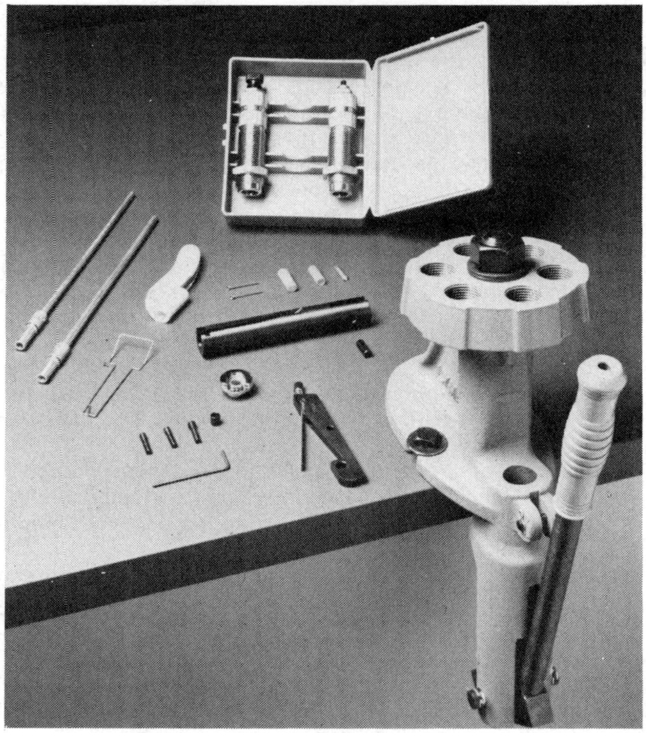

Whatever your interest, write for a free copy of the current Lyman catalog. This catalog is issued annually in December January and fully describes all the products offered by Lyman.

Our technical staff will be happy to offer assistance. If questions or problems arise, please write including all details.

COMPONENTS

Cartridge Cases

The chief function of a cartridge is to seal off the breech at the time of firing. To accomplish this, the case walls must expand freely so that they are tight against the sides of the chamber. This sealing action prevents the hot powder gases from leaking back around the cartridge and out through the action. Along with this, the cartridge case must withstand the chamber pressure that is built up during firing. To achieve this, the case requires a structural strength of its own—plus the additional supporting strength supplied by the breech face and chamber walls. In essence, the case functions as an intrinsic part of the gun. **A firearm is no stronger than the case that is used in it—nor is the case stronger than the firearm.**

Cartridge brass is carefully tempered in its final manufacture. The head of the case is thick and tough which gives it the strength and rigidity necessary to resist the force of the chamber pressure. The forward section of the case (neck, shoulder and body) is considerably thinner than the head section. In manufacture, these portions are given an anneal which leaves them soft and ductile. The obvious advantage is that the case walls and neck will now expand freely to release the bullet and seal the chamber while the cartridge is fired.

As shooters, we may have been rather casual in our regard for empty brass cases, but as reloaders we soon come to think differently. Without a quantity of strong and serviceable cases, we would not get far in reloading ammunition. The usual way for a reloader to obtain serviceable cases is to purchase factory loaded ammunition. After this "store bought" ammo has been fired, the empty cases are retained for future reloading. The reloader may also purchase new cases from his component dealer.

To make sure your cases are in prime condition, start with either new or once-fired cases. Never use brass of unknown origin such as that found on a shooting range.

Each firing and resizing has an influence on the serviceability of the case. The battering of chamber pressure and the forces applied by the resizing die eventually work-harden the forward portion of the case and destroy its usefulness.

Once a cartridge has cracked from fatigue it is past saving and must be discarded. The remaining cases in the lot may be saved by annealing.

Annealing Cases

Annealing is the process of altering the structure of any metal so as to relieve its working stresses and increase its ductility. A simple form of brass annealing enables the reloader to re-soften the neck and shoulder of his cases after they have become work-hardened.

The only special equipment required for this job is a concentrated heat source, such as a propane torch, and a metal dish or pan. Fill the pan to **half the case length** with water and stand a series of cases on their heads in the pan of water.

Individually, heat the neck of each case. When the neck reaches a cherry red color quickly tip the case over into the water. The water serves as a quenching medium and leaves the neck and shoulder quite soft. It also protects the head of the case, keeping it cool so that it may retain its hardness.

Overall Case Length and Case Fatigue

The battering of chamber pressure, plus the forces applied by the resizing die affect the serviceability of our cases. Visualize what would happen were we to place a chunk of brass stock on an anvil and pound on it with a heavy hammer. The repeated blows from the hammer would forge the brass into a different shape. It would become thinner and longer depending upon how hard or how often it was struck. A cartridge case is no different. Its brass material flows forward in direct relation to the pressure to which it is subjected and to the number of times it is fired and resized. As more and more of the brass is forced forward, the case neck thickens and eventually lengthens until it exceeds the maximum tolerance allowed by the chamber.

Not only do repeated firing and resizing alter the overall dimensions of the case, but cartridge brass has the characteristic of becoming harder as it is worked. In other words, the forward portion of the case loses its ductility and tends to crack rather than expand properly. This is called "fatigued brass" and it is recognized by small cracks that begin to appear in the neck, shoulder or body of the case.

The changes that occur in the structure and length of the cartridge brass work to the detriment of the handloader. By careful inspection, he must learn to recognize these changes before they become critical, and take the necessary steps to correct them. If, as recommended, the reloader separates his cases into lots and keeps a record of the number of times they have been reloaded and fired, he will know that all cases in that lot are in the same general condition. If one case shows signs of lengthening or fatiguing, then the entire lot of cases is suspect.

Carefully inspect your cases before each reloading. If your cases are new, or once fired, they will not reveal fatigue at the first reloading. However, fatigue signs may show up in subsequent loading, so you must learn to look for them. **Check your cases for splits or cracks in the neck, shoulder or body.** Reject all cases that show signs of defects, but flatten them with a pair of pliers before discarding to prevent their reuse. We suggest you separate your cartridge cases into lots and keep a record of their history.

Trimming is necessary when your cases have lengthened after numerous firings. Check your cases after resizing and never allow them to exceed the maximum listed measurement. How often you will need to trim the cases depends largely on the type of case you are using and the pressure of the load. Bottle-neck cases take more abuse from pressure than straight sided cases and require trimming more often.

Case length is of vital importance to the auto pistol reloader since most of those cartridges headspace on the case mouth; rather than a projecting rim at the base.

Typically, rimmed straight-wall cases require little or no trimming. Case length directly affects overall cartridge length and is of particular importance to crimped cartridges. Heavy rifle and handgun loadings should be crimped into a groove to ensure the bullets will not back out under recoil and bind the action or cylinder.

Cases which are notably short will require the bullet to make a longer "jump" to reach the rifling. Usually this affects accuracy adversely.

The trimming of any case more than four times is not recommended. After this amount of trimming, it may be assumed that the case walls are now too thin and the case should be discarded.

Cartridge Case Nomenclature

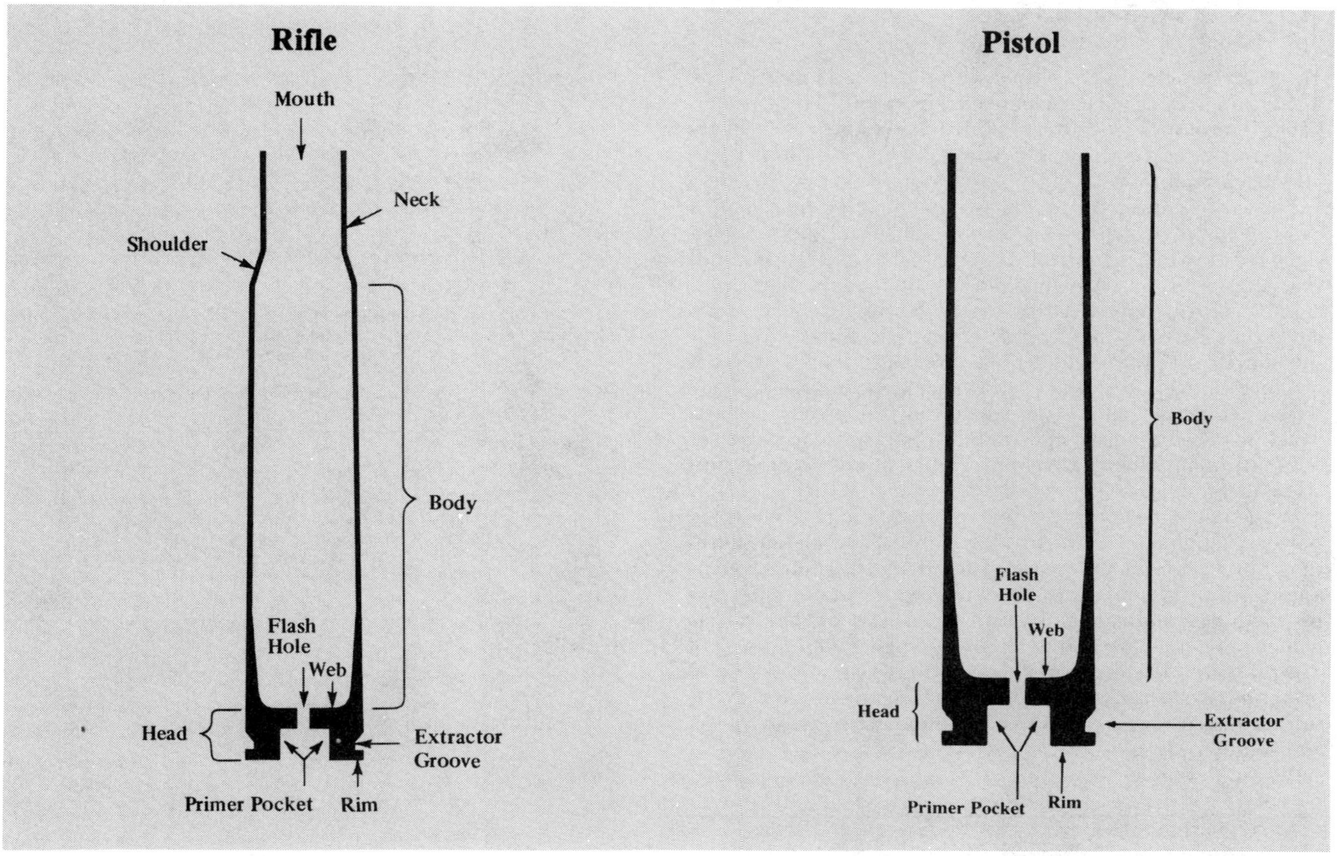

While cases vary in shape and size, the same basic nomenclature applies to all center fire cartridges.

Case Styles

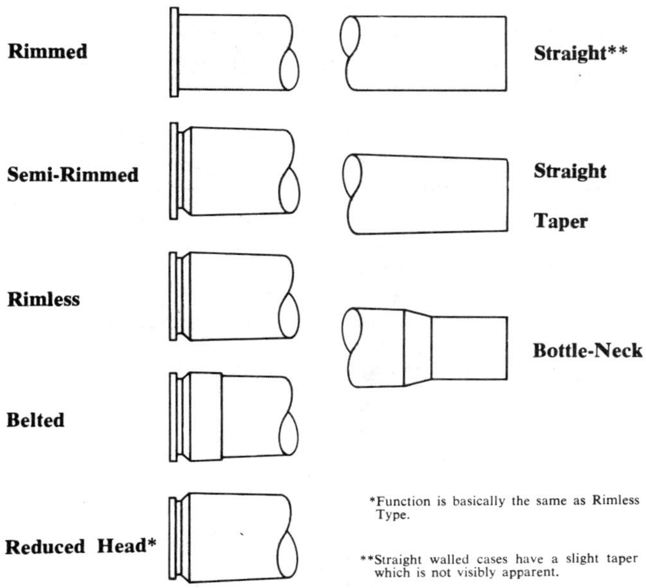

*Function is basically the same as Rimless Type.

**Straight walled cases have a slight taper which is not visibly apparent.

This illustration shows the head forms and body shapes standard to cartridge design. As you will see later, the design of the cartridge case governs the headspace dimensions. Case examples for almost any combination of the above shapes may be encountered. For example: the 30/30 uses the rimmed bottle-neck, the 30/06 uses the rimless bottle-neck, the .458 Winchester uses the belted head and the straight body.

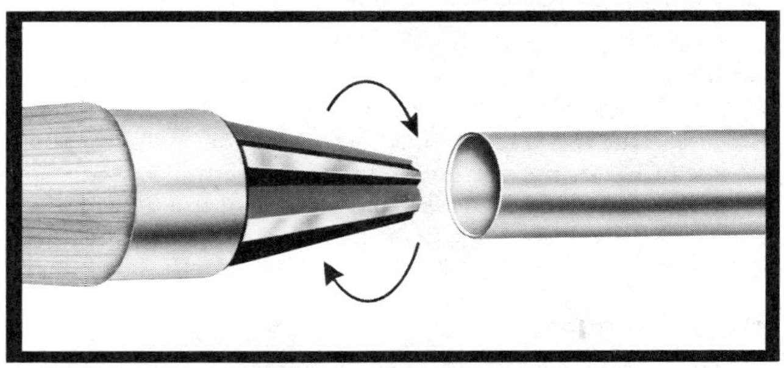

When loading new or once-fired cases, it is necessary to remove the sharp inside edges of the case mouth. This operation is called **chamfering** and its purpose is to ease insertion of the new bullet. Chamfering is required only for the first reloading of a new or once-fired case. The inexpensive Lyman hand reamer chamfers a case easily and with uniformity. Hold the case in one hand, while you lightly turn the reamer in the case mouth with the other hand. Remove very little material and **do not** cut a sharp knife edge on the case.

Propellant

Many people are concerned with the dangers of handling and loading modern smokeless powders. To the uninitiated, gun powder is often regarded as " explosive" something to be feared such as dynamite. This is quite a natural presumption.

The various modern powders used in reloading are classified as propellants, which means they are chemical mixtures designed to burn under controlled conditions, and to propel a shot charge or projectile. A high explosive, on the other hand, is a completely different breed of cat. These mixtures are quite dangerous because they are designed to detonate. When a substance such as dynamite or blasting gelatin, is detonated, it produces intense heat and violent shock waves. These shock waves exert tremendous pressure on anything they contact, which makes it almost impossible to vent away the effects of detonation involving any appreciable quantity of explosive material.

While not to be compared to explosives, modern reloading powders are nonetheless highly flammable. They not only burn, they burn vigorously. In case of accidental ignition, a great amount of gas at high temperature will be formed. If the powder is stored in its original factory canister, **as it should be**, this gas will create a pressure on the relatively fragile sides of the can. The pressure will split open the seams of the container, or pop off the lid. In such an event, the pressure remains at a low level **if sufficient space is provided to accommodate the escaping gas and vent it away**.

Never smoke—or have an open flame—in the presence of smokeless or black powder.

Handling and Storage

Smokeless Powder

For safe reloading, large quantities of reloading powder should never be stock-piled. The few cans that you do store must be handled with the care and caution due a flammable substance. In this way you control the situation so that it cannot get out of hand. Even in the case of a total house fire arising from other sources, **smokeless powders will not produce the effect of an explosion** if the proper storage precautions have been taken.

Black Powder

Unlike smokeless powder, black powder is classified as a Class "A" Explosive. Also, black powder is a good deal touchier than smokeless powder. Its ignition does not necessarily require a direct flame. It can be ignited by percussion (dropping container, etc.) or by static electricity. We recommend that black powder be handled with extreme care. Quantities should be kept to a minimum and should not be stored in the home or with other reloading powders. A clean storage area in an outbuilding (away from the home) should be used. This area should comply with the regulations set for smokeless powder. The out-building should be **locked** to protect children.

Black Powder/Pyrodex Useage Chart

Note: This chart is intended as a guide to show the appropriate uses of black powder and Pyrodex. It is not necessary to follow them exactly.

Pyrodex is not suitable, at present, for use in flintlocks as either the pan priming or the main charge.

FFFFG

Commonly called "Four F", this is the finest granulation and is used for priming flintlocks. Due to its rather limited use, it is usually somewhat difficult to obtain. When necessary, FFG may be substituted.

There is no Pyrodex equivalent.

FFFG/Pyrodex "P"

Commonly called "Triple F", this powder is used in most single shot pistols and all percussion revolvers. It is also popular for all smaller caliber rifles up to and including 50 caliber. When FFFFG is not available, FFFG may be used to prime a flintlock.

FFG/Pyrodex "R&S"

Commonly called "Double F", this is a popular powder for rifles over 50 caliber and up to 75 caliber. Also used in the larger caliber single shot pistols and most shotguns.

FG/Pyrodex "Ctg"

Commonly called "Single F", this is the coarsest granulation used for small arms. Use is pretty much restricted to rifles over 75 caliber and large bore shotguns.

Alcan Propellants

AL-5—A dense, progressive burning shotshell powder that is suitable for a variety of handgun loads. Available in ½-pound tins, 4½-, 12- and 24-pound kegs.

AL-7—A dense, very progressive-burning powder commonly used for magnum shotshells but also in some revolver loads. Packaged in ½-pound tins, 3-, 6- and 15-pound kegs.

AL-8—A coarse-grain powder used for some .44 magnum and 30 carbine loads but usually associated with shotshell reloading. Available in ½-pound tins, 3-, 6- and 15-pound kegs.

AL-120—The fastest burning powder in the Alcan line. Available in ½-pound tins, 4½, 12- and 24-pound kegs.

Dupont Propellants

SR-4756—Basically a magnum shotshell powder with limited application in heavy handgun and rifle loads. Available in ½-pound canisters, 4- and 12-pound kegs.

SR-4759—Basically a rifle powder with limited handgun applications. Probably the most useful propellant for cast bullets, because of its remarkable stability with reduced loads. Available in 4-pound caddy or 12-pound keg.

SR-7625—A versatile single-base powder which lends itself to a wide range of loading applications and is suitable for a variety of handgun and rifle cartridges. Available in ½-pound canisters, 4- and 12-pound kegs.

IMR-4227—The fastest burning powder of the IMR series, this is a versatile powder suitable for heavy handgun loads and many rifle loads. Available in 1-pound canisters, 8-pound caddies and 20-pound kegs.

IMR-4198—A single-base, fast-burning powder useful for a variety of handgun and rifle loads. Second fastest burning of the IMR series. Available in 1-pound canisters, 8-pound caddies and 20-pound kegs.

"HI-SKOR" 700X—A clean-burning, double-base powder useful for a wide range of handgun and rifle cartridges.

PB—Originally developed as a universal shotshell powder, it has a wide range of applications as a handgun and rifle propellant. This porous-base powder is available in ½-pound canisters, 4-pound caddies and 25-pound kegs.

IMR-3031—This single base powder, normally recommended by the manufacturer for medium capacity cartridges, is very versatile and performs well with cast bullets in large, straight cases such as 45-70.

IMR-4064—A single base powder for medium to large capacity cases, this propellant is useful when high velocities are desired with cast bullets.

IMR-4320—A relatively slow burning, single base powder for heavy bullets in medium to large capacity cases.

IMR-4350—A slow burning propellant useful with cast bullets when high loading density in large cases is desired.

Hercules Propellants

Reloader-7—Fastest burning of the Hercules rifle powders, this double-base propellant has some handgun uses. It is becoming increasingly popular with cast bullet rifle shooters. Available in 1-pound canisters.

Red Dot—A shotshell powder that is very useful for handgun and rifle loads. Available in 1-pound canisters, 3- and 12-pound kegs.

Green Dot—Basically designed as a 12-gauge shotgun powder, it is useful for a wide range of handgun and rifle loads. Available in 1-pound canisters, 4- and 15-pound kegs.

Blue Dot—Designed for magnum waterfowl shotshells, it is used in some handgun loads with varying degrees of success. This slow-burning double-base powder is available in 1-pound canisters and 5-pound kegs.

Herco—A slow-burning, coarse-grain, double-base powder that is effective in some handgun and rifle loads. Available in 1-pound canisters, 4-, 8- and 15-pound kegs.

Unique—An all-round double-base powder appealing to handgun and rifle shooters. This powder is very effective for reduced cast bullet, rifle loads. Packaged in 1-pound canisters and 4-, 8- and 15-pound kegs.

2400—A fine-grain, double-base powder useful in heavy handgun loads and some rifle cartridges. Available in 1-pound canisters, 4-, 8- and 15-pound kegs.

Bullseye—A high-energy, quick burning powder, perhaps the most widely used pistol powder available. It is capable of excellent accuracy in a wide range of pistol cartridges. Available in 1-pound canisters, 3- and 15-pound kegs.

Hodgdon Propellants

Trap 100—A fast burning spherical-ball-type double-base powder. It is excellent for target loads in handguns. Available in ½-pound canisters, 8- and 1-pound kegs.

HS-5—A spherical-ball-type powder. Available in 1-pound can, 8- and 12-pound kegs.

HS-6—A spherical double-base slow-burning powder. Available in 1-pound canisters and 8- and 12-pound kegs.

HS-7—the slowest burning of the Hodgdon spherical double-base powders. Gives uniform ballistics. Available in 1-pound canisters, 8- and 12-pound kegs.

H-110—A spherical-ball-type power adaptable to heavy pistol loads and 30 carbine. Available in 1-pound canisters, 8-and 20-pound kegs.

H-335—A spherical powder adaptable to a number of cartridges from small to medium case capacities. Has exhibited good performance with cast bullets in 30-30 Win. bolt action rifles.

H-4895—An extremely versatile powder in almost all rifle cases from .222 Rem. to .458 Win.

HP-38—A fast-burning powder for most handgun reloading requirements, particularly .38 Special loads. Available in 12-ounce canisters and 8-pound kegs.

BL-C(2)—This spherical-ball-type powder is popular with a large number of handgun shooters and performs well in the .222 Rem. Available in 1-pound canisters, 20-, 50- and 100-pound kegs.

Winchester Propellants

231—A very fast high-energy pistol powder designed for target and standard velocity loads in handguns. Available in 1-pound canisters, 3-, 8- and 12-pound kegs.

296—This is the factory powder used in .357 magnum and .44 magnum loads. The manufacturer cautions against using reduced loads with this powder. A reduction in powder charge from listed weights can cause dangerous pressures. Available in 1-pound canisters, 3- and 8-pound kegs.

630—A moderately slow but high-energy pistol powder which is also very effective in cast bullet rifle loads. Available in 1-pound canisters and 8-pound kegs.

748—Basically a rifle powder that has proven to be well suited for cast bullets.

760—A rifle powder covering a broad range of applications where a higher loading density with cast bullets may be desirable.

Projectiles and Lubricant

Rifle

The one firm rule for a cast rifle bullet is that it *must* fit the bore of the rifle in which it will be used. These lead alloy bullets, even the harder ones of Linotype or heat-treated wheelweights, are much easier to damage in firing than their jacketed descendants. They must fit perfectly or they cannot perform at their best.

There is, even now, detectable variation in the bore and groove measurements of modern centerfire rifles. Cast bullets which do not fit properly are inaccurate and contribute to leading.

Bullets designed by Guy Loverin work well in most rifles. Their short ogive and long groove-filling bodies produce excellent alignment with the bore's centerline. Perhaps their only shortcoming is that lube grooves are sometimes exposed when the bullet is seated to touch the rifling. It is then desireable to adjust the sizer to stop the lubricant below that point.

Examples 1 and 2 show bullet designs which intend for the body to fill the grooves and the nose to ride atop the lands, perhaps engraving slightly. The crucial feature of these designs is the diameter of the nose. If it does not contact the tops of the lands, then accuracy will suffer.

However, when the bullets are correctly mated to a barrel, their performance rivals that of jacketed bullets.

Example 4 represents a cast bullet type that tends to perform less well than the Loverin and Ideal designs. The reason is that there is so little bearing surface to align the bullet in the barrel. These bullets can shoot well *if* you work at it.

From time to time we receive letters or telephone calls wherein the customer states he's quite happy with his #280473 or #311413 and can't understand why others shy away from them.

We have prepared the following chart showing our most popular bullet designs in a wide range of chamberings.

Please note that some of these designs may not be available from year-to-year. A current Lyman catalog will carry the mould listing. These catalogs are available in January of each year, free for the asking.

These four shapes are basic to cast rifle bullet design. Actual bullets may vary to some extent, but, generally, they fall into one of these four categories.

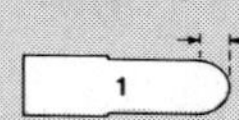

SHORT OGIVE

(IDEAL DESIGNS) Bullets that conformed to either of these general shapes performed very well in our testing. Due to their short ogive the greatest weight mass bears on some portion of the rifling. The driving bands bear directly on the groove, while the nose acts as a pilot to align the bullet in the throat and bore.

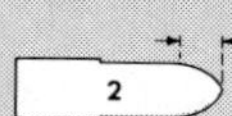

SHORT OGIVE

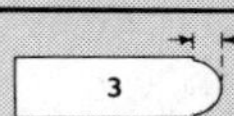

SHORT OGIVE

(LOVERIN DESIGN) Bullets having this general shape performed extremely well in all of our testing. Due to their short ogive, the greatest weight mass bears directly on the groove. While the design does not always provide for a section of the bullet to bear on the lands, the many driving bands of groove diameter keep them well aligned.

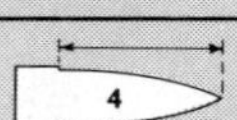

LONG OGIVE

Bullets of this general shape proved to be the worst performers in our testing. Their long flowing ogives leave a large portion of the bullet unsupported by the rifling. Cast bullets of this general shape can easily misalign themselves as they enter the throat and bore.

Pistol

Reloaders of pistol cartridges have, in the past, tended to use more jacketed bullets—on the average. Only target shooters and those who sought the most economical form of projectile used lead bullets to any great degree.

Today, the trend is shifting. The retail price of a box of jacketed bullets has reached the point where more and more reloaders (always economy-minded) are becoming bullet casters. Today we know much more about various alloys than we did 30 years ago. New lubricants, like Alox/beeswax, have permitted maximum pistol velocities without leading. Now the handgunner can efficiently use cast bullets for most of his needs without sacrifice.

Basically, bullet shapes are similar for both cast lead and jacketed pistol bullets; an exception being the jacketed rifle projectiles useable in pistol cartridges like the .221 Fireball and .256 Winchester.

There are several factors to consider when selecting a bullet besides having carefully tested load data for a given projectile weight:

1. Weight— For optimum potential accuracy the bullet must fall within a range peculiar to a given caliber. Also, the velocity at which that bullet will be driven is a factor. The whole idea is to properly blend the three (weight, rate of twist and velocity) to produce the results satisfying you, the reloader. Weight is important to the target shooters for its effect upon sectional density; hunters are concerned with weight for its implications to penetration potential on game.

2. Nose Design— Target shooters usually select a wad cutter-design so that the target paper will be cut cleanly for easier scoring. Although the aerodynamic shape of these bullets is poor, it is more than adequate for the limited (25 & 50 yards) distances involved.

Silhouette shooters using the big .357 Magnum, .41 Magnum and .44 Magnum often choose the heavy semi-wad cutter designs of Keith and Thompson. These bullets have the weight to drop the rams at 200 meters.

Hunters use much the same criteria as do the silhouette shooters. Additionally, hunters can use the appropriate Lyman composite bullet to obtain improved terminal ballistics on game.

3. Intended Use— There's no point in using a big heavy bullet if you intend to do some informal plinking—a lighter one will do. This reduces the cost of the bullet and powder charge and produces a nice savings to you.

Conversely, don't load a light bullet to run the silhouette course or hunt deer. You need the bigger bullets for their ballistic efficiency.

Note: Space considerations in this book limited our trajectory and wind drift tables to cast rifle bullets. Cast pistol bullets are covered in Lyman's PISTOL AND REVOLVER HANDBOOK. Muzzleloading projectiles—roundballs and conicals—are covered in Lyman's BLACK POWDER HANDBOOK.

Cast Bullet Suitability Chart

Rifle

Chambering	Bullet #
.22 Hornet	225415
.218 Bee	225415
.222 Remington	225415 225462
.223 Remington	225415 225462
.222 Remington Magnum	225415 225462
.225 Winchester	225415 225462
.224 Weatherby Magnum	225415 225462
.22/250 Remington	225415 225462
.220 Swift	225415 225462
.243 Winchester	245496
.244 Remington	245496
.25/20 Winchester	257312
.256 Winchester Magnum	257312
.250 Savage (.250/3000)	257312
.257 Roberts	257312
.25/06 Remington	257312 257464
.257 Weatherby Magnum	257312
.270 Winchester	280473
.270 Weatherby Magnum	280473
7mm Mauser (7 x 57mm)	287405 287308
.280 Remington	287405 287308
.284 Winchester	287405 287308
7 x 61mm Sharpe & Hart	287405 287308
7mm Remington Magnum	287405 287308
7mm Weatherby Magnum	287405 287308
.30 Carbine	311359 311410
.30 Remington	311291
.303 Savage	311291 311334
.30/30 Winchester	311291 31141
.300 Savage	311359 311466 311291 311467
.308 Winchester	311359 311465 311466 311291 31141 311467 311334 301620 301618
.30/40 Krag	311359 311466 311291
30/06 Springfield	311359 311466 311291 31141 311334 311299 301618 301620
.300 H&H Magnum	311466 311291 311334
.300 Winchester Magnum	311466 311291 311467 311334
.308 Norma Magnum	311466 311291 311334
.300 Weatherby Magnum	311291 311334
7.62 Russian	311466 311291
7.65 Argentine Mauser	311466 311299
.303 British	311466 311299
7.7mm Japanese	311466 311299
.32/20 Winchester	311316
8 x 57mm Mauser	323470
8mm Remington Magnum	323470
.35 Remington	358430 358315
.350 Remington Magnum	358430 358315
.358 Winchester	358430 358315
.358 Norma Magnum	358315
.375 Winchester	375248 375449
.375 H&H Magnum	375248
.378 Weatherby Magnum	375248 375449
.38/55 Winchester	375248
.38/40 Winchester	40143
.44/40 Winchester	42798 429434
.44 Remington Magnum (rifle)	429348 429215 429360 429421 429244
.444 Marlin	42798 429215 429360 429421 429244
.45/70 Government	
(a) Trap door Springfield actions	457191 457122 457124 457193
(b) 1886 Winchester 1895 Marlin	457191 457122 457193
(c) Ruger #1&3 Mauser actions	457122 457193 457406 457125
.458 Winchester Magnum	457191 457406
.50/70 Government	515141
.50/90-2½"-Sharps	515141
.50/140-3¼-Sharps	515141

Pistol

Chambering	Bullet #
.221 Remington Fireball	225415
.25 ACP	252435
.256 Winchester	257312
.30 Luger	313249 313226
.30 Mauser	313249
.30 M1 Carbine	311359 311410
.32 ACP	311252 313249
.32 Smith & Wesson	311252 313249
.32 Smith & Wesson Long	311252 313226 3118
9mm Luger	358242 358345 356402
.38 Smith & Wesson	358242 (92 & 121 gr.) 358480 358495 358311
.38 Super Auto	358242 356402 358242 358480 358311
.380 ACP	358242 (92 & 121 gr.)
.38 Special	358242 358242 (92 & 121 gr.) 358495 358212 35863 35891 358156 358311 357446 358429
.357 Magnum	358242 (92 & 121 gr.) 358345 356402 358480 358495 358477 358156 358311 358429
.38/40 Winchester	40143
.41 Magnum	410610 410459
.44 Smith & Wesson Special	42798 429215 429360 429421 429244
.44 Remington Magnum	429348 429215 429360 429421 429244
.44/40 Winchester	42798
.45 ACP	452389 452488 452460 452374
.45 Auto Rim	452389 452460 452374 452423
.45 Colt	454190 452424 454190

Lubricant

Lubricant, applied to the grooved body of a given bullet design, provides a film between the bullet metal and the walls of the bore. This film greatly reduces friction as the bullet travels down the barrel and either eliminates or minimizes leading.

When this film breaks down, for whatever reason, leading occurs and reduced accuracy results.

Modern muzzleloaders can successfully use a tremendous variety of concoctions for lubricants. Today there are a number of commercial substances, for use with patched balls and conical bullets, in addition to the traditional mixtures of beeswax, tallow and that modern favorite, crisco.

Demands on lubricant performance increased when the black powder cartridges of the late 1800's were developed. The bullets in these cartridges were a groove-filling mechanical fit in the bore and thus needed greater protection from friction. Velocities of these cartridges were relatively low—to 1500 fps—and little work was required to develop satisfactory lubricants for those soft alloy (or pure lead) bullets.

The development of smokeless powder, and the resulting "modern" cartridges, drove ammunition companies to the use of jacketed bullets to realize the performance potential of the new propellant. These jacketed bullets eliminated the leading problem and provided additional resistance in the bore which caused the smokeless powder to burn better.

Today there are more sophisticated lubricants available:

The Alox-beeswax (or synthetic replacement recently developed by Alox Corp.) mixture, when mixed 50/50 to the NRA formula, is the best high-velocity lubricant available today. This mixture has proven itself in recent years and is widely available from a number of firms, including Lyman. Be sure the product states it is made "to the NRA formula" or that it is comprised of 50/50 Alox/beeswax. Otherwise, you may be buying a different formula—and there are some on the market.

The Alox Corporation has produced a synthetic beeswax for incorporation in NRA formula. All reports of its performance, to date, indicate it is equal to the original formulation.

The NRA formula of Alox/beeswax can take cast bullets up to almost 3,000 fps under optimum conditions. Typically it is not necessary on loads under 1,000 fps and another type may be used.

Lyman has sold its Ideal Lubricant for years. This substance is ideal (pardon the pun) for handgun target loads and other low velocity pistol applications. Riflemen can successfully use this substance on loads to about 1600 fps.

Note: Alloy hardness is an important—and equally vital—consideration in terms of lubricant performance.

There are various liquid lubricants now on the market which appear to have potential. The nice thing about these lubes is that the bullet can be dipped and allowed to dry; then sized and loaded normally. The lubricant dries to a hard, non-tacky surface and does not collect lint, grit, etc. as might bullets lubed with the more conventional substances.

These "dip 'n dry" lubricants might prove especially valuable to riflemen using bore-riding bullet designs, like #311334, which might benefit from lube on the bullet's nose.

A ripe area for experimentation with the new lubricants lies in the various formulas (as produced by thinning) possible and the combination of both the Alox/beeswax and liquid lubricant.

Miscellaney

As we researched and experimented, in the course of preparing this Handbook, certain questions came up and our technical staff took little side-trips here and there to investigate potentially interesting topics.

Since some of these forays might be general interest, we have reproduced the results below. Where firing data is shown, the specifics of the test match those set forth in the Data Section proper.

Effect of Alloy Variance on Pressure and Velocity

No. 2 Alloy vs. Linotype
Bullet #311291, 30-06 Springfield:
Sized .308
29.0 gr. SR4759

	No. 2 Alloy	Linotype
Velocity, Avg. fps	2173	2196
Max.	2195	2219
Min.	2165	2158
Spread	30	61
Std. Dev.	9.94	18.36
Pressure Avg. CUP	37,100	36,800
Max.	41,100	41,100
Min.	30,900	30,500
Spread	10,200	10,600
Std. Dev.	2693	3702

Effect of Wad vs No Wad

#311291 in 30-06 & #457122 in 45-70 (Ruger Chamber)
Dacron Wads, 1/4" Thick, Cut in 5/8" Squares

	#311291 30-06 With Wad	#311291 30-06 Without Wad	#457122 45-70 With Wad	#457122 45-70 Without Wad
Velocity, Avg. fps	2140	2130	1893	1852
Max.	2155	2146	1911	1915
Min.	2124	2117	1859	1681
Spread	31	29	52	234
Std. Dev.	11.5	10.2	18.1	80.2
Pressure Avg. CUP	37,000	35,600	37,700	32,600
Max.	38,600	38,100	40,200	39,200
Min.	33,700	33,200	33,200	17,200
Spread	4900	4900	7000	22,000
Std. Dev.	1660	1409	2353	8135

When developing a load, if a wad is desired, it should be used from the beginning as the charge weight is increased. It should never be added as an afterthought once a maximum load has been established since its presence could result in a pressure increase of 2,000 CUP or more.

Effect of Bullet Alloy on As-Cast Weight

	Lead	Wheel-weight	#2 Alloy	Lino-type
Avg. weight as cast grains	176.2	173.4	167.5	163.3

Primers

For a cartridge to function properly, the successful operation of each component is required. Primarily from a safety standpoint, we have stated that the cartridge case is the most important component. However, if a primer does not ignite, we will have a misfire and all of our concern for case inspection, powder selection, etc., will be for naught. Therefore, every component, including the primer, is critical to the performance of the load.

Actually, the primer is responsible for only a small percentage of the accuracy of a given load, but its burning characteristics will add or detract from the overall pressure. How much these burning characteristics affect pressure is relative. It depends primarily upon the application of a specific primer to a specific set of conditions. Generally, the primer's influence on pressure is minimal when compared to the possible effects of other conditions and components. This influence, however, is not so slight that it can be overlooked.

In testing metallic cartridges we noted pressure variations (due to primer change) of approximately 2,000 C.U.P. Further testing could indicate still greater variations. The point is that while metallic cartridge primers do not seem to vary greatly, enough variation exists to require us to re-work a load when changing a primer type or brand. The reloader should stick with the same brand and type of primer when working up and using a load. If he changes the primer brand or type, he must then go back to the "starting load" and begin again.

If you examine a "Boxer type" primer carefully you will note that the anvil protrudes slightly beyond the end of the primer cup. This is not an oversight in manufacture! The anvil is supposed to protrude so that it will seat solidly against the bottom of the primer pocket and firmly resist the force of the firing pin blow. If the primer is seated incorrectly (leaving space between anvil and bottom of pocket) then the entire primer will move forward when struck by the firing pin. Such a condition retards the blow of the firing pin and causes inconsistent ignition.

When seating primers, make sure that the primer bottoms in the primer pocket. However, the primer should not be crushed. Normally, when a primer is seated correctly, it will be flush with the head of the case, or perhaps a few thousandths below. Under no circumstances should the primer stick out beyond the case head. Such a condition can prove dangerous, particularly in auto loading firearms, for the primer is in an exposed position. If such an improperly seated primer were to receive a blow from the bolt face, it could fire before the round was safely chambered.

Boxer primers having either rough or flat dome construction may be encountered and the correctly shaped priming punch is available for either contour. Use the flat punch for the flat primer and the concave punch for the round primer. Before seating a primer it is wise to examine it visually to make sure the anvil is not missing. Although this rarely occurs, the resultant misfire could be embarrassing if it should come at the wrong moment (such as when that trophy buck shows himself from behind the hemlocks).

Handling Primers

A word of caution on handling primers. Primers are detonated by percussion (a sharp blow). As packaged by the manufacturer, primers are quite safe, but they should not be tossed about or handled in a careless manner. Keep your primers in the factory container until they are to be used. Never store primers in a makeshift container. An old glass jar full of primers could become lethal if dropped or accidentally knocked onto the floor.

Crimped Primers

Some military cases are loaded in a manner which employs a slight crimp over the primer to hold it snugly in place. These cases can be decapped without difficulty, but the crimp must be removed before the case can be primed again. Various commercial tools, which include the Lyman primer pocket reamer, are available to do this job. Primer pocket reamers come in two sizes, large and small.

One last thought—never decap live primers. If for some reason live ammunition must be disassembled, pull the bullet, pour out the powder, chamber the primed case in a firearm, and snap the primer before decapping.

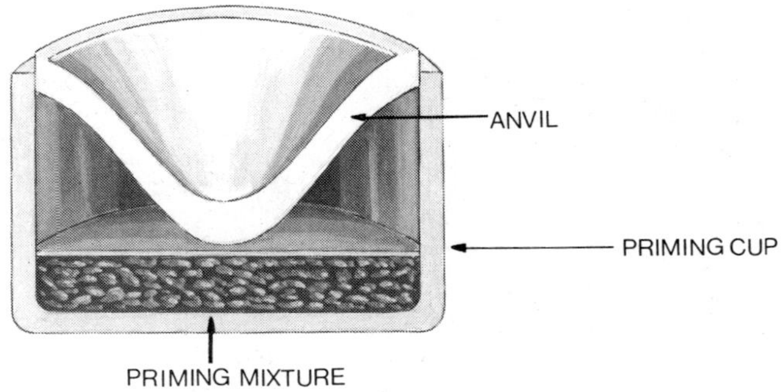

NOMENCLATURE OF "BOXER TYPE" PRIMER

The boxer type primer is used in all American metallic cartridges. The battery cup primer used in shotshells is of the same basic design, but contains an extra cup to facilitate its use in the less rigid shotshell base. The Berdan type primer, still popular in Europe, functions in the same manner as the Boxer type, but with this primer the anvil is actually part of the case.

NOTE: This section refers only to rifle and pistol cartridge primers. For information pertaining to shotshell primers, see the *Lyman Shotshell Handbook, 2nd Edition.*

U.S. Boxer Primer Chart

	Large Rifle	Large Rifle Magnum	Small Rifle	Small Rifle Magnum	Large Pistol	Large Pistol Magnum	Small Pistol	Small Pistol Magnum
Remington	9½	9½M	6½	7½	2½		1½	5½
Winchester Western	8½-120		6½-116		7-111	7M-111F	1½-108	1½M-108
Federal	210	215	*200		150	100		
	†210M		**205					
			††205M					
Alcan	Large Rifle Max-Fire		Small Rifle Max-Fire		Large Pistol Max-Fire		Small Pistol Max-Fire	
CCI	200	250	400	450	300	350	500	550
		BR2		BR4				

*Also may be used for magnum, small pistol loads.
**Thick cup design, especially for .17 Rem. and .22 centerfire loads.
†Match version of No. 210.
††Match version of No. 205.

RELOADING RIFLE & PISTOL CARTRIDGES

Reloading a cartridge requires a series of SIX basic mechanical operations. So that you will understand the fundamentals of each operation, we will first treat them graphically and then explain why each operation is necessary. Four out of the six operations are performed by the reloading dies.

1. Full Length Resizing:

When a cartridge is fired, the side walls of the case expand to the chamber size of the gun. This is necessary for the case to function properly and seal the chamber. These walls remain pretty much at their expanded size and do not snap back to original dimensions. Since all chambers are not identical, cases fired in one gun may not chamber in another. Even the chambers of a revolver cylinder will, in fact, vary from one to another. For this reason, it is necessary to compress the walls of the case to a standard diameter that is acceptable to all firearms of the same caliber. This operation is called resizing.

2. Decapping:

This operation consists of simply removing the old or fired primer.

3. Inside Neck Expanding:

After the case has been resized, the inside diameter of the neck will be too small to accept the bullet. Inside neck expanding enlarges the inside diameter of the neck to a size which will receive and hold the bullet securely. For almost all pistol cases and optionally for rifle cases, a two-step expanding plug is used to open up the inside of the case neck. The first step on this plug is slightly smaller than bullet diameter, while the second step is a few thousandths larger. The idea behind this is to allow the bullet to enter the case freely without shaving lead. The actual difference between the two steps is not visually apparent. The illustrations have been exaggerated for purposes of clarification.

4. Priming:

This operation consists of inserting a new primer into the primer pocket.

5. Charging Powder:

This operation consists of carefully weighing out and pouring the appropriate powder charge into the case.

6. Bullet Seating:

The last operation in the reloading process is seating a new bullet into the case.

In the preceding text we covered reloading graphically and have given the reader a general idea of what is required. Now by employing photographs we will explain the actual reloading of a cartridge. You will note that we are using a set of three reloading dies to perform four of the six operations. Further along in reloading you will hear of two-die sets and even four-die sets. The difference is that two- and three-die sets combine some of the operations, while a four-die set performs each operation separately. Due to their shape, most pistol cartridges require the use of a three-die set.

Cast bullet loads in rifle cartridges are greatly improved by augmenting the normal two-die set with the Lyman "M" die. The slight flare produced by the "M" die ensures no bullet metal will be shaved as the cast bullet is seated.

Making Cartridges

We now assume that your cases have been properly inspected, that you have selected a load and purchased the necessary components. Your reloading press should be assembled and mounted according to the instructions supplied with it. Many

Rifle:

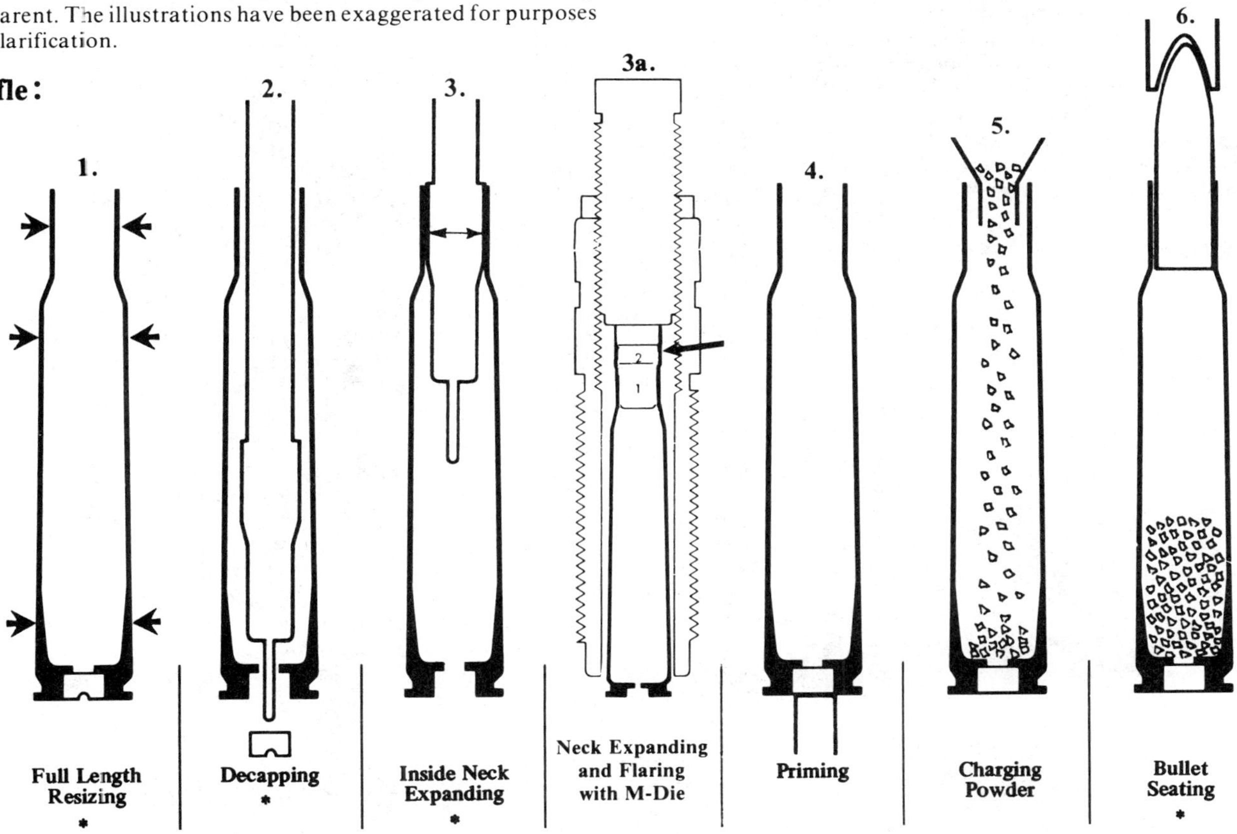

*Operations 1, 2, 3, and 6 are performed by reloading dies.

Pistol:

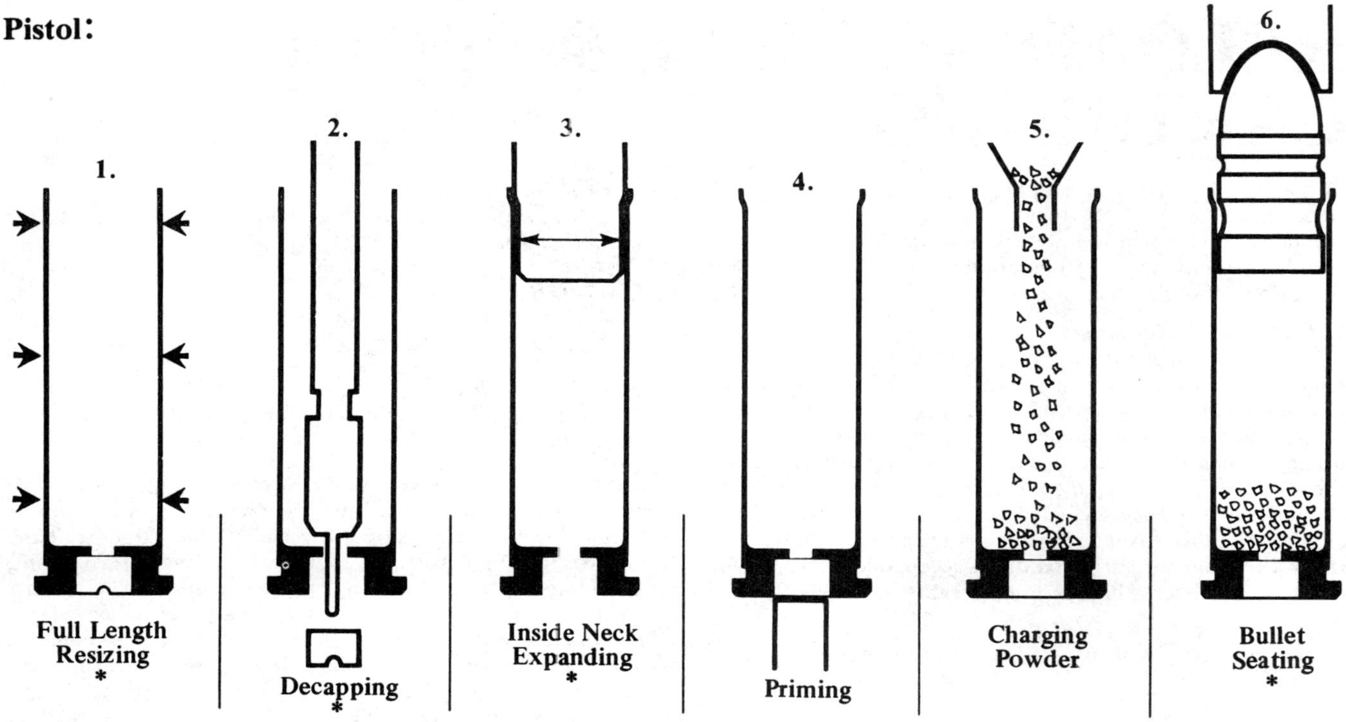

*Operations 1, 2, 3, and 6 are performed by reloading dies.

reloading presses may be assembled to function on either the up-stroke or down-stroke of the handle. The presses illustrated here are operating on the down-stroke.

Cleaning the inside of the case mouths with a bore brush (dry) of the appropriate caliber, then dipping the necks into powdered graphite, will greatly reduce the effort required to work the necks—and thus extend case life.

Unless you are reloading pistol cases with a tungsten carbide (T/C) F/L die, which requires no special lubricant, you must lubricate your cases by wiping them with a cloth—or rolling them on a pad—sparingly wet with Lyman Sizing Lubricant. This special lubricant will cut friction to a minimum and ease the sizing operation Apply a very thin coat, for too much grease will trap air in the die and cause "lube dents." Although cases dented in this manner may be used for reloading, as the dents are ironed out in firing, it is not considered good reloading practice.

Screw your Full Length Resizing Die into the head of the press, adjust it according to the instructions furnished, and you are ready to commence loading.

Reloading: Step by Step

This section of the Handbook concerns itself with the basics of rifle and pistol cartridge reloading. For the benefit of the novice, the text and illustrations have been intentionally simplified, but not over-simplified. Still, there is considerably more to learn, for reloading is that kind of a hobby.

Stay with the recommended "starting load" until you are familiar with the fundamentals of reloading. Progress slowly and never load over your head. If you don't understand something, ask. Seek the advice of knowledgeable reloading friends, or the Lyman Technical Staff who gladly offer assistance.

The reloading of rifle cartridges will be demonstrated on the new Lyman O-Mag Press; pistol cartridges on our popular Spar-T Press.

Reloading On The O-Mag Press

Step One *(Full-Length Resizing and Decapping)*

Slide the head of your cartridge case into the shell holder and pull the press handle down all the way. If the die is adjusted properly, the entire cartridge case will enter the die flush to the shell holder. Note in the cutaway drawing how two of the original six reloading operations (full-length resizing and decapping) are accomplished by this step.

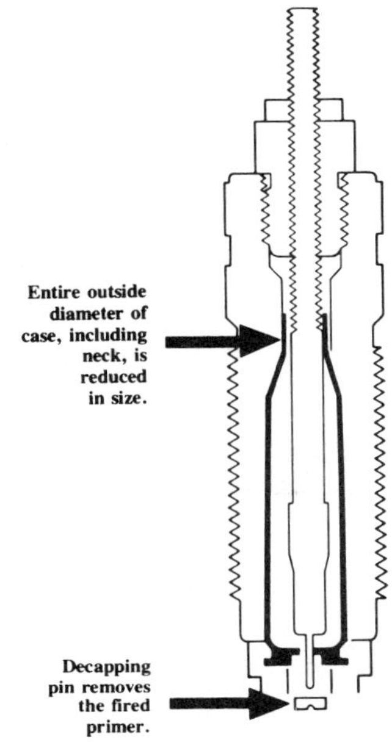

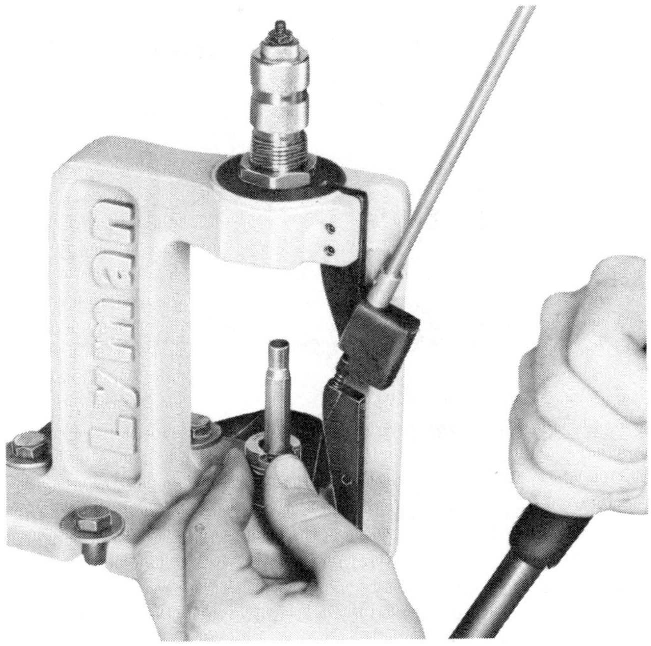

Step Two *(Inside Neck Expanding and Priming)*

As your case is withdrawn from the resizing die, two further operations are accomplished. The expanding button will automatically enlarge the neck, as shown in the cutaway drawing, and the priming punch will seat the new primer. As the expanding action of the button is automatic, you need not be concerned with it. You must, however, place the new primer (cup side up) into the priming punch sleeve. Push the priming arm forward (toward the press) and pull up on the press handle. As the ram is lowered, the priming arm will enter the slot in the side of the ram and seat the primer.

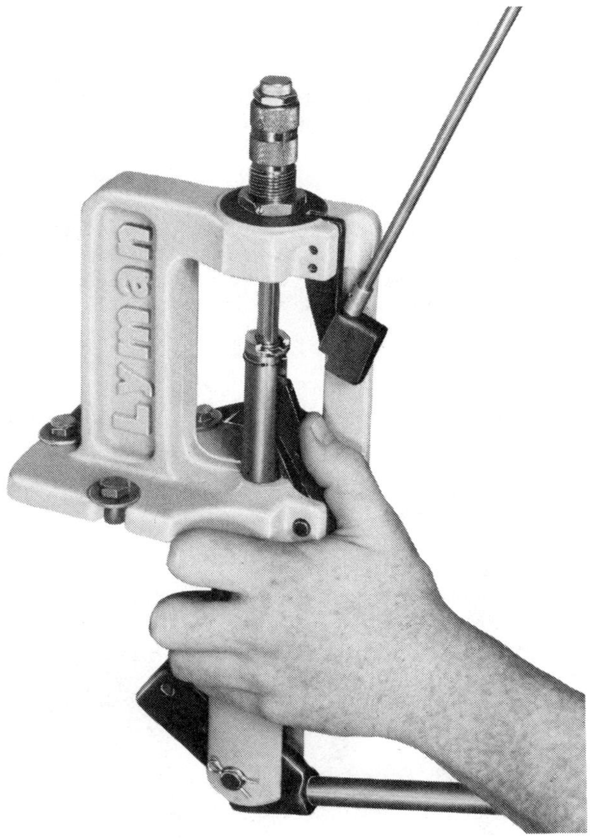

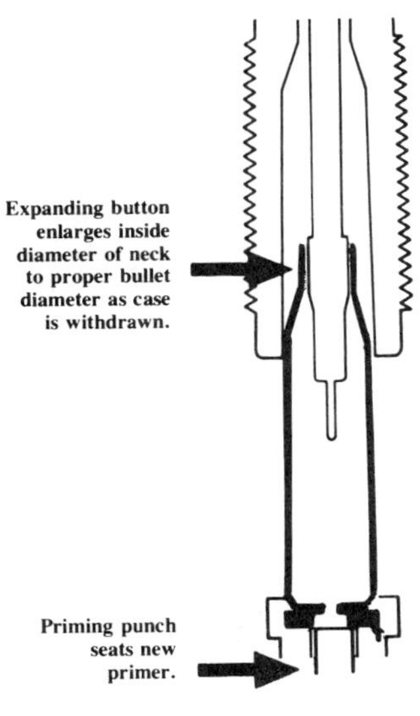

Editor's Note:
Clean the interior of your dies periodically. Grit can accumulate which will scratch both the die walls and the cartridge cases.

Step Two (Continued)

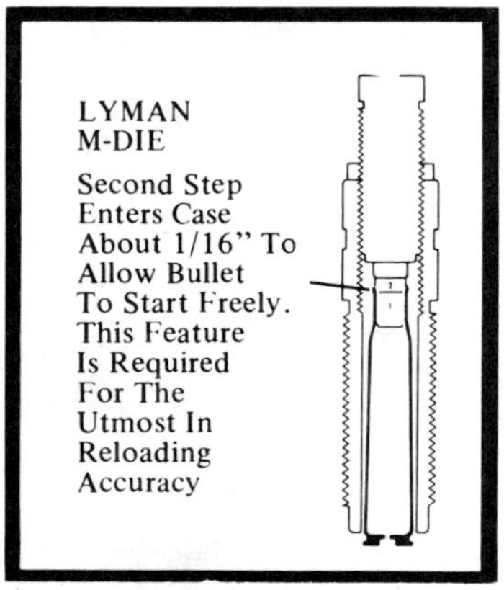

For the utmost in cast bullet reloading accuracy, the use of three reloading dies is mandatory. A Three-Die Set differs from a Two-Die Set in that it contains a separate neck-expanding die. This die (Lyman "M" die) is equipped with a two-step expanding plug (see Lyman Catalog). The "M" Die offers several advantages: (1) The bullet can be started into the case freely with the fingers and visually aligned before it enters the seating die; (2) It insures a more precise "bullet pull" from shot to shot. Consistent bullet pull (tension of neck on seated bullet) is highly important to accuracy as each bullet must release with exactly the same amount of pressure if shots are to remain consistent.

Step Three *(Charging Powder)*

You will need an accurate powder scale such as the Lyman D-7. The data section specifies the powders appropriate for your particular cartridge. It also lists a suggested weight of the powder charge in grains and in fractions of grains. For example, 9.5 grains would read as NINE and FIVE TENTHS grains. 10.0 grains would be read as TEN grains. Carefully level the powder scale as explained in the scale instructions and set it to weigh your required charge.

Slowly sprinkle small amounts of powder into the scale pan until the beam comes into balance. The beam is in balance when the pointed end (extreme left) is exactly on the zero mark.

Carefully remove the pan and pour its contents into the cartridge case. Use a powder funnel to make sure all the powder enters the case. To avoid the possibility of accidentally "Double Charging" a cartridge, you should develop a foolproof system of loading. A suggested method is to place all the uncharged cases on your left. As you pick up each case for charging, turn it up-side-down and shake it. This will insure that the case is empty. Turn the case right-side-up, charge it and place it carefully on your right. Take care when removing or replacing the scale pan that the poise are not accidentally moved.

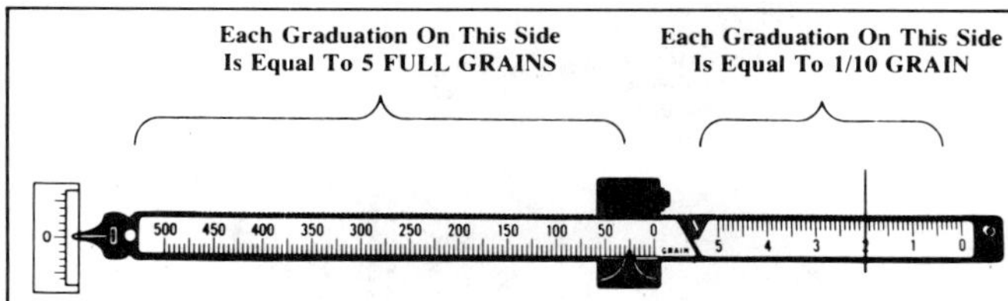

This illustration shows the beam of a modern reloading scale. Note how it is graduated on both sides of the pivot point. The scale is set by moving the two weights (poise) to the proper graduations. The large poise (on the left) is used to obtain multiples of FIVE GRAINS, while the small poise (on the right) is used for 1/10 FRACTIONS of a grain, or SINGLE grains from one to five.

EXAMPLE: The illustration shows a setting of 27.0 grains. If you wanted to decrease this 1/10 grain, you would simply move the small poise one notch to the right.

CAUTION

Technicians in the Lyman Lab have observed a potentially serious phenomena involving powder scales and plastic loading blocks and/or styrofoam packaging. It seems these substances sometimes retain a certain amount of static electricity, enough to create a electro-static field of varying radius.

This electro-static field has proven capable of radical deflection of uncharged and "zeroed" scales of all brands (available to us at the time). Of course, powder in the pan will tend to dampen the deflection but some still occurs depending on the charging level. Generally, the heavier the charge the less error... assuming the scale was first "zeroed" correctly.

We suggest you clear the surface of your reloading bench and make very sure the scale is set up accurately. Then move your equipment back piece by piece, paying particular attention to the plastic or styrofoam mentioned earlier. Please note this caution applies to all forms of reloading.

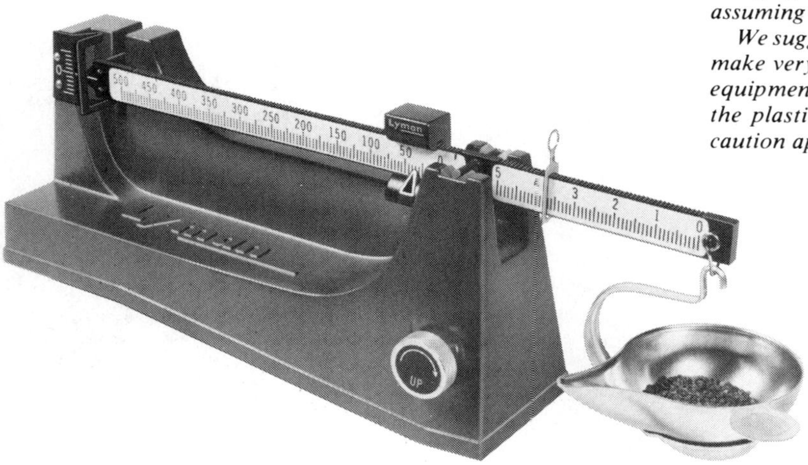

Step Four *(Bullet Seating)*

The last step in reloading a cartridge is seating the new bullet. Make certain that the overall length of the finished round is not longer than the maximum overall length listed in the data section.

This illustration shows how a bullet is seated. Screw the bullet seating die into the head of the press and adjust it according to the instructions supplied with the die. Place a primed, charged cartridge case in the shell holder and a bullet on the mouth of the case. Hold the bullet in place as you pull the press handle all the way down. As the case enters the die, the bullet will be pushed firmly into the neck of the case. Adjusting the seating screw controls the depth to which the bullet is seated. Adjusting the die body controls the crimp.

Crimping is a matter of choice and the seating die may be adjusted to crimp, or not to crimp as you desire. If you are loading hunting loads that will see hard usage in the magazine, it is wise to crimp-in the bullet. This prevents the bullets from unseating when the rifle is under recoil. Best accuracy, however, is usually obtained by not crimping-in the bullet. Target, or varmint loads, are best left uncrimped.

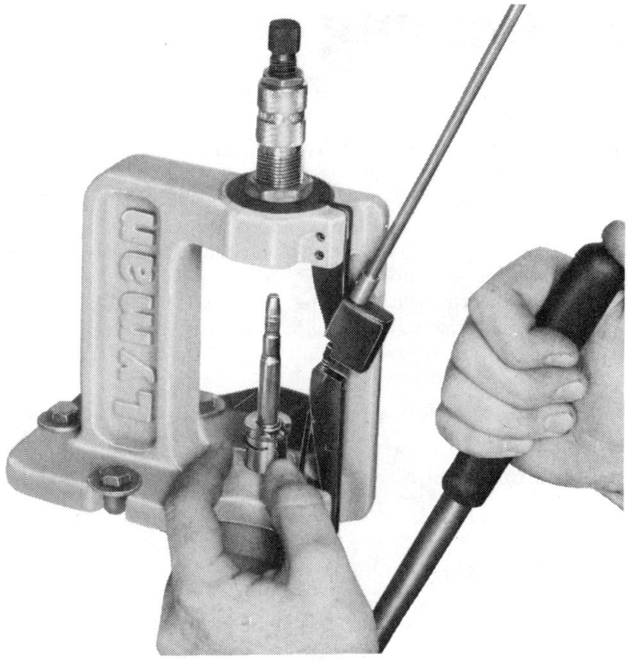

Built-in crimp shoulder affords crimping-in of bullet when desired.

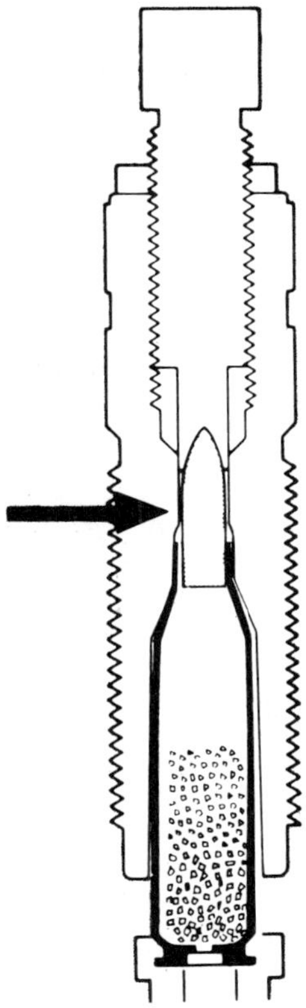

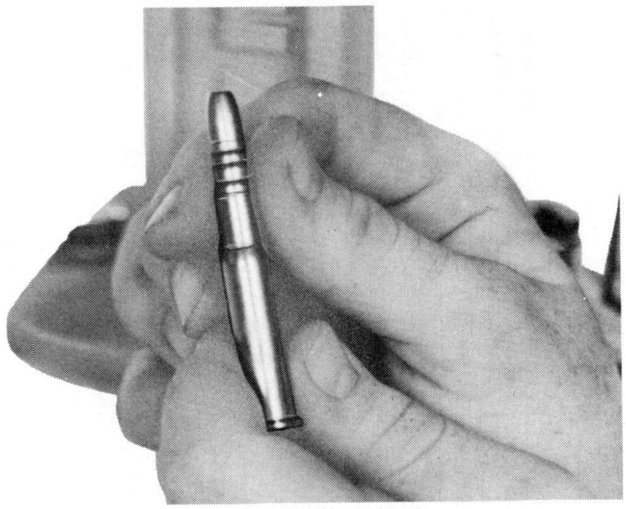

The bullet should be visually aligned in the neck before it enters the seating die. The operator should make certain that the seating die is locked tightly in adjustment and he should operate the press handle slowly and with care.

For accuracy loading, bullets should be seated friction tight, but **the case should not be crimped**. Crimping can be harmful to consistent bullet pull for two reasons: (1) It is nearly impossible to crimp each case exactly the same each time; (2) The pressure used in crimping has a tendency to bulge the neck away from the bullet. In some instances, crimping actually loosens the bullet in the neck and lightens bullet pull (see illustration).

However, loads intended for field use in other than single shot rifles or pistols must be crimped to ensure the bullets don't back out under recoil and jam the gun.

CRIMPED

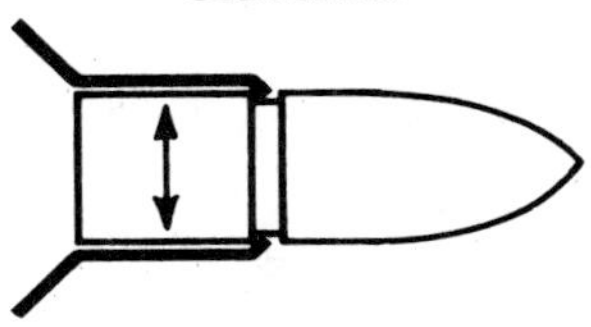

While bullet will not move forward or backward, it may be rocked sideways, out of alignment.

UNCRIMPED

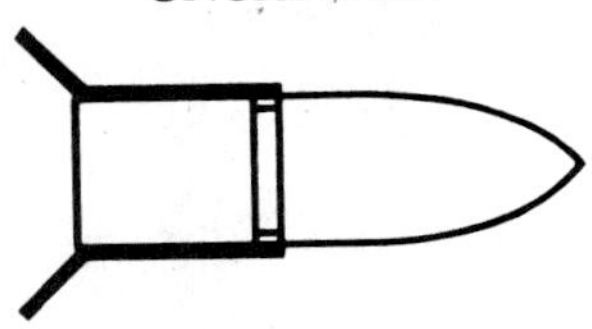

Most uniform bullet pull is afforded when the neck walls are tight against the bullet.

CAUTION: After reloading and before firing, wipe your cases to remove all sizing lubricant. The presence of oil or grease on a cartridge may dangerously increase thrust on the bolt face.

Editor's Note:

After every 300-500 rounds, remove the seating screw and clean accumulated lubricant from both the die interior and the seating screw.

Lubricant build-up can cause misaligned bullets and a gradual increase in seating depth which often escapes detection.

Reloading on the Spar-T Press

Step One *(Full-Length Resizing and Decapping)*

Install and adjust the f/l resize and decap die according to the manufacturer's instructions.

Slide the head of your cartridge case into the Shell Holder as illustrated, and pull your press handle down all the way. If the die is adjusted properly, the entire cartridge case will enter the die flush to the shell holder.

Note, in the cutaway drawing, how two of the original six reloading operations (full-length resizing and decapping) are accomplished by this step. Pull up on the press handle to remove the case from the die.

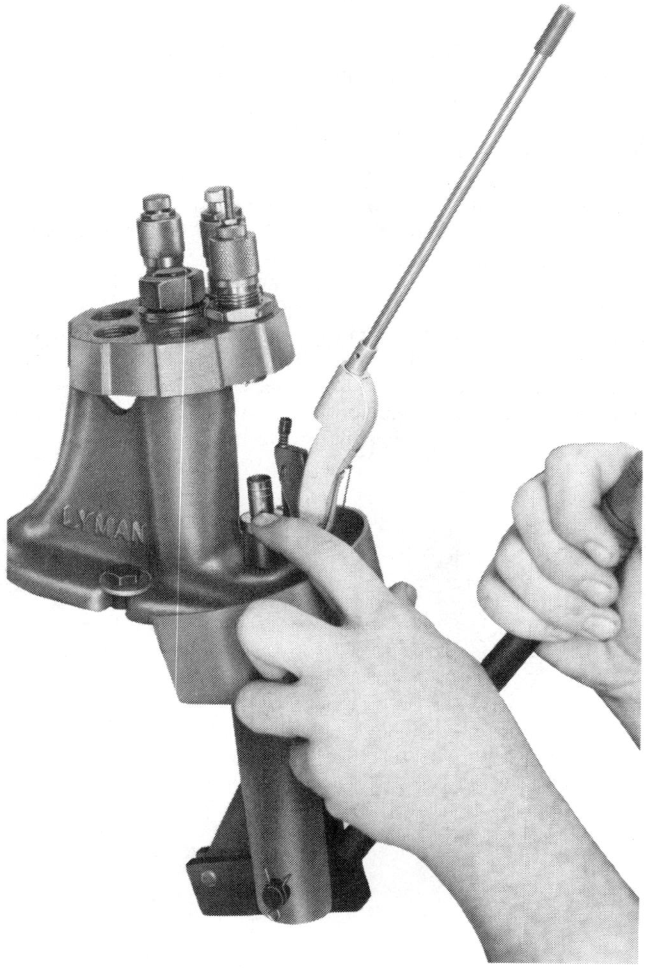

Note—The entire cartridge case enters this die, flush to the shell holder.

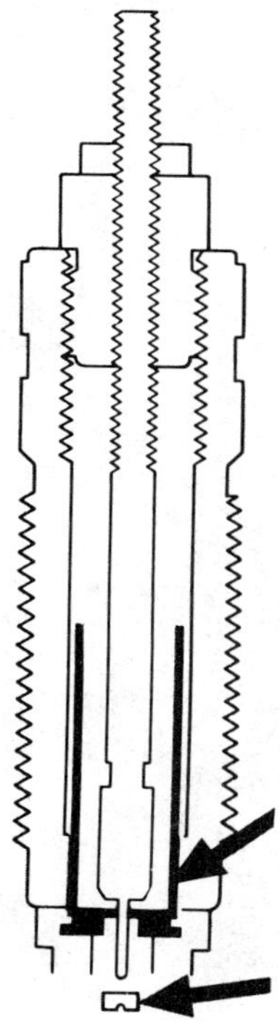

Entire outside diameter of case is reduced in size.

Decapping pin removes the fired primer.

Editor's Note:

Clean the interior of your dies periodically. Grit can accumulate which will scratch both the die walls and the cartridge cases.

Step Two *(Inside Neck Expanding)*

Screw the Neck Expanding Die into your press and adjust it according to the instructions supplied with the die. Place the resized cartridge case into the shell holder and pull down on the press handle. Note, in the drawing, how the two-step plug enters and expands the case neck. Actually, there is only a few thousandths difference in diameter between the first and second steps on the plug. This difference is so slight that it is not visually apparent. The illustration has been exaggerated for clarification.

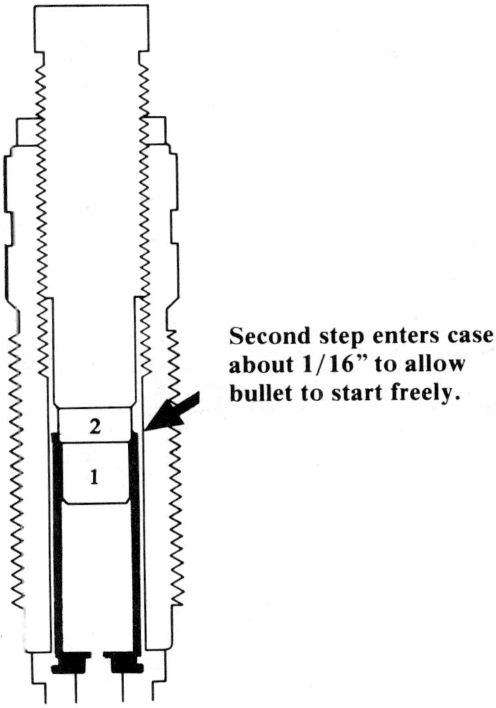

Step Three *(Priming)*

The priming operation takes place as your case is withdrawn from the Neck Expanding Die. Place the new primer (cup side up) into the priming punch sleeve. Push the primer arm forward (toward the press) and pull up on the press handle. As the ram is lowered, the priming arm will enter the slot in the side of the ram and seat the primer.

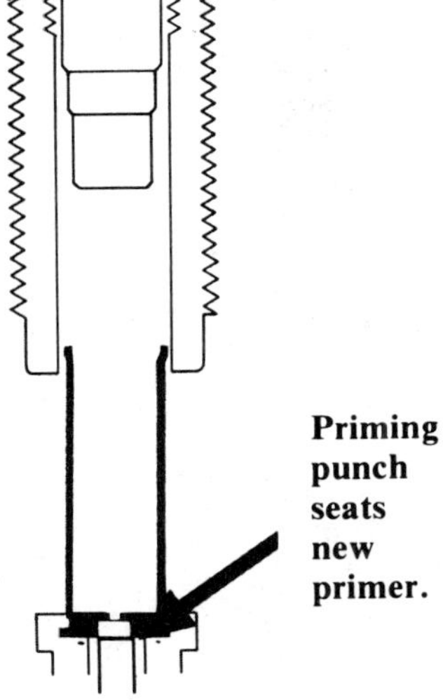

Seating Primers:

Primers are seated mainly by feel. The bottom of the anvil **must bottom** in the primer pocket. Depending on the brand of case and primer being used, this usually works out so that the primer is fully seated when the top of the primer is flush with the head of the case, or a few thousandths below the head. Under no circumstances should primers protrude. Use care not to crush the primer. Crushed primers give erratic ignition, or fail to fire.

Step Four *(Charging Powder)*

For the weighing of powder, you require an accurate powder scale such as the Lyman D-7 shown. The Data Section of the Handbook specifies the powders appropriate for your particular cartridge. It also lists a suggested weight of the charge in grains and in fractions of grains. For example, 2.2 would be read as TWO and TWO TENTHS grains. 3.0 would be read as THREE grains.

Carefully level the powder scale as explained in the instructions and set it to weigh your required charge. See the accompanying illustration for scale adjustment.

Slowly trickle small amounts of powder into the scale pan until the beam comes into balance. The beam is in balance when the pointed end (extreme left) is exactly on the zero mark.

Carefully remove the pan and pour its contents into the cartridge case. Use a powder funnel to make sure all the powder enters the case. Because pistol powders are comparatively fast burning, most normal charges take up very little room in the cartridge case. In other words, it is possible to **accidentally double charge**, or even triple charge many pistol cases. This, of course, would prove extremely dangerous and a foolproof system of loading must be developed. A suggested method is to place all the uncharged cartridge cases on your left. As you pick up each case for charging, turn it up-side-down and shake it. This will insure that the case is empty. Turn the case right-side-up, charge it and place it carefully on your right. Take care, when removing or replacing the scale pan that the poise are not accidentally moved.

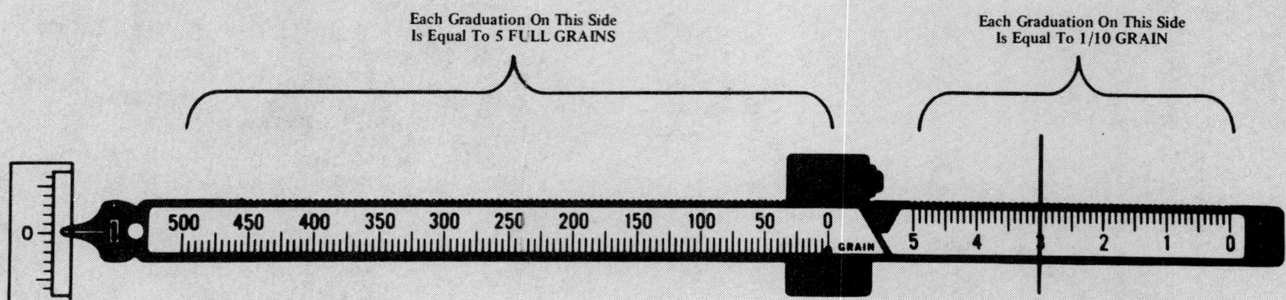

The illustration shows the beam of a modern reloading scale. Note how it graduated on both sides of the pivot point. The scale is set by moving the two weights (poise) to the proper graduations. The large poise (on the left) is used to obtain multiples of FIVE GRAINS, while the small poise (on the right) is used for FRACTIONS of a grain or SINGLE grains from one to five.

Example: The illustration shows a setting of 3.0 grains. If you wish to increase this to 8.0 grains, simply move the large poise one notch to the left.

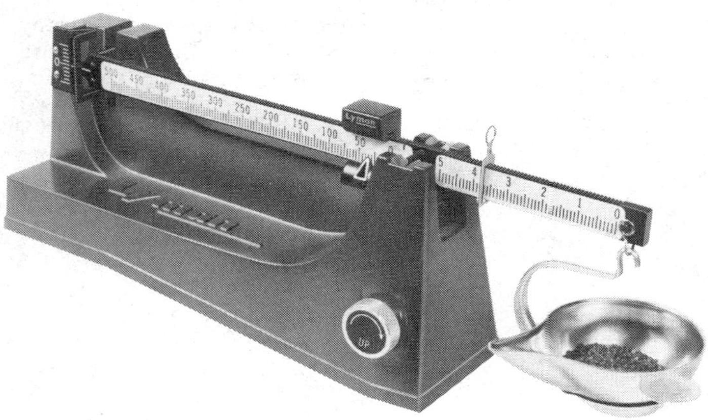

CAUTION

Technicians in the Lyman Lab have observed a potentially serious phenomena involving powder scales and plastic loading blocks and/or styrofoam packaging. It seems these substances sometimes retain a certain amount of static electricity, enough to create a electro-static field of varying radius.

This electro-static field has proven capable of radical deflection of uncharged and "zeroed" scales of all brands (available to us at the time). Of course, powder in the pan will tend to dampen the deflection but some still occurs depending on the charging level. Generally, the heavier the charge the less error... assuming the scale was first "zeroed" correctly.

We suggest you clear the surface of your reloading bench and make very sure the scale is set up accurately. Then move your equipment back piece by piece, paying particular attention to the plastic or styrofoam mentioned earlier. Please note this caution applies to all forms of reloading.

Step Five *(Bullet Seating)*

The last operation in reloading a cartridge case is seating the new bullet. Be sure the overall length of the finished round is not longer than the MAXIMUM OVERALL LENGTH listed in the Data Section. Adhering to this measurement will make certain that the finished round will function properly in your magazine or cylinder.

This illustration shows how a bullet is seated. Screw the bullet seating die into the head of the press and adjust it according to the instructions supplied. Place a primed, charged cartridge case in the shell holder and start a bullet in the mouth of the case. Pull your press handle all the way down, so when the case enters the die, the bullet will be pushed firmly into the neck of the case. Adjusting the seating screw controls the depth to which the bullet is seated. Adjusting the die body controls the crimp.

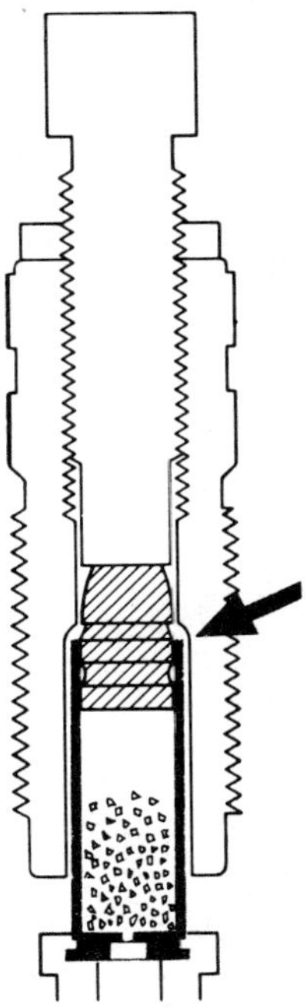

Note—The entire cartridge case does not enter this die.

Built-in crimp shoulder affords crimping-in of bullet when desired. Do not crimp on rimless autoloading cartridges which headspace on the case mouth.

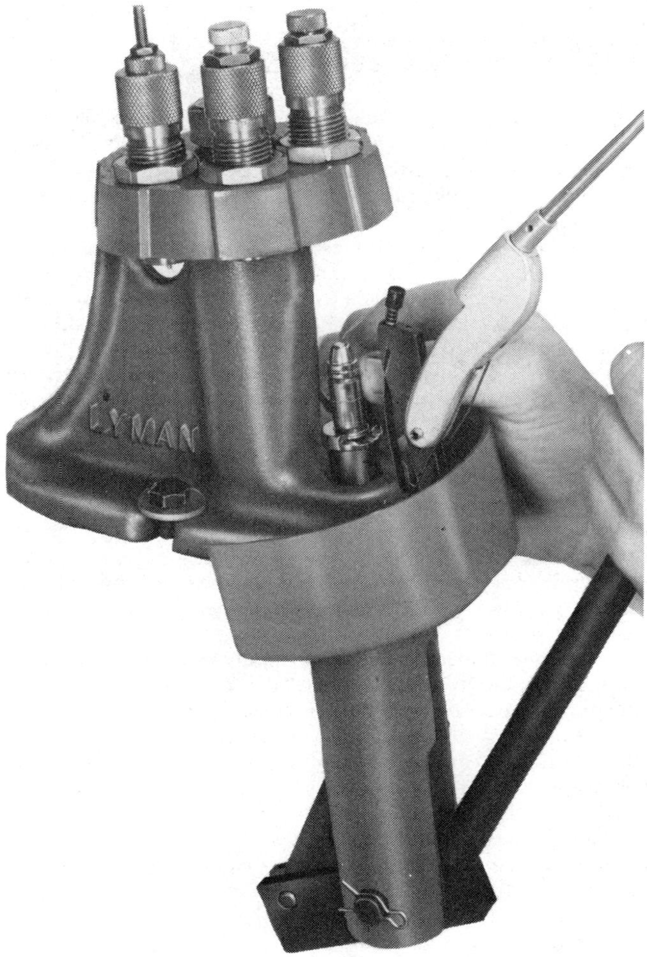

CAUTION: After reloading and before firing, wipe your cases to remove all sizing lubricant. The presence of oil or grease on a cartridge may dangerously increase thrust on the bolt face.

Editor's Note:

After every 300-500 rounds, remove the seating screw and clean accumulated lubricant from both the die interior and the seating screw.

Lubricant build-up can cause misaligned bullets and a gradual increase in seating depth which often escapes detection.

ACCURACY TIPS & TECHNIQUES

Introduction

Beyond owning a well-made, correctly bedded/fitted firearm, precision-made dies and related equipment, the cast bullet shooter can improve his shot-to-shot accuracy by paying attention to details during the various stages of casting and sizing.

Unlike the reloader who uses only commercial jacketed bullets, most cast bullet shooters opt to cast their own projectiles for a given cartridge from a myriad of bullet designs and weights. Accordingly, selection of the bullet design should be one of your early considerations.

Generally speaking, the more bearing surface you can find, the better your end results will be. Cast bullets, rifle applications in particular, aren't quite as tough as are their jacketed counterparts. Improperly sized bullets, misaligned cartridges, or faulty bore/chamber alignment can create a less than optimum projectile attitude within the barrel. In other words, a canted bullet; the centerline of which is not tracking along the centerline of the bore.

Aside from a bit of leading which may occur from related scuffing or gas-cutting, the real problems begin when the bullet leaves the barrel.

Unless the bullet base leaves the barrel simultaneously around its full diameter, the bullet is often tipped to varying degrees. The part of the bullet still in contact with the barrel drags, tipping the nose towards the sector of contact. Propellant gases, bursting from confinement on the far side of the bullet base, aggravate the tendency to tip. This phenomena is graphically illustrated with a Minie bullet in our Introduction to Muzzle-loading. The bullet base, alloy or gas-check, should be true and square—and precisely loaded. Sometimes the culprit is not the bullet's base, but the crown in the gun barrel.

By now, most shooters understand that the barrel vibrates as the projectile travels to the muzzle from the firing chamber. In handguns, this factor is not a major consideration, with any exceptions occuring in the long barreled silhouette specialty guns. Rifles, however, are affected tremendously.

Hardly a month goes by without a discussion of action and barrel bedding appearing somewhere in the shooting sports press. The focus of these discussions (which range from column commentary to full feature story) is on the bedding of barrel and receiver in the stock. The whole idea being to not interfere inconsistently in the barrel's vibrational cycle so that, given uniform ammunition, the projectile will exit the barrel at the same point in the cycle. Assuming a "perfect" projectile, the results on target are gratifying.

However, let's assume our rifle is well-bedded and gives minute-of-angle groups with jacketed bullets. This rifle should do nearly as well with cast bullets—not necessarily at similar velocities, though.

A cast projectile of short bearing surface—a spitzer, for example—may not shoot well. The spitzer point is not the culprit by itself; the key, again, is bearing surface alignment.

Now is a good time to clarify a point and return to the alignment variables mentioned early in this section. First, "bearing surface" on a cast bullet is a desirable feature primarily as an aid to alignment once the bullet is fired. A cast bullet is not as resilient as a jacketed bullet and needs that extra bearing surface to enhance bore alignment. Cast bullets which resemble jacketed match bullets tend to be least accurate in most rifles since relatively little of their body length is in contact with the bore.

Within the consideration of bullet design lies the matter of gas check designs versus non-check designs. This matter is one of application. Use of gas checks on hard (#2 alloy or harder) bullet metal isn't usually necessary until velocities exceed 1600 fps.

This means that most handgun projectiles don't necessarily require gas checks when full-house loads are used. However, use of gas check pistol bullets is very popular and thousands of shooters swear by them. One reason for this ties to bullet metal hardness. By using a softer (than #2 alloy) alloy and a gas check, a reloader can (a) shoot less expensive bullets (b) use projectiles which will tend to expand in game, transferring energy to the animal, rather than drilling on through cleanly or shattering on bone. Another consideration is that the gas checks on pistol bullets serve, to some degree, as bore scrapers. More importantly, gas checks offer protection to a bullet's base as it jumps the barrel/cylinder gap in a revolver. Finally, handgunners using the specialty silhouette guns—like the T/C Contender and Merrill—are able to approach, and even exceed, the hypothetical 1600 fps limit.

Orient Components

Everything mechanical is assembled from parts which are processed within certain dimensional specifications. Each specification has both a "maximum" and a "minimum" allowable dimension. This holds true for every rifle barrel and chamber, reloading die, bullet mould and sizing die and all related equipment. Generally, this factor is called "tolerance."

Advanced benchrest shooters eliminate (or at least minimize) the tolerance by using very expensive equipment machined with the highest possible precision. Also, these same shooters use jacketed bullets which help tremendously.

Does this mean the average shooter, with his off-the-shelf (or nearly so) rifle or pistol, must accept lack-lustre cast bullet accuracy? The answer is a resounding "No."

By identifying bullets from each mould cavity and processing them—and cartridge cases—with each being positioned just like its predecessor, groups may be halved, literally. This "orientation" must be carried through the chambering and firing, with the loaded round loaded in the same relative position each time. The best way to employ this technique is to mark mould cavities, as noted in the next page, and to index chambering on the cartridge case's caliber or brand marking. This works well with jacketed bullets, too.

Casting for Accuracy

In casting for maximum accuracy, the by-word is "consistency". Know the materials and procedures with which you are dealing—and be consistent in every aspect. To experiment with the variables of alloy, lubricant and sized diameter, alter *only one* of those variables at a time. In this way, any changes in performance can be linked to their cause. (The same caution applies to cartridge cases, primers and propellants, too—but that's another matter).

Single vs. Multi-Cavity Moulds:

Over the years, many cast bullet shooters who have been interested in pure accuracy experimentation have advocated only the use of single cavity blocks on the premise that multi-cavity blocks do not produce truly identical bullets from their individual cavities.

In a literal sense this is quite true. In a practical sense, however, there's more to be said, particularly regarding the double cavity mould. Lyman's, and most other brands, double cavity moulds are cut with the same cutter in a sequential operation. Cutter wear, affecting as-cast diameter, is not a factor between cavities of a given block going through production.

After the first cavity is cut, the blocks are opened and the machining fixture is traversed to the index point for the second cavity. The blocks are then closed over the spinning cherry cutter to produce the second cavity.

Where can variance occur? First, the blocks may not be compressed equally for each cavity with resulting *minor* diameter variance. Second, a worn block fixture could shift minutely as the blocks close the second time, producing a slight variance in the out-of-round dimension. Finally, depth of cut may shift for the same reason or because of spindle wear in the machine head.

It is our opinion that the vast majority of bullet casters are best served by well-made multi-cavity moulds—particularly double cavity. The only shooters who might truly benefit from a single cavity mould are those who are shooting state-of-the-art benchrest guns using advanced reloading techniques. Even in this instance a double cavity block can make sense.

How? Very simply. A double cavity block provides two choices of as-cast bullet. If the shooter can truly establish that the bullet from one cavity shoots better and/or to a different point of impact than the bullet from the other cavity, he has several very palatable courses of action open to him. First, he can cast from only the one cavity he chooses. Second, he can mark each cavity to clearly identify the bullets during post-casting visual inspection. Typically this marking is accomplished by placing light punch or file marks in the nose area of a given projectile. Placement of marks in both cavities, in the same positions, allows the bullets to be uniformly oriented, if desired, during reloading.

Some casters advocate placement of the marks on the mould parting line and state that the "fin" of metal produced can be easily knocked off with a fingernail. However, too heavy a blow to the marking punch could easily displace block metal into the parting area and prohibit correct closure of the block.

Lyman, by the way, does not endorse that practice since placement of punch marks constitutes a deliberate marring of the cavity and invalidates the product warranty.

By mechanically identifying bullets from the cavities with a punch mark system or carefully segregating them into two cooling piles as the blocks are emptied each time, the shooter produces a substantial number of good bullets. If he chooses not to use the product of both cavities in his pet benchrest rig, he can accumulate projectiles for his over-the-course big bore rifle or sporter, at very least. More common is the practice of treating the two bullets as a component change which may require minor adjustments to sights or propellant change—much like shifting from one lot/brand of case, primer or powder to another. Accuracy is virtually identical.

Smoking Mould Cavities

Some advanced bullet casters find that applying a coat of soot to the mould cavity enables them to produce good bullets faster and easier.

The technique is called "smoking" and is best done with a grease/oil-free flame, such as match rather than a candle. Good results have been reported from users of carbide lamps, also.

To brush or not to brush the soot from the cavity before casting becomes the question. Since moulds and casting technique vary, we feel that experimentation will provide the answer.

Additionally, some casters report that their start-up time is reduced and better bullets result from applying a good cold blue to the mould cavity. Again, experimentation will provide the answer for your situation. Please note this technique applies only to iron or steel blocks.

Remember, you must start with clean, degreased blocks whatever you do.

Experiment with Sizing Diameters

Reloaders using cast bullets must deal with several variables which do not arise with jacketed bullets. Among these, sizing diameter is very important.

Most cast bullet shooters want to size their projectiles to match the jacketed bullet diameter appropriate for a given chambering. This is understandable—and also wrong—in many instances.

The cast bullet shooter should be prepared to experiment with sizing diameters, beginning with *his measured* groove diameter and ascending in increments of one or two thousandths a time.

This will require an investment in several extra sizing dies but will prove well worth it. Pressures won't be affected since the bullet metal is so soft compared to gilding metal and bullet weight remains unchanged.

Willingness to experiment (always upward from your measured groove diameter) in this area will provide the following benefits: reduced sizing of as-cast bullet; tighter case neck fit in chamber and, last but not least, a *sure fit* of bullet to barrel.

For the best possible accuracy and performance, you *must* experiment with sizing diameters other than those which match appropriate jacketed bullet diameters.

Bench-Testing Rifles and Pistols

In order to evaluate the performance of a given load, it is necessary to fire a series of targets and note the resultant groups. "That is obvious—tell me something I don't know" you might be thinking.

There's more to it, though, than sitting on a hillside, leaning against a rock and sending a few shots into a dirt clod at the foot of the hill.

The best way to evaluate the performance of a given load is from a solid benchrest, at known-distance targets over a level range. The benchrest relieves your body of the task of holding the gun absolutely still; known-distance targets show you exactly how the load performs at known—and useful—distance increments; and the level range ensures there is no trajectory distortion. Gravity exerts a reduced influence on a projectile when it is fired either uphill or downhill. That is why hunters tend to overshoot game either above or below them.

You must, of course, have a rifle or pistol capable of delivering every possible bit of accuracy. This is not to say that the only possible testing has to be done with sophisticated and specialized target guns—just that you understand your gun and any shortcomings it may have.

Benchrest shooters have developed a fairly universal approach to shooting the bench guns wherein they have the least possible body contact with their rifle.

- The stock fore-end is not gripped by hand; that hand, instead, controls the butt bag support.
- While the shoulder does touch the butt plate, it does so very, very gently.
- And, if the trigger hand does wrap around the stock wrist, it does so very lightly. Another, and perhaps more common, trigger release technique is to place the forefinger on the trigger, the thumb behind the triggerguard and "pinch" off the shot.

Well, those techniques are well and good for the benchrester with his special stocks, rests and related paraphernalia. Those techniques, however, are probably not ideal for a shooter equipped in a more modest or all-around fashion.

After some consideration, we submit the following bench test techniques as being the most practical for general accuracy testing involving cast bullets:

1. **Handgunners should get good results by sitting behind a shooting bench, gripping their pistol firmly in two hands. Next, extend your arms and bring them down on a firm rest which should make contact from the wrists back no more than six inches. Experiment for the position offering the greatest degree of comfort and consistency—then return to that position for each shot.**
2. Riflemen should seat themselves behind a sturdy bench and adjust the heights of both stool, bench and rest to permit an erect sitting position.

   Remove both sling and sling swivels (if they are of the quick-detach sort). Lay the rifle on the front rest and snuggle in behind the butt. Grip the forearm and pull it firmly into your shoulder. Grip the wrist firmly and position your hand for a good controlled trigger squeeze. This position is the easiest to master. However, some practice is necessary.

   The rests for both the handgunner and rifleman should be of a firm consistency—but not hard and solid. A rifle fired from a rock or fence post will shoot to a different point than one fired from a padded rest. Handgunners will notice something similar—plus some pain—if they shoot off an unpadded rest.

Whether your interest lies in rifles or pistols, the important consideration in bench testing is consistency. Furthermore, don't grip your piece so tightly that you tremble for that will defeat your purpose.

Relax. Enjoy. And squeeze that trigger....

Neck-Turning

This is a component preparation procedure used by benchrest rifle shooters to produce truly uniform wall thickness in the walls of their cartridge case necks. Goals are consistent neck tension on the bullets, conducive to uniform shot-to-shot combustion and thus accuracy, and a close fit of case neck to chamber to enhance bore-bullet-case alignment.

Neck-turning is desirable to the advanced cast bullet rifle shooter for similar reasons.

However, this procedure may be necessary to develop the most accurate combinations in your sporter or target rifle and here is why: Fit of bullet to your gun's bore is the single most important component consideration. Pursuit of that goal may well lead you to sized cast bullets that are several thousandths of an inch larger in diameter than the appropriate jacketed bullet. This oversize projectile does not expand the case neck by the same amount, but there is some degree of case neck diameter increase.

Thus, in some chambers the neck area may not accept the loaded cast bullet round. The answer, if you wish to continue with the larger bullets, is to reduce wall thickness in the case neck.

What may seem just another problem to overcome often can be a blessing in disguise. Neck-turning not only produces additional space for oversize bullets but makes neck wall thickness uniform—greatly enhancing bullet/bore alignment.

The variation in case wall thickness may surprise you. Buy or borrow a tubing micrometer and measure some of your brass. Surprised? Variations of a thousandth or two are not uncommon. This is another reason to orient components and the loaded cartridge in the chamber—if you cannot be "perfect", be consistent.

Gas Checks

There are two types of gas checks commonly used: The simple brass cup with nominally uniform wall thickness, as sold by Lyman; the gliding metal crimp-on cup sold by Hornady. Both are satisfactory if they fit the bullet properly. However, both types have been criticized for systemic shortcomings as follows:

1. The press-fit gas check has been accused of falling off the bullet after it leaves the muzzle, disturbing the bullet by its departure (inconsistent departure, at that) and thus contributing to inaccuracy. It is true the Lyman check often separates from the bullet after firing and, in fact, was designed to do so. This may or may not contribute to reduced accuracy. Comparison testing, changing only the type of gas check, is the best way to resolve the matter for your particular application.
2. The crimp-on gas check is designed to actually bite into the bullets' shank to ensure retention in place after firing. However, the key to success with these gas checks is to be sure they are seated square with the bullet base before the crimp is pressed into the bullets' gas check shanks. Criticism of the crimp-on check centers on misalignment during the sizing/application operation which produces a projectile with an irregular base. A true, square base is imperative for accuracy.

Additionally, some shooters feel that they must exert excessive force during the sizing operation when the gas check must crimp onto a very hard alloy bullet; linotype alloy, for example. However, the crimp-on gas checks have a Brinnell hardness of about 100 contrasted to a 22 BHN rating for linotype metal. The gas check can take the pressure, the questionable factors are the bullet itself and the sizing/lubricating machinery.

Difficulty of installation is a problem common to both types of gas check for the same reason—diameter of the cast bullet shank is too large to allow the bullet base to bottom out inside the gas check cup. The strength of the gas check material prevents it from uniformly expanding to handle the increased diameter.

The other side of the coin is the situation wherein the gas check shank is too small to retain the slip-fit gas check. There are several remedies open:

1. Use a dab of thick lubricant on bullet base to retain the gas check during loading operation. This is best done on straight cases or those in which the bullet's base is still within the grip of the case neck, otherwise the gas check could fall off into the powder charge.
2. Use a harder alloy to increase the as-cast diameter of the bullet and thus the gas check shank.
3. Anneal the gas checks, check for fit.
4. Switch to the crimp-on gas check.

A possible solution to this problem, for both types of gas checks, lies in annealing the checks to soften the material. The softened material can be moved to a greater degree than the unannealed while still protecting the bullets' base in a totally adequate manner.

For optimum accuracy, gas checks should be weight-segregated into groups. The criteria for this segregation is up to you, but an allowable variance similar to that used in inspection your cast bullets is logical.

For load development and related broad spectrum experiments, a half-grain increment should prove sufficient. "Fine tuning" for accuracy may lead you to increments as low as three-tenths of a grain. As always, best results are obtained by uniformity and consistency.

There is a certain benefit to seating gas checks squarely before the sizing operation. The crimp-on style of checks are sometimes prone to premature crimping which usually produces an untrue base—not good.

To aid us in our experiments, we fabricated a simple gas check seating fixture which fits inside the sizing die retainer nut and aligns a gas check cavity with the top punch to ensure square seating.

Once the gas check is firmly seated, the fixture is removed; sizing and lubrication proceed normally.

Seating Depth

This factor is a consideration just as much for the cast bullet shooter as for the jacketed bullet reloader. Both bullet types benefit from reducing the "jump" from case mouth to engagement with the rifling.

It is fairly standard procedure, with jacketed bullets, to determine the point at which the bullet's ogive firmly bears against the rifling and then back off the die a turn to seat the bullet about 1/16" back from the rifling. The reason for this is to allow the bullet to start moving before it meets the resistance of the rifling. Failure to allow this fraction of an inch has been proven to substantially boost chamber pressures, a potentially dangerous situation which is always detrimental to accuracy.

Cast bullets, because they are "softer" (20 BHN vs 100 BHN) than jacketed bullets, suffer more from any free travel before rifling engagement. Seating the bullet to engrave from the rifling often improves accuracy. Try both techniques, even if the actual overall length exceeds that listed in our tables.

Keep in mind that the finished cartridge must function in your gun's clip or magazine and feed reliably into the chamber.

Single shot target rifles usually pose no problems. Other rifles and pistols, however, often limit usable overall length by their clip and magazine—or cylinders.

Ammunition assembled for field use must be utterly reliable with no possibility of the bullet wedging into the barrel and being pulled from the case as the action is opened for clearing.

Technical Ramblings *by Ken Mollohan*

In a previous newsletter, the effects of ignition variables on cast bullet loads were considered, and the conclusion reached that (within the limits of reliable ignition), the milder the primer, the lower the pressures would be, and the better the results with cast bullets. Now let us consider why milder pressures mean better accuracy.

The immediate reaction is to say "Because they don't upset the bullet and distort it so much when it's fired. High pressures deform the base in the throat, before the nose has time to start moving. Also, the high pressures have high temperature to match, and melt the base of the bullet. Naturally, the misshapen, molten blob leads the bore, and is inaccurate".

That's all very reasonable and very logical. Unfortunately, there is reason to believe it is quite wrong. Let's examine it point-by-point.

1. ***Melting of the Base by Propellent Gases:* It's true that propellent gases are quite hot—many times hotter than is required to melt lead. However, their heat capacity (specific heat) is very low: they transfer very little thermal energy as they cool. By contrast, the specific heat of lead is very high—it takes a lot of thermal energy to raise it's temperature. Also, transfer of heat is very slow; while only miliseconds elapse before the bullet is beyond them. Pass your hand quickly through a small flame. Though exposure time will be far greater than a bullet's, you won't feel the heat. Then put a bullet-sized chunk of lead in the flame. Does it melt instantly? No—several seconds are needed to even melt edges. Then think about the common practice of exposing bare lead in jacketed bullets—Noslers and military slugs, for example. Do they cause leading from melting at the base? Finally, consider plain-based slugs at *full power* in cases like the M-1 Carbine and 44 Magnum rifles: no melting, no leading, good accuracy. While leading and inaccuracy do occur, it seems most unlikely that gross melting is responsible, or even likely.**
2. *Pressure Deformation in the Throat and First Part of the Barrel:* Dr. Mann credited this as a major source of inaccuracy, and ran extensive tests with shortened barrels. Beyond question, he obtained misshapen, deformed bullets which were not capable of accuracy. Later tests produced similar results with jacketed bullets: even tough, hard military hardball would expand, under the pressure, to resemble an umbrella. There seemed to be no question: the tests with short barrels proved conclusively that the pressures in the throat and first part of the bore *could* deform bullets drastically, and destroy their accuracy potential.

   If enough pressure is applied to turn a military hardball into an "umbrella", the base must upset to fill the bore and throat perfectly: the upsetting force will see to that. Then it will be squirted down, into the bore, and on it's way. Only the fact that the copper jacket will take such reforming enables it to survive. A soft, plastic lead alloy has no chance unless the load is reduced...or so they say.

   Trouble is, digging into the backstop of almost any range will produce an awful lot of jacketed bullets that do not show any bearing (sliding marks) in the corners. Some will show only slight marks in the center of the grooves, and a few (from oversize barrels) will show no bearing on the grooves at all.—And this is not a result of anemic handloads either, because the same thing is found with 30-60 military hardball. Those slugs didn't deform or expand one 10th to fill bore, throat or anything else. And if they didn't, maybe—must maybe—hard cast lead alloys don't, either. In fact, it's hard to see how deformation in the barrel can affect accuracy to any great extent (assuming properly designed bullets). First of all, if they upset in the throat, and are reswaged going into the bore, the upset and swaging cannot be perfectly uniform. Some portion will receive more upset and/or swaging than another. Yet recovered fragments from factory-velocity loads show uniform lube grooves and uniform band widths; no significant upset. And, if we accept the theory that perfect upsetting and swaging occurs, it means that a perfect bullet is delivered into the bore; in either case, great loss of accuracy does not occur due to upsetting in the throat.

   Does it occur in the bore? Theoretically it can: under pressure from behind from the hot gasses, and with the body prevented from deforming (by the walls of the bore), the nose can sag, or collapse. This collapse is unlikely to be uniform, and accuracy results. Col. Harrison of the NRA showed this to be a major problem with poorly supported, sharp designs.

   But well-designed bullets like 311291 are another story. Since it is already full bore diameter in the nose, it cannot upset more than a few thousandths of an inch to groove diameter, where the expansion will be stopped by the barrel. Even if the worst possible case is assumed, wild inaccuracy cannot result: bullets with lead shaved from one side to duplicate the imbalance that would occur if *only* one side sagged (which is highly improbable) shoot poorly, but not wildly. And Loverin designs have almost no nose at all left unsupported to sag. Yet they, too, will shoot wildly with maximum loads. Thus nose upset does not seem to be a major source of inaccuracy. And since the body is totally supported and cannot collapse, the problem does not seem to be there, either. Granted this is not in accord with many theories, as mentioned above. But if you are unconvinced that pressure deformation in the throat and bore are not problems, consider the following experiment.

   Load Lyman's 311291 in a 30-30 case at about 40,000 psi. In a good rifle, it will shoot alongside jacketed ammo of the best quality, if properly made and loaded. Now load it to 40,000 psi in a .300 Magnum. The only thing it will shoot along side of is a blunderbus—if you're lucky.

   Since in each case, you can have the same bullet, same alloy, same sizing, same lube, same gas check, same pressure—same everything but accuracy. You can even get cute and rechamber a 30-30 barrel (on a suitable action) to 300 Mag. without changing the fact that results will go from great to lousy.

   Then load the 30-30 at 40,000 psi with a plain base bullet and note the loss of accuracy. Yet plain base bullets give good results at 40,000 psi in the M1 carbine.

Reprinted by permission of the Cast Bullet Association, Inc., from the *Fouling Shot*, 3/77.

Speaking "Frankly" *by Frank Marshall, Jr.*

On .30-30 Loads...

Ed Harris recently gave me some W-W .375 Win. cases to play with. The .375 would make a pretty good cast bullet rifle as it's, using a 270-gr. or so bullet, or it could be necked down to .35 for a good dual purpose hunting/target rifle with better ballistics than either the .375 or .30 cal. rifles on cases of that capacity.

Measuring the .375 cases, I found the neck and body walls are thicker. When necked down in a .30-30 die the cartridge neck with a .308 jacketed bullet seated ran .331", compared to .327" max. for the standard .30-30. This would be dangerous in a close neck rifle, but I happened to have a sloppy chambered Savage bolt-action .30-30 I use for a rough duty hunting rifle, and these .375 cases provide in effect a CBC with this rifle. This rifle normally only shoots cast bullets well if sized to .312".

For hunting with this rifle I use the 311284 Lyman bullet with 30 grs. of 4350 with any large rifle primer. After lubing I run it deep in the lube die with a flat top punch, using washers around the punch as adjustable stops to control the amount of upset. You can use these to bump bullets nicely in standard lubricators. To get best concentricity, however, don't lock down the set screw which holds the top punch, for this may force it crooked. Simply put a dab of beeswax or bullet lube on the top of the shank on the punch and push it in. The wax will make it stay put while it seeks its center. Use the same trick on the washers too. Using this technique I take the 311284 bullet and form a flat-nosed bullet with .303" bore riding section in one pull. The exact size, of course can be controlled to fit any barrel. My Savage 840, however is a 6 groove 12" twist with .302" bore and .309 groove. The barrel is smooth and uniform, but just a bit odd as to dimensions. The rifle is plain as hell, but is a real workhorse and is effective on Blue Ridge bears and white tails.

I seat this reshaped bullet with flat-nose, .303" bore riding portion and .312" body to just feed reliably from the box magazine on the Savage. I estimate its velocity with 30 grs. of 4330 as about 1750 f.p.s. from the 22" barrel. This sounds mild, but considering this bullet weighs over 220 grs. and this velcity exceeds the old Super-X load for the .32-40 with 165-gr. bullet, it is no popgun. In fact, its energy just about duplicates what you got with the .303 Savage with 190-gr. bullet. You must remember, of course, that while the .303 claimed 1950 with the 190-gr. bullet, those figures were based on a 26" barrel and most of the short 1899 Savage carbines never saw the far side of 1800 f.p.s. with factory loads.

This bullet is cast of the same 90-7-3 lead, antimony, tin alloy I use, only sized to .3095" for my Winchester Model 70 target .30'06. It is my best and favorite .30 cal. bullet. The bullets I reject for match use usually wind up in the .30-30 Savage, or in my Remington 788 .30-30, which I don't knock around with. These bullets are already weight segregated, having been rejected only for minor visual defects. I think the heavy deformation from reshaping them helps compress the voids, as they shoot with equal accuracy to my selected match bullets, in the 788 .30-30. I used these same bullets with .301 nose, and bands sized .3095" to shoot the under m.o.a. group at Wappwallopen, without reshaping them.

The Rem. 788 with 10" twist and the Savage 840 with 12" twist both stabilize the shortened flat-nosed 311284 without any yaw noticeable on the target down to about 1500 f.p.s., though this may change at longer ranges than 100 yds. or when the weather gets cold.

I shot a whitetail buck quartering away with this load, which entered the right ham and exited the opposite shoulder up close to the neck, and knocked a groove 8" long on a pine tree it hit. The exit hole indicated fair mushrooming and a good wound channel, but no lead chips were found, just the Hornady gas check about half way through the hindquarter. This 90-7-3 alloy isn't real hard, and gives just about optimum performance. I have seen deer hit the same way with .30-30 jacketed 170-gr. loads where the bullets never got past the diaphragm. The deer ran on and on, gutshot, a hell of a mess, due to poor penetration. This heavy bullet, in contrast, is very deadly on these tough going away shots.

For an optimum load in .30-30 I need a new .30 cal. 200-gr. bullet with three bands like the #311284, but with a shorter nose and blunt shape like the #311440. This would be more for a hunting than as a target bullet, and the blunt shape would be an advantage for the woods and brush shooting I do.

With modern powders in a .30-30 you could get 2000 f.p.s. with safe pressures for a Model 94 lever gun, for instance using about 36 grs. of W-W 760 with the bullet seated to feed through the magazine.

I have a thing about wadcutter bullets, in that I feel they are exceptional in terms of accuracy and game effect, when they have a wide, flat-nose and a bore riding area at least 1 cal. long. I have never been able to match groups shot with nose bore riding bullets using the all-body, non-nose bearing bullets of the Loverin or Pope form. The nose riding form was probably never tested fairly by the old masters, although Pope's multi-groove tapered bullet was in effect the same in purpose as the snug nose we seek today.

If Pope, Zischang and their breed were around today they would see what accurate really is, with a cast lead bullet in a breech loading rifle as we now do in our CBA matches.

The old boys shot some good, great groups, granted, but they could not consistently plunk them in under 1 m.o.a. which is now commonplace among our top shots on a good day, even with the light rifles. We have caught up to the old masters and are equalling their performance with far lighter rifles.

We are getting to the point where to win we must shoot great even under the lousy conditions and fellows must be able to deliver the goods on demand, not just wait for the perfect time. We are learning to be riflemen. I predict that the performance of cast bullets will soon equal the best of any bullet type for either match use or hunting if we continue to develop at our present rate. Seeing the way our scores have increased since our first national, that doesn't seem very far fetched at all.

Speaking "Frankly"

The .30-'06' Is An Oldie, But a Goodie

by Frank Marshall, Jr.

When Fairfax Rod & Gun Club decided to shoot at a 1" (gulp) 10 ring at 100 yds. with cast bullets for record, under CBA rules, I had to rustle through my "retired" rack for a suitable rifle. I also dug out the old notebook in which I had recorded over a half-century of cast load skullduggery.

Looking through the old notes of my Uncle Will, I refreshed the memories of the one-holer groups he used to shoot with his 10" twist star-gaged Springfield with 28" heavy barrel on a standard Springfield Sporter stock of dense, well-fitted walnut. All pertinent factors were well-documented, and the rifle showed amazing consistency in over two pages of legible data using the Ideal 311284 bullet. The loads were light by today's standards, giving an estimated velocity of 1600 f.p.s. using old FA26 corrosive primers with Unique, Hi-Vel #2 or No. 80 powder, plus a smattering of a few others, like Sharpshooter and Lightning, which few reading this will remember.

I distinctly remember my uncle Will saying, "with this load the powder's not the thing, it's the velocity. It has to be enough to stabilize, but not enough to lead. For low disturbance from recoil, minimal heat buildup over a long run of shots, the loads must be fairly light, the alloy only medium hard, so it will still seal the bore over the whole 28", while being strong enough to handle the 10" twist. All are essential."

Will, using a 5X on that Springfield, shooting from sandbags at 100 yds., would bet and win every time you could cover all or part of every one with a quarter. I saw this many, many times. This is equal to a 10X possible on the A-15 Smallbore Rifle Target. Not many of the pre-WWI masters could do this consistently, though they shot a few better. Will could do it on demand any time you asked him. His reaction, when asked how he could do it when legendary shots of single-shot rifles couldn't, was a casual, "they don't have boltguns." He knew the inherent superiority of the bolt action.

Few of the old masters left in the post-WWI period really understood the tricks or understood the principles of good cast bullet load development. They didn't understand why, they just experimented till they found something that worked. Once they found a "trick", it was a well-kept secret. There was no free exchange of information like the CBA enjoys today; for back then a trick meant the difference between winning a healthy wager and losing hard-earned money.

Today, of course, those same old masters wouldn't stand a chance against our top ten CBA record holders. Our guns are far better today, and we really do know and understand more. The long passage of time simply makes the occasional very good scores stand out as legendary, whereas the average standard of rifle accuracy at that time was really pretty abysmal, by today's standards.

Strange as it may sound, however, the old pros knew there was no such thing as a perfect bullet, case, powder, primer, chamber, throat, barrel, sights, bedding jobs or loading tools. Most of all, there was no perfect shooter, handloader or gunsmith. I learned early that once I accepted this and reacted accordingly, I'd crossed the big barrier.

The solution-now and then-has been to orient the imperfections to make their effect uniform on the target. Consistency of imperfection is the key to accuracy. It is amazing the fine results possible with simple loading tools like a Lee Loader or Lyman 310 dies in a 7/8-14 adapter if you concentrate on uniform position of the dies in the press, indexing them to a reference mark, the same with the bullet, orienting to a mould mark throughout sizing and loading and shooting, and the case with the sizing and seating die, and so forth.

In bullet sizing it is especially important to keep the top punch oriented and concentric to the sizing die. Your sizing die should be oriented in the tool also.

You strive for perfection by realizing your equipment is not perfect. Bullet selection boils down to the best one out of three, using the old plumber's pot, gas stove and dipper. My casting conditions are most primitive by most of you guys standards.

However, last year at Wapwallopen Ed Harris and I both won aggregates and had to choose awards. Ed agreed to let me have a Saeco electric furnace, if I would agree to try it; chiding me in a friendly way. I'll now go modern, thanks to him. I still feel, however, the old pot and dipper can produce truly excellent bullets, but perhaps with more rejects. I'm not sure.

I keep at least nine lbs. of lead in the pot (that holds 11 lbs.) and preheat the mould on the top edge of the pot; never in gas flame which may warp it, then cast until the bullets are well-fitted without any wrinkles. I then start keeping bullets for later inspection, and regulate my casting rate to stay just shy of frosty bullets. I don't work fast, but hold the spout or dipper to the mould for a few seconds to keep it warm, then wait a few seconds for the sprue to completely harden before opening the mould.

I flux every twenty bullets, and prior to coming out of the pot with a full dipper I stir back and forth twice through the alloy and bring the dipper up from the bottom uniformly each time. Dennis Marshall has remarked that lead alloys don't gravity-segregate and mixing is not necessary, and since he's the engineer, I won't argue with him, but I feel the stirring has other benefits, such as dislodging impurities and helping to prevent getting dross or flux in the cavity by always dipping from the bottom of the pot. The system works, I won't change.

I place the dipper spout centered on the pourhole of the mould, held horizontally, and roll them vertically together in a smooth and uniform motion. I get fine bullets by this method, but I must be very careful to select and weigh them for match loads and maybe only 20 or so out of 100 are of a quality I would take to Wapwallopen to shoot against the like of Sears, Ardito, Rollins, Sarty, Musselman and others of similar ability.

The long winter months of '78 found me with plenty of time and all my moulds cleaned, as no trick could be left unturned to face that 1" ten ring come spring. I ran batches of 311467, 31141, 311440, and my all-time .30 cal. favorite, the long, lean, wind-bucking 220-gr. #311284. After honing out a .309" die to give a .3095" bullet for a snug fit in my .3095" throat, I was ready for the serious work to begin.

All bullets were constantly inspected visually for any defects, in every phase from mold to final seating in the case. All were weighed into three lots, plus or minus 0.1 gr., as segregated to (-); (on) and (+). Those outside the + 0.1 grain scope went into the plinker/practice/fouler box. I throw away many bullets others would shoot, but I don't have very many unexplained fliers, either.

Using my pre-war Winchester Model 70 target rifle with factory 10" twist bull barrel, I've always sought to duplicate the ideal conditions my uncle Will found worked in his Springfield. They have worked for me, too. I worked on his indexing theory throughout casting, sizing, loading and shooting.

I concentrated on his "ideal" 1600 f.p.s. velocity, mostly with the #311284, but it gave very good results in my usual 87-10-3 lead-antimony tin alloy with all the above bullets (old, good Ideal moulds casting bullets of correct dimensions requiring little or no sizing). Early on I tried various powders to get this velocity, and in turn tried 15 grs. of Unique (one of Will's favorites), 20 grs. of #2400 or IMR-4227; and 24 grs. of IMR-4198. RL-7 is similar to 4198 and should work OK with the same approximate charge.

When SR-4759 was reintroduced a few years ago, I ordered a caddy with high hopes, since it was designed specifically for the purpose of making reduced loads in the .30 cal. '06 case, for downloading military-type bullets to simulate downrange impacts in testing armor plate, helmets, etc. 21 grs. of SR-4759 turned out right with the favorite #311284.

Of several accurate loads I had, I found the 311284 bucked the wind best and had, overall, the most consistent performance. This was just as my uncle Will had found in his Springfield. This isn't surprising, as my Model 70 has the same gov't 4-groove rifling form using in Springfield .30 cal. barrels, the same 10" twist and is internally similar. The Model 70 is essentially a refined Mauser or Springfield action anyway, with a faster locktime, better trigger, etc/. but the barrel weight and stock, handling qualities on the bags, etc. were almost identical to the Type T Springfield.

My Model 70 could be nitpicked on fine points; by modern standards it isn't what benchresters today would consider a super rifle. It's groove diameter is a fat .3085" plus, and the bore is .3005", requiring a nose no less than .301" diameter. Some shooters go a bit larger. The leade is worn somewhat, but smooth, a condition I find favorable towards use of lead bullets. At the front of the short throat, it measures .3095"; not with a minimum chamber, but not a bad one either. It is good and concentric, the bolt closing square on the case and both locking lugs bearing evenly.

Although when I started out, I realized there were much hotter rifles around, my hopes were tempered liberally by my memories of what uncle Will had done years ago with a very similar, and probably no better rifle. Most of all, though, I was limited by what I had, since when you're on half-rations, that's what you go with. I have more time than money, and I make up for in care to details what I may lack in fancy substance. Speed loading, high volume shooting and super cast bullet accuracy are not compatible.

Once I decided on the #311284 as best, I never toyed with the load after that. I use 21 grs. of SR-4759, weighed, no filler, the #311284 cast of 87-10-3 Pb-Sb-Sn, which weighs 220-grs., mixed from old hard wheel weights and bar solder. Hornady gaschecks are seated and all grooves filled with Alox and a thin layer is allowed to run onto the .301" nose. This .3095" bullet is seated snug in the neck so the gascheck is at its base, which presses the front band hard on chambering against the front leade cone. I use FA59 match cases, but there's nothing magic about them. What's important is to have a good uniform batch of brass; all the same maker, preferably the same lot.

The old .30-'06 seems to have fallen from favor among the younger shooters, but I think they are selling it short. True, the .30-30 and .308 are good cast bullet rounds, but the '06 is too. The small cases, like the .30 Johnson, .30 Herett, and 308x1½, I'm not sure about. I think the bigger case is OK provided you have a powder which will ignite easily. A small case develops more pressure to get the bullet up to velocity and this hurts more than helps. A high velocity may buck the wind better but is not as consistent. Above 1800 f.p.s. you cannot get as consistent performance, due to leading, as Ed Harris stated in No. 20; but at 1600 f.p.s. I never clean my rifle, because I don't need to. I use plenty of lubricant and the bore doesn't lead at all.

The performance of this combination speaks well for it. The first time we fired at Fairfax under CBA rules I fired a 196-4X in a light gale, and knew I had a combination capable of sub-moa, even though I admit I was holding deep into those gusts when there was no time to wait any longer. The next match was a fair day; 198-8X which set a CBA record. The next match I fired a 197-9X, which while not a record, was great for the conditions. In the August match before we went to Wapwallopen last year, I held a shot wrong for a nine, but finished with a 199-4X with deceptively calm, varying mirage conditions. At Wapwallopen, of course, I had to contend with 200 yd. shooting, but the combination held up well there too, giving me a 295-14X for the 100-200 yd. aggregate and a 200-yd. record 1.894" group. The old .30-'06 with 10" twist, using an ancient, ornery base-pour bullet with a light load giving it only 1200 f.p.s. or so remaining velocity at 200 yards, delivered the goods fine.

After returning to Fairfax we had one more match to go to complete our schedule. Rather than casting more bullets, I used selected "rejects" for the last match, which hung in for a 198-9X and a season aggregate of 595-21X out of a possible 600.

The rest of the story is significant; from the consistency of the record scores fired in club matches and nationals, plus doing 100 and 200 yard score and group shooting all with the same combination. I feel this proves the worth of the slow heavy bullet in bucking the wind without leading the barrel. I never had an eight for record all season, nor did I use my cleaning rod once.

The fact that a mediocre rifle could produce a 99+ season average with a 40% X count should make you ponder what a really good rifle could do with such a load. It would take a hell of a salesman to convince me that a good .30-'06 with 10" twist isn't in the running. Being that we are still a ways from knowing all the answers, this may not be it, but it's competitive with the alternatives. When I can shoot 10X possibles all day I'll give up shooting and try fishing or frisbees.

Reprinted by permission of the Cast Bullet Association, Inc., from the *Fouling Shot*, 8/79.

A Clean Barrel is an Accurate Barrel

Leading

Mention "cast bullets" in an assemblage of shooters and, chances are, visions of dull, streaked bores—woefully inaccurate—will cross most minds. This, however, is not usually the case. Lead alloy bullets, of reasonably suitable hardness, properly sized, lubed and loaded, rarely lead bores to such a degree.

A real culprit has been factory pistol ammunition, most visibly the .357 Magnum and .44 Magnum, which has been loaded with swaged lead bullets. By its very nature the swaging process dictates a bullet metal far too soft for magnum—perhaps even normal service—velocities. Many shooters, unfamiliar with bullet casting techniques, see all non-jacketed bullets as "lead bullets" and draw incorrect conclusions.

The fact is that, usually, lead deposits from properly constructed alloy reloads are of a minor nature, at worst, and annoy the accuracy buff far more than the hunter or casual shooter.

But, once lead—in whatever degree—occurs, how do we rid ourselves of it?

Many leaded bores are thoroughly cleaned by the vigorous and repeated use of bronze bore brushes and nitro powder solvent.

Handguns, revolvers in particular, seem to collect a disproportanate amount of lead in chamber mouths and forcing cones. Once damaged at the onset of its travel, said bullet leaves liberal swathes of lead as it streaks down the bore. Bore brushes sometimes cannot make a dent in this mess and more drastic means must be used.

Steel wool is drastic, to be sure. Although judicious scrubbing usually causes no visible effect, most shooters shy away from its use.

There's a device called the Lewis Lead Remover which has served handgunners well over the years. Basically it employs a tightly fitted brass or bronze patch to scrub out the lead.

Perhaps the easiest remedy for both handgunners and riflemen is to fire several rounds of jacketed ammunition. This removes the leading very effectively, leaving only the jacket fouling to be removed before the barrel is ready for alloy bullets again.

Jacket Fouling

It is a fact of life that your rifle will not shoot alloy bullets as well as it might until the bore has been scrubbed clean, removing the particles of gilding metal left from the passage of jacketed bullets. This problem doesn't seem to be noticeable in handguns; probably because of the difficulty in isolating variables.

While a good scrubbing with stiff bronze bore brushes and solvent may do the trick, many shooters follow that with a polishing with a mild abrasive paste like Brobst JB Bore Cleaner. Others have used toothpaste, an even milder abrasive.

Pastes like JB have been used, with good effect, for some years by benchrest shooters to remove jacket fouling. Proper usage does not damage a bore.

For those determined to rid their barrel of the last particle of jacket material, a second round of solvent and bore brush is in order. Wipe the bore clean with patches; then run a patch wet with clean solvent down the bore. Allow the rifle to stand several days, then run a tight clean patch down the bore. If it comes out bearing some greenish-black streaks, you have not yet reached Utopia.

Just how far you carry the cleaning, beyond bore brush and powder solvent, is largely a matter of personal inclination. Serviceable accuracy can be had after a good scrubbing.

At the conclusion of any cleaning process, there will be a film of solvent—or whatever—left in the bore. For best results this should be removed. Knowledgeable shooters employ a carburetor cleaner like GUMOUT, wetting the bore with the substance. This cleaner is removed with clean patches when it becomes tacky. A film-free bore results.

By now, most of us realize the value in using chamber guides for our rifles or using cleaning rods with muzzle alignment bushings to eliminate uneven wear.

HUNTING WITH CAST BULLETS

Introduction

Over the years, shooters have been bombarded with new high-intensity centerfire cartridges which go farther, flatter and faster than the cartridges which preceded them. These jacketed bullet loadings are fine ballistic products, to be sure, and do an excellent job in the field.

When the utmost in projectile performance is required, riflemen have turned to the jacketed bullet to deliver the goods. This is logical enough, since the cartridges which most of us fire today were designed for jacketed bullets in high intensity loadings and, understandably, are at their best when capped with a jacketed projectile.

But there is a place for cast bullets in hunters' rifles, pistols and muzzleloaders when you consider the average distance at which North American game is taken is around 50-60 yards. Cast bullets do a darn good job of harvesting game—plus Africa's Big Five—using poured projectiles of varying hardness.

Cast bullet shooters adapt hunting technique to their chosen firearm—which means they must close to within 100 yards of their quarry—and then shoot straight. Hunting with handguns and muzzleloaders interests us, in this Handbook, because of the projectiles used.

Handgunners can do just as well with hard alloy bullets as they can with jacketed hollow-points. Commercial pistol cartridges suitable for hunting small to medium game include the .357 Magnum, .41 Magnum .44 Magnum and the .45 Colt. The best cast bullet designs are heavy, with a semi-wad cutter profile. It really doesn't matter if these large bullets expand since they, typically, punch a full-diameter hole right through game. This ensures vitals will be reached if your shot is true and that bleed-out will be fast.

Muzzleloaders hunt deer and black bear with long guns of .45 caliber or larger. The projectile a given rifle can deliver can vary from a patched roundball to a conical bullet of twice the weight. The practical hunting range of muzzleloaders is about 100 yards. Beyond that, roundballs drop off radically and the conicals—while retaining their energy with much greater efficiency—become harder to place properly due to their trajectory.

This nice bull moose was taken by Jim Henry of Waitsfield, VT, during a canoe hunt along the Missinabi River in northern Ontario in September, 1979.
Henry used a .50 caliber Lyman Plains Rifle and #504617 ahead of 80 grains G-O 3fg blackpowder. He made a broadside shot at about 30 yards. The pure lead bullet, a nominal 370 grains, penetrated both lungs and was recovered, moderately expanded and intact, under the skin on the off-side. The bull traveled less than 100 yards, allowing follow-up shots to prevent it's wandering into the bush.

This New England whitetail buck was taken, by your editor, with a Super Black Hawk .44 magnum. The load was #429421 (lino) sized .429" over 24 grains of IMR 4227. The shot was taken broadside, at about 50 yards; the animal went another 50 yards, and collapsed. The Keith bullet had zipped right through, cutting a full-caliber wound. Meat loss was negligible.

Centerfire riflemen are another and, in a sense, newer participant in hunting with cast lead bullets. A hundred years ago—and less—lead bullets were all there were for riflemen. More recently, jacketed bullets, allowing optimum ballistics, have held sway.

There are many applications of cast rifle bullets to hunting not limited to those "old" workhorse cartridges like the .45/70 and .50/70. Modern cartridges, like the .222, .30-30, .30/06 and .308 can effectively use alloy bullets for a variety of needs.

Bullets cast of the hardest alloy may shatter if bone is struck. This means penetration will be reduced and meat damage increased—let alone the increased probability of wounding a game animal. Hunters seeking medium game—deer, bear, etc.—should trade off some velocity to allow use of a softer bullet metal which doesn't tend to shatter. #2 Alloy, or a metal in that range, is a good choice. Also, hollowpoint designs are often not practical except in the larger bullets cast in softer metal.

Varmint hunters, on the other hand, can benefit from the tendency of hard (BHN 22+) alloy to shatter to achieve explosive performance on their quarry coupled with the flattest possible trajectory.

Whatever your application of cast lead projectiles, it is your obligation, as a sportsman, to harvest game cleanly. The only way to accomplish that, with surety, is to develop your loads and practice, practice, practice.

CAST BULLETS FOR HUNTING

by C.E. Harris and Dennis Marshall

From the muzzle-loading period through the end of the blackpowder era, cast bullets were commonly used for hunting. Although swaged bullets were loaded in factory ammunition, outdoorsmen in rural or frontier areas often cast their own bullets and handloaded their own cartridges for reasons of easy resupply or economy. Soft lead alloys were entirely adequate for the modest velocities of the day, which averaged about 1300 and rarely exceeded 1600 f.p.s. Blackpowder cartridges relied mainly on their large caliber and substantial bullet weight for killing power, though at the highest velocities attainable with blackpowder, soft lead-tin alloys provided excellent expansion and weight retention for game loads. Blackpowder "express" cartridges used light bullets at higher than usual velocities, and were said to be more effective on thin-skinned game, due to the shock caused from this "mushrooming." Relatively light-weight, soft, hollow-point cast bullets such as the Lyman #457122, designed by A.C. Gould, were found to enhance expansion in blackpowder cartridges. Metal such as 1:20 tin-lead is still unsurpassed for hunting or target shooting, using blackpowder or mild smokeless loads in calibers such as the .32-40, .38-55 and .45-70.

Modern cartridges easily obtain over 1600 f.p.s. with gas check bullets, and require stronger alloys than can be obtained by hardening lead with tin alone. Antimony is essential in bullet metal, once velocity is increased beyond the capability of plain-based bullets.

Hard bullets cast of linotype metal are best for high velocity, cast bullets target loads, because of its superior strength and casting quality. However, hard bullets do not give reliable performance on game, even with maximum loads. At low velocities, with light loads or at excessive ranges, hard bullets do not expand, but simply penetrate clean through, causing little tissue damage. At high velocities, particularly at close range, the front portion of the bullet fragments severely upon impact, causing severe surface damage and excessive meat loss. Hollow-point cast bullets of hard alloy perform similarly, but in the extreme. The entire hollow-point portion shatters, leaving only the base of the bullet intact to provide minimal penetration. This is in marked contrast to hollow-point bullets cast of soft alloy and fired at blackpowder velocities, where they provide classic "mushroom" expansion with good weight retention. This difference in performance is explained by differences in the physical properties of the alloys used, and the velocity with which they strike the game target.

Regardless of alloy composition, cast rifle bullets expand and subsequently fracture in two distinct stages. Expansion is a relatively smooth, continuous process in which the nose metal deforms, first by slight axial compression, and then by shear wherein metal flows toward and over the perimeter of the bullet to form the "mushroom." Fracture of the expanded nose occurs periodically. As expansion continues, a crack initiates beneath the mushroom at the perimeter of the bullet and propagates in a curved path (much like an ogive) toward the nose of the bullet. If the bullet retains adequate velocity as this point is reached, the expanded metal sloughs off by a ductile fracture mechanism, and the process repeats itself until force against the nose diminishes below a level needed to sustain the deformation.

Soft lead alloys with only small percentages of antimony are more ductile than harder ones, and can withstand greater deformation before fracturing occurs. The presence of tin, within the normal range used for bullet metal, has little effect on retarding expansion, and does not contribute to the fracturing mechanism. Soft alloys, however, are inadequate for high velocity loads. Because of their reduced strength, they can neither deliver suitable accuracy, nor withstand the high velocities of heavier loads and remain intact (unless contained by a jacket).

Use of a moderately hard alloy permits a compromise of

Bullet cast of half and half wheelweights and linotype shed its nose at high velocity. Arrows point to locations of stress cracks caused by shearing of deformed metal, which lead to its fracturing. Light colored region near nose is soft, recrystallized metal caused by plastic deformation, whereas dark region below retains the harder, as-cast structure. Deformed corner of bullet base resulted when bullet yawed in the recovery medium and shed its gascheck. Note also that while a stress crack on the left side of the bullet nose indicates incipient fracture of some nose material, the bullet's yawing reduced the stress on this part so that it was not shed, hence the assymetrical shape of the recovered bullet.

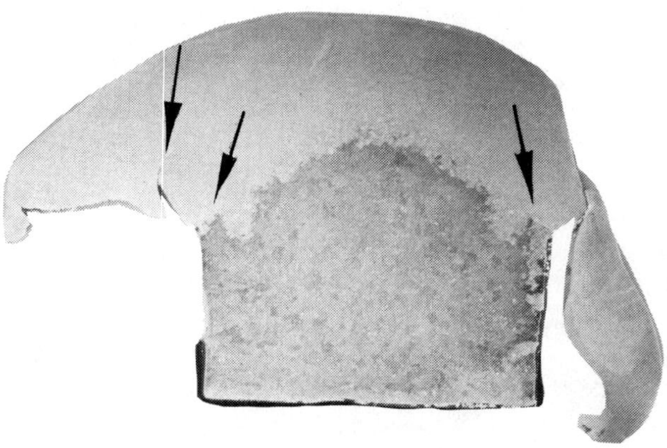

Bullet cast of wheelweight metal with 2% tin added shows good expansion and weight retention. Arrows indicate stress cracks caused by shear of displaced nose material. Note incipient fracture of large fragment at right which is barely attached to the rest of the bullet.

reasonably high velocity and good hunting accuracy, with adequate striking energy, penetration and bullet performance. With correct loads having properly dimensioned bullets, good lubrication and suitable powder charge, cast bullets can equal the accuracy of factory loads, providing penetration and energy equal to .30-30 Win. or .30-40 Krag factory loads; in cartridges such as the .30-30, .308 Win., or .30-'06. This is entirely adequate for the majority of deer hunting, which rarely requires shooting beyond 150 yds.

Alloys not harder than Lyman No. 2 metal (90% lead, 5% tin, 5% antimony, approximately 16 BHN), will give best results. Softer alloys are preferable, if they can provide suitable accuracy from the particular rifle used. We have used wheel-weight metal with 2% tin added, about 13.5 BHN, with very good results in several .30 cal. rifles at velocities up to about 2100 f.p.s., although it requires a smooth barrel, careful lubrication, and thorough cleaning at regular intervals to maintain good accuracy, comparable to factory loads.

Optimum cast bullet performance on game requires the striking velocity not be excessive, otherwise violent fracturing of the bullet, poor weight retention and excessive meat damage will occur. If alloy and velocity are controlled to provide at least 60% weight retention, adequate expansion and penetration will be available even for quartering and raking shots on deer-sized animals. In firing tests, .30 cal. cast bullets weighing 175-190 grs., cast of alloys similar to No. 2 metal and fired at velocities from 2050 to 2100 f.p.s., retained 60% of their original weight, gave penetration equal to factory 180-gr. .30-'06 loads, and expanded to about 1½ times their original diameter.

Available reports from hunters and firing experiments suggest meat destruction and bullet fragmentation become excessive at striking velocities above about 2100 f.p.s. Alloys harder than No. 2 metal exhibit increased fracturing and greater weight loss at somewhat lower velocities than softer bullets. This becomes evident upon examination of the data in the accompanying table.

Firing tests were conducted using conventional lubricated, gaschecked bullets in .30 cal. rifles, using .30-30 and .30-'06 factory loads as controls. Each shot was chronographed, the penetration measured, and the bullets recovered. Averages of velocity, penetration, expansion and weight retention are based on no less than four bullets in each case. Two alloys used represent two simple formulations which can be readily mixed from common materials, being slightly softer, and slightly

CAST BULLET EXPANSION VS. ALLOY AND STRIKING VELOCITY
Averages of Not Less Than Four Rounds Shot Into Wet Paper at 25 Yards

Bullet Type and Alloy	Bullet weight* (grs.)	Vel @ 15 ft. (f.p.s.)	Pene-tration (ins.)	Expan-sion (cals.)**	% Weight Retained	Remarks
Remington .30-30 Core Lokt	170	2074	12	1.58	80	CONTROL "A"
Winchester .30-30 Power Pt.	170	2081	12	1.64	77	CONTROL "B"
Winchester .30-'06 Power Pt.	180	2635	20	2.14	77	CONTROL "C"
LYMAN #31141 new wheelweights +4% tin (BHN 13.5)	175.5	1434	19	1.24	96	marginal expansion
	175.5	1679	20	1.98	91	excellent expansion
	175.5	2037	18	1.42	66	good expansion, shed nose in fragments
	175.5	2108	16	1.49	61	as above but more flattened
LYMAN #31141 the equivalent of old wheelweights plus 2% tin (BHN 16.5)	173	1724	30	1.04	96	no significant expansion
	173	1771	24	1.17	84.8	marginal expansion some fragmenting
	173	2020	18	1.11	58.4	much fragmenting shed nose
Darr #308170XQ (much like 31141 but nose-pour) new linotype (BHN 21)	179	1725	28	1.04	97	slight flattening minimal expansion
	179	1964	16	1.05	58	much fragmenting shed nose
	179	2130	17	1.00	53	as above, but more severe fracturing

*Bullet weights include lubricant in all grooves + gascheck.

**Expansion is average of recovered bullet diameter at largest and smallest points, divided by unfired diameter, i.e. .308″ for jacketed control, .310″ for cast bullets.

harder than Lyman No. 2 metal. Virgin linotype metal was used for comparison purposes with a bullet of similar shape, to illustrate the greater fracturing which accompanies harder alloys.

Telephone books which have been soaked thoroughly overnight, and stacked without squeezing out the excess water, were used as an expansion medium. Penetration, expansion and weight retention of factory loads compare closely with actual bullets recovered from game, or shot into gelatine blocks. Actual penetration of the factory loads in wet paper was about 15-20% greater than factory data for the same loads fired in gelatine. The relative penetration of the .30-30 and .30-'06 loads was the same, and expansion and weight retention of bullets recovered in the wet paper were almost identical with factory experience for the same loads fired in gelatine. The difference in penetration is partly explained by the fact that gelatine blocks are a homogeneous mass, while the stacked phone books, stood on edge, contained some airspace despite efforts to press them tightly together. The effect is analogous to comparing penetration in spaced pine boards vs. one solid log.

Such firing is necessarily subjective, but because all bullets were fired into the same medium, at the same 25-yd. range and approximate velocity levels, using factory ammunition as a standard, it is possible to make a judgment on their relative performance. It should be noted, however, that the rotational velocity of a bullet diminishes less slowly than its striking velocity, and reduced load firings at short range do not necessarily produce the same results that actual firing at long range would. Expansion and weight loss at longer ranges would probably be somewhat greater than shown by the lower velocities here. Nonetheless, the experimental data show good agreement with the available reports from hunters who have used cast bullets on game, most of which is shot at close range.

Factors determining load choice are the anticipated range at which game will be shot, and the penetration and striking energy required. Moderately hard alloys in the range of 13-16 BHN will permit good performance over a useful range of striking velocities.

At a muzzle velocity of 2100 f.p.s., a 175-gr. #31141 bullet retains 1649 f.p.s. and 1057 ft.-lbs. of energy at 150 yds. This will provide reasonable expansion of correct alloys and adequate penetration for deer-sized game. Heavier bullets will permit greater retained energies, but expansion fails at long range before striking energy become inadequate.

Exterior ballistics of typical cast bullet loads are shown in the back of this book. It is recommended that for deer hunting, the bullet retain at least 1700 f.p.s. striking velocity to insure expansion, while having not less than 1000 ft.-lbs. of energy. These criteria will help determine maximum useful range of .30 cal. cast bullet loads for hunting purposes.

For bullets of the same alloy and shape, the percentage of weight loss at a given striking velocity is relatively unaffected by sectional density, within the normal range of bullet weights for a given cartridge.

Optimum penetration, striking energy and bullet performance will be obtained using the heavier available cast bullets for a given caliber. These should be cast of metal not harder than No. 2 alloy, and loaded to provide striking (rather than muzzle) velocities not exceeding about 2100 f.p.s. When anticipated shooting ranges are likely to exceed 100 yds., higher velocity cast bullet loads may be useful to obtain greater retained energies, but they will be unnecessarily destructive at close ranges. For short-range woods hunting, lighter loads, around 1800-2000 f.p.s., will give good results.

Because the upper practical muzzle velocity limit for cast bullets, consistent with good hunting accuracy in most rifles, is around 2100 f.p.s., the striking energy of such loads in .30 cal. rifles is about equal to the .30-30 Winchester with bullets such as the #31141, or the .30-40 Krag when heavier bullets like the #311284 are used. This generally limits their use to deer-sized animals, except when used by a very experienced hunter who can get within close range, pick his shots carefully, and shoot well. Although reports have been received of hunters killing moose and elk at short range with a.30 cal. cast bullet loads, this cannot be recommended to the average hunter, any more than you could recommend .30-30 factory loads for such game. Their energy is marginal for game larger than deer.

When greater striking energy is needed than is provided by .30 cal. cast bullet loads, hunters should either use jacketed bullets, or resort to larger calibers offering greater energy with suitable cast bullet loads. The above principles are completely applicable to larger calibers, such as the .35 Rem., .358 Win., or .375 H&H. Correct cast bullet loads in these cartridges can equal or exceed the performance of .35 Rem., .358 Win. or .350 Rem. Mag. factory loads. In proper hands, they would be adequate for any North American game at moderate ranges. As for the larger bores, over .40 cal., today's hunters will find they perform as well with cast bullets today as they did during the blackpowder era when they were the principal meat gatherers for a growing nation.

AN INTRODUCTION TO MUZZLELOADING

A Historical Review of Muzzleloading

Records describing the early use of guns are often vague, conflicting and incomplete. The defenders of Seville in 1247 supposedly used a rock-throwing cannon to defend their city. This is one of the earliest references to the use of cannon. Next, a fresco painting dated 1340 by Paola Neri includes the image of a single person firing a hand cannon.

As primitive as these weapons were they did sometimes fire and were capable of causing considerable damage. Whether or not various records are accurate it is still safe to assume that by 1350 the use of cannon and hand cannon was growing rapidly.

Early hand cannon were crude affairs consisting of a long iron tube closed at one end and attached to a pole by which the cannon was held. A touch hole was drilled near the closed end of the tube to allow the shooter to ignite the main charge with a match or ember. To load, the shooter poured an undetermined amount of impure, unpredictable powder down the barrel. Choice of projectiles was limited only by his surroundings. The guns were often overloaded and burst in the shooter's hands. When the gun did fire properly the only sure effect was the beating taken by the man holding the pole.

The hand cannon was slowly modified from the pole-mounted crude tube of the fourteenth century to a more or less bona-fide handgun. The barrel was fastened to a sturdy wooden stock styled with a curved butt permitting rudimentary sighting. The Germans are credited with developing the matchlock—the first lock for the hand cannon—in the early fifteenth century. Combination of the two produced the most successful firearm to date. A matchlock consists of an S-shaped lever holding the fuse (or match) at one end with the hand-operated trigger lever at the other end. This lever pivots around a pin through the gunstock. Clasping the trigger sweeps the burning end of the match into position over the priming charge, ignites same and—hopefully—sets off the main charge.

There were many faults inherent in the matchlock system. The wind could move the match and ignite the priming before the shooter was prepared. Also, the wind could blow the powder out of the pan or even extinguish the fuse. Obviously, the weather had to be perfect—no rain, snow, hail or high winds—before the firearms had the remotest chance of effectiveness. These were the days, however, when armies marched only for brief periods between the spring thaw and the fall harvest season. Arduous campaigns in inclement weather were yet to come.

By the end of the fifteenth century many of these mechanical problems were solved. Fuses were soaked in a mixture of saltpeter and water and then dried. The result was an even-burning fuse not easily extinguished. A spring was developed to hold the match away from the powder. The touch hole was moved from the top of the gun to the side and covered to protect it—and the priming—from the elements.

Further development of the matchlock system and guns in general, was given impetus from flow social change. Previously feudal armies defended the land owned by their masters. However, as the feudal system waned and nationalism grew, wars were fought on other lands some distance from hearth and home. Economics of the time required the serfs, the bulk of any given army, to keep up the agricultural effort ensuring the lord would have sustenance for his people. The increasing length of campaigns plus the rise of nationalism saw the decline of feudal forces and the birth of the standing army concept. Defense, and armament in general, became increasingly important and the gun developed a bit faster than it might have had the feudal system contined to exist.

Even with these social and mechanical developments the guns had one large fault: They were rarely accurate at anything but point-black range. The science of ballistics was unknown and many times it was more luck than skill that resulted in a successful shot. The problem was partially solved by rifling the barrel, a process which appeared at the end of the fifteenth century.

This development changed shooting drastically. The inventor of rifling, like that of gunpowder, is not conclusively known.

The Austrian Gaspard Killner is now linked to the discovery in the 1400's. Later, Augustus Kotter of Nuremburg was credited with the invention in 1520. More than 100 years later an English patent covering the invention of rifling methods was issued to Arnold Rotsipen in 1635.

While rifling was a milestone in the development of accurate firearms its practical application was questionable. In the fifteenth and sixteenth centuries shooters used mechanically tight-fitting projectiles without patching material. When fouling accumulated in the grooves loading became extremely difficult—if not impossible. After a few shots it was necessary to pound the ball down the barrel with a mallet. Loading was terribly slow and rifling was not used on military long guns for some time.

The sixteenth century also saw the development of the wheellock . . . a self-activated ignition system. However, the wheellock was expensive to make and easily damaged. For these reasons it never fully replaced the matchlock system and had only a few years of acceptance before being bumped aside by the flintlock.

Sixteenth century tactics leaned increasingly towards the use of firearms; notably the arquebus combined with a force of pikemen. The arquebus was usually a light wheellock fired without a rest; offering the troops additional mobility. Gradually the arquebus was replaced by the more efficient, but heavier, musket. Popular military acceptance of the musket was long in coming since it was heavy to carry and took fifty-six drill movements to reload.

However, the musket was so much more effective than the arquebus that its succession was inevitable. The growing use of muskets among the ranks of infantry pikemen made cavalry shock charges even more hazardous than they had been and so the horsemen armed themselves with handguns.

The matchlock was difficult to use on horseback, but the invention of the wheellock provided troopers with a satisfactorily manageable handgun. While too delivate for use on muskets the wheellock did just fine on pistols and cavalrymen usually carried at least two in saddle holsters.

The Swedish king, Gustavus Adolphus, was not only a brilliant tactician, but a pioneer of military technological development in the 1520's. He replaced the heavy wooden musket rest with a lighter iron spike providing the musketeer not only with a lighter load but also a substitute close-quarter weapon. Eventually, Gustavus adopted a lighter musket and, an innovation for its time, the paper cartridge. The foregoing led to a major change in tactics resulting in the infantry becoming the prime fighting force by the 1700's.

An improved lock system, known as the snaphaunce, appeared later in the sixteenth century in the Netherlands. Early snaphaunces had an arm holding iron pyrites which struck the toothed steel of the frizzen and created sparks. However, iron pyrites crumbled easily and eventually flint was used in its place.

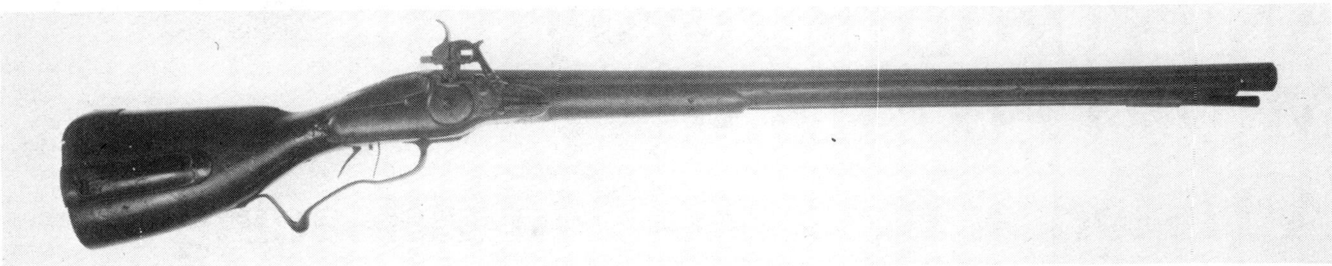

English wheellock rifle of .61 caliber circa 1680. The unusual round/octagonal barrel is marked "R. Rowland Londino." *Courtesy Winchester Gun Museum.*

As the sparks were being created the frizzen moved forward and brought the priming pan cover with it, exposing the priming charge to the shower of sparks. Later this lock was simplified and became the flintlock we know so well today.

The flintlock's biggest improvement over previous lock systems was that the cover of the flashpan opened and sparks were delivered to the charge all in one motion . . . meaning the gun would be reliable even in rainy weather. This improvement, and the simple operation and durability of the flintlock, added to its success.

Firearms took a big leap forward with the emergence of the flintlock. The long gun became a reliable piece of equipment and, more often than not, fired when the trigger was pulled!

As soon as the flintlock system appeared and its advantages were known, many governments began switching from matchlock to flintlock muskets. In England, the 1600's found the British Army with a motley collection of muskets including some matchlocks, matchlocks converted to flint and even a few modern flintlocks. The arrival of William and Mary of Orange, in 1660, brought many changes to the British Isle and one of these was the speed-up of the switch from matchlock to flintlock firearms. The new rulers decided the Brown Bess flintlock should be the standard military firearm and arranged their purchases accordingly.

In 1660 the Brown Bess musket was a .78 caliber smooth-bore with a 48″ barrel. Revisions in design occurred from time to time and by 1713 the musket had a 46″ barrel and weighed ten pounds. The lock and barrel were browned by artificial oxidation; hence the name Brown Bess. A lug atop the barrel, just back from the muzzle, fitted a detachable steel-sleeved triangular bayonet. The bayonet, by the way, which first appeared when men plugged knives into the muzzles of their muskets at Ypres in 1647, was given a permanent place in modern armies by the invention of the "ring" or "socket" bayonet in 1678.

Paper cartridges could be used with the Brown Bess for ease and speed in loading. To load his Brown Bess the British soldier bit open the end of the cartridge opposite the bullet, poured a little powder into the priming pan, poured the remainder of the charge down the muzzle, then loaded the bullet and remnants of the cartridge which served as crude patching. Use of the cartridges reduced loading time and British soldiers using paper cartridges could shoot about six shots a minute. This rate was reduced to about three shots a minute when modest accuracy was required.

The French armies under Louis XIV carried flintlocks by 1660. Francis I initiated government manufacture of firearms in 1535 and in 1669 the French government began the first national armory in St. Etienne. By 1717 the French government had standardized production of all guns according to a specific pattern. This pattern called for a .69 caliber musket with a 47″ barrel and an overall length of 62½″. This standardization made repairs and replacement parts the same throughout the armies and allowed all the troops to use the same ammunition. The French musket of 1763 was used as a model for the first muskets produced at the American armories in Springfield, Massachusetts, and Harper's Ferry, Virginia.

Long Guns

Firearms in early American history were transplants from Europe. But conditions in the New World were much different from those in Europe and adaptation was necessary. America was thickly wooded and sparsely settled with few roads. Most travel was by foot and a long gun had to be light enough to carry long distances. Conflict with the Indians was common and the new settlers soon discovered the Indians hid behind trees and fought from ambush. It was impossible to fight them using the parade ground European tactics. The Jaeger rifle was brought to Pennsylvania Dutch country by the Germans. They quickly began to adapt their rifle to the new country and thus evolved the Pennsylvania Rifle.

Around 1740 an American development did a great deal to change the course of shooting history. This was the development of loading an under bore-size ball inside a cloth patch. Grease on the patch kept the powder fouling soft and its tight fit pushed fouling from the previous shot before it. It also kept the ball from coming into direct contact with the lands and thereby eliminated leading. The Pennsylvania rifle, using this system, took a mere twenty seconds to load, compared to the Jaeger which needed fifteen minutes.

The New World demanded a different breed of men and they needed a dependable rifle. By 1760 modifications on the Pennsylvania rifle had evolved into another distinct species, the Kentucky (or American) Long Rifle. The Kentucky Rifle was lighter than the Pennsylvania and weighed between nine and ten pounds. Bore size was reduced to get more bullets per pound of lead and the barrel was lengthened to get the most velocity out of a given powder charge.

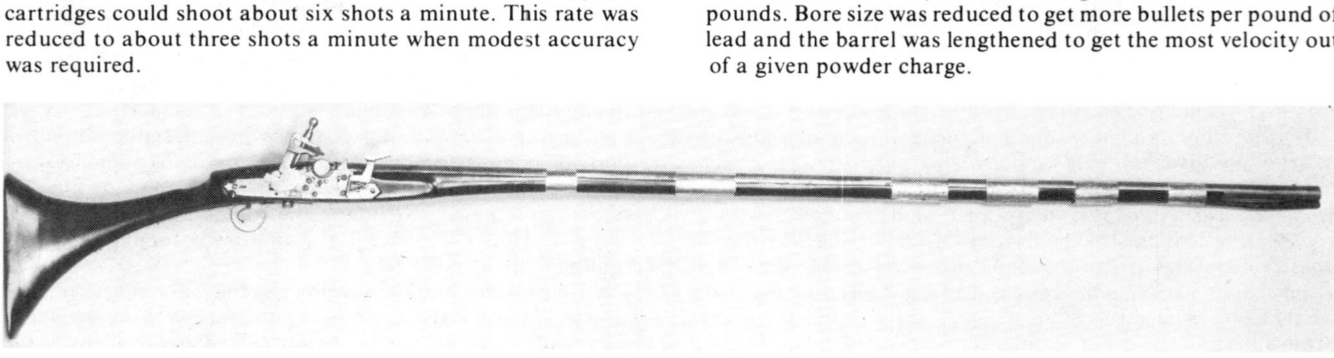

Moorish snaphaunce with .62 caliber barrel held to stock with six brass bands. *Courtesy Winchester Gun Museum*

Contrary to popular belief, riflemen were not a major factor in the winning of the Revolutionary War. The long-barrelled weapon was certainly superior in accuracy and range to the short rifle used by the Hessian Haegers and the service smoothbore musket. The rifle took less time to load since riflemen used a cloth patch around an under bore-size round ball. However, in the broad view, riflemen were few and far between and it was the Brown Bess and muskets like it that fired the short-range volleys and made the bayonet rush which decided the day.

Crack rifleman Henry Morgan conceded rifles were effective only when supported by muskets and bayonets. At Brooklyn Heights (1776), riflemen firing from cover were bayonetted by British and Hessians who drew their fire and then rushed before the riflemen could reload for another volley.

Britain developed a gun called the Blunderbuss in the eighteenth century and this firearm was manufactured into the early nineteenth century. It featured a barrel about fifteen inches long with a smoothly "belled" muzzle. These unique pieces were used for protection against highwaymen but were never used for hunting.

Gunsmiths were always trying to improve their firearms and in the late eighteenth century the first chamber plugs were invented. Previously, when the priming charge exploded, the flame touched only a small corner of the main powder charge. With the chamber plug the flame exploded directly into the center of the main powder charge increasing the power to be realized within a given barrel-length.

However, with the good comes the bad and, as beneficial as the plug was, it did have one serious drawback: The flame had to pass through the chamber plug to reach the main charge and this increased the ignition time and thereby lessened accuracy. Henry Nock, a British gunsmith, came to the shooter's rescue in 1787. Nock created a larger ante-chamber which was connected to the main powder charge by a small cylindrical hole. When the hammer fell the flame from the priming charge entered the ante-chamber and was forced through a small aperture, speeding ignition and overcoming the greatest drawback of the earlier chamber plugs. Nock's plug allowed shorter, and more easily removable, barrels.

By 1800 America was becoming quite heavily populated and the westward move was on. Both firearms and ammunition were in great demand. Eleuthere Irenee DuPont arrived in America from Paris with his father and brother in 1800, as political refugees, bent on developing a tract of land in Virginia. Eleuthere's plans changed when he ran out of gunpowder on a hunting trip and purchased some of American manufacture. Unimpressed with its quality he decided to begin manufacturing powder. This was the humble beginning of what is today one of the country's largest manufacturing firms.

The DuPont plant was a great success. The quality of the powder was much better than anything available in this country. Gunpowder had, at this time, many uses other than for shooting. People believed that drinking a mixture of gunpowder and rum made them courageous. Gunpowder was also used as a seasoning for meat, and as a cauterizing agent on open wounds. An interesting belief was that burning gunpowder purified the air . . . a questionable benefit given the acrid aroma. Nonetheless, the troops at Valley Forge were ordered to burn the powder of one rifle charge per night in each tent to purify the air.

In the nineteenth century there were two great changes in the firearms industry: Breechloading and the Percussion Cap.

The concept of breechloading had intrigued gunsmiths and shooters alike for many years. All attempts at designing a workable breechloader failed because the propellant gases always managed to escape around the breech seal—diminishing the velocity of the projectile as well as affecting the shooter.

In 1664 Abraham Hall developed one of the first breechloading rifles but the design was not successful. It ws not until 1811 that John H. Hall (no relation to Abraham) perfected a breechloading system and applied for a United States patent that same year. Hall submitted his ideas to the American government and the government, impressed, sent Hall to the Harper's Ferry Armory in 1816 to implement production of the design. Shortly afterward, the U.S. became the first nation to adopt a breechloader as a military firearm. Muzzleloaders remained the backbone of military armament, though, as there was still great difficulty in keeping the breech tightly closed.

Development of the percussion cap was the second great nineteenth century firearms innovation. In 1805 Reverend Alexander Forsyth, an Englishman, discovered the explosive properties of fulminate of mercury. With this discovery came the development of the percussion cap. Joshua Shaw, another Englishman, is credited with the actual invention of the cap. Although he finished the development in 1813 he did not release the invention until he completed his relocation to the U.S. in 1814. Then he applied for—and received—an American patent.

Shaw's cap contained a small amount of fulminate and was placed over a "nipple" through which a channel ran leading to a bolster or "drum" and finally into the main powder charge. When the hammer fell on the cap, the fulminate ignited and a jet of flame was directed into the main charge, igniting it with a surity never before attainable.

By 1840 the percussion cap was widely accepted. Between 1820 and 1830 quite a few flintlock guns were adapted to the percussion system but guns produced after this period usually were designed for the cap. Flintlocks continued to be manufactured for many years as there was some reluctance—as there usually is—in making a change.

American rifles continued to change and the Plains Rifle emerged in the United States around 1840. Its chief characteristics were the thick-wristed halfstock and the big-bored short barrel. The needs of western pioneers for a stronger-barreled gun to fire greater charges behind heavier bullets not only led to the foregoing design development but also to the use of "cast steel" barrels in the mid-1800's.

This new cast steel replaced, and was superior to, the existing welded iron barrels. This strength allowed the westener to increase his charge safely. The range of his shots tended to be double that of his Eastern brother and the game twice—or even greater—the size of Eastern animals. The westener had to of the Kentucky Rifle is sometime absent but the western rifle usually with the bison, moose and grizzly.

The Plains Rifle is seldom found in flintlock since that ignition system was making its exit by 1830. The familiar patchbox of the Kentucky Rifle is sometimes absent but the

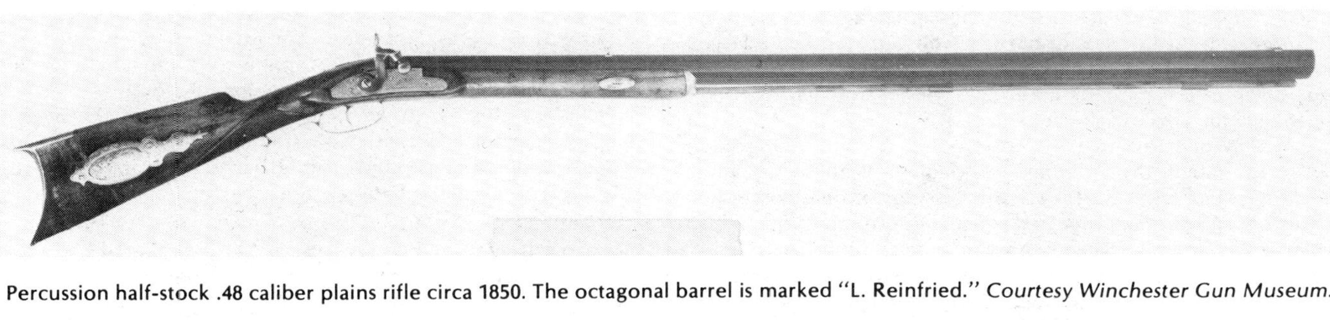

Percussion half-stock .48 caliber plains rifle circa 1850. The octagonal barrel is marked "L. Reinfried." *Courtesy Winchester Gun Museum.*

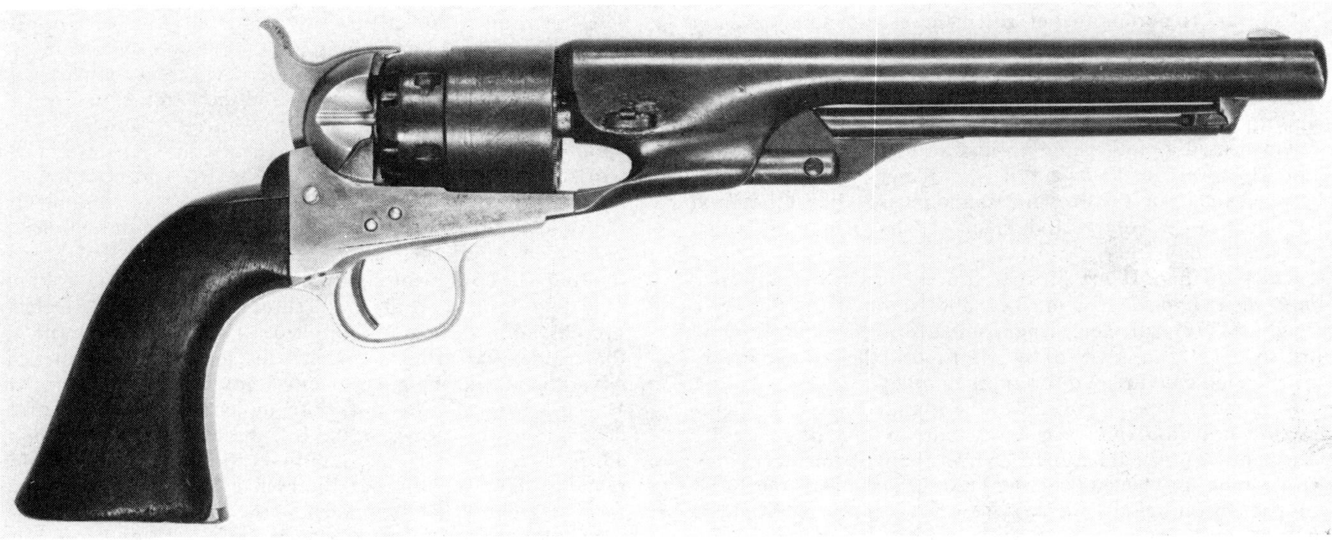

Colt's 1860 Army was the most widely used Colt percussion revolver. More than 200,000 of these guns were made with production halting in 1872. During the Civil War 34 percent of all pistols ordered by the Union were the Colt 1860 Army .44. *Courtesy NRA Gun Museum.*

western rifle usually features double set triggers and brass or iron furniture.

Handguns developed in much the same manner as long guns—first the hand cannon, the matchlock, the wheellock and finally the flintlock and percussion models. Single shot handguns were prominent for many years but not until the early 1800's was the multi-shot revolver patented. The idea of a multi-shot handgun intrigued gunsmiths and shooters alike—and many pistols were developed which allowed more than one shot before reloading.

The first multi-shot pistols had two barrels, one over the other, and one trigger for each barrel. A second model, known as the turnbarrel pistol, featured two barrels arranged vertically which pivoted on a central base pin. After firing the first barrel the shooter rotated the barrels to bring the unfired barrel in line with the lock. While this model had a slower action than the fixed over/under model it required only one lock—which made it less expensive to produce.

In the early years of the nineteenth century four-barrel turn pistols were quite common. However, the multi-shot pistol known today as the revolver did not emerge until the 1830's.

Handguns

In 1835 Samuel Colt applied for a British patent for a revolver and, in the following year, applied for an American patent on the same gun. It is interesting to note that while Sam Colt is most famous for his revolvers it was his rifles that initially brought him success. His first plant was in Paterson, New Jersey, and his first government contract was for rifles.

His first pistol, the Paterson, was available in three models and featured a folding trigger. Colt tried to sell the government this revolving handgun but failed. Failure to get the contract, plus a ruinously expensive lawsuit, closed the Paterson factory.

Soon after, as fate would have it, Colt was awarded a large government contract for his revolvers. He accepted the contract but no longer had manufacturing facilities and had to find outside help. This outside help was Eli Whitney.

Whitney had invented the cotton gin but before he could secure a patent the machine was stolen. Whitney became embroiled in a legal battle over the missing cotton gin and the many copies of it that appeared throughout the country. In his despair Whitney accepted a government contract for 10,000 handguns and spent the next year developing a system to produce the guns. This system incorporated interchangeable parts—a manufacturing innovation—and was an important step in the development of mass production. In the early 1840's Whitney had the two items that Sam Colt needed most desperately: Manufacturing know-how and a manufacturing plant. The Colt-Whitney partnership seemed ideal. The first fruits of this partnership was the .44 caliber Whitneyville-Walker revolver. Colt had modified his original design for more appeal to the government and the Walker had a fixed trigger and brass triggerguard which fathered the now-familiar revolver silhouette.

With the success of the Whitneyville-Walker Sam Colt began to get the orders he had needed a few years earlier. In 1847 Colt opened a factory in Hartford, Connecticut, and began producing a lighter version of the Walker. Since that day the Hartford factory has been busy producing Colt firearms.

In 1851 Sam Colt was invited to the Great Exhibition in London. He stirred such great interest among the English that he was invited to read a paper on the Colt revolver before the Institute of Civil Engineers. Encouraged, despite understandable resentment from British firearms manufacturers, Colt began producing revolvers in Britain during 1853. His hopes for a fat government contract were squashed, however, when a revolver of British design and manufacture was chosen as the standard military firearm. The Colt facility closed shortly afterwards.

In the late 1840's Colt again sought outside assistance to improve his business and teamed up with Elisha King Root. Root designed much of the machinery used in the Hartford facility and, following the .36 caliber 1851 Navy revolver. Root's streamlining of the basic revolver design is easy to spot.

In 1860 Colt and Root manufactured a revolver to replace the various Dragoon models and fill the performance gap between the heavyweight .44's and the lightweight .36 Navy. The new revolvers was called the Army Holster Pistol, Model of 1860 and it arrived just in time to play the major handgun role in our Civil War. The .44 caliber six-chambered revolver weighed two pounds, eleven ounces and featured a round rebated cylinder and the new Root-designed creeping loading lever.

The Navy Pistol, Model 1861, followed the Model 1860 and closely resembled it. The Navy was a smaller six-chambered pistol of .36 caliber with a round barrel (as opposed to the Model 1851 which had an octagonal barrel).

1862 was a year of finalities for Sam Colt. His last percussion revolver design, the Police Model of 1862, appeared that year. Hereafter, all new designs would be engineered for the metallic cartridge. However, Sam Colt, himself, would not see them since 1862 was his last year. His life had been a series of successes and failures but he left a legacy of firearms excellence which lives to this day. The name Colt was, and still is, symbolic of quality and dependability in firearms.

The closest Colt competitor, and the producer of what has been called the finest percussion revolver design, was the firm of E. Remington & Sons who began producing revolvers as soon as the early Colt patents expired. The 1858 New Model Army and Navy models were the most popular Remington percussion designs and those which represent the peak of percussion revolver design.

The 1858 New Model Army .44 saw service in our Civil War and was favored by many because of the integral topstrap which provided more frame support than did the open-top Colt design. The 1858 also had a one-piece frame and grip assembly contrasted to Colt's three piece design. The sturdiness of this design was carried on into the cartridge era via the 1875 Remington revolver.

Hundreds of years have elapsed between the initial discovery of gunpowder and the development and acceptance of the metallic cartridge. In this chapter we have not attempted to treat the various eras as thoroughly as they deserve—only to provide an outline of how things happened. Since the scope of this book is intended to cover only muzzle-loading black-powder guns it is with the decline of the percussion era that we close this discussion.

Shooting the Minie Ball

At one time or another nearly everyone tries to shoot something other than a patched round ball out of his muzzle loader. This means either a solid or hollow-base conical bullet. What most shooters don't fully realize is the relative complexity of making the switch to conicals...and accomplishing worthwhile performance.

We all know a Minie ball has a hollow base and is designed for an easy sliding fit down the bore. The Minie was designed just before our Civil War and relies on the igniting powder charge to spread the skirts to seal the bore and firmly engage the rifling. Usually there are several scraping grooves around the bullet which collect fouling from the preceding shot as the bullet travels towards the muzzle.

The other conical slug has a solid base and relies on a different technique for sealing the bore and locking into the rifling. These bullets, like the Lyman #454612, have a top band dimension several thousandths of an inch *larger* than bore diameter. The center and bottom bands are sized at, or a thousandth of an inch below, bore diameter. As the powder charge ignites the bullet's base is forced down the bore; beginning to move before the bullet's tip. This situation compresses the bullet's overall length and the soft lead spreads out, filling the grooves and engaging the rifling. For bullets of this design perhaps the most critical dimension is that of the bore into which they are loaded. These solid conicals *must* be at least bore diameter and ought to engrave slightly on loading for best results.

Aside from the two basic bullet patterns there are at least six closely integrated variables which affect performance. They are: Depth of rifling, rate of twist, barrel length, velocity, fouling, and bullet design.

Depth of Rifling: Conicals work best in a barrel with shallower grooves than those typically found in roundball barrels. Where the roundball works best with .010" to .012" grooves the conical barrel needs .004" to .005" to do its best.

There's another basic difference, too. Often the conical barrel will have fewer lands and grooves than its round ball counterpart. For example, the Zouave musket sports a three groove barrel as do a number of other muskets. These broad shallow grooves make it easier for the Minie bullet to effectively expand, seal the bore and clean it with the scraping grooves on each slug as it is fired down the barrel. Deep rifling with numerous lands and grooves requires the bullet to deform itself more to engage the rifling, increases the liklihood of blow-by and increases drag both in the barrel and in the air.

Rate of Twist: For some reason this factor seems to be overlooked when shooters consider a conical bullet. To many people the solution is simple — "If I can get it down the barrel I'll shoot it." Without a second thought, the shooter expects his roundball barrel with the slow twist to produce equal accuracy with an elongate bullet, two or three times heavier than the patched ball. There is much more to the situation as you will see.

Within a given caliber, the faster the twist the better a conical (either solid or hollow base) will tend to perform in terms of accuracy. It is also true, however, that the faster the twist the more fouling will remain in the bore...affecting accuracy and ease of reloading. More about that later.

The suitability of a rate of twist varies with the velocity at which the bullet is fired. For example, the replica Zouave has a 1-72" twist as does nearly every other .58 musket on the market today. Firing the Zouave and either the #575213-OS or the #575213 with "G.I." charges of 60 grains of 2Fg blackpowder produces good accuracy. For precise target work skirmishers and match shooters often back down to 40 or 50 grains of 2Fg.

However, when the charge is upped beyond 80 grains of 2Fg, accuracy usually begins to fall off. Heretofore this has been chalked off to skirt deformation—if not to "blown" skirts—and little if any conjecture arose concerning the velocity/twist relationship.

Photographic examination of several of these bullets in mid-air shows the skirts are in good shape at charges up to 100 grains of 2Fg. So, with skirts intact the culprit has to be stabilization... these faster-moving Minies need a faster twist.

In the last couple of years two new rates of twist (for .58 conicals) have arrived on the market. The 1-60" twist is offered by Navy Arms in its "magnum" Hawken .58. This gun has good high velocity accuracy and has proven itself in the field, taking the largest African game. Next, the 1-48" twist appears in the Enfield replicas offered by Jana International and Euro Arms of America. Although the Enfield appeared around the time of our Civil War (as did the Zouave and 1861 Springfield) with the faster twist, the charges it consumed were similar to the 60 grain 2Fg Zouave loads. That this twist has high performance potential may be deduced from the fact that more and more nationally—ranked NMLRA musked competitors are using—and winning with—custom musket barrels with the 1-48" twist.

The two guns discussed so far are clearly defined as Minie—launchers by the respective manufacturers. There are two other faster twists than 1-72"—1-66" and 1-60"—which appear in .58 roundball barrels offered by Douglas, Large, Bauska and Green River Rifleworks. With their numerous and deep lands and grooves these barrels are best suited for the patched roundball but will shoot a Minie with acceptable hunting accuracy.

Stepping down to .50 caliber the only currently available conical twist is 1-48". Lyman, Thompson/Center and Navy Arms offer this twist in their respective Hawkens. Other.50 caliber twists are available—usually 1-66" or 1-72"—from barrelmakers servicing the muzzleloading market.

Next comes the .45's which include production guns with *bore* dimensions of .450"-.453". Within this bore-size category there is only one conical twist on the market today: 1-48". This seems to be a "compromise" twist allowing the shooter to produce good groups with either the roundball *or* the conical within certain velocity constraints. Slower roundball twists are available from a number of barrelmakers.

The .44's are right behind with their nominal bore specifications of .445-.446" and offer the purchaser of a production gun two conical twists: 1-22" and 1-56". The faster twist is the best, corresponding with the specifications for the 45/70 Government projectile, and offers near-optimum stabilization of conical slugs. However, offsetting some of the benefits of the excellent stabilization is the 1-22"s tendency to retain more fouling than a barrel with a slower rate of twist. By the same token, a 1-48" twist will retain more "soot" than a 1-56" or 1-66" twist if all other conditions are equal. Although seemingly unlikely the 1-56" twist will do a reasonably good job with Lyman #445369 as well as with the patched roundball. Guns within this group are produced by Harrington and Richardson, Navy Arms and Numrich Arms.

Barrel Length: This factor reflects judgements on handling and carrying ease, interior ballistics and the degree of specialization of the particular long gun design. The projectile's "barrel time" and propellant burning rate are perhaps the most important considerations as in the following example.

A 300 grain Minie ahead of 175 grains of 2Fg in a 28" barrel will burn most, if not all, of the charge. Substitute a 500 grain Minie and more complete combustion results. The reason? The heavier bullet requires more energy to get it started and, once started, takes longer than a lighter bullet to travel the same distance. This means more time for the powder charge to burn within the barrel.

Cut that barrel back a couple inches and, chances are, the 300 grain bullet becomes noticeably unable to consume the charge. The 500 grain slug, however, will continue to be the most efficient. When the barrel/bullet weight combination becomes inefficient with 2Fg, the shooter's logical next step is to reduce the charge weight level and move to the next finer granulation of sporting powder. However, this is not to be done lightly since chamber pressure and bullet design limitations impose practical limits.

Velocity: This factor is entwined with both the rate of twist and bullet design and coverage here may seem redundant. However, the point needs to be made that "faster is not always better" for, in fact, usually it is not. You must experiment with your gun and the various brands and granulations of powder to produce a load satisfying *your* accuracy and energy needs. Keep in mind that a conical can be propelled beyond the stabilization range of your gun's twist and also beyond the limits of the individual bullet design.

Fouling: This is an especially important consideration for the conical bullet user. Residue from the preceding shot(s) lies between the walls of the bore and the bullet loaded atop the fresh charge in your barrel. It's not too difficult to understand how the fouling can not only keep a bullet from fully expanding but can deform it as large amounts of fouling build up ahead of a departing projectile and are"overrun" by it. The solid and hollow-base bullets are vulnerable to fouling-related problems in different ways.

Solid-base bullets, designed to lightly engrave the upper bearing band during the seating process, are very difficult to load if fouling is heavy. Fortunately, they seem less subject to inferior performance after being (almost literally) hammered down a heavily fouled bore.

Traditional hollow-base Minies are not so rugged and are affected by rough handling. However, because they are sized for a sliding fit—and not to engrave—they can be successfully loaded in bores too dirty for similar solid-base conicals. When the time comes for forceful seating the most serious potential consequences are the deformation fo the skirt *and* canting of the projectile in the barrel.

Fouling is an individual matter. On a hot dry day fouling is very hard and dry; conversely, a damp (or humid) day tends to leave fouling soft. Brands and granulation of powder produce residue in varying degrees ... the coarser the granulation the more fouling will be left in the bore. Lubricants react differently with the igniting powder charge and contribute, in one way or another, to the fouling condition.

Finally, fouling is always heaviest in the chamber area and for several inches up the barrel. Heavy charges invariably produce very hard dry residue and the faster the rate of twist the greater the buildup will be regardless of caliber. Since it is of great importance to firmly seat these heavy conicals on the charge, safe loading and shooting are somewhat jeopardized by having the greatest obstruction in the most critical area. Today there is only a uniform twist rate available in the long guns suited for conicals and it is interesting to speculate on the utility of shallow-groove gain twist barrels which would have the slow roundball twist at the breech-end to minimize fouling in the chamber area and ease the attendant projectile loading and seating problems.

Bullet Design: Perhaps the only thing common to all muzzle-loading conical bullets is the pure lead of which they are cast. The following review of the Lyman bullets should assist you in selecting the proper conical for your muzzle-loader.

#445599— A 250 grain minie with three scraping grooves and a reasonably sturdy skirt. Originally designed for 1-56" rates of twist this bullet does a bit better in the faster 1-22" rate of twist. If used in the 1-56" twist start with a fairly light charge and increase it by slight increments until the desired accuracy is attained.

#445369—The big brother of #445599, this Minie is longer and weighs in at 291 grains. Moderate charges won't affect the skirt and best accuracy is found with the 1-22" twist. However, the bullet does very well in the 1-56" twist and produces good hunting accuracy.

Note: The #445599 and #445369 case a nominal .445" bullet and are suitable only for rifles with a .445"-.446" bore. If you own one of the H&R .45's, Numrich .45's or a Navy Arms Hqeken .45, these are the conicals for you. There ar other suitable guns on the market and the best practice is to slug the barrel and determine the bore size by measuring the lead slug with a micrometer.

#454616—This solid-base 225 grain bullet was designed specifically for use in the Thompson/Center Hawken .45 with the 1-48" twist. Just how it might shoot in a slower roundball twist would be an interesting subject for experimentation. This bullet is too large for the .445" bores but could be sized down in a device like Lyman's 450 Lubricator/Sizer. Since this bullet was designed to engrave a couple thousandths of an inch on the top bearing band during loading, any sizing would make all three bands the same diameter.

#454613— A 265 grain Minie designed specifically for the Lyman Plains Rifle with the 1-48" twist. Combining characteristics of both the solid and hollow-base designs, the #454613's top scraping band is sized to engrave lightly while loading, while the base cavity features a heavy skirt capable of handling heavy charges.

#454612—This 300 grain solid-base conical was also designed for the 1-48" twist Lyman Plains Rifle. Typical of solid-base designs, the top bearing band is a couple thousandths over bore size and the lands will lightly engrave the top band as the bullet is loaded. This bullet, and the others like it, become very difficult to load unless the bore is cleaned every couple shots.

#504617— A 375 grain solid designed for a 1-48" twist. The upper bearing band is designed to be lightly engraved as the bullet is loaded into a .500" bore found on the Thompson/Center Hawken and the Navy Arms Hawken. Performance of this design is limited chiefly by the pressures encountered in the higher charges.

#533476— One of the oldest Minie designs in the Lyman line, it is also the only one currently available for .54 caliber. Although there is no .54 musket presently on the market, the #533476 does a good job when used in a Douglas 1-66" twist roundball barrel. Experiement with your rifle and adjust the powder charge until the best combination is found.

#542622—A newly-developed .54 Minie (similar to No. 577611) featuring a heavy skirt. This bullet is designed for maximum loads in the Lyman Plains Rifle. T/C Renegade or other .54 rifles with a 1-48″ twist and nominal .540″ bore.

#557489—This is the original "Christmas Tree" bullet for the Sharps .54 breechloading percussion rifle and carbine. Today this 460 grain bullet is an excellent choice for both the U.S.-made Shiloh replicas and Italian replicas of the percussion

#575494—This is the lightest .58 Minie available today—weighing in at just 315 grains. Just 55 grains heavier (in pure lead) than the .560″ roundball the #575494 seems to give its best performance when fired at low velocities from a replica Zouave with a 1-72″ twist. Performance in faster twists such as the 1-60″ found in the Navy Arms Hawken or the 1-48″ found in the Enfields has yet to be evaluated. However, the basically thin-skirted design of the bullet logically indicates the use of only moderate charges to avoid deformation of the skirt and canting of the bullet.

Wide flat shoulders on the uppermost bearing band give wadcutter performance on targets and small game. Fifty yard accuracy from a Zouave runs around 1″-1½″ with charges up to 50 grains of G-O 2Fg. Increasing the charge opens the group.

#575602—Another hollow-base design best suited to lighter charges. This 400 grain bullet was the only one to "come apart" during the Pressure and Velocity testing in the Lyman pressure guns. A notable feature of the 602 is its wide and relatively shallow square-shouldered grease/scraping grooves. While these grooves will retain a stiff lubricant, such as a beeswax and Crisco mixture, a looser lube like pure Crisco or Lyman minie lubricant is best placed in the hollow base.

#575213-OS—An old style design directly descended from Civil War projectiles and one of the favorites of modern North-South skirmishers. The 213-O looks like a Minie should ... a some what streamlined configuration with three adequate scraping grooves. Light to moderate (40 to 60 grains of G-O 2Fg) produces good accuracy for most musket shooters. This is a good hunting bullet at standard musket velocities.

#575213—Probably the most popular and accurate Minie available to today's musket shooter. The 213 is a bit longer and heavier than the Old Style and weighs in at 505 grains. Musket shooters usually find their most accurate load somewhere between 40 and 50 grains of G-O 2Fg or slightly greater charges of Curtis & Harvey 2Fg. Accuracy holds until standard musket charges (and thus velocities and pressures) are exceeded.

#577611—A new hollow-base design weighing 540 grains. This bullet is intended for use ahead of especially heavy charges for hunting medium to big game. Special features include an extra-thick skirt and two sharply defined scraping grooves. Examination of this bullet in mid-air confirms the design's performance since the skirt shows no deformation.

#57730—This 570 grain slug is the heaviest standard design offered by Lyman. While the #57730 hasn't enjoyed a great reputation for accuracy in 1-72″ twist barrels the availability of 1-60″ and 1-48″ twist barrels may prove to be the answer and encourage use of a fine heavyweight Minie design.

Conclusion:

You have just read a fairly comprehensive discussion of muzzle-loading conical bullets. Their function and final performance is somewhat more complex and complicated than most people realize. The shooter is best advised to relax and enjoy the get-acquainted period as he experiments with a variety of loads and lubricants. Relaxation, after all, is one of the major byproducts of muzzle-loading.

Loading the Minie in a Rifled Musket

As in every muzzle-loading gun there are several operations the wise shooter performs before pouring the first powder charge down the bore. First he dries the bore and chamber area with clean patches and removes any oil accumulation visible in the nipple vent. Next the shooter will snap one or two caps on the nipple to *make sure* the channel from the nipple through the barrel wall is open. For a quick visual verification place the muzzle near a leaf, blade of grass or similar object—the cap blast will noticeably move it around if the vent is clear. Finally, run the patch down the bore one last time to collect any new debris. Now, you're ready to load and here's how that goes.

1. Set the musket's butt on the ground with the barrel angled so the muzzle is well away from your body. This skirmisher is reaching into his cartridge pounch for a prepared load.
2. Pour the measured charge down the barrel. Some shooters use pre-measured charges loaded into carboard tubing—a common technique employed by skirmishers. For "civilian" shooters the important thing to remember is *not* to load directly from a flask. There have been instances where an ember from the preceding shot has remained alive long enough to ignite the next charge as it dropped down the barrel. The resulting flash touched off the powder within the flask causing serious injury to the shooter. Use a separate measure.
3. Push the lubricated Minie ball skirt-down into the bore, place the recessed ramrod head over the Minie and smoothly ram it home. Strive for a smooth motion that leaves the Minie ball seated atop the powder without air space or undue compression.
4. Bring the hammer to full cock, dip a cap out of the belt pouch (suitable for civilians as well as skirmishers, this pouch is probably the easiest way to handle musket caps) and press it firmly over the nipple—you're ready to fire. If the shot is not to be made immediately, carefully lower the hammer to half-cock.

After firing leave the hammer down over the exploded cap as you reload. This restricts air circulation and helps smother any sparks left behind by your preceding shot. **Remember—avoid having your hands or face directly over the muzzle during the loading operation. After the gun is loaded follow the safety rules used for modern firearms.**

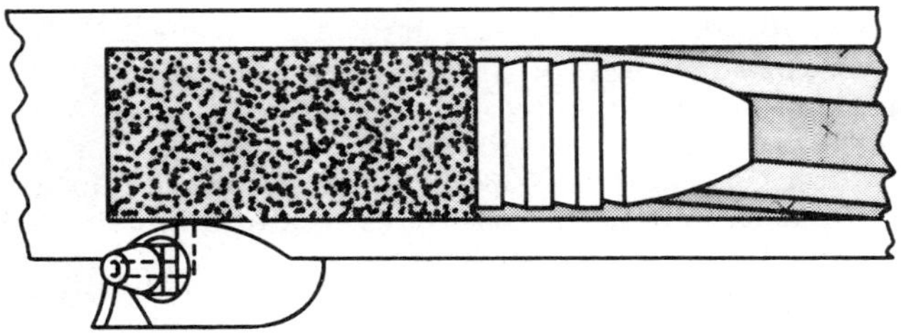

Loaded and ready to go, this Minie (obviously undersize) awaits the cap's flash to ignite the main powder charge.

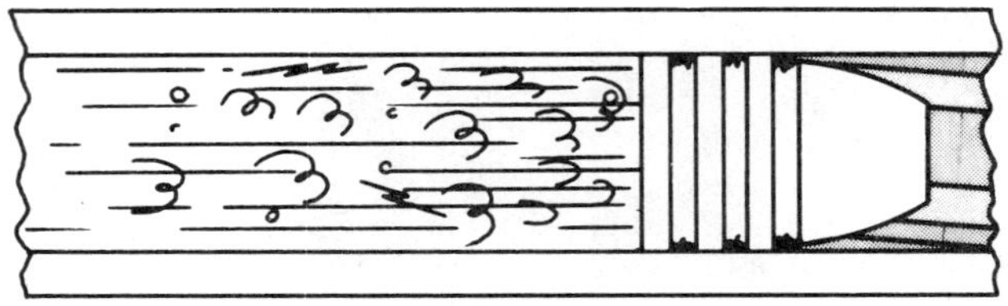

Speeding down the bore, this Minie is now shorter and fatter than it was just before the trigger was pulled. The grooves, or rings, around Minies are scraping grooves and here you see debris/fouling from preceding shots being scraped off and blown out of the barrrel.

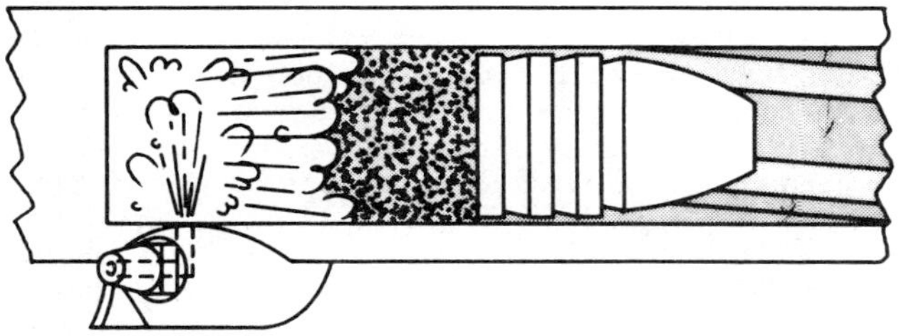

The rear of the charge is ignited first and pushes the front portion of the charge, as well as the Minie, up the barrel. The skirt of the bullet starts to expand but some propellant gas does get around the spreading base.

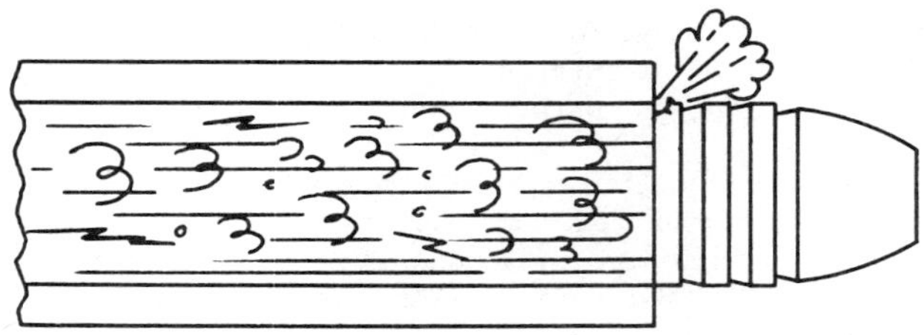

The most fragile, and critical, part of a Minie is its skirt. If the bullet is cast from a cold mould a flaw can result which will give way just as the bullet clears the muzzle as you see here. This results in the Minie tipping a bit and losing part of its accuracy. Skirt damage, with the same results, can also occur if a bullet is forced down a heavily fouled bore.

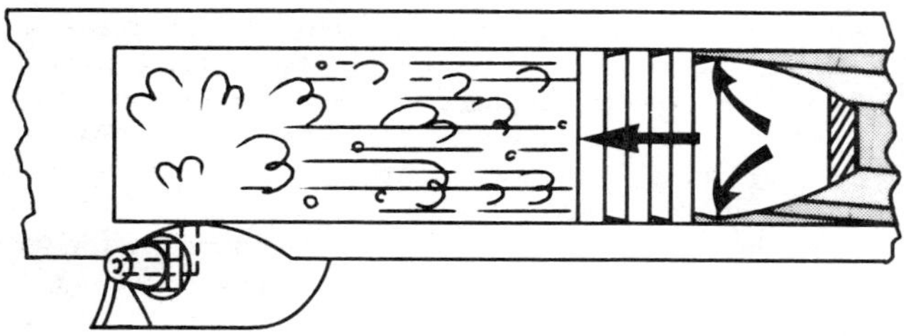

Within a short distance of its starting point, the bullet has filled the bore as much as it is going to, the entire charge of powder has been ignited (providing we are dealing with standard Zouave charges of 60 grains of 2Fg) and the Minie has not only spread out but foreshortened within the bore as the arrows indicate.

If all goes well, your Minie will exit the bore and go on its way looking like this one. However, too heavy a powder charge can turn that nice square base into something resembling the big end of a shuttlecock. Remember to tailor your charges to the design limits of the bullets. Minies, of necessity, must be cast of soft lead and are fairly easy to damage.

Shooting the Patched Ball

The Patched Roundball:

Without a doubt, the most popular and widely used muzzle-loading projectile is the patched roundball. This combination of a soft lead sphere and a scrap of lubricated cloth is peculiarly American and has been with us ever since its worth was proven prior to and during our Revolutionary War.

Over the years different styles of rifling have been devised and used with some degree of success. Characteristics ran from extremely deep grooves to relatively shallow ones; from two or three lands to eight or more. Throughout this period and up through today the most constant single factor has been the patch around the ball—that hasn't changed ... nor has the lead ball, for that matter.

A roundball, whether fired from a rifle or a smoothbore, needs the patch to take up the slack, called "windage", between the ball and barrel walls. A properly fitted ball and patch will seal the bore and keep propellant gases *behind* the bullet where they do the most good. At the same time the ball is held rigidly in position and not allowed to wander from side to side as it speeds towards the muzzle. In a smoothbore this means the ball leaves the muzzle the same way each shot; it does the same in a rifle *plus* the ball is stabilized by the rifling and given a high degree of accuracy.

While patching is the common ingredient, along with the lead ball, of roundball shooting, there are several other considerations which we, the modern shooter, must understand if we are to obtain maximum performance from our guns. These are rate of twist, velocity, patching and lubricant.

Rate of Twist: Experience taught our forefathers that a very slow twist was the best for stabilizing the roundball. Unlike some of the other traditional tales that have come down to us the slow twist has once again proven itself.

Suitability of twist is directly related to the length and weight of a bullet. Closely related is the length and positioning of bearing surface on any given bullet. The roundball is the smallest single projectile which can completely fill a particular bore from groove to groove, and thus has the lowest sectional density and bearing surface of any projectile possible in that caliber. The physics of such a ball require a very slow rotation for optimum stabilization—providing the gun barrel is of rifle length, over (approximately) 20 inches.

For barrels of single shot pistols or revolvers the most appropriate twist is one notably faster than that found in a rifle of the same caliber. For example, the best roundball twist for a .45 caliber rifle is, let's say, one turn in 66 inches. However barrel a single shot pistol with a piece that slow twist tube and check the results. Performance at 25 yards might be reasonable but at 50 yards the groups will open notably. The best twist for that .45 pistol is a faster one: 1-19″ or 1-22″. The fast twist will handle light to heavy pistol charges (up to around 49 grains 3Fg) and keep the groups nice and tight.

Velocity: A patched roundball usually can be propelled within broad velocity bounds without causing deterioration of accuracy beyond acceptable hunting standards. This generality is most true of barrels with a 1-72″ twist, the fastest twist used in roundball rifles.

This phenomena is a direct function of velocity interacting with the ball's stabilization needs. Lower velocity balls will perform well in short pistol barrels of fast twist but not in longer verisions of the same barrel at higher velocity. The shooter's best course is to ascertain the rate of twist of his rifle either from the manufacturer's literature or by measuring the rate with a tightly-patched cleaning rod. Velocity, by itself, is generally not a major factor to the roundball shooter using one of the slower twist barrels. Just as important to performance—perhaps even more so—is the cloth in which the ball is wrapped.

Patching: The soft lead ball should never touch the lands and grooves of the bore. Cloth patching is used to make up the difference between ball diameter and groove diameter.

This cloth serves not only to wedge the ball tightly into the rifling but also to protect the bore from leading and prevent the escape of propellant gases. Obviously the patch takes a real beating. Pillow and mattress ticking, denim and other durable hardweave fabrics produce the best results for most people and the shooter can usually find something adequate at his neighborhood fabric center. Of course, old 'jeans out of the family rag bag can be used, as can nearly any material if you try hard enough!

Use of a looser weave material of proper thickness might seem to result in a nice tight fit of the ball atop the powder charge...but that's not the true test of patching material. The cloth must prove its merit during the split second it travels to the muzzle wrapped around the lead ball.

The softer cloth is more prone to damage from the hot propellant gases, friction of the barrel walls and the potential cutting effect of the rifling's lands. Perhaps the worst offense the shooter can commit is to load an excessively loose ball and patch combination. Although there is no perfect combination, the shooter must experiment and find the best possible pairing which allows him to load and ram the ball with reasonable ease even with a small amount of fouling in the bore. The loose-fitting ball and patch allows the propellant gases to by-pass the patch, burning away all but the most substantial cloth. With its patch partially torn or burned away, the ball is free to wander from one side of the bore to the other and the resulting accuracy leaves much to be desired. There are, however, a couple of things the shooter can do to minimize damage to the patch: Ensure the muzzle is lightly chamfered and use a good lubricant.

Safety Note: Lyman suggests you use only patching material of 100% natural fiber. Synthetics or blends are not good. Further, plastic wadding should not be used.

Lubricants: A general description of a lubricant is a substance which reduces the friction between two bearing surfaces and expedites the motion of one body against another. Patch lubricant plays a crucial role not only in softening fouling but in softening the patch, allowing the cloth to more closely mould itself to the bore of the rifle. Substances used to impregnate the patching range from saliva, powder solvent and various oils to commercially prepared lubricants designed for the black-powder shooter—such as those marketed by Blue and Gray Products, Hodgdon's and Lyman. Best results with most of the substances (except saliva) are obtained by working the lubricant into the material and storing the cloth a day or two, allowing the substance to more fully permeate the fibers.

Use of the saliva-moistened patch is most practical at the range when the shot will be fired in short order. In a hunting situation it should be considered *only* a field expedient to be used if nothing else is available. The reason is that the saliva will rust the bore, particularly the area of the chamber region against which the patched ball bears. Repeated use of the spit patch can easily cause pitting; and has, to the dismay of some modern shooters.

Choice of the proper lubricant is a matter for experimentation by the individual shooter. He should try to protect his patching from the sharp edges of the bore's lands and the sometimes rough interior finish on some new muzzle-loading barrels.

Your selection of the proper cloth patching should be based on an understanding of the relationship between the bore of *your* rifle and the roundball which will be used. Ball diameter must be less than that of the bore and the cloth must not only fill the grooves but also allow a tight sliding fit between the lands and the patched ball. Follow the suggestions of your rifle's manufacturer concerning projectile diameter. Several companies, Lyman, CVA, and Thompson/Center, make specific diameters available either with the purchase of the gun or as a component in a valuable accessory kit. Most other guns will have standardized bore sizes and will be suited for one of the standard roundball diameters such as .440″, .445″, .490″ or .560″. To make your final judgement on ball and patch you must measure the bore.

The best way to measure the bore of a gun is to drive an oversize lead slug through it from muzzle to breech. Take the barrel from the stock, remove the breechplug and clamp the barrel in a vise with padded (non-marring) jaws. Lightly lubricate the bore and start the oversize slug on its way using a long brass rod and a hammer. When the now-engraved slug drops from the open breech you will have a perfect print of your bore's dimensions. Now, with a micrometer, measure both the bore and groove diameter. Here's a sample: Groove—.526″; Bore—.503″.

We know the ball must be smaller than the bore diameter so let's select a .498″ diameter ball as the best choice for this bore.

With the ball adequately undersize (.005″) to fit into the bore we now determine the needed patching thickness to seal the grooves.

Groove—.526″
Ball— .498″
.028″ difference

Now divide the difference by two and the minimum patching thickness is determined: .014″. Remember, there is a thickness of patching on each side of the ball and the difference between the ball and groove diameters must be halved to determine the thickness of cloth needed. Usually it is better to buy cloth that is several thousandths thicker since the lubricated material will compress upon loading.

Now that you're on the way—keep experimenting with your rifle. Vary the powder charge, cleaning technique, patching or whatever. That's part of the fun of muzzle-loading. But remember to vary only one condition at a time so you can easily keep track of cause and effect.

Loading the Patched Roundball:

Loading the patched roundball requires the same techniques regardless whether the gun is rifled or smoothbore, flintlock or caplock. The following preliminaries should be performed before the first charge of the day is poured down the barrel.

Since oil and any other form of moisture is the enemy of the successful blackpowder shooter, the bore and chamber area should receive a good cleaning just before the gun is loaded. Run fresh patches down the barrel until they come out clean and dry. Clean the flash channels of both flintlocks and caplocks with pipe cleaners—pushing the flexible stem on into the barrel.

Next, place a cap on the nipple of the caplock, hold the muzzle near a blade of grass, bit of dirt, etc. and drop the gun's hammer on the cap. Detonation of the fulminate will cause a small but noticeable blast to emanate from the muzzle—moving the blade of grass or bit of dust. If the blast does not manifest itself, the shooter must go back over the gun to clear away the obstruction.

Finally, after all is clear, run a last clean patch down the barrel to catch any freshly dislodged lubricant. Now the gun is ready to be loaded.

1. Set the rifle's butt on the ground with the muzzle inclided in a safe direction—well away from your body. Measure and pour the powder down the barrel using an adjustable measure or pre-weighed charges.
2. Lay your strip of lubricated patching cloth (pre-cut patch) over the muzzle. Center the ball and press it into the bore until it is flush with the muzzle. Tight-fitting combinations can be seated by reversing your ball starter and "rolling" it over the ball.
3. Cut the excess patching from around the ball. Specially designed patch knives are ideal for this task although nearly any sharp object will suffice.
4. Push the ball into the bore using your ball starter.
5. Seat the ball firmly on the powder with the ramrod. The desired ramming stroke is smooth and uninterrupted. Jabbing or tamping the ball down the bore may result in serious deformation or uneven seating force upon the charge. Make sure the ball is firmly seated since an air space could cause a bulged barrel—or worse. Marking your ramrod at the appropriate level is a handy trick. Return the ramrod to the thimbles.
6. Place the hammer on full cock and prime your piece—either with powder or percussion cap. You are ready to fire. Lower the hammer to half-cock if the shot will not be made right away.

Safety Notes

Caplocks: After firing leave the hammer down over the exploded cap as you reload. This helps smother any sparks left from the preceding shot. Keep your hands and face away from the muzzle.

Flintlocks: Before squeezing the trigger at the range, check to your right and make sure your buddy is not standing in line with your barrel's touch-hole. When a fully loaded flintlock goes off there is a jet gas that shoots straight out from the vent and leaves its mark on unwary bystanders. Warn your companions and take extra care before squeezing the trigger.

All Misfires: Should your gun fail to fire ... keep the muzzle pointed in a safe direction until the chance for a hangfire has passed and you are satisfied the charge is truly "dead". Next, inspect the nipple and/or vent, remove any obvious obstruction, reprime and try the shot again. If the charge continues to balk, you may have to work some fine powder into the nipple or vent with your pick, reprime and shoot. At worst, it may be necessary to dismantle the rifle, unbreech the barrel and drive out the load.

Balls Seated Without Powder: This seems to happen to everyone at one time or another. It may be necessary to use a "worm" or similar device or even unbreech the barrel—but before you go to those extremes try this: Work some fine powder into the flash channel, prime and shoot. Work more powder into the channel and barrel, *seat the ball,* prime and shoot. This should do it.

Loading and Shooting the Percussion Revolver

For guns which dominated the firearms scene for such a short time—roughly thirty years—the caplock revolvers have proved a tremendous attraction to the American shooter. These guns appeal not only to skirmishers and other Civil War buffs but also to anyone who has ever been the least interested in the settling of the West. Replicas of the early Colt and Remington revolvers launched the current interest in modern muzzleloaders and continue to be the most popular single type of blackpowder firearm.

Today these revolvers are made of modern steels and are virtually impossible to damage using sporting blackpowder due to their limited chamber capacity. For just a couple of pennies a shot, the plinker or target shooter can fill in a session at the range with big-bore shooting of either the .36 or .44 revolvers and obtain performance close to that of centerfire guns of the same caliber.

Unlike their centerfire counterparts the caplock revolvers are fun throughout the loading, shooting and cleaning stages of a day in the field or at the range. The successful shooter is the one who lavishes the most love and attention upon his "jewel." Now here are some tips on loading and shooting which can provide the basis for hours of enjoyable blackpowder shooting.

Loading:

Blocked flash channels and accumulated grease or solvent are the chief causes for misfires and "weak" shots. The key to success is to ensure these factors don't interfere with your shooting. So, before pouring that first charge into a chamber, dry the bore and each chamber with clean dry patches. Hold the gun up to a strong light source and look through the nipple channels—a strong glow means a clear channel. This cleaning and examination is best done with the cylinder out of the revolver but can be accomplished without disassembly if necessary. The last step prior to loading is to snap a cap or two on each nipple to completely dry the flash channels and chamber areas. Holding the muzzle near a bit of dust, scrap of patching etc. will visibly prove the arm's readiness when the object in front of the muzzle responds to the cap's blast.

Hold the revolver upright in your left hand and pour a measured amount of blackpowder into each chamber. If the gun is to be fired right away at a target you may choose to load all six chambers. If, on the other hand, the pistol will be carried in the field, the safest course is to load only five chambers and let down the hammer on the nipple of the empty sixth.

Place a ball (sprue up and centered for uniformity) over each chamber mouth, rotate the cylinder to position the ball under the rammer and smoothly seat the ball firmly over the charge . . . and below the chamber mouth. Try to exert the same force while loading each ball. Remember—watch out for that powderless safety chamber. The chamber mouth should shave a thin ring of lead from the ball—this is your best insurance against multiple discharge.

Seal and lubricate each loaded chamber by filling the space remaining above each bullet with your favorite lubricant; such as Crisco, Hodgdon's or Lyman Lubricant. These lubricants soften powder fouling and help prevent multiple discharges.

Finally, point the muzzle in a safe direction (as you should have been doing all along) and cap each loaded chamber. If the caps are a bit loose squeeze the skirts together a bit for a snug fit on the nipple—otherwise recoil will "de-cap" your pistol in one or two shots. Now, with all the loaded chambers capped, lower the hammer onto the empty chamber. Don't rely on safety pins or notches. Your pistol is ready to cock and fire. Handle it carefully—just as you would a modern cartridge revolver.

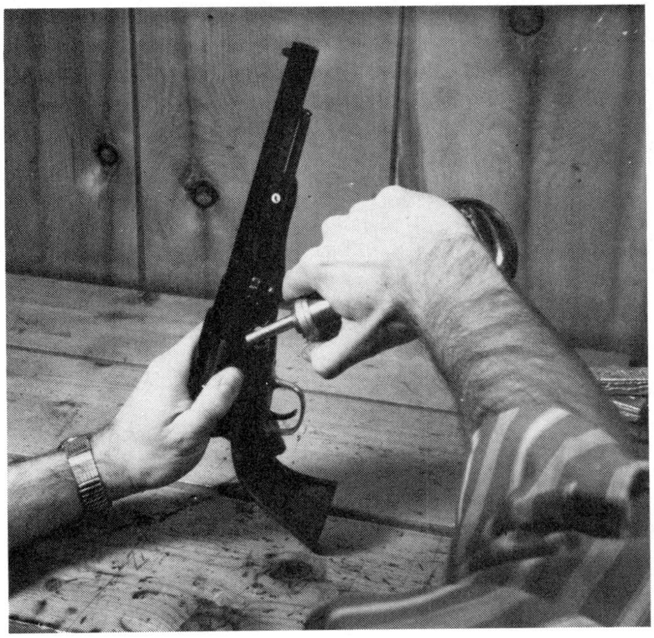

Carefully pour a measured amount of blackpowder into each chamber.

Release the loading lever from its retaining latch and smoothly ram each ball home.

Lubricant over each chamber mouth not only softens fouling but lubricates the cylinder pin and internal workings.

The last step—with the muzzle pointed in a safe direction—is to cap the nipple of each loaded chamber.

REFERENCE

Pressure

Every shooter has a general idea of what happens when a cartridge is fired in the chamber of a firearm. The powder charge is ignited by the primer flash. As the powder burns, it generates an enormous amount of gas which creates a force in all directions. This force, which we identify as chamber pressure, presses against the case head and drives the cartridge back against the bolt face. It pushes the case walls out tight against the sides of the chamber. Seeking to escape the confines of the case, the pressure force builds to the point where it unseats the bullet and drives it with increasing speed down the barrel. What concerns the reloader is how high this pressure builds and how safely it is controlled during firing.

Variables Affecting Pressure

The pressure for a given load varies considerably with the tolerance of the individual firearm. The condition of the throat, bore diameter, groove diameter, and chamber dimensions, etc., all have a substantial bearing on the pressure generated in a gun. When we add to this list such other variables as primer brand and type, case capacity, bullet type and hardness, bullet weight, etc., we see there is more to this business of pressure than simply pouring the proper amount of powder into the case.

Conditions That Can Raise Pressure

Powder: When we add more powder to the case, we naturally increase the pressure of the load. All other conditions being equal, fast burning powders create higher pressures than slow burning powders when used in equal amounts. The powders shown in this Handbook were carefully chosen for the applications and loads listed. Use these powders as recommended and never exceed the maximum load listed.

Under certain conditions, reduced loads can also cause trouble. This phenomenon, popularly termed "detonation", has never been satisfactorily explained. Yet, evidence does exist that under certain conditions reduced loads can jump pressures enormously. The reader is, therefore, cautioned against reducing the suggested listed starting loads.

Bullets: As previously indicated, slight variations in bullet style can influence the pressure of the load. All else being equal, a heavier or a harder bullet will boost pressure due to its increased weight mass and/or its greater resistance to the rifling. Oversize bullets and bullets with more bearing surface will also up pressure. Your bullet diameter should fit your rifle or handgun. When working up a load, use a specific style and weight of bullet. Do not assume that you can change from one bullet style, etc., to another without altering the pressure of the load.

Cases: The inside capacity (volume) of the cartridge case has a direct relationship to the pressure of the load. The experienced reloader sticks with one specific brand of case when working up his loads. He realizes that assorted brand cases, when used with the same powder charge, will give variations in pressure.

Primers: Primer characteristics will affect the pressure of the load. When working up a load, use the same brand and type of primer.

Chamber and Bore Dimensions: The dimensions of your particular rifle or pistol do have a direct relationship to the pressure generated by your reloaded cartridge. The tolerance range for domestic firearms has been taken into consideration when determining the suggested loads. If your gun is imported, (sporter or military), we suggest you slug the barrel before attempting to reload. Your bullet diameter should correspond to the groove diameter of the firearm. Cast bullets can vary .002″ over the groove diameter. In all instances, the load should be worked up gradually.

Overall Length: To conform to chamber dimensions, the total measurement of the seated bullet and case normally should not exceed the listed overall length in the Data Section. This accomplishes two things: (1) it ensures the loaded round will feed properly through the magazine and (2) it keeps the bearing surface of the bullet back from the rifling.

Case Length: Repeated firing and resizing affects the overall length and thickness of the cartridge brass. This combination of forces constantly pushes the case material forward, gradually making the case neck longer and somewhat thicker than it was originally. When the case has exceeded the maximum tolerance allowed by the chamber, an increase in pressure results. The overall case length which is listed in the Data Section will inform you as to the proper length for your cartridge brass. A suggested "Trim-to-Length" is also specified.

Crimp: Crimping the case does have some influence on pressure. If a crimp is to be used, then all loads should be crimped in a uniform manner.

Deep Seating Bullets: Our testing indicates that excessive (below the junction of the neck and shoulder) seating of the bullet does increase pressure. Most shorter bullets, in a given caliber, will conform to maximum overall length without excessive seating. Longer bullets should be seated exactly to overall length and in all cases, the load should be worked up gradually.

Visual Signs of Excess Pressure

Lacking pressure testing equipment, the experienced reloader confines himself to the guidelines set down in his reloading manual. In other words, he begins with the suggested starting load and works slowly upwards until he finds his best performing load. When working up a load, he holds the variables that affect pressure to a minimum by using components of the same brand and style. His cases are either new or once fired.

After each firing he inspects the cases carefully for signs of excess pressure. If all is in order, he continues by adding another increment of powder to the total charge. He never exceeds the maximum load, for he knows that the moment he does, he is eliminating all margins of safety. He concerns himself with accuracy and performance rather than "block-busting energy". In most cases, if he is concentrating on accuracy, his loads will probably remain a little on the mild side.

The following text deals with pressure signs as they might appear when working with new cases:

Excessive Headspace: When this condition is present, the bolt face is not properly supporting the case head. Before reloading, we must be sure that our firearm is completely free from headspace problems.

At low pressure (the starting load) the first sign of excessive headspace is usually the backing out of primers. As the pressure increases slightly, the case moves back in the chamber and reseats the primer. At this stage, the case appears quite normal. As the pressure of the load builds, the case will no longer move backward but will tend, instead, to cling to the chamber wall. Lacking necessary support in the head area, the case will stretch and eventually crack around its circumference.

Primers: We feel that flattened or cratered primers, if interpreted correctly, **are a positive indication of extremely high pressure**. Working with pressure guns in our laboratory, we have never experienced cratered primers when the pressures

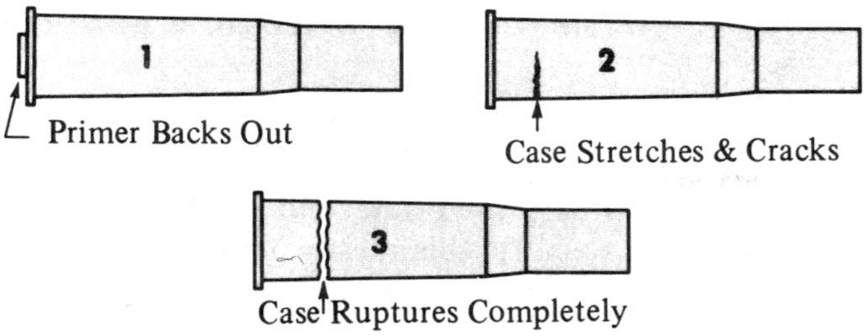

The above illustrations are indications of excessive headspace. Do not use any firearm which shows such signs of headspace problems.

were reasonable. When the pressure climbs the primer starts to show it.

The following illustration shows several changes in primer contour which may be interpreted as signs of excess pressure. The reloader should stick with the same brand and type of primer when developing his load. If all is going well and he then suddenly experiences one of these changes in contour, **he should question the safety of his load**.

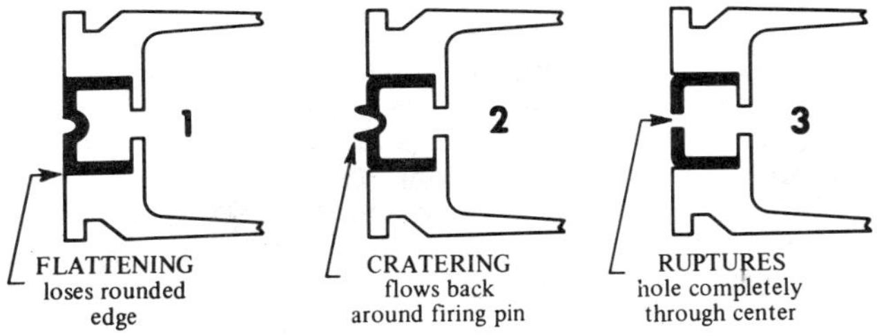

To interpret these signs correctly, common sense must be used. For example, soft (or thin) primers will flatten more than hard (or thick) primers with the same amount of pressure. Also, a long firing pin can cause the condition shown in illustration (3). Each of these conditions, however, would be obvious by its consistency. In other words, they would not occur abruptly or half-way through a test.

Primer Pockets: As the pressure of the load increases, the head of the case is subjected to more and more stress. Eventually, when the pressure is severe enough, it will open the primer pocket, forcing the primer out. The signs shown below can also be caused by case fatigue and it is for this reason that the reloader is instructed to use only new or once fired cases when working up a load. When these signs are experienced with **new strong cases**, they definitely indicate extremely high pressure. Illustration (1), (2), and (3) show the same condition in three stages of severity.

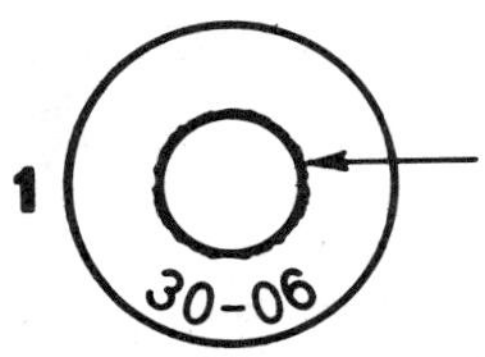

Dark smudge line around primer indicates gas leakage.

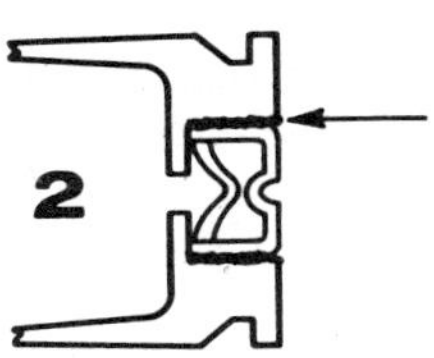

Pocket opens, leaving primer loose. A dark smudge, indicating gas leakage, is usually present.

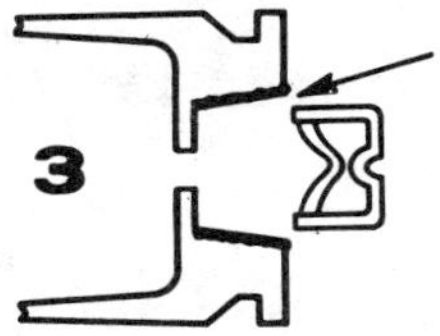

Pocket opens completely, freeing the primer. A dark smudge, indicating heavy gas leakage, is usually present.

Hard Extraction: When experienced suddenly, hard extraction is a definite sign of high pressure. Rough chambers can also create extraction problems, but when such a condition is caused by a rough chamber it remains constant.

Flowing of Brass in Head Area: When the cartridge brass flows back into the extractor port, or the head stamping flattens, or the head takes on the irregularities of the bolt face, we have positive indications of very high pressure. Indications so severe, in fact, that they are usually accompanied by other signs of pressure such as blown primers, stretched pockets, visible gas leaks, etc.

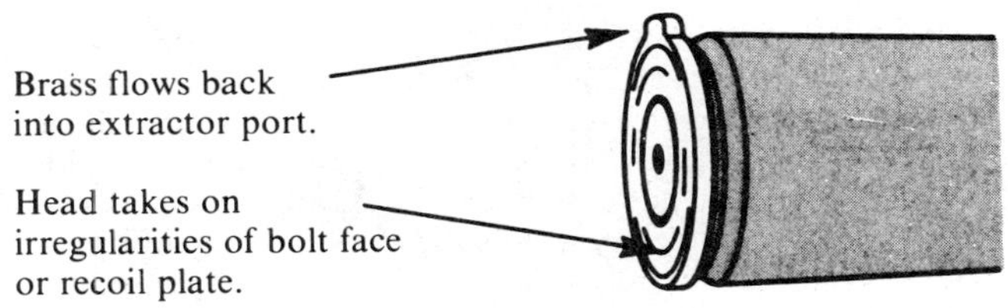

Head Expansion: The side walls and neck of the cartridge case are designed to expand freely and to seal the chamber at the time of firing. This is necessary to prevent gas from leaking back through the action. When a cartridge case is badly smudged by gas on its outer walls and neck (without other indications) it is actually a sign that the pressure is too low, and that the case walls are not expanding properly.

The head of the case, designed to resist pressure and expansion, may be viewed for signs of pressure. However, it is one of the most difficult signs to interpret correctly and a good micrometer is required. If the chamber dimensions of your firearm are slightly on the large side, you will always experience some head expansion (on the first firing) with new brass or with factory loaded rounds. The best method is to fire a few factory loaded rounds in your gun and measure the expansion of the brass. This measurement can then be accepted as normal and any expansion over it should be considered excessive. Normally, expansion of .001" is considered as evidence of very high pressure.

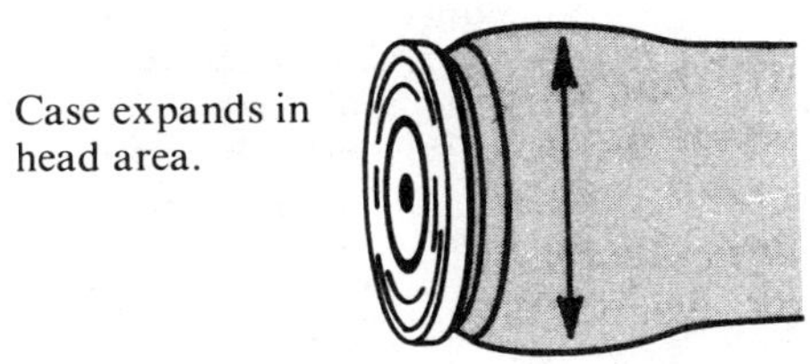

Visible Gas Leaks: At very low pressure, when the case walls are not expanding, gas leakage may be visible. Usually it is accompanied by other signs such as blown primers, stretched pockets, etc.

Special Purpose Projectiles

A. Composite Cast Pistol Bullets

Lyman's new composite cast pistol bullets are a combination of an old idea, new technique and technology. The desire to create an expanding lead hunting bullet which can be driven at maximum velocity without leading is not a new one.

Composite bullets, of varying sorts for various purposes, have been used by American shooters for about 150 years. One type of projectile, used in muzzleloading slug guns, may be seen even today at any of the larger muzzleloading shoots. These slug gun bullets consist of a soft lead jacket and hard lead nose/core.

Some years back, Lyman offered a special mould line which would cast a nose/core of soft lead. Design of the nose portion was a direct "lift" from given popular rifle bullets and the resultant soft nose was intended to be identical to the nose of the same design cast full length in hard alloy. The mould system worked this way: After preparing an ample supply of the soft lead noses, the caster turned to the full length mould and a pot of harder alloy. Before each cast, a soft lead nose was set into the blocks and hard alloy poured in on top of it.

The first of these bullets was introduced in Ideal Handbook #11. The soft nose feature was made only for #457124 and #457193—both intended for the .45/70 Govt. round. Ideal Handbooks #17 through #31 show an expanded composite rifle bullet offering to include designs for .30, .32 and .38 caliber—as well as the .45/70 bullet, now called #457194. It is interesting to note that after Ideal Handbook #11, the design of the smaller bullets featured a rather short stem and the accompanying copy touted the advantages of the gas check feature (just introduced) and said not a word about terminal performance on game!

Reports of actual performance have eluded us but we may speculate about the system itself. In all probability, these bullets suffered from fabrication problems which, in the extreme, caused rejection; in a lesser state (assuming the caster didn't weigh each projectile) caused poor accuracy.

In the first place, the nose/core mould had to produce a near-perfectly fitting piece. If too small, the hard alloy would flow around and over its sides; if too large, the mould would not close completely and molten metal could flow between the blocks, causing a severe "flashing" problem at very least.

In either case, the chance of a nose/core misalignment was fairly good. However, these two-piece bullets were offered in the larger bore chamberings like the .38-55 Winchester, a rather low intensity cartridge which operated at modest velocity and pressure levels. The .38-55 was, and still is, an excellent cast bullet cartridge which gracefully made the transition from blackpowder to smokeless loading.

A problem almost certainly encountered simultaneously with the foregoing was that of casting the hard alloy down over a cool lead stem and having the mould fill properly with no voids. Again, specific information documenting performance of this mould system has not been available. However, anyone who has ever cast a bullet with a hollow point or hollow base knows a cold nose or base plug is not conducive to the production of void-free bullets. Those plugs must be brought up to temperature, just like the mould blocks. The caster of those early composite bullets was faced with what amounted to a cold plug on each cast.

The current Lyman composite system for pistol bullets has, we believe, circumvented the inherent problems of its predecessor. Furthermore, most bullet casters will have the necessary lubricator/sizer and need only buy the composite blocks set from their dealer and two-part epoxy from a local store to be in business! There is no swaging involved which, as we know, requires expensive dies.

Performance of the New Lyman Bullets

Lyman does not make any spectacular claims regarding composite bullet performance. We have satisfied ourselves that these projectiles, with proper load development, are as accurate as most other equivalent cast or jacketed pistol bullets driven at maximum safe velocity.

Externally these bullets are designed to be virtually identical to three proven plain-base pistol bullets designed by Elmer Keith. We did slightly alter the location of the crimping groove on #358624 to permit a practical crimping depth when loaded in .357 Magnum brass. Here's the list of comparisons:

#358429 (168 gr.) = Composite #358624 (170 gr.) Both use same top punch & seating screw.
#429421 (245 gr.) = Composite #429625 (232 gr.) Both use same top punch & seating screw.
#454424 (255 gr.) = Composite #452626 (245 gr.) Both use same top punch & seating screw.

Note: Load data for the composites is the same as that for the appropriate one-piece bullet. The nominal weights are, practically speaking, the same and reflect the kind of variation a cast bullet shooter experiences when shifting from one lot of alloy to another; or even additional ingots within the same lot.

Terminal Ballistics

Terminal performance is better than that of a hard alloy one-piece bullet of the same design. The hard bullet typically punches right through and (probably) exits on the far side, imparting only a portion of its energy to the target.

The composite bullets' soft nose begins to mushroom on impact. This nose expansion increases the frontal area, slowing the bullet much faster than its hard alloy counterpart. Part of the soft lead nose might break free; sheared off as it "washes" down over the Linotype metal jacket mouth. When this occured in the Lyman tests, the fragments were found right with the spent bullet. There was never a trail of fragments through the Duc Seal expansion medium. Practically speaking, the bullets did their job.

However, should a bullet's nose mushroom, wash down and be sheared off, the shooter is left with a flat-nosed slug of at least 70% of original weight. This flat nose, coupled with a lighter overall weight, will cause continued rapid deceleration and transmission of energy to the target. Probability of the composite bullet expending its energy and remaining in the target, compared to its hard-cast counterpart, is much greater.

Thus, for the same basic reasons, the Lyman composite pistol bullet system will yield terminal performance approaching that of a jacketed bullet of the same weight. Interestingly, a broad systemic comparison can be made between the Lyman composite system and that of the famous Nosler Partition Rifle Bullet. If the Nosler's nose should be lost, a specially-designed jacket keeps a lead-filled base unit intact, ensuring continued penetration and energy transmission.

Lyman Tests

These three projectiles (#358624, #429625 and #452626) were tested for accuracy and expansion at 50 and 100 yards on the Lyman factory range. "Full-house" loads were used in all tests to duplicate probable hunting conditions. We learned two things: a) The composite pistol bullets are more comparable to a jacketed bullet than a hard alloy cast bullet in terms of accuracy/expansion. b) The proper choice of propellant can halve, literally, your group size! Be sure to try several different powders before settling on a load.

Assembly, Sizing and Lubrication

Separate moulds (sold as a set) are used to cast the nose/core and the jacket/base. These moulds are intended for pure lead and Linotype alloy, respectively. #2 Alloy may be substituted for Lino if loads are reduced somewhat and no tolerance problem develops with the nose/core piece. The shooter can experiment, as well, with slightly harder nose/core metal but should be aware of two potential problem areas:

1. Tolerances for each mould cavity are based on a known shrinkage factor for a given alloy. Shrinkage differs notably between pure lead and Lino; much less difference occurs between #2 Alloy and Lino. Changing the composition of metal for either part could impair fit of the two parts through increase/decrease in part dimensions.
2. We did, in fact, experiment with a stiffer nose/core metal but found it caused some fragmentation in the base/jacket of brittle Lino alloy. In short, staying with the pure lead nose/core seems a practical choice to us. Now, on to actual assembly.

- After casting an adequate quantity of each component, set up the lubricator/sizer with the proper sizing (H&I) die and the special "G" punch furnished in each composite mould set.
- Back off pressure on the lubricant reservoir so that no lube will flow into the sizing die.
- Size the jacket/base unit (only) without application of lubricant. The special "G" punch closely fits the jacket cavity and serves to ensure accurate alignment during this phase of the operation.
- Mix the two-part epoxy according to product directions. Use of this type bonding agent allows plenty of time to process bullets, correct early mistakes and keep equipment clean and free from epoxy.

Prepare clean hard surface upon which to mix your epoxy. Most two-part formulas call for a 50-50 mix of resin and hardener. A small flat applicator of some sort is handy for spreading the epoxy on the core sidewalls.

- Dip the core end of the nose into the epoxy and evenly coat the base and sides. Set the nose/core into the sized jacket and put both aside. Using the same technique, quickly process the rest of the components. Note: Don't apply excessive epoxy. Clean off any overflow before going on to the next bullet.
- Allow the epoxy in the loosely-assembled bullets to begin to stiffen. This could be a matter of 30 to 45 minutes; perhaps more or less. While you are waiting, install the appropriate regular "G" punch and load the reservoir with an Alox-type lubricant such as that sold by Lyman.
- After the foregoing time has elapsed, try one of the bullets in the lube/sizer. Don't crank in lubricant until the bullet parts stay together as they come out of the sizer. If the parts separate, leaving the nose in the top punch, reassemble that one bullet and let the whole batch sit another ten minutes or so—then repeat the test.
- Once the epoxy has adequately set up, crank in the lubricant and process the bullets as you would a regular cast bullet. The lube/sizer positively aligns the two parts by centering the nose in the jacket. Keep the sizer and bullet free of excess epoxy.

During last stage of assembly, some additional pressure can be exerted on the lube/sizer downstroke to firmly press the nose into the jacket. A firm fit is all that's needed—don't overdo it. Neither the bullet nor your lubricator/sizer can handle much swaging-type pressure.

- Set aside the finished bullets to cure completely; usually overnight will do the job. See directions from manufacturer of the epoxy. Afterwards, load as you would any cast pistol bullet.

We would appreciate hearing from shooters who use the bullets in the field. While these three bullets have been pretty well tested to the point where there'll be no surprises, there is no real substitute for information based on actual field use on game. Please write to the attention of the editor. Thanks . . . and good shooting.

B. Paper Patch Bullets

With the accuracy potential of cast .30 caliber rifle bullets pretty well established around 2200-2300 f.p.s., Colonel Harrison, of the NRA Technical Staff, then turned his attention of a means by which these rifle bullets could be propelled at a greater rate while retaining reasonable accuracy.

The answer to this inquiry came from the past—as they often do. With the goal of preventing contact between alloy and bore wall and, perhaps, a reduction in metal cost, i.e. softer alloy than was normally used for maximum loads, Colonel Harrison turned to paper patching.

He performed a great many experiments, recording them in THE AMERICAN RIFLEMAN, and his works resulted in #301618 and #301620; both of which are covered in this Handbook.

Testing of these bullets occurred, primarily, in match-grade rifles chambered for the .30/06 and .308 Winchester. Later, additional work was done in a .300 Winchester Magnum indicating these bullets were accurate up to 3,000 f.p.s.

Certainly Colonel Harrison achieved his purpose and bullet casters now have the means to deliver high velocity .30 caliber alloy bullets which are cast in a somewhat softer alloy.

The implications to the hunter are great. Now his cast reloads can approach or match the ballistics of factory ammunition with the improved terminal performance of a softer bullet metal.

Casting of these bullets follows standard procedure but there are a number of special steps required — and we have outlined them for you:

Paper Patch Layout

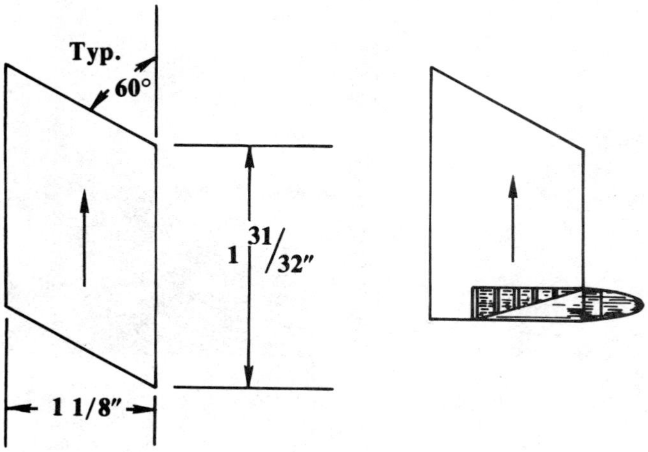

NRA PAPER PATCHED BULLETS #301618 AND #301620 ASSEMBLY INSTRUCTIONS

General:

Although these bullets are designed with a gas check shank, gas checks do not appear to be necessary and were not used in our load development. If the bullets cast greater than .3016 in diameter, they should be sized without lubricant through a degreased .301 die. Bullet metal should have a hardness of 12-15 BHN for most applications but, for full loads in excess of 2500 f.p.s., harder bullets in the range of 16-20 BHN are advisable. This would be an alloy similar to linotype (22 BHN) reduced slightly with lead. Paper should be high grade bond with a 25% cotton fiber content and will measure about .0033 in thickness. The wrapping operation is best performed on a thick, hard rubber pad which can be marked to indicate the location of the bullet point with reference to the paper.

Assembly Details: See Loading/#2 "Note" before proceeding.

1. Wet paper patch with saliva on both sides.
2. Lay paper down pointing away from you.
3. Lay bullet on paper and start wrap near corner of patch.
4. Holding the started corner against the bullet, lift the bullet and pull the patch tight while restraining the far end of the paper.
5. Set the bullet down and, while still pressing hard on the near patch corner, roll the bullet back a little to be certain the point is stuck down; then roll it forward all the way while continuing to press down. Adjust the direction of rolling as required to keep it straight.
6. When the patch is all rolled on, hold the exposed edge down and gather the projecting base paper into as fine a tail as possible, then twist to tighten the patch.
7. Allow the patched bullet to dry. An overnight drying period is usually necessary, at which time the patch will have stuck tightly on the bullet.
8. Cut the dried tail off close to the base with side-cutting pliers, leaving a small twisted stub to prevent unwinding.

Sizing and Lubrication:

1. **Spray with a teflon spray and dry**
2. **Run the bullet through a .308 sizer-luber, (#413 top punch) with Alox lube and wipe off excess lube.**

Loading:

1. Expand the case mouth enough to accept the patched bullet without tearing the paper.
2. Seat the bullet to bring the patch against the barrel forcing cone. (Seating screw #413).

Note: Prior to patching a major quantity of castings, be sure you have established two dimensions:

a. Overall cartridge length with bullet in case. The nose of the bullet should touch-even engrave on-the rifling.
b. Height of proper wrap up sides of bullet. The leading edge of the paper patch should just enter the chamber's forcing cone.

Determining the foregoing may seem a "chicken or egg" quandry. However, a few experimental bullet wraps—one to establish overall seating length and perhaps several afterward to establish height of wrap of the paper patch—should do the trick. Please note that the paper doesn't have to wrap with any relationship to the grooves in the bullet.

Determining Alloy Hardness

One of the most difficult tasks for the advanced bullet caster is the accurate determination of his alloy's hardness. Lyman's experiences in receiving bullet metal other than what was ordered (discussed in the Bullet Casting Section) concur with the general experience of other bullet casters.

As bullet casters, we are aware that certain levels of metal hardness are necessary for good performance at various velocity levels. The greater the desired velocity, the harder the bullet metal needs to be.

To aid us in our research, the Technical Staff fabricated a hardness tester which works quite well, being based on the relationship of depression diameters in sample material--against known reference--when all other variables are equal.

This system--and formula--has been around; although implementation has varied. Of the various methods and techniques which have appeared over the years, this one seems practical and easy to use.

Hardness Tests (Brinell)

In order to measure bullet hardness "standard" samples were cast using Lyman bullet #429348 because of its large diameter, flat-nose. A prototype tester was devised, which could be mounted in the reloading press, and featured a load indicator calibrated at 100 lbs. (45.4 Kg.) A 7/16" diameter (11.1mm) steel ball penetrator, installed in a 7/8" diameter x 14 die body, was placed in the threaded die mounting hole. Bullet metal samples were placed in a holder and subjected to the 100 lb. load with a dwell time of 3 seconds.

Some samples, such as pure lead, received only one indent while the harder alloys were rotated to obtain four or five impressions. Using lead, with a known Brinell hardness number (BHN) of 5, the following formula was used to determine hardness of the unknown samples.

$$BHN = 5\left(\frac{\text{Diameter of Indent in lead}}{\text{Diameter of Indent in sample}}\right)^2$$

Indent diameters were measured with dial calipers under a 3 power illuminated magnifier.

Heat Treating Details:

Wheelweight bullets were sized, after casting, with gas checks but no lube. The bullets were placed in an oven at 450° F. for one hour and quenched in room temperature water. The bullets were dried overnight at room temperature and lubed in oversized dies; .225" for .22 caliber and .309" for .30 caliber.

Hardness figures obtained by the foregoing technique were as follows:

Pure lead ---------------------------- 5.0 BHN
Wheelweights - no heat treat----- 8.9 BHN
Wheelweights - heat treated---- 27.1 BHN
Linotype ----------------------------21.8 BHN

Samples processed through our hardness tester were, from the left, linotype and pure lead.

These metal samples are, from the left, heat-treated wheelweight metal and pure lead.

Mould Reference Table

This is a listing of Lyman/ Ideal mould designs for reference only. Many of these designs are discontinued; please do not order from this reference listing.

The current annual catalog, available in Dec./ January, carries the mould designs in production—if it isn't in the catalog we are not making it.

Bullet No.	Grs. Weight #2 Alloy	Style	Top Punch	Seating Screw
22536	* 46	F.N., L	415	415
22596	50	R.N., G.C.	415	415
• 225107	37	F.N., G.C.	415	415
225209	* 38	R.N., L	438	438
225230	47	R.N., L	438	438
225322	55	F.N.	415	415
225353	43	F.N., L	415	415
• 225415	49	F.N.	415	415
• 225438	44	R.N., G.C.	438	438
225450	48	SPT, G.C.	450	450
• 225462	57	R.N., G.C.	438	438
225487	43	SPT,	450	450
• 22835	* 44	SPT, L	450	450
• 228367	59	SPT, G.C.	450	450
244203	* 60	R.N.	203	203
• 245496	84	R.N.L, G.C.	203	203
• 245497	93	R.N.L, G.C.	203	203
• 245498	99	F.N., G.C.	498	498
• 252435	51	R.N.	203	203
25716	*100	F.N., L	420	415
25719	* 80	SPT, L	418	450
25720	* 67	F.N., L	420	415
25721	87	F.N.	420	415
257205	88	F.N., L	420	415
257231	* 88	F.N.	420	415
257283	85	F.N.	420	415
257285	87	F.N.	420	415
257306	115	F.N.	420	415
• 257312	89	F.N.	420	415
• 257325	113	R.N., G.C.	325	438
257382	88	F.N.	420	415
257388	80	SPT, G.C.	418	450
• 257418	98	SPT, G.C.	418	450
• 257420	65	F.N., G.C.	420	415
257454	92	R.N., L, G.C.	463	438
• 257463	75	R.N., L, G.C.	463	438
• 257464	89	R.N., L, G.C.	463	438
263314	90	R.N.	325	438
• 266305	101	R.N., G.C.	305	305
• 266324	119	R.N.	455	438
266386	105	SPT	386	386
• 266455	127	R.N., L, G.C.	455	438
• 266469	141	R.N., L, G.C.	455	438
280411	138	SPT	411	377
• 280412	136	R.N., G.C.	468	438
• 280468	*109	R.N., L, G.C.	468	438
• 280473	125	SPT, G.C.	473	377
287129	124	R.N., G.C.	346	438
287202	* 84	R.N.	346	438
• 287221	*120	F.N., L	420	415
• 287308	164	R.N., G.C.	468	438
• 287346	135	R.N., G.C.	346	438
287373	122	F.N.	420	415
287377	148	SPT, G.C.	377	377
• 287405	150	R.N., L, G.C.	346	438
287442	139	SPT, G.C.	377	377
• 287448	118	R.N., G.C.	438	438
291379	138	SPT, G.C.	379	379
294380	189	PT, G.C.	473	377
299153	90	R.N.	465	465
299155	80	R.N.	465	465
• 300136	146	R.N., G.C.	465	465
• 301618	165	PT, PP	413	413
• 301620	190	PT, PP	413	413
3111	*200	F.N., L	8	8
3112	*165	R.N., L	465	465
3113	*158	F.N., L	8	8
• 3118	115	F.N.	8	8
31110	100	R.N.	465	465
31114	*100	F.N., L	8	8
31123	125	F.N.	8	8
• 31125	240	R.N.	465	465
31133	100	F.N.	8	8
• 31141	173	F.N., G.C.	8	8
311157	165	F.N., L	8	8
311206	*125	R.N.	465	465
311224	195	R.N.	465	465
311227	90	R.N.	465	465
311234	120	F.N., L	8	8
311240	92	R.N.	465	465
311241	*125	R.N.	465	465
311245	87	R.N.	465	465
• 311252	77	R.N.	465	465
311255	115	R.N.	465	465
311257	110	R.N.	465	465
311259	*150	R.N.	467	467
311264	*125	R.N.	8	8
311274	195	R.N.	346	438
311278	197	PT	278	278
311280	196	PT	278	278
311281	175	F.N.	8	8
311282	217	R.N.	467	467
• 311284	214	R.N., G.C.	467	467
311288	175	R.N.	467	467
• 311290	210	PT, G.C.	467	467
• 311291	170	R.N., G.C.	291	465
• 311299	202	PT, G.C.	467	467
• 311316	112	F.N., G.C.	8	8
311329	185	SPT, G.C.	329	329
311331	218	R.N., G.C.	278	278
311332	180	PT, G.C.	413	413
311333	197	R.N., G.C.	278	278

*Bullet offered in more than one weight.
•Bullet listed in the loading tables of this handbook.
**Weight in lead.

Bullet No.	Grs. Weight #2 Alloy	Style	Top Punch	Seating Screw
• 311334	190	PT, G.C.	413	413
311335	206	PT, G.C.	467	467
311340	188	PT	329	413
311351	158	R.N.	329	329
• 311359	115	PT, G.C.	359	359
311365	198	SPT, G.C.	329	329
311375	168	PT, G.C.	413	413
311397	153	SPT, G.C.	329	329
311403	167	F.N., L	8	8
• 311407	180	F.N., L, G.C.	445	445
• 311410	130	PT	467	467
311413	169	PT, G.C.	413	413
311414	150	PT, G.C.	329	329
• 311419	88	F.N., G.C.	8	8
• 311440	147	F.N., G.C.	8	8
• 311441	117	F.N., G.C.	8	295
• 311465	122	R.N., L, G.C.	465	465
• 311466	152	R.N., L, G.C.	467	467
• 311467	177	PT, L, G.C.	467	467
• 311576	120	R.N., G.C.	465	465
31355	85	R.N.	465	465
31356	125	F.N., L	8	8
31357	100	F.N.	8	8
• 313226	93	R.N.	226	226
• 313249	84	R.N.	226	226
313260	*100	F.N., L	445	445
313307	220	R.N.	226	226
313445	95	S.W.C.	445	445
313492	93	W.C.	445	445
313493	104	F.N.	8	8
316204	170	R.N.	470	470
316275	200	R.N.	470	470
316475	155	R.N., L, G.C.	470	470
31947	*120	F.N., L	295	295
31948	*120	R.N.	470	470
31950	**105	PT	295	295
31952	196	F.N., L	295	295
319162	185	F.N., L	295	295
319247	165	F.N.	295	295
319261	150	F.N.	295	295
319273	185	F.N.	295	295
319289	185	F.N.	295	295
319295	174	F.N., G.C.	295	295
319323	181	SPT	467	467
319350	195	F.N., L	295	295
32115	*125	F.N.	295	295
321232	170	F.N.	295	295
321265	153	F.N.	295	295
• 321297	181	F.N., G.C.	295	295
321298	151	F.N.	295	295
• 321317	161	R.N., G.C.	470	470
• 321427	134	R.N., G.C.	470	470
• 32359	115	F.N.	295	295
32360	*125	F.N.	295	295
32361	98	PT	366	366
32362	83	R.N.	470	470
323236	188	R.N., L	467	467
323357	183	F.N., L	295	295
• 323366	182	SPT, G.C.	366	366
• 323378	243	R.N., G.C.	470	470
• 323470	165	R.N., L, G.C.	470	470
• 323471	215	PT, L, G.C.	366	278
• 323481	185	PT, G.C.	278	471
328371	151	PT	366	366
• 33889	247	F.N., G.C.	320	320
338237	*145	F.N.	320	320
• 338320	201	F.N., G.C.	320	320
338337	224	R.N.	447	447
350293	165	R.N.	293	293
• 350319	167	R.N., G.C.	447	447
• 350447	183	F.N., G.C.	447	447
350457	250	F.N., L, G.C.	447	447
• 350482	251	R.N., G.C.	320	320
354433	158	PT	447	447
• 356402	121	C	402	402
• 356404	95	F.N.	495	495
356472	140	W.C.	429	429
357443	158	F.N.	395	395
• 357446	162	S.W.C.	429	429
357453	151	S.W.C.	429	429
357511	107	S.W.C.	203	203
357512	130	S.W.C.	203	203
3589	282	R.N., G.C.	430	430
35842	195	C	395	395
• 35863	148	W.C.	344	344
35864	140	R.N.	311	311
35870	150	R.N.	311	311
35871	146	R.N.	311	311
35872	115	R.N.	430	430
35873	105	R.N.	430	430
35875	*160	F.N.	429	429
35887	*125	W.C.	311	395
• 35891	148	W.C.	495	495
35893	125	C, G.C.	93	93
35897	232	R.N., G.C.	97	97
358101	75	W.C.	495	495
• 358156	155	S.W.C., G.C.	429	429
358160	150	PT, Heel	311	311
• 358212	146	R.N.	311	311
• 358242	* 92	R.N.	311	311
358250	156	R.N.	311	311
358269	129	R.N.	311	311
358271	150	S.W.C.	429	429
358302	112	C	302	302
• 358311	158	R.N.	311	311
358313	170	F.N.	449	449
• 358315	204	R.N., G.C.	311	311
• 358318	245	R.N., G.C.	311	311
358339	136	R.N.	430	430
358344	150	W.C.	344	344
• 358345	115	S.W.C.	429	429
358356	108	R.N., S.W.C.	430	430
358363	70	C	395	395
358385	150	R.N.	311	311
358394	87	R.N., W.C.	430	430
358395	148	H.B.	395	395
358416	158	R.N.	311	311
358425	*112	W.C.	402	402

*Bullet offered in more than one weight.
•Bullet listed in the loading tables of this handbook.
**Weight in lead.

Bullet No.	Grs. Weight #2 Alloy	Style	Top Punch	Seating Screw
• 358429	168	S.W.C.	429	429
• 358430	*150	R.N.	430	430
358431	160	S.W.C.	429	429
358432	*148	W.C.	429	429
• 358477	150	S.W.C.	429	429
• 358480	133	S.W.C.	429	429
• 358495	141	W.C.	495	495
358500	149	S.W.C.	203	203
358502	119	R.N.	203	203
358503	125	W.C.	203	203
358624	170	S.W., COM	429	429
364615	**128	C, MX	N/A	N/A
366408	245	F.N., L	295	295
37578	260	R.N.	167	167
37579	*170	F.N.	449	449
37580	250	C	80	97
37581	253	R.N.	167	167
37582	170	F.N.	449	449
37583	**145	F.N.	449	449
37584	*176	F.N., L	449	449
37586	265	F.N.	449	449
375164	190	F.N.	449	449
375165	250	F.N.	449	449
375166	330	F.N., L	449	449
375167	267	R.N.	167	167
• 375248	249	F.N.	449	449
375272	310	R.N.	449	449
• 375296	265	F.N., G.C.	449	449
375355	238	F.N., L	449	449
• 375449	264	F.N., G.C.	449	449
386177	196	R.N., G.C.	167	167
386178	200	R.N.	178	178
• 40143	172	F.N.	43	43
• 40188	170	F.N.	43	88
401452	196	S.W.C.	452	452
40392	*190	F.N., L	43	43
40395	*145	C	95	95
403168	200	F.N.	43	43
403169	245	F.N.	43	43
403171	370	F.N.	88	88
403172	285	F.N.	88	88
403173	*250	F.N.	43	43
406150	330	R.N.	43	43
41026	199	W.C.	402	402
41027	217	W.C.	402	402
• 41028	212	F.N.	43	43
41032	*200	S.W.C.	429	429
410214	101	R.N., W.C.	43	43
410219	375	F.N.	43	43
• 410426	240	R.N.	263	263
• 410459	220	S.W.C.	610	429
• 410610	215	S.W.C.	610	429
412174	260	F.N.	43	43
412263	288	R.N.	263	263
415175	300	F.N.	43	43
419180	200	C	263	263
419181	285	F.N.	43	43
• 42798	205	F.N.	98	98
427100	168	R.N.	263	263

Bullet No.	Grs. Weight #2 Alloy	Style	Top Punch	Seating Screw
427103	400	R.N.	263	263
429106	161	R.N.	251	251
429184	235	R.N.	251	251
429185	210	PT	303	303
• 429215	210	S.W.C., G.C.	421	421
429220	175	S.W.C.	220	220
• 429244	245	S.W.C., G.C.	421	421
429251	255	R.N.	251	251
• 429303	200	S.W.C.	303	303
429336	250	S.W.C.	421	421
• 429348	180	W.C.	348	348
429352	245	W.C.	348	348
• 429360	232	S.W.C.	360	360
429383	245	R.N.	251	251
429384	241	R.N.	251	251
429398	248	W.C.	98	98
• 429421	245	S.W.C.	421	421
429422	235	S.W.C., H.B.	421	421
• 429434	215	F.N., G.C.	98	98
429436	214	R.N., W.C.	251	251
429478	210	R.N.	251	251
429508	220	S.W.C.	203	203
429509	170	S.W.C.	449	449
429510	190	S.W.C.	203	203
429513	150	S.W.C.	203	203
429518	245	S.W.C.	203	203
• 429625	232	S.W.C. COM	421	421
436218	*205	F.N., L	360	360
439186	370	R.N.	187	187
441267	360	F.N., L	421	421
445369	**291	C,MN	421	421
445599	**250	C,MN	190	190
446109	340	F.N.	421	421
446110	340	F.N.	421	421
446187	*330	R.N.	187	187
450225	170	R.N.	374	374
450229	**155	C, MN	460	460
451112	*275	F.N.	449	449
451113	*305	R.N.	187	187
451114	450	F.N.	421	421
45266	215	F.N.	460	460
• 452374	225	R.N.	374	374
• 452389	185	R.N., W.C.	374	374
452400	240	R.N.	374	374
• 452423	238	S.W.C.	424	424
452424	255	S.W.C.	424	424
452428	230	W.C.	348	348
• 452460	200	S.W.C.	460	460
452484	225	R.N., G.C.	374	374
452486	193	S.W.C.	460	460
• 452488	195	S.W.C.	374	374
452490	230	S.W.C., G.C.	424	424
452491	220	S.W.C., G.C.	424	424
452505	190	S.W.C.	203	203
452626	245	S.W.C., COM	424	424
45467	**200	C,MN	374	374
• 45468	**175	C,MN	460	460
• 454190	250	F.N.	190	190
454309	235	W.C.	374	374

*Bullet offered in more than one weight.
•Bullet listed in the loading tables of this handbook.
**Weight in lead.

Bullet No.	Grs. Weight #2 Alloy	Style	Top Punch	Seating Screw
• 454424	255	S.W.C.	424	424
454485	250	F.N.	190	190
454506	190	S.W.C.	203	203
454612	**300	C	N/A	N/A
454613	**265	C, MN	N/A	N/A
454616	**220	C, MX	N/A	N/A
• 457102	445	F.N., G.C.	191	191
457121	*475	F.N., L	191	191
• 457122	*290	F.N., H.P.	191	191
• 457124	385	R.N., L	374	374
• 457125	500	R.N., L	374	374
• 457126	455	R.N., L	374	374
457127	210	R.N.	374	374
457130	145	C	130	130
457131	285	F.N.	190	190
457132	*535	R.N.	374	374
• 457191	292	F.N., L	191	191
457192	350	F.N.	191	191
• 457193	405	F.N., L	191	191
457195	225	F.N., H.B.	191	191
457196	290	R.N., H.B.	191	191
457235	505	R.N.	374	374
457401	193	PT, W.C.	401	401
• 457406	475	R.N., G.C.	374	374
457483	378	R.N., G.C.	191	191
• 462560	545	R.N., G.C.	424	424
465451	488	F.N.	191	191

Bullet No.	Grs. Weight #2 Alloy	Style	Top Punch	Seating Screw
470216	*230	F.N., L	460	460
500294	447	R.N.	374	374
504617	**370	C, MX	N/A	N/A
509133	*465	F.N.	191	191
509134	360	R.N.	191	191
• 512138	450	F.N., L	191	191
• 515139	*330	F.N., H.P.	191	191
• 515141	450	F.N.	141	141
• 515142	515	F.N., L	141	141
518144	285	F.N.	141	141
518145	350	F.N.	141	141
53344	**415	C, MN	N/A	N/A
533476	**410	C, MN	N/A	N/A
538146	**347	C	N/A	N/A
540619	**405	C, MX	N/A	N/A
542622	**425	C, MN	N/A	N/A
557456	**475	C, MN	N/A	N/A
557489	**440	PT	N/A	N/A
575213	**505	C, MN	N/A	N/A
575213OS	**460	C, MN	N/A	N/A
575387	**565	R.N.	N/A	N/A
575494	**315	C, MN	N/A	N/A
575602	**400	C, MN	N/A	N/A
57730	**570	R.N., MN	N/A	N/A
577611	**530	C, MN	N/A	N/A
68569	**730	C, MN	N/A	N/A

*Bullet offered in more than one weight.
•Bullet listed in the loading tables of this handbook.
**Weight in lead.

Bullet Abbreviations

PT	Pointed	W.C	Wad Cutter	PP	Paper Patch	MX	MAXI
SPT	Spire Point	S.W.C.	Semi-Wad Cutter	L	Loverin Design	MN	MINIE
F.N.	Flat Nose	H.P.	Hollow Point	COM	Composite (Lead Nose/Lino Body)	N/A	Not Applicable
R.N.	Round Nose	C	Conical	G.C.	Gas Check	H.B.	Hollow Base

Rates of Twist Tables

A. Rifle

Chambering Twist	Mfgr./Model
.22 Hornet – 1 in 16":	Savage 219, 340; Winchester 70. Ruger #3
.219 Zipper – 1 in 14":	Marlin 336.
.222 Remington – 1 in 14":	Browning; Colt; Husqvarna; Marlin; Remington 722, 725, 700, 600, 40XB, 760, 788, 660; Savage 24V; 112V; 340; Sako; Winchester 70, 770.
.222 Remington – 1 in 16":	J.C. Higgins 52.
.222 Remington Magnum – 1 in 12":	Sako.
.222 Remington Magnum – 1 in 14":	Browning; Remington 722, 700, 40XB; Husqvarna.
.223 Remington – 1 in 12":	Colt Bolt Action, Colt AR-15; Remington 760, 700; H&R; Sako.
.223 Remington – 1 in 14":	Remington 40XB, Husqvarna. Savage – 24V
.224 Weatherby Magnum – 1 in 14":	Weatherby.
.225 Winchester – 1 in 14":	Winchester 70, 670; Savage 340.
.22-250 – 1 in 14":	Browning; Remington 700, 788, 40XB; Savage 110; H&R; Husqvarna; Ruger; Winchester 70, 770.
.220 Swift – 1 in 14":	Winchester 70; Ruger 77
.220 Weatherby Rocket – 1 in 14":	Weatherby.
.243 Winchester – 1 in 10":	Browning; Colt; FN; Harrington and Richardson – models 300, 308, 360; Husqvarna; Mannlicher-Schoenauer; Musketeer; Mossberg 800; Remington 700; Savage 99, 110; Sako Bolt Action and Lever Action; Schultz and Larson; Winchester 70, 88, 100, 670, 770; J.C. Higgins 51-L; Ruger.
.243 Winchester – 1 in 9":	Remington 660, 700, 788 (1969).
.244 Remington – 1 in 10":	Remington 700.
.244 Remington – 1 in 12":	Remington 722, 760, 740, 725; 40XB; Sako.
6MM Remington – 1 in 9":	Remington 600, 700, 742, 760, 660.
6MM Remington – 1 in 10":	Remington 40XB; Schultz & Larsen; Ruger.
6MM International – 1 in 12":	Remington 40XB.
6 x 67MM – 1 in 12":	Remington 40XB.
.256 Winchester – 1 in 14":	Marlin 62.

Chambering Twist	Mfgr./Model
.250 Savage - 1 in 10":	Savage 99, Ruger 77.
.257 Roberts - 1 in 10":	Remington 722, 760; Winchester 70; Ruger 77.
.25/06 - 1 in 10":	Remington 700; Savage 112-R Winchester 70.
.257 Weatherby Magnum - 1 in 12":	Weatherby.
.257 Weatherby Magnum - 1 in 14":	Mannlicher-Schoenauer.
6.5 Mannlicher-Schoenauer - 1 in 8¼":	Mannlicher-Schoenauer.
6.5 x 55MM - 1 in 8":	Husqvarna.
6.5 x 55MM - 1 in 8¼":	Schultz & Larsen.
6.5 x 55MM - 1 in 9":	Remington 40XB.
6.5MM Remington Magnum - 1 in 9":	Remington 600, 660, 700; Ruger.
6.5MM x 68MM - 1 in 11":	Mannlicher-Schoenauer.
.264 Winchester Magnum - 1 in 9":	Browning; FN; Musketeer; Remington 700; Savage 110; Sako; Schultz & Larsen; Winchester 70, 670, 770.
.264 Winchester Magnum - 1 in 10":	Colt; Mannlicher-Schoenauer.
.270 Winchester - 1 in 9":	Mannlicher-Schoenauer.
.270 Winchester - 1 in 9½":	Husqvarna.
.270 Winchester - 1 in 10":	Browning; FN; Harrington & Richardson 300; High Standard; J.C. Higgins 50, 51, 51-L; Husqvarna; Musketeer; Remington 721, 760, 725, 700; Savage 110; Sako; Schultz & Larsen; Winchester 70, 670, 770; Ruger.
.270 Weatherby Magnum - 1 in 12":	Weatherby.
7 x 57MM - 1 in 8.7":	FN.
7 x 57MM - 1 in 9":	Mannlicher-Schoenauer.
7 x 57MM - 1 in 9½":	Ruger.
7 x 57MM - 1 in 10":	Winchester 70.
.280 Remington - 1 in 9¼":	Remington 760, 740, 742, 700, 721, 725.
.280 Remington - 1 in 9½":	Ruger.
.284 Winchester - 1 in 10":	Browning; Savage 99; Winchester 88, 100.
7 x 61 Sharpe & Hart - 1 in 10":	Schultz & Larsen
7MM Remington Magnum - 1 in 9":	Husqvarna; Remington 40XB, 700 (1969).
7MM Remington Magnum - 1 in 9¼":	Remington 700.
7MM Remington Magnum - 1 in 9½":	Savage 110; Sako; Winchester 70, 770; Ruger.
7MM Remington Magnum - 1 in 10":	Browning; FN; Harrington & Richardson 300; Musketeer; Schultz & Larsen.
7MM Weatherby Magnum - 1 in 12":	Weatherby.
7MM Weatherby Magnum - 1 in 10":	Weatherby (1965).
.30 M1 Carbine - 1 in 20":	Marlin 62.
.30-30 Winchester - 1 in 12":	Savage 170, 340, 219; Winchester 64, 94, 66.
.30-30 Winchester - 1 in 10":	Marlin 366; Remington 788.
.30 Remington - 1 in 12":	Remington 81, 141.
.300 Savage - 1 in 12":	Remington 81, 722, 760; Savage 99.
.308 Winchester - 1 in 10":	Colt; J.C. Higgins 51-L; Mannlicher-Schoenauer; Marlin; Musketeer; Remington 722, 740, 742, 760, 700, 40XB, 788, 660, 600; Schultz & Larsen; Ruger.

Chambering Twist	Mfgr./Model
.308 Winchester - 1 in 12":	Browning Bolt Action; FN; Harrington and Richardson 300, 308; Husqvarna; Mossberg 800; Savage 99, 110; Sako Bolt Action and Lever Action; Winchester 70, 88, 100, 670, 770.
.30-06 - 1 in 10":	Browning; Colt; FN; High Standard; J.C. Higgins 50, 51, 51-L; Harrington & Richardson 300; Mannlicher-Schoenauer; Marlin; Musketeer; Remington 721, 760, 740, 742, 725, 700, 40XB; Savage 110; Sako; Schultz & Larsen; Winchester 70, 670, 770; Ruger.
.30-06 - 1 in 12":	Husqvarna.
.300 H & H Magnum - 1 in 10":	Browning; Remington 721; Sako; Winchester 70.
.308 Norma Magnum - 1 in 10":	Musketeer; Schultz & Larsen.
.308 Norma Magnum - 1 in 12":	Browning; Husqvarna.
.30-.388 - 1 in 10":	Remington 40XB.
.300 Winchester Magnum - 1 in 10":	Browning; FN; Harrington & Richardson 300; Musketeer; Remington 700; Sako; Savage 110; Winchester 70, 670, 770; Ruger.
.300 Winchester Magnum - 1 in 12":	Husqvarna.
.300 Weatherby Magnum - 1 in 12":	Weatherby.
.32 Remington - 1 in 14":	Remington 81, 141.
.32 Winchester Special - 1 in 16":	Marlin 336; Winchester 64, 94.
8 x 68S - 1 in 11":	Mannlicher-Schoenauer.
.338 Winchester Magnum - 1 in 10":	Mannlicher-Schoenauer; Savage 110; Winchester 70; Ruger.
.338 Winchester Magnum - 1 in 12":	Browning; Sako.
.340 Weatherby Mangnum - 1 in 10":	Weatherby.
.348 Winchester - 1 in 12":	Winchester 71.
.35 Remington - 1 in 16":	Marlin 336; Remington 81, 141, 600, 760.
.35 Remington - 1 in 12":	Savage 170.
.358 Winchester - 1 in 10":	Mannlicher-Schoenauer.
.358 Winchester - 1 in 12":	Savage 99; Schultz & Larsen; Winchester 70, 88, 100.
.350 Remington Magnum - 1 in 16":	Remington 600, 660, 700; Ruger.
.358 Norma Magnum - 1 in 12":	Husqvarna; Schultz & Larsen.
.375 Winchester - 1 in 12":	Winchester 94.
.375 H & H Magnum - 1 in 12":	Remington 700; Winchester 70; Ruger.
.375 H & H Magnum - 1 in 14":	Browning.
.375 Weatherby Magnum - 1 in 12":	Weatherby.
.378 Weatherby Magnum - 1 in 12":	Weatherby.
.44 Magnum - 1 in 38":	Ruger Carbine; Marlin 336; Remington 788; Winchester 94.
.444 Marlin - 1 in 38":	Marlin 336.
.45-70 - 1 in 20":	Ruger; Marlin.
.458 Winchester Magnum - 1 in 14":	Remington 700; Winchester 70; Ruger.
.458 Winchester Magnum - 1 in 15":	Mannlicher-Schoenauer.
.458 Winchester Magnum - 1 in 16½":	Browning.
.460 Weatherby Magnum - 1 in 16":	Weatherby.

B. Pistol

Chambering Twist	Mfgr./Model
.22 Jet – 1 in 14″	Thompson – Contender
.22 Jet – 1 in 15″	Smith & Wesson
.22 Hornet – 1 in 14″	Thompson – Contender
.221 Fireball – 1 in 12″	Remington XP-100
.221 Fireball - 1 in 14″	Thompson - Contender
.222 Remington - 1 in 14″	Thompson - Contender
.256 Winchester - 1 in 14″	Ruger; Thompson - Contender
.30 Carbine - 1 in 12″	Thompson - Contender
.30 Carbine - 1 in 20″	Ruger
.30-30 Winchester - 1 in 14″	Thompson - Contender
.30 Herrett - 1 in 14″	Thompson - Contender
.32 Colt New Police - 1 in 14″	Colt
.32 ACP - 1 in 16″	Mauser
.32 Smith & Wesson, Smith & Wesson Long - 1 in 18¾″	Smith & Wesson
.357 Magnum - 1 in 14″	Colt
.357 Magnum - 1 in 16″	Ruger
.357 Magnum - 1 in 18″	Thompson - Contender
.357 Magnum - 1 in 18¾″	Smith & Wesson
.357 Herrett - 1 in 14″	Thompson - Contender
.35 Remington - 1 in 14″	Thompson - Contender
.38 Smith & Wesson - 1 in 18¾″	Smith & Wesson
.38 Special - 1 in 14″	Colt
.38 Special - 1 in 17″	Charter
.38 Special - 1 in 18¾″	Smith & Wesson
.38 Super - 1 in 16″	Colt
9mm Parabellum - 1 in 10″	Smith & Wesson
9mm Parabellum - 1 in 16″	Colt
.41 Magnum - 1 in 18¾″	Smith & Wesson
.41 Magnum - 1 in 20″	Ruger; Thompson - Contender
.44 Special - 1 in 16″	Colt
.44 Magnum - 1 in 20″	Ruger; Smith & Wesson
.44 Magnum - 1 in 22″	Thompson - Contender
.45 Auto Rim - 1 in 15″	Smith & Wesson
.45 Auto - 1 in 16″	Colt; Thompson - Contender
.45 Winchester Magnum - 1 in 16″	Thompson - Contender
.45 Colt - 1 in 24″	Thompson - Contender

Conversion Factors

Multiply	By	To Obtain
Atmospheres	14.70	Pounds per square inch
Kilograms per square centimeter	14.23	Pounds per square inch
Pounds per square inch	0.07032	Kilograms per cm^2
Drams	1.772	Grams
Drams	0.0625	Ounces
Grains (Troy)	1	Grains (Avoirdupois)
Grains	0.0648	Grams
Grams	15.43	Grains
Grams	0.03527	Ounces
Kilograms	1000	Grams
Kilograms	2.205	Pounds (Avoirdupois)
Ounces	16	Drams
Ounces	437.5	Grains
Ounces	0.0625	Pounds (Avoirdupois)
Ounces	28.35	Grams
Pounds	7000	Grains
Pounds	453.6	Grams
Pounds	16	Ounces
Centimeters	0.3937	Inches
Centimeters	0.01	Meters
Centimeters	10	Millimeters
Inches	2.540	Centimeters
Meters	100	Centimeters
Meters	3.281	Feet
Meters	39.37	Inches
Meters	1.094	Yards
Feet per Second	0.3048	Meters per Second
Feet per Second	0.6818	Miles per Hour
Meters per Second	3.281	Feet per Second
Miles per Hour	88	Feet per Minute
Miles per Hour	1.467	Feet per Second
Liters	0.2642	Gallons
Liters	1.057	Quarts
Liters	1000	Milliliters
Square Centimeters	0.1550	Square Inches
Square Inches	6.452	Square Centimeters

Decimal-Fraction Table

Drill Size	Decimal	Drill Size	Decimal	Drill Size	Decimal	Drill Size	Decimal
80	.0135	42	.0935	13/64	.2031	X	.3970
79	.0145	3/32	.0938	6	.2040	Y	.4040
1/64	.0156	41	.0960	5	.2055	13/32	.4062
78	.0160	40	.0980	4	.2090	Z	.4130
77	.0180	39	.0995	3	.2130	27/64	.4219
76	.0200	38	.1015	7/32	.2188	7/16	.4375
75	.0210	37	.1040	2	.2210	29/64	.4531
74	.0225	36	.1065	1	.2280	15/32	.4688
73	.0240	7/64	.1094	A	.2340	31/64	.4844
72	.0250	35	.1100	15/64	.2344	1/2	.500
71	.0260	34	.1110	B	.2380	33/64	.5156
70	.0280	33	.1130	C	.2420	17/32	.5312
69	.0292	32	.1160	D	.2460	35/64	.5469
68	.0310	31	.1200	1/4	.2500	9/16	.5625
1/32	.0312	1/8	.1250	E	.2500	37/64	.5781
67	.0320	30	.1285	F	.2570	19/32	.5938
66	.0330	29	.1360	G	.2610	39/64	.6094
65	.0350	28	.1405	17/64	.2656	5/8	.6250
64	.0360	9/64	.1406	H	.2660	41/64	.6406
63	.0370	27	.1440	I	.2720	21/32	.6562
62	.0380	26	.1470	J	.2770	43/64	.6719
61	.0390	25	.1495	K	.2810	11/16	.6875
60	.0400	24	.1520	9/32	.2812	45/64	.7031
59	.0410	23	.1540	L	.2900	23/32	.7188
58	.0420	5/32	.1562	M	.2950	47/64	.7344
57	.0430	22	.1570	19/64	.2969	3/4	.7500
56	.0465	21	.1590	N	.3020	49/64	.7656
3/64	.0469	20	.1610	5/16	.3125	25/32	.7812
55	.0520	19	.1660	O	.3160	51/64	.7969
54	.0550	18	.1695	P	.3230	13/16	.8125
53	.0595	11/64	.1719	21/64	.3281	53/64	.8281
1/16	.0625	17	.1730	Q	.3320	27/32	.8438
52	.0635	16	.1770	R	.3390	55/64	.8594
51	.0670	15	.1800	11/32	.3438	7/8	.8750
50	.0700	14	.1820	S	.3480	57/64	.8906
49	.0730	13	.1850	T	.3580	29/32	.9062
48	.0760	3/16	.1875	23/64	.3594	59/64	.9219
5/64	.0781	12	.1890	U	.3680	15/16	.9375
47	.0785	11	.1910	3/8	.3750	61/64	.9531
46	.0810	10	.1935	V	.3770	31/32	.9688
45	.0820	9	.1960	W	.3860	63/64	.9844
44	.0860	8	.1990	25/64	.3906	1	1.000
43	.0890	7	.2010				

RELOADING DATA INTRODUCTION

The preparation of reloading data for this, our third edition of the CAST BULLET HANDBOOK, took several years of planning, research and shooting. With well over 5,000 tested loads, this is our most ambitious technical project to date.

In some cases, assuming the propellant was currently available, we "picked up" cast bullet data from our second edition, published in 1973. This is particularly true for the more esoteric/obsolete chamberings and for the "fast powder" loadings in the popular chamberings.

This enabled us to focus much of the available test time on expanded loading tables for popular rifle and pistol cartridges. The rifle data, for example, contains a tremendous amount of new reloading data featuring the slower rifle powders. The second edition featured, primarily, the faster pistol and shotshell propellants. Had the two test programs been combined, a total of about five years would have been required to generate the data.

Although there is a muzzle-loading section in this Handbook, we did not publish any load data...preferring to devote the maximum space to centerfire cartridge data. Shooters seeking loading data for modern blackpowder rifles, muskets, shotguns and pistols should read our BLACKPOWDER HANDBOOK, the most authoritative text on the subject.

Usually we impose a logical—but somewhat stringent—condition on candidate components from which we assemble loads. That condition being that they be currently available at the time of publication.

It won't take a veteran bullet caster long to notice that we show data for cast bullet designs which are not currently available. Others might make this discovery when they try to buy one of those moulds. This wasn't done to mislead or confuse anyone; rather to provide data for these currently discontinued designs for which there are many moulds still in use.

One final comment on this topic: the annual Lyman catalog, available December/January each year, contains the listing of available mould designs. If you don't see a given number there, then it is not available from the factory. You must refer to the current catalog—not last year's.

The data listed in this section have been tested by our technicians and found to be safe when loaded with our test components and fired (under our laboratory controlled conditions) in our testing equipment. Since Lyman Products Corporation has no control over the manufacture of the various components listed, the actual loading, choice or condition of the firearms and components used, no responsibility for use of this data is implied or assumed.

Components:

The reader should bear in mind that the components listed are not of Lyman manufacture. Therefore, it is possible that production changes affecting ballistic performance can occur at any time without our knowledge. If there is ever a question as to the correctness of the component specified, write to its manufacturer whose address is listed in the front of this publication.

Starting Load:

It is essential that the reader begin with the suggested weight of powder listed in this bracket and work up slowly (following load development precautions) to his best performing load. The novice should use only the "starting load" for a period of time until he builds confidence and experience. Never decrease this charge as an increase in pressure could be encountered.

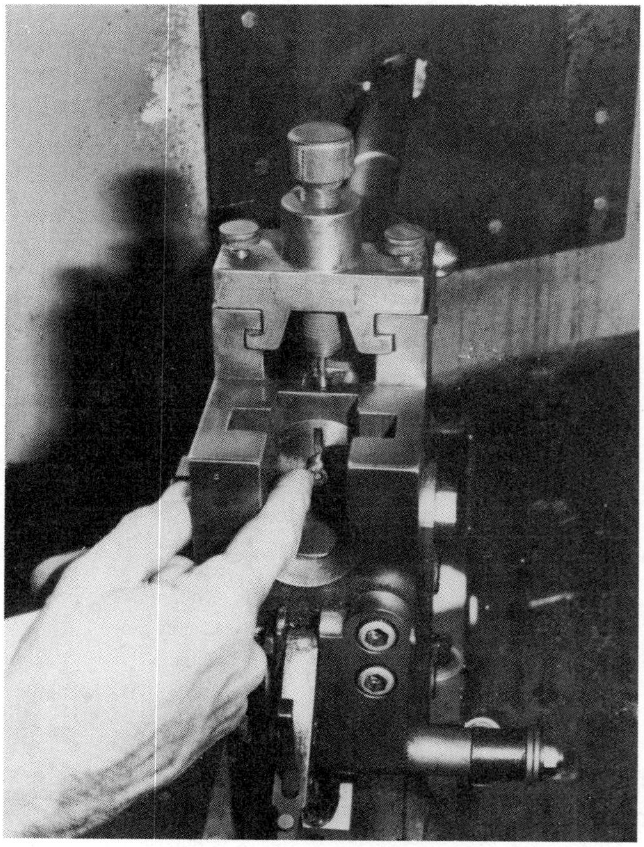

The Universal Bond Receiver operates very much like a falling-block rifle action. These receivers can be used for any centerfire cartridge or shotshell testing; requiring an appropriately chambered and collared pressure barrel.

Maximum Load:

All loads which are listed as maximum were tested and classified as maximum by our technicians in accordance with our laboratory standards. **Under no circumstances should these loads be exceeded**, nor should they be quickly accepted by the reader as a safe working maximum for his particular rifle or pistol.

Many reloaders misinterpret the meaning of the "maximum load". They wrongly assume that if a high pressure load proved safe in a test laboratory then it is equally safe under any and all conditions. This is not true. The reader must start with the "starting load" and work up his loads carefully. Working with his particular firearm and component combination, he may encounter signs of excess pressure before he reaches the maximum charge listed.

The technician classifies a load as maximum after carefully considering many aspects of its ballistic performance. **The maximum average pressure of the load is not the only criteria.** Often a load having an acceptable maximum average pressure will be rejected (or reduced) due to its erratic performance. Accuracy must also be considered, particularly when dealing with cast lead alloy bullets. In all instances, the maximum listing represents what our technicians consider to be the maximum working combination for the bullet, powder and caliber listed. These loads do not exceed SAAMI standards.

Accuracy Loads:

When a load is noted as such in the data tables proper, it means that the given combination of components produced the most uniform internal ballistics of any load tested utilizing that particular bullet design.

Unless noted in "Comments", the accuracy load was not fired at targets. The load, however, does have a high potential - assuming all external factors are optimum - for producing outstanding accuracy since uniform internal ballistics are critical to accuracy on target. You cannot have one without the other.

Test Parameters:

Velocities shown were taken at fifteen feet and not corrected to the muzzle.

Each test string began with a clean dry barrel and consisted of ten shots.

Loads exhibiting erratic internal ballistics were not pursued.

We had no problem with leading in any of our testing.

Bullets:

Bullet numbers are listed in the introductory specifications for each cartridge and in the headline above the appropriate data block - along with an illustration of that particular bullet.

Please note these bullets are artists' renderings. Comparing your bullet against the drawing could reveal minor differences. Furthermore, minor changes are sometimes made to bullets. These drawings, which appear throughout the data sections, are for general reference only and are not intended to be a precise representation.

Bullet alloy is noted as is the exact weight of each tested bullet.

Not all cast bullets within a given caliber are intended to perform equally. We have used them in the most appropriate chamberings.

Powders:

We have limited our testing to those powders which are manufactured in the United States and which are readily available to the consumer. The following brands are listed: Dupont, Winchester, Hercules, Alcan, Hodgdon and Gearhart-Owen.

Compressed Loads:

All compressed loads are indicated with a +. Depending upon the volume of the specific cartridge case used by the reader, he may, or may not, have difficulty starting bullets in such loads. If the bullet will not start, reduce the load sufficiently so that 1/10" of space remains in the case neck. Start the bullet into the case and use whatever additional pressure is required to fully seat the bullet. Failure to comply could result in a bulged case.

Filler Wads:

Dacron filler wads in the form of 1/4 inch thick batting were used in conjunction with cast bullet loads, where indicated. This material can be purchased in most yard-goods stores. It should be cut into squares, which seal the case.

When developing a load, if a wad is desired, it should be used from the beginning as the charge weight is increased. It should never be added as an afterthought, once a maximum load has been established, since its presence could result in a pressure increase of 2,000 CUP or more.

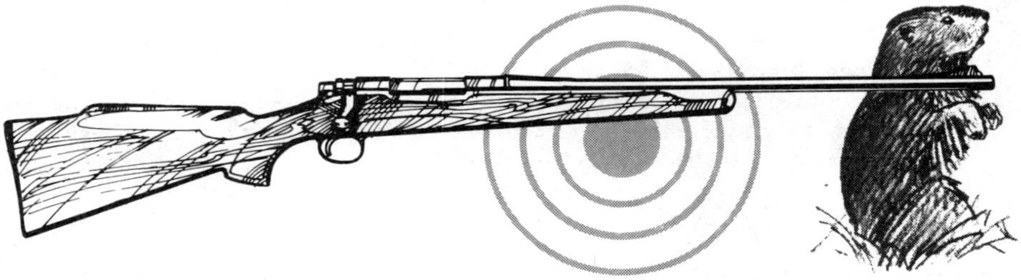

.22 Hornet

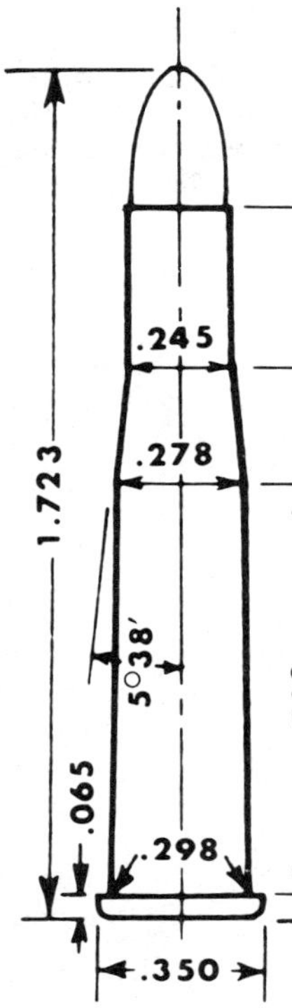

TEST COMPONENTS:

Cases .... Winchester/Western
Trim-to Length .... 1.393″
Primers .... Winchester 6½-116
Primer Size .... Small Rifle
Lyman Shell Holder .... No. 4
Cast Bullets Used .... (size to .223″ dia.)
*Gas Check Bullet .... *#225107, 37 gr.
*#225438, 41 gr.
*#225415, 45 gr.

TEST SPECIFICATIONS:
(Velocity)

Firearm Used .... Winchester Model 54
Barrel Length .... 24″
Twist .... 1-16″
Groove Diameter .... .223″

COMMENTS:

Two distinct groove diameters are encountered in rifles chambered for this cartridge (.223″ or .224″). Some of the earlier rifles make use of a .223″ groove barrel while all of the more current models have the .224″ groove. The reloader must insure that his bullet diameter corresponds to the groove diameter of his particular rifle. All of the data listed were compiled in a .223″ groove barrel.

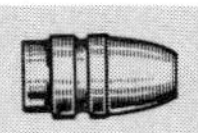

#225107
37 gr., (#2 Alloy) 1.642″ OAL

Powder	Sug. Starting Grains	Velocity F.P.S.	Pressure C.U.P.	Max. Grains	Velocity F.P.S.	Pressure C.U.P.
Red Dot	2.0	1210	—	4.0	1945	—
700X	2.0	1225	—	4.0	1965	—
Green Dot	2.5	1350	—	4.5	2040	—
PB	2.5	1350	—	4.5	2000	—
Unique	3.0	1400	—	5.0	2115	—
SR-7625	3.0	1490	—	5.0	2155	—

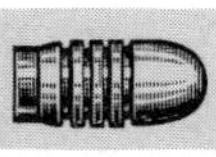

#225438
41 gr., (#2 Alloy) 1.675″ OAL

Powder	Sug. Starting Grains	Velocity F.P.S.	Pressure C.U.P.	Max. Grains	Velocity F.P.S.	Pressure C.U.P.
Red Dot	2.0	1095	—	3.8	1730	—
700X	2.0	1105	—	3.8	1730	—
Green Dot	2.5	1245	—	4.3	1835	—
PB	2.5	1225	—	4.3	1780	—
Unique	3.0	1365	—	4.7	1845	—
SR-7625	3.0	1380	—	4.7	1875	—

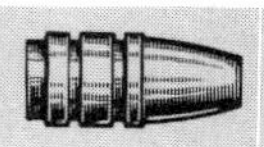

#225415
45 gr., (#2 Alloy) 1.694″ OAL

Powder	Sug. Starting Grains	Velocity F.P.S.	Pressure C.U.P.	Max. Grains	Velocity F.P.S.	Pressure C.U.P.
Red Dot	2.0	1060	—	3.7	1645	—
700X	2.0	1050	—	3.7	1665	—
Green Dot	2.5	1230	—	4.2	1765	—
PB	2.5	1185	—	4.2	1705	—
Unique	3.0	1320	—	4.6	1805	—
SR-7625	3.0	1350	—	4.4	1745	—

Note: Loads shown in shaded panels are maximum.

.218 Bee

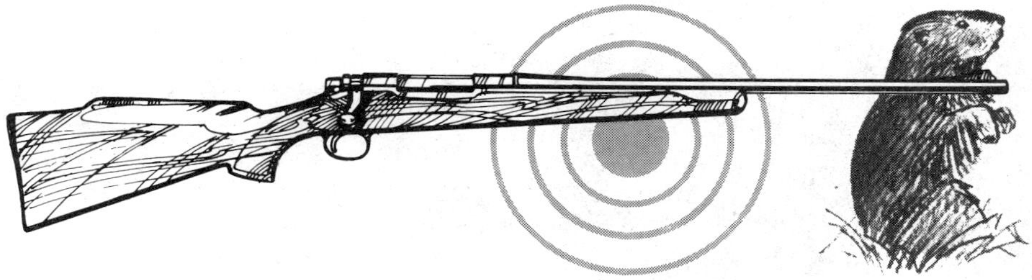

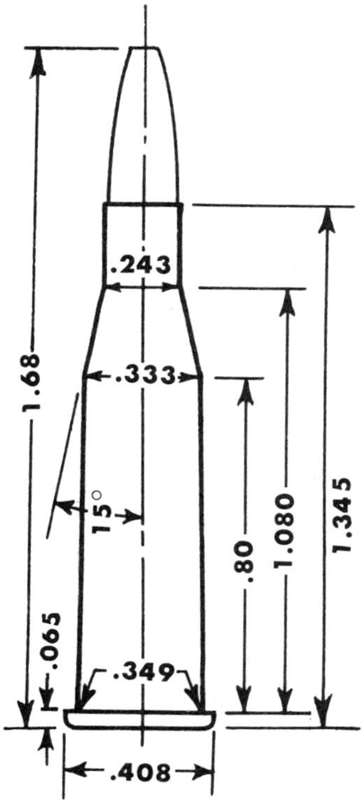

TEST COMPONENTS:

Cases .... Winchester/Western
Trim-to Length .... 1.335″
Primers .... Winchester 6½-116
Primer Size .... Small Rifle
Lyman Shell Holder .... No. 10
Cast Bullets Used .... (size to .223″ dia.)
*Gas Check Bullet *#225107, 37 gr.
*#225438, 41 gr.
*#225415, 45 gr.

TEST SPECIFICATIONS:
(Velocity)

Firearm Used .... Winchester Low Wall
Barrel Length .... 20″
Twist .... 1-16″
Groove Diameter .... .223″

COMMENTS:

Do not use pointed bullets in those rifles which have tubular magazines.

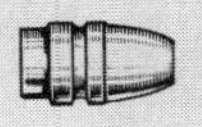

#225107
37 gr., (#2 Alloy) 1.532″ OAL

Powder	Sug. Starting Grains	Velocity F.P.S.	Pressure C.U.P.	Max. Grains	Velocity F.P.S.	Pressure C.U.P.
Red Dot	3.0	1505	—	5.0	2100	—
700X	3.0	1555	—	5.0	2165	—
Green Dot	3.5	1635	—	5.5	2230	—
PB	3.5	1615	—	5.5	2175	—
Unique	4.0	1685	—	6.0	2260	—
SR-7625	4.0	1735	—	6.5	2375	—

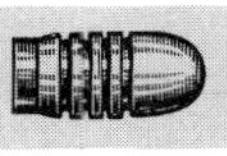

#225438
41 gr., (#2 Alloy) 1.550″ OAL

Powder	Sug. Starting Grains	Velocity F.P.S.	Pressure C.U.P.	Max. Grains	Velocity F.P.S.	Pressure C.U.P.
Red Dot	2.8	1345	—	4.8	1885	—
700X	2.8	1360	—	4.8	1910	—
Green Dot	3.3	1460	—	5.3	2005	—
PB	3.3	1420	—	5.3	1930	—
Unique	3.8	1500	—	5.8	2035	—
SR-7625	3.8	1550	—	6.2	2135	—

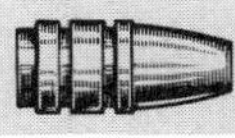

#225415
45 gr., (#2 Alloy) 1.595″ OAL

Powder	Sug. Starting Grains	Velocity F.P.S.	Pressure C.U.P.	Max. Grains	Velocity F.P.S.	Pressure C.U.P.
Red Dot	2.8	1295	—	4.7	1805	—
700X	2.8	1310	—	4.7	1805	—
Green Dot	3.3	1410	—	5.2	1915	—
PB	3.3	1365	—	5.2	1845	—
Unique	3.7	1450	—	5.7	1960	—
SR-7625	3.8	1480	—	6.0	2005	—

Note: Loads shown in shaded panels are maximum.

.222 Remington

(5.7X43mm)

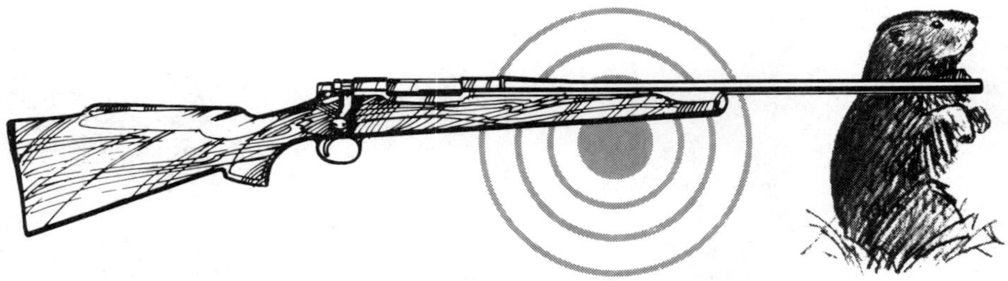

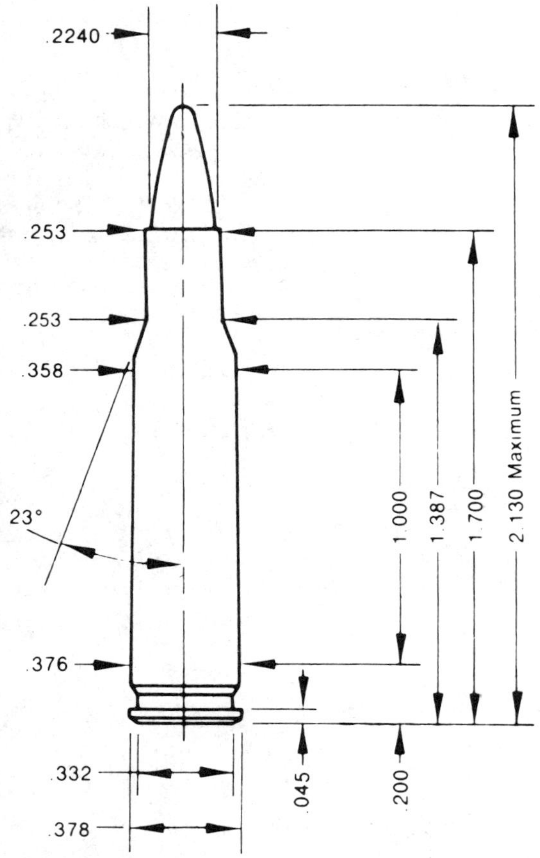

TEST COMPONENTS:

Cases .... Remington
Trim-to Length .... 1.690″
Primers .... Remington 7½
Primer Size .... Small Rifle Magnum
Lyman Shell Holder .... No. 26
Case Bullets Used .... (size to .224″ dia.)
*Gas Check Bullet *#225107, 37 gr.
*#225438, 41 gr.
*#225415, 45 gr.
*#225462, 54 gr.

TEST SPECIFICATIONS:
(Velocity & Pressure)

Firearm Used .... Universal Receiver
Barrel Length .... 26″
Twist .... 1-14″
Groove Diameter .... .224″

COMMENTS:

Best accuracy can be expected at velocities of 2000 f.p.s. to 2300 f.p.s. Based on our results at 50 yards the best high velocity bullet would be #225107 at 2600 f.p.s. to 2700 f.p.s.

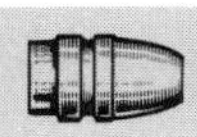

#225107
37 gr., (#2 Alloy) 2.005″ OAL

Powder	Sug. Starting Grains	Velocity F.P.S.	Pressure C.U.P.	Max. Grains	Velocity F.P.S.	Pressure C.U.P.
Red Dot	5.5	2057	25,800	7.5	2481	39,900
700X	5.0	1915	24,000	6.9	2375	39,900
Green Dot	6.0	2118	27,000	8.2	2544	40,700
PB	5.5	1915	23,400	7.7	2392	40,700
Unique	6.0	2132	21,600	7.5	2481	29,400
SR-7625	6.0	2070	27,000	8.0	2409	40,700
SR-4756	6.5	2044	21,000	9.0	2570	39,900
630	8.0	1988	18,600	10.5	2421	25,800
SR-4759	11.1	2094	11,400	14.5	2858	27,100
IMR-4227	10.4	2034	12,500	13.7	2769	27,300
748	20.2	2184	9,300	23.5	2866	20,900
BL-C(2)	*20.3	2127	9,200	24.1	2870	22,900

Note: Loads shown in shaded panels are maximum.
*Designates potentially most accurate load.

#225438
41 gr., (#2 Alloy) 1.972″ OAL

Powder	Sug. Starting Grains	Velocity F.P.S.	Pressure C.U.P.	Max. Grains	Velocity F.P.S.	Pressure C.U.P.
Red Dot	5.3	1934	27,000	7.4	2309	40,700
700X	5.0	1872	27,600	6.6	2178	39,900
Green Dot	5.5	1908	27,000	7.8	2331	40,700
PB	5.5	1845	27,000	7.2	2141	39,000
Unique	6.0	2053	24,600	8.0	2444	36,000
SR-7625	6.0	1915	30,000	7.2	2169	39,900
Herco	6.5	2044	27,000	9.1	2481	40,300
SR-4756	6.3	1988	23,400	8.7	2380	36, 400
630	8.0	1956	20,400	11.0	2403	30,000
SR-4759	11.6	2157	15,400	15.0	2797	27,600
IMR-4227	12.0	2327	17,000	15.6	2816	22,100

#225415
45 gr., (#2 Alloy) 2.107″ OAL

Powder	Sug. Starting Grains	Velocity F.P.S.	Pressure C.U.P.	Max. Grains	Velocity F.P.S.	Pressure C.U.P.
Red Dot	5.1	1785	25,200	7.4	2217	39,000
700X	4.7	1709	27,000	6.4	2053	39,900
Green Dot	5.3	1779	26,400	7.7	2232	40,700
PB	5.2	1718	25,800	7.0	2024	39,400
Unique	6.0	1972	25,800	8.0	2358	36,000
SR-7625	5.7	1785	24,000	7.0	2016	35,500
Herco	6.5	1956	27,000	8.8	2336	40,700
SR-4756	6.0	1769	21,600	8.7	2267	40,300
630	8.0	1919	22,200	11.0	2364	32,000
2400	11.0	2153	15,500	15.8	2835	31,300
SR-4759	12.2	2185	15,300	16.5	2895	36,700
IMR-4227	13.0	2314	19,700	15.9	2751	24,200
748	17.0	2019	12,100	25.0	2881	25,400

#225462
54 gr., (#2 Alloy) 2.031″ OAL

Powder	Sug. Starting Grains	Velocity F.P.S.	Pressure C.U.P.	Max. Grains	Velocity F.P.S.	Pressure C.U.P.
Red Dot	4.7	1590	27,000	7.0	2044	40,300
700X	4.3	1510	24,600	6.2	1855	39,400
Green Dot	5.0	1595	27,000	7.4	1980	40,700
PB	4.8	1510	25,800	6.5	1828	40,700
Unique	5.7	1795	25,800	8.0	2222	39,900
SR-7625	5.1	1565	27,600	6.5	1815	39,900
Herco	6.2	1808	27,000	8.6	2227	39,400
SR-4756	6.0	1666	26,400	8.1	2044	40,300
630	8.0	1851	24,000	11.0	2298	36,000
2400	11.0	2154	19,100	14.8	2820	43,700
SR-4759	11.8	2167	20,000	16.0	2805	43,500
748	*19.2	2163	14,200	24.0	2836	30,700

Note: Loads shown in shaded panels are maximum.
*Designates potentially most accurate load.

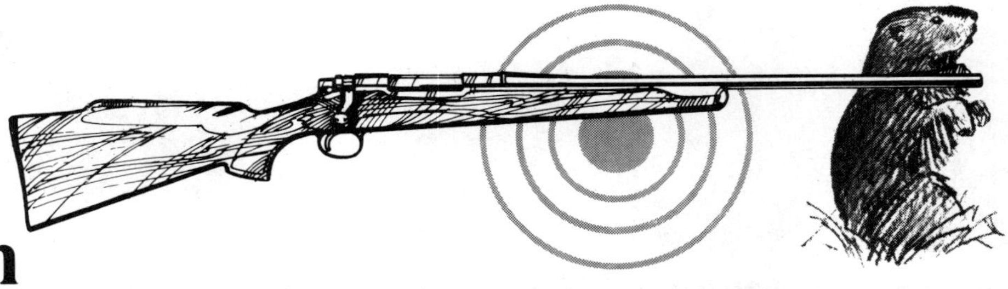

.223 Remington

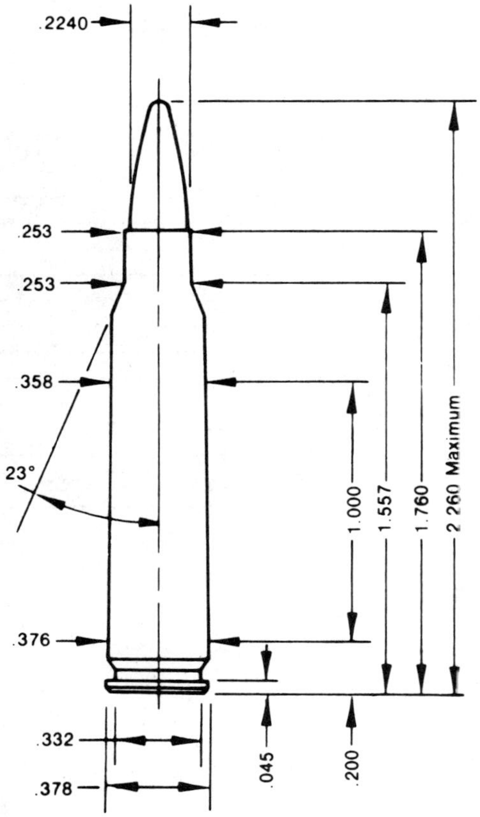

TEST COMPONENTS:

Cases .... Federal & Remington
Trim-to Length .... 1.750″
Primers .... Remington 7½ & CCI 450
Primer Size .... Small Rifle Magnum
Lyman Shell Holder .... No. 26
Cast Bullets Used .... (size to .224″ dia.)
*Gas Check Bullet *#225107, 37 gr.
*#225438, 41 gr.
*#225415, 45 gr.
*#225462, 54 gr.

TEST SPECIFICATIONS:
(Velocity & Pressure)

Firearm Used .... Universal Receiver
Barrel Length .... 24″
Twist .... 1-12″
Groove Diameter .... .224″

COMMENTS:

Best accuracy can be expected at velocities between 2000 f.p.s. to 2300 f.p.s. The most accurate bullet was #225462 at 2100 f.p.s. Based on our results at 50 yards the best high velocity bullet would be #225107 at 2600 f.p.s. to 2700 f.p.s.

When loading for the Ruger Mini-14 a positive crimp must always be provided to avoid bullet set-back during the feed cycle and also to prevent a loose bullet from lodging in the throat of the bore as the result of a "squib" load (a cartridge case which contains an inadequate amount of propellant powder.)

Note: Loads shown in shaded panels are maximum.
**Signifies Remington cases and CCI 450 primers used.

#225107
37 gr., (#2 Alloy) 1.992″ OAL

Powder	Sug. Starting Grains	Velocity F.P.S.	Pressure C.U.P.	Max. Grains	Velocity F.P.S.	Pressure C.U.P.
Red Dot	6.0	2060	23,400	9.5	2590	40,700
700X	6.0	2065	24,600	8.5	2460	39,900
Green Dot	6.5	2075	22,200	9.5	2560	36,000
PB	6.5	2005	21,600	9.0	2400	36,400
Unique	7.0	2070	19,200	10.0	2690	33,500
SR-7625	7.0	2060	22,200	9.5	2460	37,700
630	10.0	2195	19,200	12.5	2530	24,600

#225438
41 gr., (#2 Alloy) 2.034" OAL

Powder	Sug. Starting Grains	Velocity F.P.S.	Pressure C.U.P.	Max. Grains	Velocity F.P.S.	Pressure C.U.P.
Red Dot	5.7	1870	22,800	9.2	2370	40,700
700X	5.8	1875	24,000	8.2	2245	41,100
Green Dot	6.3	1905	22,800	9.3	2335	37,700
PB	6.5	1860	23,400	9.2	2250	40,300
Unique	7.0	1945	21,000	10.0	2530	36,400
SR-7625	6.8	1895	24,600	9.3	2260	40,300
630	10.0	2120	20,400	12.5	2430	27,000
**SR-4759	12.0	1878	18,100	15.0	2612	31,400
**748	22.0	2578	24,300	26.2	2809	28,400
**IMR-4895	21.0	2337	24,800	23.5	2675	28,200

#225415
45 gr., (#2 Alloy) 2.060" OAL

Powder	Sug. Starting Grains	Velocity F.P.S.	Pressure C.U.P.	Max. Grains	Velocity F.P.S.	Pressure C.U.P.
Red Dot	5.6	1795	23,400	9.0	2280	40,700
700X	5.6	1790	23,400	7.7	2080	39,400
Green Dot	6.2	1855	23,400	9.2	2270	39,000
PB	6.3	1775	24,000	8.7	2120	39,400
Unique	6.8	1900	20,400	9.5	2300	35,000
SR-7625	6.6	1820	24,000	8.8	2120	39,900
630	10.0	2080	21,000	12.5	2375	28,200
**SR-4759	12.0	1853	18,200	16.0	2640	35,900

#225462
54 gr., (#2 Alloy) 2.090" OAL

Powder	Sug. Starting Grains	Velocity F.P.S.	Pressure C.U.P.	Max. Grains	Velocity F.P.S.	Pressure C.U.P.
Red Dot	5.4	1675	22,800	8.6	2100	40,700
700X	5.4	1680	25,200	7.5	1970	39,000
Green Dot	6.0	1735	23,400	8.8	2110	38,100
PB	6.1	1675	24,000	8.5	1980	40,300
Unique	6.7	1805	22,000	9.0	2110	30,000
SR-7625	6.4	1715	25,800	8.3	1950	39,400
630	10.0	2020	22,200	12.5	2285	29,400
**H110	13.7	2120	23,700	18.0	2699	41,600
**SR-4759	11.3	1762	20,700	17.7	2611	43,600
**748	19.3	1902	21,600	26.0	2697	32,000

Note: Loads shown in shaded panels are maximum.
**Signifies Remington cases and CCI 450 primers used.

.222 Remington Magnum

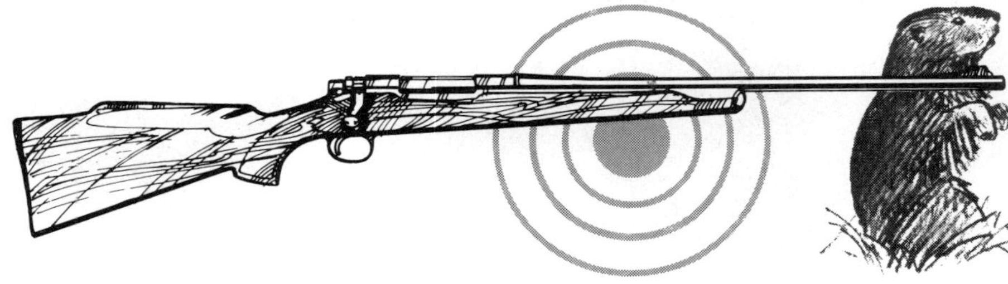

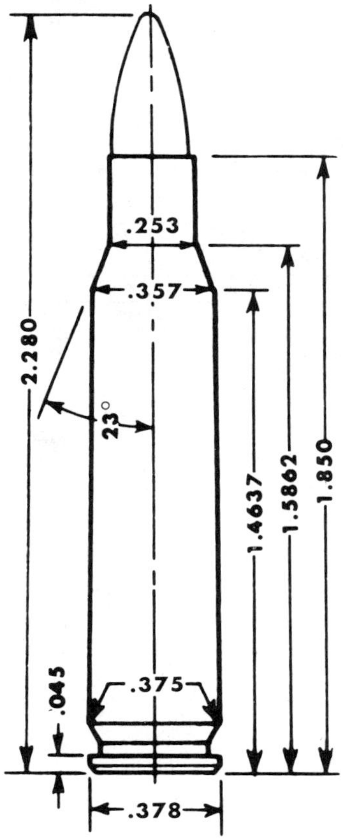

TEST COMPONENTS:

Cases .... Remington-Peters
Trim-to Length .... 1.840″
Primers .... Remington 7½
Primer Size .... Small Rifle, Magnum
Lyman Shell Holder .... No. 26
Cast Bullets Used .... (size to .224″ dia.)
*Gas Check Bullet — *#225107, 37 gr.; *#225438, 41 gr.; *#225415, 45 gr.; *#225462, 54 gr.

TEST SPECIFICATIONS:
(Velocity)

Firearm Used .... Remington Model 722
Barrel Length .... 24″
Twist .... 1-14″
Groove Diameter .... .224″

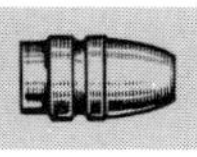

#225107
37 gr., (#2 Alloy) 2.105″ OAL

Powder	Sug. Starting Grains	Velocity F.P.S.	Pressure C.U.P.	Max. Grains	Velocity F.P.S.	Pressure C.U.P.
Red Dot	6.0	2070	—	9.5	2610	—
700X	6.0	2105	—	8.5	2510	—
Green Dot	6.5	2110	—	9.5	2585	—
PB	6.5	2050	—	9.5	2525	—
Unique	7.0	2125	—	10.0	2755	—
SR-7625	7.0	2130	—	9.5	2510	—
630	10.0	2280	—	12.5	2620	—

#225438
41 gr., (#2 Alloy) 2.132″ OAL

Powder	Sug. Starting Grains	Velocity F.P.S.	Pressure C.U.P.	Max. Grains	Velocity F.P.S.	Pressure C.U.P.
Red Dot	5.8	1895	—	9.3	2415	—
700X	5.8	1885	—	8.2	2265	—
Green Dot	6.4	1945	—	9.4	2390	—
PB	6.5	1890	—	9.3	2285	—
Unique	7.0	1975	—	10.0	2545	—
SR-7625	6.9	1940	—	9.3	2285	—
630	10.0	2150	—	12.5	2430	—

Note: Loads shown in shaded panels are maximum.

#225415
45 gr., (#2 Alloy) 2.150" OAL

Powder	Sug. Starting Grains	Velocity F.P.S.	Pressure C.U.P.	Max. Grains	Velocity F.P.S.	Pressure C.U.P.
Red Dot	5.6	1800	—	9.1	2335	—
700X	5.6	1800	—	8.0	2175	—
Green Dot	6.2	1870	—	9.2	2295	—
PB	6.3	1800	—	9.0	2190	—
Unique	6.8	1925	—	9.7	2325	—
SR-7625	6.7	1855	—	9.1	2205	—
630	9.8	1940	—	12.3	2365	—

225462
54 gr., (#2 Alloy) 2.075" OAL

Powder	Sug. Starting Grains	Velocity F.P.S.	Pressure C.U.P.	Max. Grains	Velocity F.P.S.	Pressure C.U.P.
Red Dot	5.4	1675	—	8.9	2175	—
700X	5.4	1660	—	7.8	2010	—
Green Dot	6.0	1750	—	9.0	2165	—
PB	6.1	1675	—	8.8	2045	—
Unique	6.6	1800	—	9.5	2195	—
SR-7625	6.5	1720	—	8.9	2030	—
630	9.8	1910	—	12.3	2310	—

Note: Loads shown in shaded panels are maximum.

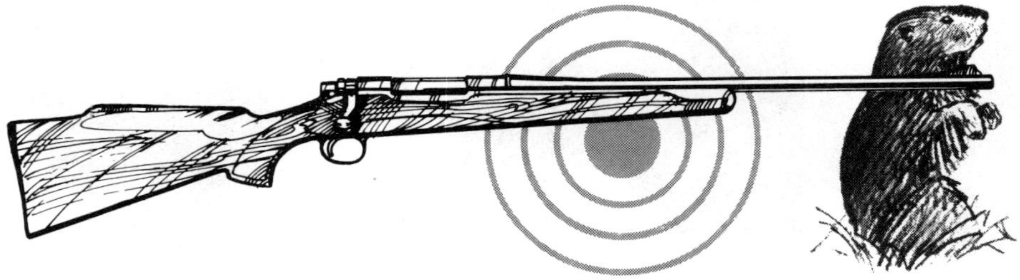

.225 Winchester

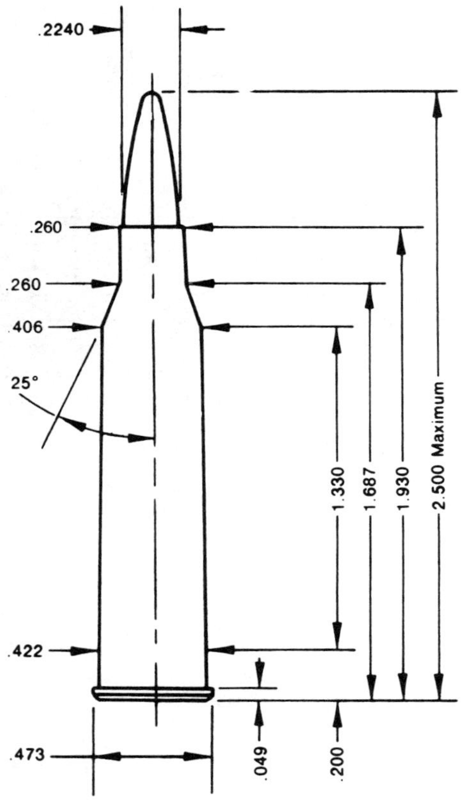

COMMENTS:

Velocity readings in this caliber tend to be very erratic from one gun to the next. Data taken in one gun will not necessarily be uniform with results in another gun.

TEST COMPONENTS:

Cases .... Winchester/Western
Trim-to Length .... 1.920″
Primers .... Winchester 8½-120
Primer Size .... Large Rifle
Lyman Shell Holder .... No. 5
Cast Bullets Used .... (size to .224″ dia.)
*Gas Check Bullet *#225107, 37 gr.
*#225438, 41 gr.
*#225415, 45 gr.
*#225462, 54 gr.

TEST SPECIFICATIONS:
(Velocity)

Firearm Used .... Winchester Model 70
Barrel Length .... 22″
Twist .... 1-14″
Groove Diameter .... .224″

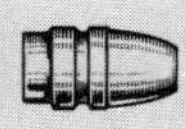

#225107
37 gr., (#2 Alloy) 2.190″ OAL

Powder	Sug. Starting Grains	Velocity F.P.S.	Pressure C.U.P.	Max. Grains	Velocity F.P.S.	Pressure C.U.P.
Red Dot	6.5	2090	—	9.5	2600	—
700X	6.5	2130	—	9.5	2600	—
Green Dot	7.0	2145	—	10.0	2620	—
PB	7.0	2065	—	10.0	2540	—
Unique	7.5	2165	—	10.5	2640	—
SR-7625	7.5	2145	—	10.5	2600	—
630	10.0	2085	—	13.0	2460	—

#225438
41 gr., (#2 Alloy) 2.210″ OAL

Powder	Sug. Starting Grains	Velocity F.P.S.	Pressure C.U.P.	Max. Grains	Velocity F.P.S.	Pressure C.U.P.
Red Dot	6.5	1875	—	9.5	2345	—
700X	6.5	1910	—	9.5	2360	—
Green Dot	7.0	1975	—	10.0	2410	—
PB	7.0	1890	—	10.0	2325	—
Unique	7.5	1990	—	10.5	2410	—
SR-7625	7.5	1920	—	10.5	2355	—
630	10.0	1935	—	13.0	2295	—

Note: Loads shown in shaded panels are maximum.

.225 Winchester (Continued)

#225415
45 gr., (#2 Alloy) 2.235" OAL

Powder	Sug. Starting Grains	Velocity F.P.S.	Pressure C.U.P.	Max. Grains	Velocity F.P.S.	Pressure C.U.P.
Red Dot	6.0	1770	—	9.0	2235	—
700X	6.0	1775	—	9.0	2185	—
Green Dot	6.5	1845	—	9.5	2280	—
PB	6.5	1760	—	9.5	2170	—
Unique	7.0	1880	—	10.0	2315	—
SR-7625	7.0	1825	—	10.0	2195	—
630	9.5	1825	—	12.5	2195	—

225462
54 gr., (#2 Alloy) 2.150" OAL

Powder	Sug. Starting Grains	Velocity F.P.S.	Pressure C.U.P.	Max. Grains	Velocity F.P.S.	Pressure C.U.P.
Red Dot	5.5	1610	—	8.5	1965	—
700X	5.5	1590	—	8.5	1995	—
Green Dot	6.0	1660	—	9.0	2060	—
PB	6.0	1565	—	9.0	1970	—
Unique	6.5	1695	—	9.5	2045	—
SR-7625	6.5	1640	—	9.5	2025	—
630	9.5	1820	—	12.5	2120	—

Note: Loads shown in shaded panels are maximum.

.224 Weatherby Magnum

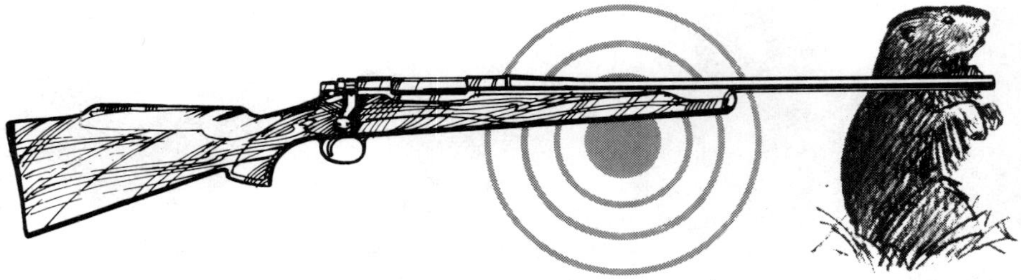

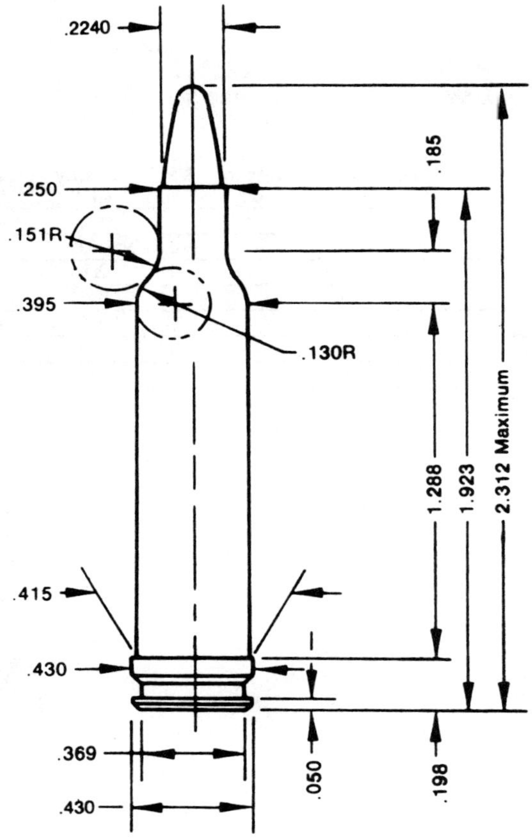

TEST COMPONENTS:

Cases .... Weatherby
Trim-to Length .... 1.915″
Primers .... Remington 9½
Primer Size .... Large Rifle
Lyman Shell Holder .... No. 3
Cast Bullets Used .... (size to .224″ dia.)
*Gas Check Bullet *#225415, 45 gr.
*#225462, 54 gr.

TEST SPECIFICATIONS:
(Velocity)

Firearm Used .... Weatherby Varmint Master
Barrel Length .... 26″
Twist .... 1-14″
Groove Diameter .... .224″

COMMENTS:

In this case the use of a magnum type primer will result in very high and dangerous pressures. Use only standard large rifle primers for this cartridge.

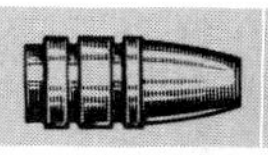

#225415
45 gr., (#2 Alloy) 2.312″ OAL, Max.

Powder	Sug. Starting Grains	Velocity F.P.S.	Pressure C.U.P.	Max. Grains	Velocity F.P.S.	Pressure C.U.P.
Unique	6.0	1818	—	7.0	2008	—
2400	11.0	1964	—	13.0	2314	—

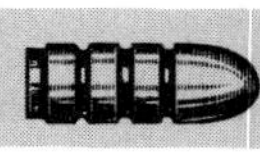

#225462
54 gr., (#2 Alloy) 2.312″ OAL, Max.

Powder	Sug. Starting Grains	Velocity F.P.S.	Pressure C.U.P.	Max. Grains	Velocity F.P.S.	Pressure C.U.P.
Unique	6.0	1727	—	8.0	2074	—
IMR-4227	13.0	2092	—	15.0	2386	

Note: Loads shown in shaded panels are maximum.

.22/250 Remington

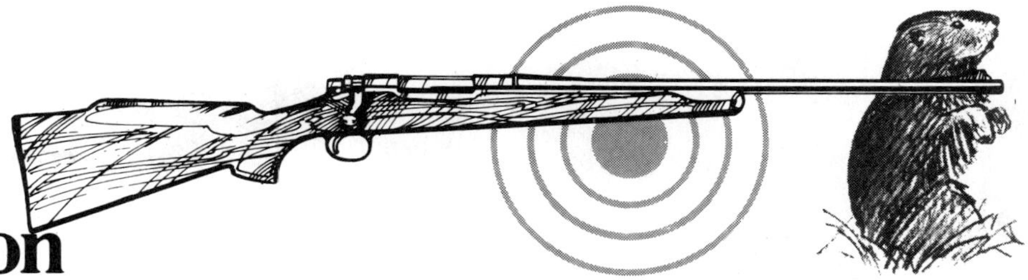

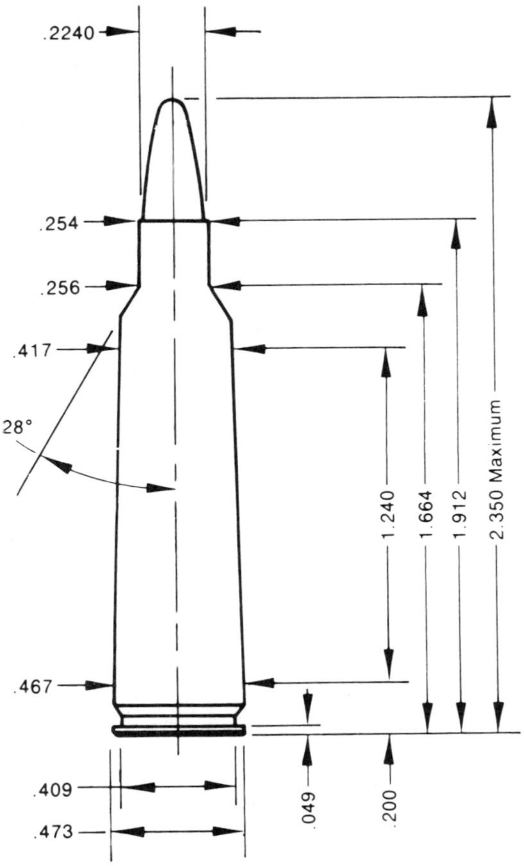

TEST COMPONENTS:

Cases .... Winchester
Trim-to Length .... 1.902″
Primers .... Winchester 8½-120
Primer Size .... Large Rifle
Lyman Shell Holder .... No. 2
Cast Bullets Used .... (size to .224″ dia.)
*Gas Check Bullet *#225107, 37 gr.
*#225438, 41 gr.
*#225415, 45 gr.
*#225462, 54 gr.

TEST SPECIFICATIONS:
(Velocity & Pressure)

Firearm Used .... Universal Receiver
Barrell Length .... 24″
Twist .... 1-14″
Groove Diameter .... .224″

COMMENTS:

Best accuracy can be expected at velocities between 2000 f.p.s.and 2300 f.p.s. The most accurate bullet was #225462 at 2100 f.p.s. Based on our results at 50 yards the best high velocity bullet would be #225107 at 2600 f.p.s. to 2700 f.p.s. Due to the wide variations in "custom" chambers, maximum loads should be approached with caution.

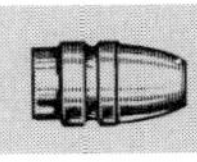

#225107
37 gr., (#2 Alloy) 2.195″ OAL

Powder	Sug. Starting Grains	Velocity F.P.S.	Pressure C.U.P.	Max. Grains	Velocity F.P.S.	Pressure C.U.P.
Red Dot	7.2	2260	19,800	9.5	2634	33,500
700X	7.0	2222	21,600	9.0	2527	32,500
Green Dot	7.5	2250	17,100	10.0	2624	32,500
PB	7.5	2211	22,200	9.5	2485	32,000
SR-7625	7.5	2158	20,400	10.0	2586	36,000
SR-4756	8.0	2064	15,400	11.0	2646	31,000
630	12.4	1975	7,900	15.5	2693	16,300
IMR-4227	15.0	2044	6,900	20.3	2748	15,800
IMR-4198	*17.7	2107	7,900	23.3	2749	14,300

Note: Loads shown in shaded panels are maximum.
*Designates potentially most accurate load.

.22/250 Remington (Continued)

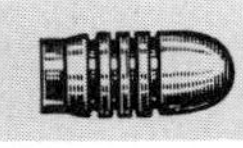

#225438
41 gr., (#2 Alloy) 2.266" OAL

Powder	Sug. Starting Grains	Velocity F.P.S.	Pressure C.U.P.	Max. Grains	Velocity F.P.S.	Pressure C.U.P.
Red Dot	7.2	2079	17,100	9.5	2404	32,000
700X	7.0	2021	17,100	9.0	2354	32,000
Green Dot	7.5	2050	15,400	10.0	2413	30,000
PB	7.5	1997	17,100	9.5	2277	30,000
SR-7625	7.5	1982	15,400	10.0	2362	35,000
SR-4756	8.0	1936	15,400	11.0	2445	31,000
630	9.0	1840	—	11.0	2090	15,400
SR-4759	15.2	2290	14,400	18.5	2807	22,300
IMR-4227	14.0	2151	12,300	20.1	2945	23,500
IMR-4198	16.9	2173	10,700	21.1	2739	20,200
748	21.4	2174	11,200	26.8	2764	18,800

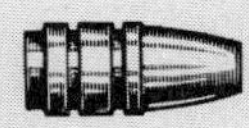

#225415
45 gr., (#2 Alloy) 2.325" OAL

Powder	Sug. Starting Grains	Velocity F.P.S.	Pressure C.U.P.	Max. Grains	Velocity F.P.S.	Pressure C.U.P.
Red Dot	7.2	1981	22,800	9.5	2323	35,000
700X	7.0	1850	18,000	9.0	2268	33,500
Green Dot	7.5	1992	21,000	10.0	2356	34,500
PB	7.5	1920	18,000	9.5	2188	31,000
SR-7625	8.0	1981	20,400	10.0	2237	36,900
SR-4756	8.0	1915	18,000	11.0	2308	32,000
630	9.0	1825	15,400	11.0	2086	18,000
SR-4759	14.5	2176	13,000	19.0	2820	26,900
IMR-4198	16.9	2173	12,100	22.1	2867	23,900
748	22.3	2175	13,600	28.5	2814	21,500

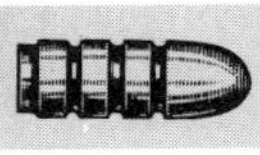

#225462
54 gr., (#2 Alloy) 2.347" OAL

Powder	Sug. Starting Grains	Velocity F.P.S.	Pressure C.U.P.	Max. Grains	Velocity F.P.S.	Pressure C.U.P.
Red Dot	7.0	1828	21,000	9.5	2184	36,400
700X	7.5	1937	29,400	9.0	2143	36,900
Green Dot	7.0	1828	21,000	10.0	2236	36,900
PB	8.0	1880	27,000	9.5	2088	36,400
SR-7625	8.5	1955	30,500	10.0	2157	38,600
SR-4756	8.5	1935	18,600	10.5	2207	35,500
630	9.0	1837	17,100	11.5	2143	21,000
SR-4759	14.5	2154	15,700	20.0	2833	33,600
IMR-4227	*15.6	2156	13,800	20.1	2858	35,300
IMR-4198	16.8	2167	14,500	23.0	2883	30,200
748	22.9	2127	12,900	28.6	2811	25,800

Note: Loads shown in shaded panels are maximum.
*Designates potentially most accurate load.

.220 Swift

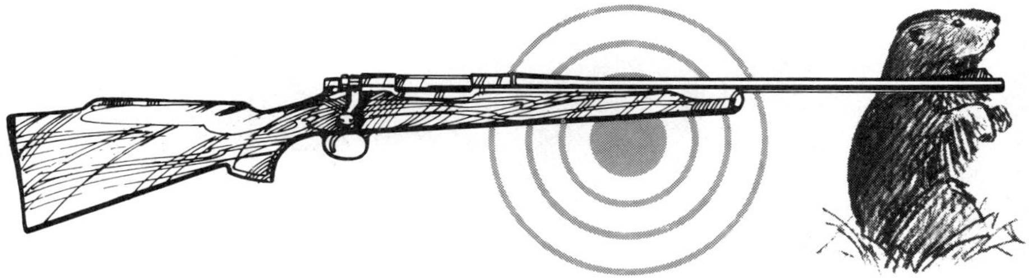

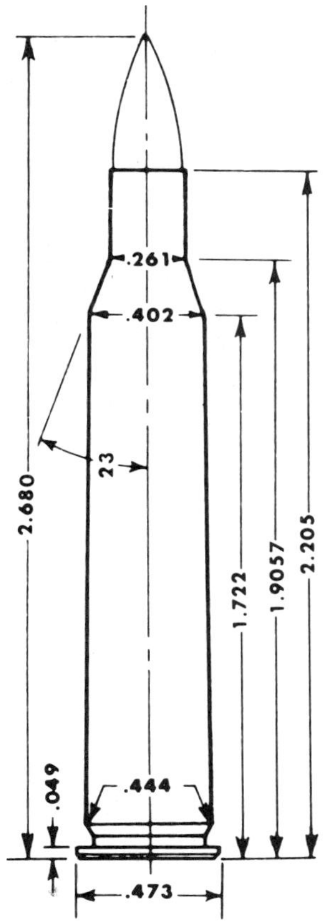

TEST COMPONENTS:

Cases .... Winchester
Trim-to Length .... 2.195″
Primers .... Winchester 8½-120 & CCI 200
Primer Size .... Large Rifle
Lyman Shell Holder .... No. 5
Cast Bullets Used .... (size to .224″ dia.)
*Gas Check Bullet .... *#225438, 41 gr.
*#225415, 45 gr.
*#225462, 54 gr.

TEST SPECIFICATIONS:
(Velocity & Pressure)

Firearm Used .... Universal Receiver
(Velocity barrel used where no pressures are listed)
Barrel Length .... 24″
Twist .... 1-14″
Groove Diameter .... .224″

COMMENTS:

Best accuracy can be expected at velocities between 2000 f.p.s. and 2300 f.p.s. The most accurate bullet was #245462 at 2100 f.p.s.

#225438
41 gr., (#2 Alloy) 2.520″ OAL

Powder	Sug. Starting Grains	Velocity F.P.S.	Pressure C.U.P.	Max. Grains	Velocity F.P.S.	Pressure C.U.P.
Red Dot	6.5	1795	—	10.0	2350	—
700X	6.5	1795	—	10.0	2335	—
Green Dot	7.0	1800	—	10.5	2365	—
PB	7.0	1750	—	10.5	2290	—
Unique	7.5	1835	—	11.0	2390	—
SR-7625	7.5	1800	—	11.0	2345	—
**630	*10.0	1819	16,000	16.2	2665	27,900
**2400	10.3	1780	16,100	16.6	2682	28,800
**SR-4759	12.3	1806	15,900	18.0	2701	29,100

Note: Loads shown in shaded panels are maximum.
*Designates potentially most accurate load.
**Signifies use of CCI 200 primers.

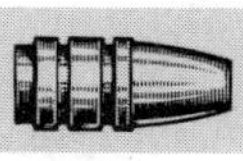

#225415
45 gr., (#2 Alloy) 2.546" OAL

Powder	Sug. Starting Grains	Velocity F.P.S.	Pressure C.U.P.	Max. Grains	Velocity F.P.S.	Pressure C.U.P.
Red Dot	6.5	1730	—	10.0	2270	—
700X	6.5	1710	—	10.0	2275	—
Green Dot	7.0	1800	—	10.5	2305	—
PB	7.0	1680	—	10.5	2210	—
Unique	7.5	1805	—	11.0	2320	—
SR-7625	7.5	1765	—	11.0	2280	—
**630	*9.9	1816	16,500	17.1	2674	31,000
**2400	10.5	1772	16,500	17.4	2723	31,500
**SR-4759	12.2	1748	15,500	18.5	2683	30,300

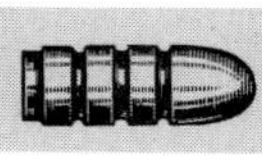

#225462
54 gr., (#2 Alloy) 2.525" OAL

Powder	Sug. Starting Grains	Velocity F.P.S.	Pressure C.U.P.	Max. Grains	Velocity F.P.S.	Pressure C.U.P.
Red Dot	6.0	1605	—	9.5	2035	—
700X	6.0	1615	—	9.5	2040	—
Green Dot	6.5	1660	—	10.0	2130	—
PB	6.5	1610	—	10.0	2060	—
Unique	7.0	1705	—	10.5	2165	—
SR-7625	7.0	1635	—	10.5	2115	—
**630	*10.5	1819	19,400	17.4	2670	39,300
**2400	11.0	1724	18,000	18.5	2706	38,600
**SR-4759	12.4	1759	18,000	19.2	2685	40,000

Note: Loads shown in shaded panels are maximum.
*Designates potentially most accurate load.
**Signifies use of CCI 200 primers.

.22 Savage

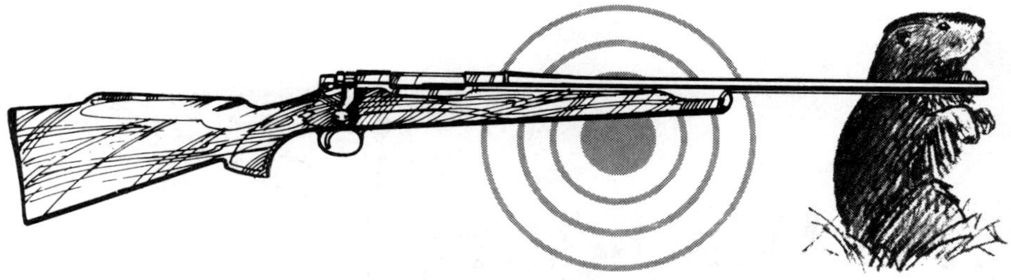

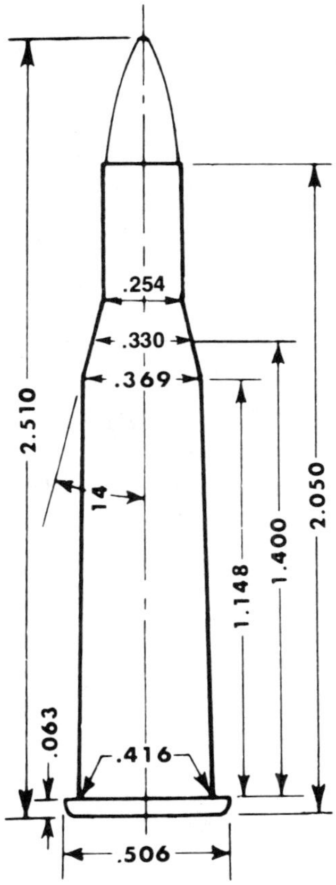

TEST COMPONENTS:

Cases .... Winchester
Trim-to Length .... 2.040″
Primers .... Winchester 8½-120
Primer Size .... Large Rifle
Lyman Shell Holder .... No. 6
Cast Bullets Used .... (size to .228″ dia.)
*Gas Check Bullet .... *#228365, 60 gr.
#22835, 78 gr.

TEST SPECIFICATIONS:
(Velocity)

Firearm Used .... Savage Model 99
Barrel Length .... 20″
Twist .... 1-12″
Groove Diameter .... .229″

COMMENTS:

Variations in groove diameters are sometimes encountered in rifles chambered for this cartridge. Most barrels have a groove diameter of .228″ while a small percentage are found with a larger groove diameter of .229″. For oversize barrels, we recommend the use of cast lead bullets sized to groove diameter.

#228367
60 gr., (#2 Alloy) 2.510″ OAL, Max.

Powder	Sug. Starting Grains	Velocity F.P.S.	Pressure C.U.P.	Max. Grains	Velocity F.P.S.	Pressure C.U.P.
Unique	6.5	1754	—	7.7	1908	—
2400	11.0	1811	—	13.0	2074	—
IMR-4227	12.0	1893	—	14.0	2145	—

#22835
78 gr., (# 2 Alloy) 2.510″ OAL, Max.

Powder	Sug. Starting Grains	Velocity F.P.S.	Pressure C.U.P.	Max. Grains	Velocity F.P.S.	Pressure C.U.P.
Unique	6.0	1501	—	7.0	1625	—
2400	9.0	1562	—	11.5	1730	—
IMR-4227	10.0	1590	—	12.5	1792	—

Note: Loads shown in shaded panels are maximum.

.243 Winchester

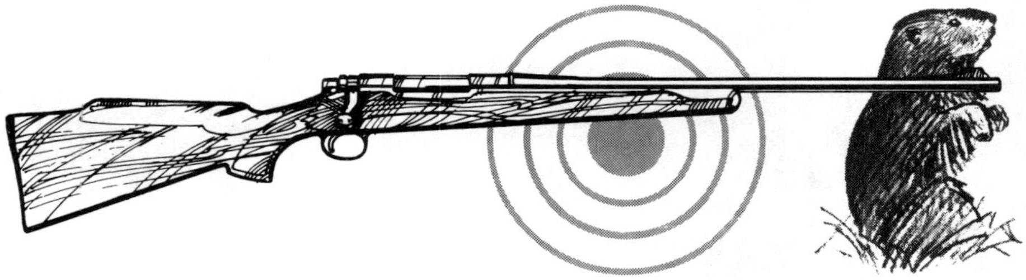

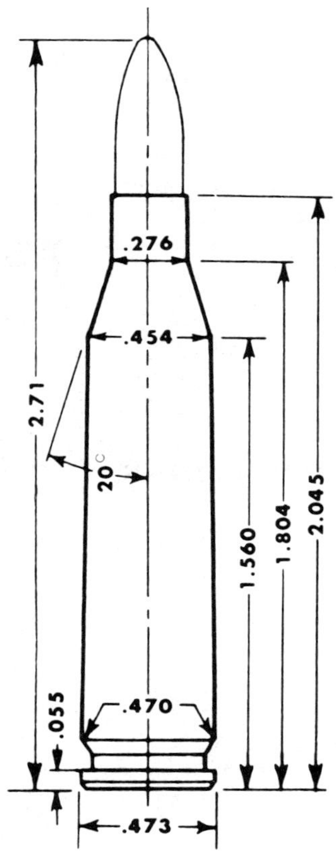

TEST COMPONENTS:

Cases .... Remington
Trim-to Length .... 2.035″
Primers .... Remington 9½
Primer Size .... Large Rifle
Lyman Shell Holder .... No. 2
Cast Bullets Used .... (size to .243″ dia.)
*Gas Check Bullets *#245496, 83 gr.
*#245497, 90 gr.
*#245498, 95 gr.

TEST SPECIFICATIONS:
(Velocity & Pressure)

Firearm Used .... Universal Receiver
Barrel Length .... 26″
Twist .... 1-10″
Groove Diameter .... .243″

COMMENTS:

If your .243 rifle is imported, we recommend that you carefully check its groove diameter. We have encountered rifles with groove diameters as small as .239″. When fired in these tight groove rifles, factory loaded .243 cartridges blew primers and even our starting loads proved too hot.

Accuracy seems to fall off rapidly with velocities over 1750 f.p.s. Best results were obtained with bullet #245497 at 1700 f.p.s.

#245496
83 gr., (#2 Alloy) 2.480″ OAL

Powder	Sug. Starting Grains	Velocity F.P.S.	Pressure C.U.P.	Max. Grains	Velocity F.P.S.	Pressure C.U.P.
Red Dot	10.0	1845	33,000	12.0	2032	40,300
700X	8.5	1684	30,500	10.5	1904	39,000
PB	9.0	1653	29,400	11.0	1836	36,400
SR-7625	9.5	1703	30,500	11.5	1909	39,400
SR-4756	11.0	1846	31,500	13.0	2013	38,600
630	*12.5	1626	13,300	17.5	2134	24,300
SR-4759	14.9	1668	12,800	20.0	2156	21,400
IMR-4227	14.5	1623	11,900	19.7	2156	23,200
IMR-4198	16.0	1638	10,800	22.7	2175	19,200
RX7	16.0	1676	12,400	21.0	2092	20,200
748	26.1	2114	18,200	39.0	2906	43,800
H335	25.8	2191	21,200	37.8	2926	46,000
H4895	25.0	2126	17,300	37.2	2970	47,400
IMR-4320	28.2	2178	20,800	40.2	2996	45,700

Note: Loads shown in shaded panels are maximum.
*Designates potentially most accurate load.

#245497
90 gr., (#2 Alloy) 2.455" OAL

Powder	Sug. Starting Grains	Velocity F.P.S.	Pressure C.U.P.	Max. Grains	Velocity F.P.S.	Pressure C.U.P.
Red Dot	8.5	1611	27,000	11.5	1934	39,400
700X	7.5	1513	26,400	10.0	1762	37,300
PB	8.0	1477	23,400	10.5	1815	39,900
SR-7625	8.5	1521	25,800	11.0	1768	36,400
SR-4756	9.5	1586	26,400	12.5	1869	36,400
630	18.8	2146	28,400	23.8	2556	47,300
SR-4759	21.0	2182	28,000	26.2	2605	46,700
IMR-4227	20.9	2197	28,100	25.5	2537	44,500
IMR-4198	23.9	2198	23,500	30.7	2757	45,300
RX7	22.0	2157	24,200	29.6	2695	46,300
748	26.0	2077	19,300	38.0	2828	46,300
H335	25.0	2128	20,200	36.3	2790	47,300

Note: Loads shown in shaded panels are maximum.

#245498
95 gr., (#2 Alloy) 2.590" OAL

Powder	Sug. Starting Grains	Velocity F.P.S.	Pressure C.U.P.	Max. Grains	Velocity F.P.S.	Pressure C.U.P.
Red Dot	8.0	1482	22,800	10.0	1699	33,000
700X	7.0	1386	22,200	9.5	1653	34,500
PB	8.0	1424	24,600	10.0	1611	32,000
SR-7625	8.5	1463	27,000	10.5	1652	35,500
SR-4756	10.0	1550	26,400	12.0	1790	35,000
630	16.0	1803	19,700	23.5	2426	43,100
SR-4759	20.8	2112	29,800	25.8	2475	47,600
IMR-4227	19.3	2036	26,500	24.5	2456	46,600
IMR-4198	20.0	1826	16,500	30.2	2572	42,900
RX7	20.0	1874	20,000	32.0	2608	47,100

6mm Remington
.244 Remington

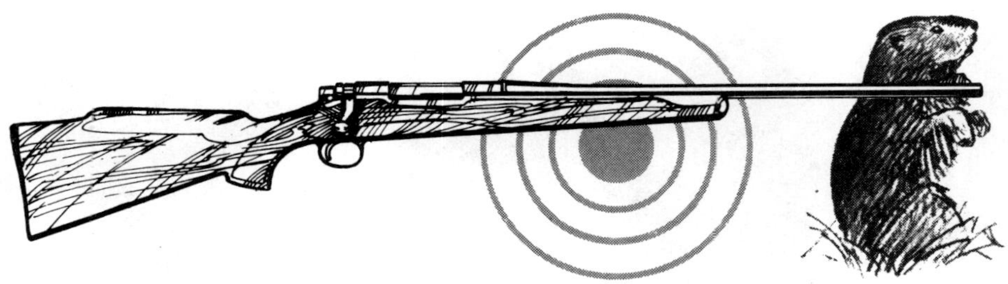

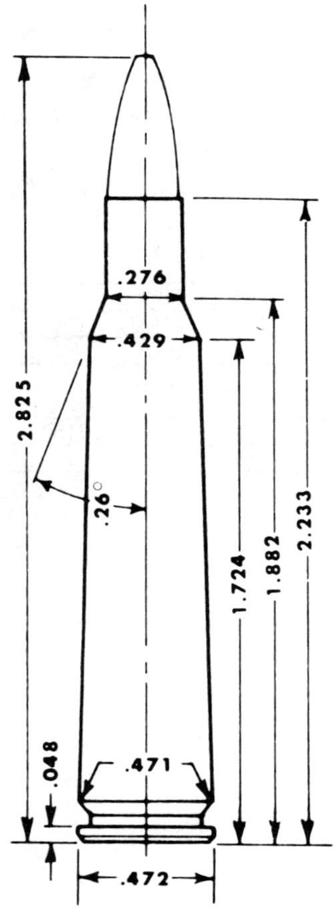

TEST COMPONENTS:

Cases .... Remington
Trim-to Length .... 2.225″
Primers .... Remington 9½
Primer Size .... Large Rifle
Lyman Shell Holder .... No. 2
Cast Bullets Used .... (size to .243″ dia.)
*Gas Check Bullets *#245496, 83 gr.
*#245497, 90 gr.
*#245498, 95 gr.

TEST SPECIFICATIONS:
(Velocity & Pressure)

Firearm Used .... Universal Receiver
Barrel Length .... 24″
Twist .... 1-9″
Groove Diameter .... .243″

COMMENTS:

The only difference between the 6mm Remington and the .244 Remington cartridge is in the factory bullet weights which are loaded into these two cases. The cases themselves are identical except for the headstamps.

Normally, .244 rifles have a 1-12″ twist and the cartridge is factory loaded with 75 and 90 grain bullets to accommodate this rifling. The 6mm Remington cartridge is factory loaded with a 100 grain bullet to better accommodate the 1-9″ twist of the 6mm rifles.

Accuracy seems to fall off rapidly with velocities over 1750 f.p.s. Best results were obtained with bullet #245497 at 1700 f.p.s.

#245496
83 gr., (#2 Alloy) 2.690″ OAL

Powder	Sug. Starting Grains	Velocity F.P.S.	Pressure C.U.P.	Max. Grains	Velocity F.P.S.	Pressure C.U.P.
Red Dot	9.0	1726	30,500	11.5	1973	40,700
700X	8.5	1635	29,400	10.7	1879	40,700
Green Dot	9.0	1703	27,600	11.5	1947	39,900
PB	9.0	1642	27,600	11.0	1829	38,600
SR-7625	9.0	1628	29,400	11.0	1818	38,600
SR-4756	10.0	1725	29,400	12.5	1974	40,700
630	*12.0	1736	20,200	15.5	2041	26,300
SR-4759	13.0	1661	16,100	19.5	2202	34,000
IMR-4227	13.7	1711	17,300	19.7	2181	28,200
IMR-4198	16.8	1828	17,000	21.0	2144	22,500
RX7	14.5	1651	17,100	20.0	2117	25,800
748	20.8	1772	14,700	26.0	2173	23,400
H4895	19.6	1760	16,300	24.5	2132	23,300

Note: Loads shown in shaded panels are maximum.
*Designates potentially most accurate load.

#245497
90 gr., (#2 Alloy) 2.687" OAL

Powder	Sug. Starting Grains	Velocity F.P.S.	Pressure C.U.P.	Max. Grains	Velocity F.P.S.	Pressure C.U.P.
Red Dot	8.5	1582	28,800	11.0	1831	40,300
700X	8.0	1551	24,600	10.2	1776	39,000
Green Dot	8.5	1581	27,000	11.0	1826	38,600
PB	8.5	1540	21,000	10.6	1729	39,000
SR-7625	8.5	1523	25,200	10.6	1739	39,400
SR-4756	9.5	1627	28,800	12.0	1845	38,600
630	14.0	1817	22,300	17.5	2111	30,700
SR-4759	15.8	1840	22,100	19.8	2134	30,700
IMR-4227	15.4	1812	22,200	19.2	2101	30,500
IMR-4198	16.9	1749	17,000	21.1	2094	25,500
RX7	16.5	1779	19,800	21.5	2138	27,900
748	20.0	1726	15,500	26.0	2151	25,200
H335	19.0	1760	17,400	24.5	2152	26,900
H4895	19.0	1679	16,800	24.0	2109	25,700

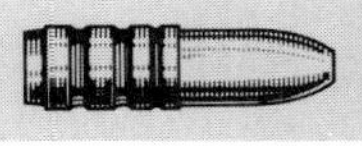

#245498
95 gr., (#2 Alloy) 2.793" OAL

Powder	Sug. Starting Grains	Velocity F.P.S.	Pressure C.U.P.	Max. Grains	Velocity F.P.S.	Pressure C.U.P.
Red Dot	8.1	1491	25,800	10.6	1756	38,600
700X	7.6	1456	25,200	9.9	1699	38,100
Green Dot	8.2	1489	24,000	10.7	1732	36,400
PB	8.2	1438	25,800	10.3	1630	37,300
SR-7625	8.2	1448	24,600	10.3	1639	36,900
SR-4756	9.0	1498	27,600	11.7	1751	37,700
630	*13.7	1739	21,500	18.0	2079	31,900
SR-4759	15.0	1698	19,600	19.8	2064	30,400
IMR-4227	14.5	1642	18,300	19.2	2052	30,700
IMR-4198	16.5	1684	17,100	21.1	2056	25,800
RX7	16.0	1629	17,300	21.7	2092	29,400
748	20.4	1679	16,400	26.0	2101	26,900
H335	19.0	1675	16,700	25.0	2149	30,600
H4895	19.0	1682	19,300	25.0	2154	29,500

Note: Loads shown in shaded panels are maximum.
*Designates potentially most accurate load.

.25/20 Winchester

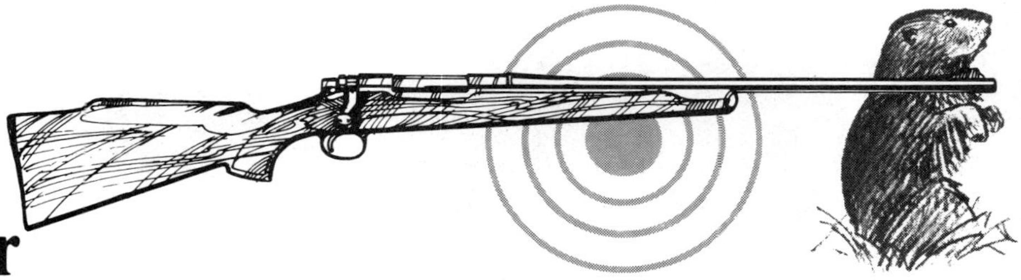

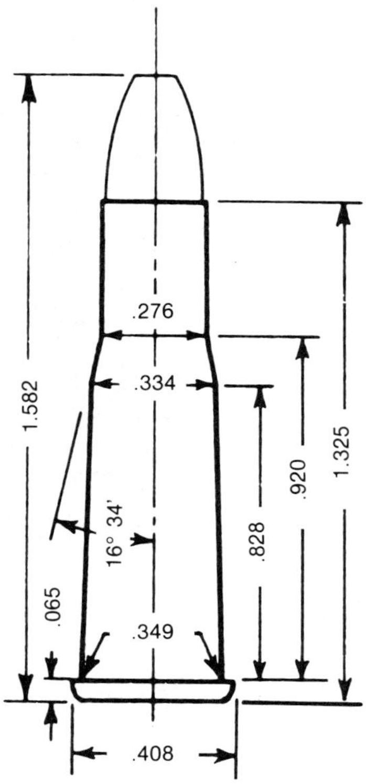

TEST COMPONENTS:

Cases .... Winchester/Western
Trim-to Length .... 1.312″
Primers .... Winchester 6½-116
Primer Size .... Small Rifle
Lyman Shell Holder .... No. 10
Cast Bullets Used .... (size to .257″ dia.)
*Gas Check Bullets *#257420, 68 gr.
*#257312, 88 gr.

TEST SPECIFICATIONS:
(Velocity)

Firearm Used .... Winchester Model 92
Barrel Length .... 24″
Twist .... 1.14″
Groove Diameter .... .256″

COMMENTS:

Use only round, or flat nose bullets, in those rifles which have tubular magazines. Caution: these loads are not intended for rifles which were designed for black powder cartridges.

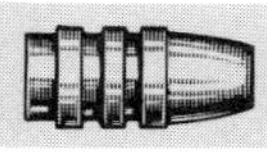

#257420
68 gr., (#2 Alloy) 1.560″ OAL

Powder	Sug. Starting Grains	Velocity F.P.S.	Pressure C.U.P.	Max. Grains	Velocity F.P.S.	Pressure C.U.P.
Red Dot	3.5	1320	—	4.5	1560	—
700X	3.5	1345	—	4.5	1580	—
Green Dot	4.0	1400	—	5.0	1665	—
PB	4.0	1360	—	5.0	1605	—
Unique	4.5	1455	—	6.0	1805	—
SR-7625	4.5	1480	—	5.5	1700	—

#257312
88 gr., (#2 Alloy) 1.555″ OAL

Powder	Sug. Starting Grains	Velocity F.P.S.	Pressure C.U.P.	Max. Grains	Velocity F.P.S.	Pressure C.U.P.
Red Dot	3.2	1125	—	4.2	1355	—
700X	3.2	1160	—	4.0	1320	—
Green Dot	3.7	1220	—	4.7	1425	—
PB	3.6	1175	—	4.5	1345	—
Unique	4.2	1270	—	5.5	1555	—
SR-7625	3.9	1230	—	4.7	1390	—

Note: Loads shown in shaded panels are maximum.

.256 Winchester Magnum

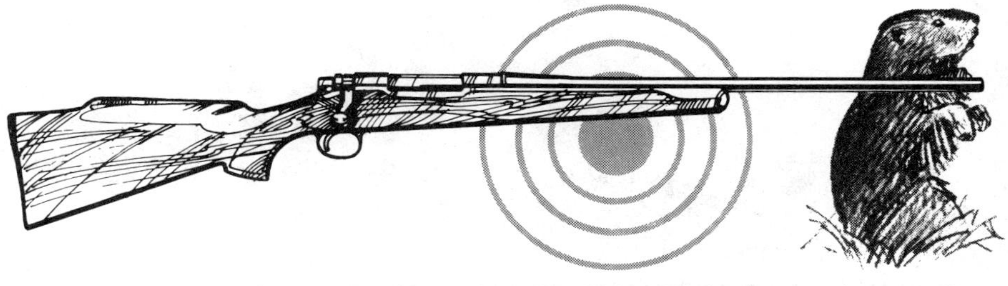

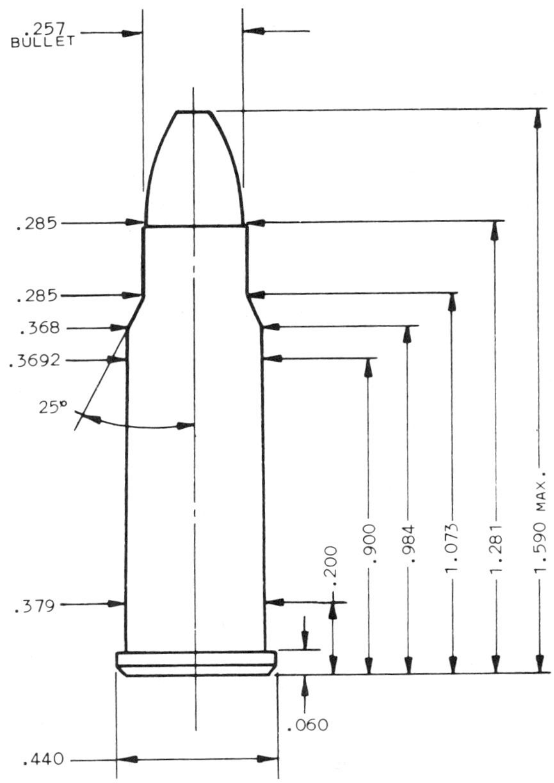

TEST COMPONENTS:

Cases .... Winchester/Western
Trim-to Length .... 1.275″
Primers .... Remington 6½
Primer Size .... Small Rifle
Lyman Shell Holder .... No. 1
Cast Bullets Used .... (size to .257″ dia.)
*Gas Check Bullets *#257420, 68 gr.
*#257312, 88 gr.

TEST SPECIFICATIONS:
(Velocity)

Firearm Used .... Marlin Model 62
Barrel Length .... 24″
Twist .... 1-14″
Groove Diameter .... .257″

COMMENTS:

Many of the rifles chambered for this cartridge have slightly oversize chambers which create a problem for the reloader. These chambers cause excessive case stretching which makes it impossible to resize brass after two or three firings.

The small case capacity of the cartridge limits suitable reloading powders to a very few. Ballistics and accuracy, however, are good.

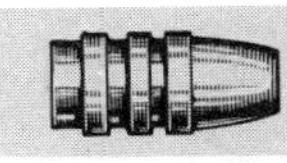

#257420
68 gr., (#2 Alloy) 1.590″ OAL, Max.

Powder	Sug. Starting Grains	Velocity F.P.S.	Pressure C.U.P.	Max. Grains	Velocity F.P.S.	Pressure C.U.P.
Unique	5.0	1669	—	6.0	1883	—
2400	9.0	1745	—	11.0	2105	—
IMR-4227	10.0	1779	—	12.0	2096	—

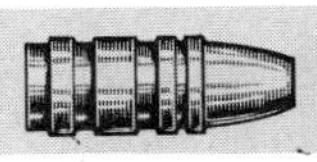

#257312
88 gr., (#2 Alloy) 1.590″ OAL, Max.

Powder	Sug. Starting Grains	Velocity F.P.S.	Pressure C.U.P.	Max. Grains	Velocity F.P.S.	Pressure C.U.P.
Unique	5.0	1492	—	6.0	1697	—
2400	10.0	1808	—	12.0	2123	—
IMR-4227	11.0	1824	—	13.0	2123	—

Note: Loads shown in shaded panels are maximum.

.250 Savage

(.250/3000 Savage)

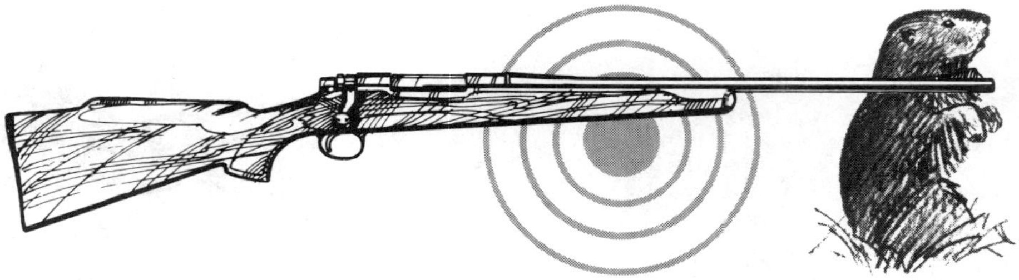

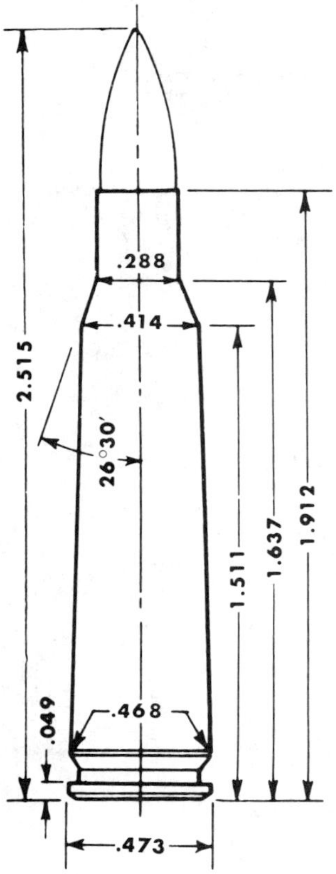

TEST COMPONENTS:

Cases .... Norma
Trim-to Length .... 1.902″
Primers .... Remington 9½
Primer Size .... Large Rifle
Lyman Shell Holder .... No. 2
Cast Bullets Used .... (size to .257″ dia.)
*Gas Check Bullets .... *#257463, 75 gr.
*#257312, 88 gr.
*#257418, 105 gr.
*#257325, 112 gr.

TEST SPECIFICATIONS:
(Velocity)

Firearm Used .... Savage
Barrel Length .... 20″
Twist .... 1-14″
Groove Diameter .... .257″

#257463
75 gr., (#2 Alloy) 2.225″ OAL

Powder	Sug. Starting Grains	Velocity F.P.S.	Pressure C.U.P.	Max. Grains	Velocity F.P.S.	Pressure C.U.P.
Red Dot	7.0	1585	—	10.0	2000	—
700X	7.0	1595	—	10.0	2015	—
Green Dot	7.5	1615	—	10.5	2015	—
PB	8.0	1600	—	10.5	1945	—
Unique	8.0	1625	—	11.0	2030	—
SR-7625	8.5	1625	—	11.0	1980	—

COMMENTS:

A good many rifles which are chambered for this cartridge have 1-10″ twist barrels. In these rifles, bullets less than 75 grains will seldom give accuracy.

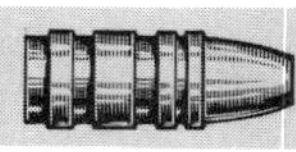

#257312
88 gr., (#2 Alloy) 2.205″ OAL

Powder	Sug. Starting Grains	Velocity F.P.S.	Pressure C.U.P.	Max. Grains	Velocity F.P.S.	Pressure C.U.P.
Red Dot	7.0	1510	—	9.5	1800	—
700X	6.5	1460	—	9.0	1750	—
Green Dot	7.0	1495	—	10.0	1835	—
PB	7.5	1500	—	10.0	1770	—
Unique	7.5	1525	—	10.5	1865	—
SR-7625	8.0	1550	—	10.5	1815	—

Note: Loads shown in shaded panels are maximum.

(.250/3000 Savage) (Continued)

#257418
105 gr., (#2 Alloy) 2.417″ OAL

Powder	Sug. Starting Grains	Velocity F.P.S.	Pressure C.U.P.	Max. Grains	Velocity F.P.S.	Pressure C.U.P.
Red Dot	6.0	1335	—	9.0	1665	—
700X	6.0	1345	—	8.5	1625	—
Green Dot	6.5	1375	—	9.5	1720	—
PB	7.0	1375	—	9.5	1645	—
Unique	7.0	1420	—	10.0	1745	—
SR-7625	7.5	1435	—	10.0	1680	—

#257325
112 gr., (#2 Alloy) 2.375″ OAL

Powder	Sug. Starting Grains	Velocity F.P.S.	Pressure C.U.P.	Max. Grains	Velocity F.P.S.	Pressure C.U.P.
Red Dot	6.0	1255	—	8.5	1515	—
700X	5.5	1195	—	8.0	1475	—
Green Dot	6.0	1230	—	9.0	1560	—
PB	6.5	1235	—	9.0	1495	—
Unique	6.5	1270	—	9.5	1590	—
SR-7625	7.0	1280	—	9.5	1530	—

Note: Loads shown in shaded panels are maximum.

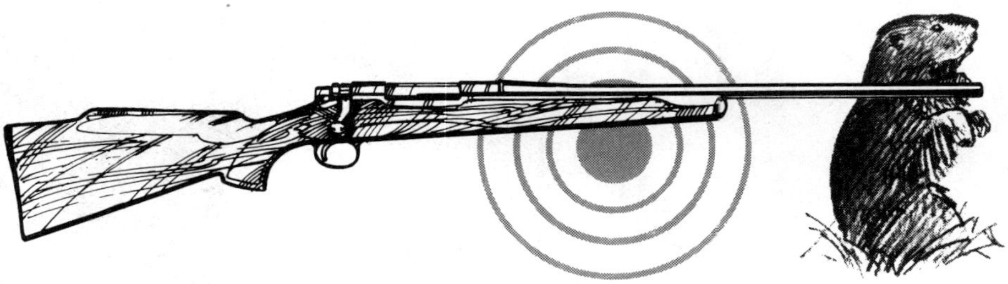

.257 Roberts

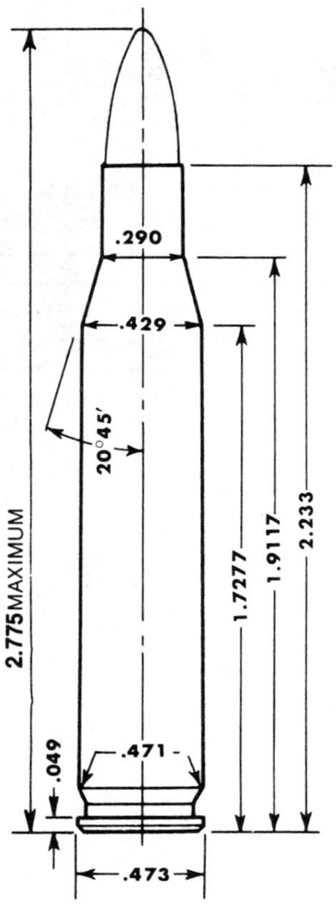

TEST COMPONENTS:

Cases .... Remington Peters
Trim-to Length .... 2.223"
Primers .... Remington 9½
Primer Size .... Large Rifle
Lyman Shell Holder .... No. 8 or 2
Cast Bullets Used .... (size to .257" dia.)
*Gas Check Bullets .... *#257463, 75 gr.
*#257312, 88 gr.
*#257418, 105 gr.
*#257325, 112 gr.

TEST SPECIFICATIONS:
(Velocity)

Firearm Used .... Ruger Model 77
Barrel Length .... 24"
Twist .... 1-12"
Groove Diameter .... .257"

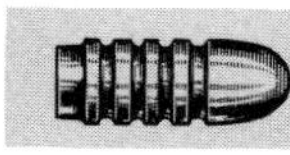

#257463
75 gr., (#2 Alloy) 2.555" OAL

Powder	Sug. Starting Grains	Velocity F.P.S.	Pressure C.U.P.	Max. Grains	Velocity F.P.S.	Pressure C.U.P.
Red Dot	7.0	1525	—	12.0	2115	—
700X	7.0	1545	—	12.0	2125	—
Green Dot	8.0	1605	—	13.0	2170	—
PB	8.0	1555	—	13.0	2090	—
Unique	9.0	1680	—	14.0	2230	—
SR-7625	9.0	1650	—	13.5	2135	—
630	12.0	1770	—	13.0	2300	—

#257312
88 gr., (#2 Alloy) 2.620" OAL

Powder	Sug. Starting Grains	Velocity F.P.S.	Pressure C.U.P.	Max. Grains	Velocity F.P.S.	Pressure C.U.P.
Red Dot	7.0	1430	—	12.0	1960	—
700X	7.0	1435	—	11.5	1925	—
Green Dot	8.0	1515	—	12.5	1985	—
PB	8.0	1450	—	12.5	1905	—
Unique	9.0	1590	—	13.5	2040	—
SR-7625	9.0	1535	—	13.0	1940	—
630	12.0	1705	—	18.0	2190	—

Note: Loads shown in shaded panels are maximum.

COMMENTS:

Most commericial rifles for this cartridge used a short Magazine which limits overall cartridge length to 2.775". Prevalent also was the use of a long chamber throat which gives best accuracy results when the bullet is seated just short of touching the rifling. This combination of a short magazine and long throat does not allow for best performance.

Cartridge brass varies in the contour of the extractor cut and hence one shell holder will not always fit all cases. The use of a Lyman No. 8 shell holder is suggested wherever possible. A No. 2 shell holder can be used whenever a No. 8 proves unsatisfactory.

#257418
105 gr., (#2 Alloy) 2.760" OAL

Powder	Sug. Starting Grains	Velocity F.P.S.	Pressure C.U.P.	Max. Grains	Velocity F.P.S.	Pressure C.U.P.
Red Dot	7.0	1365	—	11.5	1840	—
700X	7.0	1380	—	11.0	1820	—
Green Dot	8.0	1445	—	12.0	1870	—
PB	8.0	1395	—	12.0	1800	—
Unique	9.0	1510	—	13.0	1925	—
SR-7625	9.0	1470	—	12.0	1800	—
630	12.0	1635	—	18.0	2090	—

#257325
112 gr., (#2 Alloy) 2.710" OAL

Powder	Sug. Starting Grains	Velocity F.P.S.	Pressure C.U.P.	Max. Grains	Velocity F.P.S.	Pressure C.U.P.
Red Dot	6.5	1240	—	11.0	1700	—
700X	6.5	1245	—	10.5	1665	—
Green Dot	7.5	1325	—	11.5	1730	—
PB	7.5	1255	—	11.5	1655	—
Unique	8.5	1400	—	12.5	1785	—
SR-7625	8.5	1340	—	11.5	1645	—
630	11.0	1500	—	17.0	1970	—

Note: Loads shown in shaded panels are maximum.

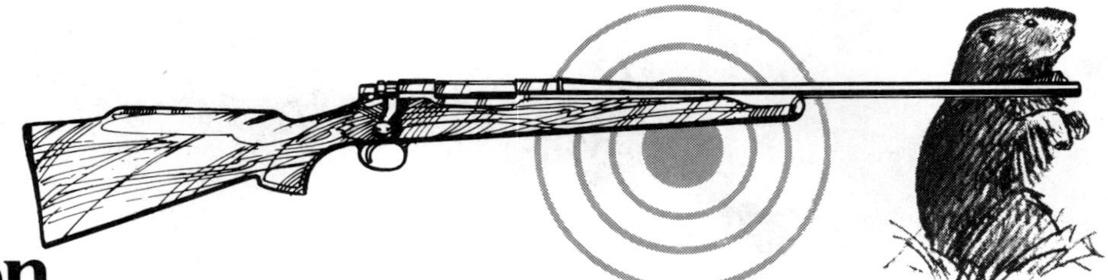

.25/06 Remington

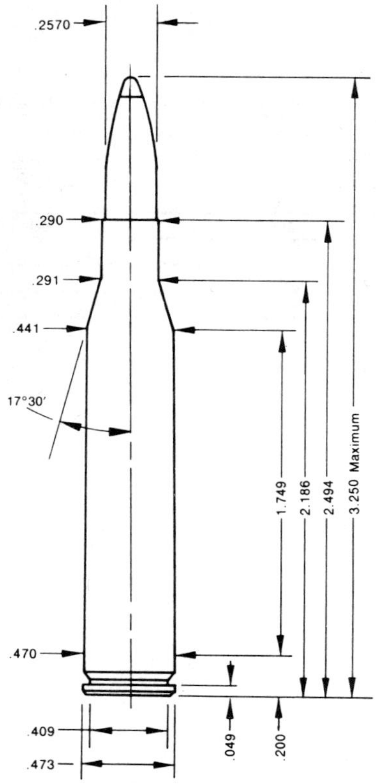

TEST COMPONENTS:

Cases .... Remington
Trim-to Length .... 2.484″
Primers .... Winchester 8½-120 & CCI 200
Primer Size .... Large Rifle
Lyman Shell Holder .... No. 2
Cast Bullets Used .... (size to .257″ dia.)
*Gas Check Bullets *#257420, 68 gr.
*#257463, 75 gr.
*#257312, 88 gr.
*#257464, 90 gr.
*#257418, 105 gr.
*#257325, 112 gr.

TEST SPECIFICATIONS:
(Velocity & Pressure)

Firearm Used .... Universal Receiver; used as a velocity barrel where no pressures are listed
Barrel Length .... 26″
Twist .... 1-10″
Groove Diameter .... .257″

COMMENTS:

Accuracy above 2000 f.p.s. proved to be erratic. Best results were obtained with bullet #257312 at 1600 f.p.s. and bullet #257418 at 1800 f.p.s.

#257420
68 gr., (#2 Alloy) 2.897″ OAL

Powder	Sug. Starting Grains	Velocity F.P.S.	Pressure C.U.P.	Max. Grains	Velocity F.P.S.	Pressure C.U.P.
Red Dot	11.0	2087	—	14.0	2379	39,400
700X	10.0	1989	—	13.0	2269	38,100
Green Dot	12.0	2149	—	15.5	2400	41,600
PB	11.0	1967	—	14.5	2304	39,400
SR-7625	11.0	1978	—	15.0	2322	41,600
SR-4756	12.0	2008	—	16.0	2389	40,700
**630	18.5	2160	14,500	24.0	2622	25,400
**2400	19.0	2123	15,200	27.0	2875	36,100
**IMR-4227	21.0	2196	15,300	25.5	2691	27,600
**IMR-4198	20.0	2073	14,700	29.5	2635	22,300
**H4895	29.5	2193	14,600	36.9	2685	24,100

Note: Loads shown in shaded panels are maximum.
**Signifies use of CCI 200 primers.

.25/06 Remington (Continued)

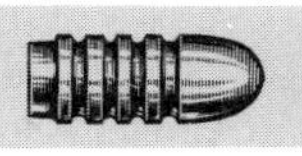

#257463
75 gr., (#2 Alloy) 3.250" OAL, Max.

Powder	Sug. Starting Grains	Velocity F.P.S.	Pressure C.U.P.	Max. Grains	Velocity F.P.S.	Pressure C.U.P.
Red Dot	11.0	1997	—	12.5	2171	33,000
700X	10.0	1898	—	13.0	2192	36,000
Green Dot	12.0	2080	28,800	13.5	2226	36,900
PB	11.0	1907	—	14.0	2174	37,700
SR-7625	11.0	1867	—	14.5	2220	41,600
SR-4756	12.0	1937	—	15.5	2265	36,400

#257312
88 gr., (#2 Alloy) 2.867" OAL

Powder	Sug. Starting Grains	Velocity F.P.S.	Pressure C.U.P.	Max. Grains	Velocity F.P.S.	Pressure C.U.P.
Red Dot	11.0	1856	31,000	12.5	1970	35,500
700X	10.0	1785	27,000	12.5	2006	38,600
Green Dot	12.0	1906	33,500	13.0	1992	36,000
PB	11.0	1784	—	13.5	1978	38,600
SR-7625	11.0	1747	—	14.0	2011	41,100
SR-4756	12.0	1816	—	15.0	2040	35,500
**630	18.0	2164	21,600	29.0	2874	44,200
**2400	20.7	2179	23,500	29.0	2838	47,100
**SR-4759	20.5	2185	21,600	31.5	2897	48,600
**748	29.0	2150	15,700	43.5	2840	34,900

#257464
90 gr., (#2 Alloy) 2.911" OAL

Powder	Sug. Starting Grains	Velocity F.P.S.	Pressure C.U.P.	Max. Grains	Velocity F.P.S.	Pressure C.U.P.
**630	20.0	2184	22,100	29.0	2826	45,600
**2400	20.0	2162	22,000	29.0	2781	47,800
**SR-4759	22.2	2193	22,100	32.8	2859	47,900
IMR-4227	22.0	2225	23,600	30.0	2753	46,300
**IMR-4198	24.5	2182	19,000	33.0	2770	35,900
**748	30.0	2127	16,800	46.3	2960	39,700

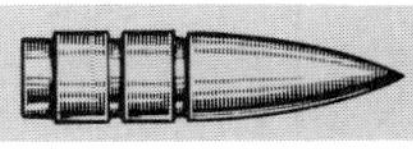

#257418
105 gr., (#2 Alloy) 2.912" OAL

Powder	Sug. Starting Grains	Velocity F.P.S.	Pressure C.U.P.	Max. Grains	Velocity F.P.S.	Pressure C.U.P.
Red Dot	10.5	1741	25,200	12.5	1904	35,000
700X	10.0	1695	—	12.0	1876	36,000
Green Dot	11.5	1806	28,200	13.0	1926	36,000
PB	11.0	1709	—	13.0	1865	35,500
SR-7625	11.0	1689	25,200	13.5	1899	37,300
SR-4756	12.0	1730	—	15.0	1998	37,300
**630	20.0	2124	—	28.0	2623	48,200
**2400	22.0	2152	26,300	30.8	2677	47,400
**SR-4759	24.0	2171	29,800	32.0	2652	48,700
**IMR-4198	25.0	2125	27,000	38.0	2845	46,200

#257325
112 gr., (#2 Alloy) 2.984" OAL

Powder	Sug. Starting Grains	Velocity F.P.S.	Pressure C.U.P.	Max. Grains	Velocity F.P.S.	Pressure C.U.P.
**630	21.0	2149	32,000	28.0	2578	49,000
**2400	23.5	2164	31,300	32.0	2660	49,000
**SR-4759	24.5	2184	30,400	32.0	2575	48,000
**IMR-4198	27.5	2174	28,100	38.0	2739	43,600

Note: Loads shown in shaded panels are maximum.
**Signifies use of CCI 200 primers.

.257 Weatherby Magnum

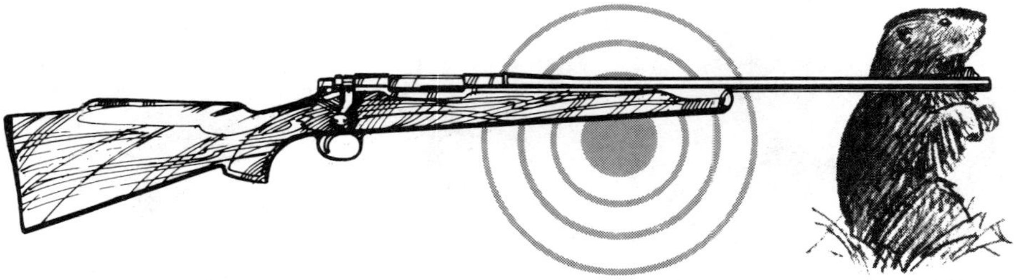

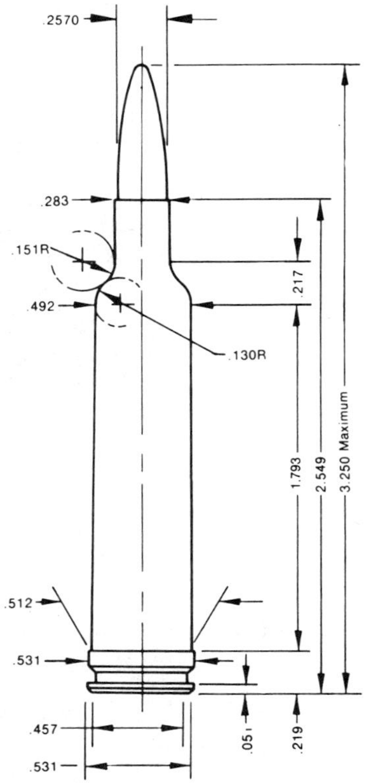

TEST COMPONENTS:

Cases .... Weatherby
Trim-to Length .... 2.540″
Primers .... Remington 9½
Primer Size .... Large Rifle
Lyman Shell Holder .... No. 13
Cast Bullets Used .... (size to .257″ dia.)
*Gas Check Bullets .... *#257312, 88 gr.
*#257418, 105 gr.

TEST SPECIFICATIONS:
(Velocity)

Firearm Used .... Weatherby Mark V
Barrel Length .... 26″
Twist .... 1-12″
Groove Diameter .... .257″

COMMENTS:

The data listed for this cartridge was obtained in a Weatherby rifle and is intended for Weatherby rifles only. The free-boring constructed into these firearms allow higher velocities at safe working pressures. For custom rifles which are not freebored, maximum loads should be reduced a full 5%. Even then, they should be approached with caution.

Do not use the magnum primers with cast bullet data. Cast bullet data were obtained with Remington 9½ primers.

#257312
88 gr., (#2 Alloy) 3.250″ OAL, Max.

Powder	Sug. Starting Grains	Velocity F.P.S.	Pressure C.U.P.	Max. Grains	Velocity F.P.S.	Pressure C.U.P.
Unique	13.0	1915	—	19.0	2386	—
2400	19.0	1984	—	24.0	2309	—

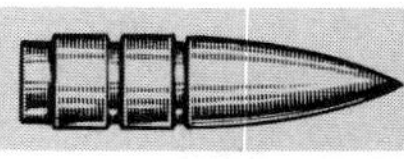

#257418
105 gr., (#2 Alloy) 3.250″ OAL, Max.

Powder	Sug. Starting Grains	Velocity F.P.S.	Pressure C.U.P.	Max. Grains	Velocity F.P.S.	Pressure C.U.P.
Unique	12.0	1763	—	16.0	2079	—
2400	18.0	1808	—	23.0	2145	—

Note: Loads shown in shaded panels are maximum.

6.5 Japanese

(6.5 x 50mm Arisaka)

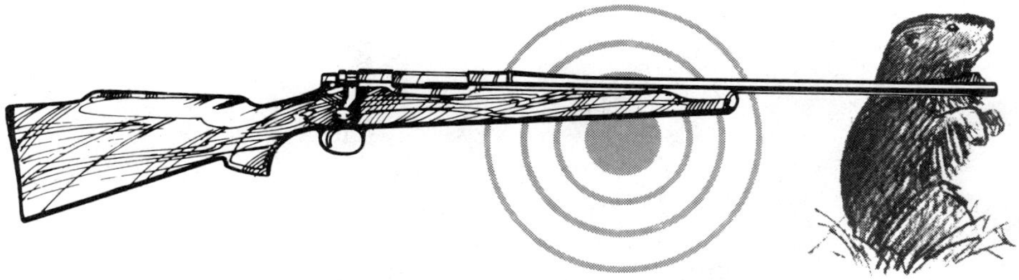

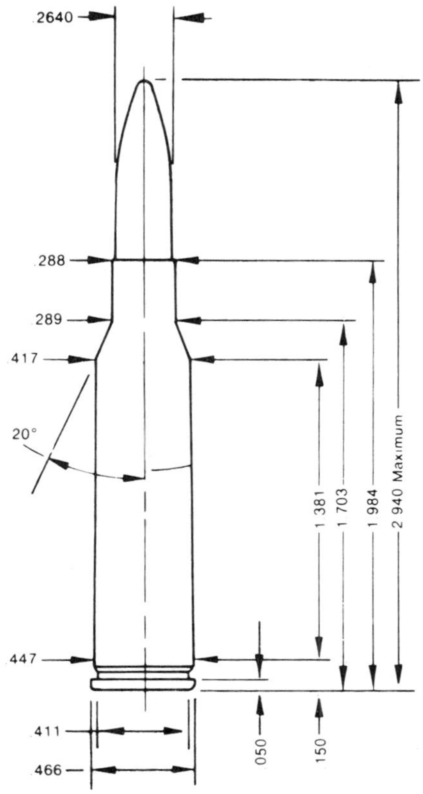

TEST COMPONENTS:

Cases .... Norma
Trim-to Length .... 1.978″
Primers .... Remington 9½
Primer Size .... Large Rifle
Lyman Shell Holder .... No. 5
Cast Bullets Used .... (size to .264″ to .266″ dia.)
*Gas Check Bullets .... *#266305, 103 gr.
#266324, 119 gr.
*#266455, 129 gr.
*#266469, 143 gr.

TEST SPECIFICATIONS:
(Velocity)

Firearm Used .... Japanese Service Carbine
Barrel Length .... 19″
Twist .... 1-9″
Groove Diameter .... .267″

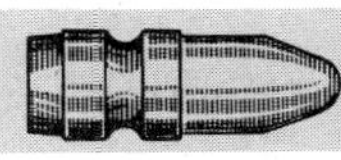

#266305
103 gr., (#2 Alloy) 2.940″ OAL, Max.

Powder	Sug. Starting Grains	Velocity F.P.S.	Pressure C.U.P.	Max. Grains	Velocity F.P.S.	Pressure C.U.P.
2400	16.5	1824	—	19.0	2044	—
IMR-4227	21.0	2028	—	23.0	2207	—
IMR-4198	24.0	2114	—	27.0	2352	—
IMR-4895	28.0	2016	—	33.0	2375	—

COMMENTS:

Norma brass was used exclusively for our testing of this cartridge. Bullets weighing 119 grains or more gave the best accuracy. A wide variation in groove diameter is common with these military rifles and we recommend that you slug your bore before reloading.

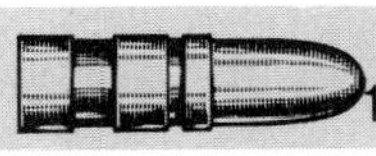

#266324
119 gr., (#2 Alloy) 2.940″ OAL, Max.

Powder	Sug. Starting Grains	Velocity F.P.S.	Pressure C.U.P.	Max. Grains	Velocity F.P.S.	Pressure C.U.P.
2400	15.0	1615	—	18.0	1868	—
IMR-4227	19.5	1893	—	22.0	2083	—
IMR-4198	22.0	1919	—	24.0	2057	—
IMR-4895	28.0	1976	—	31.0	2092	—

Note: Loads shown in shaded panels are maximum.

#266455
129 gr., (#2 Alloy) 2.940" OAL, Max.

Powder	Sug. Starting Grains	Velocity F.P.S.	Pressure C.U.P.	Max. Grains	Velocity F.P.S.	Pressure C.U.P.
2400	14.0	1488	—	19.0	1960	—
IMR-4227	19.0	1818	—	21.0	1988	—
IMR-4198	23.0	1980	—	27.0	2145	—
IMR-4895	27.0	1996	—	31.0	2242	—

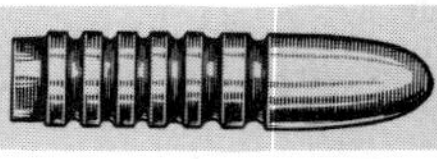

#266469
143 gr., (#2 Alloy) 2.940" OAL, Max.

Powder	Sug. Starting Grains	Velocity F.P.S.	Pressure C.U.P.	Max. Grains	Velocity F.P.S.	Pressure C.U.P.
2400	14.0	1515	—	18.0	1845	—
IMR-4227	18.0	1683	—	21.0	1956	—
IMR-4198	22.0	1893	—	26.0	2159	—
IMR-4895	26.0	1858	—	32.0	2272	—

Note: Loads shown in shaded panels are maximum.

6.5 Italian

(6.5 x 52mm Mann. Carcano)

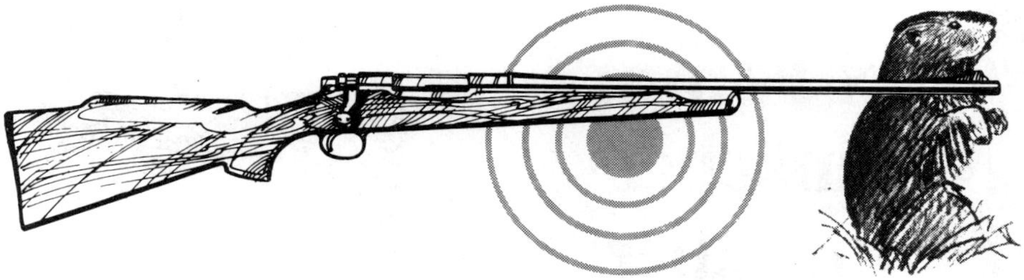

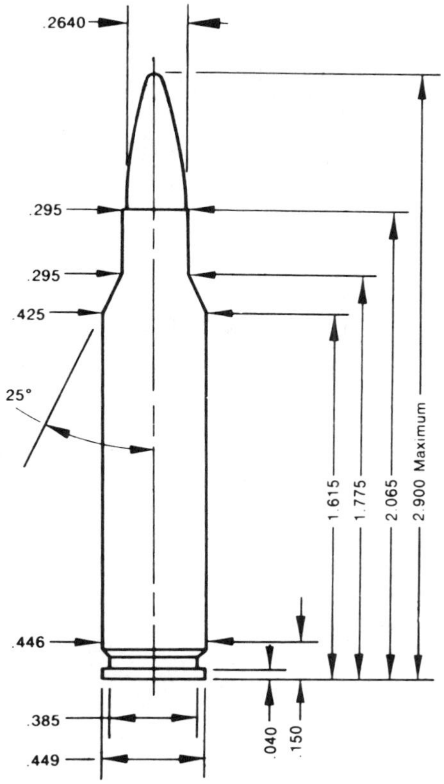

TEST COMPONENTS:

Cases .... Norma
Trim-to Length .... 2.055″
Primers .... Remington 9½
Primer Size .... Large Rifle
Lyman Shell Holder .... No. 28
Cast Bullets Used .... (size to .264″ to .266″ dia.)
*Gas Check Bullets .... #266324, 119 gr.
*#266455, 129 gr.
*#266469, 143 gr.

TEST SPECIFICATIONS:
(Velocity)

Firearm Used .... Italian Service Carbine
Barrel Length .... 21″
Twist .... Gain Twist
Groove Diameter .... .267″

COMMENTS:

Due to the wide variations in groove diameters which exist in these surplus military rifles, we suggest that you proceed with caution. Slug your bore to determine the correct bullet diameter for your rifle. See the reference section for information on slugging your bore. If your groove diameter is larger than .264″, you could experience gas blow-by (gas from breech due to low pressure) when using the listed starting loads with the standard bullet size.

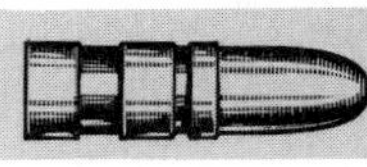

#266324
119 gr., (#2 Alloy) 2.900″ OAL, Max.

Powder	Sug. Starting Grains	Velocity F.P.S.	Pressure C.U.P.	Max. Grains	Velocity F.P.S.	Pressure C.U.P.
Unique	9.0	1445	—	11.0	1615	—
2400	21.0	1960	—	23.0	2114	—

#266455
129 gr., (#2 Alloy) 2.900″ OAL, Max.

Powder	Sug. Starting Grains	Velocity F.P.S.	Pressure C.U.P.	Max. Grains	Velocity F.P.S.	Pressure C.U.P.
Unique	8.0	1351	—	10.0	1529	—
2400	19.0	1901	—	21.0	2040	—

#266469
143 gr., (#2 Alloy) 2.900″ OAL, Max.

Powder	Sug. Starting Grains	Velocity F.P.S.	Pressure C.U.P.	Max. Grains	Velocity F.P.S.	Pressure C.U.P.
Unique	8.0	1296	—	10.0	1481	—
2400	15.0	1545	—	18.0	1763	—

Note: Loads shown in shaded panels are maximum.

6.5 x 54mm Mannlicher-Schoenauer

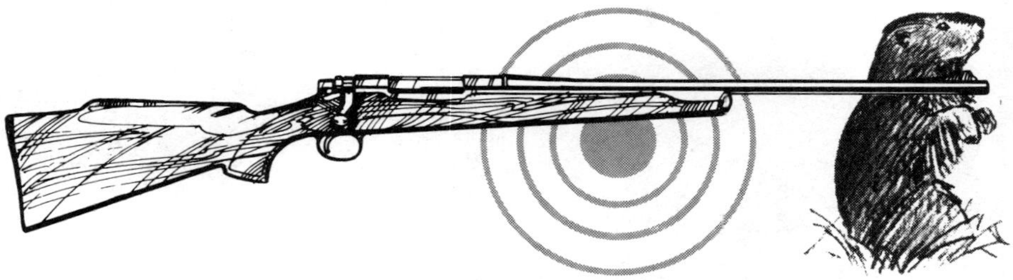

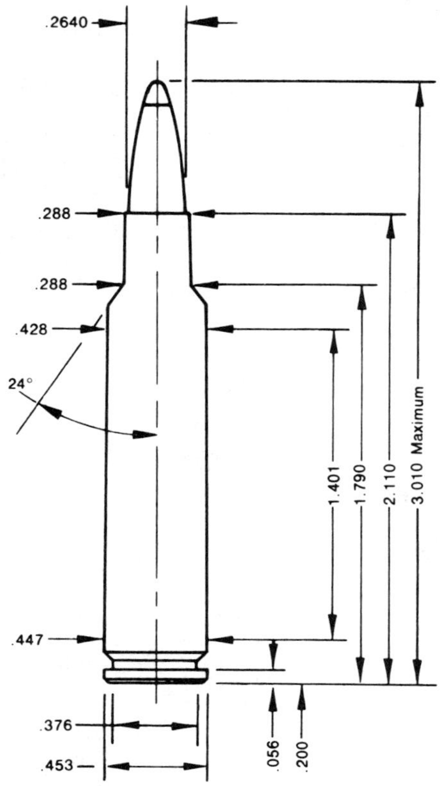

TEST COMPONENTS:

Cases	Norma
Trim-to Length	2.100″
Primers	Remington 9½
Primer Size	Large Rifle
Lyman Shell Holder	No. 28
Cast Bullets Used	(size to .264″ dia.)
*Gas Check Bullets	*#266305, 103 gr.
	#266324, 119 gr.
	*#266455, 129 gr.
	*#266469, 143 gr.

TEST SPECIFICATIONS:
(Velocity)

Firearm Used	Custom Mauser
Barrel Length	20″
Twist	1-7½″
Groove Diameter	.265″

COMMENTS:

More than half of the rifles which we have encountered in this caliber show signs of excessive headspace. Before using a 6.5 x 54mm rifle, we recommend that its headspace be checked by a competent gunsmith.

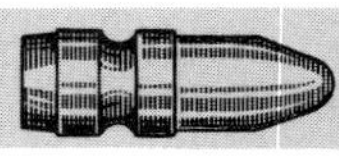

#266305
103 gr., (#2 Alloy) 3.010″ OAL, Max.

Powder	Sug. Starting Grains	Velocity F.P.S.	Pressure C.U.P.	Max. Grains	Velocity F.P.S.	Pressure C.U.P.
2400	12.0	1535	—	19.0	2114	—
IMR 4227	13.0	1526	—	22.0	2257	—
IMR 4198	16.0	1655	—	24.0	2183	—
IMR 4895	20.0	1524	—	31.0	2232	—

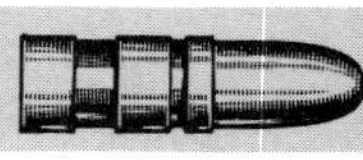

#266324
119 gr., (#2 Alloy) 3.010″ OAL, Max.

Powder	Sug. Starting Grains	Velocity F.P.S.	Pressure C.U.P.	Max. Grains	Velocity F.P.S.	Pressure C.U.P.
2400	12.0	1466	—	19.0	2004	—
IMR 4227	13.0	1508	—	21.5	2092	—
IMR 4198	16.0	1592	—	24.0	2127	—
IMR 4895	20.0	1519	—	31.0	2178	—

Note: Loads shown in shaded panels are maximum.

6.5 Mannlicher-Schoenauer (Continued)

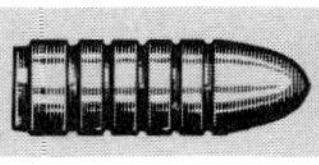

#266455
129 gr., (#2 Alloy) 3.010" OAL, Max.

Powder	Sug. Starting Grains	Velocity F.P.S.	Pressure C.U.P.	Max. Grains	Velocity F.P.S.	Pressure C.U.P.
2400	12.0	1466	—	19.0	1996	—
IMR 4227	13.0	1474	—	21.5	2074	—
IMR 4198	16.0	1639	—	23.0	2066	—
IMR 4895	20.0	1602	—	31.0	2257	—

#266469
143 gr., (#2 Alloy) 3.010" OAL, Max.

Powder	Sug. Starting Grains	Velocity F.P.S.	Pressure C.U.P.	Max. Grains	Velocity F.P.S.	Pressure C.U.P.
2400	12.0	1390	—	19.0	1923	—
IMR 4227	13.0	1382	—	21.0	1953	—
IMR 4198	16.0	1510	—	22.0	1915	—
IMR 4895	19.0	1400	—	30.0	2132	—

Note: Loads shown in shaded panels are maximum.

6.5mm Swedish Mauser

(6.5 x 55)

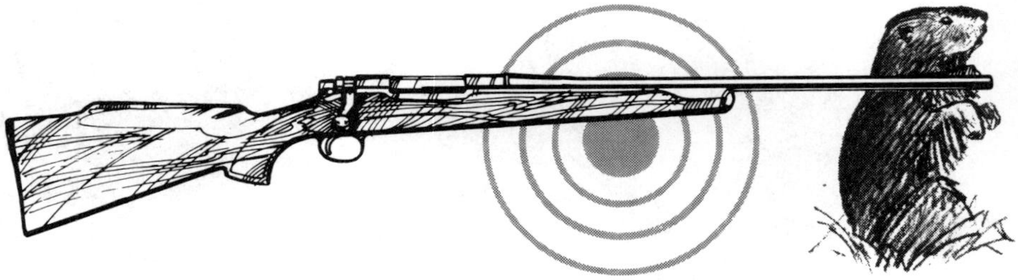

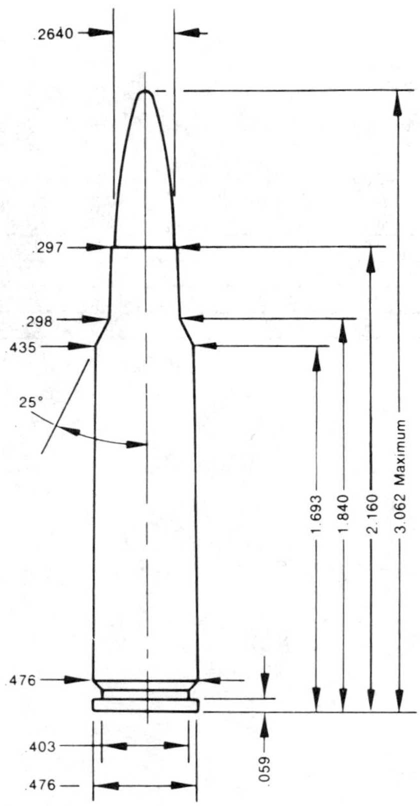

TEST COMPONENTS:

Cases .... Norma
Trim-to Length .... 2.150″
Primers .... Remington 9½
Primer Size .... Large Rifle
Lyman Shell Holder .... No. 27
Cast Bullets Used .... (size to .266″ dia.)
*Gas Check Bullets .... #266324, 119 gr.
*#266455, 129 gr.
*#266469, 143 gr.

TEST SPECIFICATIONS: (Velocity)

Firearm Used .... HVA Bolt Action
Barrel Length .... 23½″
Twist .... 1-7½″
Groove Diameter .... .266″

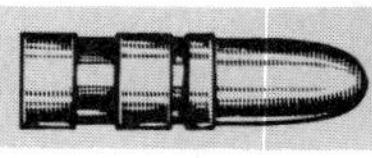

#266324
119 gr., (#2 Alloy) 2.625″ OAL.

Powder	Sug. Starting Grains	Velocity F.P.S.	Pressure C.U.P.	Max. Grains	Velocity F.P.S.	Pressure C.U.P.
Red Dot	7.0	1235	—	10.0	1575	—
700X	7.0	1240	—	10.0	1570	—
Green Dot	8.0	1320	—	11.0	1640	—
PB	8.0	1240	—	12.0	1640	—
Unique	9.0	1390	—	12.0	1690	—
SR-7625	9.0	1295	—	13.0	1685	—
630	10.0	1365	—	16.0	1820	—

COMMENTS:

Due to the wide variations in groove diameters in firearms chambered for the 6.5 x 55 cartridge, we suggest you slug your bore to determine the correct bullet diameter for your rifle.

#266455
129 gr., (#2 Alloy) 2.710″ OAL.

Powder	Sug. Starting Grains	Velocity F.P.S.	Pressure C.U.P.	Max. Grains	Velocity F.P.S.	Pressure C.U.P.
Red Dot	7.0	1175	—	10.0	1515	—
700X	7.0	1200	—	10.0	1525	—
Green Dot	8.0	1270	—	11.0	1585	—
PB	8.0	1215	—	12.0	1580	—
Unique	9.0	1340	—	12.0	1640	—
SR-7625	9.0	1290	—	13.0	1645	—
630	10.0	1315	—	16.0	1800	—

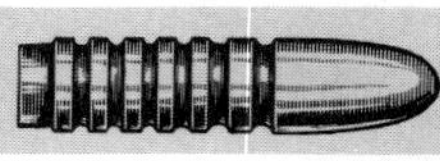

#266469
143 gr., (#2 Alloy) 2.825″ OAL.

Powder	Sug. Starting Grains	Velocity F.P.S.	Pressure C.U.P.	Max. Grains	Velocity F.P.S.	Pressure C.U.P.
Red Dot	7.0	1135	—	9.5	1335	—
700X	7.0	1135	—	9.5	1330	—
Green Dot	8.0	1205	—	10.5	1465	—
PB	8.0	1145	—	11.5	1465	—
Unique	9.0	1275	—	11.5	1525	—
SR-7625	9.0	1220	—	12.5	1525	—
630	10.0	1300	—	15.5	1710	—

Note: Loads shown in shaded panels are maximum.

6.5mm Remington Magnum

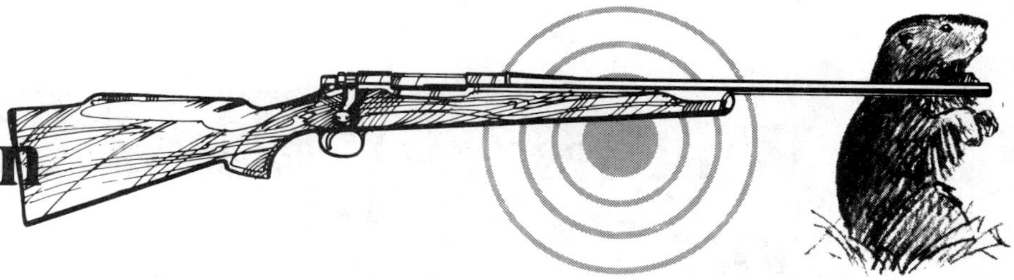

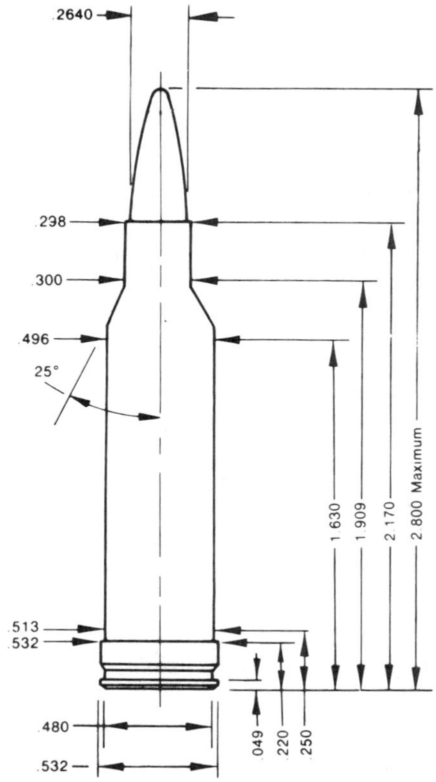

TEST COMPONENTS:

Cases	Remington Peters
Trim-to Length	2.160″
Primers	Remington 9½
Primer Size	Large Rifle
Lyman Shell Holder	No. 13
Cast Bullets Used	(size to .264″ dia.)
*Gas Check Bullets	#266324, 119 gr. *#266455, 129 gr. *#266469, 143 gr.

TEST SPECIFICATIONS:
(Velocity)

Firearm Used	Remington Model 700
Barrel Length	24″
Twist	1-9″
Groove Diameter	.264″

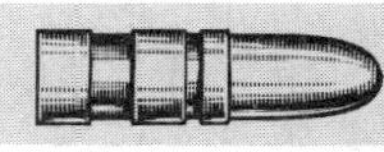

#266324
119 gr., (#2 Alloy) 2.635″ OAL.

Powder	Sug. Starting Grains	Velocity F.P.S.	Pressure C.U.P.	Max. Grains	Velocity F.P.S.	Pressure C.U.P.
Red Dot	9.0	1340	—	14.5	1785	—
700X	9.0	1345	—	14.5	1780	—
Green Dot	10.0	1485	—	15.5	1880	—
PB	10.0	1405	—	15.5	1815	—
Unique	11.0	1545	—	16.5	1930	—
SR-7625	11.0	1470	—	16.5	1860	—

#266455
129 gr., (#2 Alloy) 2.605″ OAL.

Powder	Sug. Starting Grains	Velocity F.P.S.	Pressure C.U.P.	Max. Grains	Velocity F.P.S.	Pressure C.U.P.
Red Dot	9.0	1275	—	14.0	1725	—
700X	9.0	1300	—	14.0	1730	—
Green Dot	10.0	1445	—	15.0	1835	—
PB	10.0	1370	—	15.0	1765	—
Unique	11.0	1500	—	16.0	1885	—
SR-7625	11.0	1435	—	16.0	1810	—

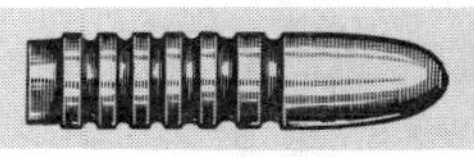

#266469
143 gr., (#2 Alloy) 2.615″ OAL.

Powder	Sug. Starting Grains	Velocity F.P.S.	Pressure C.U.P.	Max. Grains	Velocity F.P.S.	Pressure C.U.P.
Red Dot	8.5	1215	—	13.5	1620	—
700X	8.5	1200	—	13.5	1615	—
Green Dot	9.5	1275	—	14.5	1655	—
PB	9.5	1205	—	14.5	1595	—
Unique	10.5	1390	—	15.5	1755	—
SR-7625	10.5	1335	—	15.5	1700	—

Note: Loads shown in shaded panels are maximum.

.264 Winchester Magnum

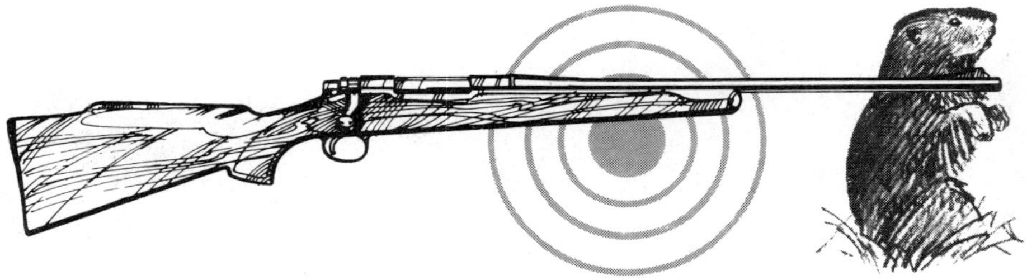

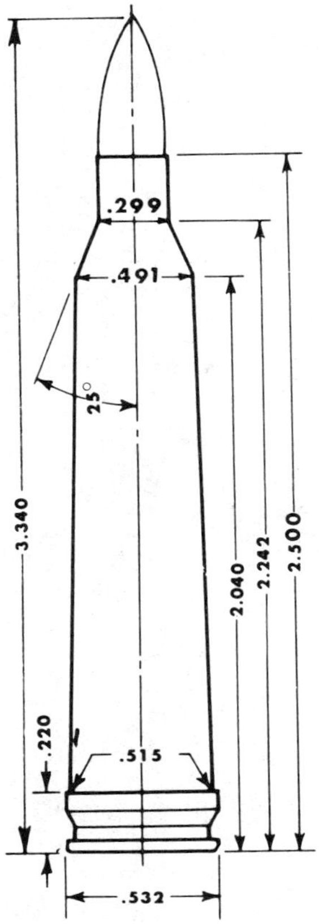

TEST COMPONENTS:

Cases .... Winchester/Western
Trim-to Length .... 2.490″
Primers .... Winchester 8½-120
Primer Size .... Large Rifle
Lyman Shell Holder .... No. 13
Cast Bullets Used .... (size to .264″ dia.)
*Gas Check Bullets #266324, 119 gr.
*#266455, 129 gr..
*#266469, 143 gr.

TEST SPECIFICATIONS:
(Velocity)

Firearm Used .... Universal Receiver Velocity Barrel
Barrel Length .... 24″
Twist .... 1-9″
Groove Diameter .... .264″

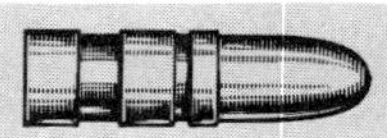

#266324
119 gr., (#2 Alloy) 2.991″ OAL

Powder	Sug. Starting Grains	Velocity F.P.S.	Pressure C.U.P.	Max. Grains	Velocity F.P.S.	Pressure C.U.P.
Red Dot	12.0	1565	—	17.0	1885	—
700X	12.0	1565	—	17.0	1875	—
Green Dot	13.0	1620	—	18.0	1925	—
PB	13.0	1535	—	18.0	1855	—
Unique	14.0	1670	—	19.0	1960	—
SR-7625	14.0	1605	—	19.0	1905	—
630	17.0	1695	—	22.0	1985	—

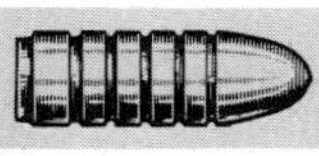

#266455
129 gr., (#2 Alloy) 2.983″ OAL

Powder	Sug. Starting Grains	Velocity F.P.S.	Pressure C.U.P.	Max. Grains	Velocity F.P.S.	Pressure C.U.P.
Red Dot	12.0	1535	—	16.5	1835	—
700X	11.5	1490	—	16.5	1840	—
Green Dot	13.0	1590	—	17.0	1850	—
PB	12.5	1470	—	17.5	1815	—
Unique	13.5	1600	—	18.5	1930	—
SR-7625	13.5	1510	—	18.5	1850	—
630	16.5	1685	—	21.5	1965	—

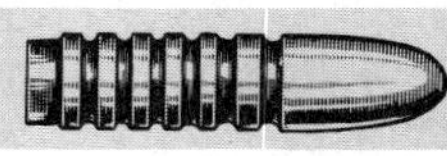

#266469
143 gr., (#2 Alloy) 3.068″ OAL

Powder	Sug. Starting Grains	Velocity F.P.S.	Pressure C.U.P.	Max. Grains	Velocity F.P.S.	Pressure C.U.P.
Red Dot	11.0	1390	—	16.0	1715	—
700X	11.0	1380	—	16.0	1715	—
Green Dot	12.5	1470	—	16.5	1735	—
PB	12.0	1375	—	17.0	1690	—
Unique	13.0	1495	—	18.0	1795	—
SR-7625	13.0	1430	—	18.0	1730	—
630	16.0	1585	—	21.0	1865	—

Note: Loads shown in shaded panels are maximum.

.270 Winchester

(6.9 x 64mm)

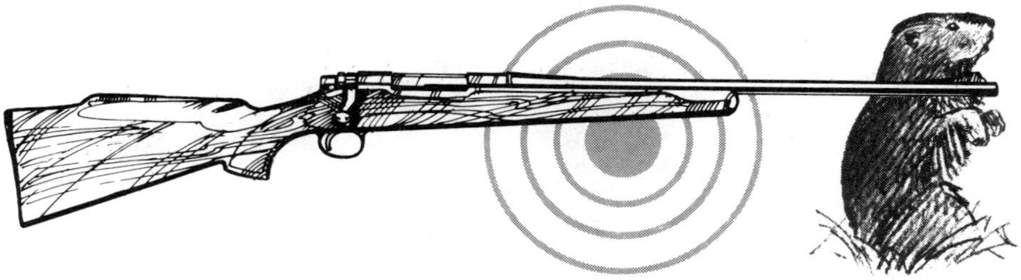

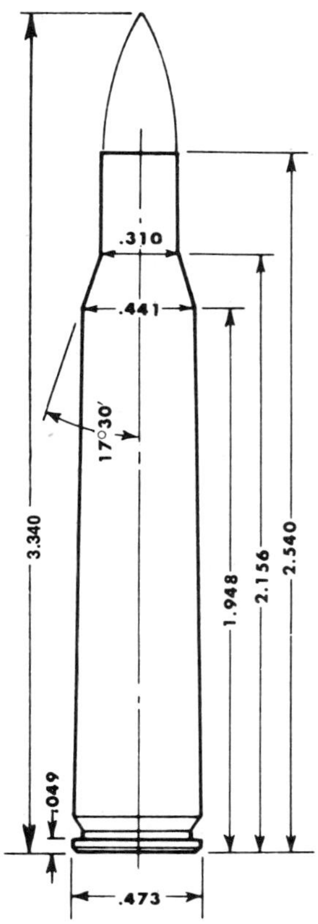

TEST COMPONENTS:

Cases .... Winchester
Trim-to Length .... 2.530″
Primers .... Winchester 8½-120
Primer Size .... Large Rifle
Lyman Shell Holder .... No. 2
Cast Bullets Used .... (size to .277″ dia.)
*Gas Check Bullets *#280468, 122 gr.
*#280473, 124 gr.
*#280412, 136 gr.

TEST SPECIFICATIONS:
(Velocity & Pressure)

Firearm Used .... Universal Receiver
Barrel Length .... 26″
Twist .... 1-10″
Groove Diameter .... .277″

COMMENTS:

The lighter bullets gave the best performance, with #280468 showing the best 50 yard accuracy at 2000 f.p.s.

#280468
122 gr., (#2 Alloy) 3.069″ OAL

Powder	Sug. Starting Grains	Velocity F.P.S.	Pressure C.U.P.	Max. Grains	Velocity F.P.S.	Pressure C.U.P.
Red Dot	12.0	1716	33,500	14.0	1890	40,700
700X	10.0	1555	19,000	13.0	1796	39,400
PB	11.5	1609	27,600	14.3	1822	41,100
Unique	13.0	1836	31,000	15.7	2057	39,900
SR-7625	11.0	1560	22,000	14.0	1784	40,700
SR-4756	12.6	1677	27,000	16.0	1912	41,600
SR-4759	23.0	2189	29,400	31.5	2647	49,000
RX7	25.0	2163	24,300	37.5	2784	47,800
IMR 3031	35.2	2184	20,900	44.0	2892	43,000
748	*36.0	2162	20,200	45.0	2875	42,600

Note: Loads shown in shaded panels are maximum.
*Designates potentially most accurate load.

.270 Winchester (Continued)

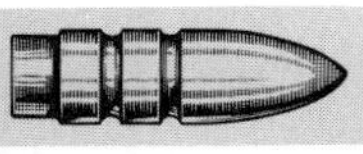

#280473
124 gr., (#2 Alloy) 3.150" OAL

Powder	Sug. Starting Grains	Velocity F.P.S.	Pressure C.U.P.	Max. Grains	Velocity F.P.S.	Pressure C.U.P.
Red Dot	11.0	1587	29,400	13.5	1780	39,400
700X	10.0	1489	25,200	13.0	1729	39,900
Green Dot	11.5	1607	31,500	14.5	1839	40,300
PB	11.0	1499	27,000	13.5	1695	39,400
Unique	12.0	1698	28,200	15.3	1962	40,300
SR-7625	11.0	1505	25,800	13.5	1683	39,000
Herco	12.5	1647	30,500	15.5	1894	39,400
SR-4756	12.0	1554	24,600	15.5	1833	39,900
SR-4759	24.0	2134	27,000	32.0	2598	49,200
RX7	25.0	2065	19,900	39.0	2778	49,200
IMR 3031	31.0	2185	19,900	42.8	2914	48,900
748	34.0	2237	20,200	49.5	2955	45,600

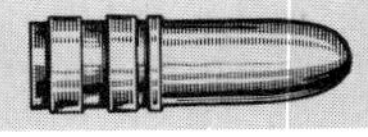

#280412
136 gr., (#2 Alloy) 3.095" OAL

Powder	Sug. Starting Grains	Velocity F.P.S.	Pressure C.U.P.	Max. Grains	Velocity F.P.S.	Pressure C.U.P.
Red Dot	10.5	1495	28,200	13.0	1700	36,400
700X	10.0	1462	26,400	12.5	1662	38,100
Green Dot	10.5	1542	31,000	13.0	1646	35,500
PB	10.5	1431	23,400	13.0	1614	36,000
Unique	11.5	1612	27,000	14.8	1860	37,700
SR-7625	10.5	1408	22,800	13.0	1593	36,400
Herco	12.0	1568	28,200	15.0	1796	36,900
SR-4756	11.5	1490	20,400	15.0	1731	37,700
SR-4759	23.5	2077	29,200	30.0	2414	47,900
RX7	25.4	2051	23,500	39.3	2609	42,200
IMR-3031	29.2	2104	21,500	40.0	2733	49,800
748	32.0	2099	19,400	49.0	2808	44,700

Note: Loads shown in shaded panels are maximum.

.270 Weatherby Magnum

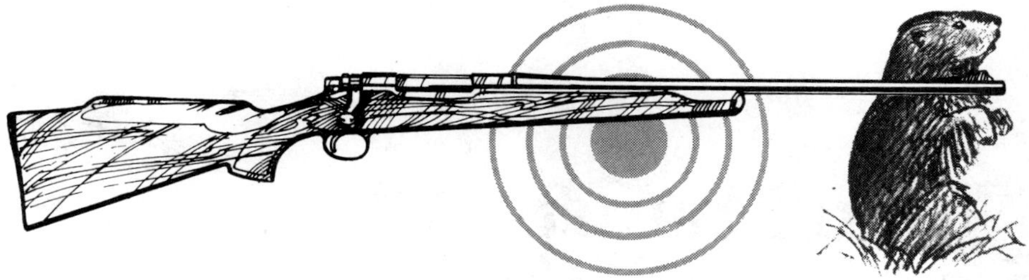

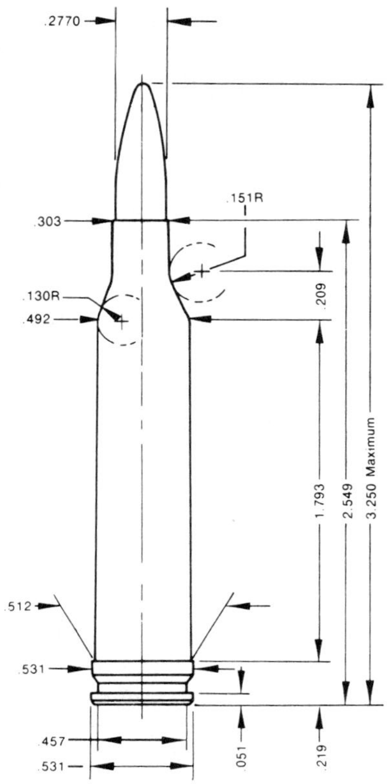

TEST COMPONENTS:

Cases	Weatherby
Trim-to Length	2.540″
Primers	Remington 9½
Primer Size	Large Rifle
Lyman Shell Holder	No. 13
Cast Bullets Used	(size to .277″ dia.)
*Gas Check Bullets	*#280468, 122 gr. *#280473, 124 gr. *#280412, 136 gr.

TEST SPECIFICATIONS:
(Velocity)

Firearm Used	Weatherby Mark V
Barrel Length	26″
Twist	1-12″
Groove Diameter	.277″

COMMENTS:

The data for this cartridge was obtained in a Weatherby rifle and is intended for Weatherby rifles only. The free-boring constructed into these firearms allow higher velocities at safe working pressures. For custom rifles which are not free-bored, maximum loads should be reduced a full 5%. Even then, they should be approached with caution.

Do not use the magnum primers, however, with cast bullets. Only standard large rifle primers should be used with cast bullet loads.

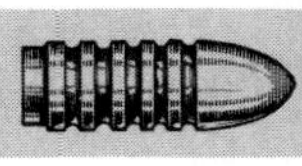

#280468
122 gr., (#2 Alloy) 3.250″ OAL, Max.

Powder	Sug. Starting Grains	Velocity F.P.S.	Pressure C.U.P.	Max. Grains	Velocity F.P.S.	Pressure C.U.P.
Unique	14.0	1875	—	18.0	2169	—
2400	22.0	1956	—	28.0	2364	—
IMR 4227	27.0	2096	—	29.0	2232	—

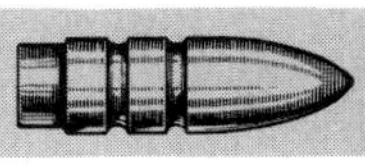

#280473
124 gr., (#2 Alloy) 3.250″ OAL, Max.

Powder	Sug. Starting Grains	Velocity F.P.S.	Pressure C.U.P.	Max. Grains	Velocity F.P.S.	Pressure C.U.P.
Unique	14.0	1756	—	16.0	1882	—
2400	20.0	1785	—	26.0	2118	—
IMR 4227	25.0	1964	—	27.0	2053	—

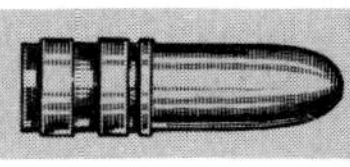

#280412
136 gr., (#2 Alloy) 3.250″ OAL, Max.

Powder	Sug. Starting Grains	Velocity F.P.S.	Pressure C.U.P.	Max. Grains	Velocity F.P.S.	Pressure C.U.P.
Unique	13.0	1639	—	16.0	1838	—
2400	20.0	1762	—	25.0	2040	—
IMR 4227	24.0	1879	—	26.0	1988	—

Note: Loads shown in shaded panels are maximum.

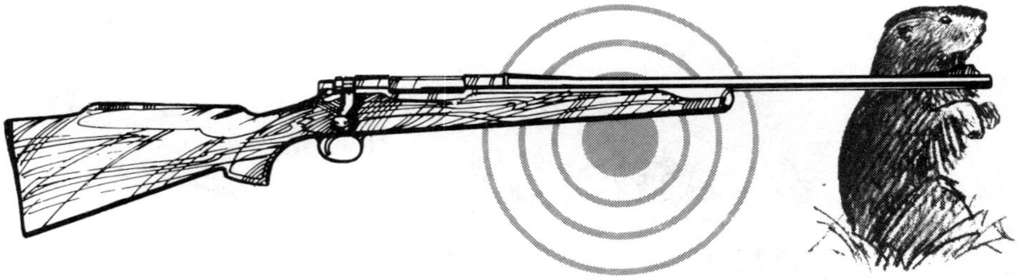

7mm Mauser

(7 x 57)

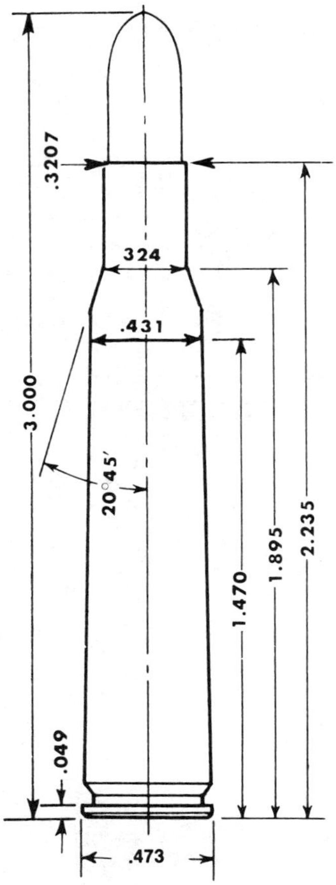

TEST COMPONENTS:

Cases .... Federal
Trim-to Length .... 2.225″
Primers .... Federal 210 & CCI 200
Primer Size .... Large Rifle
Lyman Shell Holder .... No. 2
Cast Bullets Used .... (size to .284″ dia.)
*Gas Check Bullets *#287448, 119 gr.
*#287346, 135 gr.
*#287405, 150 gr.
*#287308, 162 gr.

TEST SPECIFICATIONS:
(Velocity & Pressure)

Firearm Used .... Universal Receiver
Barrel Length .... 24″
Twist .... 1-8¾″
Groove Diameter .... .284″

COMMENTS:

Most 7mm's are imported rifles and their groove diameters can vary considerably. We recommend that you slug your barrel to determine its exact groove diameter.

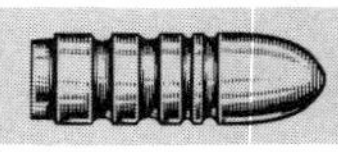

#287448
119 gr., (#2 Alloy) 2.690″ OAL

Powder	Sug. Starting Grains	Velocity F.P.S.	Pressure C.U.P.	Max. Grains	Velocity F.P.S.	Pressure C.U.P.
Red Dot	9.0	1405	24,000	14.0	1845	41,600
700X	9.0	1425	24,600	12.5	1755	41,600
Green Dot	10.0	1550	28,200	14.0	1835	39,900
PB	10.0	1490	28,800	12.5	1665	39,400
Unique	11.0	1610	25,800	15.5	1985	41,100
SR-7625	10.5	1515	28,800	13.0	1690	39,000
630	15.0	1735	24,000	21.0	2155	34,500

Note: Loads shown in shaded panels are maximum.

#287346
135 gr., (#2 Alloy) 2.685" OAL

Powder	Sug. Starting Grains	Velocity F.P.S.	Pressure C.U.P.	Max. Grains	Velocity F.P.S.	Pressure C.U.P.
Red Dot	9.0	1350	26,400	12.0	1625	38,100
700X	8.5	1310	24,000	11.5	1590	40,300
Green Dot	9.5	1380	24,600	12.5	1640	39,400
PB	9.5	1305	—	12.0	1540	37,700
Unique	10.5	1490	25,200	15.0	1850	39,900
SR-7625	10.0	1405	30,000	12.0	1540	36,900
630	15.0	1675	25,800	21.0	2050	37,300
**2400	17.0	1753	20,800	22.3	2059	30,300
**SR-4759	19.5	1763	18,200	*24.7	2158	36,400
**RX7	20.5	1691	15,000	28.0	2145	28,100
**748	28.0	1882	16,100	35.0	2120	21,200

#287405
150 gr., (#2 Alloy) 2.800" OAL

Powder	Sug. Starting Grains	Velocity F.P.S.	Pressure C.U.P.	Max. Grains	Velocity F.P.S.	Pressure C.U.P.
Red Dot	8.5	1245	22,800	11.5	1525	38,100
700X	8.5	1240	24,600	11.0	1485	39,000
Green Dot	9.0	1315	25,200	12.0	1540	37,700
PB	9.5	1245	25,200	12.0	1475	38,600
Unique	10.0	1405	27,000	14.5	1685	39,900
SR-7625	9.5	1230	24,000	12.0	1465	38,600
630	14.5	1535	24,000	20.5	1990	37,300
**SR-4759	20.0	1778	23,300	24.3	2039	37,700
**RX7	23.0	1755	18,700	27.0	2006	27,700
**748	27.2	1685	13,700	34.0	2066	24,100
**H335	26.4	1734	16,100	33.0	2153	31,200
**H4895	*25.6	1772	18,300	32.0	2156	31,300
**IMR-4350	31.0	1771	17,400	38.0	2178	28,900

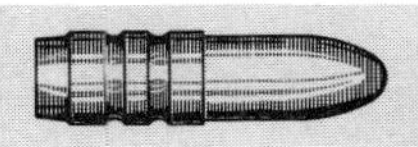

#287308
162 gr., (#2 Alloy) 2.800" OAL

Powder	Sug. Starting Grains	Velocity F.P.S.	Pressure C.U.P.	Max. Grains	Velocity F.P.S.	Pressure C.U.P.
Red Dot	8.5	1200	25,200	11.5	1465	38,100
700X	8.5	1190	26,400	10.5	1400	36,900
Green Dot	9.0	1215	24,000	12.0	1485	36,900
PB	9.5	1195	25,200	12.0	1415	38,100
Unique	10.0	1335	25,800	14.5	1610	38,600
SR-7625	9.5	1185	22,800	11.5	1385	37,700
630	14.5	1465	22,800	20.5	1905	37,700
**SR-4759	20.0	1740	25,800	25.0	2016	41,200
**IMR-4227	19.0	1660	23,400	24.0	1984	41,200
**RX7	29.0	2037	31,800	32.0	2235	41,900
**748	33.0	2019	24,800	41.0	2417	42,600
**760	36.0	1937	20,800	45.0	2415	39,100
**IMR-4350	35.0	1965	22,800	45.0	2449	39,500

Note: Loads shown in shaded panels are maximum.
*Designates potentially most accurate load.
**Signifies CCI 200 primers used.

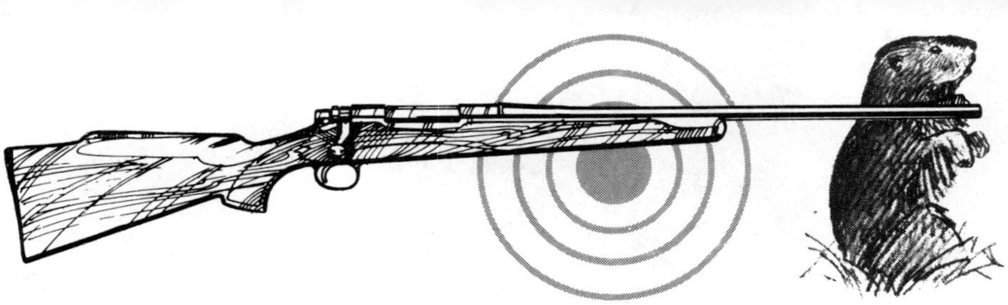

.280 Remington

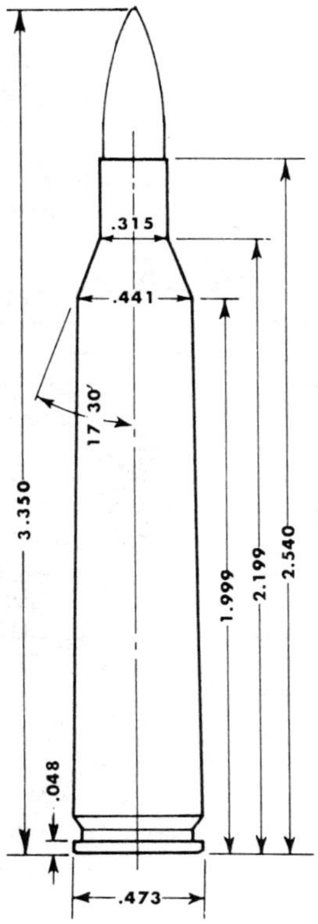

COMMENTS:

This cartridge gave very uniform velocities and excellent accuracy with all the powders and bullets tested.

TEST COMPONENTS:

Cases .... Remington/Peters
Trim-to Length .... 2.540″
Primers .... Remington 9½
Primer Size .... Large Rifle
Lyman Shell Holder .... No. 2
Cast Bullets Used .... (size to .284″ dia.)
*Gas Check Bullets *#287448, 119 gr.
*#287346, 135 gr.
*#287405, 150 gr.
*#287308, 162 gr.

TEST SPECIFICATIONS:
(Velocity)

Firearm Used .... Remington Model 742
Barrel Length .... 22″
Twist .... 1-9½″
Groove Diameter .... .284″

#287448
119 gr., (#2 Alloy) 2.612″ OAL

Powder	Sug. Starting Grains	Velocity F.P.S.	Pressure C.U.P.	Max. Grains	Velocity F.P.S.	Pressure C.U.P.
Red Dot	11.0	1585	—	16.0	1955	—
700X	10.5	1545	—	15.0	1890	—
Green Dot	11.5	1600	—	16.5	1980	—
PB	11.0	1500	—	15.5	1840	—
Unique	12.0	1625	—	17.0	1985	—
SR-7625	11.5	1535	—	16.0	1880	—
630	15.0	1660	—	20.0	1960	—

#287346
135 gr., (#2 Alloy) 2.765″ OAL

Powder	Sug. Starting Grains	Velocity F.P.S.	Pressure C.U.P.	Max. Grains	Velocity F.P.S.	Pressure C.U.P.
Red Dot	11.0	1495	—	15.5	1815	—
700X	10.5	1465	—	14.5	1750	—
Green Dot	11.5	1525	—	16.0	1840	—
PB	11.0	1420	—	15.5	1750	—
Unique	12.0	1535	—	17.0	1890	—
SR-7625	11.5	1460	—	16.0	1775	—
630	15.0	1585	—	20.0	1895	—

Note: Loads shown in shaded panels are maximum.

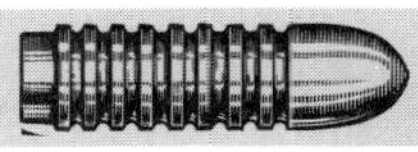

#287405
150 gr., (#2 Alloy) 2.794" OAL

Powder	Sug. Starting Grains	Velocity F.P.S.	Pressure C.U.P.	Max. Grains	Velocity F.P.S.	Pressure C.U.P.
Red Dot	10.5	1405	—	15.0	1710	—
700X	10.0	1365	—	14.0	1645	—
Green Dot	11.0	1425	—	15.5	1735	—
PB	10.5	1330	—	15.0	1640	—
Unique	11.5	1440	—	16.5	1785	—
SR-7625	11.0	1370	—	15.5	1670	—
630	14.5	1485	—	19.5	1815	—

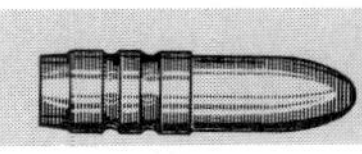

#287308
162 gr., (#2 Alloy) 2.700" OAL

Powder	Sug. Starting Grains	Velocity F.P.S.	Pressure C.U.P.	Max. Grains	Velocity F.P.S.	Pressure C.U.P.
Red Dot	10.0	1305	—	14.5	1600	—
700X	9.5	1250	—	13.5	1535	—
Green Dot	10.5	1325	—	15.0	1630	—
PB	10.0	1225	—	14.5	1525	—
Unique	11.0	1345	—	16.0	1695	—
SR-7625	10.5	1240	—	15.0	1550	—
630	14.0	1415	—	19.0	1725	—

Note: Loads shown in shaded panels are maximum.

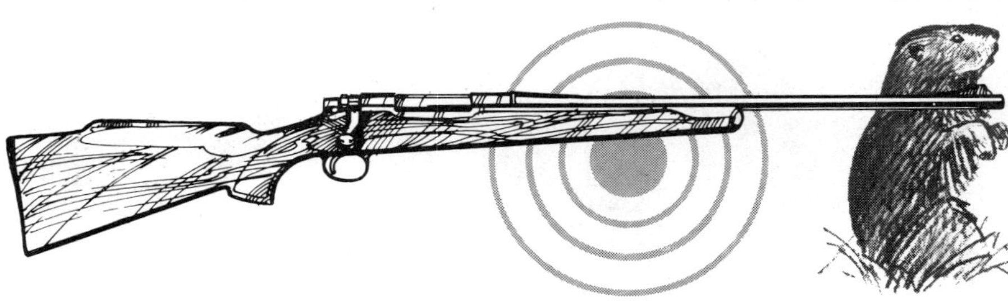

.284 Winchester

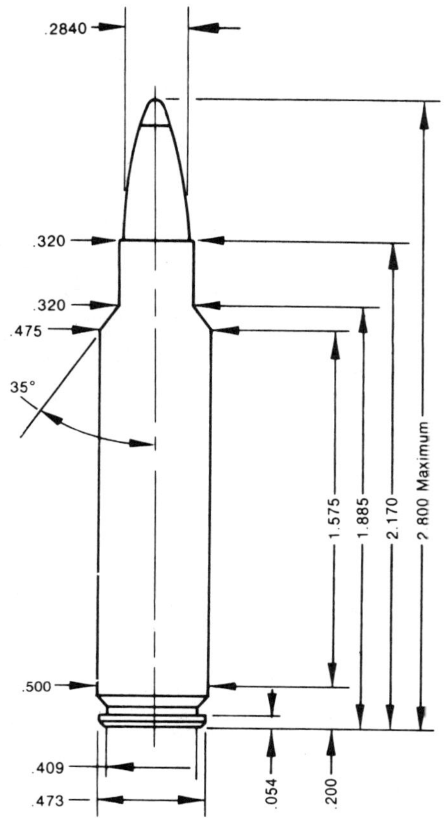

TEST COMPONENTS:

Cases	Winchester/Western
Trim-to Length	2.160″
Primers	Winchester 8½-120
Primer Size	Large Rifle
Lyman Shell Holder	No. 2
Cast Bullets Used	(size to .284″ dia.)
*Gas Check Bullets	*#287448, 119 gr.
	*#287346, 135 gr.
	*#287405, 150 gr.
	*#287308, 162 gr.

TEST SPECIFICATIONS:
(Velocity)

Firearm Used	Ruger Model 77
Barrel Length	22″
Twist	1-10″
Groove Diameter	.284″

COMMENTS:

Maximum loads for individual rifles seem to vary quite a bit in this caliber. We recommend that maximum loadings be approached with caution.

#287448
119 gr., (#2 Alloy) 2.612″ OAL

Powder	Sug. Starting Grains	Velocity F.P.S.	Pressure C.U.P.	Max. Grains	Velocity F.P.S.	Pressure C.U.P.
Red Dot	9.0	1415	—	15.5	1950	—
700X	9.0	1425	—	15.5	1955	—
Green Dot	10.0	1510	—	16.0	1965	—
PB	10.0	1440	—	16.0	1835	—
Unique	11.0	1565	—	17.0	2015	—
SR-7625	11.0	1510	—	17.0	1955	—
630	15.0	1650	—	20.0	1995	—
IMR-4227	17.0	1550	—	21.0	1815	—

Note: Loads shown in shaded panels are maximum.

.284 Winchester (Continued)

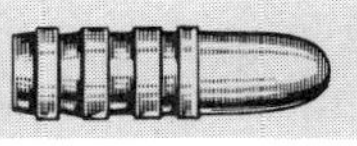

#287346
135 gr., (#2 Alloy) 2.765" OAL

Powder	Sug. Starting Grains	Velocity F.P.S.	Pressure C.U.P.	Max. Grains	Velocity F.P.S.	Pressure C.U.P.
Red Dot	9.0	1340	—	15.5	1830	—
700X	9.0	1335	—	15.5	1830	—
Green Dot	10.0	1410	—	16.0	1855	—
PB	10.0	1360	—	16.0	1780	—
Unique	11.0	1475	—	17.0	1885	—
SR-7625	11.0	1415	—	17.0	1840	—
630	15.0	1590	—	20.0	1915	—
IMR-4227	17.0	1525	—	21.0	1785	—

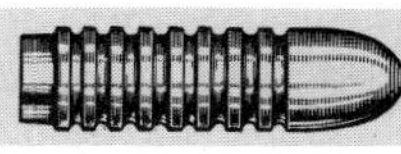

#287405
150 gr., (#2 Alloy) 2.794" OAL

Powder	Sug. Starting Grains	Velocity F.P.S.	Pressure C.U.P.	Max. Grains	Velocity F.P.S.	Pressure C.U.P.
Red Dot	10.0	1370	—	15.5	1760	—
700X	10.0	1380	—	15.0	1735	—
Green Dot	11.0	1435	—	16.0	1780	—
PB	11.0	1375	—	16.0	1715	—
Unique	12.0	1500	—	17.0	1840	—
SR-7625	11.5	1410	—	16.5	1740	—
630	15.5	1580	—	20.0	1870	—
IMR-4227	17.5	1535	—	21.0	1745	—

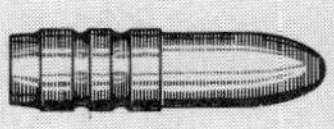

#287308
162 gr., (#2 Alloy) 2.700" OAL

Powder	Sug. Starting Grains	Velocity F.P.S.	Pressure C.U.P.	Max. Grains	Velocity F.P.S.	Pressure C.U.P.
Red Dot	10.0	1310	—	15.5	1690	—
700X	10.0	1310	—	15.0	1660	—
Green Dot	11.0	1370	—	16.0	1705	—
PB	11.0	1305	—	15.5	1610	—
Unique	12.0	1430	—	17.0	1755	—
SR-7625	11.5	1330	—	16.0	1625	—
630	15.5	1505	—	19.5	1745	—
IMR-4227	17.5	1350	—	21.0	1620	—

Note: Loads shown in shaded panels are maximum.

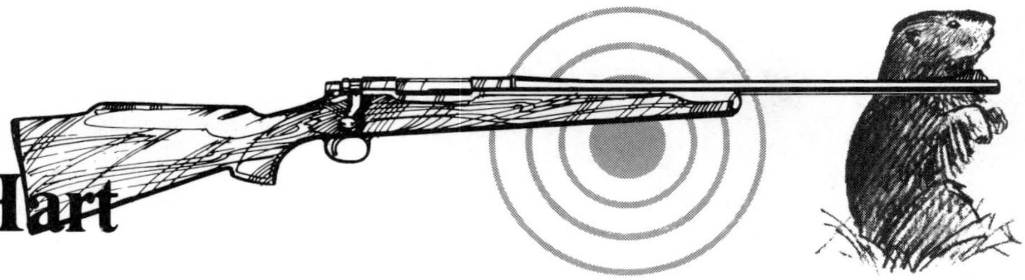

7mm Sharpe & Hart
(7 x 61mm)

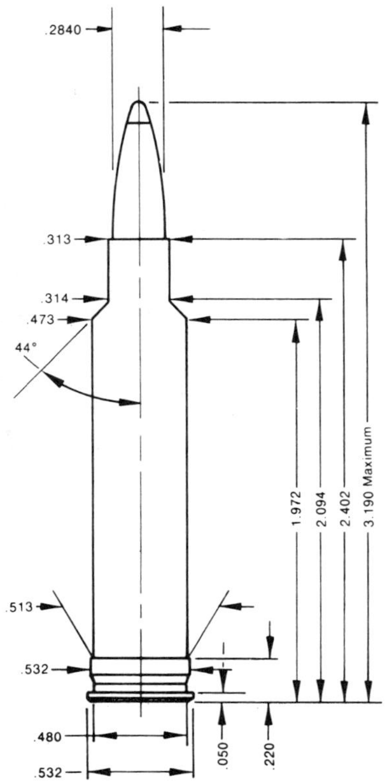

TEST COMPONENTS:

Cases .... Norma
Trim-to Length .... 2.392″
Primers .... Remington 9½
Primer Size .... Large Rifle
Lyman Shell Holder .... No. 13
Cast Bullets Used .... (size to .284″ dia.)
*Gas Check Bullets *#287346, 135 gr.
*#287405, 150 gr.
*#287308, 162 gr.

TEST SPECIFICATIONS:
(Velocity)

Firearm Used .... Shultz and Larsen 65 DL
Barrel Length .... 24″
Twist .... 1-10″
Groove Diameter .... .285″

COMMENTS:

The 7 x 61mm proved to be one of the most efficient cartridges we tested. Accuracy was good and velocities were extremely uniform. Compared to other magnum cases, this cartridge produced very high velocities with minimum amounts of powder. The rifle used for testing was free-bored. For rifles which do not have a free-bored throat, this data should be reduced a full 5%. Even then, maximum loading should be approached with caution.

#287346
135 gr., (#2 Alloy) 3.190″ OAL, Max.

Powder	Sug. Starting Grains	Velocity F.P.S.	Pressure C.U.P.	Max. Grains	Velocity F.P.S.	Pressure C.U.P.
Unique	15.0	1805	—	18.0	2016	—
2400	18.0	1658	—	24.0	2070	—
IMR-4227	22.0	1828	—	26.0	2092	—

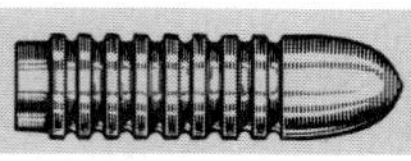

#287405
150 gr., (#2 Alloy) 3.190″ OAL, Max.

Powder	Sug. Starting Grains	Velocity F.P.S.	Pressure C.U.P.	Max. Grains	Velocity F.P.S.	Pressure C.U.P.
Unique	14.0	1683	—	17.0	1876	—
2400	17.0	1564	—	23.0	2008	—
IMR-4227	22.0	1811	—	25.0	1988	—

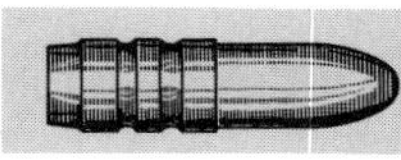

#287308
162 gr., (#2 Alloy) 3.190″ OAL, Max.

Powder	Sug. Starting Grains	Velocity F.P.S.	Pressure C.U.P.	Max. Grains	Velocity F.P.S.	Pressure C.U.P.
Unique	13.0	1557	—	16.0	1754	—
2400	16.0	1508	—	22.0	1858	—
IMR-4227	20.0	1655	—	24.0	1872	—

Note: Loads shown in shaded panels are maximum.

7mm Remington Magnum

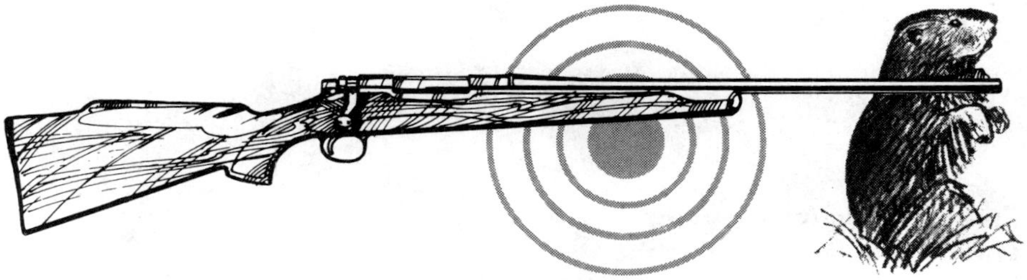

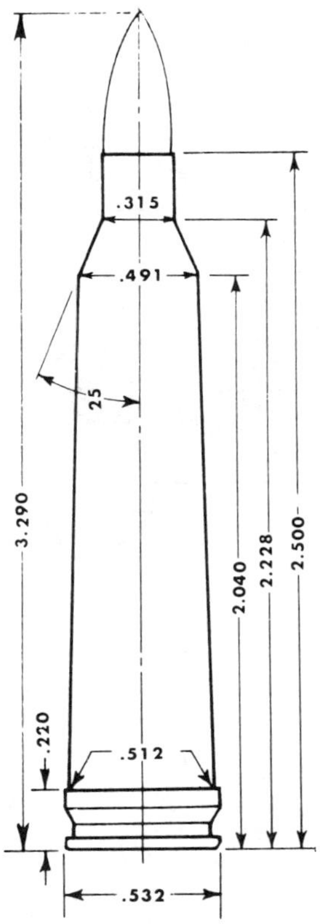

TEST COMPONENTS:

Cases .... Winchester
Trim-to Length .... 2.490″
Primers .... Remington 9½ M
Primer Size .... Large Rifle, Magnum
Lyman Shell Holder .... No. 13
Cast Bullets Used .... (size to .284″ dia.)
*Gas Check Bullets *#287448, 119 gr.
*#287346, 135 gr.
*#287405, 150 gr.
*#287308, 162 gr.

TEST SPECIFICATIONS:
(Velocity & Pressure)

Firearm Used .... Universal Receiver
Barrel Length .... 26″
Twist .... 1-9½″
Groove Diameter .... .284″

COMMENTS:

Accuracy with this cartridge was erratic at velocities in excess of 1800 f.p.s. Bullet #287448 seemed to give the best performance.

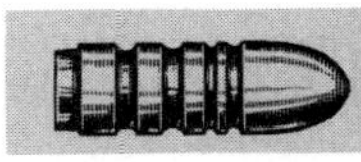

#287448
119 gr., (#2 Alloy) 3.290″ OAL, Max.

Powder	Sug. Starting Grains	Velocity F.P.S.	Pressure C.U.P.	Max. Grains	Velocity F.P.S.	Pressure C.U.P.
Red Dot	18.0	2084	31,500	21.0	2270	40,700
700X	14.0	1793	25,200	18.5	2112	38,100
Green Dot	16.0	1924	24,600	18.0	2042	29,400
PB	16.0	1857	25,200	20.5	2141	40,300
SR-7625	16.0	1844	25,200	21.0	2168	40,300
SR-4756	20.0	2077	25,800	24.5	2321	39,900

Note: Loads shown in shaded panels are maximum.

7mm Remington Magnum (Continued)

#287346
135 gr., (#2 Alloy) 2.967" OAL

Powder	Sug. Starting Grains	Velocity F.P.S.	Pressure C.U.P.	Max. Grains	Velocity F.P.S.	Pressure C.U.P.
Red Dot	17.0	1903	31,500	20.0	2078	40,700
700X	14.5	1722	27,600	17.5	1929	37,700
Green Dot	15.0	1731	24,600	18.0	1930	31,000
PB	16.5	1756	27,600	19.5	1939	38,600
SR-7625	17.0	1795	31,000	20.0	1972	39,900
SR-4756	20.0	1931	30,000	23.5	2153	40,700
SR-4759	29.7	2164	29,500	37.1	2564	49,200
IMR-4198	32.4	2187	26,900	44.0	2776	49,800
RX7	32.0	2159	26,100	43.5	2674	46,700
748	32.0	2163	26,800	54.0	2683	41,200
H4895	40.0	2158	25,000	52.0	2815	47,400

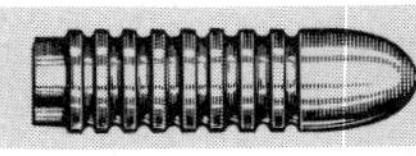

#287405
150 gr., (#2 Alloy) 3.052" OAL

Powder	Sug. Starting Grains	Velocity F.P.S.	Pressure C.U.P.	Max. Grains	Velocity F.P.S.	Pressure C.U.P.
Red Dot	16.0	1782	30,000	19.0	1943	38,100
700X	13.5	1601	25,800	16.5	1798	34,000
Green Dot	15.0	1675	25,800	18.0	1849	31,000
PB	15.5	1652	26,400	18.5	1820	36,000
SR-7625	15.5	1635	25,800	19.0	1851	37,700
SR-4756	19.0	1813	25,800	22.5	2004	36,900
630	18.0	1730	18,000	21.0	1903	20,400
SR-4759	28.4	2164	33,500	35.5	2396	47,100
IMR-4198	33.3	2186	27,500	42.0	2574	46,900
RX7	34.0	2157	29,000	42.5	2560	47,800
748	41.4	2174	27,400	51.8	2676	46,100
H4895	40.0	2171	28,600	50.0	2739	49,500

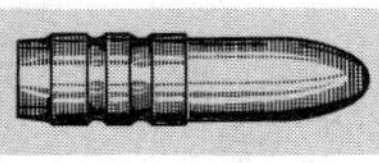

#287308
162 gr., (#2 Alloy) 3.192" OAL

Powder	Sug. Starting Grains	Velocity F.P.S.	Pressure C.U.P.	Max. Grains	Velocity F.P.S.	Pressure C.U.P.
Red Dot	14.5	1604	26,400	18.0	1807	33,500
700X	13.0	1504	23,400	16.0	1701	33,000
Green Dot	14.5	1577	23,400	18.0	1786	31,500
PB	15.0	1553	25,800	18.0	1723	34,500
SR-7625	15.0	1541	25,200	18.5	1749	35,500
SR-4756	18.5	1728	26,400	22.0	1912	36,400
630	17.5	1643	19,200	21.0	1816	21,000
SR-4759	33.0	2174	35,900	37.0	2371	48,700
IMR-4198	33.6	2157	32,100	42.0	2526	49,200
RX7	35.0	2177	35,000	41.5	2377	44,700
748	41.5	2181	32,600	53.0	2559	47,900
H4895	38.1	2190	31,300	48.0	2528	48,100

Note: Loads shown in shaded panels are maximum.

7mm Weatherby Magnum

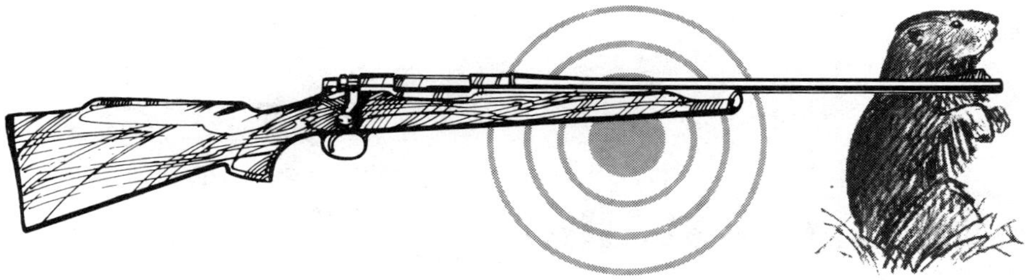

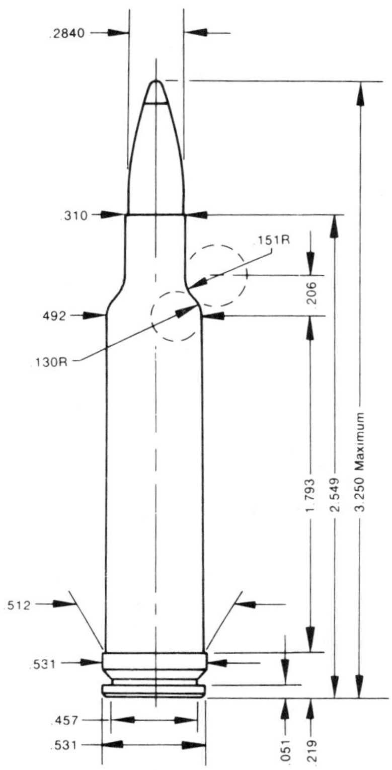

TEST COMPONENTS:

Cases .... Weatherby
Trim-to Length .... 2.540″
Primers .... Remington 9½
Primer Size .... Large Rifle
Lyman Shell Holder .... No. 13
Cast Bullets Used .... (size to .284″ dia.)
*Gas Check Bullets .... *#287405, 150 gr.
*#287308, 162 gr.
#287221, 179 gr.

TEST SPECIFICATIONS:
(Velocity)

Firearm Used .... Weatherby Mark V
Barrel Length .... 26″
Twist .... 1-12″
Groove Diameter .... .2825″

COMMENTS:

The data listed for this cartridge were obtained in a Weatherby rifle and are intended for Weatherby rifles only. The free-boring constructed into these firearms allows higher velocities at safe working pressures. For custom rifles which are not free-bored, maximum loads should be reduced a full 5%. Even then, they should be approached with caution.

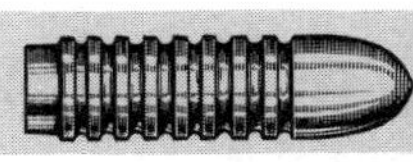

#287405
150 gr., (#2 Alloy) 3.260″ OAL, Max.

Powder	Sug. Starting Grains	Velocity F.P.S.	Pressure C.U.P.	Max. Grains	Velocity F.P.S.	Pressure C.U.P.
Unique	15.0	1683	—	18.0	1872	—
2400	18.0	1559	—	24.0	1930	—
IMR-4227	23.0	1760	—	26.0	1930	—

#287308
162 gr., (#2 Alloy) 3.260″ OAL, Max.

Powder	Sug. Starting Grains	Velocity F.P.S.	Pressure C.U.P.	Max. Grains	Velocity F.P.S.	Pressure C.U.P.
Unique	14.0	1572	—	17.0	1763	—
2400	17.0	1501	—	23.0	1828	—
IMR-4227	22.0	1686	—	25.0	1838	—

#287221
179 gr., (#2 Alloy) 3.260″ OAL, Max.

Powder	Sug. Starting Grains	Velocity F.P.S.	Pressure C.U.P.	Max. Grains	Velocity F.P.S.	Pressure C.U.P.
Unique	13.0	1481	—	16.0	1674	—
2400	16.0	1412	—	22.0	1763	—
IMR-4227	21.0	1592	—	24.0	1742	—

Note: Loads shown in shaded panels are maximum.

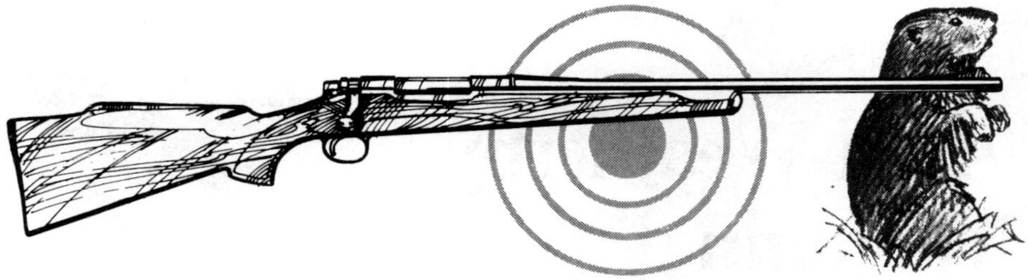

7.35mm Italian
(Terni)

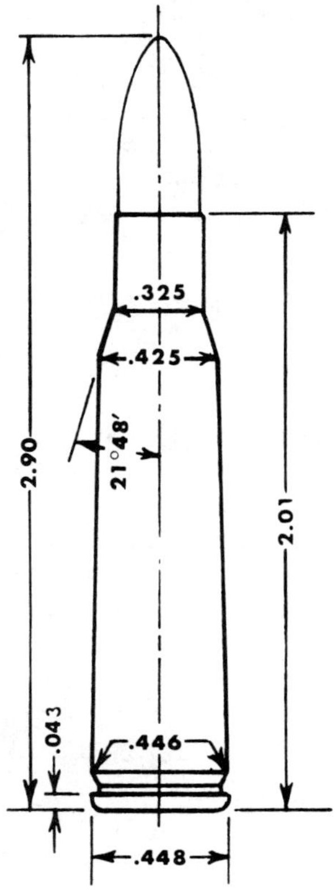

TEST COMPONENTS:

Cases .... Formed from Norma 6.5 x 54 Brass
Trim-to Length .... 2.000″
Primers .... Winchester 8½-120
Primer Size .... Large Rifle
Lyman Shell Holder .... No. 28
Cast Bullets Used .... (size to .299″ & .301″ dia.)
*Gas Check Bullet *#300136, 150 gr.

TEST SPECIFICATIONS:
(Velocity)

Firearm Used .... Italian Military Carbine
Barrel Length .... 21″
Twist .... Gain
Groove Diameter .... .302″

COMMENTS:

Here again a large variation in groove diameters exist. Match the bullet diameter as closely as possible to the groove diameter.

Lead bullets can be sized .299″ or .301″. Most actual groove diameters run between .300″ and .302″. Due to the relatively poor design and workmanship of these rifles, we do not recommend their use except for lead bullet shooting.

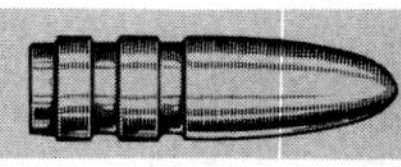

#300136
150 gr., (#2 Alloy) 2.755″ OAL, Max.

Powder	Sug. Starting Grains	Velocity F.P.S.	Pressure C.U.P.	Max. Grains	Velocity F.P.S.	Pressure C.U.P.
2400	16.0	1650	—	22.0	2070	—
IMR-4227	16.0	1535	—	25.0	2136	—
IMR-4198	19.0	1595	—	30.0	2262	—
IMR-3031	24.0	1625	—	33.0	2247	—

Note: Loads shown in shaded panels are maximum.

.30 Carbine

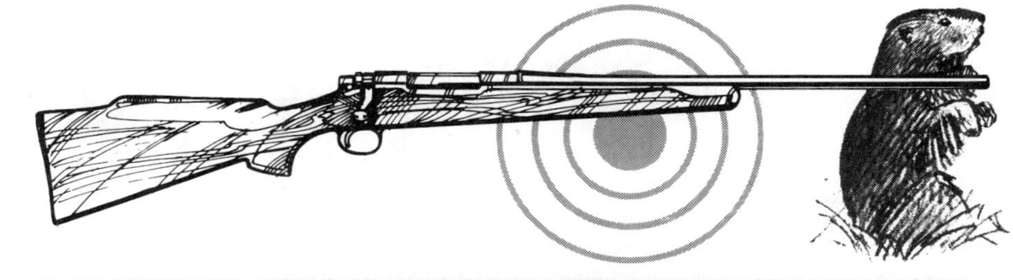

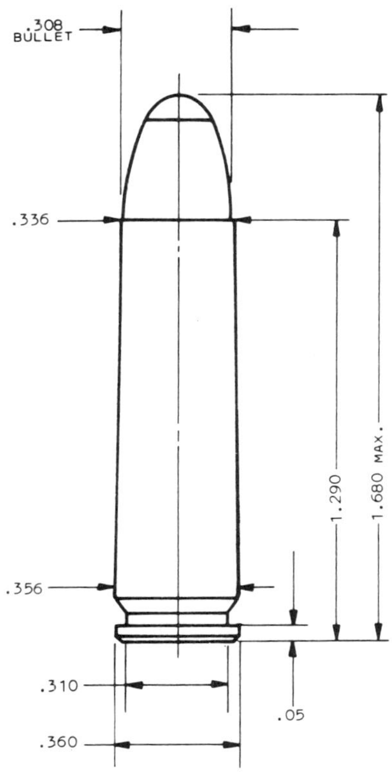

TEST COMPONENTS:

Cases .... Remington
Trim-to Length .... 1.286″
Primers .... Remington 6½
Primer Size .... Small Rifle
Lyman Shell Holder .... No. 19
Cast Bullets Used .... (size to .308″ dia.)
*Gas Check Bullets .... *#311359, 113 gr.
*#311576, 120 gr.
#311410, 130 gr.

TEST SPECIFICATIONS:
(Velocity & Pressure)

Firearm Used .... Universal Receiver
Barrel Length .... 20″
Twist .... 1-20″
Groove Diameter .... .308″

COMMENTS:

Best ballistic uniformity was obtained with bullet No. 311410 using AL-8 powder. Not all carbines will function semi-automatically with all loads. Bullet No. 311410 with AL-8 or IMR 4227 will function all carbines we have tested.

This cartridge headspaces on the mouth and case length is, therefore, critical. Never trim cases shorter than the trim-to length shown and never crimp bullets.

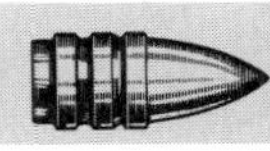

#311359
113 gr., (#2 Alloy) 1.680″ OAL, Max.

Powder	Sug. Starting Grains	Velocity F.P.S.	Pressure C.U.P.	Max. Grains	Velocity F.P.S.	Pressure C.U.P.
Unique	5.5	1388	27,600	6.5	1555	38,600
SR-7625	4.6	1194	22,800	5.6	1368	39,000
SR-4756	5.5	1298	23,400	6.5	1463	38,100
AL-8	9.5	1577	32,000	11.0+	1728	39,400
2400	11.0	1633	31,500	12.5+	1828	40,300
IMR-4227	11.5	1497	27,000	13.3+	1733	39,000

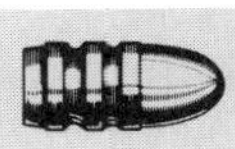

#311576
120 gr., (#2 Alloy) 1.680″ OAL, Max.

Powder	Sug. Starting Grains	Velocity F.P.S.	Pressure C.U.P.	Max. Grains	Velocity F.P.S.	Pressure C.U.P.
Unique	5.0	1300	22,200	6.3	1526	39,000
SR-4756	5.1	1232	23,400	6.1	1407	39,400
AL-8	9.2	1510	31,500	10.8+	1727	40,700
2400	10.8	1623	29,400	12.3+	1795	38,100
IMR-4227	11.0	1426	25,800	13.1+	1709	40,300

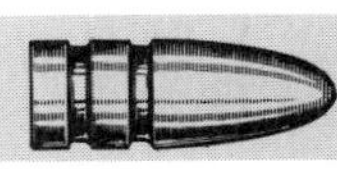

#311410
130 gr., (#2 Alloy) 1.680″ OAL, Max.

Powder	Sug. Starting Grains	Velocity F.P.S.	Pressure C.U.P.	Max. Grains	Velocity F.P.S.	Pressure C.U.P.
Unique	4.9	1260	23,400	6.1	1468	35,000
SR-4756	5.0	1180	22,000	6.0	1371	37,300
AL-8	9.0	1497	30,500	10.4+	1675	40,300
2400	10.5	1564	28,800	12.0+	1733	38,600
IMR-4227	10.6	1366	24,600	12.7+	1647	38,600

Note: Loads shown in shaded panels are maximum.
+Designates a compressed powder charge.

.30 Remington

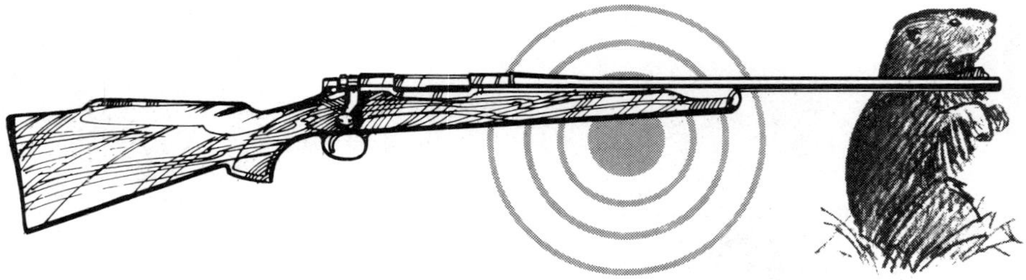

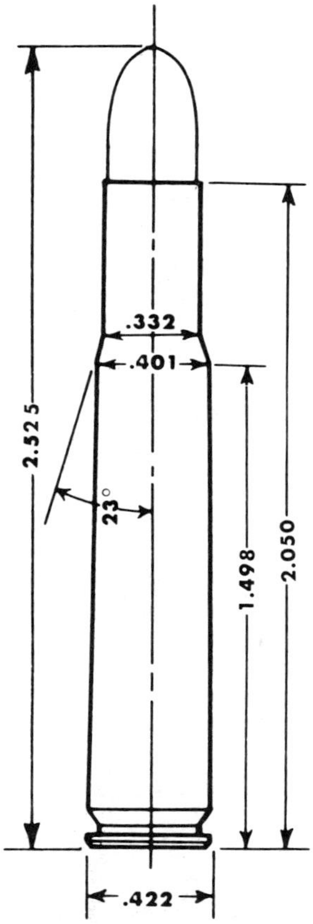

TEST COMPONENTS:

Cases .... Remington
Trim-to Length .... 2.040″
Primers .... Remington 9½
Primer Size .... Large Rifle
Lyman Shell Holder .... No. 15
Cast Bullets Used .... (size to .307″ & .308″ dia.)
*Gas Check Bullet *#311291, 169 gr.

TEST SPECIFICATIONS:
(Velocity)

Firearm Used .... Remington Model 81
Barrel Length .... 22″
Twist .... 1-12″
Groove Diameter .... .307″

COMMENTS:

Groove diameters of rifles in this caliber may run a bit tight (as small as .306″ dia.). We recommend sizing to groove diameter.

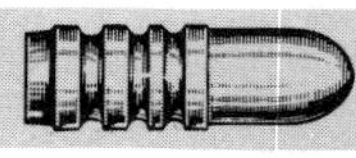

#311291
169 gr., (#2 Alloy) 2.525″ OAL, Max.

Powder	Sug. Starting Grains	Velocity F.P.S.	Pressure C.U.P.	Max. Grains	Velocity F.P.S.	Pressure C.U.P.
2400	16.0	1730	—	20.0	2000	—
IMR-4227	17.5	1754	—	22.0	2024	—
IMR-4198	19.0	1697	—	25.5	2118	—

Note: Loads shown in shaded panels are maximum.

.303 Savage

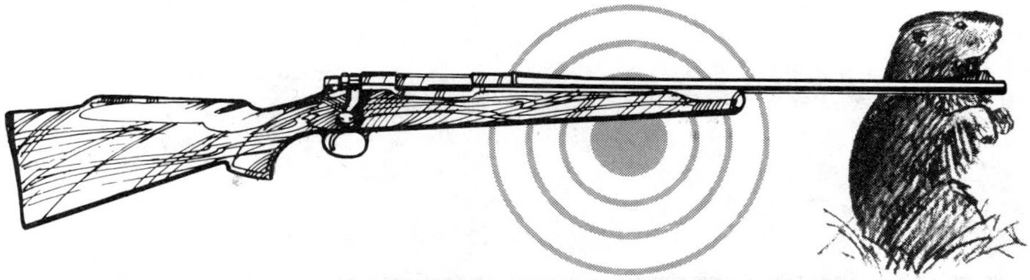

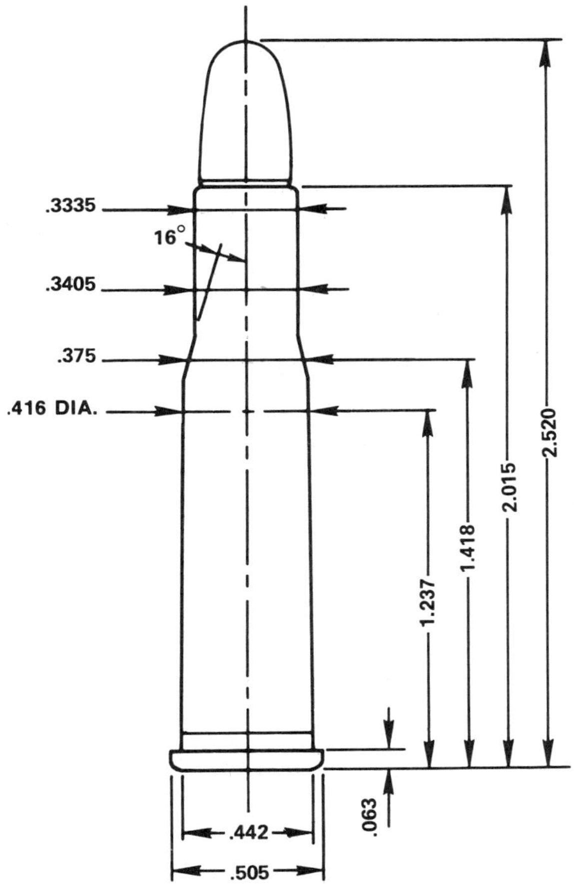

TEST COMPONENTS:

Cases .... Remington
Trim-to Length .... 2.010″
Primers .... Remington 9½
Primer Size .... Large Rifle
Lyman Shell Holder .... No. 7
Cast Bullets Used .... (size to .308″ dia.)
*Gas Check Bullets *#311291, 169 gr.
*#311334, 187 gr.

TEST SPECIFICATIONS:
(Velocity)

Firearm Used .... Savage Model 99
Barrel Length .... 26″
Twist .... 1-12″
Groove Diameter .... .308″

#311291
169 gr., (#2 Alloy) 2.520″ OAL, Max.

Powder	Sug. Starting Grains	Velocity F.P.S.	Pressure C.U.P.	Max. Grains	Velocity F.P.S.	Pressure C.U.P.
Unique	8.0	1326	—	11.0	1930	—
2400	15.0	1625	—	20.0	1980	—
IMR-4227	16. 0	1618	—	22.0	2008	—

COMMENTS:

The rifles chambered for this cartridge which we have tested, show no appreciable variation in standard .308 groove diameter. Our consumer correspondence, however, indicates that some grooves may run a little on the large side. We suggest that you slug your barrel before reloading, and use a cast bullet sized to your groove diameter.

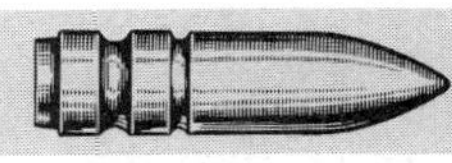

#311334
187 gr., (#2 Alloy) 2.520″ OAL, Max.

Powder	Sug. Starting Grains	Velocity F.P.S.	Pressure C.U.P.	Max. Grains	Velocity F.P.S.	Pressure C.U.P.
Unique	7.0	1273	—	10.0	1413	—
2400	14.0	1411	—	19.0	1811	—
IMR-4227	16.0	1492	—	22.5	1934	—

Note: Loads shown in shaded panels are maximum.

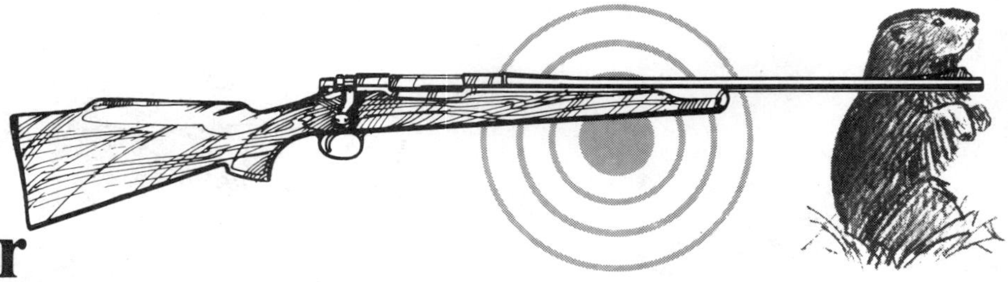

.30/30 Winchester

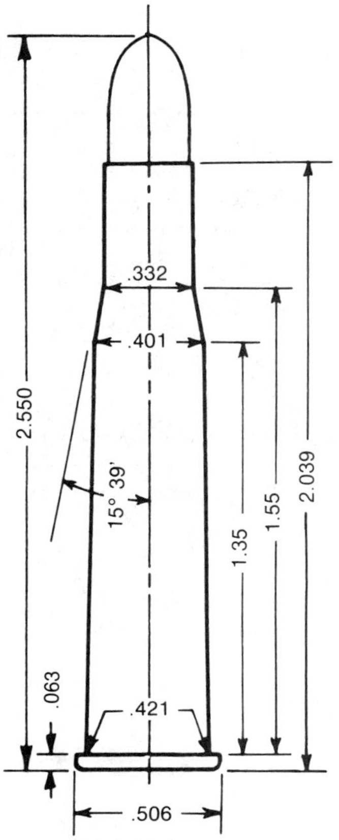

TEST COMPONENTS:

Cases .... Winchester
Trim-to Length .... 2.028″
Primers .... Winchester 8½-120 & CCI 200
Primer Size .... Large Rifle
Lyman Shell Holder .... No. 6
Cast Bullets Used .... (size to .308″ dia.)
*Gas Check Bullets
*#311441, 115 gr.
*#311465, 122 gr.
*#311440, 151 gr.
*#311291, 169 gr.
*#31141, 170 gr.
*#311407, 173 gr.

TEST SPECIFICATIONS:
(Velocity & Pressure)

Firearm Used .... Universal Receiver
Barrel Length .... 26″
Twist .... 1-12″
Groove Diameter .... .308″

COMMENTS:

Rifles with tubular magazines require the use of flat or blunt-nosed bullets and a crimp to keep the bullet in place. Bullet #311465 is an excellent performer at higher than normal cast bullet velocities (2400 f.p.s. - 2500 f.p.s.) but can be used only in bolt action rifles because of the rather pointed nose configuration.

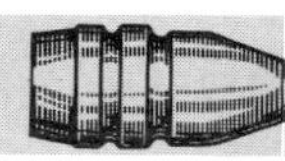

#311441
115 gr., (#2 Alloy) 2.370″ OAL

Powder	Sug. Starting Grains	Velocity F.P.S.	Pressure C.U.P.	Max. Grains	Velocity F.P.S.	Pressure C.U.P.
Red Dot	7.0	1400	25,200	10.0	1747	38,600
700X	7.0	1415	27,600	8.9	1643	38,600
Green Dot	7.5	1430	25,200	10.5	1738	38,600
PB	7.0	1320	22,200	9.7	1602	38,600
Unique	7.5	1480	21,600	11.4	1913	36,400
SR-7625	7.5	1379	23,400	10.0	1638	36,400
Herco	10.0	1645	27,000	12.5	1907	36,900
SR-4756	8.5	1474	23,400	11.9	1793	37,700
**630	13.5	1598	15,600	*20.0	2289	34,700
**2400	13.5	1553	15,700	21.2 *	2282	34,300
**SR-4759	15.5	1600	15,200	22.5	2203	34,600
**RX7	20.0	1626	10,400	33.5	2585	29,400

Note: Loads shown in shaded panels are maximum.
*Designates potentially most accurate load.
**Signifies CCI 200 primers used.

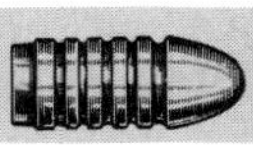

#311465
122., (#2 Alloy) 2.447″ OAL

Powder	Sug. Starting Grains	Velocity F.P.S.	Pressure C.U.P.	Max. Grains	Velocity F.P.S.	Pressure C.U.P.
**630	12.8	1631	19,100	19.7	2213	34,300
**SR-4759	15.7	1516	11,900	22.8	2263	35,300
**RX7	20.3	1828	15,600	32.8	2551	29,700
**748	25.1	1714	15,000	37.5+	2368	27,500
**H4895	24.3	1661	12,800	*35.4+	2460	30,300

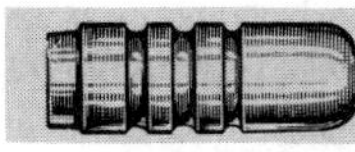

#311440
151 gr., (#2 Alloy) 2.480″ OAL

Powder	Sug. Starting Grains	Velocity F.P.S.	Pressure C.U.P.	Max. Grains	Velocity F.P.S.	Pressure C.U.P.
Red Dot	6.8	1257	25,800	9.0	1481	38,100
700X	6.5	1235	27,000	8.0	1401	36,400
Green Dot	7.0	1231	23,400	9.5	1478	36,900
PB	7.0	1222	26,400	9.0	1422	38,100
Unique	7.5	1333	21,600	11.0	1691	37,300
SR-7625	7.5	1283	27,600	9.2	1413	36,900
Herco	9.0	1386	25,800	12.0	1660	36,900
SR-4756	8.0	1282	22,200	11.1	1573	36,400
**SR-4759	14.4	1529	16,700	18.0	1887	33,500
**RX7	22.5	1971	27,000	30.0	2318	35,600
**748	28.0	1939	22,300	37.5+	2433	35,000
**H335	25.5	1983	23,800	33.0	2328	31,700

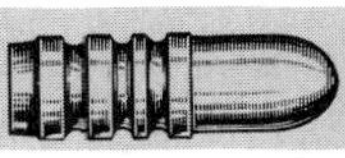

#311291
169 gr., (#2 Alloy) 2.521″ OAL

Powder	Sug. Starting Grains	Velocity F.P.S.	Pressure C.U.P.	Max. Grains	Velocity F.P.S.	Pressure C.U.P.
Red Dot	6.5	1145	24,600	8.5	1363	34,500
700X	6.0	1101	26,400	7.5	1253	34,000
Green Dot	7.0	1126	23,400	9.0	1348	33,500
PB	6.5	1093	24,000	8.7	1311	36,900
Unique	7.0	1211	19,800	10.6	1568	36,000
SR-7625	6.5	1107	22,800	9.0	1348	36,400
Herco	9.0	1328	28,200	11.5	1538	38,100
SR-4756	7.0	1123	19,800	10.5	1431	35,000
**SR-4759	14.8	1530	19,900	18.5	1862	36,100
**RX7	19.0	1635	20,200	28.0	2152	34,400
**IMR-3031	22.5	1599	18,300	28.5	2095	32,500
**748	23.8	1604	16,900	37.3+	2355	35,100
**H335	20.3	1638	18,500	32.5	2230	32,800

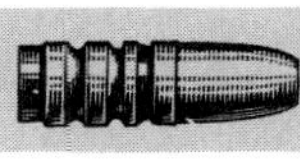

#31141
170 gr., (#2 Alloy) 2.510″ OAL

Powder	Sug. Starting Grains	Velocity F.P.S.	Pressure C.U.P.	Max. Grains	Velocity F.P.S.	Pressure C.U.P.
Red Dot	6.5	1159	24,600	8.5	1348	34,500
700X	6.0	1124	24,600	7.5	1269	34,000
Green Dot	7.0	1192	25,200	9.0	1373	33,500
PB	6.5	1100	23,400	8.7	1286	36,900
Unique	7.0	1240	21,600	10.6	1555	35,500
SR-7625	6.5	1118	22,800	9.0	1337	36,900
Herco	8.5	1281	25,800	11.0	1505	33,000
SR-4756	7.0	1102	19,800	10.5	1423	36,400
**SR-4759	15.5	1613	23,700	17.7	1803	32,600
**RX7	20.0	1632	19,800	28.6	2165	34,500
**IMR-3031	21.6	1555	20,100	27.0	2016	31,300
**748	24.8	1613	22,500	35.0+	2270	32,900
**H335	22.0	1596	23,200	30.0	2254	33,800

Note: Loads shown in shaded panels are maximum.
*Designates potentially most accurate load.
+Designates compressed powder charge.
**Signifies CCI 200 primers used.
Bullet #311465 should not be used in tubular magazines.

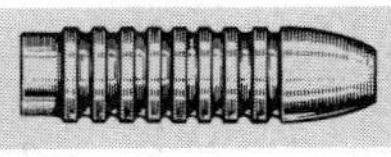

#311407
173 gr., (#2 Alloy) 2.484" OAL

Powder	Sug. Starting Grains	Velocity F.P.S.	Pressure C.U.P.	Max. Grains	Velocity F.P.S.	Pressure C.U.P.
700X	6.0	1102	26,400	7.5	1273	37,700
Green Dot	6.5	1131	25,800	8.5	1306	35,500
PB	6.5	1085	24,000	8.5	1279	38,100
Unique	7.0	1203	24,000	10.4	1519	36,900
SR-7625	6.5	1087	22,800	9.0	1302	38,900
Herco	9.0	1301	27,600	11.0	1472	36,400
**SR-4759	15.0	1596	24,300	16.7	1733	32,900
**RX7	19.8	1632	20,800	26.0	2055	32,800
**IMR-3031	22.5	1628	20,800	26.0	1943	33,700
**748	23.3	1631	19,700	33.0	2184	33,400
**H335	22.5	1628	22,800	29.0	2181	32,700
**760	*28.4	1625	18,000	35.5+	2134	33,800

Note: Loads shown in shaded panels are maximum.
*Designates potentially most accurate load.
+Designates a compressed powder charge.
**Signifies CCI 200 primers used.

.300 Savage

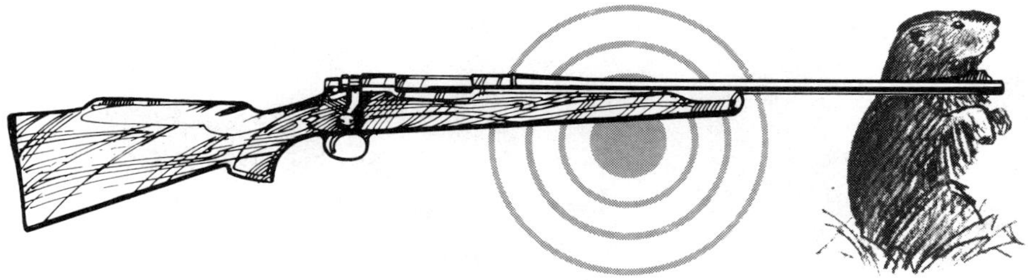

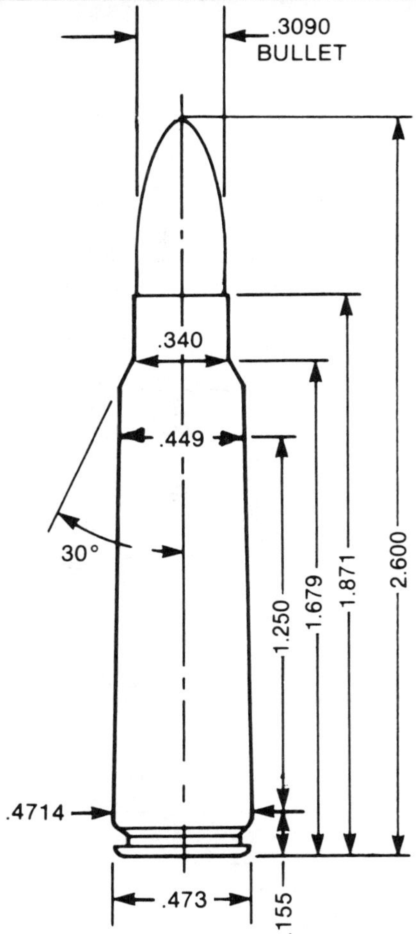

TEST COMPONENTS:

Cases .... Federal & Remington
Trim-to Length .... 1.865″
Primers .... Federal 210
Primer Size .... Large Rifle
Lyman Shell Holder .... No. 2
Cast Bullets Used .... (size to .309″ dia.)
*Gas Check Bullets — *#311359, 113 gr.; *#311576, 120 gr.; *#311440, 151 gr.; *#311466, 151 gr.; *#311291, 169 gr.; *#311467, 178 gr.

TEST SPECIFICATIONS:
(Velocity & Pressure)

Firearm Used .... Universal Receiver
Barrel Length .... 25″
Twist .... 1-12″
Groove Diameter .... .309″

COMMENTS:

Best results were obtained with the 151 grain and heavier cast bullets.

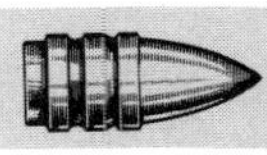

#311359
120 gr., (#2 Alloy) 2.280″ OAL

Powder	Sug. Starting Grains	Velocity F.P.S.	Pressure C.U.P.	Max. Grains	Velocity F.P.S.	Pressure C.U.P.
Red Dot	8.0	1490	25,200	11.0	1845	39,900
700X	8.0	1475	25,200	11.0	1775	42,000
Green Dot	8.5	1525	22,800	12.0	1850	39,900
PB	9.0	1490	25,800	12.0	1800	41,600
Unique	9.0	1530	—	13.0	1960	36,400
SR-7625	9.5	1550	28,200	12.5	1815	42,900
SR-4756	10.0	1555	23,400	14.0	1920	41,600

Note: Loads shown in shaded panels are maximum.

#311576
120 gr., (#2 Alloy) 2.244" OAL

Powder	Sug. Starting Grains	Velocity F.P.S.	Pressure C.U.P.	Max. Grains	Velocity F.P.S.	Pressure C.U.P.
Red Dot	8.0	1480	25,200	11.5	1825	39,400
700X	8.0	1470	26,400	11.0	1780	42,000
Green Dot	8.5	1505	24,600	12.0	1845	38,600
PB	9.0	1495	26,400	12.0	1770	40,700
Unique	9.0	1530	—	13.5	1955	37,700
SR-7625	9.0	1485	25,200	12.0	1760	42,400
SR-4756	10.0	1560	25,200	14.0	1900	39,400

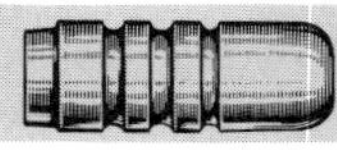

#311440
151 gr., (#2 Alloy) 2.315" OAL

Powder	Sug. Starting Grains	Velocity F.P.S.	Pressure C.U.P.	Max. Grains	Velocity F.P.S.	Pressure C.U.P.
Red Dot	8.0	1335	26,400	11.0	1600	41,600
700X	8.0	1315	28,800	10.0	1515	39,000
Green Dot	8.5	1370	27,000	11.5	1625	38,100
PB	9.0	1335	28,200	11.0	1550	40,700
Unique	9.0	1385	23,400	12.5	1700	37,300
SR-7625	9.0	1320	25,200	10.0	1505	38,600
SR-4756	10.0	1385	23,400	13.0	1635	39,000
**630	*18.5	1912	35,400	22.5	2219	38,100
**SR-4759	20.0	1992	29,200	23.2	2378	41,800
**RX7	24.7	1906	18,400	33.0	2453	40,700
**IMR-3031	26.5	1853	17,600	36.5	2544	39,900
**748	30.5	1955	18,500	43.0	2594	37,200

#311466
151 gr., (#2 Alloy) 2.435" OAL

Powder	Sug. Starting Grains	Velocity F.P.S.	Pressure C.U.P.	Max. Grains	Velocity F.P.S.	Pressure C.U.P.
Red Dot	8.0	1295	23,400	11.0	1570	39,000
700X	8.0	1310	24,600	10.5	1535	40,300
Green Dot	8.5	1330	25,200	11.5	1595	38,100
PB	9.0	1320	25,200	11.0	1500	37,300
Unique	9.0	1360	22,800	13.0	1705	37,700
SR-7625	9.0	1305	24,600	11.0	1495	38,100
SR-4756	10.0	1375	24,000	13.5	1665	41,100
**630	19.0	1943	23,800	22.7	2228	38,200
**SR-4759	21.0	1983	25,400	24.5	2236	41,500
**RX7	27.0	1961	17,500	33.4	2455	40,100
**IMR-3031	28.0	1866	16,400	37.1	2522	39,200
**748	32.0	1879	13,300	43.0	2625	37,000

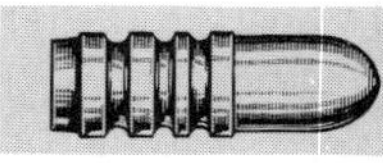

#311291
169 gr., (#2 Alloy) 2.358" OAL

Powder	Sug. Starting Grains	Velocity F.P.S.	Pressure C.U.P.	Max. Grains	Velocity F.P.S.	Pressure C.U.P.
Red Dot	7.5	1185	24,000	10.5	1460	37,700
700X	8.0	1245	26,400	10.0	1420	39,000
Green Dot	8.0	1205	23,400	11.0	1485	36,400
PB	9.0	1250	30,500	11.0	1410	42,000
Unique	8.5	1235	22,200	13.0	1625	39,400
SR-7625	9.0	1240	30,500	11.0	1415	41,600
SR-4756	10.0	1310	25,800	12.5	1520	39,900
**630	18.0	1815	26,400	*20.8	2044	38,200
**SR-4759	20.0	1888	28,800	22.5	2047	40,200
**RX7	22.3	1812	19,300	32.4	2339	39,500
**IMR-3031	26.7	1823	18,700	35.0	2388	38,400
**748	29.4	1817	17,000	41.4	2456	36,600

Note: Loads shown in shaded panels are maximum.
*Designates potentially most accurate load.
**Signifies use of Remington cases.

.300 Savage (Continued)

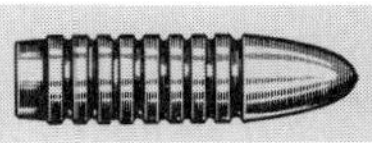

#311467
178 gr., (#2 Alloy) 2.525" OAL

Powder	Sug. Starting Grains	Velocity F.P.S.	Pressure C.U.P.	Max. Grains	Velocity F.P.S.	Pressure C.U.P.
Red Dot	7.5	1185	24,600	10.5	1425	38,100
700X	8.0	1230	28,800	10.0	1400	39,900
Green Dot	8.0	1210	25,200	11.0	1470	38,100
PB	8.5	1205	27,000	10.5	1370	38,100
Unique	8.5	1240	22,800	13.0	1600	38,100
SR-7625	8.5	1195	25,800	10.5	1360	36,400
SR-4756	10.0	1290	24,600	12.0	1450	36,000
**630	16.8	1699	20,500	21.3	2018	37,100
**SR-4759	18.0	1685	22,400	20.8	1920	36,900
**RX7	26.0	1746	15,800	32.0	2198	33,900
**IMR-3031	26.0	1689	15,100	30.0	2103	29,700
**748	28.0	1737	15,700	*36.0	2113	23,200

Note: Loads shown in shaded panels are maximum.
*Designates potentially most accurate load.
**Signifies use of Remington cases.

.308 Winchester

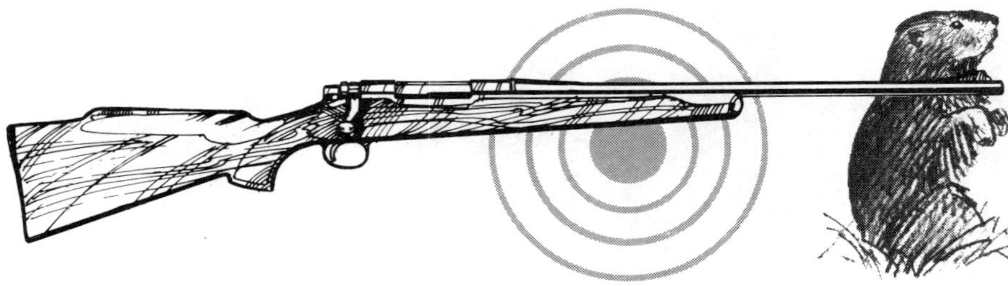

(7.62mm NATO)

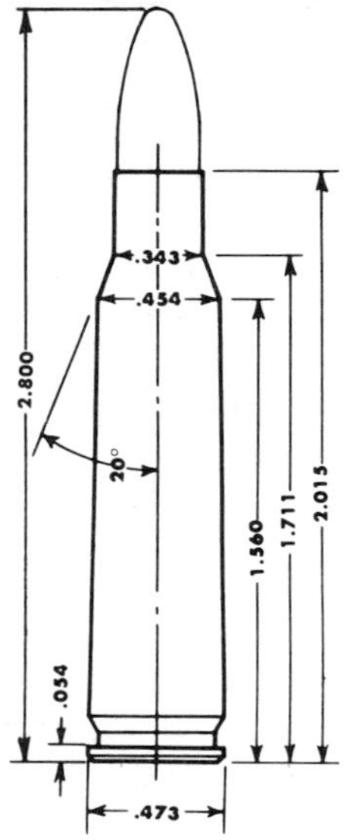

TEST COMPONENTS:

Cases .... Winchester
Trim-to Length .... 2.005″
Primers .... Winchester 8½-120
Primer Size .... Large Rifle
Lyman Shell Holder .... No. 2
Cast Bullets Used .... (size to .308″ dia.)
*Gas Check Bullet *#311359, 113 gr.
*#311576, 120 gr.
*#311465, 122 gr.
*#311440, 151 gr.
*#311466, 151 gr.
NRA Paper Patched #301618, 160 gr.
*#311291, 169 gr.
*#31141, 170 gr.
*#311467, 178 gr.
*#311407, 173 gr.
*#311334, 187 gr.
NRA Paper Patched 301620, 200 gr.

COMMENTS:

If military cases are used, maximum loads should be reduced by 2 full grains. This cartridge is capable of exceptionally fine accuracy with most component combinations.

The best bullet for higher than normal cast bullet velocities (2500 f.p.s. to 2600 f.p.s.) seems to be #311465, based on our 50 yard accuracy tests.

Refer to the section on techniques for paper patching with regard to bullets #301618 and #301620.

TEST SPECIFICATIONS:
(Velocity & Pressure)

Firearm Used .... Universal Receiver
Barrel Length .... 26″
Twist .... 1-12″
Groove Dia. .... .308″

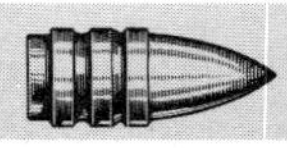

#311359
113 gr., (#2 Alloy) 2.800″ OAL, Max.

Powder	Sug. Starting Grains	Velocity F.P.S.	Pressure C.U.P.	Max. Grains	Velocity F.P.S.	Pressure C.U.P.
Red Dot	10.0	1623	24,000	14.0	2000	39,400
700X	9.0	1540	24,600	12.5	1879	40,300
Green Dot	11.0	1692	25,200	15.0	2049	40,300
PB	10.5	1601	24,600	13.5	1879	38,600
Unique	12.0	1831	24,600	17.0	2272	40,700
SR-7625	12.0	1704	26,400	15.0	1953	39,400
Herco	14.0	1886	25,800	17.0	2159	38,600

#311576
120 gr., (#2 Alloy) 2.800″ OAL, Max.

Powder	Sug. Starting Grains	Velocity F.P.S.	Pressure C.U.P.	Max. Grains	Velocity F.P.S.	Pressure C.U.P.
Red Dot	10.0	1617	25,200	14.0	1968	38,600
700X	9.0	1508	24,600	12.3	1845	37,700
Green Dot	11.0	1683	25,800	15.0	2012	40,300
PB	10.5	1575	25,800	13.5	1838	39,900
Unique	12.0	1810	24,000	17.0	2252	40,700
SR-7625	11.0	1600	27,000	14.0	1851	39,000
Herco	14.0	1879	28,800	17.0	2136	39,400

#311465
122 gr., (#2 Alloy) 2.450″ OAL

Powder	Sug. Starting Grains	Velocity F.P.S.	Pressure C.U.P.	Max. Grains	Velocity F.P.S.	Pressure C.U.P.
630	15.0	1755	15,700	*25.6	2575	45,400
RX7	20.0	1761	12,100	34.7	2605	32,200
748	25.0	1735	12,300	41.2	2596	31,900
H4895	24.2	1735	11,700	39.7	2595	30,000
IMR-4064	27.2	1791	13,500	40.0	2594	31,300

Note: Loads shown in shaded panels are maximum.

#311440
151 gr., (#2 Alloy) 2.325″ OAL

Powder	Sug. Starting Grains	Velocity F.P.S.	Pressure C.U.P.	Max. Grains	Velocity F.P.S.	Pressure C.U.P.
Red Dot	10.0	1445	26,400	13.5	1742	40,300
700X	9.0	1364	26,400	12.0	1630	39,900
Green Dot	10.5	1472	27,000	13.5	1715	39,900
PB	10.5	1432	30,000	13.0	1609	39,900
Unique	11.5	1616	28,200	15.0	1920	39,900
SR-7625	11.5	1481	30,500	14.0	1661	39,400
Herco	13.5	1694	30,000	16.0	1858	38,600
SR-4756	12.5	1524	26,400	16.0	1795	39,000
630	18.5	1942	30,100	24.4	2301	47,300
2400	17.0	1788	26,000	24.0	2239	46,100
SR-4759	18.0	1790	24,000	26.8	2332	49,300
IMR-4227	18.0	1753	25,300	25.7	2237	46,100
IMR-4198	24.1	1970	26,300	31.6	2479	47,400
RX7	24.1	2006	26,900	36.0	2611	44,600

#311466
151 gr., (#2 Alloy) 2.525″ OAL

Powder	Sug. Starting Grains	Velocity F.P.S.	Pressure C.U.P.	Max. Grains	Velocity F.P.S.	Pressure C.U.P.
Red Dot	10.0	1466	27,600	13.0	1730	40,700
700X	9.0	1407	28,200	11.5	1625	39,900
Green Dot	10.5	1461	25,200	13.5	1724	38,600
PB	10.0	1405	29,400	12.5	1613	39,400
Unique	11.0	1586	26,400	15.0	1933	41,000
SR-7625	11.0	1490	30,500	13.0	1650	37,700
Herco	13.0	1623	27,600	15.5	1824	39,000
SR-4756	12.0	1512	24,600	16.0	1826	39,900
SR-4759	17.5	1760	22,600	27.0	2401	47,800
IMR-4227	20.0	1771	18,500	31.0	2503	48,300
IMR-4198	21.8	1781	15,200	36.0+	2727	49,900
RX7	23.0	1779	14,800	39.5	2748	47,100
IMR-3031	27.0	1767	15,300	42.0+	2810	49,700
748	38.4	1767	16,500	48.0+	2920	48,000

#301618
160 gr., (#2 Alloy) 2.505″ OAL

Powder	Sug. Starting Grains	Velocity F.P.S.	Pressure C.U.P.	Max. Grains	Velocity F.P.S.	Pressure C.U.P.
RX7	20.2	1739	23,500	33.8	2594	51,600
H4895	25.0	1747	22,900	40.0+	2641	43,500
IMR-4064	26.8	1786	23,200	42.0+	2718	47,800
IMR-4320	25.8	1760	22,900	*41.0	2649	51,100
760	29.8	1797	22,100	43.5	2622	46,100

Note: Loads shown in shaded panels are maximum.
*Designates potentially most accurate load.
+Designates a compressed powder charge.

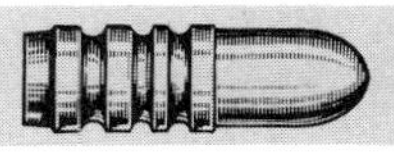

#311291
169 gr., (#2 Alloy) 2.510″ OAL

Powder	Sug. Starting Grains	Velocity F.P.S.	Pressure C.U.P.	Max. Grains	Velocity F.P.S.	Pressure C.U.P.
Red Dot	10.0	1367	28,800	12.5	1576	39,900
700X	8.0	1209	23,400	11.5	1547	39,900
Green Dot	11.0	1434	30,000	13.0	1582	37,700
PB	9.0	1220	22,200	12.5	1553	38,100
Unique	10.0	1368	23,400	15.0	1816	37,700
SR-7625	10.0	1290	28,800	13.0	1560	39,900
Herco	12.0	1479	26,400	15.5	1718	37,700
SR-4756	12.0	1416	30,000	14.5	1607	37,700
SR-4759	19.5	1847	22,300	26.0	2255	47,800
IMR-4227	21.6	1867	22,400	27.8	2270	49,500
IMR-4198	24.0	1897	22,700	30.5	2341	48,200
RX7	23.0	1870	21,200	41.5	2602	49,700
IMR-3031	*28.5	1868	20,200	39.5+	2653	49,800
748	31.7	1853	21,100	46.0+	2764	49,300

.308 Winchester (Continued)

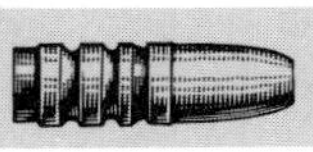

#31141
170 gr., (#2 Alloy) 2.617" OAL

Powder	Sug. Starting Grains	Velocity F.P.S.	Pressure C.U.P.	Max. Grains	Velocity F.P.S.	Pressure C.U.P.
Red Dot	10.0	1355	27,000	12.5	1565	39,000
700X	8.0	1206	24,000	11.5	1505	37,700
Green Dot	11.0	1434	28,800	13.0	1587	35,500
PB	9.0	1201	23,400	12.5	1479	40,300
Unique	10.0	1377	21,600	15.0	1802	37,700
SR-7625	10.0	1272	24,600	13.0	1499	38,600
Herco	12.0	1470	25,800	15.5	1733	37,700
SR-4756	12.0	1392	27,000	14.5	1587	37,700
SR-4759	21.5	1897	26,100	27.7	2345	50,300
IMR-4198	25.0	1898	22,700	34.0	2460	38,700
RX7	27.5	1924	24,100	39.0	2627	48,900
IMR-3031	29.0	1858	23,500	41.8	2709	51,200
748	32.0	1856	20,300	47.0	2790	50,400
H335	30.0	1847	19,800	44.2	2787	50,200

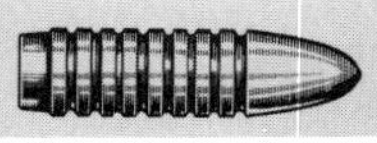

#311467
178 gr., (#2 Alloy) 2.750" OAL

Powder	Sug. Starting Grains	Velocity F.P.S.	Pressure C.U.P.	Max. Grains	Velocity F.P.S.	Pressure C.U.P.
Red Dot	10.0	1352	27,600	12.5	1555	39,900
700X	8.0	1200	22,200	11.5	1513	39,900
Green Dot	11.0	1424	30,000	13.0	1570	37,300
PB	9.5	1240	22,200	12.5	1508	39,400
Unique	10.0	1373	23,400	14.5	1730	38,100
SR-7625	10.5	1320	27,000	13.0	1508	36,900
Herco	12.0	1457	25,800	15.5	1712	37,300
SR-4756	12.0	1379	24,600	14.5	1600	40,700
SR-4759	19.0	1813	28,600	26.3	2243	47,600
IMR-4198	22.5	1782	24,400	32.0	2376	48,100
RX7	23.0	1783	25,000	34.5	2409	41,100
748	30.0	1851	23,000	44.5	2269	47,800
H335	*26.5	2007	29,500	40.0	2575	46,400

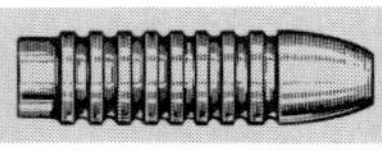

#311407
173 gr., (#2 Alloy) 2.605" OAL

Powder	Sug. Starting Grains	Velocity F.P.S.	Pressure C.U.P.	Max. Grains	Velocity F.P.S.	Pressure C.U.P.
Red Dot	10.0	1342	28,800	12.5	1538	39,400
700X	8.0	1188	24,000	11.5	1479	39,900
Green Dot	10.5	1369	29,400	12.5	1527	37,300
PB	9.0	1195	22,800	12.5	1478	39,900
Unique	10.0	1360	23,400	14.5	1715	39,900
SR-7625	10.0	1254	24,600	12.5	1470	37,700
Herco	12.5	1479	29,400	15.0	1655	37,700
SR-4756	12.0	1388	26,400	14.5	1577	36,400
2400	17.1	1702	26,900	*24.0	2134	48,300
SR-4759	18.7	1851	31,700	23.8	2146	47,400
RX7	24.8	1975	29,600	34.6	2488	48,700
IMR-3031	28.0	1958	26,700	37.1	2486	47,300
748	29.9	1981	26,700	41.0	2510	40,100
H335	26.7	1870	24,200	38.0	2488	41,600
H4895	28.0	1948	26,700	38.4+	2479	42,800

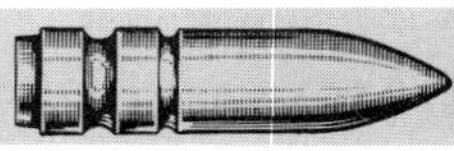

#311334
187 gr., (#2 Alloy) 2.795" OAL

Powder	Sug. Starting Grains	Velocity F.P.S.	Pressure C.U.P.	Max. Grains	Velocity F.P.S.	Pressure C.U.P.
Red Dot	9.5	1270	25,800	12.5	1492	38,000
700X	8.0	1142	23,400	11.2	1398	36,900
Green Dot	10.0	1275	25,200	12.5	1461	35,000
PB	9.0	1146	23,400	12.5	1418	39,000
Unique	9.5	1291	21,600	14.5	1680	38,600
SR-7625	10.0	1222	24,000	12.5	1402	37,000
Herco	12.0	1404	27,000	15.0	1610	37,700
SR-4756	12.0	1331	25,800	14.5	1524	35,000
2400	16.4	1602	25,100	23.6	2056	47,100
SR-4759	17.0	1605	24,000	*26.3	2137	48,200
RX7	23.9	1884	28,300	34.9	2380	43,200
IMR-3031	27.5	1802	23,700	37.0	2404	46,800
H335	29.2	1958	25,100	43.5	2500	44,700
H4895	28.3	1879	25,900	40.8	2497	45,500

Note: Loads shown in shaded panels are maximum.
*Designates potentially most accurate load.

#301620
200 gr., (#2 Alloy) 2.600" OAL

Powder	Sug. Starting Grains	Velocity F.P.S.	Pressure C.U.P.	Max. Grains	Velocity F.P.S.	Pressure C.U.P.
RX7	19.5	1643	26,600	32.2	2280	49,500
H4895	23.8	1673	25,200	38.4	2427	49,200
IMR-4064	24.6	1647	22,800	39.3+	2466	51,300
IMR-4320	24.8	1658	24,800	*36.0	2301	51,700
760	27.8	1670	24,400	41.0	2418	49,500

Note: Loads shown in shaded panels are maximum.
*Designates potentially most accurate load.
+Designates a compressed powder charge.

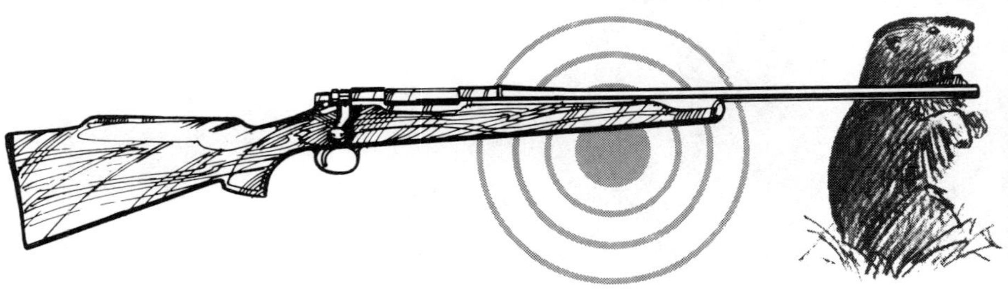

.30/40 Krag

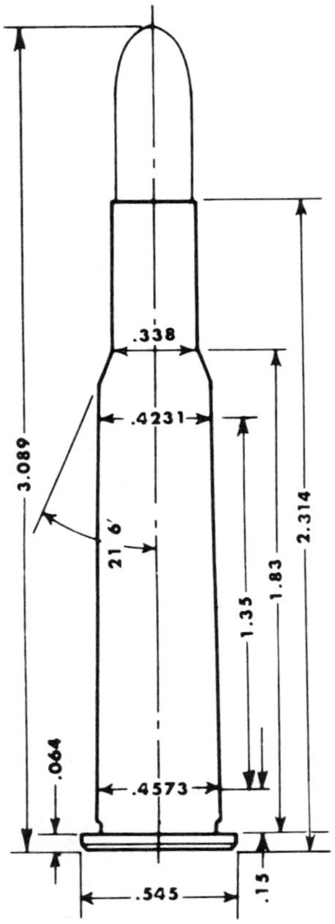

TEST COMPONENTS:

Cases .... Remington
Trim-to Length .... 2.304″
Primers .... Remington 9½
Primer Size .... Large Rifle
Lyman Shell Holder .... No. 7
Cast Bullets Used .... (size to .308″ dia.)
*Gas Check Bullets — *#311359, 113 gr.; *#311466, 151 gr.; *#311291, 169 gr.; *#311407, 173 gr.; *#311290, 208 gr.

TEST SPECIFICATIONS:
(Velocity)

Firearm Used .... Krag
Barrel Length .... 22″
Twist .... 1-10″
Groove Diameter .... .3095″

COMMENTS:

Excellent accuracy was obtained with cast bullets of 173 grains and heavier. Caution — A good many Krag rifles which we have inspected show cracks around the locking lug on the bolt. Have your rifle inspected by a gunsmith before using it.

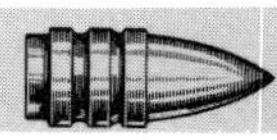

#311359
113 gr., (#2 Alloy) 3.089″ OAL, Max.

Powder	Sug. Starting Grains	Velocity F.P.S.	Pressure C.U.P.	Max. Grains	Velocity F.P.S.	Pressure C.U.P.
Unique	9.0	1492	—	13.0	1872	—
2400	19.0	1782	—	23.0	2061	—
IMR-4227	20.0	1709	—	25.0	2061	—

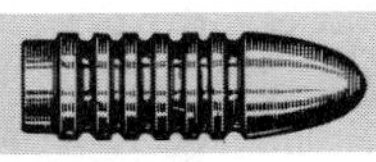

#311466
151 gr., (#2 Alloy) 3.089″ OAL, Max.

Powder	Sug. Starting Grains	Velocity F.P.S.	Pressure C.U.P.	Max. Grains	Velocity F.P.S.	Pressure C.U.P.
Unique	9.0	1381	—	13.0	1694	—
2400	19.0	1712	—	24.0	2036	—
IMR-4227	19.0	1628	—	24.0	1934	—
IMR-4198	24.0	1818	—	28.0	2032	—

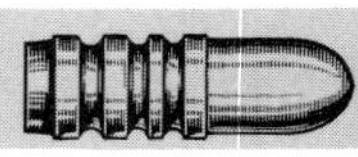

#311291
169 gr., (#2 Alloy) 3.089″ OAL, Max.

Powder	Sug. Starting Grains	Velocity F.P.S.	Pressure C.U.P.	Max. Grains	Velocity F.P.S.	Pressure C.U.P.
Unique	8.0	1220	—	12.0	1562	—
2400	18.0	1647	—	23.0	1964	—
IMR-4227	19.0	1633	—	24.0	1923	—
IMR-4198	23.0	1739	—	27.0	1980	—

Note: Loads shown in shaded panels are maximum.

.30/40 Krag (Continued)

#311407
173 gr., (#2 Alloy) 3.089" OAL, Max.

Powder	Sug. Starting Grains	Velocity F.P.S.	Pressure C.U.P.	Max. Grains	Velocity F.P.S.	Pressure C.U.P.
Unique	8.0	1204	—	11.0	1445	—
2400	17.0	1547	—	22.0	1854	—
IMR-4227	18.0	1549	—	23.0	1808	—
IMR-4198	22.0	1700	—	26.0	1901	—

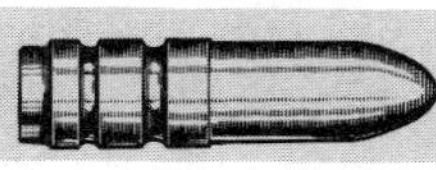

#311290
208 gr., (#2 Alloy) 3.089" OAL, Max.

Powder	Sug. Starting Grains	Velocity F.P.S.	Pressure C.U.P.	Max. Grains	Velocity F.P.S.	Pressure C.U.P.
Unique	8.0	1138	—	11.0	1355	—
2400	16.0	1415	—	21.0	1716	—
IMR-4227	17.0	1413	—	22.0	1689	—
IMR-4198	21.0	1554	—	25.0	1769	—

Note: Loads shown in shaded panels are maximum.

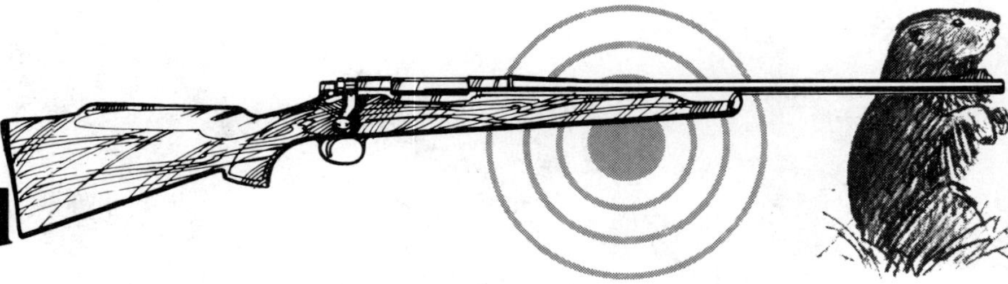

.30/06 Springfield
(7.62 x 63mm)

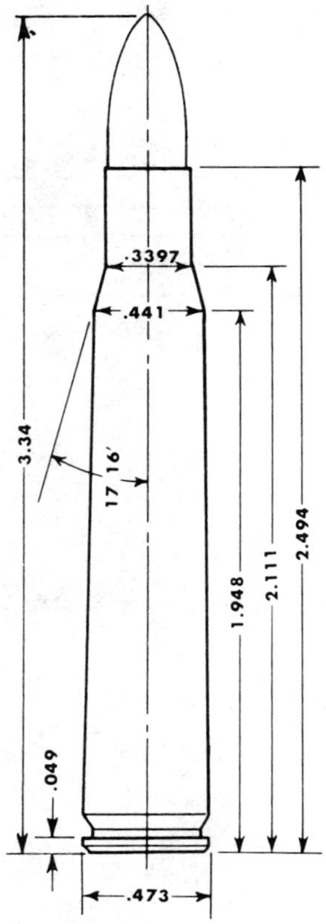

TEST COMPONENTS:

Cases .... Winchester
Trim-to Length .... 2.484″
Primers .... Winchester 8½-120
Primer Size .... Large Rifle
Lyman Shell Holder .... No. 2
Cast Bullets Used .... (size to .308″ dia.)
*Gas Check Bullets *#311359, 113 gr.
*#311465, 122 gr.
*#311440, 151 gr.
*#311466, 151 gr.
NRA Paper Patched #301618, 160 gr.
*#311291, 169 gr.
*#31141, 170 gr.
*#311334, 187 gr.
NRA Paper Patched #301620, 200 gr.
*#311299, 200 gr.
*#311290, 208 gr.
*#311284, 210 gr.

TEST SPECIFICATIONS:
(Velocity & Pressure)

Firearm Used .... Universal Receiver
Barrel Length .... 26″
Twist .... 1-10″
Groove Diameter .... .308″

COMMENTS:

Cast bullets #311291, #31141, and #311290 have proven to give outstanding accuracy. Bullet #311334 performs well in 03A3, 2 groove, issue Springfield barrels.

The best bullet for higher than normal cast bullet velocities (2500 f.p.s. to 2600 f.p.s.) seems to be #311465, based on our 50 yard accuracy tests.

Refer to the section on techniques for paper patching with regard to bullets #301618 and #301620.

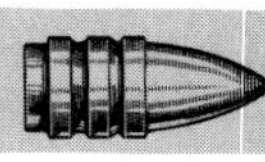

#311359
113 gr., (#2 Alloy) 3.340″ OAL, Max.

Powder	Sug. Starting Grains	Velocity F.P.S.	Pressure C.U.P.	Max. Grains	Velocity F.P.S.	Pressure C.U.P.
Red Dot	11.0	1683	23,400	15.0	2000	36,900
700X	10.0	1610	22,000	14.0	1953	38,100
Green Dot	13.0	1792	28,800	16.0	2020	38,600
PB	12.0	1689	27,000	15.5	1941	39,000
Unique	14.0	1960	27,000	19.0	2331	42,000
SR-7625	12.5	1677	21,600	16.5	2008	42,400
Herco	14.5	1876	28,200	18.0	2132	39,000

Note: Loads shown in shaded panels are maximum.

#311465
122 gr., (#2 Alloy) 2.880" OAL

Powder	Sug. Starting Grains	Velocity F.P.S.	Pressure C.U.P.	Max. Grains	Velocity F.P.S.	Pressure C.U.P.
Red Dot	11.0	1630	22,400	15.0	1953	36,400
700X	10.0	1567	20,400	15.0	1976	42,000
Green Dot	13.0	1773	31,000	16.0	1984	40,300
PB	12.0	1639	23,400	15.5	1901	38,600
Unique	13.0	1824	24,600	18.0	2207	38,100
SR-7625	12.5	1661	23,400	16.5	1976	42,900
Herco	14.5	1848	29,400	18.0	2083	40,700
630	17.5	1739	13,700	*29.0	2619	41,200
2400	18.0	1685	12,400	32.0	2617	37,600
SR-4759	20.0	1752	12,000	32.0	2602	39,300
RX7	21.0	1634	10,400	38.0	2626	31,900
748	28.0	1656	9,500	47.0	2618	28,000

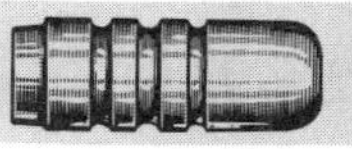

#311440
151 gr., (#2 Alloy) 2.963" OAL

Powder	Sug. Starting Grains	Velocity F.P.S.	Pressure C.U.P.	Max. Grains	Velocity F.P.S.	Pressure C.U.P.
Red Dot	11.0	1497	25,200	14.0	1724	36,400
700X	10.0	1420	25,200	13.5	1691	38,600
Green Dot	12.0	1545	29,400	15.0	1727	40,900
PB	11.5	1455	24,600	15.0	1709	40,300
Unique	12.5	1628	25,800	17.0	1964	38,100
SR-7625	12.5	1529	28,800	15.5	1736	39,000
Herco	13.5	1631	30,000	17.0	1858	39,000
SR-4756	13.5	1572	26,400	17.0	1805	36,000
630	17.0	1643	15,700	29.0	2456	47,400
2400	21.0	1739	16,100	31.0	2454	43,600
SR-4759	21.5	1728	15,300	33.0	2497	47,300
IMR-4227	23.0	1776	16,500	34.5	2493	44,800
IMR-4198	25.0	1736	13,700	36.0	2456	32,800
RX7	23.5	1669	13,700	37.0	2483	35,900
IMR-3031	*31.0	1775	13,600	42.0	2462	30,300

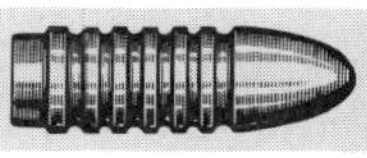

#311466
151 gr., (#2 Alloy) 3.000" OAL

Powder	Sug. Starting Grains	Velocity F.P.S.	Pressure C.U.P.	Max. Grains	Velocity F.P.S.	Pressure C.U.P.
Red Dot	11.0	1479	24,600	14.0	1715	35,500
700X	10.0	1408	22,800	13.5	1694	37,300
Green Dot	12.0	1547	29,400	15.0	1748	38,600
PB	11.5	1449	24,000	15.0	1697	38,600
Unique	12.5	1636	25,200	17.0	1941	37,300
SR-7625	12.5	1512	27,000	15.5	1733	39,400
Herco	13.5	1618	26,400	17.0	1858	39,400
SR-4756	13.5	1572	26,400	17.0	1805	37,300
630	17.0	1660	16,600	28.0	2414	45,000
2400	*20.0	1723	16,400	30.5	2446	42,000
SR-4759	19.5	1652	13,700	32.0	2477	48,100
IMR-4227	21.5	1706	14,900	34.0	2495	44,500
IMR-4198	23.0	1670	12,100	37.0	2515	38,000
RX7	23.0	1704	13,700	36.0	2476	38,500
IMR-3031	29.0	1682	12,000	42.5	2522	34,100

#301618
160 gr., (#2 Alloy) 2.923" OAL

Powder	Sug. Starting Grains	Velocity F.P.S.	Pressure C.U.P.	Max. Grains	Velocity F.P.S.	Pressure C.U.P.
RX7	23.0	1687	14,700	39.0	2543	45,200
H4895	28.0	1706	14,500	*45.0	2656	45,800
IMR-4064	29.0	1713	14,100	46.0	2731	48,300
IMR-4320	29.0	1744	15,400	46.0	2648	46,900
760	35.0	1747	13,300	50.0	2720	46,400

Note: Loads shown in shaded panels are maximum.
*Designates potentially most accurate load.

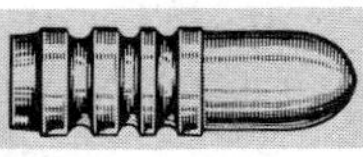

#311291
169 gr., (#2 Alloy) 3.013″ OAL

Powder	Sug. Starting Grains	Velocity F.P.S.	Pressure C.U.P.	Max. Grains	Velocity F.P.S.	Pressure C.U.P.
Red Dot	10.5	1390	24,000	13.5	1625	35,000
700X	9.5	1307	21,000	13.0	1580	36,400
Green Dot	11.5	1436	26,400	14.0	1597	35,500
PB	11.0	1340	21,000	14.5	1595	36,900
Unique	12.0	1508	24,000	16.5	1842	36,400
SR-7625	11.5	1363	23,400	14.5	1585	36,000
Herco	12.5	1488	26,400	16.0	1697	35,500
SR-4756	12.5	1418	22,200	16.5	1692	35,500
630	18.0	1638	18,600	27.8	2304	46,900
2400	18.6	1635	18,200	29.5	2296	43,100
SR-4759	20.0	1666	16,700	31.0	2325	46,900
IMR-4198	25.0	1729	15,100	38.5	2501	45,100
RX7	25.0	1733	15,100	38.3	2480	47,600
IMR-3031	29.0	1710	13,900	42.0	2469	38,200
748	30.0	1721	14,300	46.5	2531	38,700
H4895	28.0	1705	14,400	44.0	2526	43,100

#31141
170 gr., (#2 Alloy) 2.968″ OAL

Powder	Sug. Starting Grains	Velocity F.P.S.	Pressure C.U.P.	Max. Grains	Velocity F.P.S.	Pressure C.U.P.
Red Dot	10.5	1355	22,800	13.5	1585	32,500
700X	9.5	1288	22,800	13.0	1545	36,000
Green Dot	11.5	1416	25,800	14.0	1582	34,000
PB	11.0	1326	23,400	14.5	1557	37,700
Unique	12.0	1504	24,000	16.5	1811	35,500
SR-7625	11.5	1346	24,000	14.5	1570	36,900
Herco	13.0	1497	27,000	16.0	1689	36,000
SR-4756	12.5	1406	22,800	16.5	1653	36,400
630	*18.0	1662	15,900	27.8	2332	48,700
2400	20.0	1751	22,200	27.0	2141	34,500
SR-4759	20.0	1642	13,900	31.5	2329	47,400
IMR-4198	25.0	1717	13,100	38.5	2463	43,100
RX7	25.0	1661	12,800	40.5	2515	46,900
IMR-3031	30.0	1636	11,100	44.0	2522	40,800
748	31.0	1695	12,800	47.0	2503	37,500
H4895	28.5	1706	14,100	45.0	2496	40,700

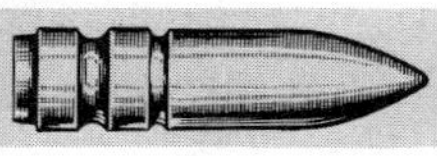

#311334
187 gr., (#2 Alloy) 3.247″ OAL

Powder	Sug. Starting Grains	Velocity F.P.S.	Pressure C.U.P.	Max. Grains	Velocity F.P.S.	Pressure C.U.P.
Red Dot	10.0	1278	22,800	13.5	1514	36,000
700X	9.5	1236	23,400	13.0	1483	36,400
Green Dot	11.5	1347	25,800	14.0	1499	35,500
PB	11.0	1275	26,700	14.5	1499	38,600
Unique	11.5	1404	24,000	16.5	1715	36,000
SR-7625	11.5	1289	25,200	14.5	1488	37,300
Herco	13.0	1432	28,200	16.0	1607	38,100
SR-4756	12.5	1353	25,200	16.5	1595	36,200
2400	20.0	1666	25,800	27.0	2040	32,500
IMR-4227	20.0	1570	19,000	28.0	1976	29,400
IMR-4198	23.0	1653	19,200	31.0	2040	28,800
SR-4759	22.0	1667	16,900	31.5	2247	49,300
RX7	24.0	1687	17,500	39.0	2329	41,600
IMR-3031	30.0	1727	16,300	45.0	2492	44,200
H4895	30.0	1708	16,100	45.0	2448	46,900
IMR-4064	32.0	1641	14,100	46.0	2454	38,800
IMR-4320	32.0	1649	14,800	46.0	2435	42,000
IMR-4350	35.0	1693	15,500	50.0	2452	42,600

#301620
200 gr., (#2 Alloy) 3.101″ OAL

Powder	Sug. Starting Grains	Velocity F.P.S.	Pressure C.U.P.	Max. Grains	Velocity F.P.S.	Pressure C.U.P.
RX7	25.0	1684	20,000	*36.5	2276	46,600
H4895	29.5	1679	18,000	42.3	2370	47,500
IMR-4064	32.0	1682	15,400	44.0	2434	47,900
760	31.0	1648	16,500	45.0	2412	48,600
IMR-4350	35.0	1664	15,700	48.0	2417	46,800

Note: Loads shown in shaded panels are maximum.
*Designates potentially most accurate load.

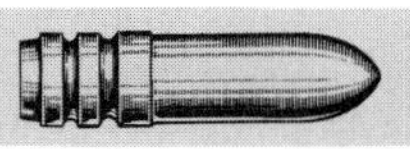

#311299
200 gr., (#2 Alloy) 3.167″ OAL

Powder	Sug. Starting Grains	Velocity F.P.S.	Pressure C.U.P.	Max. Grains	Velocity F.P.S.	Pressure C.U.P.
SR-4759	21.5	1620	18,000	*30.2	2131	47,600
RX7	22.0	1650	18,200	36.0	2228	42,500
IMR-3031	29.0	1726	15,900	42.0	2373	42,500
H4895	29.0	1637	15,400	42.5	2299	39,900
IMR-4064	30.5	1703	16,800	45.0	2423	44,400
IMR-4320	31.0	1664	16,800	45.0	2365	40,300
IMR-4350	36.0	1740	17,600	49.0	2403	44,800

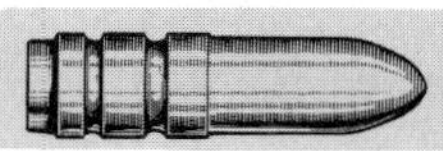

#311290
208 gr., (#2 Alloy) 3.065″ OAL

Powder	Sug. Starting Grains	Velocity F.P.S.	Pressure C.U.P.	Max. Grains	Velocity F.P.S.	Pressure C.U.P.
Red Dot	10.0	1215	24,000	13.5	1447	34,500
700X	9.5	1172	23,800	12.5	1381	35,500
Green Dot	11.5	1302	28,200	14.0	1443	39,000
PB	11.0	1218	27,000	14.0	1398	37,300
Unique	11.5	1345	25,800	16.0	1633	36,900
SR-7625	11.5	1246	27,000	14.0	1406	37,300
Herco	13.0	1379	30,500	16.0	1550	39,000
SR-4756	12.5	1307	25,800	16.0	1501	39,900
2400	19.0	1562	23,400	26.5	1968	35,000
IMR-4227	20.0	1572	24,600	26.5	1865	31,000
IMR-4198	24.5	1691	24,000	30.5	2020	33,000
SR-4759	17.5	1719	29,700	*28.6	2041	45,400
RX7	24.3	1625	19,200	35.8	2154	40,900
IMR-3031	30.0	1700	17,100	41.5	2324	46,000
H4895	30.0	1668	18,200	43.0	2313	47,700
IMR-4064	32.0	1667	16,300	44.5	2389	48,100
IMR-4320	32.0	1736	19,800	45.0	2375	47,600
IMR-4350	35.0	1697	18,000	50.0	2398	46,500

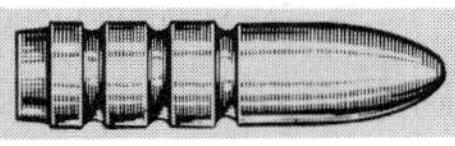

#311284
210 gr., (#2 Alloy) 3.063″ OAL

Powder	Sug. Starting Grains	Velocity F.P.S.	Pressure C.U.P.	Max. Grains	Velocity F.P.S.	Pressure C.U.P.
Red Dot	10.5	1244	27,600	13.5	1418	39,400
Green Dot	11.0	1248	26,400	13.5	1402	36,400
PB	11.0	1194	27,000	14.0	1384	37,000
Unique	11.5	1326	26,400	15.5	1577	36,400
SR-7625	11.5	1242	30,000	13.0	1320	36,400
Herco	12.5	1335	30,500	15.5	1510	39,000
SR-4756	12.5	1264	26,400	15.5	1445	35,000
2400	19.0	1592	27,000	24.0	1827	32,000
SR-4759	*22.0	1679	27,800	27.8	2004	45,300
RX7	25.0	1685	22,200	38.2	2254	47,000
H4895	29.0	1730	20,900	43.0	2335	45,600
IMR-4064	32.0	1769	22,100	44.2	2364	46,000
IMR-4350	35.8	1774	19,200	50.0	2422	47,800

Note: Loads shown in shaded panels are maximum.
*Designates potentially most accurate load.

.300 Holland & Holland Magnum

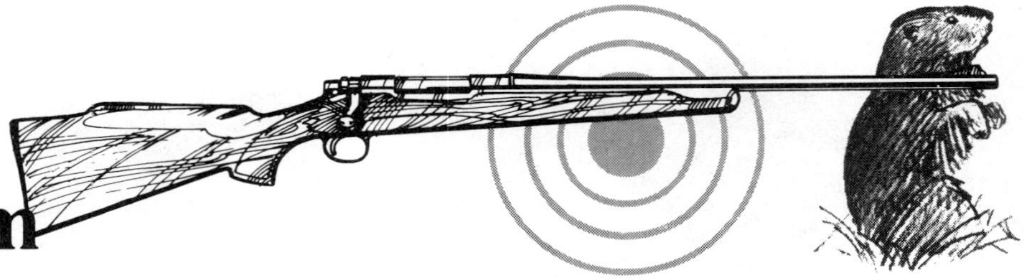

(7.63 x 72mm)

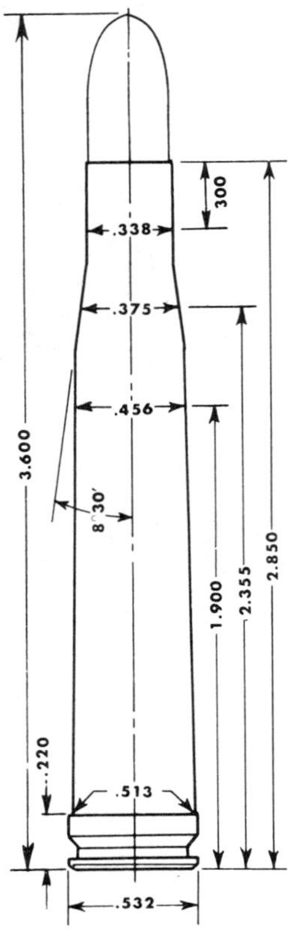

TEST COMPONENTS:

Cases .... Winchester/Western
Trim-to Length .... 2.840″
Primers .... Winchester 8½-120
Primer Size .... Large Rifle
Lyman Shell Holder .... No. 13
Cast Bullets Used .... (size to .308″ dia.)
*Gas Check Bullets *#311466, 151 gr.
*#311291, 169 gr.
*#311334, 187 gr.

TEST SPECIFICATIONS: (Velocity)

Firearm Used .... Remington Model 721
Barrel Length .... 26″
Twist .... 1-10″
Groove Diameter .... .308″

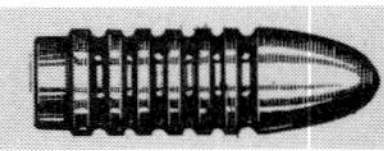

#311466
151 gr., (#2 Alloy) 3.345″ OAL

Powder	Sug. Starting Grains	Velocity F.P.S.	Pressure C.U.P.	Max. Grains	Velocity F.P.S.	Pressure C.U.P.
Red Dot	12.0	1490	—	15.0	1695	—
700X	12.0	1495	—	15.0	1700	—
Green Dot	13.0	1545	—	16.0	1735	—
PB	13.0	1485	—	16.5	1710	—
Unique	15.0	1665	—	18.0	1850	—
SR-7625	13.5	1500	—	17.0	1720	—
630	18.0	1675	—	22.0	1900	—

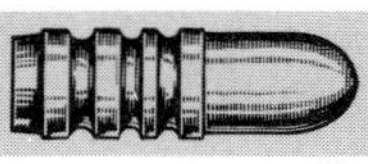

#311291
169 gr., (#2 Alloy) 3.335″ OAL

Powder	Sug. Starting Grains	Velocity F.P.S.	Pressure C.U.P.	Max. Grains	Velocity F.P.S.	Pressure C.U.P.
Red Dot	12.0	1400	—	15.0	1615	—
700X	12.0	1385	—	15.0	1590	—
Green Dot	13.0	1445	—	16.0	1640	—
PB	13.0	1380	—	16.5	1595	—
Unique	15.0	1570	—	18.0	1750	—
SR-7625	13.5	1395	—	17.0	1620	—
630	18.0	1585	—	22.0	1810	—

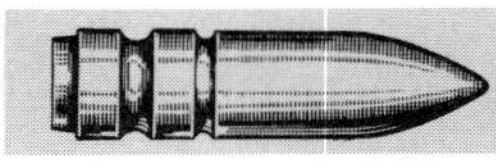

#311334
187 gr., (#2 Alloy) 3.610″ OAL

Powder	Sug. Starting Grains	Velocity F.P.S.	Pressure C.U.P.	Max. Grains	Velocity F.P.S.	Pressure C.U.P.
Red Dot	12.5	1365	—	15.0	1515	—
700X	12.5	1370	—	15.0	1525	—
Green Dot	13.5	1420	—	16.5	1595	—
PB	13.5	1360	—	16.5	1535	—
Unique	15.5	1505	—	18.0	1655	—
SR-7625	14.0	1380	—	17.0	1550	—
630	18.5	1560	—	22.0	1745	—

Note: Loads shown in shaded panels are maximum.

.300 Winchester Magnum

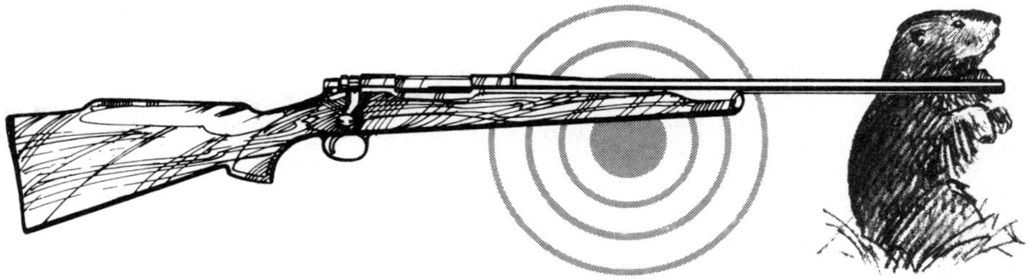

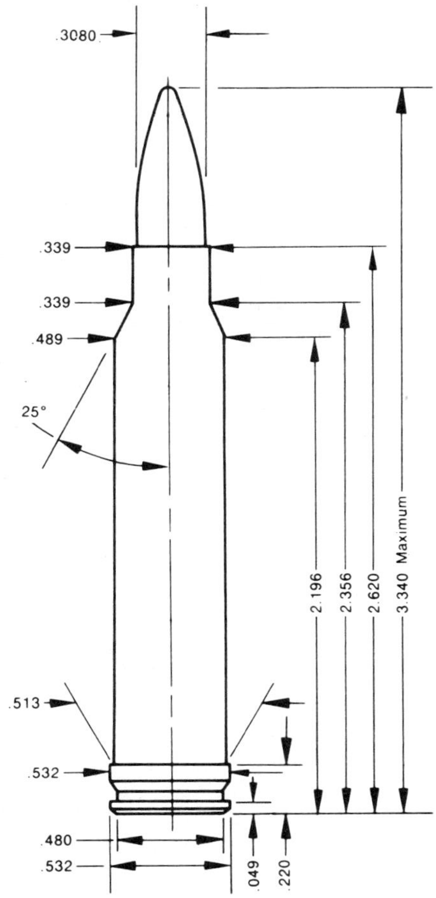

TEST COMPONENTS:

Cases .... SAKO
Trim-to Length .... 2.610″
Primers .... Winchester 8½-120
Primer Size .... Large Rifle
Lyman Shell Holder .... No. 13
Cast Bullets Used .... (size to .308″ dia.)
*Gas Check Bullets *#311440, 151 gr.
*#311466, 151 gr.
*#311291, 169 gr.
*#311467, 178 gr.
*#311407, 173 gr.
*#311334, 187 gr.

TEST SPECIFICATIONS:
(Velocity & Pressure)

Firearm Used .... Universal Receiver
Barrel Length .... 23″
Twist .... 1-10″
Groove Diameter .... .308″

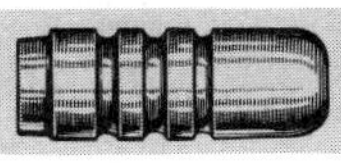

#311440
151 gr., (#2 Alloy) 3.038″ OAL

Powder	Sug. Starting Grains	Velocity F.P.S.	Pressure C.U.P.	Max. Grains	Velocity F.P.S.	Pressure C.U.P.
Red Dot	16.0	1720	28,200	19.0	1890	37,700
700X	15.0	1660	28,800	17.0	1770	34,000
Green Dot	16.0	1710	25,200	19.0	1885	35,500
PB	15.5	1600	27,000	18.5	1770	35,000
Unique	18.0	1810	25,200	22.0	2035	36,900
SR-7625	16.5	1650	30,000	19.5	1810	37,700
630	21.0	1765	20,400	26.0	2020	27,000

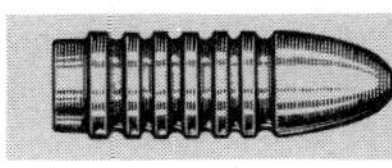

#311466
151 gr., (#2 Alloy) 3.185″ OAL

Powder	Sug. Starting Grains	Velocity F.P.S.	Pressure C.U.P.	Max. Grains	Velocity F.P.S.	Pressure C.U.P.
Red Dot	16.0	1725	29,400	19.5	1910	39,000
700X	15.0	1660	29,400	18.0	1840	38,600
Green Dot	16.0	1700	27,600	19.5	1890	35,500
PB	15.5	1605	26,400	18.5	1765	35,500
Unique	18.0	1790	26,400	22.0	2000	35,500
SR-7625	16.0	1620	28,200	19.0	1785	39,000
630	21.0	1770	21,600	26.0	2040	28,200

.300 Winchester Magnum (Continued)

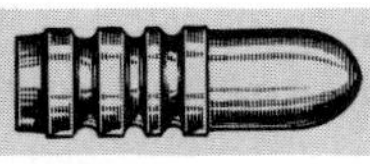

#311291
169 gr., (#2 Alloy) 3.135" OAL

Powder	Sug. Starting Grains	Velocity F.P.S.	Pressure C.U.P.	Max. Grains	Velocity F.P.S.	Pressure C.U.P.
Red Dot	16.0	1640	31,000	19.0	1790	40,700
700X	14.0	1515	28,800	17.0	1690	39,400
Green Dot	16.0	1620	30,000	19.0	1775	38,600
PB	15.0	1495	28,200	18.5	1660	36,900
Unique	18.0	1705	29,400	22.0	1910	38,600
SR-7625	15.0	1480	27,600	18.0	1650	36,000
630	21.0	1710	20,400	26.0	1965	27,600
SR-4759	30.5	2075	28,400	40.0	2472	50,900
RX7	37.0	2184	27,200	48.0	2575	46,100
748	45.0	2176	25,400	60.0	2729	46,400
H4895	40.5	2157	25,900	56.0	2748	51,200
IMR-4064	42.0	2073	21,100	60.0	2810	45,900

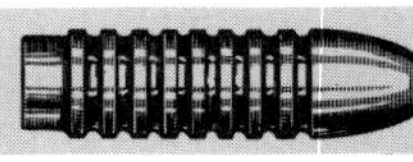

#311407
173 gr., (#2 Alloy) 3.180" OAL

Powder	Sug. Starting Grains	Velocity F.P.S.	Pressure C.U.P.	Max. Grains	Velocity F.P.S.	Pressure C.U.P.
Red Dot	15.0	1535	25,800	18.0	1680	35,000
700X	14.0	1470	24,600	17.5	1665	36,400
Green Dot	14.0	1520	23,400	18.5	1705	33,000
PB	15.0	1445	25,200	19.0	1655	37,700
Unique	17.5	1635	25,800	21.5	1840	31,500
SR-7625	15.0	1435	26,400	19.0	1650	36,400
630	21.5	1700	20,400	26.5	1925	25,800
SR-4759	29.5	1982	27,700	39.0	2366	48,300
RX7	36.0	2051	25,900	48.0	2534	48,200
748	43.5	2041	22,900	57.5	2665	47,600
	39.0	2079	25,100	52.0	2623	49,400
IMR-4064	44.0	2160	25,500	60.0	2811	50,200

#311467
178 gr., (#2 Alloy) 3.240" OAL

Powder	Sug. Starting Grains	Velocity F.P.S.	Pressure C.U.P.	Max. Grains	Velocity F.P.S.	Pressure C.U.P.
Red Dot	15.0	1565	30,500	18.0	1720	37,300
700X	14.0	1480	25,800	17.0	1640	36,900
Green Dot	15.0	1535	25,200	18.5	1720	34,500
PB	15.0	1470	26,400	18.5	1650	37,300
Unique	17.5	1650	27,000	21.5	1845	37,300
SR-7625	15.0	1460	24,600	18.5	1650	36,900
630	21.5	1675	20,400	26.5	1915	23,400
SR-4759	*30.0	2025	28,400	40.0	2403	49,500
RX7	37.0	2050	25,100	48.0	2514	47,300
748	45.3	2069	22,100	59.0	2676	46,500
H4895	41.5	2194	29,600	54.5	2680	50,200
IMR-4064	41.0	2014	21,000	60.0	2808	46,400

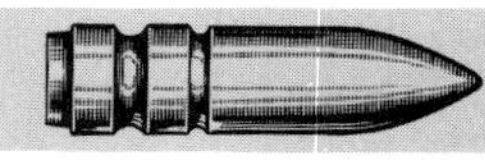

#311334
187 gr., (#2 Alloy) 3.370" OAL

Powder	Sug. Starting Grains	Velocity F.P.S.	Pressure C.U.P.	Max. Grains	Velocity F.P.S.	Pressure C.U.P.
Red Dot	15.0	1495	25,200	18.0	1650	34,000
700X	14.0	1435	25,200	17.5	1620	35,500
Green Dot	15.0	1490	24,000	18.5	1665	32,000
PB	15.0	1425	25,200	19.0	1615	36,400
Unique	17.5	1605	26,400	22.5	1845	38,600
SR-7625	15.0	1420	25,200	19.0	1615	36,400
630	21.0	1645	20,400	26.0	1870	24,600
SR-4759	32.3	1856	24,400	39.0	2319	50,100
RX7	37.0	2049	27,900	50.0	2500	49,600
H4895	39.5	2056	25,800	55.7	2567	48,500
IMR-4064	43.0	2027	21,700	61.5	2755	50,500
760	52.0	2102	24,900	70.5	2775	49,900

Note: Loads shown in shaded panels are maximum.
*Designates potentially most accurate load.

.308 Norma Magnum

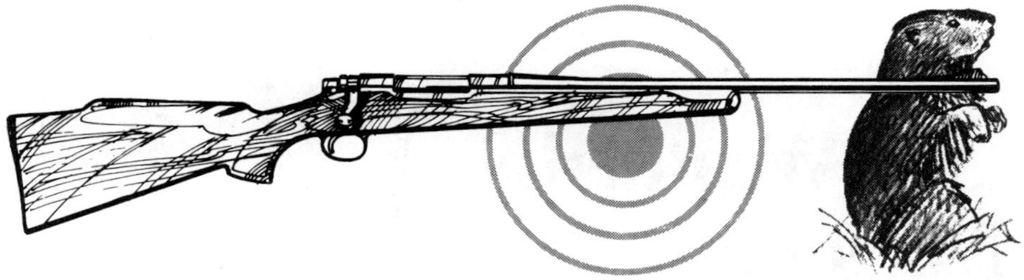

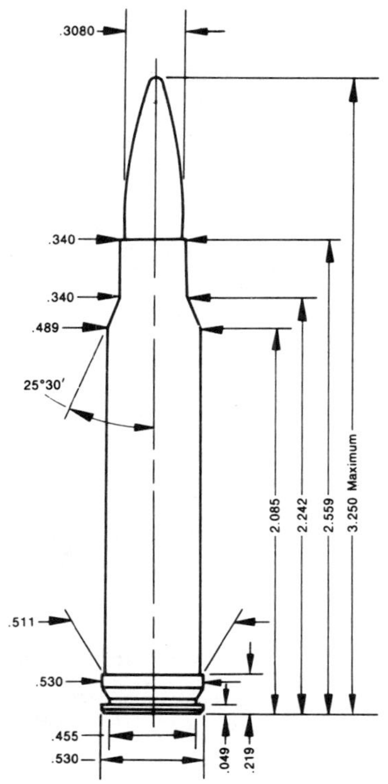

TEST COMPONENTS:

Cases	Norma
Trim-to Length	2.550″
Primers	Remington 9½
Primer Size	Large Rifle
Lyman Shell Holder	No. 13
Cast Bullets Used	(size to .308″ dia.)
*Gas Check Bullets	*#311466, 151 gr.
	*#311291, 169 gr.
	*#311334, 187 gr.

TEST SPECIFICATIONS: (Velocity)

Firearm Used	Schultz & Larsen 65 DL
Barrel Length	24″
Twist	1-10″
Groove Diameter	.308″

COMMENTS:

Our test rifle had a considerable amount of free-bore. On custom rifles which are not free-bored, maximum loads should be reduced by a full 5% from those listed. Even with this reduction, such loads should be approached with caution.

We suggest that only standard large rifle primers be used with cast bullet loads.

#311466
151 gr., (#2 Alloy) 3.250″ OAL, Max.

Powder	Sug. Starting Grains	Velocity F.P.S.	Pressure C.U.P.	Max. Grains	Velocity F.P.S.	Pressure C.U.P.
Unique	15.0	1700	—	18.0	1893	—
2400	24.0	1845	—	27.0	2012	—
IMR-4227	24.0	1715	—	28.0	1949	—

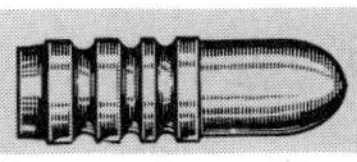

#311291
169 gr., (#2 Alloy) 3.250″ OAL, Max.

Powder	Sug. Starting Grains	Velocity F.P.S.	Pressure C.U.P.	Max. Grains	Velocity F.P.S.	Pressure C.U.P.
Unique	14.0	1569	—	17.0	1763	—
2400	22.0	1715	—	26.0	1919	—
IMR-4227	22.0	1600	—	27.0	1890	—

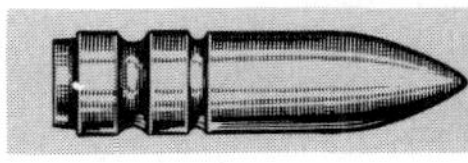

#311334
187 gr., (#2 Alloy) 3.250″ OAL, Max.

Powder	Sug. Starting Grains	Velocity F.P.S.	Pressure C.U.P.	Max. Grains	Velocity F.P.S.	Pressure C.U.P.
Unique	14.0	1488	—	16.0	1612	—
2400	20.0	1552	—	27.0	1879	—
IMR-4227	20.0	1412	—	27.0	1792	—

Note: Loads shown in shaded panels are maximum.

.300 Weatherby Magnum

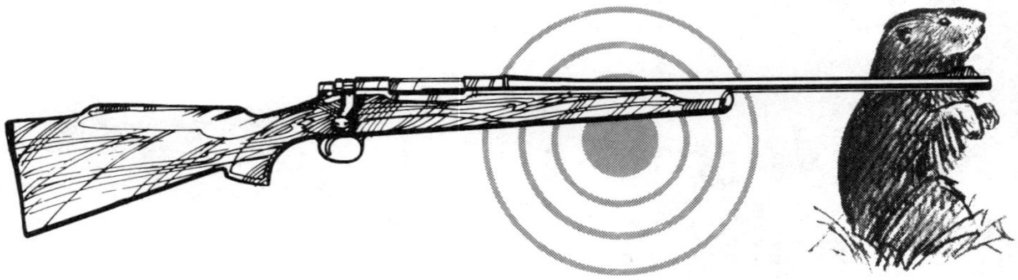

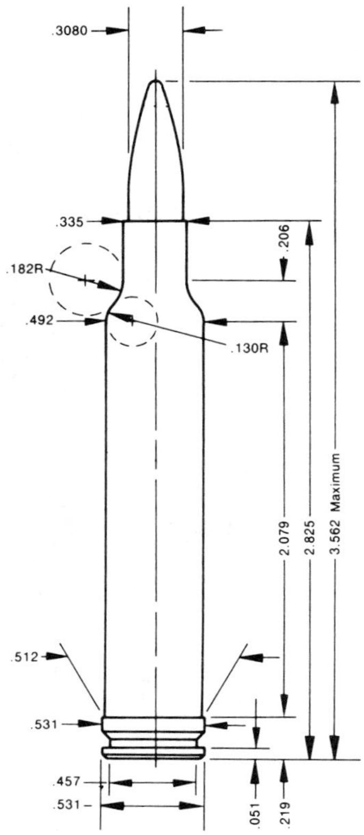

TEST COMPONENTS:

Cases .... Weatherby
Trim-to Length .... 2.815″
Primers .... Remington 9½
Primer Size .... Large Rifle
Lyman Shell Holder .... No. 13
Cast Bullets Used .... (size to .308″ dia.)
*Gas Check Bullets .... *#311291, 169 gr.
*#311334, 187 gr.
*#311290, 208 gr.

TEST SPECIFICATIONS:
(Velocity)

Firearm Used .... Weatherby Mark V
Barrel Length .... 26″
Twist .... 1-12″
Groove Diameter .... .3075″

COMMENTS:

The data listed for this cartridge were obtained in a Weatherby rifle and are intended for Weatherby rifles only. The free-boring constructed into these firearms allows higher velocities at safe working pressures. For custom rifles which are not free-bored, maximum loads should be reduced a full 5%. Even then, they should be approached with caution.

Use only standard large rifle primers with cast bullet data.

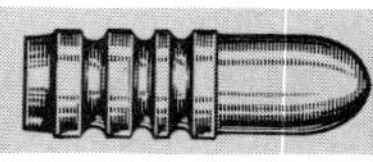

#311291
169 gr., (#2 Alloy) 3.562″ OAL, Max.

Powder	Sug. Starting Grains	Velocity F.P.S.	Pressure C.U.P.	Max. Grains	Velocity F.P.S.	Pressure C.U.P.
Unique	15.0	1572	—	19.0	1818	—
2400	23.0	1736	—	28.0	1976	—

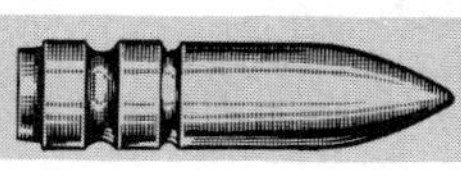

#311334
187 gr., (#2 Alloy) 3.562″ OAL, Max.

Powder	Sug. Starting Grains	Velocity F.P.S.	Pressure C.U.P.	Max. Grains	Velocity F.P.S.	Pressure C.U.P.
Unique	14.0	1449	—	18.0	1680	—
2400	22.0	1607	—	27.0	1828	—

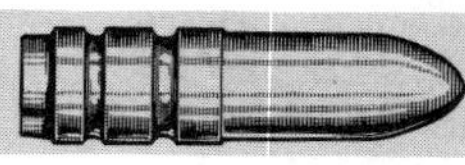

#311290
208 gr., (#2 Alloy) 3.562″ OAL, Max.

Powder	Sug. Starting Grains	Velocity F.P.S.	Pressure C.U.P.	Max. Grains	Velocity F.P.S.	Pressure C.U.P.
Unique	12.0	1265	—	16.0	1506	—
2400	20.0	1474	—	25.0	1718	—
IMR-4227	31.0	1838	—	35.0	2000	—

Note: Loads shown in shaded panels are maximum.

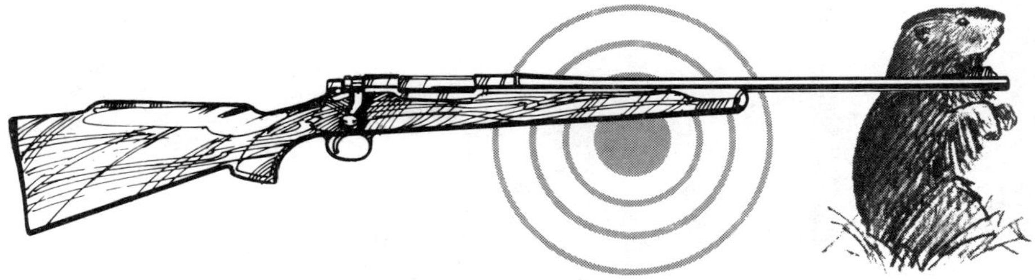

7.62 Russian

(7.62 x 54R mm)

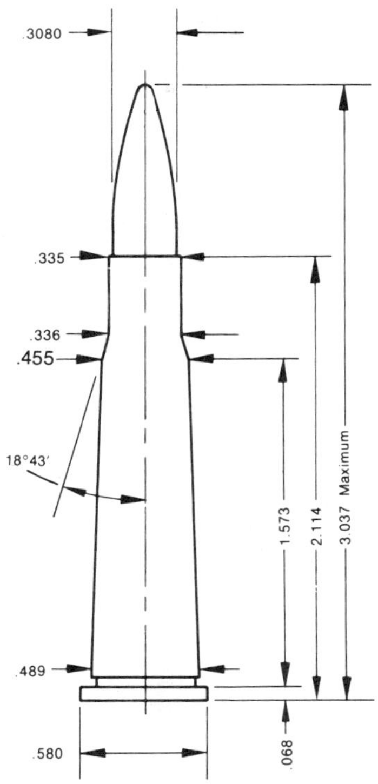

TEST COMPONENTS:

Cases .... Norma
Trim-to Length .... 2.105″
Primers .... Remington 9½
Primer Size .... Large Rifle
Lyman Shell Holder .... No. 17
Cast Bullets Used .... (size to .310″ dia.)
*Gas Check Bullets — *#311466, 151 gr.; *#311291, 169 gr.; *#311290, 208 gr.

TEST SPECIFICATIONS:
(Velocity)

Firearm Used .... Westinghouse/Russian
Barrel Length .... 24″
Twist .... 1-10″″
Groove Diameter .... .310″

COMMENTS:

Groove and bore dimensions of these Russian military rifles vary considerably. Most rifles have a groove diameter of .310″ and for these we recommend cast bullets of .310″ diameter. Due to the limitations of the chamber size, larger bullets are not recommended even in rifles which have groove diameters larger than .310″

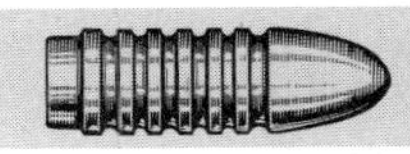

#311466
151 gr., (#2 Alloy) 3.037″ OAL, Max.

Powder	Sug. Starting Grains	Velocity F.P.S.	Pressure C.U.P.	Max. Grains	Velocity F.P.S.	Pressure C.U.P.
Unique	12.0	1592	—	15.0	1827	—
2400	23.5	1980	—	27.0	2197	—

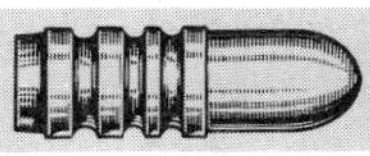

#311291
169 gr., (#2 Alloy) 3.037″ OAL, Max.

Powder	Sug. Starting Grains	Velocity F.P.S.	Pressure C.U.P.	Max. Grains	Velocity F.P.S.	Pressure C.U.P.
Unique	11.0	1461	—	14.0	1680	—
2400	22.0	1865	—	24.0	1941	—

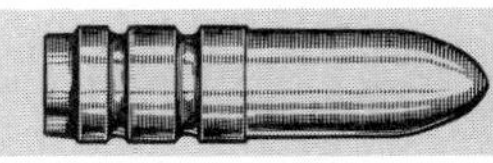

#311290
208 gr., (#2 Alloy) 3.037″ OAL, Max.

Powder	Sug. Starting Grains	Velocity F.P.S.	Pressure C.U.P.	Max. Grains	Velocity F.P.S.	Pressure C.U.P.
Unique	10.0	1261	—	13.0	1483	—
2400	17.0	1413	—	22.0	1754	—
IMR-4227	27.0	1882	—	30.0	2028	—

Note: Loads shown in shaded panels are maximum.

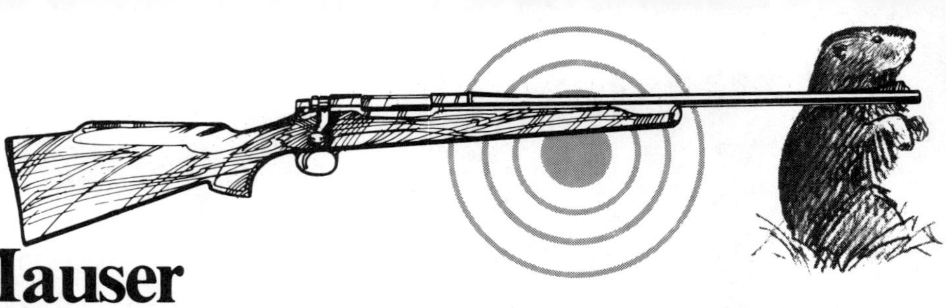

7.65 Argentine Mauser

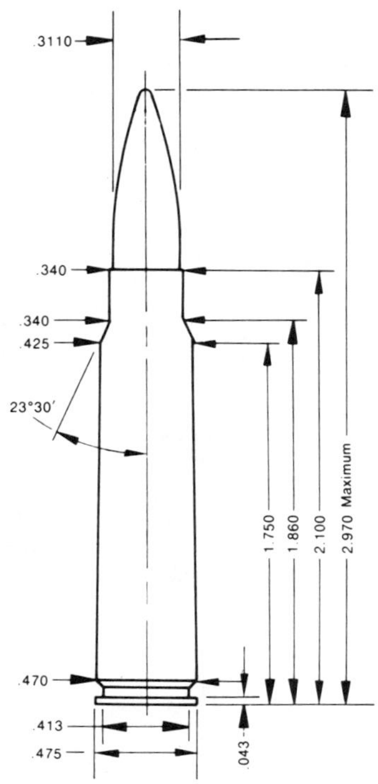

TEST COMPONENTS:

Cases .... Norma
Trim-to Length .... 2.100″
Primers .... Winchester 8½-120
Primer Size .... Large Rifle
Lyman Shell Holder .... No. 2
Cast Bullets Used .... (size to .312″ dia.)
*Gas Check Bullets *#311466, 151 gr.
*#311299, 200 gr.
*#311284, 210 gr.

TEST SPECIFICATIONS:
(Velocity)

Firearm Used .... Military Mauser Model 91
Barrel Length .... 29¼″
Twist .... 1-9 4/5″
Groove Diameter .... .312″

COMMENTS:

A wide variation of groove diameters exists with rifles chambered for this caliber. We recommend that you slug your barrel before reloading. #311284 gave best results in our test rifle.

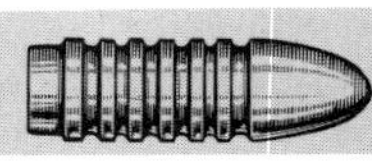

#311466
151 gr., (#2 Alloy) 2.680″ OAL

Powder	Sug. Starting Grains	Velocity F.P.S.	Pressure C.U.P.	Max. Grains	Velocity F.P.S.	Pressure C.U.P.
Red Dot	10.0	1475	—	14.0	1780	—
700X	10.0	1475	—	14.0	1785	—
Green Dot	11.0	1530	—	14.5	1795	—
PB	11.0	1470	—	15.0	1770	—
Unique	12.0	1590	—	15.0	1820	—
SR-7625	11.5	1485	—	15.5	1775	—
630	15.0	1600	—	20.0	1940	—

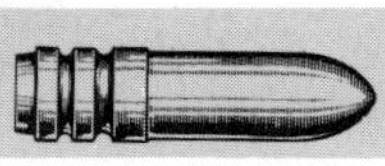

#311299
200 gr., (#2 Alloy) 2.853″ OAL

Powder	Sug. Starting Grains	Velocity F.P.S.	Pressure C.U.P.	Max. Grains	Velocity F.P.S.	Pressure C.U.P.
Red Dot	9.5	1260	—	12.5	1490	—
700X	9.0	1225	—	11.0	1380	—
Green Dot	10.0	1295	—	13.0	1515	—
PB	9.5	1200	—	12.0	1380	—
Unique	10.0	1285	—	13.0	1505	—
SR-7625	10.0	1230	—	12.5	1405	—
630	13.0	1380	—	18.0	1695	—

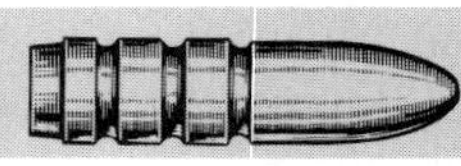

#311284
210 gr., (#2 Alloy) 2.855″ OAL

Powder	Sug. Starting Grains	Velocity F.P.S.	Pressure C.U.P.	Max. Grains	Velocity F.P.S.	Pressure C.U.P.
Red Dot	9.0	1190	—	12.0	1415	—
700X	9.0	1190	—	11.0	1345	—
Green Dot	9.5	1215	—	12.0	1400	—
PB	9.5	1160	—	12.0	1340	—
Unique	9.5	1200	—	13.0	1445	—
SR-7625	10.0	1195	—	12.5	1365	—
630	13.0	1330	—	18.0	1645	—

Note: Loads shown in shaded panels are maximum.

.303 British

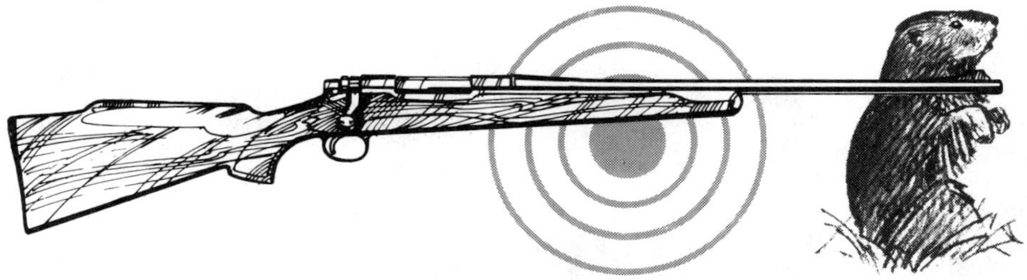

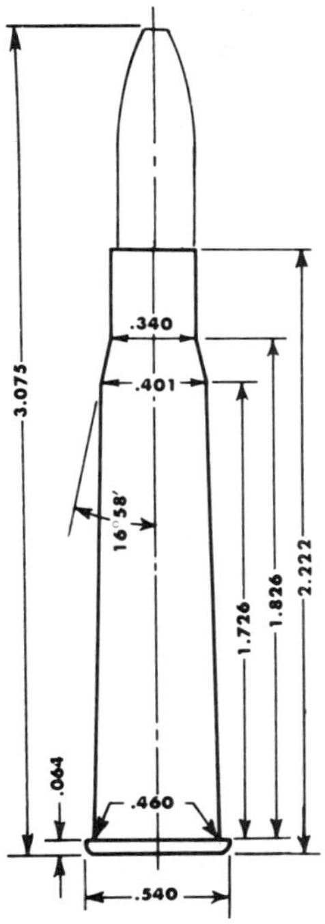

TEST COMPONENTS:

Cases .... Federal
Trim-to Length .... 2.212″
Primers .... Federal 210
Primer Size .... Large Rifle
Lyman Shell Holder .... No. 7
Cast Bullets Used .... (size to .312″ dia.)
*Gas Check Bullets .... *#311466, 151 gr.
*#311299, 200 gr.
*#311284, 210 gr.

TEST SPECIFICATIONS:
(Velocity)

Firearm Used .... Martini Single Shot
Barrel Length .... 22″
Twist .... 1-10″
Groove Diameter .... .312″

COMMENTS:

An extreme variation in groove diameters exists in rifles chambered for this cartridge. In rifles which we have checked, diameters range from .309″ to .317″. We suggest that you slug your barrel before reloading. Cast bullets should be sized as near groove diameter as possible.

#311466
151 gr., (#2 Alloy) 2.788″ OAL

Powder	Sug. Starting Grains	Velocity F.P.S.	Pressure C.U.P.	Max. Grains	Velocity F.P.S.	Pressure C.U.P.
Red Dot	10.0	1430	—	13.5	1695	—
700X	10.0	1445	—	13.0	1660	—
Green Dot	10.5	1460	—	14.0	1715	—
PB	10.5	1410	—	13.0	1585	—
Unique	11.0	1475	—	15.0	1760	—
SR-7625	11.0	1425	—	13.5	1610	—

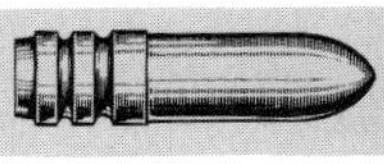

#311299
200 gr., (#2 Alloy) 2.930″ OAL

Powder	Sug. Starting Grains	Velocity F.P.S.	Pressure C.U.P.	Max. Grains	Velocity F.P.S.	Pressure C.U.P.
Red Dot	10.0	1290	—	13.0	1490	—
700X	9.5	1240	—	12.5	1460	—
Green Dot	10.0	1280	—	13.5	1515	—
PB	10.0	1220	—	12.5	1390	—
Unique	10.5	1295	—	14.0	1530	—
SR-7625	10.5	1250	—	12.5	1380	—

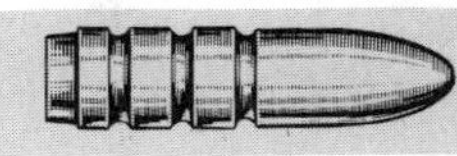

#311284
210 gr., (#2 Alloy) 3.005″ OAL

Powder	Sug. Starting Grains	Velocity F.P.S.	Pressure C.U.P.	Max. Grains	Velocity F.P.S.	Pressure C.U.P.
Red Dot	10.0	1245	—	12.5	1415	—
700X	9.0	1165	—	12.0	1375	—
Green Dot	9.5	1190	—	13.0	1425	—
PB	9.5	1130	—	12.0	1310	—
Unique	10.0	1210	—	13.5	1450	—
SR-7625	10.0	1170	—	12.0	1300	—

Note: Loads shown in shaded panels are maximum.

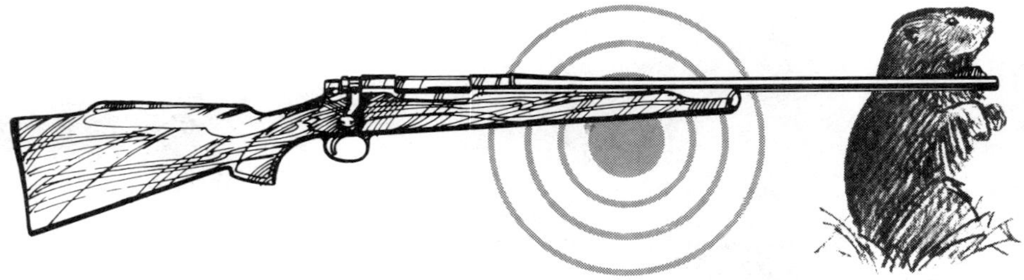

7.7mm Japanese

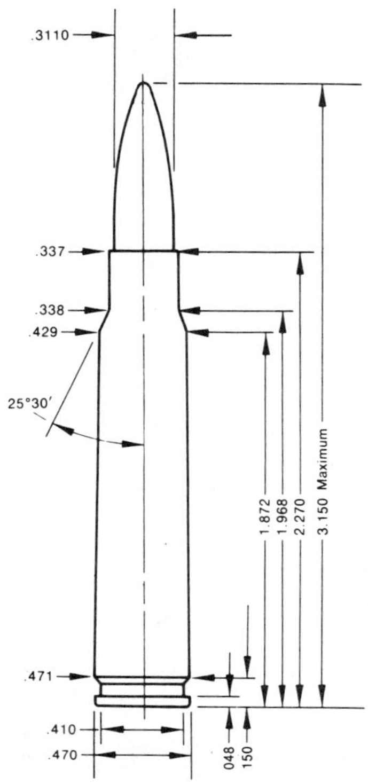

TEST COMPONENTS:

Cases .... Norma
Trim-to Length .... 2.260″
Primers .... Remington 9½
Primer Size .... Large Rifle
Lyman Shell Holder .... No. 2
Cast Bullets Used .... (size to .312″ dia.)
*Gas Check Bullets *#311466, 151 gr.
*#311299, 200 gr.
*#311284, 210 gr.

TEST SPECIFICATIONS:
(Velocity)

Firearm Used .... Japanese Arisaka
Barrel Length .... 22″
Twist .... 1-9¾″
Groove Diameter .... .312″

COMMENTS:

An extreme variation in groove diameters exists in rifles chambered for this cartridge. In rifles which we have checked, diameters range from .310″ to .317″. We suggest that you slug your barrel before reloading.

Due to wide variations in bullet and groove diameters, utmost care should be used when working up loads.

Cast bullets should be sized as close to groove diameter as possible.

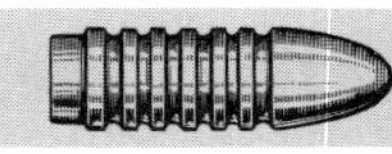

#311466
151 gr., (#2 Alloy) 2.844″ OAL

Powder	Sug. Starting Grains	Velocity F.P.S.	Pressure C.U.P.	Max. Grains	Velocity F.P.S.	Pressure C.U.P.
Red Dot	10.0	1331	—	12.0	1513	—
700X	9.5	1324	—	11.5	1466	—
PB	10.5	1235	—	13.0	1481	—
Unique	11.0	1424	—	15.0	1706	—
2400	22.0	1615	—	24.0	1801	—
IMR-4227	28.0	1838	—	30.0	1976	—

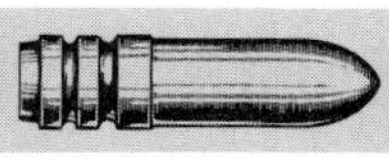

#311299
200 gr., (#2 Alloy) 3.005″ OAL

Powder	Sug. Starting Grains	Velocity F.P.S.	Pressure C.U.P.	Max. Grains	Velocity F.P.S.	Pressure C.U.P.
Red Dot	10.0	1185	—	12.0	1316	—
700X	9.5	1176	—	11.5	1290	—
PB	10.5	1166	—	12.5	1329	—
Unique	10.0	1223	—	14.0	1449	—
2400	21.0	1549	—	24.0	1763	—
IMR-4227	28.0	1805	—	30.0	1923	—

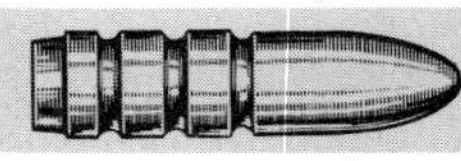

#311284
210 gr., (#2 Alloy) 3.005″ OAL

Powder	Sug. Starting Grains	Velocity F.P.S.	Pressure C.U.P.	Max. Grains	Velocity F.P.S.	Pressure C.U.P.
Red Dot	10.0	1183	—	12.0	1308	—
700X	9.5	1123	—	11.5	1274	—
PB	10.5	1142	—	12.5	1268	—
Unique	10.0	1158	—	13.0	1438	—
2400	21.0	1538	—	24.0	1751	—
IMR-4227	28.0	1760	—	30.0	1869	—

Note: Loads shown in shaded panels are maximum.

.32/20 Winchester

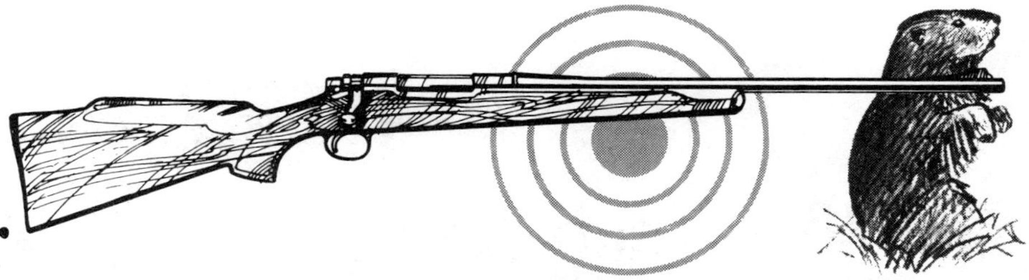

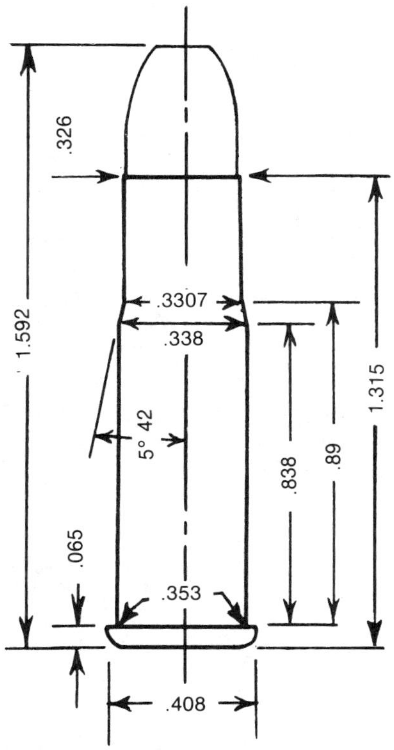

TEST COMPONENTS:

Cases .... Winchester/Western
Trim-to Length .... 1.305″
Primers .... Winchester 6½-116
Primer Size .... Small Rifle
Lyman Shell Holder .... No. 10
Cast Bullets Used .... (size to .313″ dia.)
*Gas Check Bullets .... *#311419, 91 gr.
*#311316, 115 gr.

TEST SPECIFICATIONS:
(Velocity)

Firearm Used .... Marlin Model 27S
Barrel Length .... 24″
Twist .... 1-22″
Groove Diameter .... .313″

COMMENTS:

Do not use these loads in rifles which were designed for black powder or in handguns.

Some cartridge cases may be encountered which are actually shorter than the listed "Trim-to-length." This will cause no problems in your loading, but such cases for this cartridge should be segregated according to length because of the different adjustments which will be necessary in your bullet seating die.

Some small variations in groove diameter exist in rifles chambered for this cartridge. It is advisable to slug your barrel and size cast bullets accordingly.

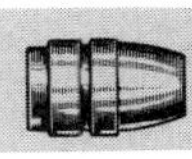

#311419
91 gr., (#2 Alloy) 1.570″ OAL

Powder	Sug. Starting Grains	Velocity F.P.S.	Pressure C.U.P.	Max. Grains	Velocity F.P.S.	Pressure C.U.P.
Red Dot	3.5	1135	—	4.5	1355	—
700X	3.0	1040	—	4.2	1340	—
Green Dot	3.8	1165	—	4.8	1380	—
PB	3.5	1050	—	5.2	1430	—
Unique	4.5	1185	—	6.0	1550	—
SR-7625	3.8	1085	—	5.8	1515	—

#311316
115 gr., (#2 Alloy) 1.555″ OAL

Powder	Sug. Starting Grains	Velocity F.P.S.	Pressure C.U.P.	Max. Grains	Velocity F.P.S.	Pressure C.U.P.
Red Dot	3.2	1030	—	4.1	1215	—
700X	2.8	940	—	3.9	1195	—
Green Dot	3.5	1045	—	4.5	1255	—
PB	3.2	955	—	4.9	1295	—
Unique	4.2	1110	—	5.5	1390	—
SR7625	3.5	985	—	5.5	1380	—

Note: Loads shown in shaded panels are maximum.

.32/40 Winchester

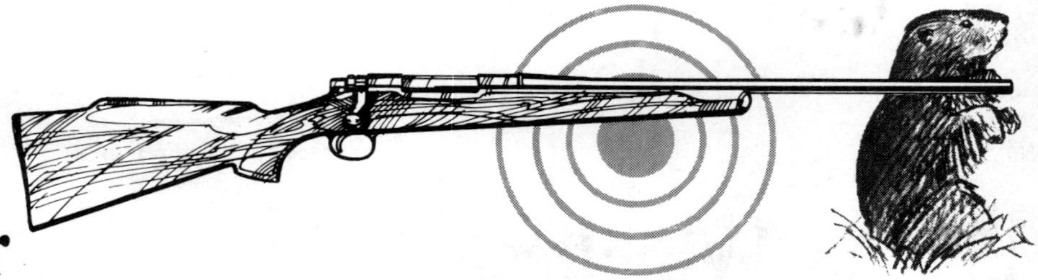

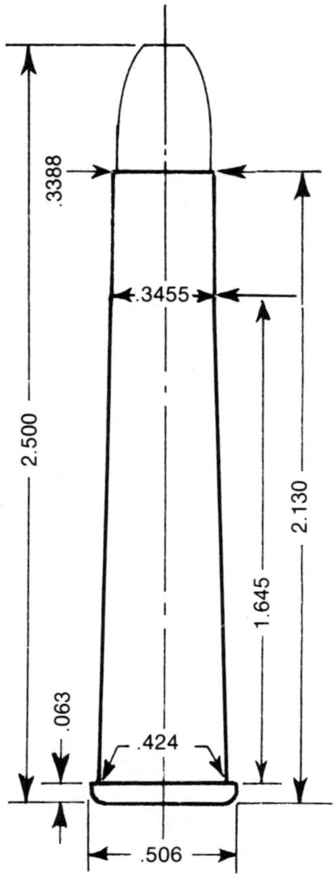

TEST COMPONENTS:

Cases .... Remington Peters
Trim-to Length .... 2.120″
Primers .... Remington 9½
Primer Size .... Large Rifle
Lyman Shell Holder .... No. 6
Cast Bullets Used .... (size to .322″ dia.)
*Gas Check Bullets *#321317, 164 gr.
*#321297, 184 gr.

TEST SPECIFICATIONS:
(Velocity)

Firearm Used .... Winchester Hi-Wall
Barrel Length .... 30″
Twist .... 1-16″
Groove Diameter .... .322″

COMMENTS:

Use only blunt or round nose bullets in rifles which have tubular magazines and crimp the bullets to prevent their movement.

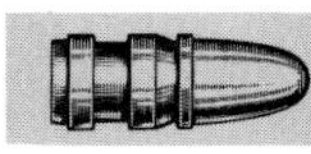

#321317
164 gr., (#2 Alloy) 2.585″ OAL

Powder	Sug. Starting Grains	Velocity F.P.S.	Pressure C.U.P.	Max. Grains	Velocity F.P.S.	Pressure C.U.P.
Red Dot	5.0	1015	—	7.5	1290	—
700X	5.0	1010	—	7.5	1280	—
Green Dot	5.5	1070	—	8.0	1305	—
PB	5.5	1000	—	8.5	1290	—
Unique	6.0	1080	—	8.5	1315	—

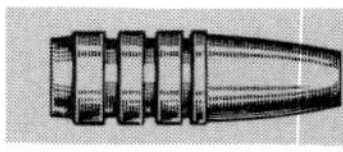

#321297
184 gr., (#2 Alloy) 2.570″ OAL

Powder	Sug. Starting Grains	Velocity F.P.S.	Pressure C.U.P.	Max. Grains	Velocity F.P.S.	Pressure C.U.P.
Red Dot	4.8	825	—	7.3	1195	—
700X	4.8	790	—	7.3	1190	—
Green Dot	5.3	980	—	7.8	1225	—
PB	5.3	910	—	8.3	1220	—
Unique	5.6	980	—	8.3	1240	—

Note: Loads shown in shaded panels are maximum.

.32 Remington

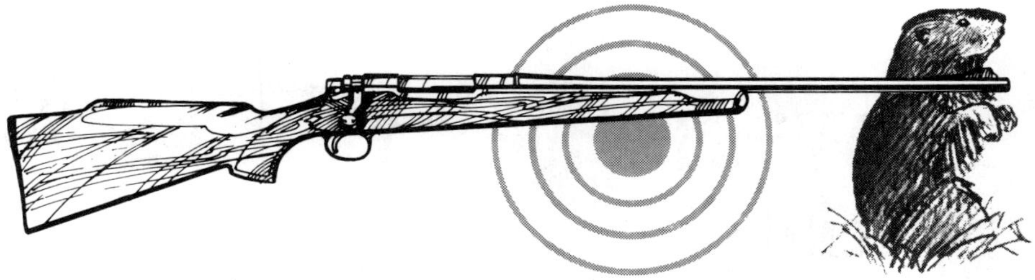

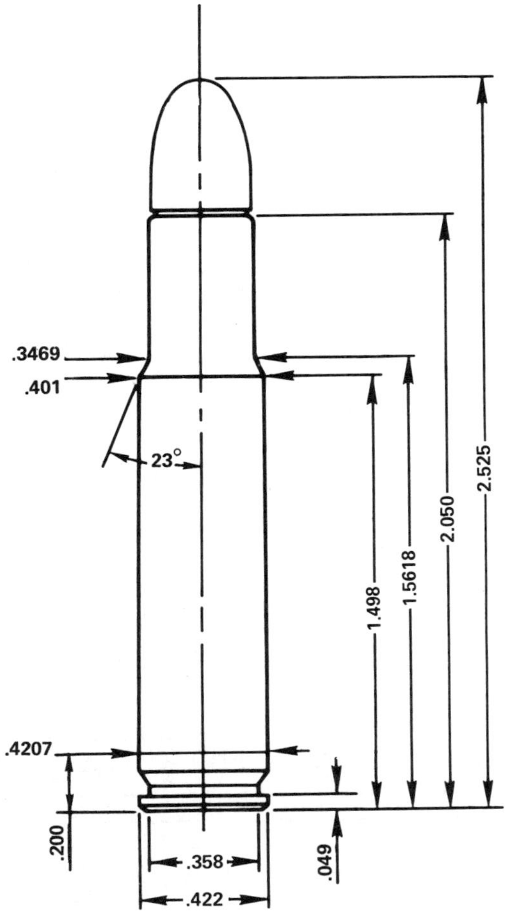

TEST COMPONENTS:

Cases .... Remington
Trim-to Length .... 2.045″
Primers .... Remington 9½
Primer Size .... Large Rifle
Lyman Shell Holder .... No. 15 or 3
Cast Bullets Used .... (size to .319″ to .321″ dia.)
*Gas Check Bullets *#321317, 164 gr.
*#321297, 184 gr.

TEST SPECIFICATIONS:
(Velocity)

Firearm Used .... Remington Model 81
Barrel Length .... 22″
Twist .... 1-14″
Groove Diameter .... .319″

COMMENTS:

New brass cases which exceed the maximum case length listed may be encountered. We suggest that you check your case length before reloading and trim your case if necessary.

The Lyman No. 15 Shell Holder is proper for all cases with the exception of Remington brand. For Remington and Peters cases use the Lyman No. 3 Shell Holder.

Data listed for Unique powder will not function the action of semi-auto rifles.

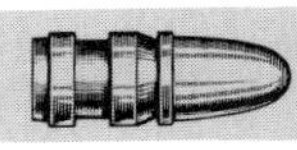

#321317
164 gr., (#2 Alloy) 2.525″ OAL, Max.

Powder	Sug. Starting Grains	Velocity F.P.S.	Pressure C.U.P.	Max. Grains	Velocity F.P.S.	Pressure C.U.P.
Unique	7.0	1234	—	12.0	1720	—
2400	18.0	1798	—	21.0	2012	—

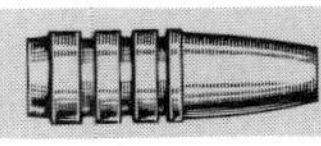

#321297
184 gr., (#2 Alloy) 2.525″ OAL, Max.

Powder	Sug. Starting Grains	Velocity F.P.S.	Pressure C.U.P.	Max. Grains	Velocity F.P.S.	Pressure C.U.P.
Unique	7.0	1159	—	9.0	1369	—
2400	16.0	1552	—	19.0	1788	—

Note: Loads shown in shaded panels are maximum.

.32 Winchester Special

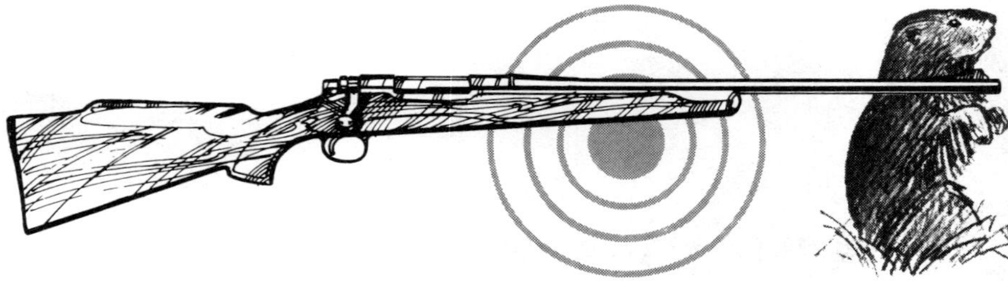

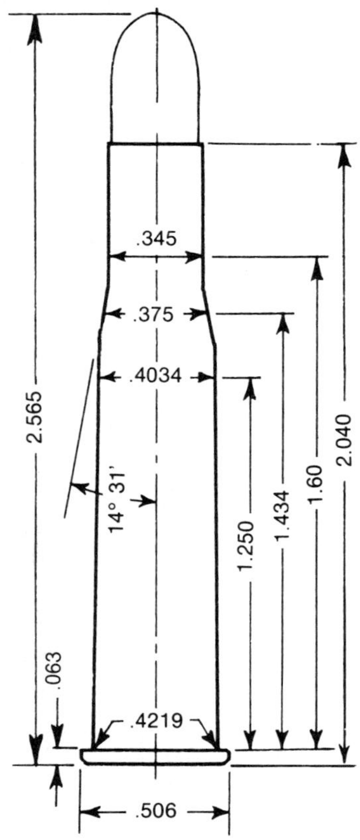

TEST COMPONENTS:

Cases .... Federal
Trim-to Length .... 2.035″
Primers .... Federal 210
Primer Size .... Large Rifle
Lyman Shell Holder .... No. 6
Cast Bullets Used .... (size to .321″ dia.)
*Gas Check Bullets *#321427, 137 gr.
*#321317, 164 gr.
*#321297, 184 gr.

TEST SPECIFICATIONS:
(Velocity)

Firearm Used .... Winchester Model 94
Barrel Length .... 20″
Twist .... 1-16″
Groove Diameter .... .321″

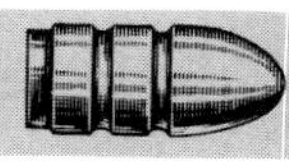

#321427
137 gr., (#2 Alloy) 2.362″ OAL

Powder	Sug. Starting Grains	Velocity F.P.S.	Pressure C.U.P.	Max. Grains	Velocity F.P.S.	Pressure C.U.P.
Red Dot	8.0	1410	—	11.0	1710	—
700X	7.5	1380	—	10.0	1640	—
Green Dot	8.5	1420	—	11.5	1710	—
PB	8.0	1325	—	10.5	1575	—
Unique	8.5	1445	—	11.5	1760	—
SR-7625	8.0	1300	—	10.5	1570	—
Herco	10.0	1495	—	12.5	1730	—
SR-4756	9.0	1345	—	12.0	1645	—

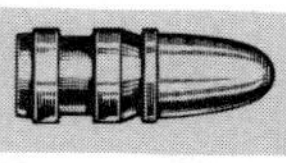

#321317
164 gr., (#2 Alloy) 2.471″ OAL

Powder	Sug. Starting Grains	Velocity F.P.S.	Pressure C.U.P.	Max. Grains	Velocity F.P.S.	Pressure C.U.P.
Red Dot	7.5	1265	—	10.5	1540	—
700X	7.2	1235	—	9.7	1480	—
Green Dot	8.0	1270	—	11.0	1555	—
PB	7.7	1185	—	10.2	1430	—
Unique	8.0	1295	—	11.0	1590	—
SR-7625	7.5	1160	—	10.0	1415	—
Herco	9.5	1340	—	12.0	1570	—
SR-4756	8.7	1230	—	11.5	1495	—

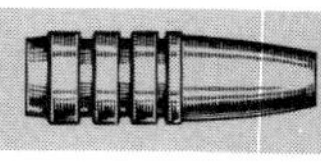

#321297
184 gr., (#2 Alloy) 2.550″ OAL

Powder	Sug. Starting Grains	Velocity F.P.S.	Pressure C.U.P.	Max. Grains	Velocity F.P.S.	Pressure C.U.P.
Red Dot	7.0	1155	—	10.0	1420	—
700X	6.7	1135	—	9.0	1340	—
Green Dot	7.5	1170	—	10.5	1440	—
PB	7.2	1100	—	9.7	1320	—
Unique	7.5	1195	—	10.5	1485	—
SR-7625	7.5	1120	—	9.5	1295	—
Herco	9.0	1265	—	11.5	1480	—
SR-4756	8.2	1140	—	11.0	1395	—

Note: Loads shown in shaded panels are maximum.

8mm Mauser

8 x 57JS

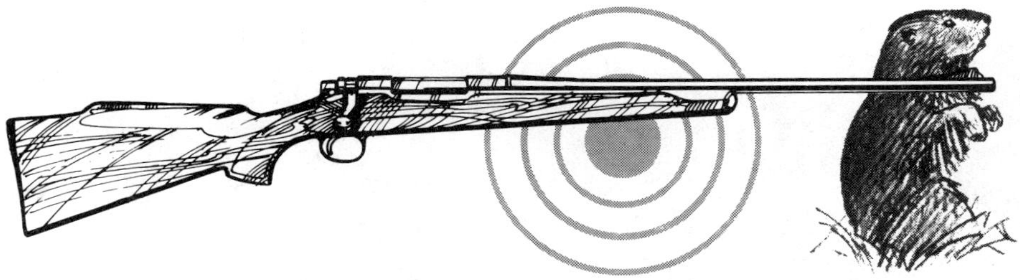

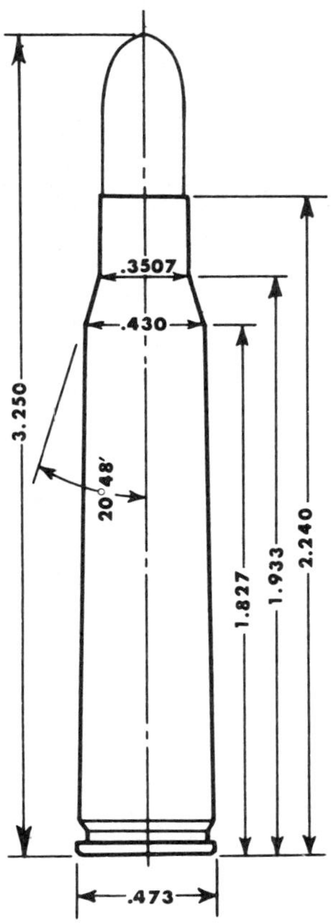

TEST COMPONENTS:

Cases .... Federal
Trim-to Length .... 2.235″
Primers .... Federal 210
Primer Size .... Large Rifle
Lyman Shell Holder .... No. 2
Cast Bullets Used .... (size to .323″ dia. Refer to comments below)
Gas Check Bullets #32359, 120 gr.
*#323470, 160 gr.
*#323366, 181 gr.
*#323471, 214 gr.

TEST SPECIFICATIONS:
(Velocity & Pressure)

Firearm Used .... Czech BRNO Mauser and Special Test Barrel
Barrel Length .... 24″ & 26″
Twist .... 1-9¼″ & 1-9½″
Groove Diameter .... .323″

COMMENTS:

This is the 8 x 57JS bore rifle.
Before loading for this cartridge, we recommend that you slug your barrel. For non-standard (small groove diameter) barrels, we suggest the use of lead alloy bullets sized as near to the exact groove diameter as possible. These data are not for use in .318″ diameter, J bore rifles.

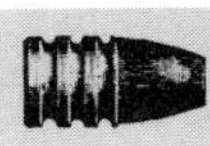

#32359
120 gr., (#2 Alloy) 2.563″ OAL

Powder	Sug. Starting Grains	Velocity F.P.S.	Pressure C.U.P.	Max. Grains	Velocity F.P.S.	Pressure C.U.P.
Red Dot	11.0	1690	—	15.0	2015	—
700X	11.0	1690	—	15.0	2000	—
Green Dot	12.0	1745	—	16. 0	2045	—
PB	12.5	1700	—	16.5	1995	—
Unique	14.0	1855	—	18.0	2150	—
SR-7625	13.0	1715	—	17.5	2050	—
630	17.0	1740	—	23.0	2120	—

Note: Loads shown in shaded panels are maximum.

8mm Mauser (8 x 57JS) (Continued)

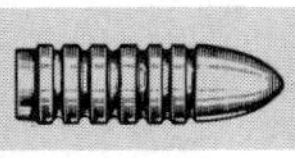

#323470
160 gr., (#2 Alloy) 2.730" OAL

Powder	Sug. Starting Grains	Velocity F.P.S.	Pressure C.U.P.	Max. Grains	Velocity F.P.S.	Pressure C.U.P.
Red Dot	10.0	1355	—	14.0	1645	—
700X	10.0	1360	—	14.0	1655	—
Green Dot	11.0	1405	—	15.0	1710	—
PB	11.5	1385	—	15.5	1670	—
Unique	13.0	1545	—	17.0	1820	—
SR-7625	12.0	1410	—	16.5	1725	—
630	16.0	1540	—	22.0	1895	—
**SR-4759	20.5	1831	18,100	25.2	2139	30,200
**IMR-4198	*27.0	1925	16,700	35.0	2482	34,900
**RX7	28.0	1917	16,900	40.0	2527	33,700
**748	36.0	2015	17,300	45.0	2524	30,600
**H4895	31.0	1957	18,600	37.5	2330	28,200

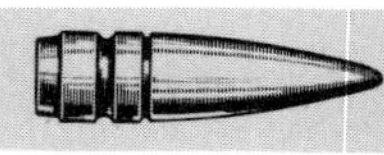

#323366
181 gr., (#2 Alloy) 3.010" OAL

Powder	Sug. Starting Grains	Velocity F.P.S.	Pressure C.U.P.	Max. Grains	Velocity F.P.S.	Pressure C.U.P.
Red Dot	10.0	1315	—	14.0	1580	—
700X	10.0	1300	—	14.0	1570	—
Green Dot	11.0	1365	—	15.0	1620	—
PB	11.5	1335	—	15.0	1565	—
Unique	13.0	1490	—	17.0	1705	—
SR-7625	12.0	1360	—	16.0	1610	—
630	16.0	1495	—	22.0	1795	—
**SR-4759	22.0	1871	25,100	25.0	2052	34,100
**IMR-4198	25.0	1802	18,800	31.0	2151	30,400
**RX7	27.5	1951	24,100	32.8	2165	30,300
**748	35.0	2009	21,700	44.0	2390	32,400
**H335	30.5	1923	21,500	38.5	2233	31,800
**H4895	29.8	1894	21,300	37.8	2310	33,700

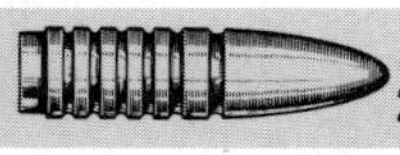

#323471
214 gr., (#2 Alloy) 3.000" OAL

Powder	Sug. Starting Grains	Velocity F.P.S.	Pressure C.U.P.	Max. Grains	Velocity F.P.S.	Pressure C.U.P.
Red Dot	10.0	1205	—	13.0	1410	—
700X	10.0	1215	—	13.0	1420	—
Green Dot	11.0	1260	—	14.0	1465	—
PB	11.5	1245	—	14.0	1405	—
Unique	13.0	1385	—	16.0	1580	—
SR-7625	12.0	1280	—	15.0	1460	—
630	16.0	1425	—	21.0	1705	—
**SR-4759	19.0	1593	23,100	23.0	1826	31,900
**IMR-4198	22.0	1654	22,700	26.6	1920	32,800
**RX7	23.0	1617	20,900	29.0	1918	30,800
**748	*30.0	1686	18,700	40.0	2178	33,900
**H4895	27.0	1645	19,900	35.5	2044	32,500

Note: Loads shown in shaded panels are maximum.
*Designates potentially most accurate load.
**Designates the use of a special test barrel.

8mm Remington Magnum

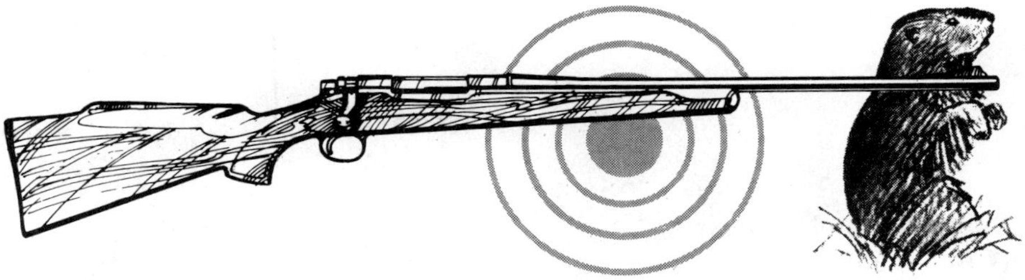

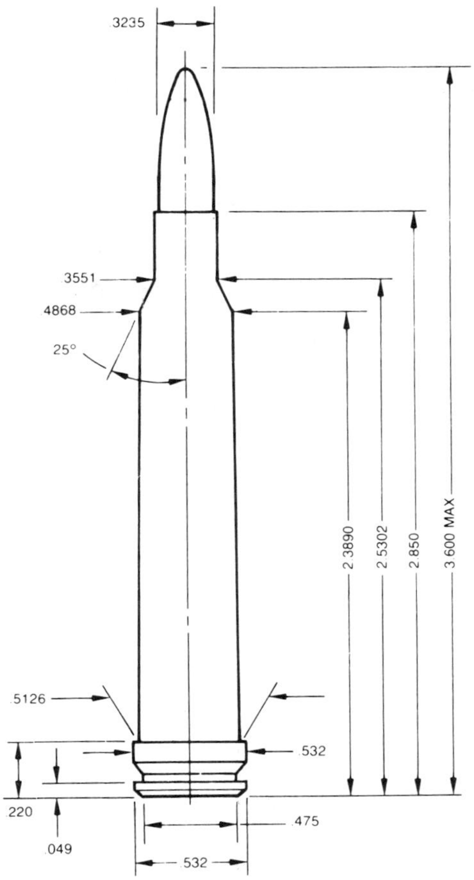

TEST COMPONENTS:

Cases .... Remington
Trim-to Length .... 2.840″
Primers .... CCI 250
Primer Size .... Large Rifle Magnum
Lyman Shell Holder .... No. 13
Cast Bullets Used .... (size to .323″ dia.)
*Gas Check Bullets *#323470, 160 gr.
*#323481, 187 gr.
*#323471, 214 gr.
*#323378, 242 gr.

TEST SPECIFICATIONS:
(Velocity & Pressure)

Firearm Used .... Remington Model 700
Barrel Length .... 24″
Twist .... 1-10″
Groove Diameter .... .323″

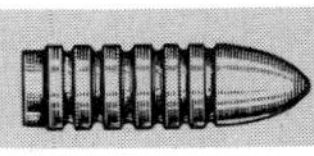

#323470
160 gr., (#2 Alloy) 3.398″ OAL

Powder	Sug. Starting Grains	Velocity F.P.S.	Pressure C.U.P.	Max. Grains	Velocity F.P.S.	Pressure C.U.P.
630	*20.0	1693	—	31.5	2292	—
2400	24.0	1714	—	35.0	2312	—
SR-4759	23.0	1712	—	39.0	2469	—
IMR-4227	24.0	1729	—	40.0	2432	—
IMR-4198	26.0	1707	—	44.0	2471	—
RX7	29.0	1655	—	49.0	2465	—
H4895	32.0	1711	—	50.0	2488	—

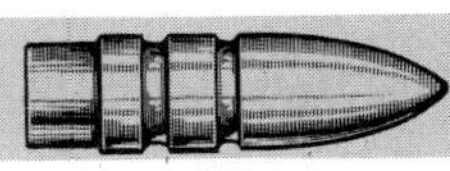

#323481
187 gr., (#2 Alloy) 3.600″ OAL

Powder	Sug. Starting Grains	Velocity F.P.S.	Pressure C.U.P.	Max. Grains	Velocity F.P.S.	Pressure C.U.P.
630	*22.5	1708	—	31.5	2132	—
2400	26.0	1740	—	34.0	2115	—
SR-4759	24.0	1664	—	37.0	2262	—
IMR-4227	23.0	1742	—	41.5	2316	—
IMR-4198	29.0	1732	—	44.0	2327	—
RX7	32.0	1752	—	49.0	2416	—
H4895	34.0	1728	—	53.0	2452	—

Note: Loads shown in shaded panels are maximum.
*Designates potentially most accurate load.

COMMENTS:

This cartridge is based on the full-length belted .375 H&H Magnum case, necked down to take the .323″ bullets, with the shoulder angle increased to 25° and the body taper decreased. Based on our 50 yard accuracy tests bullet #323378 gave the best performance at 1700 f.p.s. Accuracy deteriorated at velocities over 1900 f.p.s.

8mm Remington Magnum (Continued)

#323471
214 gr., (#2 Alloy) 3.600" OAL

Powder	Sug. Starting Grains	Velocity F.P.S.	Pressure C.U.P.	Max. Grains	Velocity F.P.S.	Pressure C.U.P.
630	*24.0	1758	—	30.0	2006	—
2400	25.0	1716	—	34.0	2047	—
SR-4759	26.0	1731	—	36.0	2112	—
IMR-4227	27.0	1702	—	37.0	2080	—
IMR-4198	27.0	1669	—	41.0	2167	—
RX7	30.0	1698	—	46.0	2247	—
H4895	33.0	1670	—	48.0	2237	—

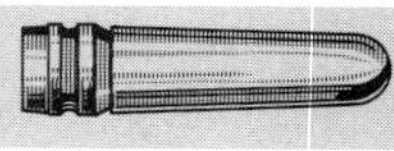

#323378
242 gr., (#2 Alloy) 3.820" OAL

Powder	Sug. Starting Grains	Velocity F.P.S.	Pressure C.U.P.	Max. Grains	Velocity F.P.S.	Pressure C.U.P.
630	*26.0	1703	—	33.0	1968	—
2400	28.0	1684	—	33.0	1879	—
SR-4759	27.5	1660	—	34.0	1901	—
IMR-4227	29.0	1633	—	36.0	1877	—
IMR-4198	31.0	1661	—	38.0	1890	—
RX7	33.0	1676	—	45.0	2099	—
H4895	37.0	1664	—	47.5	2025	—

Note: Loads shown in shaded panels are maximum.
*Designates potentially most accurate load.

.338 Winchester Magnum

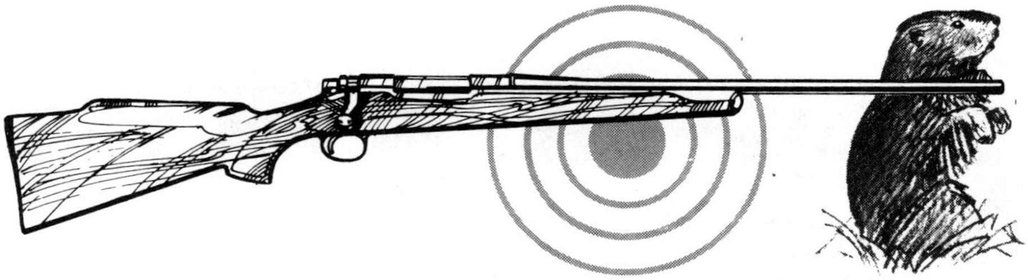

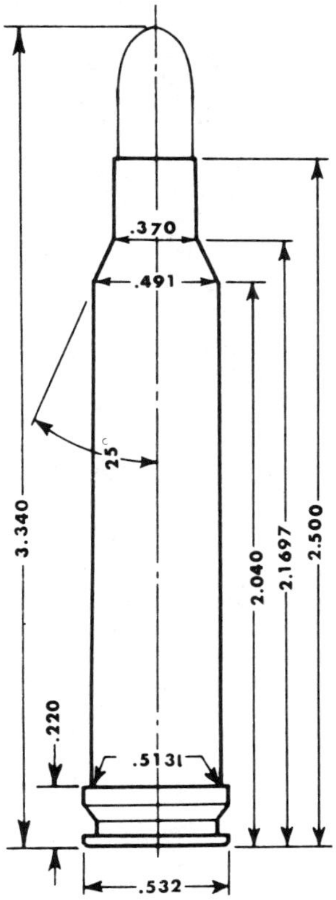

TEST COMPONENTS:

Cases .... Winchester/Western
Trim-to Length .... 2.490″
Primers .... Winchester 8½-120
Primer Size .... Large Rifle
Lyman Shell Holder .... No. 13
Cast Bullets Used .... (size to .338″ dia.)
*Gas Check Bullets *#338320, 203 gr.
*#33889, 249 gr.

TEST SPECIFICATIONS:
(Velocity)

Firearm Used .... Winchester Model 70
Barrel Length .... 24″
Twist .... 1-10″
Groove Diameter .... .336″

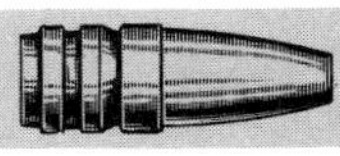

#338320
203 gr., (#2 Alloy) 3.150″ OAL

Powder	Sug. Starting Grains	Velocity F.P.S.	Pressure C.U.P.	Max. Grains	Velocity F.P.S.	Pressure C.U.P.
Red Dot	12.0	1320	—	16.0	1575	—
700X	12.0	1320	—	16.0	1580	—
Green Dot	13.0	1360	—	17.0	1615	—
PB	13.0	1300	—	18.0	1590	—
Unique	14.0	1400	—	19.5	1705	—
SR-7625	14.0	1340	—	20.0	1710	—
630	19.0	1500	—	25.0	1840	—

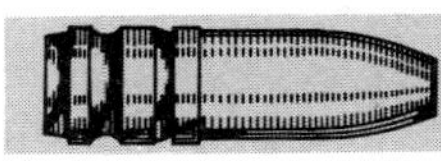

#33889
249 gr., (#2 Alloy) 3.271″ OAL

Powder	Sug. Starting Grains	Velocity F.P.S.	Pressure C.U.P.	Max. Grains	Velocity F.P.S.	Pressure C.U.P.
Red Dot	11.5	1150	—	15.5	1390	—
700X	12.0	1180	—	15.5	1390	—
Green Dot	13.0	1210	—	16.5	1420	—
PB	13.0	1155	—	17.5	1395	—
Unique	14.0	1245	—	19.0	1510	—
SR-7625	14.0	1190	—	19.5	1465	—
630	18.5	1355	—	24.5	1615	—

COMMENTS:

Fine accuracy and uniform velocities were obtained with all of the powders and bullets tested.

.340 Weatherby Magnum

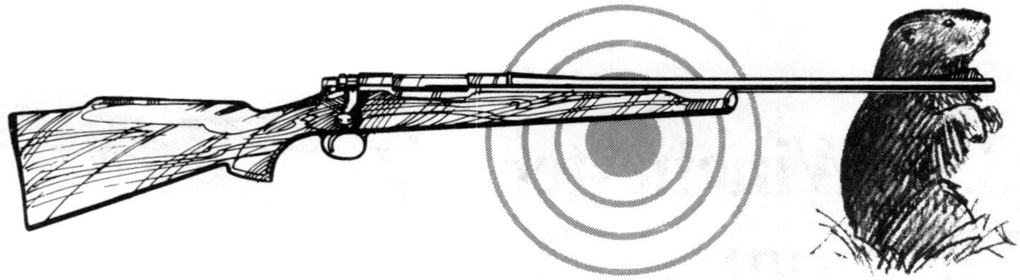

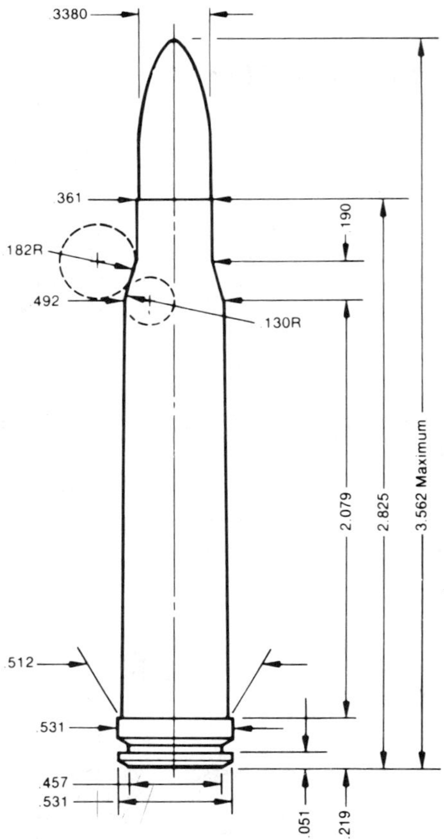

TEST COMPONENTS:

Cases .... Weatherby
Trim-to Length .... 2.815″
Primers .... Remington 9½
Primer Size .... Large Rifle
Lyman Shell Holder .... No. 13
Cast Bullets Used .... (size to .338″ dia.)
*Gas Check Bullets .... *#338320, 203 gr.
*#33889, 249 gr.

TEST SPECIFICATIONS:
(Velocity)

Firearm Used .... Weatherby Mark V
Barrel Length .... 26″
Twist .... 1-10″
Groove Diameter .... .338″

COMMENTS:

The data listed for this cartridge were obtained in a Weatherby rifle and are intended for Weatherby rifles only. The free-boring constructed into these firearms allows higher velocities at safe working pressures. For custom rifles which are not free-bored, maximum loads should be reached a full 5%. Even then, they should be approached with caution.

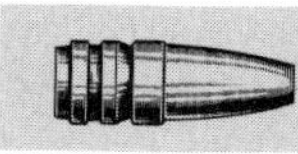

#338320
203 gr., (#2 Alloy) 3.562″ OAL, Max.

Powder	Sug. Starting Grains	Velocity F.P.S.	Pressure C.U.P.	Max. Grains	Velocity F.P.S.	Pressure C.U.P.
IMR-4198	32.0	1782	—	38.0	2044	—
IMR-3031	40.0	1949	—	44.0	2100	—
IMR-4895	42.0	1930	—	48.0	2109	—

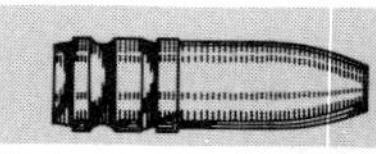

#33889
249 gr., (#2 Alloy) 3.562″ OAL, Max.

Powder	Sug. Starting Grains	Velocity F.P.S.	Pressure C.U.P.	Max. Grains	Velocity F.P.S.	Pressure C.U.P.
IMR-4198	25.0	1442	—	37.0	1923	—
IMR-3031	39.0	1768	—	43.0	1962	—
IMR-4895	41.0	1811	—	47.0	2024	—

Note: Loads shown in shaded panels are maximum.

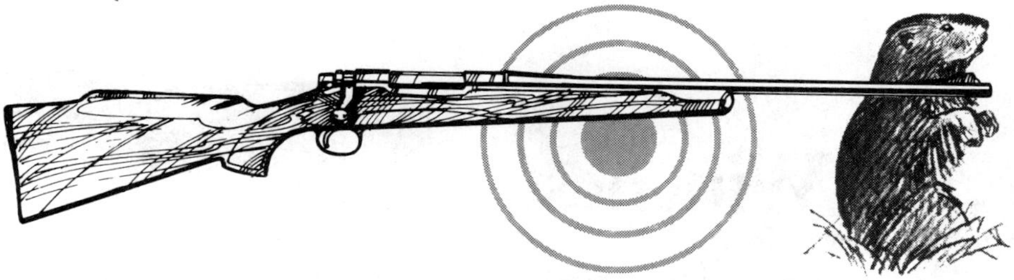

.348 Winchester

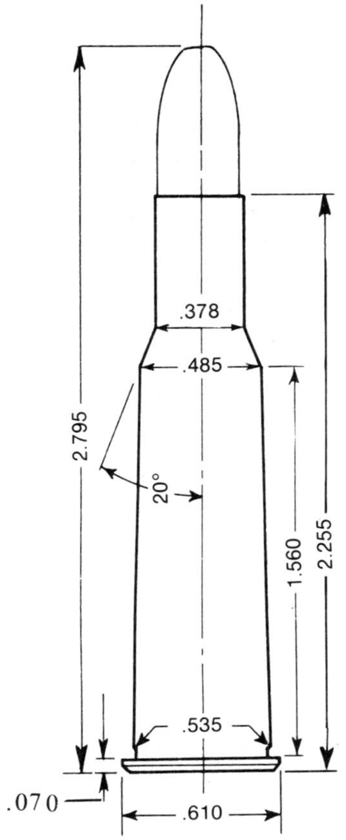

TEST COMPONENTS:

Cases .... Remington
Trim-to Length .... 2.245″
Primers .... Remington 9½
Primer Size .... Large Rifle
Lyman Shell Holder .... No. 18
Cast Bullets Used .... (size to .348″ dia.)
*Gas Check Bullets *#350447, 187 gr.
*#350482, 255 gr.

TEST SPECIFICATIONS:
(Velocity)

Firearm Used .... Winchester Model 71
Barrel Length .... 24″
Twist .... 1-12″
Groove Diameter .... .3485″

COMMENTS:

Only blunt or round nose bullets should be used in these rifles which have tubular magazines. Bullets should also be crimped in place to prevent their movement during recoil.

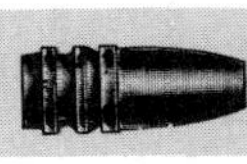

#350447
187 gr., (#2 Alloy) 2.795″ OAL, Max.

Powder	Sug. Starting Grains	Velocity F.P.S.	Pressure C.U.P.	Max. Grains	Velocity F.P.S.	Pressure C.U.P.
2400	20.0	1610	—	23.0	1779	—
IMR-4227	20.0	1510	—	30.5	2123	—
IMR-4198	25.0	1672	—	35.0	2150	—
IMR-3031	35.0	1818	—	47.0	2338	—

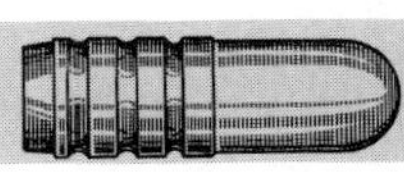

#350482
255 gr., (#2 Alloy) 2.795″ OAL, Max.

Powder	Sug. Starting Grains	Velocity F.P.S.	Pressure C.U.P.	Max. Grains	Velocity F.P.S.	Pressure C.U.P.
2400	19.0	1412	—	22.0	1560	—
IMR-4227	21.0	1428	—	30.0	1838	—
IMR-4198	27.0	1661	—	38.5	2096	—
IMR-3031	35.0	1798	—	46.0	2217	—

Note: Loads shown in shaded panels are maximum.

.351 Winchester

Self-Loading

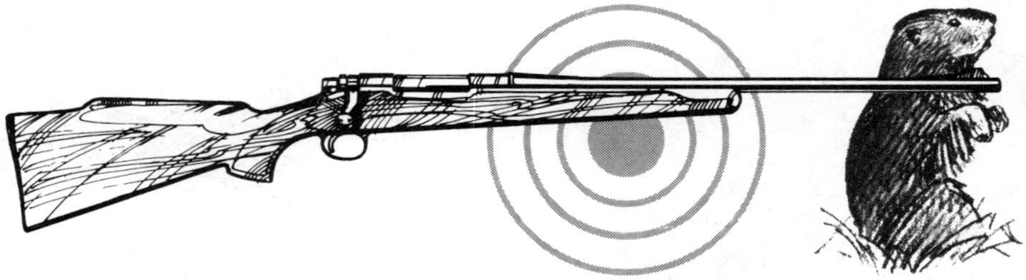

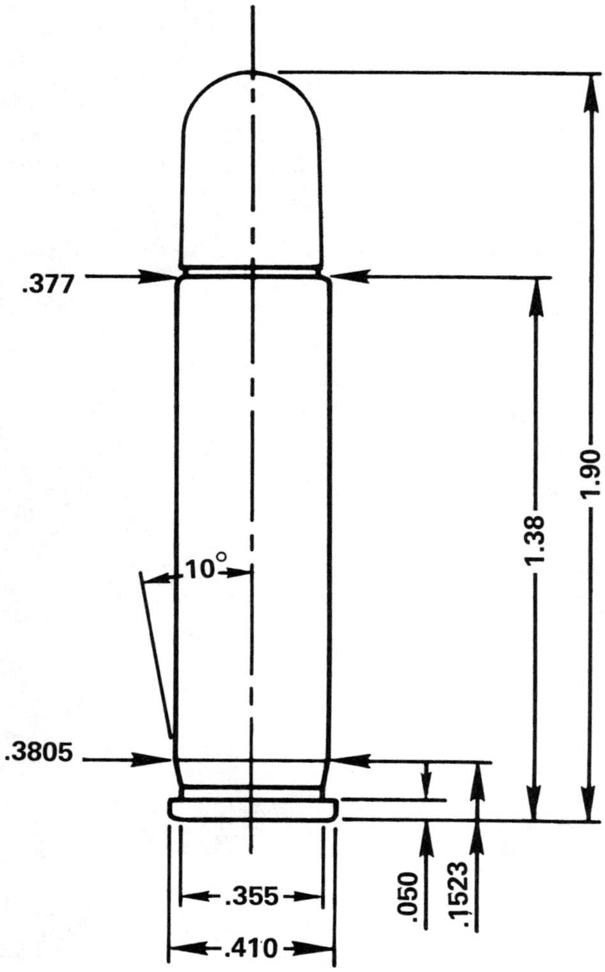

TEST COMPONENTS:

Cases .... Winchester
Trim-to Length .... 1.376″
Primers .... Remington 6½
Primer Size .... Small Rifle
Lyman Shell Holder .... No. 15
Cast Bullets Used .... (size to .352″ dia.)
*Gas Check Bullet *#350319, 171 gr.

TEST SPECIFICATIONS:
(Velocity)

Firearm Used .... Winchester Model 07
Barrel Length .... 20″
Twist .... 1-16″
Groove Diameter .... .351″

COMMENTS:

To insure positive functioning of the action, loads must be worked up to near maximum.

.348 caliber gas checks are used with cast bullets for this cartridge. Best overall results were obtained with IMR 4227 powder.

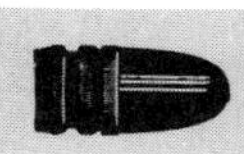

#350319
171 gr., (#2 Alloy) 1.900″ OAL, Max.

Powder	Sug. Starting Grains	Velocity F.P.S.	Pressure C.U.P.	Max. Grains	Velocity F.P.S.	Pressure C.U.P.
Unique	12.0	1692	—	12.5	1751	—
2400	17.0	1861	—	·19.0+	2020	—
IMR-4227	17.0	1658	—	19.5+	1904	—

Note: Loads shown in shaded panels are maximum.
+Designates a compressed powder charge.

.35 Remington

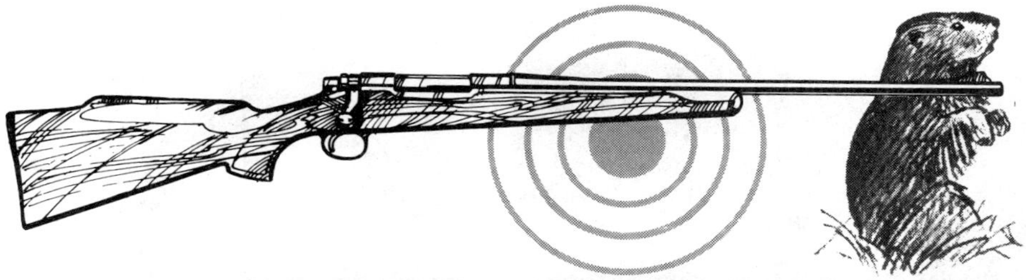

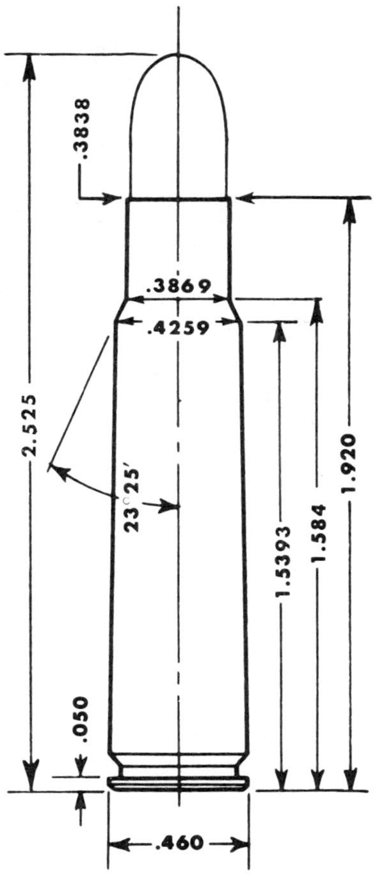

TEST COMPONENTS:

Cases .... Winchester/Western
Trim-to Length .... 1.910″
Primers .... Winchester 8½-120
Primer Size .... Large Rifle
Lyman Shell Holder .... No. 8 & 2
Case Bullets Used .... (size to .357″ dia.)
*Gas Check Bullet #358430, 150 gr.
#358430, 195 gr.
*#358315, 206 gr.

TEST SPECIFICATIONS:
(Velocity)

Firearm Used .... Marlin Model 336
Barrel Length .... 20″
Twist .... 1-16″
Groove Diameter .... .357″

COMMENTS:

Bullets should be sized to the exact groove diameter. If the rifle has multi or shallow groove rifling, then cast bullet velocities must be held to 1600 f.p.s. or less, if accuracy is to be obtained.

Use only blunt, or round nose, bullets in those rifles which have tubular magazines. Bullets are best crimped in place to prevent their moving with recoil.

The No. 8 Lyman Shell Holder works on all brass with the exception of Remington and Peters brand which requires a No. 2.

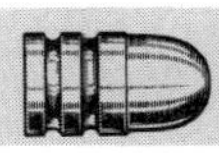

#358430
150 gr., (#2 Alloy) 2.215″ OAL

Powder	Sug. Starting Grains	Velocity F.P.S.	Pressure C.U.P.	Max. Grains	Velocity F.P.S.	Pressure C.U.P.
Red Dot	8.5	1390	—	11.0	1615	—
700X	8.0	1360	—	10.0	1545	—
Green Dot	9.0	1395	—	11.5	1625	—
PB	8.5	1310	—	11.5	1570	—
Unique	9.5	1370	—	12.5	1665	—
SR-7625	9.0	1330	—	11.5	1555	—
Herco	10.5	1425	—	13.5	1675	—
SR-4756	10.0	1355	—	13.0	1590	—
630	12.5	1340	—	16.0	1595	—

Note: Loads shown in shaded panels are maximum.

.35 Remington (Continued)

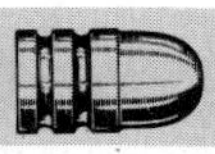

#358430
195 gr., (# 2 Alloy) 2.215" OAL

Powder	Sug. Starting Grains	Velocity F.P.S.	Pressure C.U.P.	Max. Grains	Velocity F.P.S.	Pressure C.U.P.
Red Dot	8.5	1280	—	11.0	1495	—
700X	8.0	1255	—	10.0	1415	—
Green Dot	9.0	1300	—	11.5	1515	—
PB	8.5	1225	—	11.5	1470	—
Unique	9.5	1320	—	12.0	1525	—
SR-7625	9.0	1230	—	11.5	1450	—
Herco	10.5	1360	—	13.0	1550	—
SR-4756	10.0	1275	—	13.0	1515	—
630	12.5	1320	—	16.0	1555	—

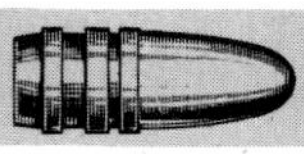

#358315
206 gr., (#2 Alloy) 2.500" OAL

Powder	Sug. Starting Grains	Velocity F.P.S.	Pressure C.U.P.	Max. Grains	Velocity F.P.S.	Pressure C.U.P.
Red Dot	8.5	1240	—	11.0	1440	—
700X	8.0	1200	—	10.0	1370	—
Green Dot	9.0	1260	—	11.5	1455	—
PB	8.5	1170	—	11.5	1405	—
Unique	9.5	1265	—	12.0	1465	—
SR-7625	9.0	1185	—	11.5	1390	—
Herco	10.5	1315	—	13.0	1490	—
SR-4756	10.0	1235	—	13.0	1445	—
630	12.5	1265	—	16.0	1505	—

Note: Loads shown in shaded panels are maximum.

.350 Remington Magnum

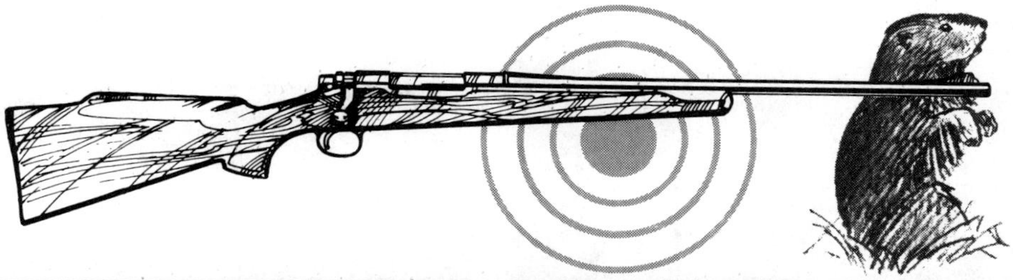

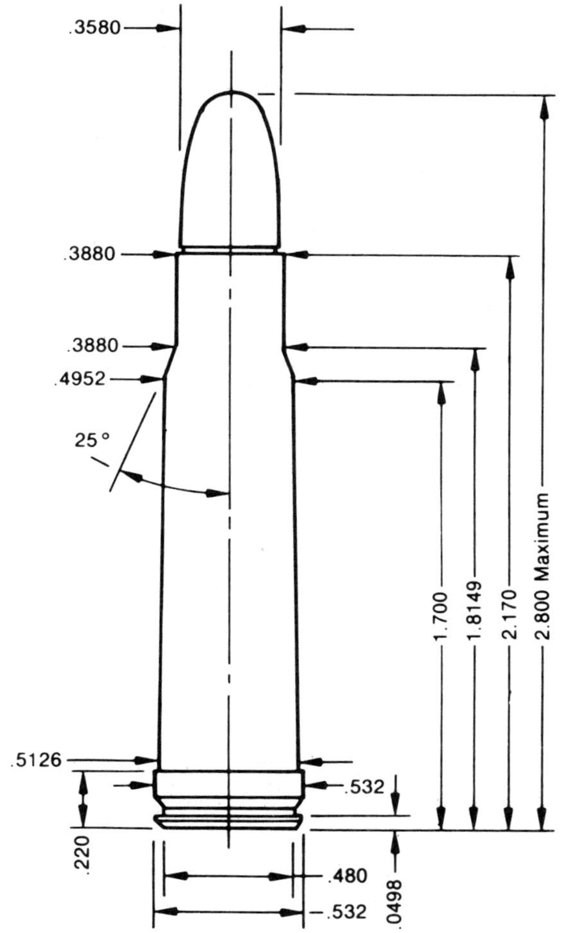

TEST COMPONENTS:

Cases .... Remington Peters
Trim-to Length .... 2.160″
Primers .... Remington 9½
Primer Size .... Large Rifle
Lyman Shell Holder .... No. 13
Cast Bullets Used .... (size to .358″ dia.)
*Gas Check Bullets #358430, 150 gr.
*#358315, 206 gr.
*#358318, 247 gr.

TEST SPECIFICATIONS:
(Velocity)

Firearm Used .... Remington Model 700
Barrel Length .... 24″
Twist .... 1-16″
Groove Diameter .... .358″

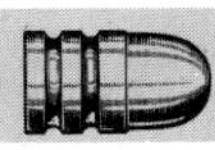

#358430
150 gr., (#2 Alloy) 2.450″ OAL

Powder	Sug. Starting Grains	Velocity F.P.S.	Pressure C.U.P.	Max. Grains	Velocity F.P.S.	Pressure C.U.P.
Red Dot	10.0	1430	—	15.0	1810	—
700X	10.0	1430	—	15.0	1800	—
Green Dot	11.0	1475	—	16.0	1840	—
PB	11.0	1390	—	16.5	1805	—
Unique	12.0	1495	—	17.5	1880	—
SR-7625	12.0	1445	—	17.5	1830	—

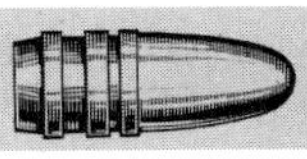

#358315
206 gr., (#2 Alloy) 2.760″ OAL

Powder	Sug. Starting Grains	Velocity F.P.S.	Pressure C.U.P.	Max. Grains	Velocity F.P.S.	Pressure C.U.P.
Red Dot	10.0	1270	—	14.5	1535	—
700X	10.0	1280	—	14.5	1540	—
Green Dot	11.0	1315	—	15.5	1620	—
PB	11.0	1250	—	16.0	1580	—
Unique	12.0	1345	—	17.0	1675	—
SR-7625	12.0	1305	—	17.0	1615	—

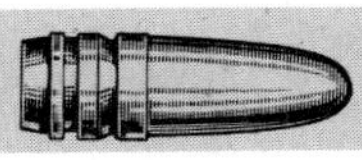

#358318
247 gr., (#2 Alloy) 2.800″ OAL

Powder	Sug. Starting Grains	Velocity F.P.S.	Pressure C.U.P.	Max. Grains	Velocity F.P.S.	Pressure C.U.P.
Red Dot	9.5	1075	—	14.0	1385	—
700X	9.5	1085	—	14.0	1400	—
Green Dot	10.5	1185	—	15.0	1470	—
PB	10.5	1130	—	15.5	1430	—
Unique	11.5	1225	—	16.5	1530	—
SR-7625	11.5	1170	—	16.5	1470	—

Note: Loads shown in shaded panels are maximum.

.358 Winchester

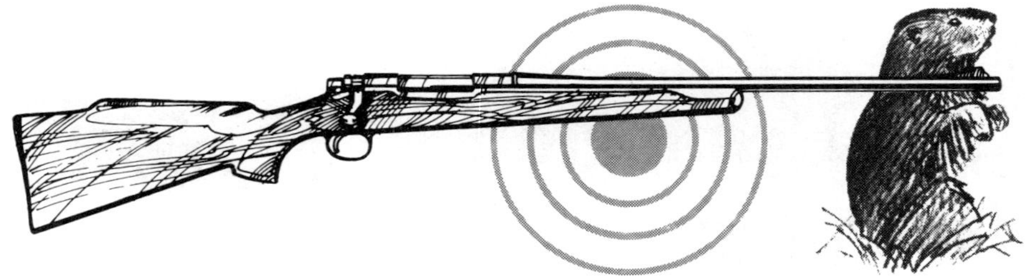

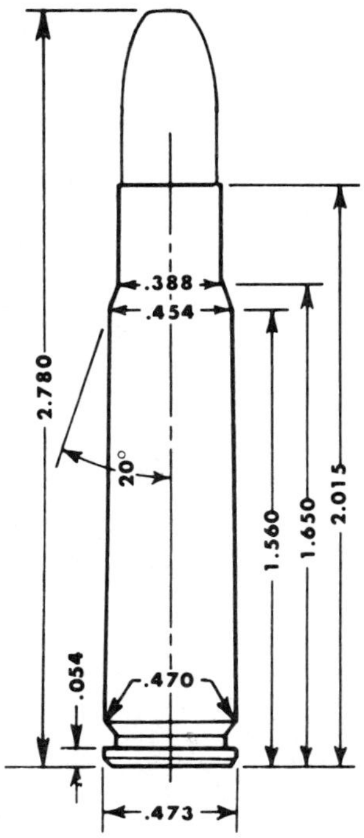

TEST COMPONENTS:

Cases .... Winchester/Western
Trim-to Length .... 2.005″
Primers .... Winchester 8½-120
Primer Size .... Large Rifle
Lyman Shell Holder .... No. 2
Cast Bullets Used .... (size to .359″ dia.)
*Gas Check Bullets #358430, 150 gr.
*#358315, 206 gr.
*#358318, 247 gr.

TEST SPECIFICATIONS:
(Velocity)

Firearm Used .... Mannlicher-Schoenauer
Barrel Length .... 20″
Twist .... 1-12″
Groove Diameter .... .362″

COMMENTS:

Accuracy was less than expected due to large groove diameter on test rifle.

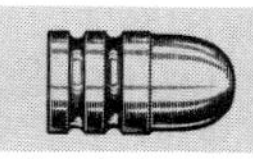

#358430
150 gr., (#2 Alloy) 2.295″ OAL

Powder	Sug. Starting Grains	Velocity F.P.S.	Pressure C.U.P.	Max. Grains	Velocity F.P.S.	Pressure C.U.P.
Red Dot	9.5	1425	—	13.5	1745	—
700X	9.5	1425	—	13.5	1745	—
Green Dot	10.5	1545	—	14.0	1765	—
PB	10.5	1500	—	14.0	1730	—
Unique	11.5	1570	—	15.0	1795	—
SR-7625	11.5	1565	—	14.5	1755	—
630	14.5	1455	—	21.0	1930	—

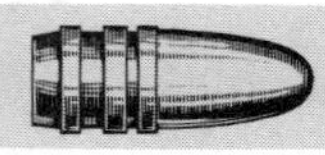

#358315
206 gr., (#2 Alloy) 2.585″ OAL

Powder	Sug. Starting Grains	Velocity F.P.S.	Pressure C.U.P.	Max. Grains	Velocity F.P.S.	Pressure C.U.P.
Red Dot	9.0	1190	—	13.0	1505	—
700X	9.0	1205	—	12.5	1490	—
Green Dot	10.0	1310	—	13.0	1495	—
PB	10.0	1265	—	13.5	1465	—
Unique	11.0	1340	—	14.5	1550	—
SR-7625	11.0	1295	—	14.0	1480	—
630	14.0	1270	—	20.0	1700	—

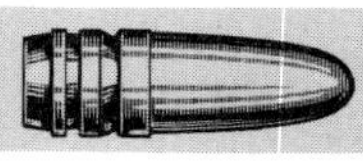

#358318
247 gr., (#2 Alloy) 2.745″ OAL

Powder	Sug. Starting Grains	Velocity F.P.S.	Pressure C.U.P.	Max. Grains	Velocity F.P.S.	Pressure C.U.P.
Red Dot	9.0	1100	—	12.5	1355	—
700X	9.0	1095	—	12.0	1330	—
Green Dot	9.5	1115	—	12.5	1340	—
PB	9.5	1070	—	13.0	1310	—
Unique	10.5	1205	—	14.0	1395	—
SR-7625	10.5	1170	—	13.5	1325	—
630	13.5	1200	—	19.0	1515	—

Note: Loads shown in shaded panels are maximum.

.358 Norma Magnum

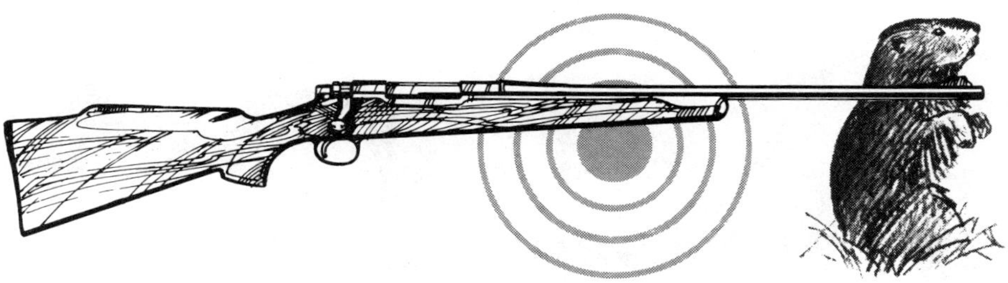

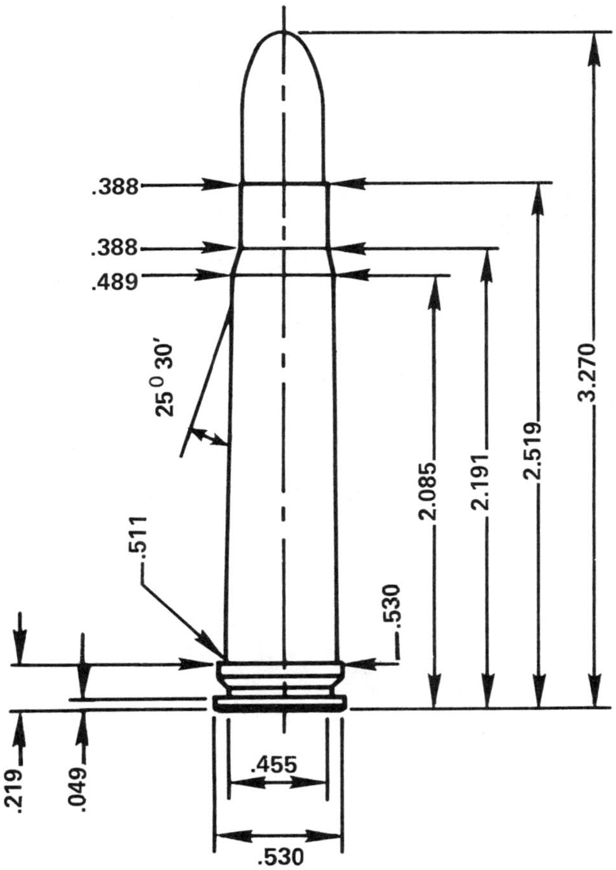

TEST COMPONENTS:

Cases .... Norma
Trim-to Length .... 2.505″
Primers .... Remington 9½
Primer Size .... Large Rifle
Lyman Shell Holder .... No. 13
Cast Bullets Used .... (size to .358″ dia.)
*Gas Check Bullets .... *#358315, 206 gr.
*#358318, 247 gr.

TEST SPECIFICATIONS: (Velocity)

Firearm Used .... Schultz & Larsen
Barrel Length .... 24″
Twist .... 1-12″
Groove Diameter .... .358″

COMMENTS:

Our test rifle had a considerable amount of free-bore. On custom rifles which are not free-bored, maximum loads should be reduced by a full 5% from those listed. Even with this reduction, such loads should be approached with caution.

#358315
206 gr., (#2 Alloy) 3.270″ OAL, Max.

Powder	Sug. Starting Grains	Velocity F.P.S.	Pressure C.U.P.	Max. Grains	Velocity F.P.S.	Pressure C.U.P.
Unique	16.0	1604	—	21.0	1865	—

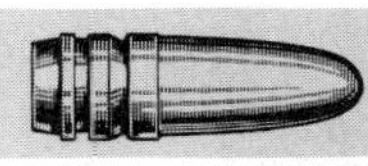

#358318
247 gr., (#2 Alloy) 3.270″ OAL, Max.

Powder	Sug. Starting Grains	Velocity F.P.S.	Pressure C.U.P.	Max. Grains	Velocity F.P.S.	Pressure C.U.P.
Unique	15.0	1436	—	19.0	1658	—

Note: Loads shown in shaded panels are maximum.

.375 Holland & Holland Magnum

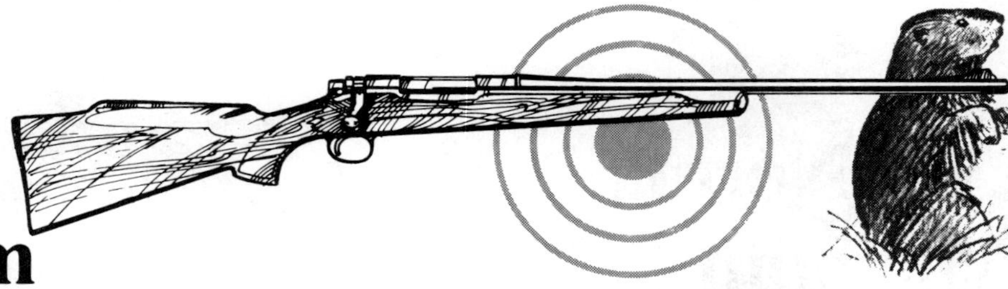

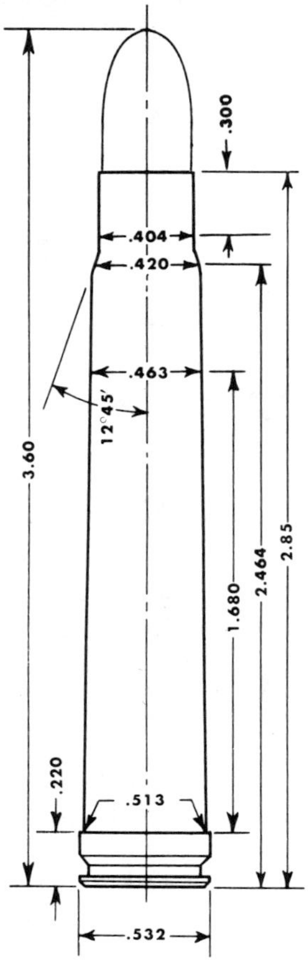

TEST COMPONENTS:

Cases .... Winchester/ Western
Trim-to Length .... 2.840″
Primers .... Winchester 8½-120
Primer Size .... Large Rifle
Lyman Shell Holder .... No. 13
Cast Bullets Used .... (size to .375″ dia.)
*Gas Check Bullet #375248, 249 gr.
*#375296, 269 gr.

TEST SPECIFICATIONS:
(Velocity)

Firearm Used .... Winchester Model 70
Barrel Length .... 24″
Twist .... 1-12″
Groove Diameter .... .376″

COMMENTS:

Due to heavy recoil we recommend that all hunting loads be crimped to prevent the bullet from moving. #375296 gave best results with all powders.

#375248
249 gr., (#2 Alloy) 3.395″ OAL

Powder	Sug. Starting Grains	Velocity F.P.S.	Pressure C.U.P.	Max. Grains	Velocity F.P.S.	Pressure C.U.P.
Red Dot	10.5	1150	—	15.0	1420	—
700X	10.0	1115	—	15.0	1410	—
Green Dot	11.5	1195	—	16.0	1450	—
PB	11.0	1105	—	16.0	1380	—
Unique	13.0	1260	—	18.0	1525	—
Herco	15.0	1335	—	20.0	1590	—
SR-4756	18.0	1430	—	24.5	1720	—
Blue Dot	17.0	1395	—	24.5	1750	—
630	20.0	1415	—	26.0	1710	—

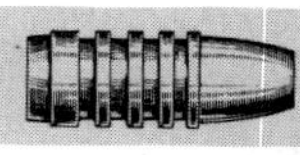

#375296
269 gr., (#2 Alloy) 3.400″ OAL

Powder	Sug. Starting Grains	Velocity F.P.S.	Pressure C.U.P.	Max. Grains	Velocity F.P.S.	Pressure C.U.P.
Red Dot	10.5	1115	—	14.5	1335	—
700X	10.5	1110	—	14.5	1335	—
Green Dot	11.5	1165	—	15.5	1370	—
PB	11.0	1075	—	15.5	1330	—
Unique	13.0	1220	—	17.5	1460	—
Herco	15.0	1295	—	19.5	1515	—
SR-4756	17.0	1350	—	23.5	1640	—
Blue Dot	17.0	1355	—	23.5	1645	—
630	19.0	1345	—	25.0	1620	—

.375 Winchester

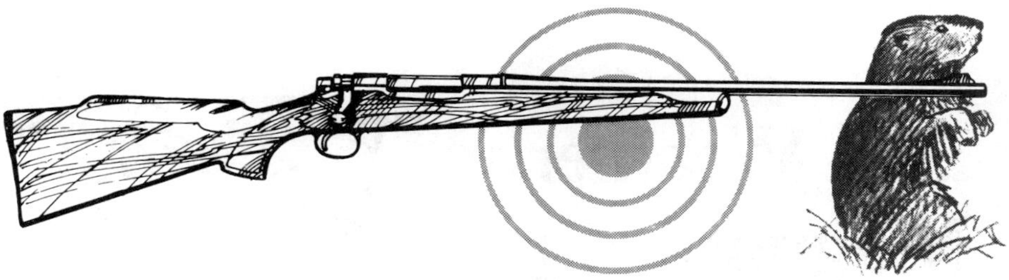

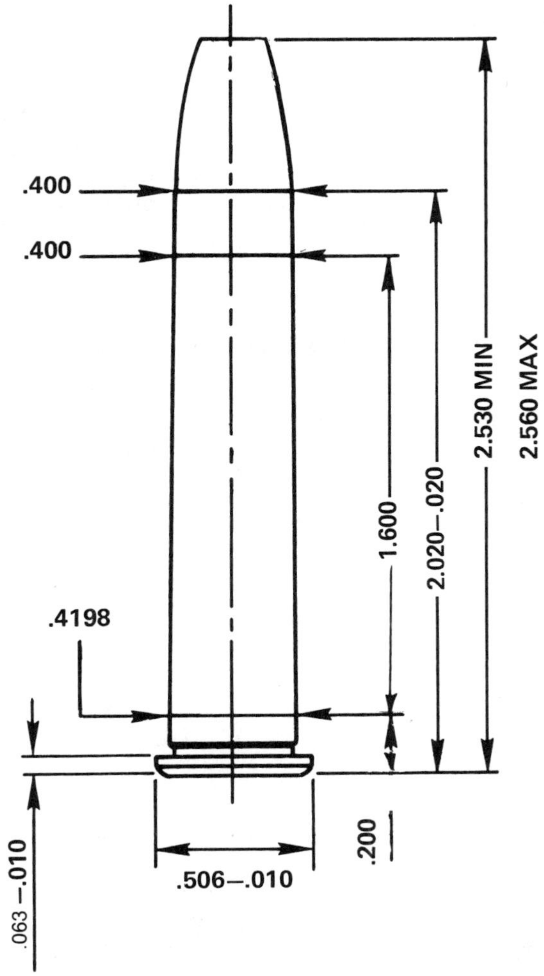

TEST COMPONENTS:

Cases .... Winchester
Trim-to Length .... 2.010″
Primers .... Winchester 8½-120
Primer Size .... Large Rifle
Lyman Shell Holder .... No. 6
Cast Bullets Used .... (size to .376″ dia.)
*Gas Check Bullet .... #375248, 248 gr.
*#375449, 264 gr.

TEST SPECIFICATIONS:
(Velocity)

Firearm Used .... Winchester Big Bore Model 94
Barrel Length .... 20″
Twist .... 1-12″
Groove Diameter .... .376″

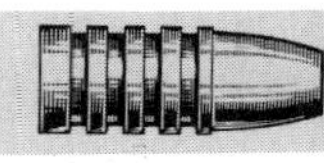

#375248
248 gr., (#2 Alloy) 2.450″ OAL

Powder	Sug. Starting Grains	Velocity F.P.S.	Pressure C.U.P.	Max. Grains	Velocity F.P.S.	Pressure C.U.P.
630	18.0	1588	—	23.5	1898	—
2400	20.0	1595	—	24.0	1897	—
SR-4759	21.0	1606	—	26.0+	1866	—
IMR-4227	22.0	1617	—	26.0	1925	—
IMR-4198	24.0	1581	—	28.5+	1907	—
RX7	27.5	1614	—	32.0+	1811	—
IMR-3031	27.0	1563	—	31.4+	1771	—

#375449
264 gr., (#2 Alloy) 2.560″ OAL

Powder	Sug. Starting Grains	Velocity F.P.S.	Pressure C.U.P.	Max. Grains	Velocity F.P.S.	Pressure C.U.P.
630	17.5	1503	—	23.5	1805	—
2400	19.5	1503	—	24.3	1806	—
SR-4759	20.0	1502	—	24.0	1738	—
IMR-4227	21.0	1520	—	24.8	1807	—
IMR-4198	23.0	1496	—	25.5	1769	—
RX7	25.0	1507	—	32.0+	1762	—
IMR-3031	26.2	1484	—	32.0+	1770	—

COMMENTS:

Do not use the .375 Winchester cartridge in any .38/55 rifles. The chambers of .38/55 rifles will accept the .375 Winchester round but it is hazardous to fire it in them. The .38/55 cartridge should not be fired in .375 Winchester rifles because the case is longer and extends into the chamber throat preventing the case mouth from expanding, which could result in the development of dangerous pressures.

Accuracy would seem to be best at 1700 f.p.s. based on our 50 yard tests.

Note: Loads shown in shaded panels are maximum.
+Designates a compressed powder charge.

.378 Weatherby Magnum

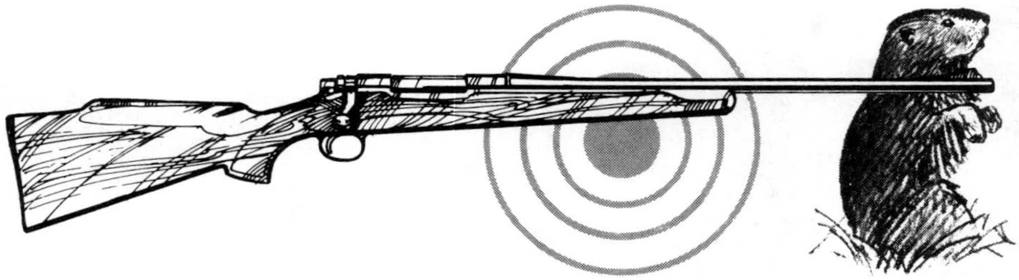

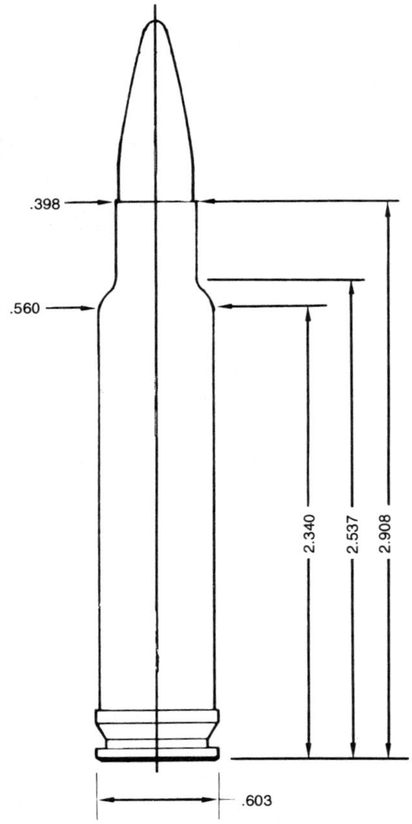

TEST COMPONENTS:

Cases .... Weatherby
Trim-to Length .... 2.903″
Primers .... Remington 9½
Primer Size .... Large Rifle
Lyman Shell Holder .... No. 17
Cast Bullets Used .... (size to .375″ dia.)
*Gas Check Bullet #375248, 249 gr.
*#375449, 267 gr.

TEST SPECIFICATIONS:
(Velocity)

Firearm Used .... Weatherby Mark V
Barrel Length .... 26″
Twist .... 1-12″
Groove Diameter .... .375″

COMMENTS:

The data listed for this cartridge were obtained in a Weatherby rifle and are intended for Weatherby rifles only. The free-boring constructed into these firearms allow higher velocities at safe working pressures. For custom rifles which are not free-bored, maximum loads should be reduced a full 5%. Even then, they should be approached with caution.

Due to heavy recoil we recommend that all bullets be crimped in place.

#375248
249 gr., (#2 Alloy) 3.690″ OAL, Max.

Powder	Sug. Starting Grains	Velocity F.P.S.	Pressure C.U.P.	Max. Grains	Velocity F.P.S.	Pressure C.U.P.
Unique	25.0	1760	—	29.0	1904	—

#375449
267 gr., (#2 Alloy) 3.690″ OAL, Max.

Powder	Sug. Starting Grains	Velocity F.P.S.	Pressure C.U.P.	Max. Grains	Velocity F.P.S.	Pressure C.U.P.
Unique	24.0	1685	—	28.0	1831	—

Note: Loads shown in shaded panels are maximum.

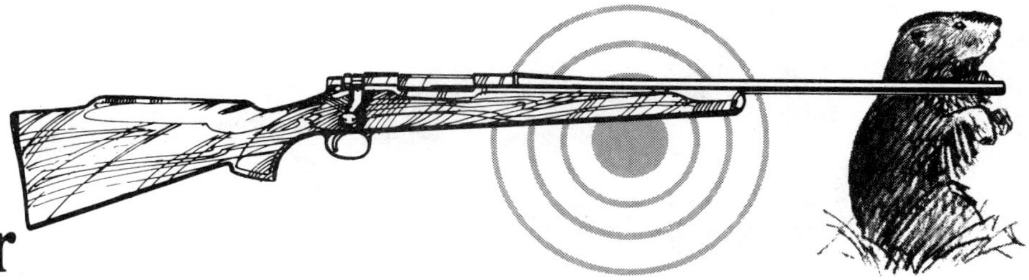

.38/40 Winchester

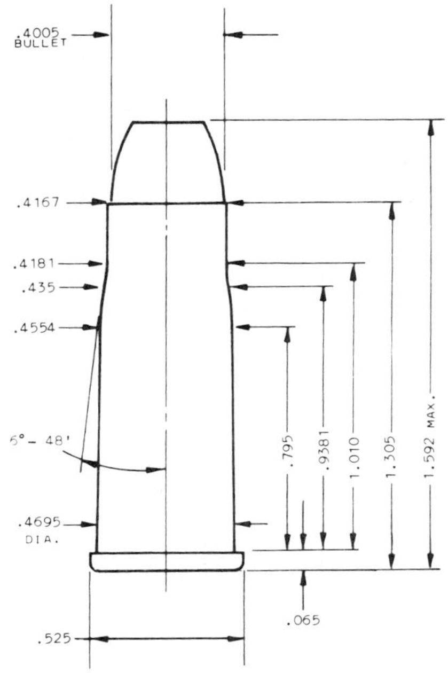

TEST COMPONENTS:

Cases .... Remington Peters
Trim-to Length .... 1.300″
Primers .... Remington 2½
Primer Size .... Large Pistol
Lyman Shell Holder .... No. 14B
Cast Bullets Used .... (size to .400″ dia.)
*Gas Check Bullets .... #40188, 170 gr.
#40143, 172 gr.

TEST SPECIFICATIONS:
(Velocity)

Firearm Used .... Winchester Model 92
Barrel Length .... 24″
Twist .... 1-36″
Groove Diameter .... .400″

COMMENTS:

These loads should not be used in handguns, or in rifles which were designed for black powder.

Individual tolerances vary greatly in rifles chambered for this cartridge. Therefore extreme care should be used in working up maximum loads.

#40188
170 gr., (#2 Alloy) 1.592″ OAL

Powder	Sug. Starting Grains	Velocity F.P.S.	Pressure C.U.P.	Max. Grains	Velocity F.P.S.	Pressure C.U.P.
Red Dot	5.5	1070	—	9.0	1465	—
700X	5.5	1075	—	9.0	1480	—
Green Dot	6.0	1100	—	9.5	1470	—
PB	7.0	1130	—	10.0	1485	—
Unique	7.0	1120	—	10.5	1515	—
SR-7625	8.0	1220	—	11.0	1565	—

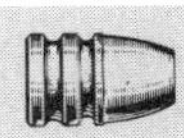

#40143
172 gr., (#2 Alloy) 1.550″ OAL

Powder	Sug. Starting Grains	Velocity F.P.S.	Pressure C.U.P.	Max. Grains	Velocity F.P.S.	Pressure C.U.P.
Red Dot	5.5	1080	—	9.0	1485	—
700X	5.5	1085	—	9.0	1505	—
Green Dot	6.0	1090	—	9.5	1480	—
PB	7.0	1155	—	10.0	1510	—
Unique	7.0	1115	—	10.5	1515	—
SR-7625	8.0	1210	—	11.0	1560	—

Note: Loads shown in shaded panels are maximum.

.38/55 Winchester

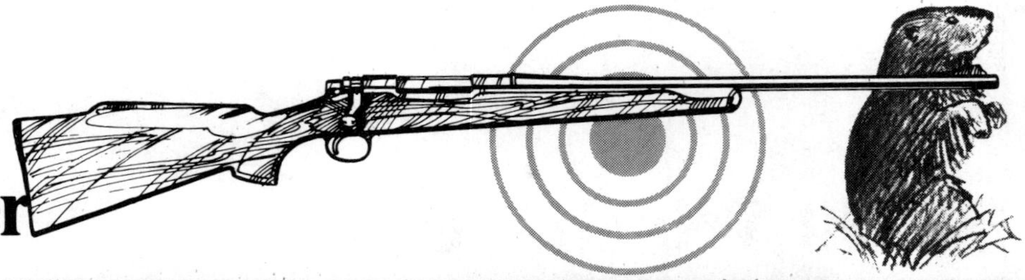

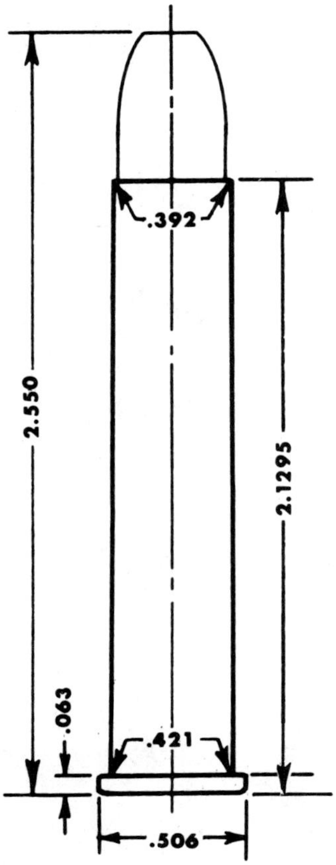

TEST COMPONENTS:

Cases .... Remington Peters
Trim-to Length .... 2.118″
Primers .... Remington 9½
Primer Size .... Large Rifle
Lyman Shell Holder .... No. 6
Cast Bullets Used .... (size to .379″ dia.)
*Gas Check Bullets #375248, 249 gr.
*#375296, 269 gr.

TEST SPECIFICATIONS:
(Velocity)

Firearm Used .... Stevens Model 44½
Barrel Length .... 29″
Twist .... 1-18″
Groove Diameter .... .379″

#375248
249 gr., (#2 Alloy) 2.550″ OAL

Powder	Sug. Starting Grains	Velocity F.P.S.	Pressure C.U.P.	Max. Grains	Velocity F.P.S.	Pressure C.U.P.
Red Dot	6.0	945	—	8.5	1145	—
700X	6.0	950	—	8.5	1150	—
Green Dot	6.5	970	—	9.0	1160	—
PB	6.5	910	—	9.5	1140	—
Unique	7.0	975	—	9.5	1170	—

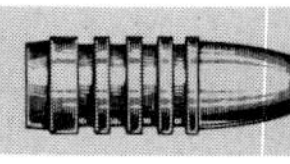

#375296
269 gr., (#2 Alloy) 2.550″ OAL

Powder	Sug. Starting Grains	Velocity F.P.S.	Pressure C.U.P.	Max. Grains	Velocity F.P.S.	Pressure C.U.P.
Red Dot	6.0	910	—	8.0	1065	—
700X	6.0	910	—	8.0	1060	—
Green Dot	6.5	940	—	8.5	1080	—
PB	6.5	880	—	9.0	1065	—
Unique	7.0	935	—	9.0	1080	—

Note: Loads shown in shaded panels are maximum.

COMMENTS:

Most of the rifles of this caliber which we have encountered show signs of excessive headspace. It is recommended that the rifle be checked by a competent gunsmith before reloading is attempted.

Slight variations in groove diameter do exist in these rifles. Where the groove diameter measures larger than .377″ the exclusive use of cast bullets is recommended for best accuracy.

.401 Winchester
Self-Loading

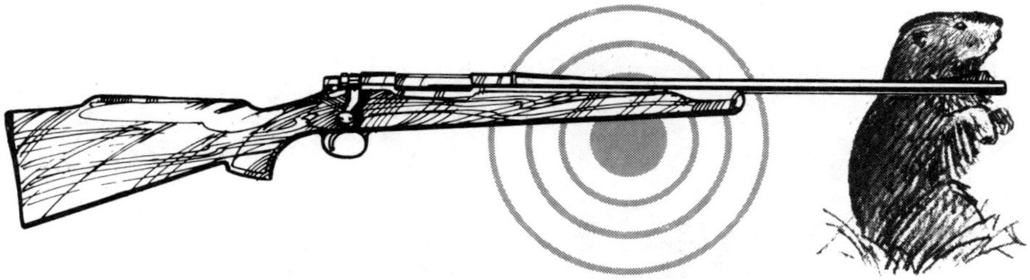

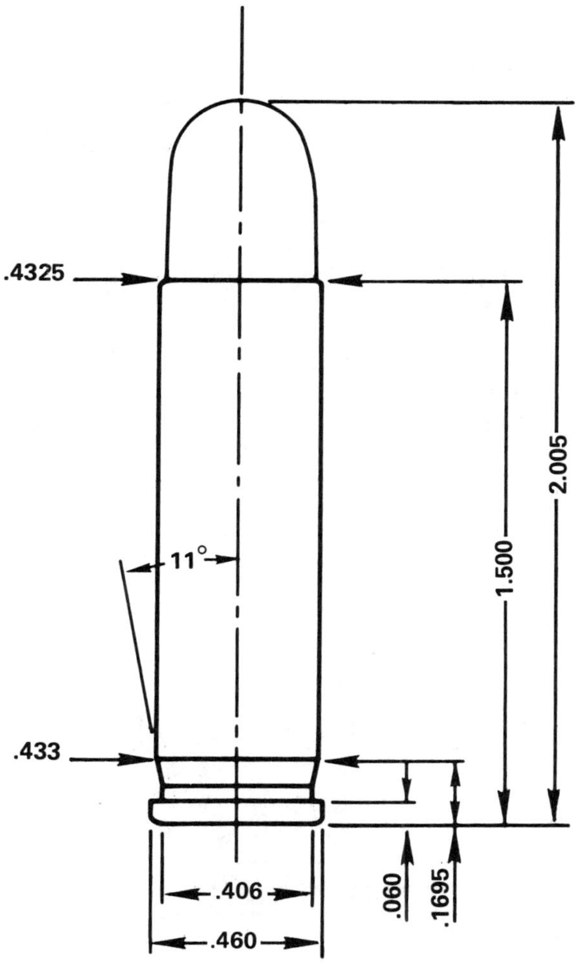

TEST COMPONENTS:

Cases .... Remington
Trim-to Length .... 1.495″
Primers .... Remington 9½
Primer Size .... Large Rifle
Lyman Shell Holder .... No. 2
Cast Bullets Used .... (size to .406″ to .407″ dia.)
*Gas Check Bullets *#41028, 212 gr.
*#410426, 240 gr.

TEST SPECIFICATIONS:
(Velocity)

Firearm Used .... Winchester Model 10
Barrel Length .... 20″
Twist .... 1-14″
Groove Diameter .... .4075″

#41028
212 gr., (#2 Alloy) 2.005″ OAL, Max.

Powder	Sug. Starting Grains	Velocity F.P.S.	Pressure C.U.P.	Max. Grains	Velocity F.P.S.	Pressure C.U.P.
Unique	12.0	1618	—	14.9	1845	—
2400	21.0	1773	—	23.5	1960	—
IMR-4227	26.0	1915	—	29.0+	2074	—

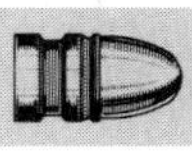

#410426
240 gr., (#2 Alloy) 2.005″ OAL, Max.

Powder	Sug. Starting Grains	Velocity F.P.S.	Pressure C.U.P.	Max. Grains	Velocity F.P.S.	Pressure C.U.P.
Unique	11.0	1470	—	13.7	1672	—
2400	20.0	1669	—	22.0	1818	—
IMR-4227	24.0	1506	—	27.5+	1968	—

Note: Loads shown in shaded panels are maximum.
+Designates a compressed powder charge.

COMMENTS:

Cases for this cartridge are often quite old and it is recommended that you keep your loads on the light side to avoid trouble with poor brass. Many cases are found which exceed the maximum length listed. Check your case length carefully and trim the cases if necessary. Our best test results were obtained with IMR 4227 powder.

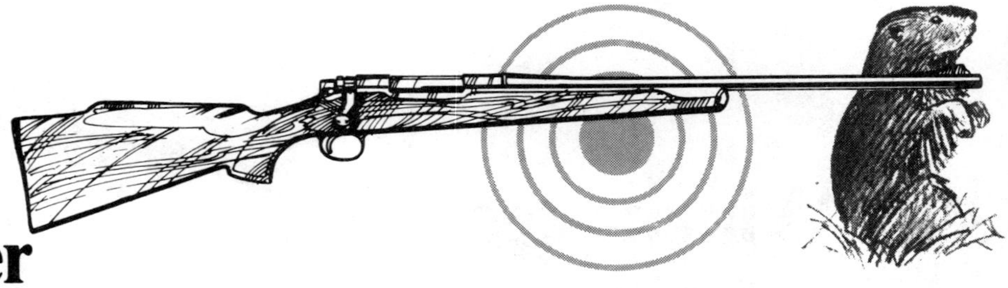

.44/40 Winchester

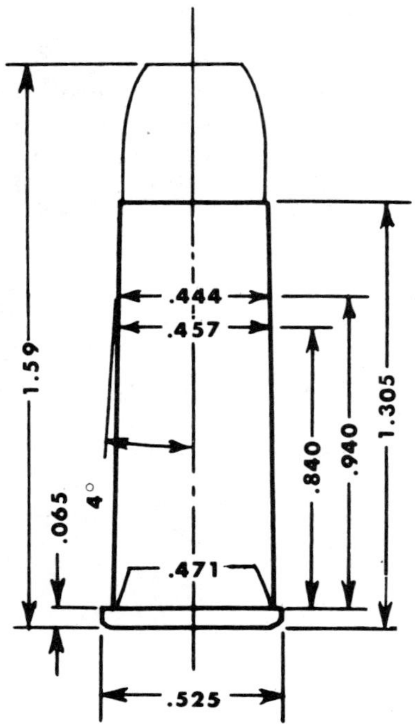

TEST COMPONENTS:

Cases .... Remington Peters
Trim-to Length .... 1.300″
Primers .... Remington 2½
Primer Size .... Large Pistol
Lyman Shell Holder .... No. 14B
Cast Bullets Used .... (size to .430″ dia.)
*Gas Check Bullets #42798, 205 gr.
*#429434, 220 gr.

TEST SPECIFICATIONS:
(Velocity)

Firearm Used .... Marlin Model 94
Barrel Length .... 24″
Twist .... 1-36″
Groove Diameter .... .436″

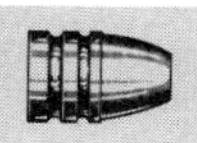

#42798
205 gr., (#2 Alloy) 1.580″ OAL

Powder	Sug. Starting Grains	Velocity F.P.S.	Pressure C.U.P.	Max. Grains	Velocity F.P.S.	Pressure C.U.P.
Red Dot	7.0	1130	—	8.0	1255	—
700X	7.0	1165	—	8.0	1270	—
Green Dot	7.3	1125	—	8.3	1230	—
PB	7.5	1075	—	9.0	1260	—
Unique	8.0	1090	—	10.5	1355	—
SR-7625	8.0	1090	—	9.5	1285	—

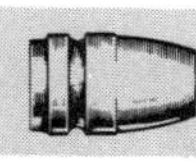

#429434
220 gr., (#2 Alloy) 1.580″ OAL

Powder	Sug. Starting Grains	Velocity F.P.S.	Pressure C.U.P.	Max. Grains	Velocity F.P.S.	Pressure C.U.P.
Red Dot	6.8	1115	—	7.8	1210	—
700X	6.8	1125	—	7.8	1225	—
Green Dot	7.1	1100	—	8.0	1185	—
PB	7.3	1060	—	8.6	1215	—
Unique	8.0	1100	—	10.2	1335	—
SR-7625	7.8	1065	—	9.0	1210	—

COMMENTS:

These loads should not be used in handguns or in rifles which were designed for black powder.

Individual tolerances vary greatly in rifles chambered for this cartridge. Therefore, extreme care should be used in working up maximum loads.

Due to variations in groove diameters, it is recommended that you slug your barrel before reloading.

Note: Loads shown in shaded panels are maximum.

.44 Remington Magnum

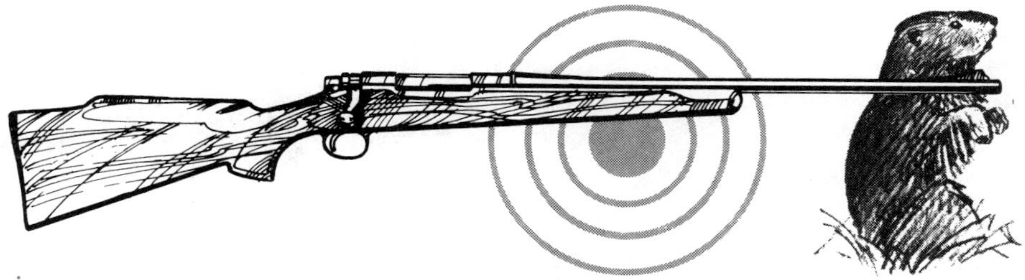

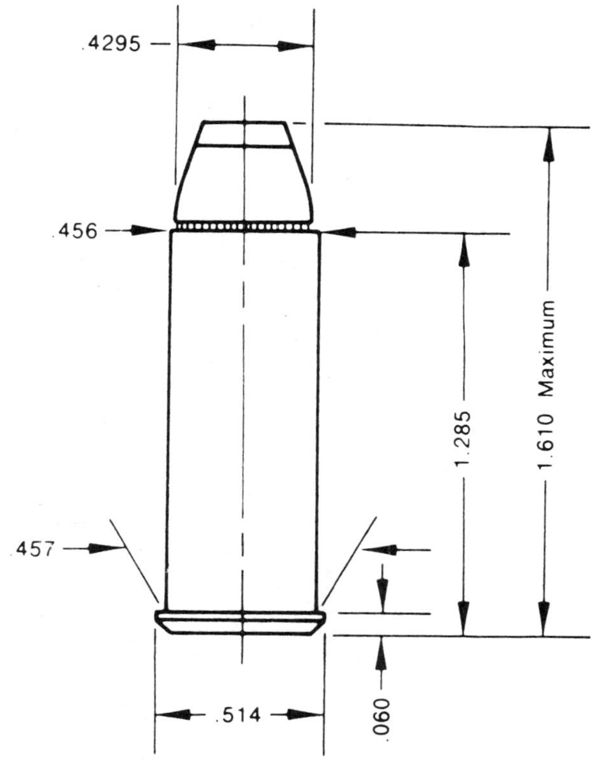

TEST COMPONENTS:

Cases .... Remington Peters
Trim-to Length .... 1.280″
Primers .... Remington 2½
Primer Size .... Large Pistol
Lyman Shell Holder .... No. 7
Cast Bullets Used .... (size to .429″ dia.)
*Gas Check Bullets #429348, 180 gr.
*#429303, 205 gr.
*#429215, 215 gr.
#429360, 232 gr.
#429421, 245 gr.
*#429244, 250 gr.

TEST SPECIFICATIONS:
(Velocity)

Firearm Used .... Ruger .44 Carbine
Barrel Length .... 18½″
Twist .... 1-38″
Groove Diameter .... .429″

COMMENTS:

Cast bullets tend to foul the gas piston and create accuracy problems in semi-automatic rifles when not cleaned regularly. Be certain that your primers are seated correctly. A high primer can fire the cartridge before the bolt is in locked position on the Ruger Carbine.

In rifles with multi or shallow groove rifling, cast bullet velocity should be held to 1600 f.p.s., or less for best accuracy.

In order to correctly seat cast bullets the maximum overall length, as listed, must be exceeded. Check overall length as listed for each specific cast bullet.

#429348
180 gr., (#2 Alloy) 1.420″ OAL

Powder	Sug. Starting Grains	Velocity F.P.S.	Pressure C.U.P.	Max. Grains	Velocity F.P.S.	Pressure C.U.P.
Bullseye	5.0	965	—	7.5	1285	—
Unique	8.0	1185	—	13.5	1770	—
AL-7	12.0	1355	—	19.0	1870	—
Herco	11.0	1420	—	15.5	1750	—
SR-4756	12.0	1485	—	17.0	1890	—
AL-8	14.0	1305	—	20.0+	1755	—
Blue Dot	14.0	1440	—	21.0	1905	—
630	15.0	1415	—	21.5	1840	—
2400	18.0	1400	—	24.0	1835	—
IMR-4227	19.0	1212	—	27.0	1820	—

Note: Loads shown in shaded panels are maximum.
+Designates a compressed powder charge.

.44 Remington Magnum (Continued)

#429303
205 gr., (#2 Alloy) 1.716" OAL

(Not For Use In Tubular Magazine)

Powder	Sug. Starting Grains	Velocity F.P.S.	Pressure C.U.P.	Max. Grains	Velocity F.P.S.	Pressure C.U.P.
Bullseye	5.0	979	—	7.5	1273	—
Unique	8.0	1207	—	13.5	1659	—
AL-7	12.0	1355	—	19.0	1855	—
Herco	11.0	1415	—	15.5	1762	—
SR-4756	12.0	1481	—	16.0	1772	—
AL-8	14.0	1290	—	20.0+	1720	—
Blue Dot	14.0	1458	—	19.5	1907	—
630	15.0	1395	—	20.5	1745	—
2400	18.0	1404	—	24.0	1841	—
IMR-4227	19.0	1291	—	26.0	1763	—

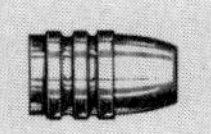

#429215
215 gr., (#2 Alloy) 1.645" OAL

Powder	Sug. Starting Grains	Velocity F.P.S.	Pressure C.U.P.	Max. Grains	Velocity F.P.S.	Pressure C.U.P.
Bullseye	5.0	935	—	7.5	1210	—
Unique	8.5	1205	—	12.5	1595	—
AL-7	13.0	1375	—	18.5	1770	—
Herco	11.0	1365	—	15.5	1685	—
SR-4756	12.0	1425	—	15.0	1660	—
AL-8	14.5	1335	—	20.0+	1656	—
Blue Dot	14.0	1370	—	19.5	1825	—
630	15.0	1320	—	20.5	1675	—
2400	18.0	1365	—	23.5	1735	—
IMR-4227	20.0	1265	—	25.5	1695	—

#429360
232 gr., (#2 Alloy) 1.655" OAL

Powder	Sug. Starting Grains	Velocity F.P.S.	Pressure C.U.P.	Max. Grains	Velocity F.P.S.	Pressure C.U.P.
Bullseye	5.0	935	—	7.5	1185	—
Unique	8.5	1200	—	12.5	1565	—
AL-7	13.0	1385	—	17.5	1685	—
Herco	11.0	1365	—	15.0	1630	—
SR-4756	11.5	1390	—	14.0	1550	—
AL-8	14.5	1300	—	20.0	1670	—
Blue Dot	14.0	1415	—	18.5	1740	—
630	15.0	1355	—	19.5	1620	—
2400	18.0	1370	—	23.0	1715	—
IMR-4227	20.0	1275	—	24.5	1605	—

#429421
245 gr., (#2 Alloy) 1.710" OAL

Powder	Sug. Starting Grains	Velocity F.P.S.	Pressure C.U.P.	Max. Grains	Velocity F.P.S.	Pressure C.U.P.
Bullseye	5.0	915	—	7.5	1180	—
Unique	8.5	1195	—	12.0	1495	—
AL-7	13.0	1355	—	17.0	1635	—
Herco	11.0	1345	—	14.5	1565	—
SR-4756	11.0	1305	—	13.0	1470	—
AL-8	14.5	1290	—	19.5	1635	—
Blue Dot	14.0	1410	—	17.5	1665	—
630	15.0	1340	—	19.5	1610	—
2400	18.0	1375	—	22.5	1680	—
IMR-4227	20.0	1310	—	24.0	1610	—

Note: Loads shown in shaded panels are maximum.
+Designates a compressed powder charge.

.44 Remington Magnum (Continued)

#429244
250 gr., (#2 Alloy) 1.680" OAL

Powder	Sug. Starting Grains	Velocity F.P.S.	Pressure C.U.P.	Max. Grains	Velocity F.P.S.	Pressure C.U.P.
Bullseye	5.0	910	—	7.5	1160	—
Unique	8.5	1180	—	12.0	1485	—
AL-7	13.0	1365	—	16.5	1580	—
Herco	11.0	1320	—	14.5	1625	—
SR-4756	11.0	1270	—	12.5	1405	—
AL-8	14.5	1290	—	19.5	1595	—
Blue Dot	14.0	1410	—	17.5	1655	—
630	15.0	1330	—	19.0	1555	—
2400	18.0	1370	—	22.0	1635	—
IMR-4227	20.0	1300	—	24.0	1595	—

Note: Loads shown in shaded panels are maximum.

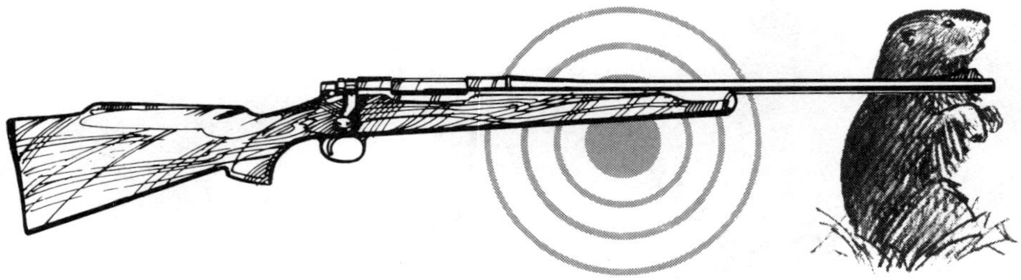

.444 Marlin

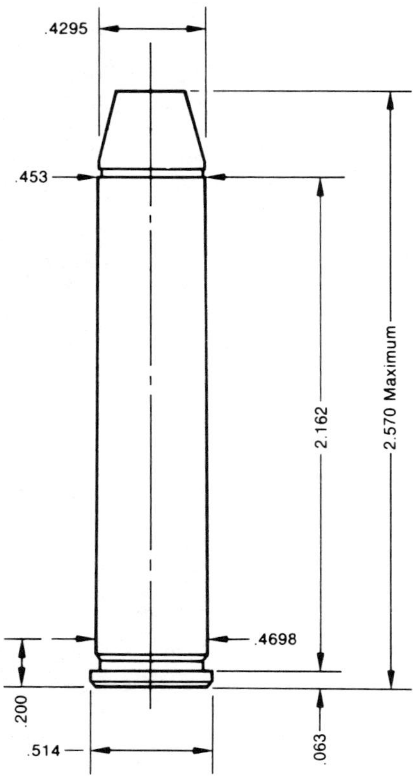

TEST COMPONENTS:

Cases .... Remington Peters
Trim-to Length .... 2.220″
Primers .... Remington 9½
Primer Size .... Large Rifle
Lyman Shell Holder .... No. 14B
Cast Bullets Used .... (size to .431″ dia.)
*Gas Check Bullets #42798, 205 gr.
*#429215, 215 gr.
#429360, 232 gr.
#429421, 245 gr.
*#429244, 250 gr.

TEST SPECIFICATIONS: (Velocity)

Firearm Used .... Marlin Model 444
Barrel Length .... 22″
Twist .... 1-38″
Groove Diameter .... .431″

COMMENTS:

In order to correctly seat cast bullets, they must exceed the maximum overall length as listed. Our data takes this into consideration and the individual overall lengths for each bullet are listed above.

#429421 should be crimped on forward edge of first driving band.

#42798
205 gr., (#2 Alloy) 2.580″ OAL

Powder	Sug. Starting Grains	Velocity F.P.S.	Pressure C.U.P.	Max. Grains	Velocity F.P.S.	Pressure C.U.P.
Red Dot	10.5	1330	—	16.0	1690	—
700X	10.0	1310	—	15.5	1690	—
Green Dot	11.0	1345	—	16.5	1700	—
Unique	12.0	1315	—	18.0	1705	—
SR-7625	13.0	1440	—	18.0	1755	—
SR-4756	14.0	1425	—	19.0	1755	—
Blue Dot	15.0	1335	—	20.0	1745	—
630	16.5	1310	—	22.0	1605	—

Note: Loads shown in shaded panels are maximum.

#429215
215 gr., (#2 Alloy) 2.595" OAL

Powder	Sug. Starting Grains	Velocity F.P.S.	Pressure C.U.P.	Max. Grains	Velocity F.P.S.	Pressure C.U.P.
Red Dot	10.5	1335	—	16.0	1685	—
700X	10.0	1295	—	15.5	1655	—
Green Dot	11.0	1325	—	16.5	1685	—
Unique	12.0	1335	—	18.0	1735	—
SR-7625	13.0	1405	—	18.0	1705	—
SR-4756	14.0	1350	—	19.0	1725	—
Blue Dot	15.0	1295	—	20.0	1690	—
630	16.5	1290	—	22.0	1605	—

#429360
232 gr., (#2 Alloy) 2.575" OAL

Powder	Sug. Starting Grains	Velocity F.P.S.	Pressure C.U.P.	Max. Grains	Velocity F.P.S.	Pressure C.U.P.
Red Dot	10.5	1295	—	15.5	1630	—
700X	10.0	1260	—	15.0	1600	—
Green Dot	11.0	1290	—	16.0	1615	—
Unique	12.0	1300	—	17.5	1660	—
SR-7625	13.0	1380	—	17.5	1650	—
SR-4756	14.0	1365	—	18.5	1680	—
Blue Dot	15.0	1310	—	19.5	1605	—
630	16.5	1275	—	22.0	1585	—

#429421
245 gr., (#2 Alloy) 2.550" OAL

Powder	Sug. Starting Grains	Velocity F.P.S.	Pressure C.U.P.	Max. Grains	Velocity F.P.S.	Pressure C.U.P.
Red Dot	10.0	1225	—	15.0	1555	—
700X	9.5	1200	—	14.5	1535	—
Green Dot	10.5	1235	—	15.5	1570	—
Unique	11.5	1270	—	17.0	1620	—
SR-7625	12.5	1310	—	17.0	1590	—
SR-4756	13.5	1305	—	18.0	1590	—
Blue Dot	14.5	1240	—	19.0	1570	—
630	16.0	1240	—	21.5	1540	—

#429244
250 gr., (#2 Alloy) 2.620" OAL

Powder	Sug. Starting Grains	Velocity F.P.S.	Pressure C.U.P.	Max. Grains	Velocity F.P.S.	Pressure C.U.P.
Red Dot	10.0	1205	—	15.0	1520	—
700X	9.5	1175	—	14.5	1500	—
Green Dot	10.5	1215	—	15.5	1525	—
Unique	11.5	1225	—	17.0	1590	—
SR-7625	12.5	1285	—	17.0	1560	—
SR-4756	13.5	1230	—	18.0	1560	—
Blue Dot	14.5	1240	—	19.0	1540	—
630	16.0	1210	—	21.5	1510	—

Note: Loads shown in shaded panels are maximum.

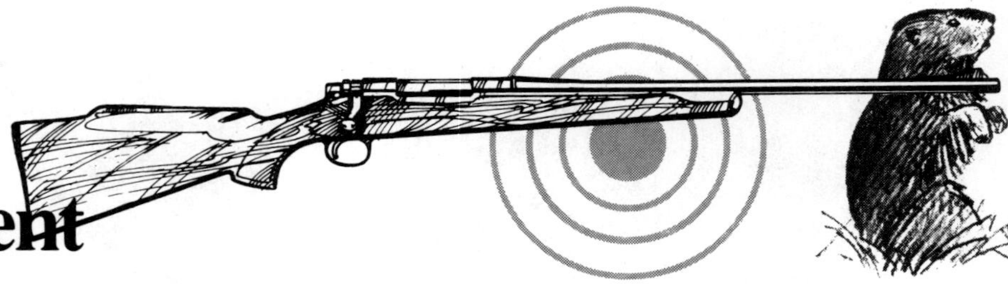

.45/70 Government

For 1873 Springfield

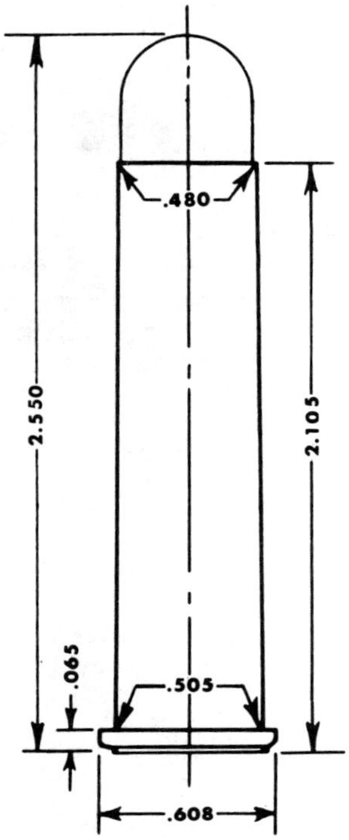

TEST COMPONENTS:

Cases .... Remington
Trim-to Length .... 2.100″
Primers .... Remington 9½
Primer Size .... Large Rifle
Lyman Shell Holder .... No. 17
Cast Bullets Used .... (size to .457″ dia.)
*Gas Check Bullets #457191, 293 gr.
#457122, 322 gr.
#457124, 366 gr.
#457193, 420 gr.

TEST SPECIFICATIONS:
(Velocity & Pressure)

Firearm Used .... Universal Receiver
(Standard Government Chamber)
Barrel Length .... 24″
Twist .... 1-18″
Groove Diameter .... .457″

COMMENTS:

Loads greater than 15,000 C.U.P. should be used only in Springfields (the so-called trapdoor Springfield) in good condition and should not exceed 17,000 C.U.P.

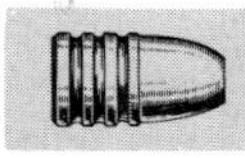

#457191
293 gr., (#2 Alloy) 2.550″ OAL

Powder	Sug. Starting Grains	Velocity F.P.S.	Pressure C.U.P.	Max. Grains	Velocity F.P.S.	Pressure C.U.P.
SR-4759	26.0	1311	9,400	30.0	1602	14,600
IMR-4198	28.5++	1327	9,900	34.5++	1613	16,600
IMR-3031	34.0++	1269	10,500	44.0++	1577	15,300

#457122
322 gr., (#2 Alloy) 2.550″ OAL

Powder	Sug. Starting Grains	Velocity F.P.S.	Pressure C.U.P.	Max. Grains	Velocity F.P.S.	Pressure C.U.P.
630	*16.5++	1117	9,600	21.5++	1378	16,300
SR-4759	19.0++	1043	8,100	25.0++	1412	16,400
IMR-4198	23.0++	1088	7,700	32.0++	1480	17,000
RX7	47.0++	1687	13,100	51.0	1813	15,800
IMR-3031	29.0++	1056	7,600	37.0++	1435	16,800

Note: Loads shown in shaded panels are maximum.
*Designates potentially most accurate load.
++Designates the use of a ½ gr. Dacron wad, ⅝″ square x ¼ thick over the powder.

.45/70 Government, For 1873 Springfield (Continued)

#457124
366 gr., (#2 Alloy) 2.540″ OAL

Powder	Sug. Starting Grains	Velocity F.P.S.	Pressure C.U.P.	Max. Grains	Velocity F.P.S.	Pressure C.U.P.
SR-4759	19.0++	1035	9,500	23.3++	1298	15,400
IMR-4198	23.0++	1039	7,400	*30.0++	1363	16,100
RX7	37.0++	1300	7,200	48.0	1764	16,200
IMR-3031	30.0++	1090	8,900	39.0++	1403	17,000

#457193
420 gr., (#2 Alloy) 2.550″ OAL

Powder	Sug. Starting Grains	Velocity F.P.S.	Pressure C.U.P.	Max. Grains	Velocity F.P.S.	Pressure C.U.P.
SR-4759	18.0++	942	9,000	*23.0++	1207	15,700
IMR-4198	21.5++	952	6,800	28.5++	1267	13,900
RX7	33.0++	1198	7,800	41.0	1467	13,300
IMR-3031	34.0	1161	11,100	38.5	1352	16,000

Note: Loads shown in shaded panels are maximum.
*Designates potentially most accurate load.
++Designates the use of a ½ gr. Dacron wad, ⅝″ square x ¼ thick over the powder.

.45/70 Government

For 1886 Winchester and 1895 Marlin

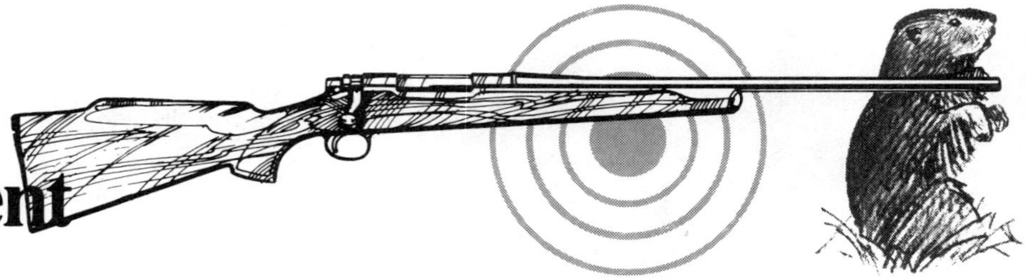

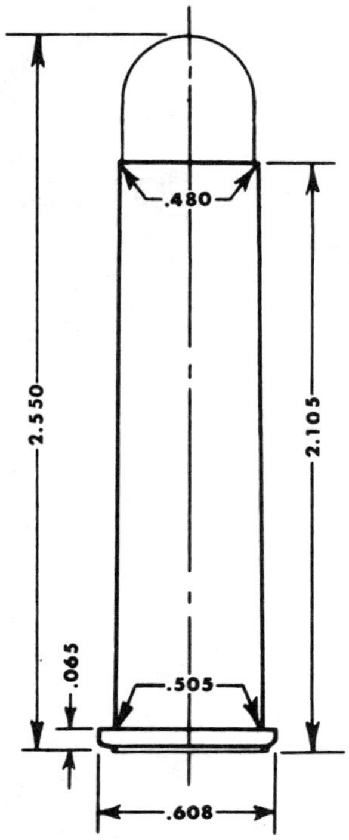

TEST COMPONENTS:

Cases .... Remington
Trim-to Length .... 2.100″
Primers .... Remington 9½
Primer Size .... Large Rifle
Lyman Shell Holder .... No. 17
Cast Bullets Used .... (size to .457″ dia.)
*Gas Check Bullets #457191, 293 gr.
#457122, 322 gr.
#457193, 420 gr.

TEST SPECIFICATIONS:
(Velocity & Pressure)

Firearm Used .... Universal Receiver (Standard Government Chamber)
Barrel Length .... 24″
Twist .... 1-18″
Groove Diameter .... .457″

COMMENTS:

Bullets should be crimped for use in tubular magazines.

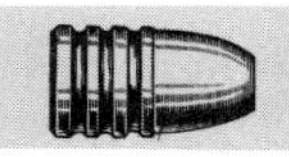

#457191
293 gr., (#2 Alloy) 2.550″ OAL

Powder	Sug. Starting Grains	Velocity F.P.S.	Pressure C.U.P.	Max. Grains	Velocity F.P.S.	Pressure C.U.P.
Unique	12.0++	1200	13,200	15.0++	1416	24,500
630	23.7++	1562	16,000	*29.6++	1847	26,900
2400	24.4++	1524	15,300	30.5++	1842	26,900
SR-4759	30.0	1602	14,600	34.0	1802	21,000
IMR-4227	25.4++	1495	11,400	31.0++	1778	24,800
IMR-4198	34.5++	1613	16,600	49.7	2065	24,600
IMR-3031	51.0	1695	16,500	59.0+	2065	27,400

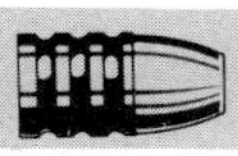

#457122
322 gr., (#2 Alloy) 2.550″ OAL

Powder	Sug. Starting Grains	Velocity F.P.S.	Pressure C.U.P.	Max. Grains	Velocity F.P.S.	Pressure C.U.P.
Unique	9.5++	963	9,600	14.5++	1295	24,400
630	20.5++	1331	14,900	*27.3++	1653	26,600
2400	21.0++	1293	14,200	28.0++	1640	25,000
SR-4759	25.0++	1412	16,400	31.5++	1719	28,000
IMR-4227	22.1++	1254	13,800	30.0++	1639	27,000
IMR-4198	33.5++	1539	18,800	43.0	1867	28,000
RX7	50.0++	1842	16,200	58.5+	2132	27,000
IMR-3031	47.0++	1723	22,000	54.0	1934	28,000

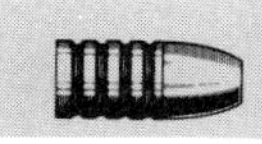

#457193
420 gr., (#2 Alloy) 2.550″ OAL

Powder	Sug. Starting Grains	Velocity F.P.S.	Pressure C.U.P.	Max. Grains	Velocity F.P.S.	Pressure C.U.P.
Unique	11.0++	1006	16,800	13.5++	1163	27,200
630	*20.0++	1244	18,600	24.5++	1445	26,000
2400	21.0++	1252	17,700	25.7++	1490	27,500
SR-4759	23.0++	1207	15,700	28.5++	1418	24,500
IMR-4227	21.0++	1163	14,700	26.2++	1418	24,200
IMR-4198	28.5++	1267	13,900	41.5	1737	28,300
RX7	40.0	1493	13,500	48.5	1828	25,000
IMR-3031	38.5	1352	16,000	47.0	1684	25,000

Note: Loads shown in shaded panels are maximum.
*Designates potentially most accurate load.
+Designates a compressed powder charge.

++Designates the use of a ½ gr. Dacron wad, ⅝″ square x ¼″ thick over the powder.

.45/70 Government

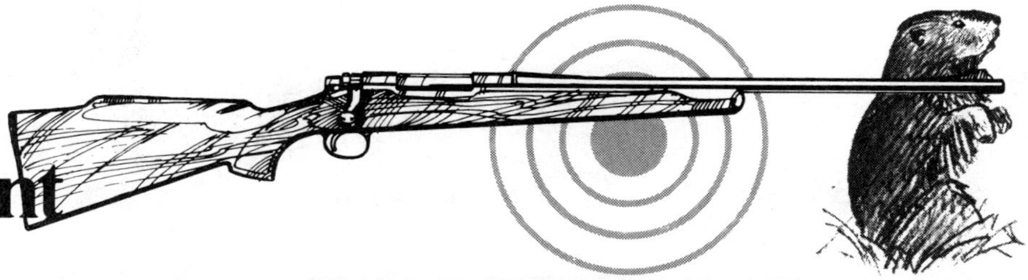

For Ruger #1 & #3

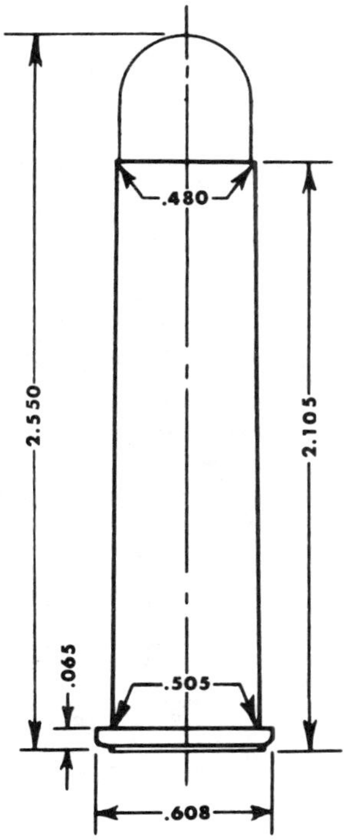

TEST COMPONENTS:

Cases .... Remington
Trim-to Length .... 2.105″
Primers .... Remington 9½
Primer Size .... Large Rifle
Lyman Shell Holder .... No. 17
Cast Bullets Used .... (size to .457″ dia.)
*Gas Check Bullets #457122, 322 gr.
#457124, 366 gr.
#457193, 420 gr.
*#457102, 426 gr.
*#457406, 451 gr.
#457125, 464 gr.

TEST SPECIFICATIONS:
(Velocity & Pressure)

Firearm Used ...... Universal Receiver, Ruger Chamber
Barrel Length .... 28″
Twist .... 1-20″
Groove Diameter .... .456″

COMMENTS:

The Ruger chamber features a throat or "ballseat" permitting overall cartridge lengths in excess of 2.550″, the maximum length usually specified for the .45-70 Government cartridge.

Duplex loads were developed for the Ruger using 10% SR-4759 over the primer, followed by 90% FFG black powder. This technique results in slightly improved velocity compared to black powder alone and is cleaner burning.

With heavily compressed powder charges it is advisable to crimp the case because of the tendency of the bullet to back out once the seating pressure is released.

#457122
322 gr., (#2 Alloy) 2.610″ OAL

Powder	Sug. Starting Grains	Velocity F.P.S.	Pressure C.U.P.	Max. Grains	Velocity F.P.S.	Pressure C.U.P.
Unique	12.5++	1171	16,900	17.5++	1472	38,700
630	29.0	1740	24,700	35.0	1999	40,000
2400	27.0	1608	21,200	34.5	1958	40,000
SR-4759	30.5	1635	20,600	39.5	1973	38,300
IMR-4227	29.0++	1635	23,500	35.0++	1893	37,700
IMR-4198	43.0++	1930	26,900	51.0	2220	37,600
IMR-3031	52.0	1865	21,500	60.0+	2179	33,700
FFg	40.0++	1041	8,400	70.0+	1399	14,700
Pyrodex CTG	46.5	1153	7,100	62.0+	1445	12,800
Duplex Load						
SR-4759	5.0	1142	7,700	7.0	1507	14,900
FFg	45.0++	—	—	63.0	—	—

Note: Loads shown in shaded panels are maximum.
*Designates potentially most accurate load.
+Designates a compressed powder charge.

++Designates the use of a ½ gr. Dacron wad, ⅝″ square x ¼″ thick over the powder.

45/70 Government, For Ruger #1 & #3 (Continued)

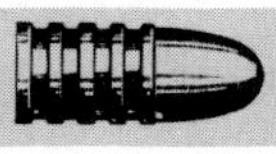

#457124
366 gr., (#2 Alloy) 2.723″ OAL

Powder	Sug. Starting Grains	Velocity F.P.S.	Pressure C.U.P.	Max. Grains	Velocity F.P.S.	Pressure C.U.P.
Unique	10.0++	939	10,300	17.5++	1411	38,500
630	25.0++	1501	19,300	31.5++	1807	36,300
2400	26.5++	1517	20,400	33.5++	1853	37,900
SR-4759	30.0	1478	16,300	39.0	1880	36,700
IMR-4227	27.0++	1458	18,300	34.0++	1769	34,300
IMR-4198	41.5++	1832	25,400	50.5	2135	37,500
IMR-3031	50.0	1714	17,800	60.0+	2117	33,000
FFg	52.5++	1087	7,700	70.0+	1328	14,400
Pyrodex CTG	39.4	1054	7,400	52.5+	1198	9,000
Duplex Load						
SR-4759	5.0	1038	7,600	6.5	1375	13,300
FFg	45.0++	—	—	58.5	—	—

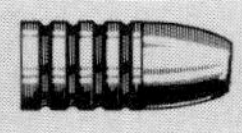

#457193
420 gr., (#2 Alloy) 2.680″ OAL

Powder	Sug. Starting Grains	Velocity F.P.S.	Pressure C.U.P.	Max. Grains	Velocity F.P.S.	Pressure C.U.P.
Unique	13.5++	1113	22,800	16.5++	1275	38,200
630	25.4++	1452	23,800	31.0++	1699	37,700
2400	26.3++	1451	23,800	32.0++	1698	37,100
SR-4759	28.0++	1370	19,300	35.0	1682	36,400
IMR-4227	27.8++	1457	24,200	34.0++	1684	37,700
IMR-4198	37.6++	1662	25,600	47.0	1932	36,600
RX7	46.0++	1687	17,800	56.0+	2058	34,700
IMR-3031	44.8	1599	20,200	56.0+	1996	35,600
FFg	63.0	1185	12,500	70.0+	1268	16,400
Pyrodex CTG	58.0+	1084	9,100	65.0+	1161	10,100
Duplex Load						
SR-4759	6.0	1259	13,700	7.0	1388	18,700
FFg	56.0++	—	—	63.0+	—	—

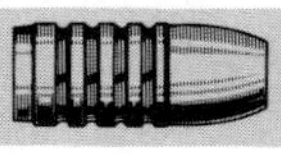

#457102
426 gr., (#2 Alloy) 2.680″ OAL

Powder	Sug. Starting Grains	Velocity F.P.S.	Pressure C.U.P.	Max. Grains	Velocity F.P.S.	Pressure C.U.P.
Unique	10.0++	857	12,300	16.5++	1277	40,300
630	22.0++	1323	19,400	30.0++	1689	40,000
2400	24.0++	1342	20,900	32.0++	1692	38,300
SR-4759	26.0++	1400	20,700	35.4	1702	37,300
IMR-4227	25.0++	1317	18,300	33.3++	1634	34,700
IMR-4198	40.0++	1741	28,200	46.3	1954	39,200
IMR-3031	48.0	1700	25,400	56.5+	2002	38,500
FFg	52.5++	1042	10,000	70.0+	1256	16,900
Pyrodex CTG	49.0	1099	9,400	65.0+	1306	16,600
Duplex Load						
SR-4759	5.3	1147	11,500	7.0	1389	22,100
FFg	47.0++	—	—	63.0	—	—

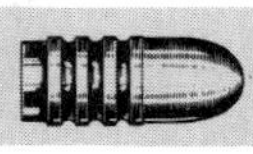

#457406
451 gr., (#2 Alloy) 2.650″ OAL

Powder	Sug. Starting Grains	Velocity F.P.S.	Pressure C.U.P.	Max. Grains	Velocity F.P.S.	Pressure C.U.P.
Unique	11.0++	950	17,100	15.2++	1209	40,000
630	20.0++	1221	19,900	27.5++	1564	39,600
2400	18.5++	1355	25,100	27.5++	1517	34,400
SR-4759	24.0++	1305	19,900	32.0	1635	39,400
IMR-4227	24.0++	1267	20,400	31.5++	1568	36,600
IMR-4198	33.0++	1445	24,100	43.0	1833	38,400
IMR-3031	36.0++	1304	15,300	44.0+	1590	29,100
FFg	47.3++	952	7,900	63.0+	1168	15,000
Pyrodex CTG	38.2++	927	7,600	51.0+	1096	10,200
Duplex Load						
SR-4759	4.7	1036	9,800	6.3	1309	21,000
FFg	42.6++	—	—	56.7	—	—

Note: Loads shown in shaded panels are maximum.
+Designates a compressed powder charge.
++Designates the use of a ½ gr. Dacron wad, ⅝″ square x ¼″ thick over the powder.

#457125
464 gr., (#2 Alloy) 2.878″ OAL

Powder	Sug. Starting Grains	Velocity F.P.S.	Pressure C.U.P.	Max. Grains	Velocity F.P.S.	Pressure C.U.P.
Unique	10.0++	888	14,100	14.8++	1153	33,700
630	23.0++	1344	22,600	30.0++	1643	40,400
2400	24.0++	1350	22,100	28.0++	1604	38,500
SR-4759	23.0++	1237	15,400	33.0	1635	36,600
IMR-4227	22.0++	1169	16,400	30.0++	1532	35,000
IMR-4198	33.0++	1477	22,100	43.5	1797	34,900
IMR-3031	42.0++	1548	22,300	52.0+	1860	34,700
FFg	45.7++	926	7,800	61.0+	1125	12,300
Pyrodex CTG	48.8++	1098	10,400	65.0+	1282	19,800
Duplex Load						
SR-4759	4.6	989	8,400	6.1	—	—
FFg	40.4++	—	—	53.9+	1261	16,500

Note: Loads shown in shaded panels are maximum.
+Designates a compressed powder charge.
++Designates the use of a ½ gr. Dacron wad, ⅝″ square x ¼″ thick over the powder.

.458 Winchester Magnum

(11.5 x 63mm)

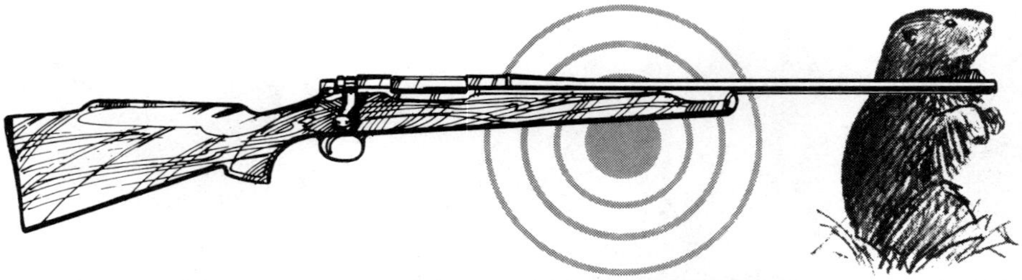

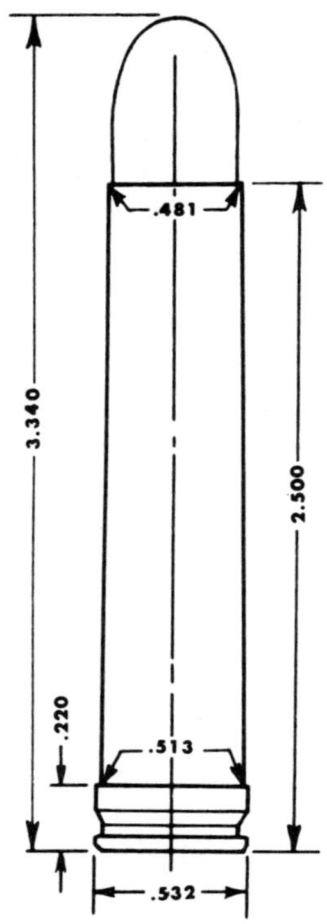

TEST COMPONENTS:

Cases .... Winchester/Western
Trim-to Length .... 2.495″
Primers .... Winchester 8½-120
Primer Size .... Large Rifle
Lyman Shell Holder .... No. 13
Cast Bullets Used .... (size to .459″ dia.)
*Gas Check Bullets #457191, 292 gr.
#457124, 385 gr.
*#457406, 482 gr.
*#462560, 552gr.

TEST SPECIFICATIONS:
(Velocity)

Firearm Used .... Ruger #1
Barrel Length .... 24″
Twist .... 1-14″
Groove Diameter .... .459″

COMMENTS:

Our best test results were obtained when using the Winchester 8½ - 120 primer. Reloading die adjustments are critical with the heavily compressed loads. If the crimping and seating screw adjustments are not exactly on the crimping groove, the case will be dented or crushed in loading.

Due to heavy recoil, all bullets should be crimped in place.

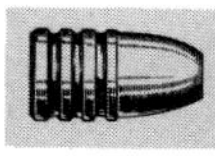

#457191
292 gr., (#2 Alloy) 2.995″ OAL

Powder	Sug. Starting Grains	Velocity F.P.S.	Pressure C.U.P.	Max. Grains	Velocity F.P.S.	Pressure C.U.P.
Unique	19.0	1460	—	23.0	1625	—
2400	28.0	1585	—	33.0	1765	—
IMR-4227	44.5	2100	—	50.5	2305	—
IMR-4198	54.0	2135	—	58.0	2205	—
IMR-3031	60.0	1885	—	69.0	2220	—
IMR-4064	64.0	1865	—	72.0	2110	—

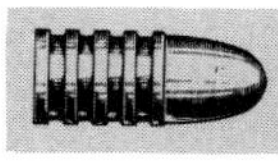

#457124
385 gr., (#2 Alloy) 3.012″ OAL

Powder	Sug. Starting Grains	Velocity F.P.S.	Pressure C.U.P.	Max. Grains	Velocity F.P.S.	Pressure C.U.P.
Unique	18.0	1315	—	22.0	1510	—
2400	25.0	1310	—	31.0	1565	—
IMR-4227	44.0	1935	—	50.0	2090	—
IMR-4198	53.0	2010	—	57.0	2125	—
IMR-3031	59.0	1900	—	68.0+	2210	—
IMR-4064	62.0	1860	—	70.0+	2115	—

Note: Loads shown in shaded panels are maximum.
+Designates a compressed powder charge.

#457406
482 gr., (#2 Alloy) 3.090″ OAL

Powder	Sug. Starting Grains	Velocity F.P.S.	Pressure C.U.P.	Max. Grains	Velocity F.P.S.	Pressure C.U.P.
Unique	17.0	1175	—	21.0	1320	—
2400	24.0	1240	—	30.0	1480	—
IMR-4227	41.0	1700	—	47.0	1855	—
IMR-4198	49.0	1830	—	53.0	1940	—
IMR-3031	57.0	1850	—	66.0+	2120	—
IMR-4064	60.0	1805	—	68.0+	2060	—

#462560
552 gr., (#2 Alloy) 3.160″ OAL

Powder	Sug. Starting Grains	Velocity F.P.S.	Pressure C.U.P.	Max. Grains	Velocity F.P.S.	Pressure C.U.P.
Unique	16.0	1050	—	20.0	1200	—
2400	23.0	1110	—	29.0	1330	—
IMR-4227	39.0	1555	—	45.0	1705	—
IMR-4198	48.0	1690	—	52.0	1810	—
IMR-3031	56.0	1745	—	65.0+	1980	—
IMR-4064	59.0	1730	—	67.0+	1955	—

Note: Loads shown in shaded panels are maximum.
+Designates a compressed powder charge.

.460 Weatherby Magnum

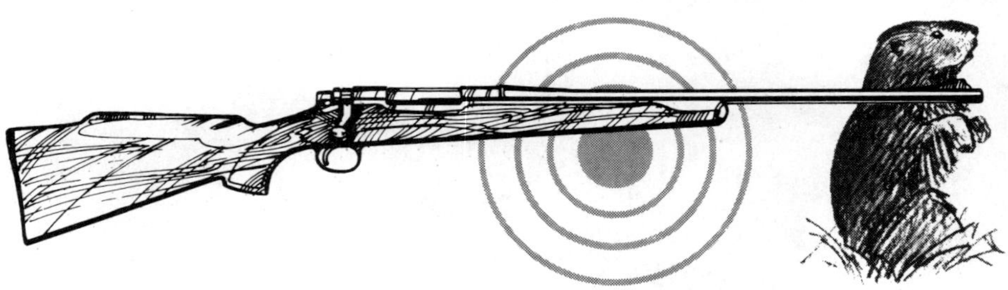

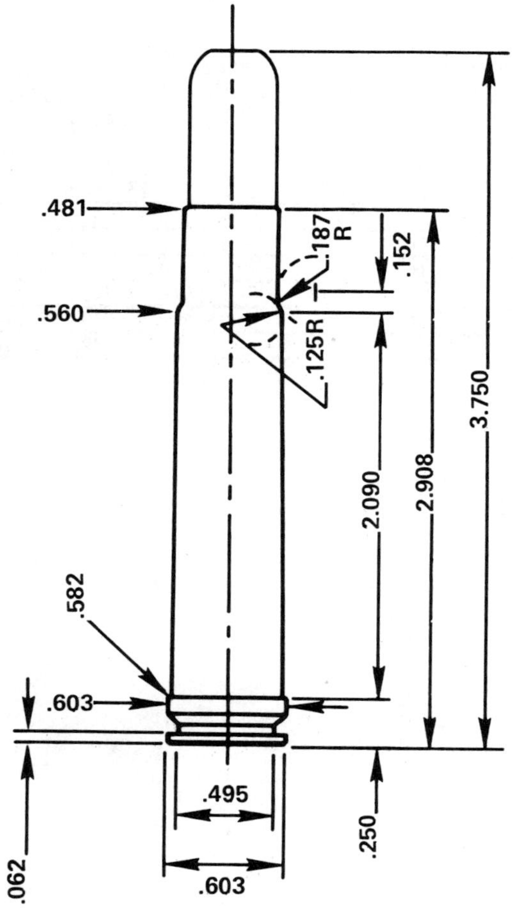

TEST COMPONENTS:

Cases .... Weatherby
Trim-to Length .... 2.903″
Primers .... Remington 9½
Primer Size .... Large Rifle
Lyman Shell Holder .... No. 17
Cast Bullets Used .... (size .457″to .459″ dia.)
*Gas Check Bullet *#462560, 552 gr.

TEST SPECIFICATIONS:
(Velocity)

Firearm Used .... Weatherby Mark V
Barrel Length .... 26″
Twist .... 1-16″
Groove Diameter .... .457″

COMMENTS:

The data listed for this cartridge were obtained in a Weatherby rifle and are intended for Weatherby rifles only. The free-boring constructed into these firearms allows higher velocities at safe working pressures. For custom rifles which are not free-bored, maximum loads should be reduced a full 5%. Even then, they should be approached with caution.

Due to heavy recoil of these rifles all bullets should be crimped in place.

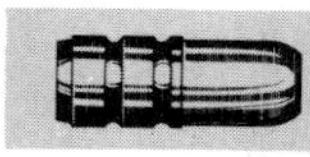

#462560
552 gr., (#2 Alloy) 3.750″ OAL, Max.

Powder	Sug. Starting Grains	Velocity F.P.S.	Pressure C.U.P.	Max. Grains	Velocity F.P.S.	Pressure C.U.P.
Unique	20.0	1150	—	28.0	1396	—

Note: Loads shown in shaded panels are maximum.

.50/70 Government

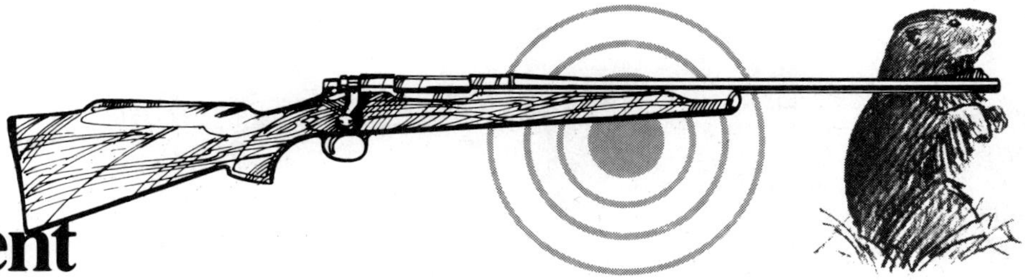

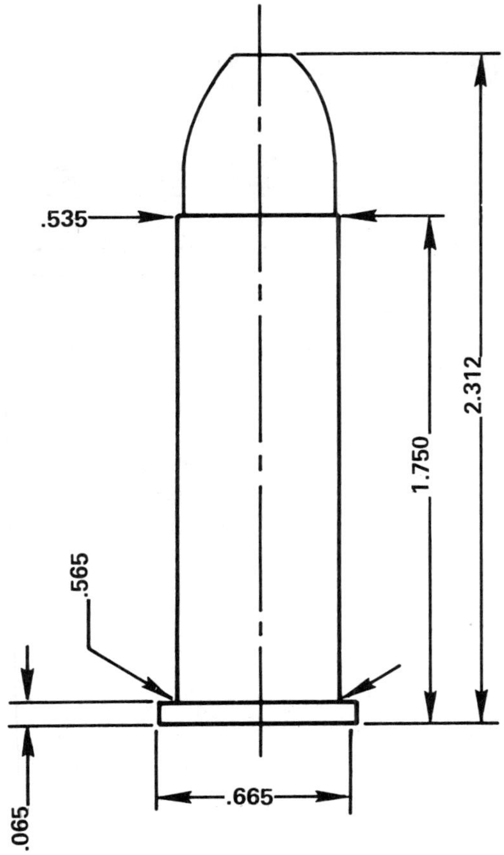

TEST COMPONENTS:

Cases	BELL
Trim-to Length	1.740″
Primers	Fed. 215
Primer Size	Large Rifle, Magnum
Lyman Shell Holder	No. 22 or 31
Cast Bullets Used	(size to .512″ dia.)
Gas Check Bullets	#515139, 334 gr.
	#515141, 422 gr.
	#512138, 440 gr.
	#515142, 498 gr.

TEST SPECIFICATIONS:
(Velocity)

Firearm Used	Shiloh Sharps
Barrel Length	22″
Twist	1-48″
Groove Diameter	.511″

COMMENTS:

The load data apply strictly to the use of modern brass cases in new, replica rifles, made with modern steel.

#515139
334 gr., (#2 Alloy) 2.125″ OAL

Powder	Sug. Starting Grains	Velocity F.P.S.	Pressure C.U.P.	Max. Grains	Velocity F.P.S.	Pressure C.U.P.
Unique	11.3++	1114	—	15.5++	1380	—
630	18.0++	1154	—	24.3++	1465	—
SR-4759	21.0++	1177	—	27.5++	1461	—
IMR-4227	21.0++	1109	—	29.0++	1464	—
IMR-4198	26.5++	1179	—	34.5++	1471	—
IMR-3031	36.0	818	—	44.0	1181	—
FFg	50.0	1113	—	70.0+	1318	—
Pyrodex CTG	38.0	1072	—	52.5	1370	—
Duplex Load						
SR-4759	3.5	1028	—	6.0	1295	—
FFg	32.5++			54.0+		

Note: Loads shown in shaded panels are maximum.
+Designates a compressed powder charge.
++Designates the use of a ½ gr., Dacron wad, ⅝″ square x ¼″ thick, over the powder.

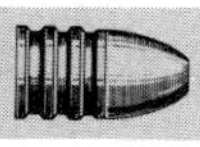

#515141
422 gr., (#2 Alloy) 2.312" OAL

Powder	Sug. Starting Grains	Velocity F.P.S.	Pressure C.U.P.	Max. Grains	Velocity F.P.S.	Pressure C.U.P.
Unique	11.0++	1003	—	14.8++	1211	—
630	18.3++	1098	—	22.7++	1306	—
SR-4759	22.0++	1114	—	26.5++	1313	—
IMR-4227	22.0++	1102	—	27.3++	1333	—
IMR-4198	25.5++	1129	—	30.0++	1305	—
FFg	50.0	1050	—	70.0+	1238	—
Pyrodex CTG	37.0	1037	—	52.5	1252	—
Duplex Load						
SR-4759	4.5	1000	—	7.0	1348	—
FFg	41.5++			63.0+		

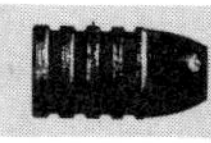

#512138
440 gr., (#2 Alloy) 2.133" OAL

Powder	Sug. Starting Grains	Velocity F.P.S.	Pressure C.U.P.	Max. Grains	Velocity F.P.S.	Pressure C.U.P.
Unique	11.0++	1006	—	14.5++	1207	—
630	18.5++	1140	—	22.5++	1318	—
SR-4759	21.7++	1144	—	26.0++	1321	—
IMR-4227	21.0++	1094	—	26.3++	1326	—
IMR-4198	26.0++	1129	—	31.0++	1304	—
IMR-3031	33.0++	1078	—	39.0++	1292	—
FFg	48.0	987	—	60.0+	1153	—
Pyrodex CTG	33.0	952	—	45.0+	1170	—
Duplex Load						
SR-4759	3.5	916	—	6.0	1244	—
FFg	32.5++			54.0+		

#515142
498 gr., (#2 Alloy) 2.223" OAL

Powder	Sug. Starting Grains	Velocity F.P.S.	Pressure C.U.P.	Max. Grains	Velocity F.P.S.	Pressure C.U.P.
Unique	10.5++	940	—	13.8++	1114	—
630	15.7++	973	—	21.0++	1218	—
SR-4759	18.5++	942	—	24.0++	1215	—
IMR-4227	19.0++	950	—	24.5++	1202	—
IMR-4198	22.3++	950	—	28.5++	1188	—
IMR-3031	27.5++	865	—	31.5++	1005	—
FFg	43.0	913	—	55.0+	1054	—
Pyrodex CTG	32.0	892	—	42.0	1074	—
Duplex Load						
SR-4759	4.3	955	—	5.5	1150	—
FFg	38.7++			49.5+		

Note: Loads shown in shaded panels are maximum.
+Designates a compressed powder charge.
++Designates the use of a ½ gr., Dacron wad, ⅝" square x ¼" thick, over the powder.

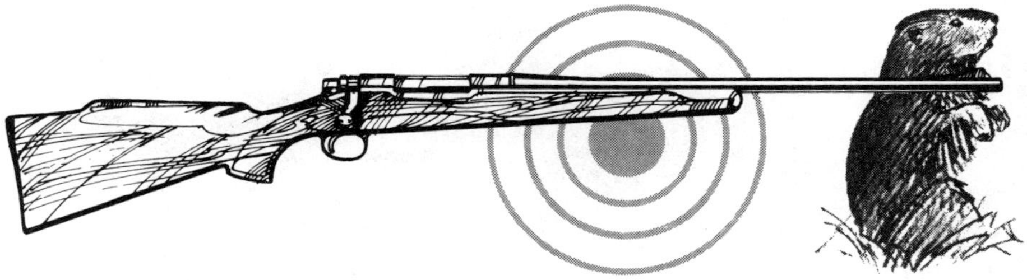

.50/90 Sharps

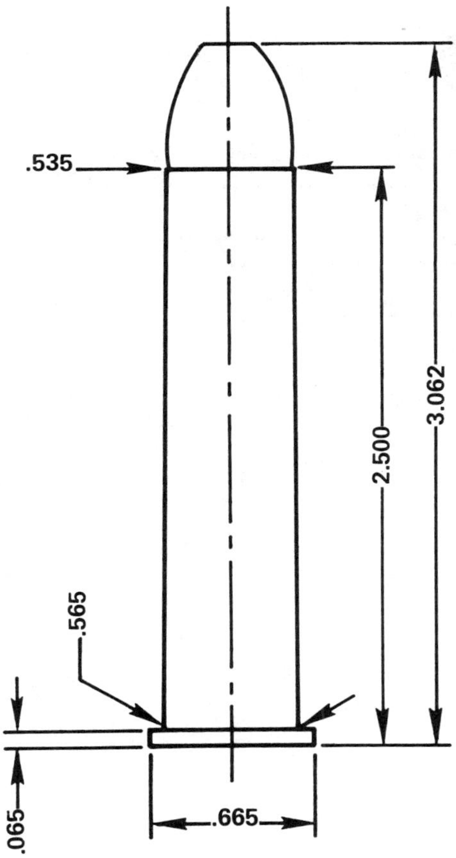

TEST COMPONENTS:

Cases .... BELL (Trimmed Back From 50-Sharps, 3¼″ Cases.)
Trim-to Length .... 2.490″
Primers .... Fed. 215
Primer Size .... Large Rifle, Magnum
Lyman Shell Holder .... No. 22 or 31
Cast Bullets Used .... (size to .512″ dia.)
*Gas Check Bullets
#515139, 334 gr.
#515141, 422 gr.
#512138, 440 gr.
#515142, 498 gr.

TEST SPECIFICATIONS:
(Velocity)

Firearm Used .... Shiloh Sharps, Sporter
Barrel Length .... 30″
Twist .... 1-36″
Groove Diameter .... .510″

COMMENTS:

The load data apply strictly to the use of modern brass cases in new, replica rifles, made with modern steel. Cases were formed by trimming 50-Sharps-3¼″ basic brass cases to 2.490″ and annealing the neck area.

#515139
334 gr., (#2 Alloy) 2.900″ OAL

Powder	Sug. Starting Grains	Velocity F.P.S.	Pressure C.U.P.	Max. Grains	Velocity F.P.S.	Pressure C.U.P.
Unique	11.5++	1112	—	18.0++	1459	—
630	17.5++	1133	—	26.7++	1528	—
SR-4759	21.0++	1151	—	30.0++	1561	—
IMR-4227	20.5++	1121	—	30.7++	1511	—
IMR-4198	23.3++	1089	—	35.8++	1510	—
FFg	58.5	1148	—	90.0	1585	—
Pyrodex CTG	30.0++	1092	—	67.5	1654	—
Duplex Load						
SR-4759	4.7	1227	—	9.0	1652	—
FFg	42.3++			81.0+		

Note: Loads shown in shaded panels are maximum.
+Designates a compressed powder charge.
++Designates the use of a ½ gr., Dacron wad, ⅝″ square x ¼″ thick, over the powder.

.50-90 Sharps (Continued)

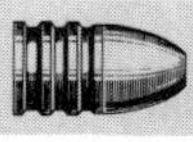

#515141
422 gr., (#2 Alloy) 3.030" OAL

Powder	Sug. Starting Grains	Velocity F.P.S.	Pressure C.U.P.	Max. Grains	Velocity F.P.S.	Pressure C.U.P.
Unique	13.0++	1062	—	20.0++	1386	—
630	15.6++	1019	—	27.0++	1446	—
SR-4759	21.0++	1062	—	30.0++	1434	—
IMR-4227	21.8++	1130	—	31.2++	1468	—
IMR-4198	23.5++	1069	—	36.3++	1460	—
IMR-3031	26.6++	944	—	43.0++	1394	—
FFg	58.5	1097	—	90.0+	1440	—
Pyrodex CTG	40.3	1052	—	62.0	1422	—
Duplex Load						
SR-4759	4.9	1179	—	7.5	1455	—
FFg	44.1++			67.5+		

#512138
440 gr., (#2 Alloy) 2.832" OAL

Powder	Sug. Starting Grains	Velocity F.P.S.	Pressure C.U.P.	Max. Grains	Velocity F.P.S.	Pressure C.U.P.
Unique	13.0++	1049	—	20.0++	1357	—
630	15.6++	991	—	24.0++	1347	—
SR-4759	17.9++	971	—	27.5++	1373	—
IMR-4227	18.3++	999	—	28.2++	1370	—
IMR-4198	21.6++	992	—	33.3++	1390	—
IMR-3031	26.7++	934	—	41.0++	1366	—
FFg	58.5	1151	—	90.0+	1459	—
Pyrodex CTG	43.9	1054	—	67.5	1441	—
Duplex Load						
SR-4759	5.2	1183	—	8.0	1440	—
FFg	46.8++			72.0+		

#515142
498 gr., (#2 Alloy) 2.972" OAL

Powder	Sug. Starting Grains	Velocity F.P.S.	Pressure C.U.P.	Max. Grains	Velocity F.P.S.	Pressure C.U.P.
Unique	12.4++	1002	—	19.0++	1247	—
630	16.2++	989	—	25.0++	1320	—
SR-4759	20.5++	1012	—	27.5++	1288	—
IMR-4227	20.0++	1005	—	28.5++	1323	—
IMR-4198	22.4++	1006	—	32.0++	1315	—
IMR-3031	29.5++	1013	—	40.0++	1299	—
FFg	59.0	1031	—	79.0+	1302	—
Pyrodex CTG	43.7	1007	—	58.3	1306	—
Duplex Load						
SR-4759	4.5	1044	—	7.1	1349	—
FFg	40.5++			63.9+		

Note: Loads shown in shaded panels are maximum.
+Designates a compressed powder charge.
++Designates the use of a ½ gr., Dacron wad, ⅝" square x ¼" thick, over the powder.

.50/140/3¼″ Sharps

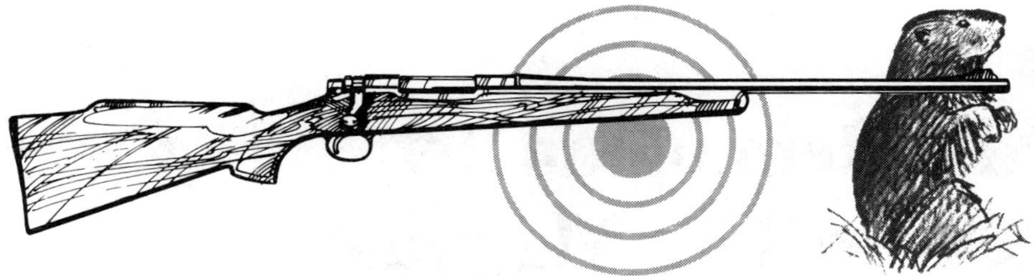

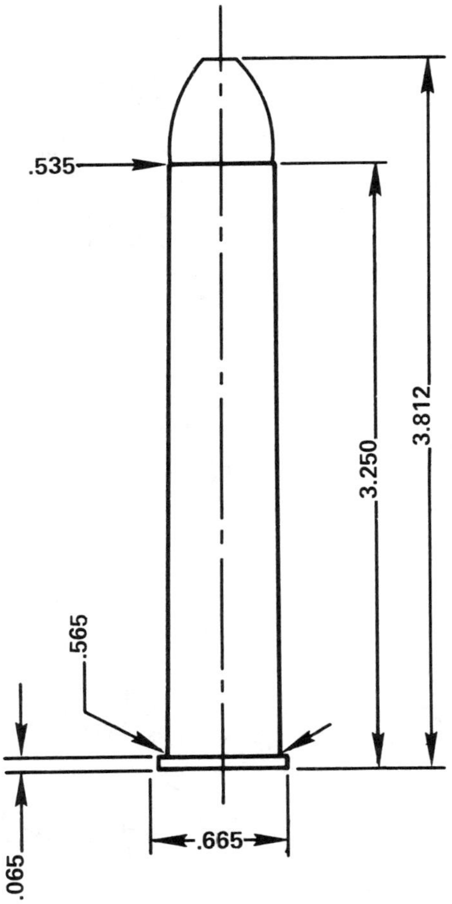

TEST COMPONENTS:

Cases .... BELL (Brass Extrusion Labs, Ltd.)
Trim-to Length .... 3.240″
Primers .... Federal 215
Primer Size .... Large Rifle Magnum
Lyman Shell Holder .... No. 22 or 31
Cast Bullets Used .... (size to .509″ dia.)
*Gas Check Bullets #515141, 422 gr.

TEST SPECIFICATIONS:
(Velocity)

Firearm Used .... Navy Arms, Remington Rolling Block
Barrel Length .... 30″
Twist .... 1-48″
Groove Diameter .... .509″

COMMENTS:

The load data apply strictly to the use of modern brass cases in new, replica rifles, made with modern steel.

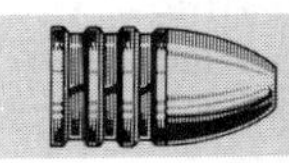

#515141
422 gr., (#2 Alloy) 3.940″ OAL

Powder	Sug. Starting Grains	Velocity F.P.S.	Pressure C.U.P.	Max. Grains	Velocity F.P.S.	Pressure C.U.P.
Unique	16.0++	1156	—	22.0++	1353	—
630	29.0++	1352	—	41.0++	1739	—
SR-4759	34.0++	1367	—	45.0++	1704	—
IMR-4227	33.0++	1343	—	46.0++	1764	—
IMR-4198	39.0++	1386	—	51.0++	1679	—
IMR-3031	49.0++	1321	—	63.0++	1681	—
FFg	107.0	1367	—	140.0+	1677	—
Pyrodex CTG	88.0	1275	—	110.0+	1608	—
Duplex Load						
SR-4759	7.5	1384	—	14.0	1755	—
FFg	67.5++	—	—	126.0+	—	—

Note: Loads shown in shaded panels are maximum.
+Designates a compressed powder charge.
++Designates the use of a ½ gr., Dacron wad, ⅝″ square x ¼″ thick, over the powder.

.22 Remington Jet Center Fire Magnum

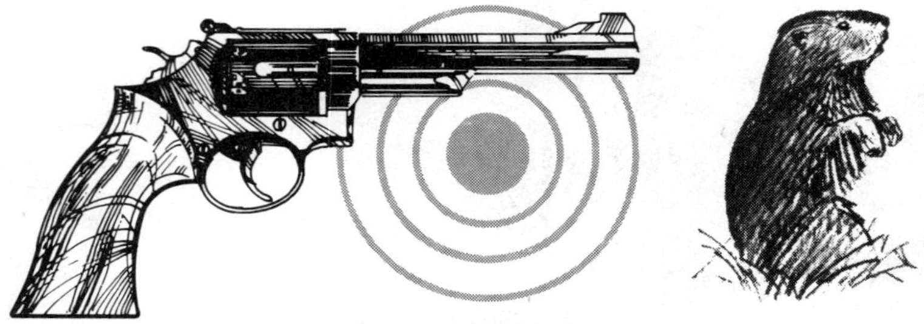

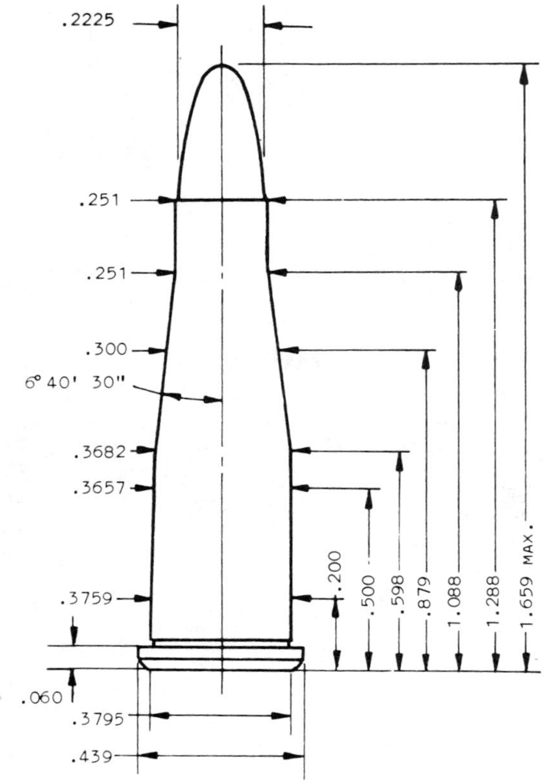

TEST COMPONENTS:

Cases .... Remington
Trim-to Length .... 1.283″
Primers .... Remington 6½
Primer Size .... Small Rifle
Lyman Shell Holder .... No. 1
Cast Bullets Used .... (size to .223″ dia.)
*Gas Check Bullets .... *#225107, 38 gr.
*#225438, 45 gr.

TEST SPECIFICATIONS:
(Velocity)

Firearm Used .... Thompson/Center Contender
Barrel Length .... 10″
Twist .... 1-15″
Groove Diameter .... .223″

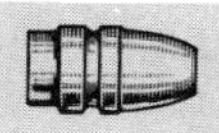

#225107
38 gr., (Linotype) 1.480″ OAL

Powder	Sug. Starting Grains	Velocity F.P.S.	Pressure C.U.P.	Max. Grains	Velocity F.P.S.	Pressure C.U.P.
2400	8.5	1848	—	12.0	2400	—
IMR-4227	9.0	1827	—	13.5	2409	—
IMR-4198	12.0	1734	—	14.5+	2125	—
630	7.5	1712	—	11.0	2370	—
HS-7	7.0	1797	—	10.0	2434	—
AL-8	7.5	1730	—	10.0	2346	—

COMMENTS:

Remove all traces of lubricant from both the cylinder and cases before firing. This will insure the proper functioning of the cylinder. Do not use .224″ diameter bullets in this cartridge. Use .222″ diameter or .223″ diameter bullets only.

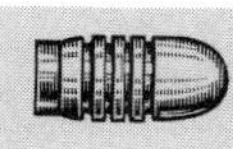

#225438
45 gr., (Linotype) 1.495″ OAL

Powder	Sug. Starting Grains	Velocity F.P.S.	Pressure C.U.P.	Max. Grains	Velocity F.P.S.	Pressure C.U.P.
2400	8.5	1704	—	11.5	2211	—
IMR-4227	9.5	1807	—	13.0	2256	—
IMR-4198	12.0	1708	—	14.5+	2126	—
630	7.5	1702	—	11.5	2446	—
HS-7	*7.5	1769	—	10.5	2370	—
AL-8	8.0	1838	—	10.5	2304	—

Note: Loads shown in shaded panels are maximum.
*Designates potentially most accurate load.
+Designates a compressed powder charge.

.221 Remington Fireball

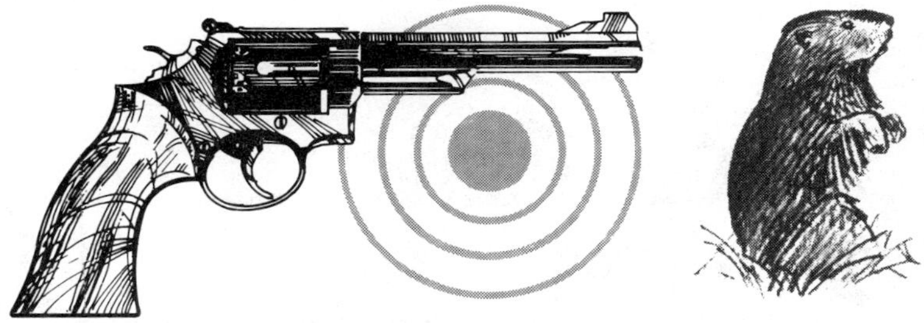

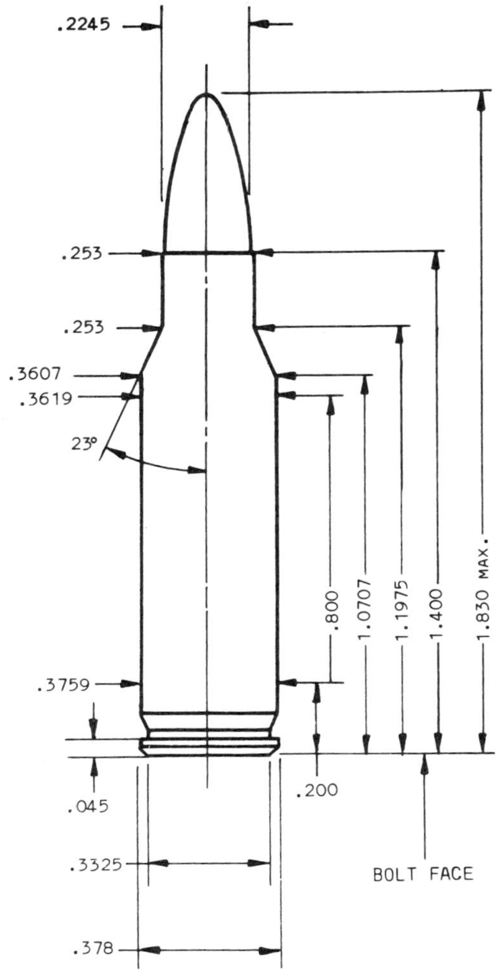

TEST COMPONENTS:

Cases .... Remington
Trim-to Length .... 1.395″
Primers .... Remington 7½
Primer Size .... Small Rifle
Lyman Shell Holder .... No. 26
Cast Bullets Used .... (size to .224″ dia.)
*Gas Check Bullets — *#225438, 45 gr.; *#225415, 50 gr.

TEST SPECIFICATIONS:
(Velocity)

Firearm Used .... Remington XP-100
Barrel Length .... 10½″
Twist .... 1-12″
Groove Diameter .... .224″

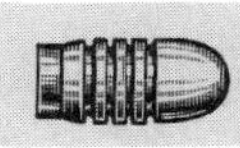

#225438
45 gr., (#2 Alloy) 1.675″ OAL

Powder	Sug. Starting Grains	Velocity F.P.S.	Pressure C.U.P.	Max. Grains	Velocity F.P.S.	Pressure C.U.P.
Red Dot	2.5	1050	—	5.0	1785	—
Green Dot	3.0	1135	—	5.5	1800	—
Unique	3.5	1180	—	6.5	1910	—
Blue Dot	7.0	1752	—	9.0	2183	—
2400	12.0	2173	—	15.0	2732	—
700X	2.5	1050	—	4.8	1570	—
PB	3.0	1090	—	5.5	1770	—
SR-7625	3.5	1180	—	5.5	1775	—
SR-4759	9.5	1725	—	*12.5	2153	—
IMR-4227	13.0	2237	—	16.0	2590	—
630	8.0	1745	—	11.0	2210	—
HS-7	7.5	1708	—	10.0	2144	—
AL-8	8.5	1776	—	11.0	2171	—

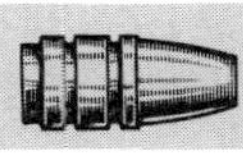

#225415
50 gr., (#2 Alloy) 1.665″ OAL

Powder	Sug. Starting Grains	Velocity F.P.S.	Pressure C.U.P.	Max. Grains	Velocity F.P.S.	Pressure C.U.P.
Red Dot	2.5	1015	—	4.5	1440	—
Green Dot	3.0	1105	—	5.0	1665	—
Unique	3.5	1110	—	6.0	1755	—
Blue Dot	7.5	1731	—	9.7	2184	—
2400	12.0	2207	—	14.5	2512	—
700X	2.5	1005	—	4.5	1460	—
PB	3.0	1050	—	5.3	1685	—
SR-7625	3.5	1115	—	5.3	1675	—
SR-4759	10.5	1706	—	13.5	2181	—
IMR-4227	13.0	2272	—	15.5	2475	—
630	8.0	1738	—	10.5	2157	—
HS-7	8.0	1742	—	10.5	2212	—
AL-8	*8.5	1699	—	11.0	2181	—

Note: Loads shown in shaded panels are maximum.
+Designates a compressed powder charge.

.25 Automatic (25 ACP)

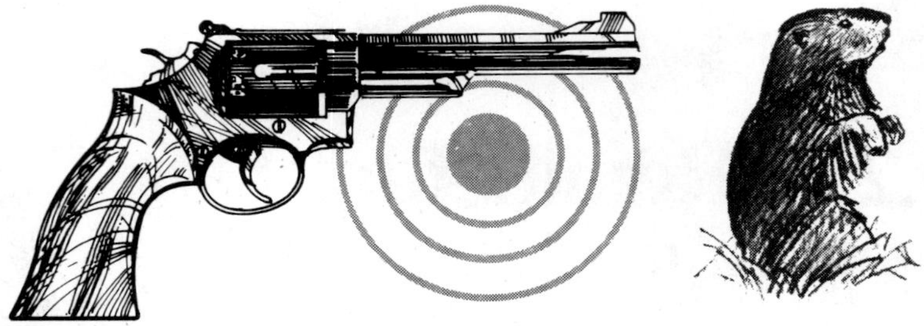

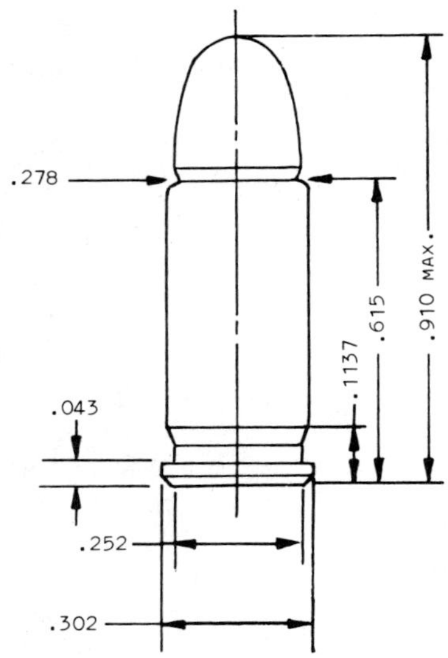

TEST COMPONENTS:

Cases .... Remington-Peters
Trim-to Length .... .610″
Primers .... Winchester 1½-108
Primer Size .... Small Pistol
Lyman Shell Holder .... RCBS No. 29
Cast Bullets Used .... (size to .251″ dia.)
#252435, 51 gr.

TEST SPECIFICATIONS:
(Velocity)

Firearm Used .... Sterling .25 Auto
Barrel Length .... 2″
Twist .... 1-16″
Groove Diameter .... .251″

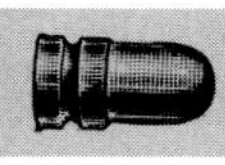

#252435
51 gr., (#2 Alloy) .910″ OAL

Powder	Sug. Starting Grains	Velocity F.P.S.	Pressure C.U.P.	Max. Grains	Velocity F.P.S.	Pressure C.U.P.
Bullseye	0.8	460	—	1.1	560	—
Red Dot	0.8	470	—	1.1	615	—
Green Dot	1.0	500	—	1.3	625	—
Unique	1.2	470	—	1.6	660	—
700X	0.8	470	—	1.1	625	—
PB	1.0	495	—	1.4	680	—
630	2.8	595	—	3.2	695	—

Note: Loads shown in shaded panels are maximum.

.256 Winchester Magnum

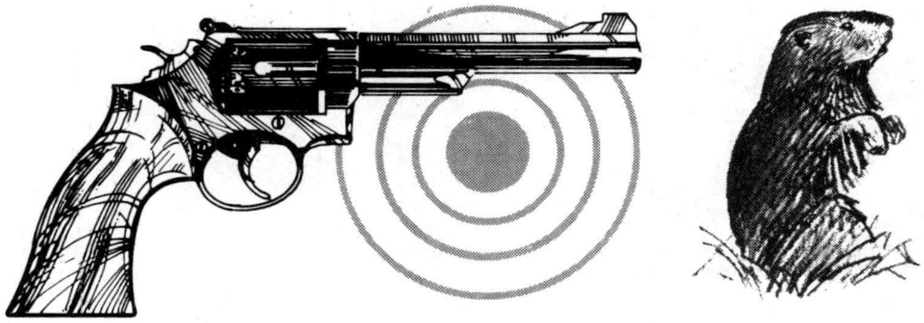

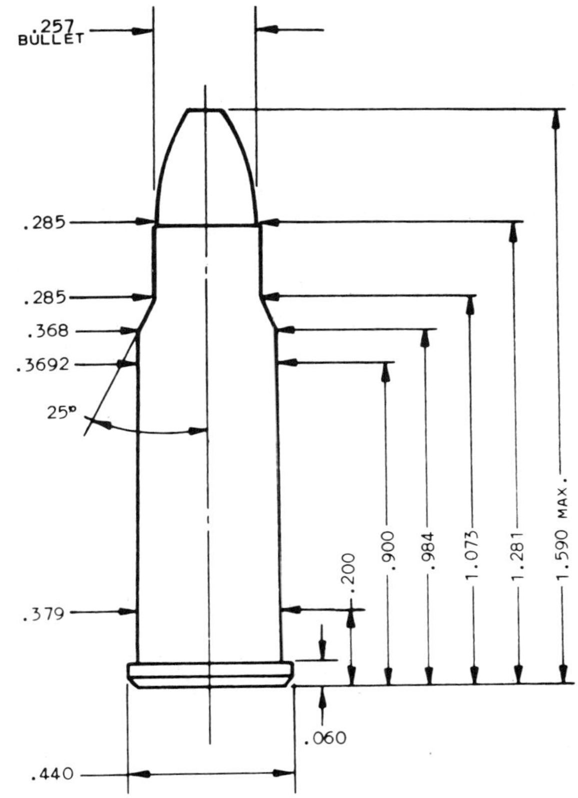

TEST COMPONENTS:

Cases .... Winchester
Trim-to Length .... 1.281″
Primers .... CCI 400
Primer Size .... Small Rifle
Lyman Shell Holder .... No. 1
Cast Bullets Used .... (size to .257″ dia.)
*Gas Check Bullets .... *#257420, 65 gr.
*#257312, 89 gr.

TEST SPECIFICATIONS: (Velocity)

Firearm Used .... Thompson/Center Contender
Barrel Length .... 10″
Twist .... 1-14″
Groove Diameter .... .257″

COMMENTS:

The small case capacity of the cartridge limits suitable reloading powders to a very few. Ballistics and accuracy, however, are good.

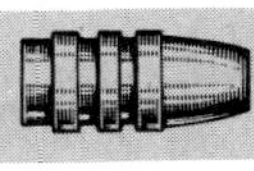

#257420
65 gr. (Linotype) 1.525″ OAL

Powder	Sug. Starting Grains	Velocity F.P.S.	Pressure C.U.P.	Max. Grains	Velocity F.P.S.	Pressure C.U.P.
SR-4756	4.0	962	—	8.0	1876	—
SR-4759	7.0	1031	—	10.0	1528	—
Blue Dot	4.3	897	—	9.0	1864	—
HS-7	4.5	920	—	10.0	1857	—
AL-8	4.3	857	—	9.0	1718	—

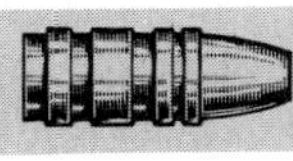

#257312
89 gr., (Linotype) 1.600″ OAL

Powder	Sug. Starting Grains	Velocity F.P.S.	Pressure C.U.P.	Max. Grains	Velocity F.P.S.	Pressure C.U.P.
SR-4756	3.8	874	—	*7.5	1624	—
SR-4759	6.8	949	—	9.5	1420	—
Blue Dot	4.0	820	—	9.5	1805	—
HS-7	4.2	829	—	9.5	1657	—
AL-8	4.3	802	—	8.5	1539	—

Note: Loads shown in shaded panels are maximum.
*Designates potentially most accurate load.

.30 Luger

(7.65mm)

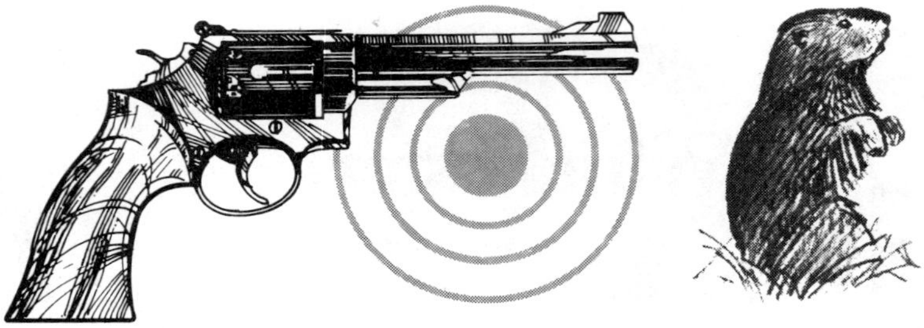

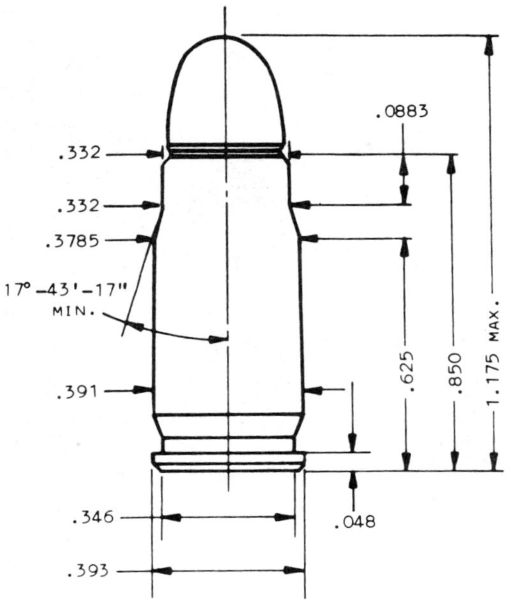

TEST COMPONENTS:

Cases .... Remington
Trim-to Length .... .845″
Primers .... Remington 1½
Primer Size .... Small Pistol
Lyman Shell Holder .... No. 12
Cast Bullets Used .... (size to .309″ dia.)
#313249, 84 gr.
#313226, 93 gr.

TEST SPECIFICATIONS:
(Velocity)

Firearm Used .... Walther P-38
Barrel Length .... 4 15/16″
Twist .... 1-9.85″
Groove Diameter .... .3085″

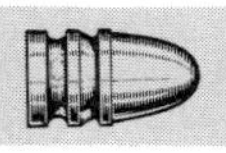

#313249
84 gr., (#2 Alloy) 1.125″ OAL

Powder	Sug. Starting Grains	Velocity F.P.S.	Pressure C.U.P.	Max. Grains	Velocity F.P.S.	Pressure C.U.P.
700X	3.7	1143	—	4.2	1204	—
PB	4.0	1105	—	4.5	1177	—
SR-4756	966	—	5.6	1182	—	
Bullseye	3.8	1106	—	4.5	1244	—
Red Dot	3.8	1116	—	4.3	1201	—
Green Dot	4.2	1111	—	4.7	1210	—
Unique	4.6	1110	—	5.2	1226	—
231	3.9	1054	—	4.4	1144	—
HS-5	5.4	1044	—	*6.0	1146	—

COMMENTS:

Variations in groove diameters will be encountered with pistols chambered for this cartridge. For best results, it is recommended that you slug your barrel. Size cast bullets as near to the exact groove diameter as possible. When using cast bullet #313249, the case should be crimped on the leading edge of the first driving band.

With some pistols, the starting loads may not function the action. In such an instance, the load should be increased, but with caution.

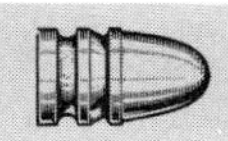

#313226
93 gr., (#2 Alloy) 1.150″ OAL

Powder	Sug. Starting Grains	Velocity F.P.S.	Pressure C.U.P.	Max. Grains	Velocity F.P.S.	Pressure C.U.P.
700X	3.5	1037	—	4.0	1128	—
PB	3.9	1047	—	4.4	1126	—
SR-4756	4.6	960	—	5.1	1094	—
Bullseye	3.6	1026	—	4.4	1159	—
Red Dot	3.7	1065	—	4.2	1152	—
Green Dot	4.1	1065	—	4.6	1150	—
Unique	4.5	1046	—	5.1	1163	—
630	7.2	1091	—	8.2	1205	—
HS-5	5.3	1062	—	*5.8	1134	—

Note: Loads shown in shaded panels are maximum.
*Designates potentially most accurate load.

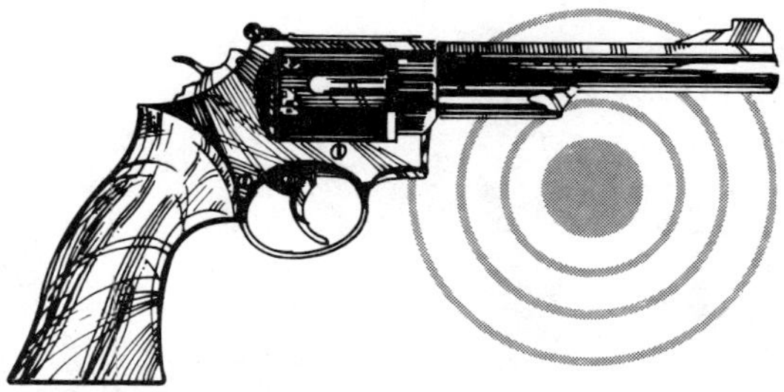

.30 Mauser

(7.63mm Mauser)

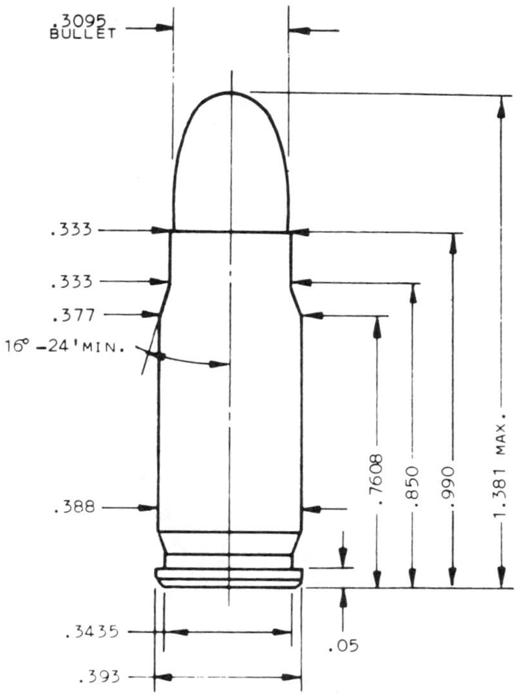

TEST COMPONENTS:

Cases .... Remington
Trim-to Length .... .985″
Primers .... Remington 1½
Primer Size .... Small Pistol
Lyman Shell Holder .... No. 12
Cast Bullets Used .... (size to .311″ dia.) #313249, 84 gr.

TEST SPECIFICATIONS:
(Velocity)

Firearm Used .... Military Mauser
Barrel Length .... 5½″
Twist .... 1-8″
Groove Diameter .... .311″

COMMENTS:

Variations in groove diameter may be encountered. Slug your barrel for best results. If your barrel is larger than .309″, we suggest that you use cast bullets sized as close to exact groove diameter as possible.

#313249
84 gr., (#2 Alloy) 1.381″ OAL

Powder	Sug. Starting Grains	Velocity F.P.S.	Pressure C.U.P.	Max. Grains	Velocity F.P.S.	Pressure C.U.P.
Bullseye	4.2	1128	—	5.0	1243	—
Red Dot	5.0	1065	—	5.6	1115	—
Unique	5.2	1018	—	6.8	1193	—

Note: Loads shown in shaded panels are maximum.

.30 M1 Carbine (Revolver)

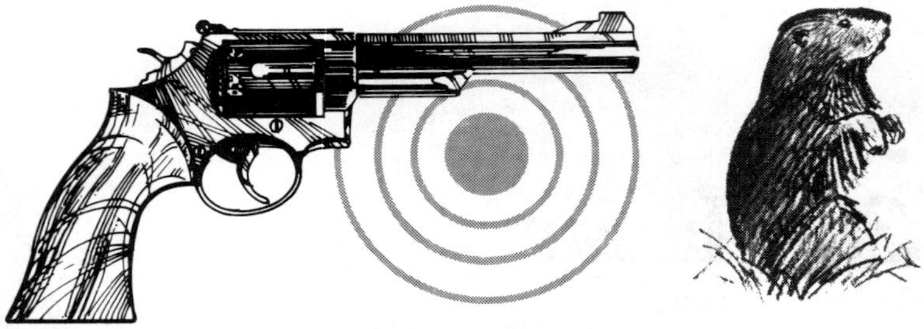

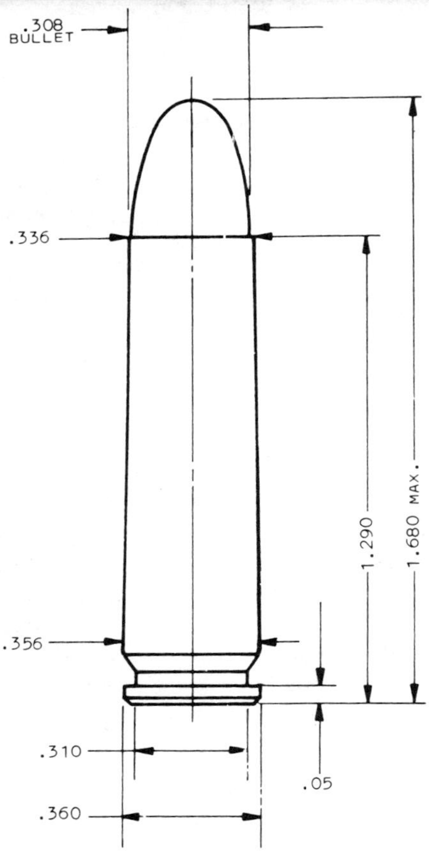

TEST COMPONENTS:

Cases .... Federal
Trim-to Length .... 1.286″
Primers .... Remington 6½
Primer Size .... Small Rifle Magnum and Small Rifle
Lyman Shell Holder .... No. 19
Cast Bullets Used .... (size to .308″ dia.)
*Gas Check Bullets *#311359, 120 gr.
*#311576, 123 gr.
#311410, 130 gr.

TEST SPECIFICATIONS: (Velocity)

Firearm Used .... Ruger Blackhawk
Barrel Length .... 7½″
Twist .... 1-20″
Groove Diameter .... .308″

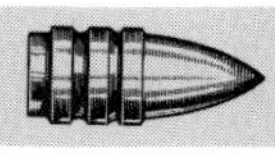

#311359
120 gr., (#2 Alloy) 1.680″ OAL

Powder	Sug. Starting Grains	Velocity F.P.S.	Pressure C.U.P.	Max. Grains	Velocity F.P.S.	Pressure C.U.P.
Unique	5.5	1138	—	6.5	1335	—
Blue Dot	6.8	1066	—	8.8	1339	—
2400	11.0	1236	—	12.5+	1436	—
SR-7625	4.6	1027	—	5.6	1162	—
SR-4756	5.5	1050	—	6.5	1220	—
IMR-4227	11.5	1169	—	13.3+	1388	—
**296	—	—	—	13.5	1415	—
H110	10.0	1055	—	13.5	1358	—
AL-8	9.5	1239	—	11.0+	1408	—

COMMENTS:

This cartridge headspaces on the mouth and case length is, therefore, critical. Never trim cases shorter than the trim-to length shown and never crimp bullets. Refer to Opening Remarks when using Winchester 296 powder.

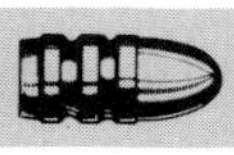

#311576
123 gr., (#2 Alloy) 1.610″ OAL

Powder	Sug. Starting Grains	Velocity F.P.S.	Pressure C.U.P.	Max. Grains	Velocity F.P.S.	Pressure C.U.P.
Unique	5.0	1017	—	6.3	1316	—
Blue Dot	6.5	997	—	8.8	1290	—
2400	10.8	1271	—	12.3+	1445	—
SR-4756	5.1	923	—	6.1	1172	—
IMR-4227	11.0	1124	—	13.1+	1342	—
**296	—	—	—	13.5	1431	—
H110	10.0	1037	—	13.5	1420	—
AL-8	9.2	1196	—	10.8+	1384	—

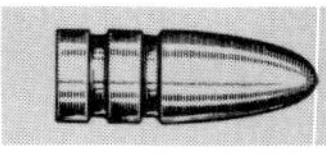

#311410
130 gr., (#2 Alloy) 1.680″ OAL,

Powder	Sug. Starting Grains	Velocity F.P.S.	Pressure C.U.P.	Max. Grains	Velocity F.P.S.	Pressure C.U.P.
Unique	4.9	1064	—	6.1	1270	—
Blue Dot	6.3	971	—	8.2	1249	—
2400	10.5	1282	—	12.0+	1414	—
SR-4756	5.0	1011	—	6.0	1182	—
IMR-4227	10.6	1116	—	12.7+	1387	—
**296	—	—	—	′12.5	1356	—
H110	9.0	979	—	12.5	1350	—
AL-8	9.0	1225	—	10.4+	1346	—

Note: Loads shown in shaded panels are maximum.
+Desginates a compressed powder charge.
**Reduced Loads not recommended by Winchester

.32 Automatic (32 ACP)
(7.65mm Browning)

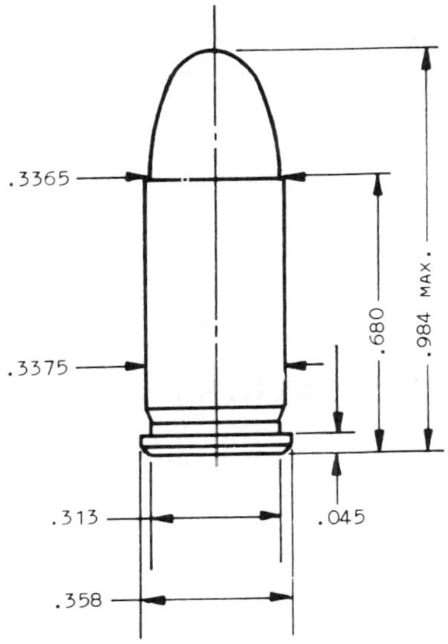

TEST COMPONENTS:

Cases .... Winchester
Trim-to Length .... .672″
Primers .... Winchester 1½-108
Primer Size .... Small Pistol
Lyman Shell Holder .... No. 9
Cast Bullets Used .... (size to .309″ dia.)
#311252, 77 gr.
#313249, 84 gr.

TEST SPECIFICATIONS:
(Velocity & Pressure)

Firearm Used .... Mauser HSc; also Special Test Barrel
Barrel Length .... 3″ (Mauser); 3¼″ (Special Barrel)
Twist .... 1-16″
Groove Diameter .... .309″

COMMENTS:

A wide variation in groove diameters is present in handguns chambered for this cartridge.

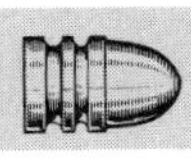

#311252
77 gr., (#2 Alloy) .975″ OAL

Powder	Sug. Starting Grains	Velocity F.P.S.	Pressure C.U.P.	Max. Grains	Velocity F.P.S.	Pressure C.U.P.
700X	1.5	750	—	1.9	895	—
PB	1.6	665	—	2.2	870	—
SR-7625	1.7	675	—	2.3	860	—
Bullseye	1.5	685	—	2.0	830	—
Red Dot	1.5	695	—	2.3	940	—
Green Dot	1.6	705	—	2.4	895	—
Unique	1.8	625	—	2.6	860	—
**231	1.5	620	7,200	2.5	888	15,000
**HP38	1.5	688	9,200	*2.0	824	13,800

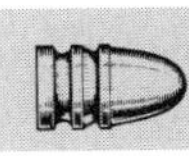

#313249
84 gr., (#2 Alloy) .984″ OAL

Powder	Sug. Starting Grains	Velocity F.P.S.	Pressure C.U.P.	Max. Grains	Velocity F.P.S.	Pressure C.U.P.
700X	1.3	685	—	1.7	830	—
PB	1.5	665	—	2.1	830	—
SR-7625	1.7	675	—	2.2	860	—
Bullseye	1.5	660	—	1.9	795	—
Red Dot	1.5	665	—	2.1	895	—
Green Dot	1.6	680	—	2.3	845	—
Unique	1.8	650	—	2.5	860	—
**630	3.5	649	7,700	5.1	882	14,400
**Trap 100	2.0	664	8,700	*2.8	814	13,700

Note: Loads shown in shaded panels are maximum.
*Designates potentially most accurate load.
**Designates special pressure test barrel.

.32 Smith & Wesson

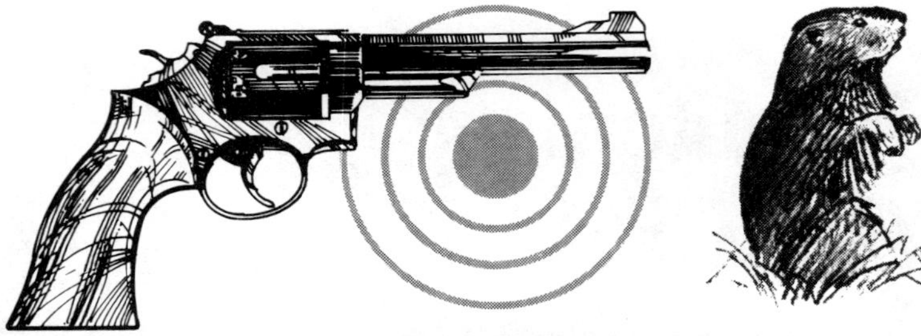

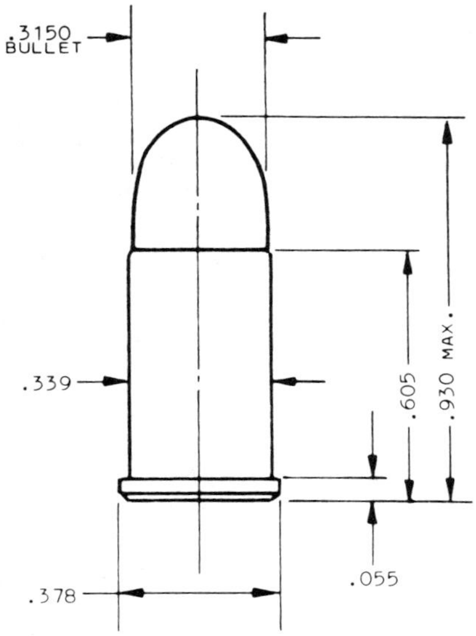

TEST COMPONENTS:

Cases .... Remington; Norma
Trim-to Length .... .600″
Primers .... Winchester 1½-108
Primer Size .... Small Pistol
Lyman Shell Holder .... No. 9
Cast Bullets Used .... (size to .310″ dia.)
#311252, 77 gr.
#313249, 84 gr.

TEST SPECIFICATIONS:
(Velocity)

Firearm Used .... Smith & Wesson Model 31
Barrel Length .... 4″
Twist .... 1-18¾″
Groove Diameter .... .312″

COMMENTS:

Use maximum loads only in solid frame revolvers. For top break models (in good condition) use only the suggested starting loads. Do not go higher as these pistols are of a relatively weak design.

Variations in groove diameter may be encountered. Slug your barrel before reloading and size your cast bullets to as near groove diameter as possible.

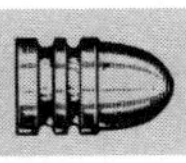

#311252
77 gr., (#2 Alloy) .860″ OAL

Powder	Sug. Starting Grains	Velocity F.P.S.	Pressure C.U.P.	Max. Grains	Velocity F.P.S.	Pressure C.U.P.
700X	1.3	598	—	1.8	758	—
PB	1.4	523	—	2.1	803	—
Bullseye	1.1	478	—	1.5	578	—
Red Dot	1.4	578	—	1.9	763	—
Green Dot	1.5	553	—	2.0	723	—
Unique	1.6	493	—	2.5	743	—
**231	1.9	641	—	*2.5	800	—
**HP38	1.4	642	—	2.2	821	—
**AL-120	1.3	537	—	1.8	721	—

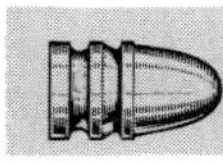

#313249
84 gr., (#2 Alloy) .860″ OAL

Powder	Sug. Starting Grains	Velocity F.P.S.	Pressure C.U.P.	Max. Grains	Velocity F.P.S.	Pressure C.U.P.
700X	1.2	613	—	1.7	798	—
PB	1.3	543	—	1.9	768	—
**SR-7625	1.6	625	—	2.1	760	—
Bullseye	1.0	468	—	1.4	578	—
Red Dot	1.3	588	—	1.7	703	—
Green Dot	‡1.4	528	—	1.9	728	—
Unique	1.5	493	—	2.4	753	—
**630	2.5	485	—	4.0	726	—
**Trap 100	1.6	558	—	2.0	645	—

Note: Loads shown in shaded panels are maximum.
*Designates potentially most accurate load.
‡Designates factory velocity duplication load.
**Designates the use of Remington cases and primers.

.32 Smith & Wesson Long

(.32 Colt New Police)

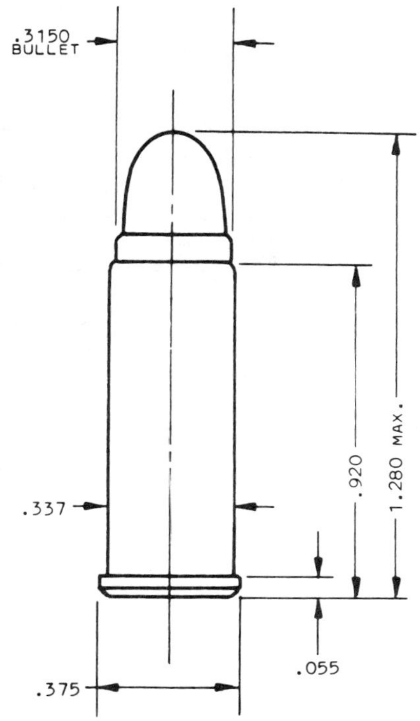

TEST COMPONENTS:

Cases .... Remington
Trim-to Length .... .915″
Primers .... Remington 1½
Primer Size .... Small Pistol
Lyman Shell Holder .... No. 9
Cast Bullets Used .... (size to .312″ dia.)
#311252, 77 gr.
#313226, 93 gr.
#3118, 115 gr.

TEST SPECIFICATIONS:
(Velocity Only)

Firearm Used .... Smith & Wesson Model 31
Barrel Length .... 4″
Twist .... 1-18¾″
Groove Diameter .... .312″

COMMENTS:

Use maximum loads only in solid frame revolvers. For top break models (in good condition) use only the suggested starting loads. Do not go higher because these pistols are of a relatively weak design.

Variations in groove diameter may be encountered. Slug your barrel before reloading and size your cast bullets to as near groove diameter as possible.

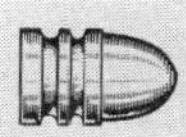

#311252
77 gr., (#2 Alloy) 1.115″ OAL

Powder	Sug. Starting Grains	Velocity F.P.S.	Pressure C.U.P.	Max. Grains	Velocity F.P.S.	Pressure C.U.P.
Bullseye	1.5	513	—	3.0	958	—
Red Dot	1.8	588	—	3.3	1028	—
Green Dot	1.9	563	—	3.4	993	—
Unique	2.5	608	—	4.0	943	—
700X	1.8	598	—	3.3	1098	—
SR-7625	2.4	583	—	3.6	988	—
231	2.0	677	—	3.5	1043	—
Trap 100	1.7	645	—	3.5	1003	—
AL-120	1.9	635	—	3.5	1071	—

Note: Loads shown in shaded panels are maximum.
*Designates potentially most accurate load.

.32 Smith & Wesson Long (Continued)

#313226
93 gr., (#2 Alloy) 1.185" OAL

Powder	Sug. Starting Grains	Velocity F.P.S.	Pressure C.U.P.	Max. Grains	Velocity F.P.S.	Pressure C.U.P.
Bullseye	1.5	518	—	2.9	898	—
Red Dot	1.7	548	—	3.0	943	—
Green Dot	1.8	528	—	3.1	893	—
Unique	2.4	543	—	3.7	883	—
700X	1.6	558	—	3.0	983	—
SR-7625	2.2	553	—	3.3	953	—
231	2.0	639	—	3.7	1064	—
HP38	1.9	738	—	3.3	1038	—
AL-120	2.0	657	—	3.5	1070	—

#3118
115 gr., (#2 Alloy) 1.180" OAL

Powder	Sug. Starting Grains	Velocity F.P.S.	Pressure C.U.P.	Max. Grains	Velocity F.P.S.	Pressure C.U.P.
Bullseye	1.4	528	—	2.0	683	—
Red Dot	1.6	553	—	2.5	783	—
Green Dot	1.7	538	—	2.6	783	—
Unique	2.1	558	—	3.5	863	—
700X	1.5	548	—	2.4	818	—
SR-7625	1.9	538	—	3.0	863	—
630	4.0	611	—	*7.0	1046	—
HS-5	3.0	671	—	5.3	1102	—
AL-5	3.0	644	—	5.1	1003	—

Note: Loads shown in shaded panels are maximum.
*Designates potentially most accurate load.

9mm Luger
(9mm Parabellum)

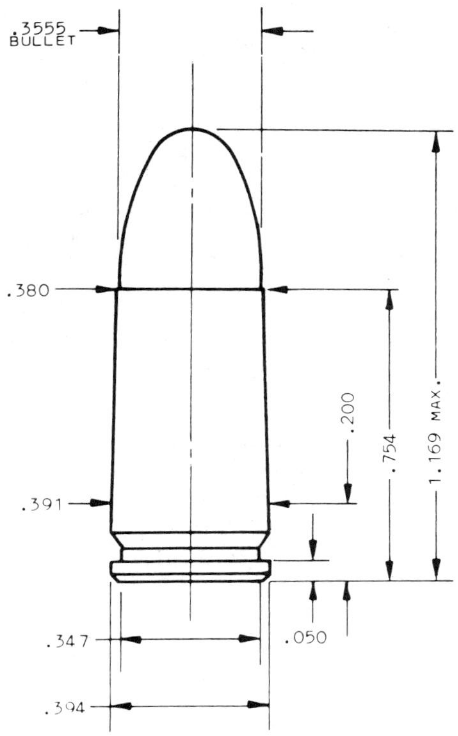

TEST COMPONENTS:

Cases .... Federal; Winchester
Trim-to Length .... .751″
Primers .... CCI 500; Winchester 1½-108
Primer Size .... Small Pistol
Lyman Shell Holder No.12
Cast Bullets Used .... (size to .358″ dia.)
#358242, 92 gr.
#358345, 115 gr.
#356402, 121 gr.

TEST SPECIFICATIONS:
(Velocity & Pressure)

Firearm Used .... Universal Receiver
Barrel Length .... 4″
Twist .... 1-10″
Groove Diameter .... .358″

COMMENTS:

Variations in groove diameter are often encountered with pistols chambered for this cartridge. However, the most common diameter is .354″. Jacketed bullets of .355″ may be used safely in all barrels ranging from .354″ to .356″. This particular cartridge headspaces on the case mouth and, therefore, the cases should not be crimped. Also, the trim-to length should be adhered to closely. A short case or crimp on the mouth can cause problems with headspace. Our most popular cast bullet for this caliber is #356402.

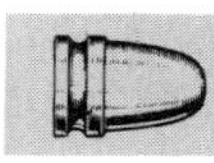

#358242
92 gr., (#2 Alloy) 1.045″ OAL

Powder	Sug. Starting Grains	Velocity F.P.S.	Pressure C.U.P.	Max. Grains	Velocity F.P.S.	Pressure C.U.P.
700X	4.1	1125	22,800	5.1	1285	31,000
PB	4.8	1115	19,800	6.0	1360	33,000
SR-7625	5.0	1180	23,400	5.9	1310	30,500
SR-4756	5.3	1168	22,400	7.2	1380	30,500
Bullseye	4.0	1060	19,800	5.2	1315	32,000
Red Dot	4.1	1120	22,800	5.5	1340	32,500
Green Dot	4.3	1095	20,400	6.0	1325	31,000
Unique	5.0	1065	19,800	6.8	1325	31,500
630	8.5	1121	20,200	12.0	1405	32,100
HS-5	5.9	1101	19,300	*7.6	1376	32,100
AL-5	5.9	1096	19,500	9.0	1405	31,500

Note: Loads shown in shaded panels are maximum.
*Designates potentially most accurate load.

9mm Luger)
(Continued)

#358345
115 gr., (#2 Alloy) .997" OAL

Powder	Sug. Starting Grains	Velocity F.P.S.	Pressure C.U.P.	Max. Grains	Velocity F.P.S.	Pressure C.U.P.
**700X	3.7	1045	25,200	4.4	1160	32,500
**PB	3.7	995	23,400	4.5	1130	32,000
**SR-7625	3.8	1005	24,000	4.5	1130	31,000
SR-4756	4.2	1005	18,800	6.1	1284	32,000
**Bullseye	3.8	1035	22,800	4.9	1200	31,000
**Red Dot	3.9	1055	24,600	4.9	1200	32,000
**Green Dot	4.0	1030	24,000	5.0	1200	32,500
**Unique	4.5	1025	22,200	5.7	1210	31,500
Blue Dot	5.8	983	16,800	8.4	1339	32,300
630	7.4	1028	18,700	*10.8+	1322	29,700
HS-5	5.0	1000	18,100	7.2	1356	32,800
AL-5	5.0	991	18,900	7.2	1269	28,700

#356402
121 gr., (#2 Alloy) 1.110" OAL

Powder	Sug. Starting Grains	Velocity F.P.S.	Pressure C.U.P.	Max. Grains	Velocity F.P.S.	Pressure C.U.P.
**700X	3.3	975	25,200	4.0	1080	33,000
**PB	3.4	915	20,400	4.3	1095	31,500
**SR-7625	3.5	935	21,000	4.3	1080	31,500
SR-4756	4.0	925	12,500	5.5	1198	32,900
**Bullseye	3.7	950	21,600	4.4	1125	32,000
**Red Dot	3.8	1020	26,400	4.5	1130	32,500
**Green Dot	3.9	970	22,200	4.8	1165	32,500
**Unique	4.3	940	21,000	5.3	1135	31,000
Blue Dot	5.9	973	18,200	8.0	1253	31,600
630	7.2	985	18,800	*10.3	1259	32,400
HS-5	5.3	1001	20,500	6.4	1176	30,200
AL-7	5.4	958	17,900	8.4+	1260	32,900

Note: Loads shown in shaded panels are maximum.
*Designates potentially most accurate load.
+Designates a compressed powder charge.
**Designates use of Winchester cases and primers.

.38 Smith & Wesson

(Colt New Police)

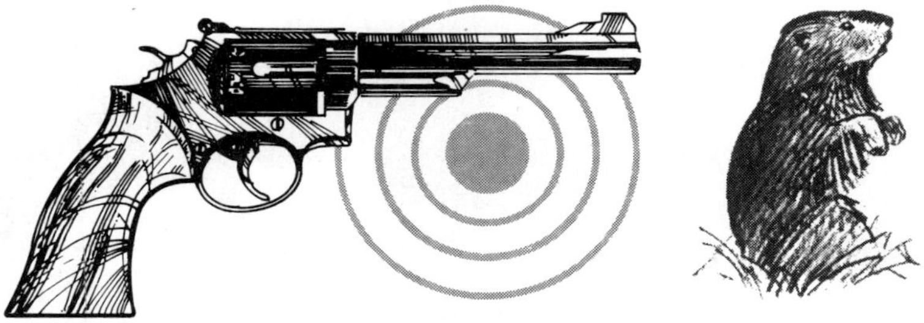

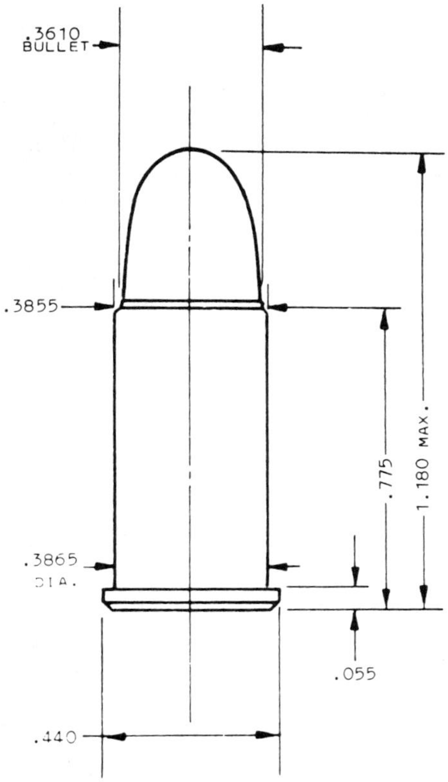

TEST COMPONENTS:

Cases .... Winchester
Trim-to Length .... .765″
Primers .... Winchester 1½-108
Primer Size .... Small Pistol
Lyman Shell Holder .... No. 21
Cast Bullets Used .... (size to .360″ dia.)
#358242, 92 gr.
#358242, 121 gr.
#358480, 133 gr.
#358495, 141 gr.
#358311, 158 gr.

TEST SPECIFICATIONS:
(Velocity & Pressure)

Firearm Used .... Special Pressure Test Barrel; Smith & Wesson Model 33
Barrel Length .... 4″
Twist .... 1-18¾″
Groove Diameter .... .361″

COMMENTS:

The loads listed are intended for solid frame revolvers which are in good condition.

Wide variations in groove diameter may be encountered. We suggest that you slug your barrel before reloading and size cast bullets as near to actual groove diameter as possible.

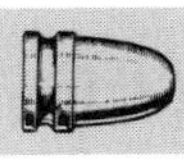

#358242
92 gr., (#2 Alloy) 1.125″ OAL

Powder	Sug. Starting Grains	Velocity F.P.S.	Pressure C.U.P.	Max. Grains	Velocity F.P.S.	Pressure C.U.P.
**Bullseye	2.4	718	6,500	3.4	965	12,800
**Unique	3.5	748	7,000	4.6	941	11,400
**Blue Dot	5.4	778	6,600	6.7	958	10,800
**700X	2.3	720	5,600	3.2	972	12,200
**SR-4756	*4.0	758	7,300	4.9	968	12,500
**231	2.7	719	6,500	3.6	944	12,100
**HS-5	4.5	782	6,600	5.3	946	10,800
**AL-5	4.3	798	7,200	5.5	980	12,300

Note: Loads shown in shaded panels are maximum
*Designates potentially most accurate load.
**Designates use of Special Pressure Test Barrel.

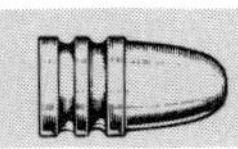

#358242
121 gr., (#2 Alloy) 1.125" OAL

Powder	Sug. Starting Grains	Velocity F.P.S.	Pressure C.U.P.	Max. Grains	Velocity F.P.S.	Pressure C.U.P.
Bullseye	2.0	560	—	3.0	805	—
Unique	3.5	700	—	4.5	945	—
700X	2.3	680	—	3.1	895	—
PB	2.6	575	—	3.4	825	—
SR-7625	2.7	580	—	3.5	830	—
**AL-5	3.5	702	9,700	4.3	827	13,000
**HS-5	3.6	676	8,700	*4.5	831	12,100
**231	2.0	610	6,900	2.8	798	12,400
Red Dot	2.4	650	—	3.2	855	—
Green Dot	2.5	615	—	3.3	815	—

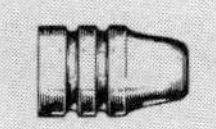

#358480
133 gr., (#2 Alloy) 1.062" OAL

Powder	Sug. Starting Grains	Velocity F.P.S.	Pressure C.U.P.	Max. Grains	Velocity F.P.S.	Pressure C.U.P.
Bullseye	2.0	595	—	2.9	820	—
Red Dot	2.4	705	—	3.1	865	—
Green Dot	2.5	680	—	3.2	830	—
Unique	3.5	790	—	4.2	920	—
700X	2.3	735	—	2.9	875	—
PB	2.6	670	—	3.2	830	—
SR-7625	2.7	675	—	3.3	850	—
**AL-5	3.0	644	8,300	3.9	799	12,400
**HS-5	3.0	620	7,500	*3.8	766	12,300
**231	1.9	568	6,400	2.8	786	12,300

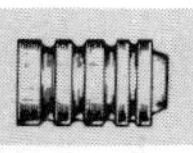

#358495
141 gr., (#2 Alloy) .990" OAL

Powder	Sug. Starting Grains	Velocity F.P.S.	Pressure C.U.P.	Max. Grains	Velocity F.P.S.	Pressure C.U.P.
Bullseye	1.6	510	—	2.4	720	—
Red Dot	1.7	560	—	2.3	715	—
Green Dot	1.8	530	—	‡2.4	680	—
Unique	3.0	710	—	3.7	885	—
700X	1.5	550	—	2.1	715	—
PB	1.8	545	—	2.4	710	—
SR-7625	1.9	535	—	2.5	725	—
**AL-5	2.5	546	6,100	3.2	702	11,500
**HS-7	3.7	619	8,200	4.5	737	11,600
**231	1.7	537	6,900	*2.3	705	11,700

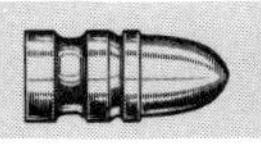

#358311
158 gr., (#2 Alloy) 1.150" OAL

Powder	Sug. Starting Grains	Velocity F.P.S.	Pressure C.U.P.	Max. Grains	Velocity F.P.S.	Pressure C.U.P.
Bullseye	1.8	545	—	2.6	730	—
Red Dot	2.1	605	—	2.8	770	—
Green Dot	2.2	595	—	2.9	745	—
Unique	3.2	715	—	3.7	805	—
700X	2.1	665	—	2.6	785	—
PB	2.3	620	—	2.8	730	—
SR-7625	2.4	660	—	2.9	770	—
**AL-7	2.7	516	5,200	4.0	754	12,200
**HS-7	*3.6	619	8,100	4.5	749	11,900
**231	1.5	486	4,600	2.5	727	12,400

Note: Loads shown in shaded panels are maximum.
*Designates potentially most accurate load.
**Designates use of Special Pressure Test Barrel.

.38 Super Automatic

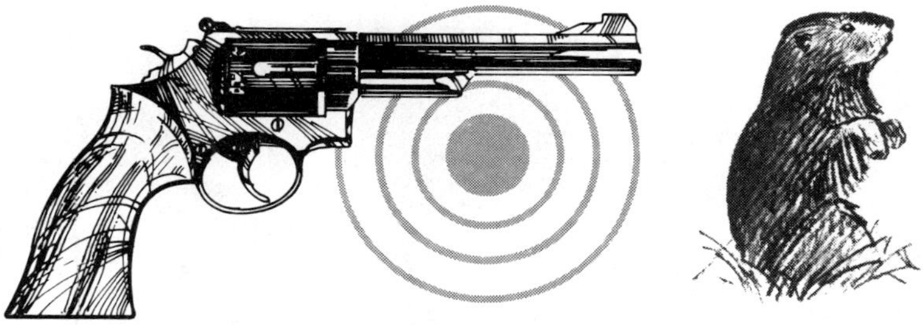

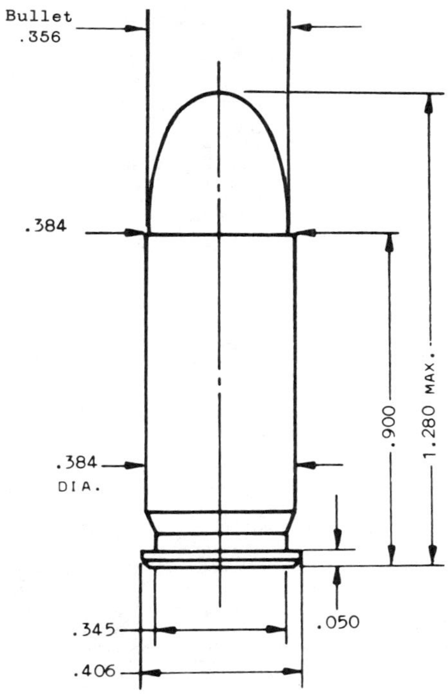

TEST COMPONENTS:

Cases .... Remington
Trim-to Length .... .895″
Primers .... Winchester 1½-108
Primer Size .... Small Pistol;
Lyman Shell Holder .... No. 12
Cast Bullets Used .... (size to .355″ dia.)
*Gas Check Bullets #358242, 92 gr.
#356404, 95 gr.
#356402, 121 gr.
#358242, 121 gr.
#358480, 133 gr.
#358311, 158 gr.

TEST SPECIFICATIONS:
(Velocity and Pressure)

Firearm Used .... Universal Receiver
Barrel Length .... 5″
Twist .... 1-16″
Groove Diameter .... .355″

COMMENTS:

These loads are intended only for those pistols which are chambered for the "Super Auto" cartridge.

For the "Standard .38 Auto" cartridge, it is recommended that you use a bullet of 133 grains or less. Reduce the starting load by ½ grain and work up to the starting load. Do not exceed this load with the "Standard Auto" cartridge.

Because this particular auto cartridge headspaces on case rim, it may be crimped if desired.

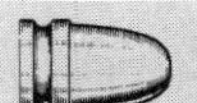

#358242
92 gr., (#2 Alloy) 1.175″ OAL

Powder	Sug. Starting Grains	Velocity F.P.S.	Pressure C.U.P.	Max. Grains	Velocity F.P.S.	Pressure C.U.P.
Bullseye	4.2	1022	19,500	5.6	1261	30,900
Unique	5.1	1061	20,400	6.3	1267	30,500
PB	4.3	1009	18,900	5.7	1222	30,500
SR-4756	5.9	1089	19,700	7.1	1263	30,300
231	4.7	1039	20,300	*6.3	1282	31,600
HP38	4.1	1026	20,600	5.8	1246	30,500
HS-5	6.3	1046	19,700	8.0	1308	31,700
AL-120	4.3	1030	20,600	5.8	1245	31,300

#356404
95 gr., (#2 Alloy) 1.130" OAL

Powder	Sug. Starting Grains	Velocity F.P.S.	Pressure C.U.P.	Max. Grains	Velocity F.P.S.	Pressure C.U.P.
Bullseye	4.5	1170	19,800	6.0	1425	31,000
Unique	5.5	1205	19,800	7.2	1460	31,000
700X	4.2	1147	20,200	6.0	1399	30,100
PB	4.9	1195	20,400	6.7	1430	31,500
SR-7625	5.4	1197	20,000	6.7	1429	31,100
231	5.2	1181	20,400	6.4	1433	30,100
HP38	*4.5	1159	20,500	6.2	1403	29,700
AL-120	4.7	1160	20,000	6.3	1395	30,200

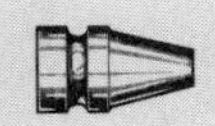

#356402
121 gr., (#2 Alloy) 1.230" OAL

Powder	Sug. Starting Grains	Velocity F.P.S.	Pressure C.U.P.	Max. Grains	Velocity F.P.S.	Pressure C.U.P.
Bullseye	4.0	1005	18,600	5.3	1235	30,000
Unique	5.1	1075	20,400	6.0	1255	31,000
700X	4.0	1025	20,600	5.5	1252	29,900
SR-7625	5.1	1099	20,600	6.6	1274	29,900
231	5.0	1092	20,700	*6.5	1307	30,800
HP38	4.4	1063	20,900	5.9	1267	29,700
HS-5	6.8	1089	19,200	8.4	1341	30,700
AL-120	4.5	1040	19,000	6.0	1235	29,800

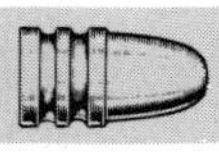

#358242
121 gr., (#2 Alloy) 1.160" OAL

Powder	Sug. Starting Grains	Velocity F.P.S.	Pressure C.U.P.	Max. Grains	Velocity F.P.S.	Pressure C.U.P.
Bullseye	3.9	970	17,600	5.4	1214	29,200
Unique	4.9	1031	20,000	6.4	1281	28,800
700X	3.7	979	19,800	5.3	1206	29,200
SR-7625	4.2	978	18,100	5.8	1230	29,300
231	4.3	988	17,700	5.8	1233	29,600
HP38	3.9	975	17,800	5.6	1230	29,400
HS-5	6.1	1030	18,200	7.6	1269	30,600
AL-120	3.8	952	17,200	5.4	1207	29,000

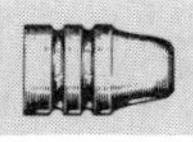

#358480
133 gr., (#2 Alloy) 1.115" OAL

Powder	Sug. Starting Grains	Velocity F.P.S.	Pressure C.U.P.	Max. Grains	Velocity F.P.S.	Pressure C.U.P.
SR-4756	5.0	975	16,900	7.0	1237	29,400
Bullseye	4.0	1000	19,800	5.3	1180	31,500
Unique	5.1	1045	21,000	6.0	1200	30,000
Blue Dot	6.8	1042	18,500	8.9	1346	30,100
630	8.4	1054	18,100	*10.6	1268	30,000
HS-6	5.9	1026	18,000	7.8	1303	30,200
AL-7	6.7	1041	18,700	8.6	1278	29,800

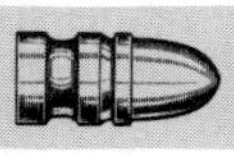

#358311
158 gr., (#2 Alloy) 1.245" OAL

Powder	Sug. Starting Grains	Velocity F.P.S.	Pressure C.U.P.	Max. Grains	Velocity F.P.S.	Pressure C.U.P.
SR-4756	4.9	927	18,900	5.8	1085	29,800
Unique	4.4	958	20,100	5.1	1058	28,100
Herco	4.5	920	21,000	5.4	1058	29,000
Blue Dot	6.5	987	20,900	7.6	1133	28,400
630	7.8	985	20,900	*9.3	1110	28,600
HS-7	6.8	956	20,200	8.3	1100	28,200

Note: Loads shown in shaded panels are maximum.
*Designates potentially most accurate load.

.380 Automatic
(9mm CORTO & 9mm SHORT)

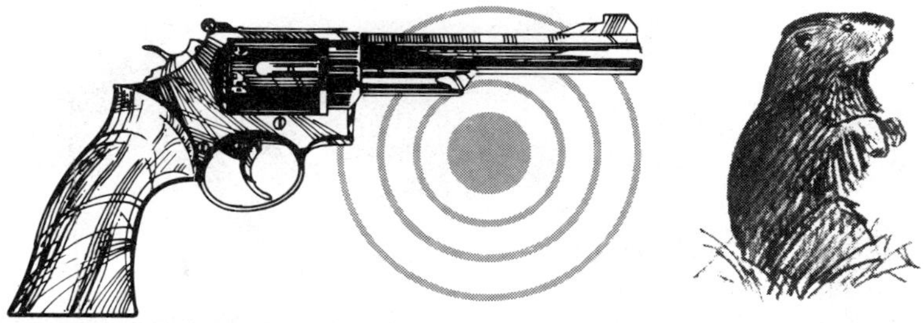

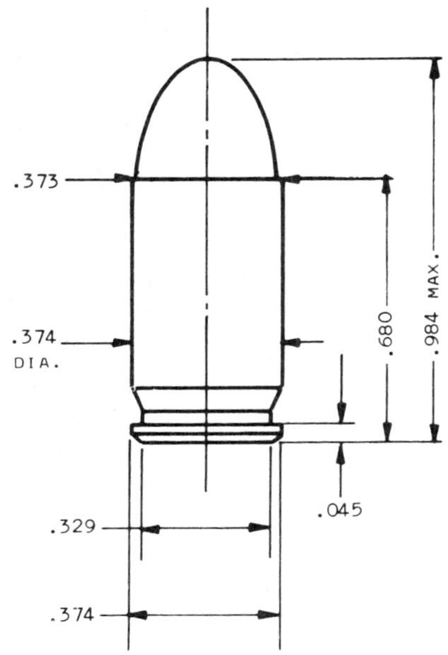

TEST COMPONENTS:

Cases .... Federal; Super Vel
Trim-to Length .... .677″
Primers .... Winchester 1½-108
Primer Size .... Small Pistol
Lyman Shell Holder .... No. 26
Cast Bullets Used .... (size to .355″ dia.)
#358242, 92 gr.
#358242, 121 gr.

TEST SPECIFICATIONS:
(Velocity and Pressure)

Firearm Used .... Universal Receiver
Barrel Length .... 3¾″
Twist .... 1-16″
Groove Diameter .... .355″

COMMENTS:

While a wide variation in groove diameter may be encountered (as large as .362″) in pistols chambered for this cartridge, the use of bullets larger than .355″ is usually not possible. This is due to chamber and case dimensions. The larger diameter bullet will bulge the case to the point where the cartridge will not chamber. If the groove diameter of your pistol runs on the large size, accuracy will be poor and there is little you can do about it.

Heavier cast bullets of .355″ diameter will sometimes cause a slight case bulge with some lots of brass. This is due to the inside taper of the case. It may be ignored as long as the cartridge will chamber and function properly.

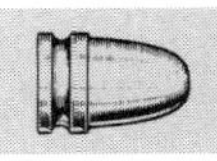

#358242
92 gr., (#2 Alloy) .980″ OAL

Powder	Sug. Starting Grains	Velocity F.P.S.	Pressure C.U.P.	Max. Grains	Velocity F.P.S.	Pressure C.U.P.
700X	1.9	695	11,100	2.5	900	15,800
PB	2.2	645	10,300	2.8	860	15,200
**SR-4756	2.8	653	6,900	4.1	970	15,600
Bullseye	2.4	750	11,800	3.0	915	16,000
Red Dot	2.0	690	11,000	2.6	880	15,200
Green Dot	2.1	655	10,000	3.0	910	16,000
Unique	3.1	755	11,800	3.7	920	16,000
**Blue Dot	4.0	667	7,000	5.5	961	14,600
**231	2.3	701	7,900	3.5	1000	15,600
**HS-5	3.1	658	6,400	*5.0	1023	15,900
**AL-5	3.0	644	6,600	5.0	1029	16,000

Note: Loads shown in shaded panels are maximum.
*Designates potentially most accurate load.
**Designates use of Federal cases.

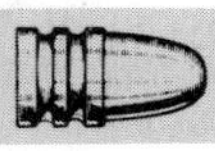

#358242
121 gr., (#2 Alloy) .980" OAL

Powder	Sug. Starting Grains	Velocity F.P.S.	Pressure C.U.P.	Max. Grains	Velocity F.P.S.	Pressure C.U.P.
700X	1.4	610	11,300	1.9	750	15,800
PB	1.7	605	10,800	2.2	750	15,800
**SR-4756	2.3	662	7,800	3.5	927	15,800
Bullseye	1.6	600	10,500	2.1	750	15,200
Red Dot	1.6	615	11,000	2.1	770	15,800
Green Dot	1.7	620	10,800	2.3	760	15,400
Unique	2.2	630	11,300	2.8	780	15,600
**Blue Dot	3.7	732	9,000	5.3	948	15,900
**231	2.1	718	9,200	3.2	946	15,800
**HS-5	2.6	619	6,300	*4.2	934	15,000
**AL-5	2.7	657	7,400	4.3	934	15,500

Note: Loads shown in shaded panels are maximum.
*Designates potentially most accurate load.
**Designates use of Federal cases.

.38 Special

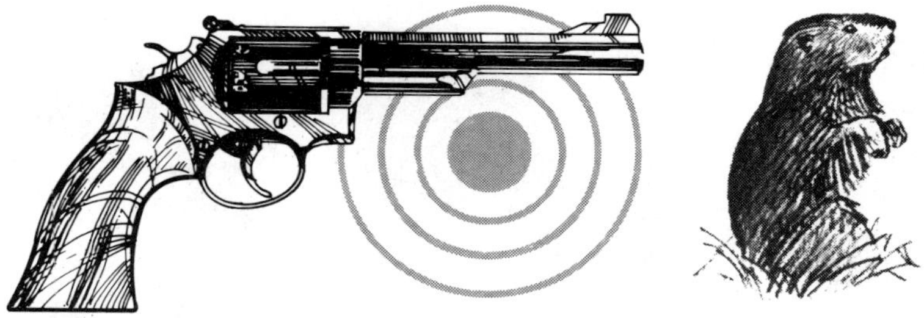

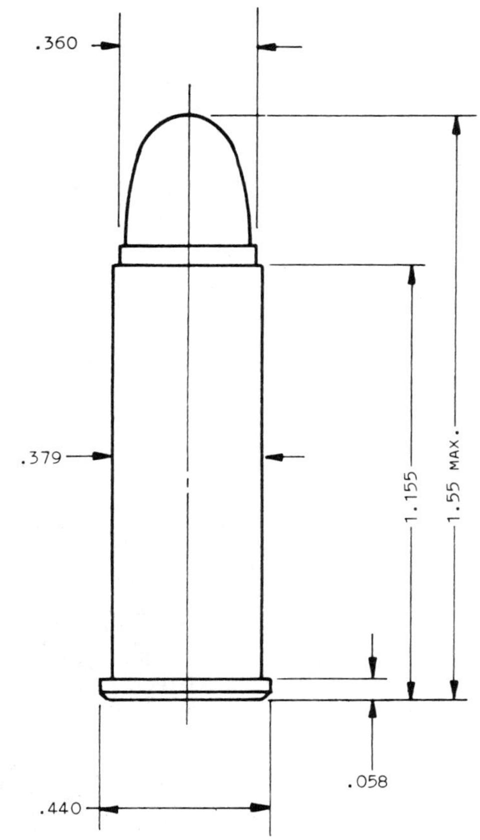

TEST COMPONENTS:

Cases .... Federal
Trim-to Length .... 1.149″
Primers .... CCI 500
Primer Size .... Small Pistol
Lyman Shell Holder .... No. 1
Cast Bullets Used .... (size to .357″ dia.)
*Gas Check #358242, 92 gr.
#358242, 121 gr.
#356402, 121 gr.
#358495, 141 gr.
#358212, 146 gr.
#35863 , 148 gr.
#35891, 148 gr.
*#358156, 158 gr.
#358311, 158 gr.
#357446, 162 gr.
#358429, 168 gr.

TEST SPECIFICATIONS:
(Velocity & Pressure)

Firearm Used .... Universal Receiver and Special Vented Barrel
Barrel Length .... 4″
Twist .... 1-18¾″
Groove Diameter .... .357″

COMMENTS:

Handguns of various makes and models will occasionally vary in groove diameter. It is wise to slug your barrel before selecting a sizing die. Size cast bullets to as near actual groove diameter as possible.

Loads designated with the symbol (+P) are higher pressure loads to give greater velocities. These loads should be used only in arms recommended for (+P) type cartridges by the gun manufacturer. Repeated use of these high pressure loads in handguns with aluminum frames or aluminum cylinders or with lightweight steel frames or cylinders, is not recommended.

For mid-range loads, we suggest bullet #358495 loaded at starting load velocities. Bullet #358311 duplicates the factory 158 grain service bullet very closely while Bullet #358429 is an excellent "Keith type" hunting bullet.

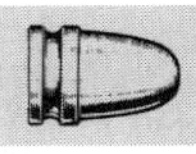

#358242
92 gr., (Linotype) 1.456″ OAL

Powder	Sug. Starting Grains	Velocity F.P.S.	Pressure C.U.P.	Max. Grains	Velocity F.P.S.	Pressure C.U.P.
700X	3.0	675	5,700	*5.0	1107	15,700
				(+P)5.3	1168	18,300
Bullseye	3.3	765	7,600	5.3	1110	15,700
				(+P)5.6	1168	17,600
Red Dot	3.8	830	8 ,800	5.2	1108	16,300
				(+P)5.5	1154	17,700
231	3.5	689	5,900	5.7	1104	16,300
				(+P)6.0	1163	17,800
HP38	3.6	836	9,100	5.0	1098	16,500
HS-5	6.5	892	9,400	8.4	1168	16,800
AL-5	6.0	911	10,400	7.9	1145	16,300

Note: Loads shown in shaded panels are maximum.
*Designates potentially most accurate load.
(+P)Designates higher pressure loads. See COMMENTS.

#358242
121 gr., (Linotype) 1.450" OAL

Powder	Sug. Starting Grains	Velocity F.P.S.	Pressure C.U.P.	Max. Grains	Velocity F.P.S.	Pressure C.U.P.
700X	2.5	647	7,600	3.9	940	16,200
				(+P)4.2	987	17,900
SR-7625	4.0	788	10,600	5.0	988	16,200
Bullseye	2.8	690	8,600	*4.6	1001	16,400
				(+P)4.9	1045	18,400
Red Dot	3.3	749	10,100	4.6	974	16,200
				(+P)5.0	1047	18,400
Unique	4.0	725	9,000	5.7	1012	16,200
231	3.4	693	8,800	5.1	986	16,600
				5.4	1036	17,900
HP38	3.0	725	9,600	4.3	961	16,700
				(+P)4.8	1045	18,300
HS-5	5.0	727	9,200	7.1	998	16,000
AL-5	4.7	732	9,600	6.4	975	16,000

#356402
121 gr., (Linotype) 1.550" OAL

Powder	Sug. Starting Grains	Velocity F.P.S.	Pressure C.U.P.	Max. Grains	Velocity F.P.S.	Pressure C.U.P.
700X	2.7	649	7,500	4.2	967	16,700
				(+P)4.5	1012	18,300
SR-7625	4.0	725	8,500	5.3	1003	16,400
Bullseye	3.8	847	11,400	*5.0	1048	16,500
				(+P)5.3	1068	18,200
Unique	3.9	774	9,700	5.0	994	15,700
				(+P)5.8	1083	17,900
231	3.5	765	9,500	4.8	1008	16,100
				(+P)5.2	1054	18,000
HP38	3.5	794	10,900	4.5	1013	16,100
				(+P)4.9	1059	17,900
HS-5	4.8	697	8,100	7.3	940	16,600
AL-5	4.8	666	9,000	7.0	1012	15,800

#358495
141 gr., (Linotype) 1.310" OAL

Powder	Sug. Starting Grains	Velocity F.P.S.	Pressure C.U.P.	Max. Grains	Velocity F.P.S.	Pressure C.U.P.
700X	2.2	636	8,300	4.0	994	16,600
SR-7625	3.1	683	9,100	4.3	905	15,600
Bullseye	3.0	778	11,000	4.1	973	16,800
Red Dot	2.8	706	9,500	4.2	961	16,300
Unique	3.8	757	10,300	5.3	1013	16,700
231	3.0	667	8,500	*4.9	1001	16,800
HP38	2.7	683	9,000	4.5	1001	16,800
HS-5	4.0	643	7,800	6.4	990	16,300
AL-5	4.0	660	8,400	6.3	991	16,300

#358212
146 gr., (Linotype) 1.515" OAL

Powder	Sug. Starting Grains	Velocity F.P.S.	Pressure C.U.P.	Max. Grains	Velocity F.P.S.	Pressure C.U.P.
700X	2.8	674	9,400	*4.0	881	15,900
SR-7625	3.5	594	6,900	5.0	899	16,100
Bullseye	2.7	612	8,000	4.3	906	15,800
Red Dot	2.8	650	8,800	4.3	917	16,400
Unique	3.7	652	9,000	5.4	928	15,900
231	3.2	617	7,900	5.0	930	16,700
HP38	3.0	654	9,300	4.4	900	16,500
HS-5	4.8	650	8,100	7.0	935	16,000
AL-5	4.8	656	8,400	6.6	902	15,500

Note: Loads shown in shaded panels are maximum.
*Designates potentially most accurate load.
(+P)Designates higher pressure loads. See COMMENTS.

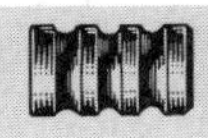

#35863
148 gr., (Linotype) 1.152″ OAL

Powder	Sug. Starting Grains	Velocity F.P.S.	Pressure C.U.P.	Max. Grains	Velocity F.P.S.	Pressure C.U.P.
700X	2.0	584	6,900	3.5	961	16,300
SR-7625	2.8	674	8,700	4.0	928	15,800
Bullseye	2.4	664	8,400	3.8	950	16,100
Red Dot	2.5	678	8,700	3.7	945	15,900
Unique	3.1	670	8,100	4.6	994	16,800
231	2.5	600	6,600	4.1	957	16,000
HP38	2.7	750	10,400	3.9	972	16,400
HS-5	4.0	778	10,900	*5.8	981	16,400
AL-5	3.3	588	6,700	5.4	988	15,700

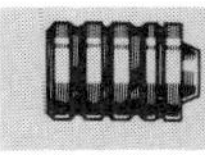

#35891
148 gr., (Linotype) 1.317″ OAL

Powder	Sug. Starting Grains	Velocity F.P.S.	Pressure C.U.P.	Max. Grains	Velocity F.P.S.	Pressure C.U.P.
700X	2.0	512	6,000	3.6	889	15,900
SR-7625	2.8	517	5,800	4.7	958	16,500
Bullseye	2.8	568	8,900	4.4	953	16,200
Red Dot	2.7	524	7,900	4.1	939	16,800
Unique	3.4	660	8,500	5.0	948	16,100
231	2.8	626	7,900	*4.6	914	15,600
HP38	2.0	635	8,700	4.1	911	16,700
HS-5	4.2	657	8,800	6.4	965	16,400
AL-5	4.3	647	8,300	6.7	986	16,800

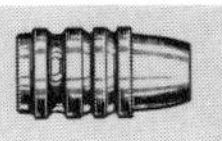

#358156
158 gr., (Linotype) 1.460″ OAL

Powder	Sug. Starting Grains	Velocity F.P.S.	Pressure C.U.P.	Max. Grains	Velocity F.P.S.	Pressure C.U.P.
PB	3.0	584	8,700	4.2	825	16,200
SR-4756	4.2	630	9,500	5.5	898	16,700
Bullseye	2.8	638	10,000	*4.0	858	16,000
				‡(+P)4.4	915	18,100
Unique	3.4	601	8,900	5.1	895	16,100
				(+P)5.4	954	18,000
Blue Dot	6.0	686	10,400	7.2	867	15,700
231	3.5	680	11,100	4.7	885	16,800
				(+P)5.0	934	18,000
630	6.3	659	10,000	8.5	896	16,500
HS-6	4.4	564	8,000	6.7	894	16,600
AL-7	5.0	649	10,400	6.7	889	16,600
HP38	2.8	642	10,200	3.9	854	16,300
				(+P)4.1	882	17,800

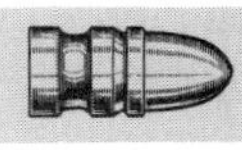

#358311
158 gr., (Linotype) 1.550″ OAL

Powder	Sug. Starting Grains	Velocity F.P.S.	Pressure C.U.P.	Max. Grains	Velocity F.P.S.	Pressure C.U.P.
PB	3.0	604	7,900	4.5	871	16,100
SR-4756	4.2	620	8,700	5.5	874	16,000
Bullseye	2.8	656	10,000	4.2	880	16,200
				‡(+P)4.5	931	18,200
Unique	3.4	611	8,300	5.1	896	16,400
				(+P)5.4	946	18,000
Blue Dot	5.9	673	9,100	7.8	919	16,400
231	3.5	675	9,600	*4.9	906	16,800
				(+P)5.2	956	18,500
630	6.3	663	9,000	8.9	895	16,000
HS-6	4.5	610	8,200	6.7	901	16,200
AL-7	5.8	716	10,000	7.0	897	15,700
HP38	2.6	621	8,800	4.3	902	16,900
				(+P)4.5	944	18,400

Note: Loads shown in shaded panels are maximum.
*Designates potentially most accurate load.
+Designates a compressed powder charge.
‡Designates factory velocity duplication load.

#357446
162 gr., (Linotype) 1.457" OAL

Powder	Sug. Starting Grains	Velocity F.P.S.	Pressure C.U.P.	Max. Grains	Velocity F.P.S.	Pressure C.U.P.
PB	3.4	646	9,500	4.6	877	16,300
SR-4756	4.6	694	10,400	5.8	921	16,800
Bullseye	3.0	680	10,100	4.2	889	16,900
Unique	3.5	639	8,900	5.2	914	16,200
Blue Dot	5.8	675	9,500	7.7	948	16,500
231	3.4	702	10,400	4.6	896	16,000
630	6.3	683	10,100	*8.6	920	15,900
HS-6	5.1	690	9,600	6.7	922	16,100
AL-7	5.2	671	9,700	6.9	909	16,200

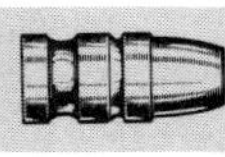

#358429
168 gr., (Linotype) 1.537" OAL

Powder	Sug. Starting Grains	Velocity F.P.S.	Pressure C.U.P.	Max. Grains	Velocity F.P.S.	Pressure C.U.P.
SR-7625	3.4	621	9,700	4.6	821	16,300
SR-4756	4.2	645	10,000	5.4	836	16,400
Bullseye	2.6	597	9,300	4.2	850	16,800
Unique	3.8	673	11,000	5.0	860	16,700
Blue Dot	5.5	647	10,000	7.3	869	16,500
231	3.0	599	9,300	4.6	855	16,900
630	6.1	660	9,900	*8.4	869	16,500
HS-6	5.0	677	11,000	6.4	856	16,800
AL-7	5.4	717	11,900	6.9	879	16,700

.357 Magnum

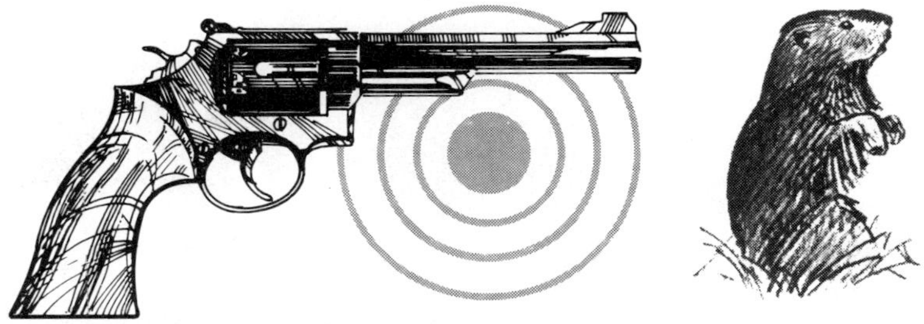

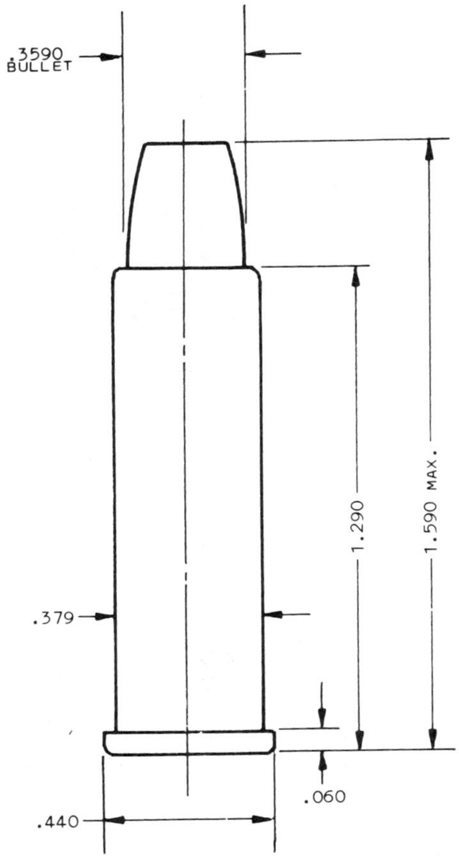

TEST COMPONENTS:

Cases .... Federal
Trim-to Length .... 1.285″
Primers .... CCI 550
Primer Size .... Small Pistol Magnum
Lyman Shell Holder .... No. 1
Cast Bullets Used .... (size to .357″ dia.)
*Gas Check Bullets #358242, 92 gr.
#358345, 115 gr.
#358242, 121 gr.
#356402, 121 gr.
#358480, 133 gr.
#358495, 141 gr.
#358477, 150 gr.
*#358156, 155 gr.
#358311, 158 gr.
#358429, 168 gr.

TEST SPECIFICATIONS:
(Velocity and Pressure)

Firearm Used .... Universal Receiver and Special Vented Barrel
Barrel Length .... 4″
Twist .... 1-18¾″
Groove Diameter .... .356″

COMMENTS:

Do not use .38 Special cases for .357 Magnum loads. Use of the smaller case will increase pressure.

Depending upon their manufacture, some variations in groove diameter exist in these handguns. Slug your bore before reloading and size cast bullets to exact groove diameter.

In order to hold the maximum overall length below 1.590″, it is sometimes necessary to crimp cast bullets on the forward edge of the first driving band.

For those desiring a mid-range load, we suggest bullet #358495 at starting load velocities. Bullet #358156 is extremely popular and an excellent choice for the heavier loads. Another great favorite is the "Keith type" bullet #358429. Refer to Opening Remarks when using Winchester 296 powder.

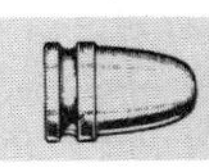

#358242
92 gr., (Linotype) 1.585″ OAL

Powder	Sug. Starting Grains	Velocity F.P.S.	Pressure C.U.P.	Max. Grains	Velocity F.P.S.	Pressure C.U.P.
700X	6.0	1206	19,700	*8.3	1581	41,500
SR-7625	7.5	1262	20,700	9.8	1613	40,900
Bullseye	7.5	1365	25,900	9.5	1630	39,900
231	8.1	1329	25,600	10.1	1629	41,200
630	14.9	1385	23,900	18.1	1717	40,700
HS-5	10.8	1404	25,700	13.6	1671	38,500
AL-5	9.2	1283	21,900	12.2	1645	39,400

Note: Loads shown in shaded panels are maximum.
*Designates potentially most accurate load.

.357 Magnum (Continued)

#358345
115 gr., (Linotype) 1.465" OAL

Powder	Sug. Starting Grains	Velocity F.P.S.	Pressure C.U.P.	Max. Grains	Velocity F.P.S.	Pressure C.U.P.
SR-7625	5.4	1036	17,600	*8.3	1446	41,900
SR-4756	6.8	1137	19,600	12.0	1626	41,900
Unique	7.5	1233	24,300	9.2	1489	41,400
Blue Dot	9.4	1134	20,400	12.8	1586	41,200
231	6.4	1166	23,900	8.4	1442	40,600
630	11.9	1226	23,100	15.5	1564	42,000
HS-5	8.0	1086	17,900	11.7	1548	41,700
AL-5	7.5	1070	18,500	10.5	1494	40,700

#358242
121 gr., (Linotype) 1.585" OAL

Powder	Sug. Starting Grains	Velocity F.P.S.	Pressure C.U.P.	Max. Grains	Velocity F.P.S.	Pressure C.U.P.
SR-7625	5.5	974	17,500	8.0	1335	41,100
Unique	6.9	1088	21,100	9.1	1409	42,000
2400	13.0	1178	25,700	16.4	1494	41,700
630	*11.4	1102	20,300	14.9	1439	39,500
HS-7	10.0	1053	18,400	13.2	1436	40,300
AL-8	9.0	925	16,300	11.9	1383	40,500

#356402
121 gr., (Linotype) 1.590" OAL

Powder	Sug. Starting Grains	Velocity F.P.S.	Pressure C.U.P.	Max. Grains	Velocity F.P.S.	Pressure C.U.P.
SR-7625	7.0	1205	23,000	*8.8	1429	41,600
Unique	7.2	1134	15,800	9.4	1482	41,300
Blue Dot	10.8	1156	17,000	13.0	1564	41,400
2400	13.5	1140	17,300	19.5+	1565	37,200
630	12.4	1230	20,800	15.5	1520	40,200
AL-8	10.3	1087	16,000	13.0	1481	40,300

#358480
133 gr., (Linotype) 1.508" OAL

Powder	Sug. Starting Grains	Velocity F.P.S.	Pressure C.U.P.	Max. Grains	Velocity F.P.S.	Pressure C.U.P.
IMR-4227	12.0	1045	21,700	17.4	1421	40,200
Unique	6.2	1006	18,300	8.6	1356	40,300
Blue Dot	10.0	1162	18,800	11.6	1426	41,000
2400	11.8	1058	15,500	16.8	1501	40,200
630	11.2	1130	17,300	*14.8	1466	40,900
AL-8	9.5	1096	19,100	12.7	1461	41,500

#358495
141 gr., (Linotype) 1.435" OAL

Powder	Sug. Starting Grains	Velocity F.P.S.	Pressure C.U.P.	Max. Grains	Velocity F.P.S.	Pressure C.U.P.
IMR-4227	12.3	1084	22,200	*16.3+	1365	39,600
Unique	5.8	960	14,600	7.9	1289	40,300
Blue Dot	9.6	1134	21,800	11.2	1382	40,600
2400	10.5	1015	18,200	14.6	1376	39,700
630	9.9	1032	16,600	13.5	1369	40,100
HS-7	9.8	1130	23,200	11.8	1356	40,700
AL-8	8.8	1012	19,800	11.6	1334	41,000

#358477
150 gr., (Linotype) 1.510" OAL

Powder	Sug. Starting Grains	Velocity F.P.S.	Pressure C.U.P.	Max. Grains	Velocity F.P.S.	Pressure C.U.P.
IMR-4227	11.8	1044	21,700	16.0+	1333	40,200
Herco	6.2	972	17,700	7.8	1225	41,900
Blue Dot	8.2	950	13,000	10.8	1356	41,200
2400	11.0	998	17,900	15.0	1362	41,400
630	10.0	1029	18,400	*13.0	1311	39,900
**296	—	—	—	17.8	1452	41,900
H110	12.4	1075	17,300	17.7	1459	41,200
AL-8	8.8	1046	23,000	10.4	1263	40,900

Note: Loads shown in shaded panels are maximum.
*Designates potentially most accurate load.
+Designates a compressed powder charge.
**Reduced loads not recommended by Winchester.

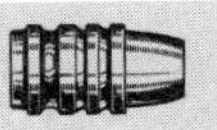

#358156
155 gr., (Linotype) 1.590" OAL

Powder	Sug. Starting Grains	Velocity F.P.S.	Pressure C.U.P.	Max. Grains	Velocity F.P.S.	Pressure C.U.P.
IMR-4227	11.4	973	24,100	*15.2	1254	41,300
Herco	5.0	785	16,300	7.5	1151	41,000
Blue Dot	8.5	975	21,500	10.5	1277	40,800
2400	10.6	999	24,900	14.0	1299	41,900
**296	—	—	—	15.6	1334	39,700
630	9.6	953	20,900	13.2	1286	42,000
H110	11.6	1037	21,800	15.7	1363	40,300
AL-8	8.2	919	22,100	10.7	1235	41,500

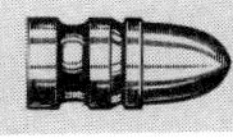

#358311
158 gr., (Linotype) 1.590" OAL

Powder	Sug. Starting Grains	Velocity F.P.S.	Pressure C.U.P.	Max. Grains	Velocity F.P.S.	Pressure C.U.P.
SR-4759	10.0	874	12,000	14.0+	1189	30,100
IMR-4227	11.8	977	19,600	*17.0+	1345	40,600
Herco	6.3	963	18,700	7.9	1203	41,000
Blue Dot	8.2	888	11,600	10.9	1316	39,200
2400	11.4	1024	20,200	15.5	1344	39,700
**296	—	—	—	18.0	1461	41,400
630	10.3	1016	18,000	13.6	1316	40,400
H110	13.0	1115	19,100	18.3	1460	40,100
AL-8	8.9	973	18,700	‡10.8	1257	39,400

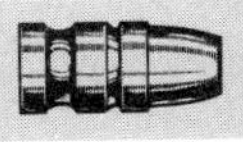

#358429
168 gr., (Linotype) 1.553" OAL

Powder	Sug. Starting Grains	Velocity F.P.S.	Pressure C.U.P.	Max. Grains	Velocity F.P.S.	Pressure C.U.P.
SR-4759	9.8	942	18,400	13.5+	1206	40,600
IMR-4227	9.8	835	14,100	14.5+	1233	40,800
Herco	5.6	885	17,900	7.1	1104	39,000
Blue Dot	8.3	970	18,400	10.0	1233	39,200
2400	9.7	879	15,900	13.5	1242	41,100
**296	—	—	—	15.2	1307	39,900
630	9.7	1012	21,400	12.5	1247	41,100
H110	*11.8	1037	18,900	15.7	1318	39,800
AL-8	8.3	882	19,500	11.2	1208	41,000

Note: Loads shown in shaded panels are maximum.
*Designates potentially most accurate load.
+Designates a compressed powder charge.
‡Designates factory velocity duplication load.
**Reduced loads not recommended by Winchester.

.38/40 Winchester

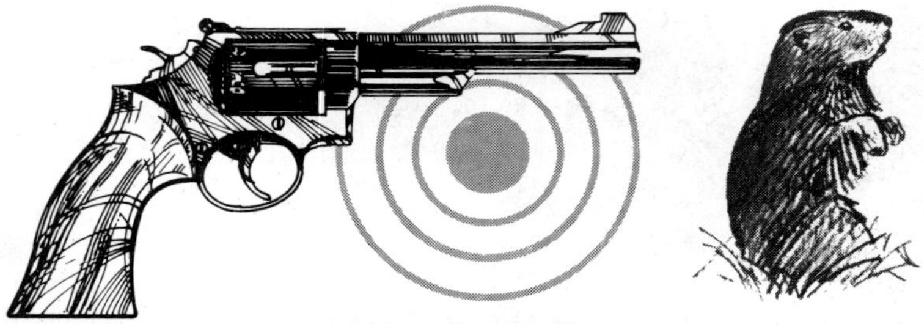

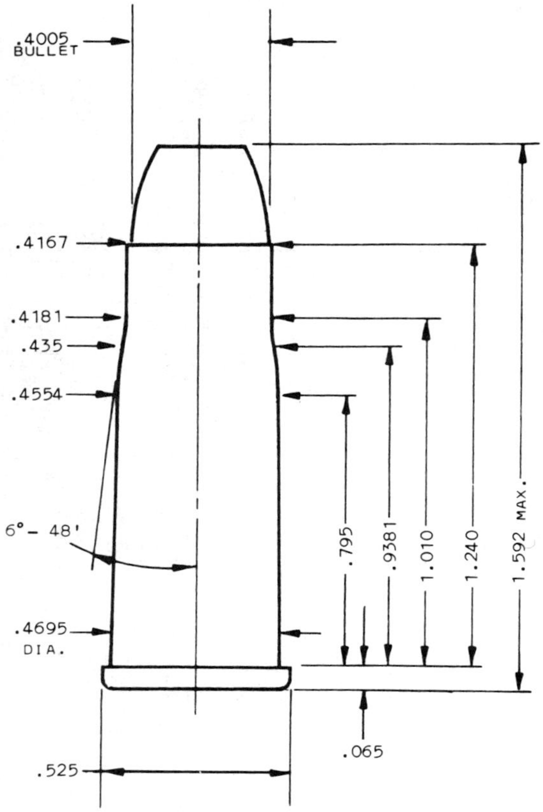

TEST COMPONENTS:

Cases .... Winchester
Trim-to Length .... 1.300″
Primers .... Winchester 7-111
Primer Size .... Large Pistol
Lyman Shell Holder .... No. 14B
Cast Bullets Used .... (size to .401″ dia.)
#40143, 172 gr.
(Can also use #40188, 170 gr.)

TEST SPECIFICATIONS:
(Velocity)

Firearm Used .... Colt S.A.A.
Barrel Length .... 7½″
Twist .... 1-36″
Groove Diameter .... .400″

COMMENTS:

These loads are not to be used in handguns which were designed for black powder. A great many revolvers which are chambered for this cartridge have groove and bore dimensions which are so far oversized that it is impossible to obtain accuracy. Because chamber dimensions limit the diameter of the bullet, nothing can be done to make these oversize barrels shoot well.

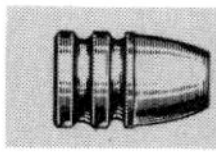

#40143
172 gr., (#2 Alloy) 1.592″ OAL

Powder	Sug. Starting Grains	Velocity F.P.S.	Pressure C.U.P.	Max. Grains	Velocity F.P.S.	Pressure C.U.P.
Bullseye	4.0	740	—	5.9	965	—
Unique	7.0	830	—	10.0	1105	—

Note: Loads shown in shaded panels are maximum.

.41 Magnum

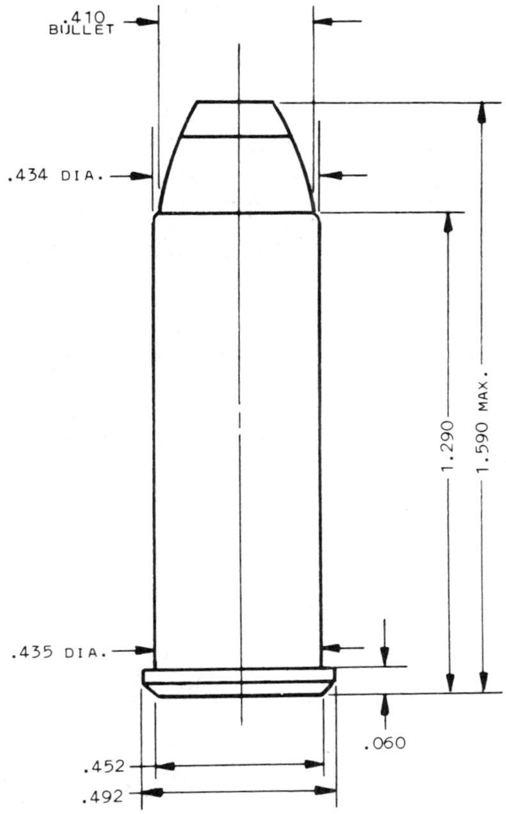

TEST COMPONENTS:

Cases .... Winchester
Trim-to Length .... 1.285″
Primers .... Remington 2½
Primer Size .... Large Pistol
Lyman Shell Holder .... No. 30
Cast Bullets Used .... (size to .410″ dia.)
*Gas Check Bullets #41028, 212 gr.
*#410610, 215 gr.
#410459, 220 gr.
#410426, 240 gr.

TEST SPECIFICATIONS:
(Velocity and Pressure)

Firearm Used .... Universal Receiver and Special Vented Barrel
Barrel Length .... 4″
Twist .... 1-18¾″
Groove Diameter .... .409″

COMMENTS:

Cast bullet #'s 410459 and 410426 list overall lengths that exceed the maximum of 1.590″ and are suitable only for the Thompson/Center Contender or similar firearms that will accept cartridges of these dimensions. The data listed for these bullets take this extra length into consideration.

Overall length with bullet #410459 - 1.700″.
Overall length with bullet #410426 - 1.707″.

#41028
212 gr., (Linotype) 1.586″ OAL

Powder	Sug. Starting Grains	Velocity F.P.S.	Pressure C.U.P.	Max. Grains	Velocity F.P.S.	Pressure C.U.P.
SR-4756	8.8	1030	21,700	10.9	1221	36,300
IMR-4227	16.2	1006	21,300	20.5	1247	35,900
Unique	8.2	998	20,300	10.7	1217	36,300
Blue Dot	12.2	1041	20,900	15.4	1333	36,500
2400	15.4	1030	21,700	20.0	1292	36,100
630	*11.9	1023	19,500	16.5	1259	35,700
HS-7	12.0	1031	21,500	15.1	1265	36,200
AL-8	13.5	1050	25,300	16.7	1264	35,500

Note: Loads shown in shaded panels are maximum.
*Designates potentially most accurate load.

#410610
215 gr., (Linotype) 1.575" OAL

Powder	Sug. Starting Grains	Velocity F.P.S.	Pressure C.U.P.	Max. Grains	Velocity F.P.S.	Pressure C.U.P.
SR-4756	8.0	909	17.100	10.8	1213	36.000
IMR-4227	14.9	956	20.100	19.2	1221	35.200
Unique	7.8	979	19.800	10.4	1214	36.100
Blue Dot	11.4	1003	20.200	14.6	1170	35.800
2400	14.2	835	19.600	18.8	1272	37.000
630	11.5	923	18.200	16.0	1251	35.800
HS-7	11.5	1007	20.900	*14.5	1249	36.800
AL-8	11.9	974	22.200	15.0+	1206	35.000

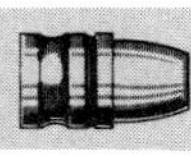

#410459
220 gr., (Linotype) 1.700" OAL

Powder	Sug. Starting Grains	Velocity F.P.S.	Pressure C.U.P.	Max. Grains	Velocity F.P.S.	Pressure C.U.P.
SR-4756	9.8	1061	20.900	*11.6	1246	33.800
IMR—4227	17.3	1048	22.100	21.5+	1280	35.500
Unique	8.1	970	18.000	11.2	1261	36.200
Blue Dot	13.0	1100	21.700	16.0	1359	36.000
2400	15.7	1022	21.100	20.1	1330	35.400
630	13.0	1012	19.000	17.5	1306	36.100
HS-7	13.0	1061	21.400	15.9	1313	36.900
AL-8	13.7	1057	23.100	16.8	1291	36.000

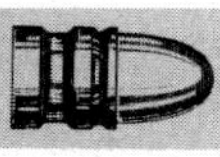

#410426
240 gr., (Linotype) 1.707" OAL

Powder	Sug. Starting Grains	Velocity F.P.S.	Pressure C.U.P.	Max. Grains	Velocity F.P.S.	Pressure C.U.P.
SR-4756	8.7	1000	23.900	10.5	1159	36.800
IMR-4227	15.0	932	20.500	19.1	1177	35.800
Unique	6.9	879	18.200	10.0	1147	36.900
Blue Dot	11.5	991	19.400	14.5	1249	36.300
2400	14.0	915	18.200	18.5	1214	36.100
630	11.1	900	18.300	*15.6	1195	36.000
HS-7	11.5	1007	21.400	14.5	1211	36.100
AL-8	12.2	936	23.000	15.5	1193	36.900

Note: Loads shown in shaded panels are maximum.
*Designates potentially most accurate load.
+Designates a compressed powder charge.

.44 Smith & Wesson Special

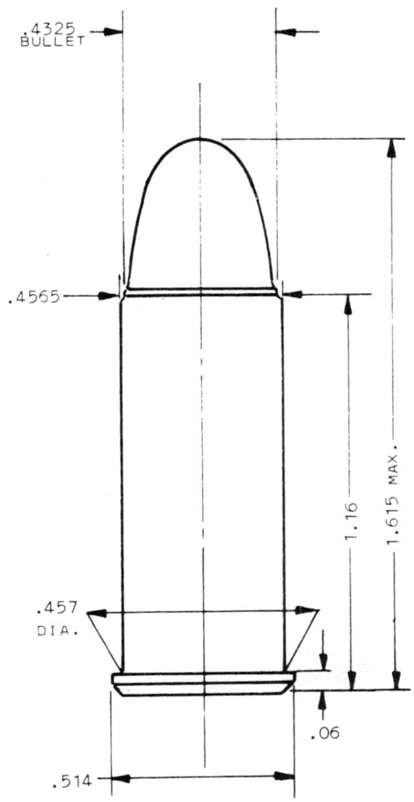

TEST COMPONENTS:

Cases .... Winchester
Trim-to Length .... 1.152″
Primers .... CCI 300
Primer Size .... Large Pistol
Lyman Shell Holder .... No. 7
Cast Bullets Used .... (size to .429″ dia.)
*Gas Check Bullets .... #42798, 205 gr.
*#429215, 215 gr.
#429360, 232 gr.
#429421, 245 gr.
*#429244, 250 gr.

TEST SPECIFICATIONS:
(Velocity and Pressure)

Firearm Used .... Universal Receiver and Special Vented Barrel
Barrel Length .... 4″
Twist .... 1-20″
Groove Diameter .... .429″

COMMENTS:

Considerable variations in groove diameters exist with handguns chambered for this cartridge. We recommend that you slug your barrel before reloading and size bullets to as near groove diameter as possible.

When loading half-jacketed bullets, do not use charge weights below suggested starting loads to avoid the danger of lodging bullets in the barrel.

Those wishing a "Keith type" bullet should choose bullet #429421.

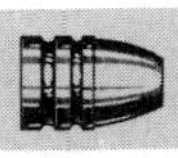

#42798
205 gr., (Linotype) 1.537″ OAL

Powder	Sug. Starting Grains	Velocity F.P.S.	Pressure C.U.P.	Max. Grains	Velocity F.P.S.	Pressure C.U.P.
SR-7625	6.8	761	9,700	7.7	838	12,600
SR-4759	13.2	651	9,200	15.2+	798	12,100
Unique	7.0	736	10,500	*8.2	869	14,000
Blue Dot	10.5	747	10,000	12.0	855	13,300
2400	13.3	734	10,800	15.2	870	13,500
630	10.8	790	12,200	12.2	876	13,800
HS-7	10.0	732	10,100	11.6	874	13,600
AL-8	11.0	767	11,400	12.5	871	13,900

Note: Loads shown in shaded panels are maximum.
*Designates potentially most accurate load.
+Designates a compressed powder charge.

.44 Smith & Wesson Special (Continued)

#429215
215 gr., (Linotype) 1.500″ OAL

Powder	Sug. Starting Grains	Velocity F.P.S.	Pressure C.U.P.	Max. Grains	Velocity F.P.S.	Pressure C.U.P.
SR-7625	6.2	732	10,900	6.8	801	13,500
Unique	6.3	729	9,500	7.7	884	13,600
Blue Dot	9.5	689	8,800	11.0	885	13,400
2400	13.0	792	11,300	14.5	905	14,000
630	7.5	534	5,500	12.1	895	13,900
HS-7	9.0	659	8,100	11.0	841	12,600
AL-8	9.4	628	8,600	10.9	828	13,700

#429360
232 gr., (Linotype) 1.525″ OAL

Powder	Sug. Starting Grains	Velocity F.P.S.	Pressure C.U.P.	Max. Grains	Velocity F.P.S.	Pressure C.U.P.
SR-4756	6.7	666	9,800	8.0	774	13,400
SR-4759	12.5	647	10,300	14.3	803	13,500
Unique	6.0	604	10,400	7.1	781	13,500
Blue Dot	9.0	646	10,300	10.0	751	13,100
2400	12.0	732	10,500	13.8	848	13,800
630	9.2	716	10,400	11.0	838	13,800
HS-7	9.5	729	10,500	11.0	853	13,900
AL-8	9.5	699	9,900	*11.0	825	13,300

#429421
245 gr., (Linotype) 1.571″ OAL

Powder	Sug. Starting Grains	Velocity F.P.S.	Pressure C.U.P.	Max. Grains	Velocity F.P.S.	Pressure C.U.P.
SR-4756	6.8	661	9,000	8.1	780	13,800
SR-4759	11.9	652	10,500	13.4	789	14,000
Unique	6.0	649	10,300	‡6.9	767	13,300
Blue Dot	9.0	710	10,800	10.0	807	13,200
2400	11.4	704	10,400	13.2	797	13,800
630	9.6	716	10,300	*10.8	819	13,700
HS-7	9.1	728	11,300	10.4	828	14,000
AL-8	8.9	609	10,000	10.3	773	12,700

#429244
250 gr., (Linotype) 1.550″ OAL

Powder	Sug. Starting Grains	Velocity F.P.S.	Pressure C.U.P.	Max. Grains	Velocity F.P.S.	Pressure C.U.P.
SR-4756	6.2	584	8,400	7.5	765	14,000
Unique	5.7	635	10,200	6.6	747	13,600
2400	10.5	602	8,500	12.3	742	12,700
630	9.3	667	10,400	10.5	769	13,200
HS-7	8.5	606	8,700	*10.0	743	13,400
AL-8	8.5	607	9,800	9.9	730	13,100
SR-4759	11.6	658	11,600	13.1	752	13,200

Note: Loads shown in shaded panels are maximum.
*Designates potentially most accurate load.

.44 Remington Magnum

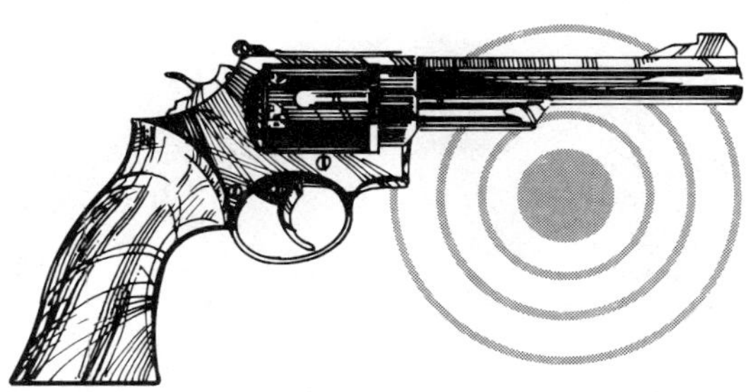

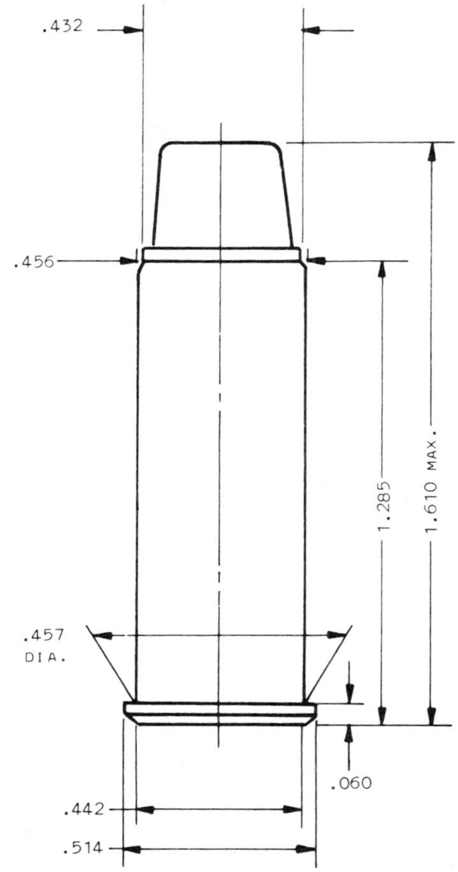

TEST COMPONENTS:

Cases .... Remington
Trim-to Length .... 1.280″
Primers .... CCI 300
Primer Size .... Large Pistol
Lyman Shell Holder .... No. 7
Cast Bullets Used .... (size to .429″ dia.)
*Gas Check Bullets .... #429348, 180 gr.
*#429303, 200 gr.
*#429215, 215 gr.
#429360, 232 gr.
#429421, 245 gr.
*#429244, 245 gr.

TEST SPECIFICATIONS:
(Velocity and Pressure)

Firearm Used .... Universal Receiver and Special Vented Barrel
Barrel Length .... 4″
Twist .... 1-20″
Groove Diameter .... .429″

COMMENTS:

To prevent bullet movement when the gun is under recoil, the cartridge requires a hard crimp on all of the heavier loads.

Cast bullet #'s 429303, 429215, 429360, 429244, 429421 list overall lengths that exceed the maximum of 1.610″ and are suitable only for the Thompson/Center Contender or similar firearms that will accept cartridges of these dimensions. The data listed for these bullets take this extra length into consideration.

Overall length with bullet #429303 - 1.692″.
Overall length with bullet #429215 - 1.645″.
Overall length with bullet #429360 - 1.660″.
Overall length with bullet #429244 - 1.680″.
Overall length with bullet #429421 - 1.710″.

Bullet #429348, loaded to starting load velocities, is recommended for mid-range use, while bullet #429215 is suggested as the best all around choice.

#429348
180 gr., (Linotype) 1.420″ OAL

Powder	Sug. Starting Grains	Velocity F.P.S.	Pressure C.U.P.	Max. Grains	Velocity F.P.S.	Pressure C.U.P.
SR-4756	11.9	1096	23,300	15.8	1359	36,900
IMR-4227	22.7	1065	21,700	27.0+	1394	35,900
Unique	11.0	1079	22,500	13.5	1343	36,500
Blue Dot	16.1	1110	22,600	20.4	1408	35,800
2400	20.3	1190	22,000	25.3+	1410	36,100
231	9.0	1002	22,000	12.6	1315	36,700
630	17.6	1144	24,500	*21.8	1390	36,000
HS-6	14.2	1082	23,200	17.3	1352	36,400
AL-7	15.7	1075	22,900	19.0	1361	36,500
Herco	11.7	1049	22,800	14.6	1344	36,400

Note: Loads shown in shaded panels are maximum.
*Designates potentially most accurate load.
+Designates a compressed powder charge.

.44 Remington Magnum (Continued)

#429303
200 gr., (Linotype) 1.692" OAL

Powder	Sug. Starting Grains	Velocity F.P.S.	Pressure C.U.P.	Max. Grains	Velocity F.P.S.	Pressure C.U.P.
SR-4756	12.3	1043	22,600	15.5	1289	36,700
IMR-4227	22.0	1038	22,100	24.5+	1201	29,600
Unique	11.0	1004	22,100	13.2	1265	36,700
Herco	11.4	1035	24,300	13.8	1245	36,700
Blue Dot	15.8	1062	22,800	19.8	1363	36,800
2400	19.0	991	20,400	24.0+	1314	34,600
231	9.1	913	19,600	*12.6	1231	36,900
630	17.5	1049	23,300	21.5	1317	36,500
HS-6	14.2	1020	22,000	17.3	1295	36,900
AL-7	16.0	999	20,400	19.0	1318	36,300

#429215
215 gr., (Linotype) 1.645" OAL

Powder	Sug. Starting Grains	Velocity F.P.S.	Pressure C.U.P.	Max. Grains	Velocity F.P.S.	Pressure C.U.P.
SR-4756	12.2	1036	24,900	15.2	1228	36,100
IMR-4227	21.8	1017	22,100	25.5+	1227	32,800
Unique	10.0	930	21,000	13.2	1221	36,000
Herco	11.4	984	22,800	13.5	1205	36,000
Blue Dot	15.7	1032	23,500	19.5	1314	37,000
2400	20.0	986	21,800	25.0	1310	37,000
231	*9.3	914	21,000	12.8	1191	37,000
630	17.8	1037	24,100	21.5	1269	36,500
HS-6	13.5	985	21,700	16.5	1238	36,000
AL-7	14.5	997	20,600	18.7	1280	34,900

#429360
232 gr., (Linotype) 1.655" OAL

Powder	Sug. Starting Grains	Velocity F.P.S.	Pressure C.U.P.	Max. Grains	Velocity F.P.S.	Pressure C.U.P.
SR-4756	9.9	887	16,000	13.3	1145	36,100
IMR-4227	20.2	961	21,900	24.5	1236	36,100
Unique	9.5	896	16,100	12.7	1159	36,200
Herco	10.7	937	23,300	13.5	1163	36,500
Blue Dot	14.4	952	20,400	18.2	1240	36,300
2400	18.5	998	22,700	23.0	1256	36,700
630	15.6	947	21,900	*19.6	1201	36,100
HS-6	12.5	921	20,800	16.2	1177	36,600
AL-8	15.0	924	21,400	20.0+	1229	36,700

#429421
245 gr., (Linotype) 1.710" OAL

Powder	Sug. Starting Grains	Velocity F.P.S.	Pressure C.U.P.	Max. Grains	Velocity F.P.S.	Pressure C.U.P.
SR-4756	11.4	927	20,500	13.1	1125	36,500
IMR-4227	20.2	938	21,300	24.0+	1196	34,600
Unique	9.8	912	20,800	13.0	1147	36,800
Herco	10.9	918	22,900	13.1	1090	34,600
Blue Dot	15.0	941	20,000	18.4	1196	35,300
2400	19.4	974	23,500	23.4	1232	36,000
630	17.1	1009	25,300	*20.7	1190	36,700
HS-7	15.2	931	20,900	18.2	1171	35,300
AL-8	14.6	856	18,700	19.6	1189	36,700

#429244
245 gr., (Linotype) 1.680" OAL

Powder	Sug. Starting Grains	Velocity F.P.S.	Pressure C.U.P.	Max. Grains	Velocity F.P.S.	Pressure C.U.P.
SR-4756	10.5	892	21,500	12.2	1119	35,300
IMR-4227	20.2	903	20,600	24.0+	1135	33,700
Unique	8.9	805	17,500	*12.1	1100	36,400
Herco	11.0	927	24,500	12.4	1085	35,400
Blue Dot	14.0	911	20,300	17.4	1160	34,800
2400	18.2	915	21,600	22.2	1165	35,300
630	16.2	942	23,700	19.3	1146	35,000
HS-7	14.5	921	22,600	17.5	1160	37,000
AL-8	14.4	901	23,200	19.5	1113	34,000

Note: Loads shown in shaded panels are maximum.
*Designates potentially most accurate load.
+Designates a compressed powder charge.

.44/40 Winchester

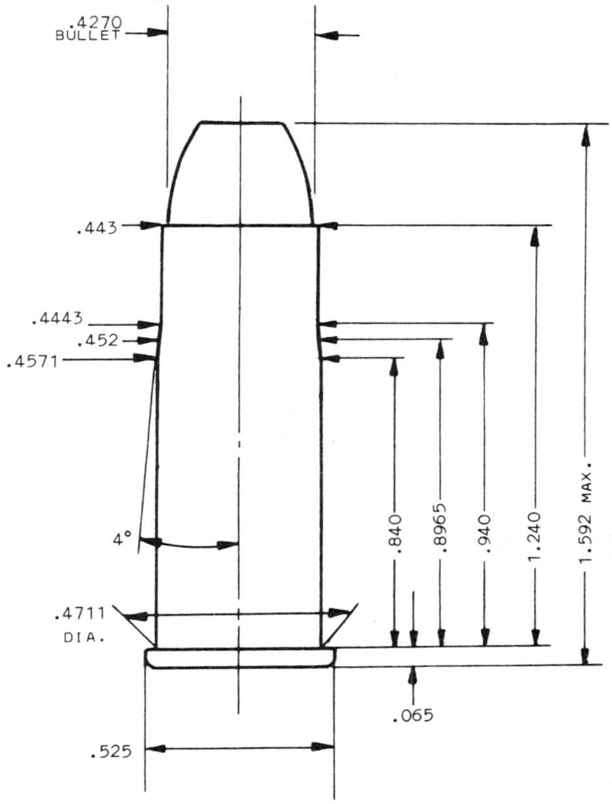

TEST COMPONENTS:

Cases .... Winchester
Trim-to Length .... 1.300″
Primers .... Winchester 7-111
Primer Size .... Large Pistol
Lyman Shell Holder .... No. 14B
Cast Bullets Used .... (size to .427″ dia.) #42798, 205 gr.

TEST SPECIFICATIONS:
(Velocity)

Firearm Used .... Colt S.A.A.
Barrel Length .... 7½″
Twist .... 1-20″
Groove Diameter .... .426″

COMMENTS:

These loads are not to be used in handguns which were designed for black powder. A wide variation in groove diameters exists with handguns chambered for this cartridge. Slug your barrel before reloading and size cast bullets as near to actual groove diameter as possible.

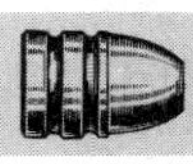

#42798
205 gr., (#2 Alloy) 1.592″ OAL

Powder	Sug. Starting Grains	Velocity F.P.S.	Pressure C.U.P.	Max. Grains	Velocity F.P.S.	Pressure C.U.P.
Bullseye	4.0	695	—	6.6	945	—
Unique	6.0	750	—	10.9	1095	—

Note: Loads shown in shaded panels are maximum.

.45 Automatic (45 ACP)

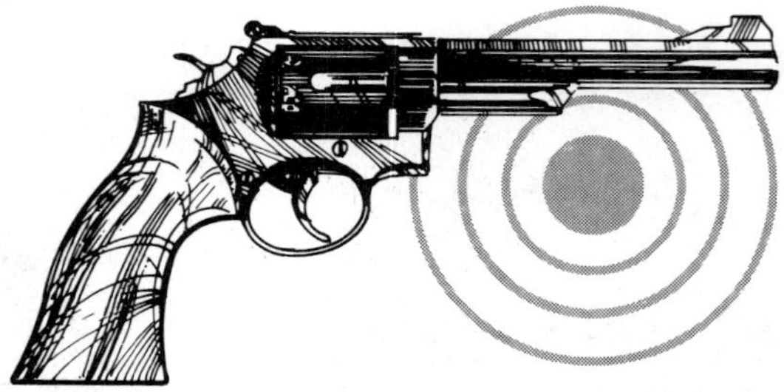

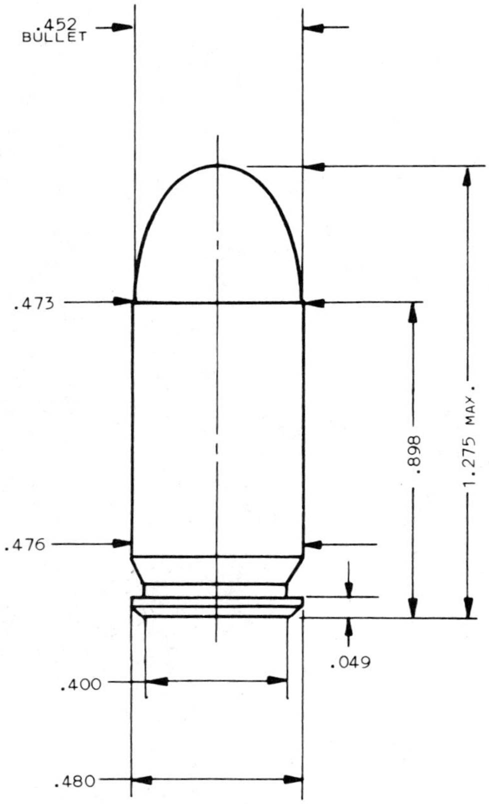

TEST COMPONENTS:

Cases .... Federal; Remington
Trim-to Length .... .895″
Primers .... Remington 2½
Primer Size .... Large Pistol
Lyman Shell Holder .... No. 2
Cast Bullets Used .... (size to .450″ dia.)
*Gas Check Bullets #452389, 185 gr.
#452488, 195 gr.
#452460, 200 gr.
#452374, 225 gr.

TEST SPECIFICATIONS:
(Velocity and Pressure)

Firearm Used .... Universal Receiver
Barrel Length .... 5″
Twist .... 1-16″
Groove Diameter .... .450″

COMMENTS:

Because this cartridge headspaces on the case mouth, the trim-to length must be adhered to closely and the case should not be crimped. A short case, or a crimp, can cause headspace problems.

While groove dimensions for these handguns will range from .450″ - .453″, the reloader may have difficulty if he uses a bullet over .451″ diameter. Case and chamber dimensions usually provide for a bullet of .451″, but larger bullets may bulge cases to the point where they fail to chamber.

Most pistols (unless they are altered) will not feed wadcutter-type bullets such as #452389. Bullet #452460 feeds well in most pistols and, when loaded to starting load velocities, is recommended for mid-range shooting.

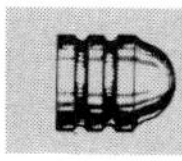

#452389
185 gr., (#2 Alloy) 1.080″ OAL

Powder	Sug. Starting Grains	Velocity F.P.S.	Pressure C.U.P.	Max. Grains	Velocity F.P.S.	Pressure C.U.P.
700X	4.0	760	8,800	5.5	1010	17,500
PB	4.6	740	8,300	5.5	920	13,500
SR-7625	5.5	825	9,800	6.5	1030	17,400
SR-4756	5.3	663	8,900	7.9	988	17,600
Bullseye	3.5	660	6,000	5.0	890	12,300
Red Dot	4.0	735	8,100	5.5	970	14,800
Green Dot	4.5	745	8,300	6.0	965	14,800
Unique	5.0	684	6,900	7.5	995	15,500
Blue Dot	7.0	636	7,600	9.2+	844	13,200
231	3.9	662	7,900	6.0	958	17,500
HS-6	5.6	613	7,500	9.1	1034	17,100
AL-5	5.1	621	7,200	*8.3	974	17,300

Note: Loads shown in shaded panels are maximum.
*Designates potentially most accurate load.
+Designates a compressed powder charge.

.45 Automatic (45 ACP) (Continued)

#452488
195 gr., (#2 Alloy) 1.120" OAL

Powder	Sug. Starting Grains	Velocity F.P.S.	Pressure C.U.P.	Max. Grains	Velocity F.P.S.	Pressure C.U.P.
700X	4.0	735	9,600	5.2	950	16,700
PB	4.6	765	10,100	5.5	910	14,800
SR-7625	5.0	780	10,300	6.2	945	16,200
SR-4756	5.3	639	8,600	7.8	977	17,500
Bullseye	3.5	680	7,700	5.0	895	13,900
Red Dot	4.0	730	9,000	5.5	950	16,200
Green Dot	4.5	740	9,400	6.0	955	15,800
Unique	5.0	695	7,900	7.5	1010	16,800
Blue Dot	7.0	646	8,200	10.7	1033	17,900
231	3.9	642	8,500	5.7	962	16,900
HS-6	6.1	652	8,300	9.0	1027	17,900
AL-5	5.8	722	8,600	*8.5	1014	17,800

#452460
200 gr., (#2 Alloy) 1.161" OAL

Powder	Sug. Starting Grains	Velocity F.P.S.	Pressure C.U.P.	Max. Grains	Velocity F.P.S.	Pressure C.U.P.
700X	4.0	745	10,300	5.2	940	17,300
PB	4.6	725	9,000	5.5	880	14,200
SR-7625	5.0	735	9,000	6.2	945	17,000
**SR-4756	5.3	704	9,100	7.3	993	17,000
Bullseye	3.5	645	6,900	5.0	860	13,400
Red Dot	4.0	695	8,400	5.3	895	14,700
Green Dot	4.5	715	9,000	5.8	895	14,400
Unique	5.0	670	7,700	7.5	980	16,600
**Blue Dot	7.1	701	8,900	10.6	1012	17,200
**231	4.0	694	9,200	6.0	987	18,000
**HS-6	6.0	690	8,600	8.7	1016	17,700
**AL-5	5.8	729	9,300	*8.4	1008	17,400

#452374
225 gr., (#2 Alloy) 1.272" OAL

Powder	Sug. Starting Grains	Velocity F.P.S.	Pressure C.U.P.	Max. Grains	Velocity F.P.S.	Pressure C.U.P.
700X	4.0	695	10,500	5.0	855	17,300
PB	4.8	705	10,300	5.5	810	14,200
SR-7625	5.0	675	9,000	6.0	850	15,200
**SR-4756	5.5	662	8,800	7.5	955	18,000
Bullseye	4.0	680	10,100	5.0	815	14,400
Red Dot	4.3	705	11,000	5.3	835	15,300
Green Dot	4.8	725	11,400	5.8	845	15,100
Unique	5.5	695	10,100	7.3	905	16,500
**Blue Dot	7.5	660	8,600	10.7	964	17,300
**231	4.0	661	9,200	*5.8	902	17,500
**HS-6	6.2	664	8,400	8.6	921	16,600
**AL-5	5.5	657	8,200	8.7	943	16,800

Note: Loads shown in shaded panels are maximum.
*Designates potentially most accurate load.
**Designates use of Federal cases.

.45 Automatic Rim

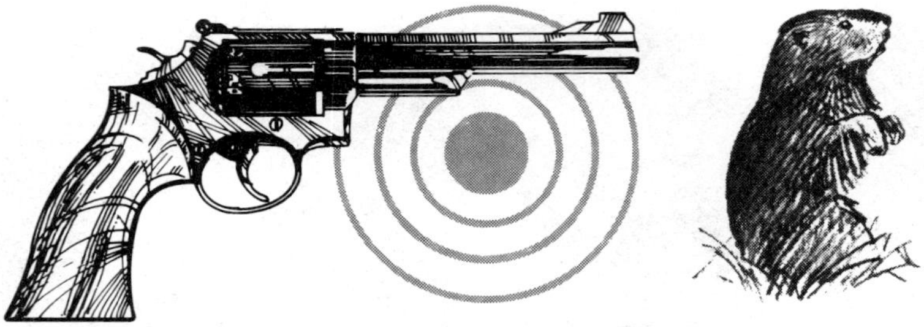

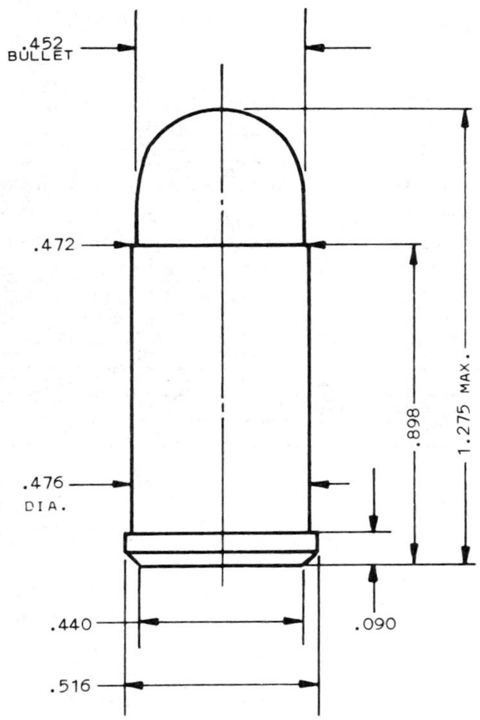

TEST COMPONENTS:

Cases .... Remington
Trim-to Length .... .894″
Primers .... Remington 2½
Primer Size .... Large Pistol
Lyman Shell Holder .... No. 14A
Cast Bullets Used .... (size to .451″ dia.)
*Gas Check Bullets #452389, 185 gr.
#452460, 200 gr.
#452374, 225 gr.
#452423, 238 gr.

TEST SPECIFICATIONS:
(Velocity & Pressure)

Firearm Used .... Special Test Barrel;
Smith & Wesson Model 25
Barrel Length .... 4½ and 6½″
Twist .... 1-16″
Groove Diameter .... .451″

COMMENTS:

Heavier loads for this cartridge should be lightly crimped to prevent bullet movement when the pistol is under recoil.

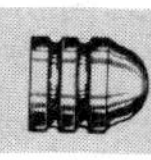

#452389
185 gr., (#2 Alloy) 1.118″ OAL

Powder	Sug. Starting Grains	Velocity F.P.S.	Pressure C.U.P.	Max. Grains	Velocity F.P.S.	Pressure C.U.P.
700X	4.0	655	—	5.3	905	—
PB	4.5	620	—	5.8	855	—
SR-7625	5.0	615	—	6.3	810	—
Bullseye	3.0	505	—	4.8	780	—
Red Dot	4.0	650	—	5.3	885	—
Green Dot	4.5	650	—	5.8	855	—
Unique	5.0	600	—	7.0	870	—
**231	4.0	735	7,200	5.7	972	13,400
**HS-6	5.5	674	5,300	*8.8	1030	14,800
**AL-5	6.5	816	8,500	8.7	1042	14,900

Note: Loads shown in shaded panels are maximum.
*Designates potentially most accurate load.
**Designates use of a special test barrel.

.45 Automatic Rim (Continued)

#452460
200 gr., (#2 Alloy) 1.161" OAL

Powder	Sug. Starting Grains	Velocity F.P.S.	Pressure C.U.P.	Max. Grains	Velocity F.P.S.	Pressure C.U.P.
700X	4.0	675	—	5.0	875	—
PB	4.5	650	—	5.5	840	—
SR-7625	5.0	660	—	6.1	850	—
Bullseye	3.0	510	—	4.7	775	—
Red Dot	4.0	645	—	5.2	845	—
Green Dot	4.5	660	—	5.6	840	—
Unique	5.0	610	—	6.8	845	—
**231	4.0	694	7,200	5.5	936	14,100
**HS-5	6.5	765	8,400	*8.2	934	12,500
**AL-5	5.7	697	6,800	8.4	977	14,400

#452374
225 gr., (#2 Alloy) 1.265" OAL

Powder	Sug. Starting Grains	Velocity F.P.S.	Pressure C.U.P.	Max. Grains	Velocity F.P.S.	Pressure C.U.P.
700X	4.0	625	—	4.8	770	—
PB	4.5	565	—	5.4	735	—
SR-7625	5.0	560	—	5.9	750	—
Bullseye	3.0	475	—	4.5	690	—
Red Dot	4.0	625	—	5.0	760	—
Green Dot	4.5	615	—	5.4	750	—
Unique	5.0	550	—	6.6	785	—
**231	4.0	660	6,900	*5.5	873	13,800
**HS-6	7.3	799	9,400	8.6	922	13,800
**AL-7	8.0	787	9,100	9.7	939	14,700

#452423
238 gr., (#2 Alloy) 1.200" OAL

Powder	Sug. Starting Grains	Velocity F.P.S.	Pressure C.U.P.	Max. Grains	Velocity F.P.S.	Pressure C.U.P.
700X	3.7	620	—	4.5	755	—
PB	3.9	535	—	4.8	690	—
SR-7625	4.1	505	—	4.9	655	—
Bullseye	3.0	500	—	4.5	720	—
Red Dot	3.5	590	—	4.3	715	—
Green Dot	4.0	605	—	4.8	715	—
Unique	4.8	585	—	6.0	750	—
**Blue Dot	6.5	654	8,100	8.5	855	13,300
**630	7.5	694	9,100	10.0	894	14,900
**HS-7	8.0	757	10,900	*9.3	864	14,200
**AL-8	7.0	628	8,300	9.6	814	13,800

Note: Loads shown in shaded panels are maximum.
*Designates potentially most accurate load.
**Designates potentially most accurate load.

.45 Colt

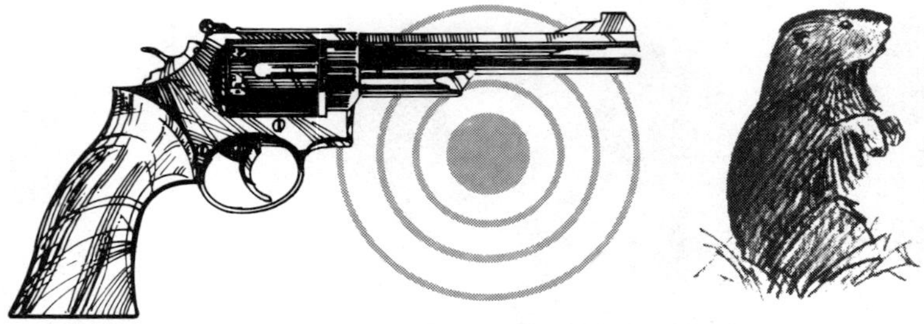

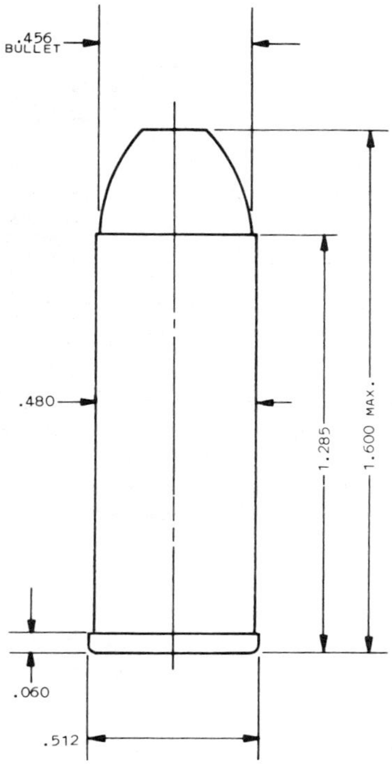

TEST COMPONENTS:

Cases .... Remington; Winchester
Trim-to Length .... 1.280″
Primers .... Winchester 7-111; Remington 2½
Primer Size .... Large Pistol
Lyman Shell Holder .... No. 11
Cast Bullets Used .... (size to .452″ dia.)
#45468, 175 gr.
#454190, 250 gr.
#454424, 255 gr.

TEST SPECIFICATIONS:
(Velocity)

Firearm Used .... Ruger Blackhawk
Barrel Length .... 7½″
Twist .... 1-16″
Groove Diameter .... .451″

Note: #452424 replaces #454424 in all data. It is the same bullet reduced slightly in diameter. Composite #452626 is also interchangeable.

COMMENTS:

Pre-World War II model revolvers normally have a groove diameter of .454″, while post-war models usually run .451″. It is wise to slug your barrel and size cast bullets to as near the exact diameter as possible.

Some sizing dies may not reduce cases enough to hold .451″ - .452″ diameter bullets tightly. An expander ball, not over .450″, should be used when loading .451″ - .452″ bullets.

Case life is very limited when using .45 Colt maximum loads, designed specifically for the Ruger and Thompson/Center Contender firearms. Many cases fired in the Lyman laboratory showed signs of longitudinal splits after only a few reloads.

In order to adhere to the maximum overall length listed, the seating depth of some cast bullets must be watched closely. Bullet #454190 should be seated to crimp on the ogive. Bullet #454424 should be seated to crimp on the forward edge of the first driving band.

Bullet #454424 loaded with Unique powder is popular for this cartridge.

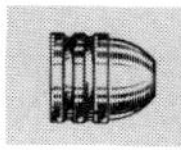

#45468
175 gr., (#2 Alloy) 1.560″ OAL

Powder	Sug. Starting Grains	Velocity F.P.S.	Pressure C.U.P.	Max. Grains	Velocity F.P.S.	Pressure C.U.P.
++Red Dot	4.0	475	—	7.0	885	—
++Green Dot	4.5	490	—	7.5	935	—
++Unique	6.0	595	—	10.0	1035	—
++700X	4.0	435	—	7.0	910	—
++PB	4.5	440	—	7.5	845	—
++SR-7625	6.0	510	—	9.5	925	—
231	7.0	770	—	9.0	1041	—
HS-5	9.5	818	—	11.0	878	—
AL-5	9.0	732	—	11.0	959	—

Note: Loads shown in shaded panels are maximum.
++Designates use of Winchester cases and primers.

#454190
250 gr., (#2 Alloy) 1.600″ OAL

Powder	Sug. Starting Grains	Velocity F.P.S.	Pressure C.U.P.	Max. Grains	Velocity F.P.S.	Pressure C.U.P.
++Red Dot	4.5	535	—	6.5	835	—
++Green Dot	4.5	505	—	7.0	835	—
++Unique	6.0	595	—	‡9.0	875	—
++700X	4.5	535	—	6.0	800	—
++PB	5.0	545	—	7.5	820	—
++SR-7625	6.0	555	—	8.5	890	—
231	*6.5	762	—	7.4	931	—
HS-5	8.0	696	—	9.8	859	—
AL-5	8.0	693	—	9.6	869	—

Note: Loads shown in shaded panels are maximum.
* Designates potentially most accurate load.
‡ Designates factory velocity duplication load.
++ Designates use of Winchester cases and primers.

#452424/#454424
255 gr., (#2 Alloy) 1.575″ OAL

Powder	Sug. Starting Grains	Velocity F.P.S.	Pressure C.U.P.	Max. Grains	Velocity F.P.S.	Pressure C.U.P.
++Red Dot	4.5	550	—	6.0	780	—
++Green Dot	4.5	500	—	6.5	765	—
++Unique	6.0	590	—	8.5	845	—
++700X	4.5	535	—	6.0	785	—
++PB	5.0	530	—	7.0	750	—
++SR-7625	6.0	555	—	8.0	835	—
630	*9.3	650	—	11.5	768	—
HS-6	7.8	630	—	9.5	761	—
AL-7	7.9	647	—	10.0	756	—

Loads for Ruger and T/C Contender only

#45468
175 gr., (#2 Alloy) 1.560″ OAL

Powder	Sug. Starting Grains	Velocity F.P.S.	Pressure C.U.P.	Max. Grains	Velocity F.P.S.	Pressure C.U.P.
Red Dot	7.0	885	—	9.0	1100	—
Green Dot	7.5	935	—	9.5	1098	—
Unique	10.0	1035	—	11.5	1192	—
700X	7.0	910	—	9.0	1159	—
PB	7.5	845	—	10.4	1140	—
SR-7625	9.5	925	—	10.5	1117	—
231	9.0	1041	—	10.0	1114	—
HS-5	11.0	878	—	12.5	1081	—
AL-5	11.0	959	—	*13.0	1107	—

#452424/#454424
255 gr., (#2 Alloy) 1.575″ OAL

Powder	Sug. Starting Grains	Velocity F.P.S.	Pressure C.U.P.	Max. Grains	Velocity F.P.S.	Pressure C.U.P.
++Red Dot	6.0	780	—	8.3	1018	—
++Unique	8.5	845	—	10.2	1081	—
++700X	6.0	785	—	8.0	1039	—
++PB	7.0	750	—	8.5	986	—
++SR-7625	8.0	835	—	9.7	1060	—
630	11.5	768	—	*15.5	1024	—
HS-6	9.5	761	—	14.0	1145	—
AL-7	10.0	756	—	14.0	1064	—

Note: Loads shown in shaded panels are maximum.
++Designates use of Winchester cases and primers.

#454190
250 gr., (#2 Alloy) 1.600″ OAL

Powder	Sug. Starting Grains	Velocity F.P.S.	Pressure C.U.P.	Max. Grains	Velocity F.P.S.	Pressure C.U.P.
++Red Dot	6.5	835	—	9.4	1097	—
++Unique	‡9.0	875	—	11.5	1185	—
++700X	6.0	800	—	9.0	1122	—
++PB	7.5	820	—	10.0	1127	—
++SR-7625	8.5	890	—	10.5	1150	—
231	7.4	931	—	10.0	1088	—
HS-5	9.8	859	—	12.0	1017	—
AL-5	9.6	869	—	*12.0	1003	—

CAST BULLET BALLISTICS

W.T. McDonald and T.C. Almgren

Historical Overview

No one knows exactly when or where the first gun was invented, but the documented history of firearms in the Western Hemisphere extends back more than six and a half centuries. The earliest written records come from both Italy and England in the year 1326. The English records shown drawings of large, vase-shaped guns about as long as a man is tall, resting on wooden platforms and firing large arrows several feet long and several inches in diameter at the head and tail. Strange as it seems, these large arrows were not unusual projectiles for that time. Large crossbow-like weapons called "ballistas" had been used at least from the time of the ancient Greeks to shoot huge arrows and these weapons, together with catapaults throwing huge stones, were the largest engines of war in the Middle Ages. They were used both to besiege and to defend fortified cities and castles in those times. The first guns in Europe seem to have been attempts to improve on the ballista. It is also interesting to note that our word "ballistics" derives from the name of that ancient siege engine.

History shows that arrows were not used very long as firearms projectiles. Small cannon tubes firing lead, iron, or stone balls weighing one or two pounds appeared in the mid 1300's. In the late 1300's the first of the huge bombards appeared. They were gigantic guns with bores many inches in diameter firing stone or iron balls weighing from several hundred pounds to a ton. They were improvements on the catapaults of the Middle Ages, and very effective in sieges against cities and castles. Firearms development proceeded in the other direction as well in that same period, and the first "handguns" (that is, guns that could be carried and fired by a single man) appeared well before the year 1400. Actually, these were hand cannons, with short barrels and bores about an inch in diameter. They were equipped with "tillers" to aid in pointing ("tiller" was the name then used for a crossbow stock), but they were designed to be fired from a city or castle wall or other fixed rest. Another full century would go by before the first rudimentary shoulder stocks would appear. Records from the late 1300's also show that handgunners had all necessary accessories, including bullet moulds for round lead balls, iron casting ladles, iron ramrods, and match materials for lighting charges through touchholes in their handguns. From these records we can conclude that cast lead bullets have been used for at least six centuries, from the time of the very first guns that could be carried and fired by a single man.

Actually, it is not surprising that early cast bullets for the handguns were round balls made of lead. Lead projectiles were used in slings long before the age of firearms. Plentiful and inexpensive, with high density and low melting point, lead was ideal compared to other possible choices (iron, stone, precious metals) in those early times. The spherical projectile shape was also a heritage from antiquity. The sling had been used for thousands of years as a weapon for both hunting and warfare, and it was well known that nearly spherical stones were more accurate than stones with irregular shape. Today we know that the round ball projectile has a relatively poor ballistic shape. Nevertheless, this basic bullet shape is still with us for muzzle-loaders, having survived more than six centuries of progress toward better ballistic shapes.

The round ball, of course, is the only practical projectile shape for smoothbores. If a nonspherical bullet is not spin-stabilized by rifling, it tumbles during flight, destroying accuracy and greatly reducing velocity and energy downrange. Round balls were used in rifles for the first two centuries of rifle development, but after that elongated pointed bullets began to prove better for use in rifled barrels. Improvements in accuracy and range were driving factors in early experiments that led to improved bullet shapes, but bullet design changes also occurred to solve some of the key problems in early rifles, like difficulty in loading, black powder fouling, and sealing the bore against powder gases escaping by the bullet.

Wads between powder and ball to seal the bore, and over the ball to prevent rolling down the bore, were used in the very earliest firearms. These measures were adequate for smoothbore cannons and handguns. When rifles appeared, though, different measures became necessary.

The first rifles came from Vienna around 1500. The idea of a spinning projectile came from the crossbow. Crossbow quarrels (short arrows) were stabilized in flight by feathered vanes which caused them to spin, and spinning quarrels were found to be more accurate than nonspinning ones. Consequently, the rifled bore was a natural idea for guns, and it was tried just about the same time that shoulder stocks were first attached to handguns. The rifle, then, is just about as old as the smoothbore gun fired from the shoulder.

It is an interesting sidenote that the first shoulder arms were given the name "arquebus" (sometimes spelled "harquebus"). At least two possible derivations of this name have been given in recent shooting literature. The one by G.W.P. Swenson in his book *Pictorial History of the Rifle* is particularly interesting. According to Mr. Swenson, "arquebus" derives from "arca bouza," meaning "bow with a hole." This explanation relates the first shoulder-fired guns to their ancestral crossbows.

Compared to the smoothbore arquebus, early rifles were very expensive and very hard to load. Rifling the barrel and holding a true bore diameter from end to end were very difficult mechanical operations required for the rifle but not for the smoothbore. In loading a rifle, the ball either matched the groove diameter or was a little larger, and it had to be driven into the barrel. Driving a ball down a fouled bore with an iron ramrod and a mallet was a slow process, and the final bullet shape seldom resembled a round ball.

These circumstances slowed the development of the rifle for more than two centuries. The greased patch materialized in the early 1600's, and this eased the loading problem. However, people in those days did not understand how rifling spins a bullet. It was believed for some time that a patch ball fired in a rifle would not spin, because the rifling lands in the barrel did not engrave the ball surface. Consequently, the greased patch, which was to become so successful in America in the next century, fell into disuse in Europe. Mainly, though, cost was the factor that kept rifles from being widely adopted for warfare in the seventeenth and eighteenth centuries. Rifles did not see limited use in a very few European armies in the 1600's and 1700's, but was primarily an expensive hunting arm for prosperous gentlemen of those periods. The smoothbore musket was the standard shoulder arm of European armies and armies here in the Americas as well until about 1850. Gunmaking machinery developed in the Industrial Revolution of the mid-nineenth century finally reduced the cost of rifles to the point where smoothbore muskets no longer had an overwhelming edge. Also, the range and accuracy of the rifle had become important military advantages. The English learned this lesson brutally in the American War of Independence, where the Pennsylvania, New York, and New England rifles with patch balls had greatly outclassed the Brown Bess musket in more than a few skirmishes, Also by 1850, the hollow base Minie bullet for rifles had completely erased any loading speed advantage of the smoothbore musket; this bullet design gave even better accuracy than the patched ball.

Experiments with nonspherical bullets in rifles began in Europe and Russia about 1700. The flat-base pointed bullet shape that we are familiar with today seems to have emerged in the early 1800's after a century of experiments. Looking back now, a hundred years seems a long time to accomplish so obvious a result. But little was known about interior or exterior ballistics in the 1700's. Bullets still had to be pounded down rifle bores, with attendent damage to point shapes. (The first breech-loaders were invented specifically to work around this problem.) Proper rifling twist rates were being developed by cut-and-try methods. And gunpowder was being improved and refined during this same period. Because so many things were changing at the same time, it wasn't until the early 1800's that pointed bullets were proved clearly superior to round balls in both range and accuracy.

After the pointed bullet, the next major development in rifle projectiles was the expanding bullet, which finally solved the loading problem for muzzle-loading rifles because the early breech-loaders just couldn't be sealed against escaping gases. The idea of the expanding bullet was simple. The bullet was cast slightly smaller than bore diameter. Then it could be pushed down even a fouled bore very easily, but it was not so small that it would slip forward if the bore were tipped down before firing. When the gun was fired, the powder gases would expand the bullet into the rifling, sealing the bore against gases escaping around the bullet. Although the idea seems simple, more than 25 years were spent developing a practical expanding bullet. The first invention by W. Greener in England in 1835 was a round ball expanded by a tapered wooden plug in a tapered hole in the ball. Delvigne in France patented an elongated hollow-base bullet in 1841. Minie added a tapered iron cup in the base of Delvigne's bullet in 1847. When the charge was ignited, this cup was supposed to be driven more deeply into the base cavity, causing the sides of the base to expand outward. Before long it was discovered that the Minie bullet worked well without the tapered cup, provided that the center of gravity of the bullet design was properly placed.

Although Minie seems to have contributed least in an invention sense, he was successful in getting his expanding bullet adopted by the French, British and United States armies. The Minie bullet as we know it today was fully developed by the time of the American Civil War. Tests in Europe and America proved it more accurate than the round ball. It retained velocity and energy downrange much better, and it shot much flatter. Hundreds of millions of Minie bullets, especially in .58 and .69 calibers, were used in the Civil War. It was a very important projectile development, but it was to be important only for a very short period of time. When the Civil War ended in 1865, Boxer-primed centerfire cartridges and breech-loading rifles and handguns to use them were a scant eight years away.

The period between 1825 and 1875 was a time of great activ ity in firearms and ammunition development in this nation and in Europe. The first metallic cartridges and the first successful breech-loading arms using them were developed during this period. In France in 1835, Casimir Lefaucheux patented his pinfire cartridge system. Lefaucheux's invention was a fully self-contained metallic cartridge with bullet, powder and a percussion cap inside. A pin extended from the side of the base of the cartridge and when this pin was struck by the falling hammer of the gun, the pin exploded the percussion cap, igniting the charge. The Lefaucheux system was popular in Europe and guns, especially shotguns and revolvers, were built for Lefaucheux cartridges until World War I. Some Lefaucheux revolvers of Belgian manufacture saw action in the American Civil War.

In America in 1848, W. Hunt patented a caseless cartridge which he called the "rocket ball". Hunt used a hollow-base lead bullet, filled the cavity with black powder, and sealed it lightly so that the flash from a percussion cap would burn through the seal and ignite the charge. A few repeating rifles were built for Hunt's cartridges. His invention was modified by Horace Smith and Daniel Wesson in the early 1850's to include a primer within each cartridge. This became the Volcanic cartridge of the mid-1850's. The hollow bullet base contained powder and primer, and was sealed by cork, brass or copper. The Volcanic Repeating Arms Company was founded in 1855 by Smith, Wesson, a shirt manufacturer named Oliver F. Winchester and some other investors to manufacture rifles and pistols based on this cartridge invention. Volcanic arms flopped, and the company went bankrupt in 1857. The main cause was weak bullet performance. The hollow bullets simply had insufficient capacity to hold a reasonable charge. It is doubtful that a Volcanic bullet ever achieved more than about 600 fps muzzle velocity, and very light bullets had to be used to achieve that much velocity. Also, the breeches of Volcanic guns could not be sealed against escaping powder gases.

Despite these failures, the Volcanic experiment was important. The cartridge was the first caseless type invented in this country. The tubular magazine Volcanic repeater was basically a sound mechanical design. And Oliver F. Winchester was able to salvage the best ideas and go on to produce the Henry rifle and a long line of Winchester repeaters.

The first rimfire cartridge, the .22BB cap, appeared in Europe about 1845. The rimfire ignition system propagated to America well before the Civil War, and breech-loading rimfire repeaters (the .56-.56 Spencer, .44 Henry, and .56-.50 Ball (Spencer cartridge)) saw action in that War. However, rimfire cartridges of that period were also weak and ballistic performance was poor, compared to the standard .58 caliber rifled musket. This muzzle-loading rifle backed a 505 grain Minie bullet with 60 grains of black powder to produce 1000 fps muzzle velocity. The Spencer rifle with the .56-.56 Spencer cartridge propelled a 450 grain bullet with 40 grains of black powder, for a muzzle velocity of about 900 fps. The .44 Henry rifled delivered a 200 grain round-nose or flat-nose bullet at about 1100 fps, using 28 grains of black powder. The .56-.50 Spencer cartridge used a 350 grain bullet and 45 grains of black powder to produce about 1075 fps muzzle velocity. The mechanical strength of the folded-head rimfire cases would not permit stronger loads.

Although the rimfire repeaters gave relatively poor ballistic performance, they had tremendous firepower compared to the muzzle-loading arms, and this advantage was well understood at the end of the Civil War. It was also obvious from Civil War experience that the metallic cartridge had solved the gastight breech problem. It was clear in 1865 that the breech-loading firearm with metallic cartridges was practical and manufacturable, but it was also clear that rimfire cartridges were not the right answer for either military or civilian needs at that time. Not only were rimfire loads weak, but rimfire cartridges could not be reloaded. This was a major drawback on the early frontier where guns meant survival and factory-loaded ammunition could be obtained only at widely separated places.

The right answer was the centerfire cartridge, which was developed in a seven or eight year period following the Civil War. The year 1873 marked a revolution in firearms development in this nation. That was the year the U.S. Army adopted the Trapdoor Springfield .45-70, Winchester introduced the Model 1873 .44-40 rifle and Colt brought out the Single Action Army .45 revolver – all using brand new centerfire black powder cartridges.

The centerfire era started with basic lead bullet shapes developed for muzzle-loaders and rimfire cartridges. The round-nose flat-base shape was the first to be used, and inside lubrication was also used from the very beginning. The flat-nose bullet shape developed very early, and this is attributed to the needs of the early tubular magazine rifles. Round-nose and flat-nose bullets helped introduce the first centerfire cartridge guns in this nation.

Within the first few years of the centerfire era, the development trend was toward large bores, very heavy bullets and huge case capacities. The .45-70-405 is the best known example, with a 405 grain bullet at about 1320 fps muzzle velocity. But the .45-70 is mild compared to the really big black powder cartridges, like the .45-120-500 Sharps (120 grains of powder, 500 grain bullet, 1520 fps muzzle velocity), the .50-140-700 Sharps (1355 fps muzzle velocity), and the .50-140-473 Winchester Express (1580 fps muzzle velocity). Basically, muzzle velocities for black powder cartridges were limited to about 1600 fps, regardless of caliber. At these velocities, large heavy bullets were necessary to retain velocity, energy, accuracy and wind-bucking capability at long ranges. The big cartridges were popular with both hunters and target shooters until smokeless powder removed the velocity barrier near the end of the century.

Hunters found, though, that the heavy solid lead bullets often had too much penetration and too little expansion for medium game animals. Hollow-point bullets then became popular for some of the big express cartridges when they were to be used on medium game. The hollow-point shape had two advantages. Hollow-point bullets were lighter than solid bullets of the same overall dimensions, and therefore had less momentum and penetration. The hollow-point also caused "mushrooming" very effectively at the low velocities of black powder cartridges.

Hollow-base bullets also made the transition to the centerfire era. Of course, it no longer was necessary to size bullets smaller than groove diameter in order to load the gun, the hollow-base bullets were used in a few factory handgun cartridges and by some handloaders for both rifles and handguns. These handloaders felt that the hollow-base shape gave a final degree of expansion and seal between bullet and bore, with a consequent improvement in accuracy. This bullet style has been especially popular with handgunners through the years. For example, Ed McGivern, the famous quick-draw, rapid-fire pistol shot of the 1920's and 30's, designed the Lyman 358395 hollow-base wadcutter and maintained that it was an excellent midrange bullet in the .38 Special. Elmer Keith designed several Lyman bullets, and among them were the 358431 and 429422 hollow-base semiwadcutters in .38 and .44 caliber.

Colt used the hollow-base bullet in at least three handgun cartridges. These were the .32 Long Colt (bullet diameter .299, groove diameter .313), .38 Long Colt (bullet diameter .357, groove diameter .375), and .41 Long Colt (bullet diameter .386, groove diameter .401). These cartridges were originally brought out in black powder days with a heeled bullet shape, with no hollow-base. The forward end of the bullet bearing surface matched the groove diameter, and a rebated heel at the base of the bullet entered the case mouth. When the cartridges were converted to smokeless powder, the hollow-base bullets were adopted so that they could be inside-lubricated. Short versions of the three cartridges (.32 Short Colt, etc.) used outside-lubricated heeled bullets even with smokeless powder loads. All these cartridges have been obsolete for years, but both hollow-base and heeled bullet moulds were available from Lyman until a short time ago.

Other basic cast bullet shapes that we recognize today were developed after the turn of this century. These are the very popular wadcutter and semiwadcutter shapes for handguns, and the long ogive spitzer shape for rifle bullets. The spitzer shape has not been as highly successful in cast bullets as in jacketed bullets for rifles, but it does perform well in medium calibers when properly loaded.

A practical smokeless powder was invented in France in 1885 by a chemist named Vielle, and in 1886 his powder was adopted for the French military 8mm Lebel cartridge. In 1892 the U.S. Army adopted the smokeless .30 Army (.30-40 Krag) cartridge and the Krag-Jorgensen bolt action magazine rifle. By 1895 the move to smokeless powder was well underway in the commercial market as well. The famous .30-30 Winchester was America's first commercial smokeless cartridge. It appeared with the Model 1894 rifle in 1895.

Smokeless powder broke the velocity barrier of black powder. Muzzle velocities over 2000 fps in medium caliber rifles became possible immediately with smokeless powder, and lead bullets were just not adequate for those velocities. Bore leading caused by the hot smokeless powder gases became very severe. Jacketed bullets were invented to solve these problems. Credit for the invention goes to Captain Rubin of the Swiss Army, in connection with the 7.5 x 54mm Swiss Army cartridge which he invented in 1889. Jacketed bullets were supposed to be the deathknell of lead bullets at the turn of this century, but that prediction turned out to be false. The lead bullet for both rifles and handguns is still going strong 80 years later, and it seems destined never to die.

Many, many cast bullets for rifles and handguns were designed between the two World Wars and in a 15 year period after World War II. Lyman entered the reloading equipment business in 1925 by purchasing the Ideal Reloading Tool Company. A selection of cast bullet designs came with that purchase, and in succeeding years the Lyman staff designed many more. Also, nearly every prominent gunwriter and many shooters designed their own cast bullets. And it seems like Lyman built the moulds for almost all of them! Lyman's *Handbook of Cast Bullets* of 1958 listed no fewer than 402 available bullet moulds. There were 71 moulds available in .30 caliber, and at least 35 moulds were listed especially for the .38 Special or .357 Magnum. Almost every possible variation and combination of bullet shapes for each caliber were represented among the moulds available at that time.

Since about 1960 there has been a general tendency to consolidate around basic, popular cast bullets in each caliber, and to eliminate a large number of rarely used bullet designs. This trend is true not only at Lyman, but within the general shooting fraternity as well. Also, the last 20 years have seen a very dedicated cast bullet research activity by Colonel E.H. Harrison, USA (Ret'd), and other staff members of the National Rifle Association. This activity has produced an NRA book entitled *Cast Bullets* by Col. Harrison. This book documents his experience with bullet designs, alloys, casting procedures, sizing, lubricants and ballistic performance. Among other experiments, Col. Harrison has reexamined paper-patched bullets. In the late 1800's paper-patched bullets were used to obtain high accuracy. While the paper-patch gripped the rifling firmly and sealed the bore effectively, it also acted as a lubricant and even cleared away black powder fouling from the previous shot. In today's smokeless cartridges, paper-patched bullets also deliver high accuracy and permit a modest increase in muzzle velocity as well. Two of Col. Harrison's paper-patched bullet designs for .30 caliber rifles are included in Lyman's present line of bullet moulds.

Recent research activities at Lyman have also led to improvements in bullet designs, lead alloys for casting, casting techniques, lubricants, and loads. In turn, these have produced improvements in muzzle velocity, accuracy and downrange ballistic performance. Today's cast bullets in modern rifles are capable of minute of angle accuracy, and muzzle velocities over 2500 fps can be achieved in medium calibers. Loading cast bullets presents an exciting challenge to the handloading shooter which just doesn't exist with jacketed bullets. This challenge to handloading skill perhaps is the chief reason why cast bullets for rifles will never disappear.

Cast Bullet Performance

In current shooting literature, cast bullet performance is almost always compared to jacketed bullet performance, and the universal standard of comparison seems to be muzzle velocity. It isn't terribly surprising that cast bullets lose out in this comparison. In a .30-06, for example, it just isn't possible to drive a 150 grain cast bullet at 3000 fps, like a 150 grain jacketed bullet can be driven, and this fact just has to be accepted. But there are other important standards, too. And, there are ways of formulating comparisons based on quantitative measures rather than qualitative opinions. Although it cannot be claimed that cast bullets outperform jacketed bullets, cast bullet performance by these quantitative measures is surprisingly good.

It is also tempting to introduce cost as a factor in these performance comparisons. At today's prices high quality jacketed bullets in small calibers cost at least five cents each, and cost goes up appreciably in larger calibers. Lead bullets cast by the handloader cost much less, even when he has to purchase the raw materials. If we could compare performance on a per dollar basis, there is little doubt that cast bullets would win. However, we'll avoid this temptation, and just look at a few pure performance factors in the paragraphs below.

Accuracy. Cast bullet accuracy depends very critically on the skill and patience of the person doing the casting, sizing, and loading. The casting process must produce bullets as nearly perfect as possible, without shape imperfections and without inclusions of the slag or dross that accumulates on the top surface of the molten lead. The bullet mould must cast bullets no more than about .003 inch oversize, so that no major reworking of the bullet shape takes place in the sizing die. Bullets must be sized and lubricated with one of the excellent modern lubricants. And the shooter may need to experiment considerably to find the best accuracy load for each gun he shoots with cast bullets.

If this seems like a long process, recall that it is no different that what our forefathers did. They didn't have all the advantages of modern technology, but they managed to shoot some pretty tight groups. Accuracy legends are legion from the regional, national, and international matches of the mid to late 1800's, and there are records of phenomenal scores by riflemen from America and Europe. Morgan James of Utica, New York, fired a nine-shot group of 0.38 inch at a range of 110 yards with a .45 caliber benchrest *muzzle-loader* in 1859. He used a composite bullet with a cast pewter point. Legend has it that Mr. Morgan was afraid to fire the tenth shot. The English specialized in long range shooting in those days. One of the best scores ever made was fired by Mr. G.C. Gibbs of Wistow in 1886. He placed 48 out of 50 consecutive shots within a 3-foot bullseye at 1000 yards with a .461 bore rifle. The other two shots were within a one-foot ring surrounding the bullseye.

These are illustrations of what lead bullets could do a century ago. Today, *good* cast bullets will shoot into a minute of angle (1 inch at 100 yards) very consistently. A little more care, selection, and load experimentation can improve cast bullet accuracy even beyond that.

Rainbow Trajectories. Cast bullets are often criticized severely because their trajectories have more drop than jacketed bullets, and the "rainbow trajectory" is supposed to be a real drawback to cast bullets. This criticism is true to some extent, but it is not as bad as it is frequently made out to be. Trajectory flatness depends upon two effects acting together. First, the bullet must be fired at a high velocity, and then it must be able to retain a high percentage of this velocity as it flies downrange. It therefore needs a high ballistic coefficient. The flattest shooting rifles in the world are the medium bore (7mm and .30 caliber) magnums because they can be loaded to very high muzzle velocities and jacketed bullets in these calibers have excellent ballistic shapes. These loads are usually the standards to which cast bullet trajectories are compared. However, there is a very practical and useful way to measure trajectory flatness. We are going to show below that cast bullets can shoot a good deal flatter than most people suspect if this method is used to get the most out of them.

Trajectory flatness really has little significance for target shooting. Target ranges are known exactly, and target guns can be sighted in well before competition begins. On the other hand, trajectory flatness is very important to the hunter. His game may step out at any range, and in the field ranges are often hard to estimate. Point blank range is a concept used for many years by hunters. The point blank range of any gun is the distance out to which a hunter can hold right on his game and be able to hit within a vital zone of the animal. As long as the animal is within the point blank range, the hunter doesn't have to hold high or low to correct for bullet trajectory. It turns out that the point blank range of any load can be maximized for a given size of game animal by simply choosing the right zero range for the gun. When this is done, the hunter gets best performance from his gun, and the point blank range is a very practical *quantitative* measure of how flat the gun can really shoot.

Point blank range depends on the size of the game animal, because the size of the vital zone varies from one animal to another. On a deer-sized animal (see **Figure 1**) the vital zone would be centered about the middle of the front shoulder, when the animal is viewed broadside. The top of the zone would be at the top of the shoulder, where a hit would break the shoulder or backbone of the animal. A hit at the bottom of the vital zone would break the front leg or be in the critical lung cavity area. The vital zone vertical dimension is then about 10 inches for medium game animals. On a varmint-sized animal the vital zone might be only 3 or 4 inches in height, and on a big game animal it might extend 15 inches.

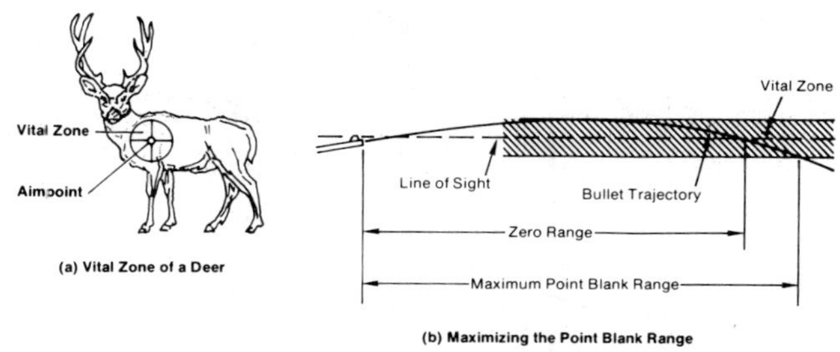

Figure 1. The Point Blank Range Concept

Figure 1 shows how to get maximum point blank range. If the hunter aims at the center of the vital zone, his bullet must not rise higher than the top edge anywhere in its trajectory. If the bullet does rise above the line marking the top of the vital zone, the hunter will shoot high if the game steps out at short range. When the bullet crosses the line marking the bottom edge of the vital zone, the crossing point is the point blank range for the load, as shown in **Figure 1**. The point blank range is maximized when the bullet rises just far enough to touch the line marking the top edge of the vital zone, and no higher or lower.

In a practical sense, a hunter maximizes point blank range for his gun and intended game by choosing a zero range that makes his bullet trajectory rise above his line of sight by an amount equal to half the vital zone vertical dimension. This can be done by using trajectory midrange height data from ballistics tables for his load. An example of how this is accomplished is given near the end of this article, but the procedure is as follows. First, the hunter figures out the vital zone vertical dimension he needs to use. Then in the ballistics table for his bullet and muzzle velocity, he finds a midrange height figure which equals half the vital zone dimension plus a correction for the height of his sights above the bore centerline. For iron sights this correction is 0.4 inch, and for a telescope sight the correction is 0.8 inch. For example, if the hunter is going after deer or other medium game, the midrange value he needs to find in the table is 5.4 inches for iron sights, or 5.8 inches for a telescope sight.

Once the right midrange height has been found in the table, the *range value* where the correct midrange occurs is the correct zero range for the load. This is the zero range that the hunter should use to sight in his rifle for the hunt. The maximum point blank range is about 40 yards farther than this zero range. Since ballistics tables list data only at certain range values, an estimation often must be made between points in the tables. The example given later explains this estimation procedure.

Figure 2 illustrates how large the maximum point blank range turns out to be for medium game at muzzle velocities appropriate for cast bullets. The three curves in the figure are for there values of ballistic coefficient which just about span the range found for cast bullets, from round balls to large caliber pointed bullets. The curves show that muzzle velocity has a large effect on maximum point blank range, since all three curves increase steeply as muzzle velocity grows. This means that, for any bullet, the higher the muzzle velocity, the longer the point blank range. Also, the curves show that for any given muzzle velocity, the higher the bullet ballistic coefficient, the longer the point blank range.

Now bear in mind that the .300 Winchester Magnum shooting the Sierra 165 grain spitzer boat tail jacketed bullet at 3200 fps muzzle velocity has a maximum point blank range of 380 yards for medium game. This same bullet in a .30-06 at 2700 fps muzzle velocity provides 330 yards maximum point blank range. The Magnum then has a 50 yard edge on the '06. The venerable .30-30 with a 150 grain jacketed flat-nose bullet at 2300 fps muzzle velocity has 230 yards maximum point blank range. These figures are from the Sierra Bullets Reloading Manual, Second Edition, and they pretty well exemplify the point blank ranges obtainable with jacketed bullets.

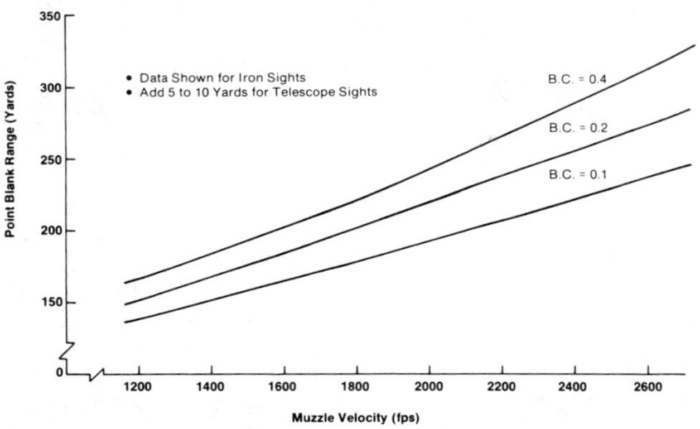

Figure 2. Maximum Point Blank Range for Cast Bullets on Medium Game (10 inch Vital Zone)

In comparison, Lyman's 301618 paper-patched 160 grain cast bullet loaded to 2400 fps at the muzzle (ballistic coefficient .310) will provide a maximum point blank range of 270 yards for medium game, as determined from **Figure 2**. Lyman's 151 grain round-nose 311466 bullet loaded to 1800 fps in a .30-30 (ballistic coefficient .250) will provide 210 yards. And the 420 grain 457193 bullet loaded to 1500 fps (ballistic coefficient .335) in a .45-70 will provide about 185 yards maximum point blank range on medium game.

These figures show pretty clearly that, while cast bullets can't quite match the point blank range performance of jacketed bullets in high power rifles, cast bullet performance is still pretty respectable. Nothing matches the flat shooting capability of the magnum cartridges with jacketed bullets loaded to maximum velocities. Nevertheless, cast bullet point blank ranges for medium game can approach 300 yards, while jacketed bullets in standard (not magnum) cartridges can approach 350 yards. So jacketed bullets have an edge, but it's not overwhelming.

Penetration and Expansion. Much controversy exists within the shooting community concerning just exactly what gives a game bullet killing power. Everyone agrees that a bullet must have the right combination of penetration and expansion within a game animal, but the controversy revolves around whether the bullet's momentum or energy is more important when it enters the animal.

Regarding penetration and expansion, it has been said that the ideal bullet for a medium to large game animal would penetrate the skin and any light bones like ribs; expand as it travels through the soft internal organs to a diameter about twice its caliber, and stop just under the outer skin on the offside of the animal. This would make sure that all the energy and momentum are spent within the animal, and minimize the meat damage. For small game or varmint animals the objectives are a little different. If neither the pelt nor the meat is valuable, a very rapidly expanding bullet which fragments within the small animal will cause instantaneous death. If damage to either meat or pelt is to be minimized, a solid bullet with little expansion is probably the best choice.

Expansion and penetration in jacketed game bullets are controlled by a combination of jacket thickness, jacket internal construction and muzzle velocity. There are thin jackets for varmint bullets, thick jackets for heavy game bullets, tapered jackets to control expansion rate, partitioned bullets to arrest expansion, etc. In a sense this is an advantage for jacketed bullets, because an optimum ballistic shape can be adopted for each bullet weight in each caliber, and then specific bullet designs can be "tailor-made" by selecting the right jacket designs. However, jacketed bullets have well known expansion problems, all of which happen because the handloader doesn't load properly for the bullet he selects. A frequent example is the magnum rifle shooter who buys a medium heavy bullet in his caliber (like a 160 grain spitzer for the 7mm Remington Magnum), loads it to maximum velocity so it will "shoot flat to 400 yards," and then manages to shoot a deer 50 or 75 yards away. Very often the bullet shatters when it enters the animal because of its super velocity; the hydrostatic shock causes instantaneous death, but about half the meat is ruined. Jacketed bullets also fail to expand at low velocities sometimes. This occurs in both rifles and handguns. An example often cited in the shooting literature is the jacketed hollow-point pistol bullet. Despite the hollow-point, these bullets do not expand reliably at velocities below about 1000 fps.

Lead bullet penetration and expansion are somewhat different. It was found a century ago that lead bullet penetration is related very closely to bullet momentum. Today it is difficult to find quantitative penetration data for any kind of commercial bullets, but a considerable number of penetration measurements using pine boards were made around the turn of the century. **Figure 3** is a plot of lead bullet penetration data listed in the *Ideal Handbook Number 17* of 1906. These data are for 31 cartridges ranging from the .22 Winchester Rim Fire with a 45 grain bullet at 1383 fps to the .50-110 Winchester with a 450 grain bullet at 1383 fps. All the cartridges had either round-nose or flat-nose bullets, and nothing is known about hardness of the bullets tested. Although there is much scatter in the data points in **Figure 3**, the curve, which is a least-squares fit to the points, shows the clear trend. Penetration increases rapidly with bullet momentum. It also turns out that heavy slow bullets have more momentum than light fast ones, as a general rule. Therefore, in any given caliber a heavy lead bullet will have more penetration that a light bullet, and larger calibers have more penetration than smaller calibers.

It also was found many years ago that lead bullets expand very well at moderate velocities, and it was found that the hollow-point is a simple remedy to underexpansion. In past years a great many cast bullet moulds were available with and without hollow-points, so that a shooter could pick a bullet design he liked and use either hollow-points or solids depending on the game he expected to shoot. Nowadays, cast bullets are fired at velocities well above those in black powder days. At modern day velocities expansion is usually adequate with solid lead bullets. Hollow-points sometimes overexpand, fragmenting within the game animal. For this reason, hollow-point bullets are becoming less popular in larger rifle calibers.

Jacketed bullets can be made to have several times the penetration that lead bullets have, but this is no advantage in most hunting situations. Cast bullets in appropriate calibers have fully adequate penetration and expansion for light and medium game, and lead bullets in large calibers have worked pretty well on heavy North American game animals as well.

Momentum and Energy. Physically, momentum is the mass of a bullet multiplied by its velocity. The mass of a bullet is its weight in pounds divided by the acceleration due to gravity. since bullet weights are given in grains, we first have to convert grains to pounds. The mass calculation is expressed by the following formula:

$$M = \frac{W}{7000 \times 32.174} = \frac{W}{225218}$$

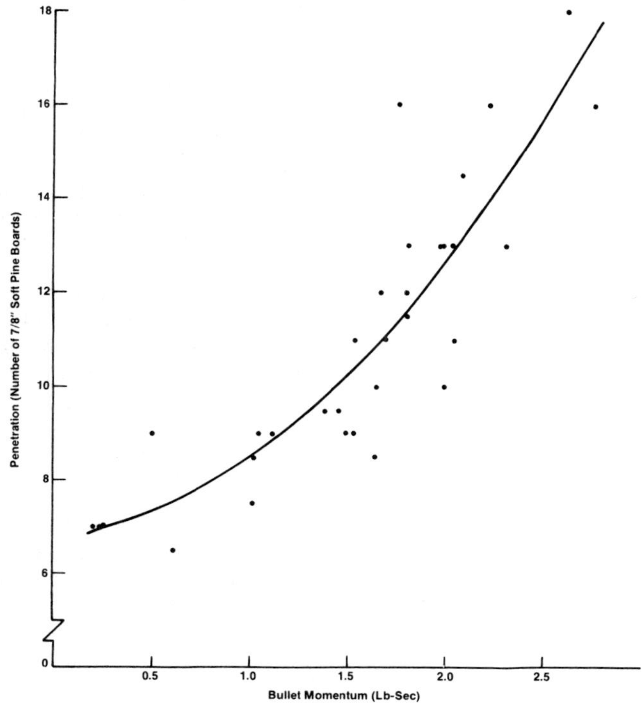

Figure 3. Lead Bullet Penetration versus Momentum

where **M** is the bullet mass and **W** is bullet weight in grains. The factor 7000 converts grains to pounds, and 32.174 ft/sec/sec is the acceleration due to gravity. After mass has been calculated, momentum is found by multiplying mass times velocity:

$$\text{Momentum} = M \times V \quad (\text{lb-sec})$$

where **V** is bullet velocity in fps. The units of momentum are pound-seconds (lb-sec).

Bullet kinetic energy is bullet momentum multiplied by half the velocity:

$$\text{Energy} = \text{momentum} \times \frac{V}{2}$$

$$= \frac{W \times V \times V}{450436} \quad (\text{ft-lbs})$$

where again **W** is bullet weight in grains and **V** is bullet velocity in fps at any point in flight where energy is to be calculated. There is also a small amount of kinetic energy associated with a bullet's spinning motion. However, this is negligible compared to the kinetic energy associated with its linear velocity, and it is almost never considered. The units of kinetic energy are foot-pounds (ft-lbs).

Momentum is sometimes described as the ability of a bullet to "push" what it strikes. This notion is not physically precise, but it is reasonably sound, because when a bullet strikes an animal it imparts a momentum to the animal. The problem with the notion is that television and motion pictures have led us to believe that raw bullet momentum can knock a man several feet through the air. This just isn't correct.

In order to get a feeling for momentum effects, consider that a baseball thrown by a professional pitcher can cross the plate at speeds near 100 mph (146.7 fps). A professional baseball must weigh between 5.0 and 5.25 ounces. The momentum of a 100 mph fastball calculates out at 1.42 lb-sec. When a batter has the misfortune to be struck by a pitched ball, he isn't knocked several feet through the air. Now, compare this baseball momentum with the muzzle momenta of some popular handgun and rifle factory loads:

.38 Special +P, 150 grains, 1090 fps	0.765 lb-sec
.44 Magnum, 240 grains, 1470 fps	1.566
.45 Auto, 230 grains, 850 fps	0.868
.243 Winchester, 100 grains, 3070 fps	1.363
.308 Winchester, 150 grains, 2860 fps	1.905
.30-06, 150 grains, 2970 fps	1.978
.45-70 Gov't, 405 grains, 1320 fps	2.374
.300 Winchester Magnum, 180 grains, 3070 fps	2.454

These figures make it clear that bullet momentum just doesn't bowl over people or game animals.

In a physical sense, energy is the bullet's ability to do work, that is, to impart kinetic energy or heat energy to the target it strikes. While bullet momenta are relatively small numbers, ranging from about 0.2 to 5.0 lb-sec, bullet energies are large numbers, ranging from around 100 to about 5000 ft-lbs.

No one knows for sure whether energy or momentum, or some combination of both, gives a game bullet killing power. More quantitative research on this subject has been done in Europe than in this country, but gathering factual experimental data is extremely difficult at best. Since a bullet cannot have energy without momentum, or vice versa, it's clear that the answer has to be an appropriate combination of the two. For instance, bullets with lots of energy and little momentum (poor penetration) are poor performers on medium and heavier game. Bullets with tremendous momentum overpenetrate many kinds of game, spending significant fractions of their energies on the surrounding countryside, and this is the other extreme of poor performance.

When we compare momentum and energy performance of cast bullets and jacketed bullets, a couple of general observations are evident immediately. First of all cast bullets can hold their own in momentum performance, at least to the limit where overpenetration becomes a problem. This is because a heavy slow bullet can have more momentum than a lighter faster one. But, since jacketed bullets have higher muzzle velocities, they tend to have higher energies than cast bullets.

Figures 4 and **5** verify these observations. Four cast bullets which are among the best performers in .270, .30 and .45 calibers have been chosen. Achievable loads with these cast

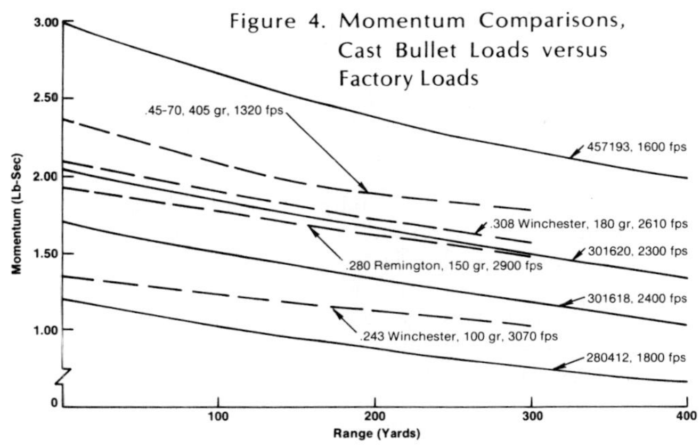

Figure 4. Momentum Comparisons, Cast Bullet Loads versus Factory Loads

bullets in cartridges like the .270 Winchester, .308 Winchester, and modern .45-70 are compared with published performance data for four factory cartridges which fairly well span the range of momentum and energy performance in standard (not magnum) loads for medium game. The two figures show performance versus range to the target, since a bullet's ability to retain energy and momentum downrange is very important for hunting. **Figure 4** shows that the heavier larger caliber cast bullet momenta equal or surpass the factory loads. It also shows clearly that a heavy slow bullet has higher momentum than a light fast bullet. **Figure 5** shows, though, that the energy performance of the best two cast bullets falls well below the performance of the best two factory cartridges. This figure also shows that light fast bullets tend to have more energy than heavy slow ones..

If we ask whether cast bullet performance is adequate for North American game animals, the answer is an unequivocal "yes" for all but perhaps the most dangerous. Certainly, lead bullets have killed every species of game on the earth. However, for the largest and most dangerous game modern day cartridges can be loaded to higher performance levels with jacketed bullets than with cast bullets. For medium game, though, a steadily growing number of hunters use cast bullets today because performance is adequate and cost is low. And within the target shooting community, cast bullets will always have an appeal because they provide the highest personal challenge to the shooter who handcrafts his own ammunition.

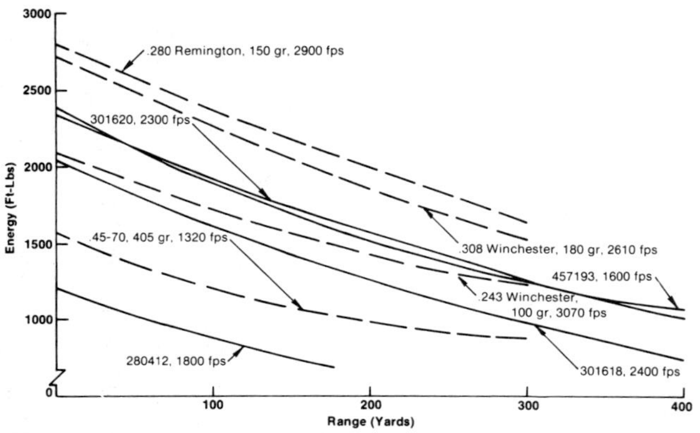

Figure 5. Energy Comparisons, Cast Bullet Loads versus Factory Loads

Ballistic Coefficients

In this handbook Lyman continues the tradition of providing shooters with measured ballistic coefficients for cast bullets, and then using these ballistic coefficients to compute the most accurate and comprehensive ballistics tables available. This tradition began with Lyman's *Black Powder Handbook* and continued through the *Pistol & Revolver Handbook* and the *Lyman Centennial Journal* to the present volume. In total, Lyman has assembled within these four handbooks the largest amount of ballistic coefficient data ever published for cast bullets, and all have been determined by actual firing tests.

As explained in the earlier handbooks, ballistic coefficient is a factor that relates the deceleration caused by drag on an actual bullet to a standard drag model established years ago for trajectory computations. Because the shape of each actual bullet differs from the "standard bullet" for which the standard drag model applies, no single value of ballistic coefficient can match actual drag deceleration to the model over the *entire range* of bullet velocity. Tests have shown, though, that a specific value of ballistic coefficient can match the drag effects over a restricted velocity range for any bullet. In this age of computers, it is easy to change the value of a bullet's ballistic coefficient as velocity changes in the trajectory computation, and very accurate ballistics can be calculated by this method for all bullets. Three values of ballistic coefficient are used for each bullet, one for high velocities, one for intermediate, and one for low velocities.

Lyman conducted a firing test program to determine the three ballistic coefficient values for each bullet. Previous experience has shown that firing tests are absolutely necessary to determine ballistic coefficients for cast bullets, because methods for calculating these coefficients from bullet shapes produce results with errors quite large for cast bullets. The ballistic coefficient for any test bullet can be determined if muzzle velocity and time of flight over a known range are measured simultaneously. Lyman set up a firing test range with three photoelectric screens and two electronic counters. Two screens, separated by 10 feet, were arranged to measure counters. Two screens, separated by 10 feet, were arranged to measure muzzle velocity at a distance of 15 feet from the muzzle of each test gun. One counter measured the time of flight of each bullet between the two muzzle screens, and the 10 ft distance divided by the measured time is the muzzle velocity of the test shot. The third screen was placed 50 yards downrange from the first screen; the second counter measured time of flight of each test bullet over that range.

Three muzzle velocity values were chosen for each cast bullet type to be tested; one high value, one intermediate and one low. Five test shots were fired with each bullet at each muzzle velocity (total of 15 shots for each bullet type). A digital computer was then used to calculate the ballistic coefficient for each test shot. This calculation involves computing a 50 yard trajectory for each test bullet fired, using the measured muzzle velocity and determining a ballistic coefficient which makes the computed time of flight match the measured time of flight. This calculation produces the ballistic coefficient value which the test bullet must have had. A few "bad flyers" were found, and these were edited out before the final ballistic coefficient assessment was made. These bad points were probably due to either bullet shape imperfections or an inappropriate rifling twist for some bullets at certain test velocities used. This editing was not arbitrary. Points were edited out only if reasonable explanations existed for the unreasonable performance.

There are two theories about determining ballistic coefficient values from such a collection of test data. One theory says that if all test bullets were perfect, all results for one bullet type at one velocity level would be identical. Then, since imperfections reduce ballistic coefficient, the correct choice is the highest test value found.

Lyman feels, though, that a shooter really wants to know the *reasonable* performance he can expect from good bullets, and not performance based on bullet perfection. Lyman personnel used the kind of reasonable care in casting the test bullets that a shooter can be expected to use. Consequently, the good test values of ballistic coefficient at each velocity level were *averaged* (in a least-squares sense since there were small shot-to-shot variations in muzzle velocity at each level).

After this analysis was completed, the trend in the average ballistic coefficient value as a function of velocity was examined for each bullet type. Appropriate velocity boundaries between high, intermediate, and low velocity ranges for ballistic coefficient values were established for each type of bullet. A ballistic coefficient value for each velocity range was then determined by interpolating or extrapolating the test data.

Table 1 shows the results of this analysis for the 45 cast bullets in this handbook. The table shows the velocity range boundaries and the ballistic coefficient value in each range for each bullet type. The values in this table have been used in Lyman's digital computer program which calculated the ballistics tables in this handbook.

Using the Ballistics Tables

A brief explanation of the terms used in the Ballistics Tables to follow is given below. Many shooters are already familiar with these terms. For those who are not, the explanations may be helpful in improving their skills. At the end of this section two procedures are explained which all shooters may find useful. The first is how to use the tables to figure where your gun will shoot if you choose a different zero range than the one in the tables. The second is the method to be used to maximize point blank range for your load.

Trajectory. The trajectory of a projectile is the actual path which that projectile follows after leaving the muzzle of a firearm. **Figure 1** below illustrates a projectile trajectory. As soon as the projectile leaves the muzzle of the gun, gravity causes it to begin to fall away from its line of departure. This causes the drop discussed later. The line of departure is an imaginary line extending along the bore in the direction that the bullet travels. Muzzle velocity, air drag, and gravity are the major contributors to shaping the trajectory. If a crosswind blows, the projectile trajectory will also be cured in the direction of the wind, that is, it will curve into the paper or out from it in **Figure 1.**

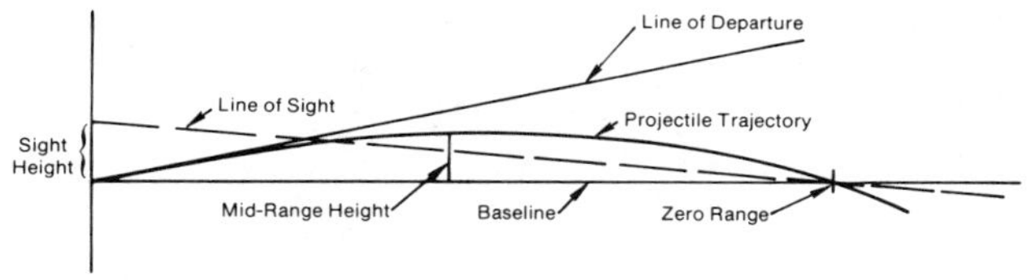

Figure 1. Parts of a Trajectory

In order to give some specific examples of the meaning of terms in the Ballistics Tables, let us suppose that we are shooting the Lyman paper-patched 301618 bullet loaded to a muzzle velocity of 2400 fps. The numerical data from the ballistics table for this bullet at this muzzle velocity will be used to illustrate the following discussion.

Velocity. Velocity is the speed of the projectile as it moves along its trajectory. Velocity is measured in feet per second (fps). The projectile moves slower and slower as it flies farther and farther from the muzzle, because air drag slows it down. The velocity of the projectile is listed in the ballistics table at each value of range from the muzzle. For example, our .30 caliber bullet fired at 2400 fps muzzle velocity has a remaining velocity of 2265 fps at 50 yards from the muzzle, 2135 fps at 100 yards, 2009 fps at 150 yards, and so forth, as shown in the ballistics table.

Energy. A moving bullet has kinetic energy, that is, energy of motion. Ordinarily, the term "kinetic" is dropped, and we speak simply about bullet energy. Energy is measured in foot-pounds (ft-lb), and it is one of the factors indicating the killing power of a bullet. The method for calculating bullet energy is explained in the preceeding section of this article.

Turning to the ballistics table for our example 301618 bullet at 2400 fps muzzle velocity, we see that the muzzle energy is 2046 ft-lb. At 50 yards it drops to 1823 ft-lb, and at 100 yards to 1619 ft-lb. At longer ranges energy continues to drop off, and this happens because bullet velocity grows progressively smaller.

Drop. When a projectile leaves the muzzle it begins to drop, just as it would if we were to hold it up and then let it go. It drops with just about the same speed when it is fired as it would if we dropped it. (There are some aerodynamic forces acting on the projectile when it is fired which keep it from falling quite as fast, and these have been taken into account in calculating the Ballistics Tables.)

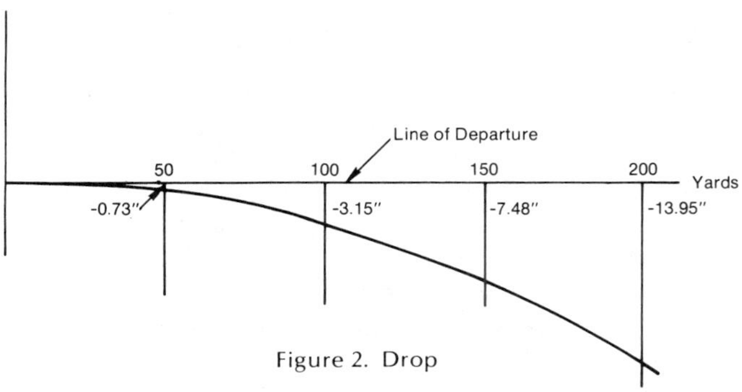

Figure 2. Drop

If we sit on the side of a hill and fire our rifle such that the line of departure (bore centerline) is level, then **Figure 2** shows how far below the line of departure our example bullet will be at each specific range. The line of departure is an imaginary line extending through the bore of the gun and is the line on which the bullet leaves the muzzle. Drop is usually measured in inches, although at long ranges from the muzzle it can grow to several feet. The minus signs on the drop values in the table signify that the trajectory passes below the line of departure.

When we zero in a rifle or handgun, we adjust the sights to tilt the barrel up (elevate the line of departure). When we have just the right amount of tilt, the trajectory is rotated upward to the point where the projectile just crosses the level baseline at the desired zero range. Then, the line of departure is above the baseline by just the amount of drop at the zero range.

Table 1. Ballistic Coefficient Values for Lyman Cast Bullets

Bullet	Weight (Grains)	Velocity Range Boundaries		Ballistic Coefficient Values		
		High to Intermediate (fps)	Intermediate to Low (fps)	High Velocities	Intermediate Velocities	Low Velocities
225107	37	2400	1800	.091	.099	.108
225438	41	2400	1800	.094	.107	.122
225415	45	2400	1800	.116	.140	.168
225462	54	2400	1800	.107	.131	.152
245496	83	1600	1200	.202	.215	.230
245497	90	1600	1200	.230	.240	.260
245498	95	1600	1200	.210	.225	.245
257420	68	2200	1800	.129	.152	.180
257464	90	1800	1400	.204	.217	.234
257312	88	1800	1400	.208	.236	.273
257418	105	1800	1400	.221	.253	.296
257325	112	1800	1400	.235	.263	.263
280468	122	2200	1600	.193	.224	.254
280473	124	1800	1400	.275	.290	.310
280412	136	1800	1400	.245	.255	.270
287448	119	2000	1600	.165	.205	.245
287346	135	1800	1400	.235	.265	.305
287405	150	1600	1200	.245	.265	.285
287308	162	1600	1200	.325	.345	.365
311359	113	2000	1600	.181	.182	.182
311441	115	2200	1600	.143	.152	.163
311576	120	1800	1400	.172	.196	.220
311465	122	2200	1600	.163	.171	.195
311410	130	1600	1200	.239	.250	.265
311440	151	2200	1600	.134	.151	.171
311466	151	2000	1600	.250	.250	.260
301618	160	2400	1800	.300	.310	.305
311291	169	2200	1600	.202	.231	.250
31141	170	2200	1600	.220	.250	.280
311467	178	2200	1600	.320	.305	.300
311407	173	1800	1400	.270	.300	.325
311334	187	1800	1400	.340	.312	.275
301620	200	2200	1800	.379	.370	.362
311299	200	1800	1400	.377	.358	.390
311290	208	1800	1400	.305	.300	.275
311284	210	1800	1400	.332	.331	.335
323470	160	2200	1600	.187	.205	.225
323471	214	1800	1400	.475	.450	.425
323378	242	1600	1200	.410	.420	.430
375248	248	1600	1200	.290	.310	.325
375449	264	1600	1200	.315	.348	.375
375167	264	1600	1200	.315	.290	.270
457191	293	1800	1400	.201	.222	.240
457193	420	1600	1200	.307	.335	.365
515141	422	1600	1200	.250	.250	.225

Figure 3 shows this situation for our example bullet. When the barrel of our rifle is tilted upward so that the line of departure passes 3.15 inches above the baseline at 100 yards, then the 301618 bullet at 2400 fps muzzle velocity will be zeroed in at 100 yards.

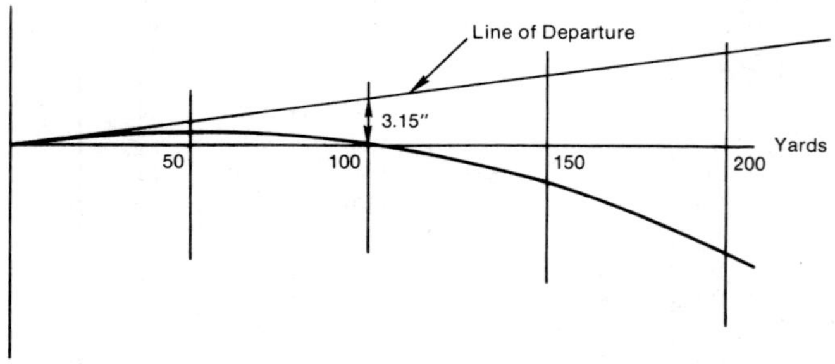

Figure 3. Zeroing In

Midrange. When a rifle or handgun is zeroed in at some particular range (as in **Figure 3**), the bullet passes above the baseline at ranges shorter than the zero range. It turns out that the projectile reaches its maximum height above the baseline at a range which is very near the midpoint between the muzzle and the zero range. We call the height of the trajectory at that midpoint the midrange height, or more frequently, the midrange. The midrange is very nearly the maximum height the projectile reaches in its flight between the muzzle and the zero range.

Figure 4 illustrates two midrange values for bullet 301618 at 2400 fps muzzle velocity. If the rifle is zeroed in at 100 yards, the midrange is 0.84 inches, and of course this midrange height occurs at a distance of 50 yards from the muzzle. If the rifle is zeroed in at 200 yards, the midrange height is 3.82 inches (at 100 yards).

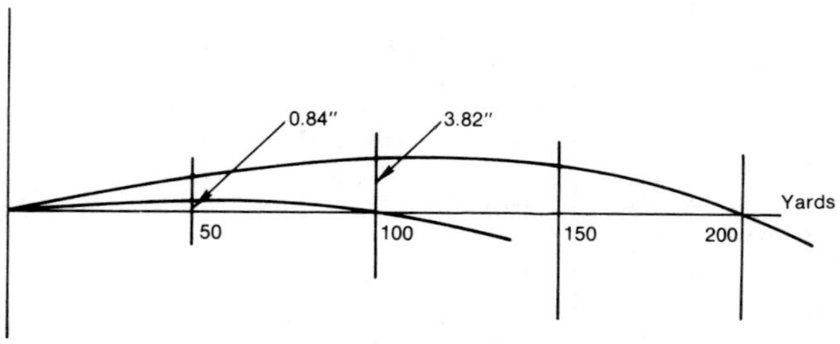

Figure 4. Midrange

Note that in the Ballistics Tables a midrange value is quoted at each range entry. In each case, this is the midrange height of the trajectory if the gun is zeroed in at the range value at the top of the column. The midrange height always occurs at a distance halfway between the muzzle and that zero range. Midrange height is important for a couple of reasons. It tells a shooter how high his bullet will rise above the line between his muzzle and target, and it is the trajectory parameter used to get the maximum point blank range performance from his rifle, as we'll explain later.

Bullet Path. The sights on a rifle or handgun are mounted above the bore. The distance between the centerline of the bore and a line through the sights (line of sight) is called the sight height. The sight height on a rifle with iron sights is about 0.75 inch. For a rifle with a telescope sight, the sight height is about 1.5 inches.

Figure 5 shows the relationship of the line of sight to the baseline and the trajectory. The numerical values given in **Figure 5** are for our example 301618 bullet at 2400 fps muzzle velocity. When the gun is zeroed in, the line of sight intersects the baseline and the trajectory at the zero range. A shooter's real interest is where his bullet is relative to his line of sight, rather than the baseline. This is called the bullet path height, usually shortened to bullet path. As shown in **Figure 5**, the 301618 bullet leaves our rifle 0.75 inch below the line of sight. So we say that the bullet path is -0.75 inch at the muzzle. Since we have our rifle zeroed in at 100 yards, the bullet is above the line of sight by 0.47 inch at 50 yards, right on at 100 yards, 2.37 inches low at 150 yards, and so forth.

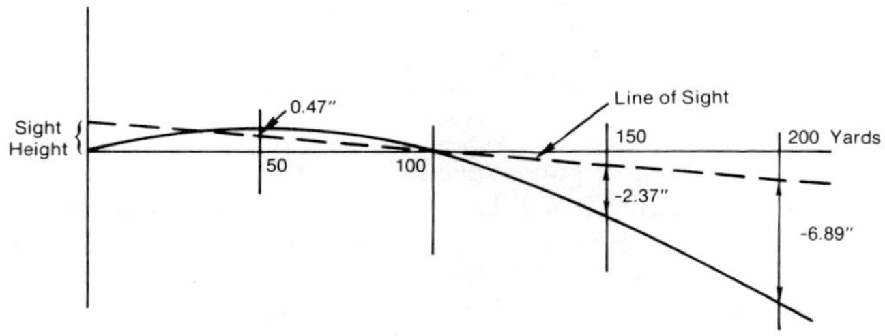

Figure 5. Bullet Path

Time of Flight. When a projectile leaves the muzzle, it requires a certain amount of time to reach a target. This is called the time of flight, and it depends on both range to the target and muzzle velocity. The farther away the target is, the more the time required to reach it. The higher the muzzle velocity, the shorter the time of flight. The Ballistics Tables list the time of flight to reach each range for each muzzle velocity case. Our example bullet which leaves the muzzle at 2400 fps requires .064 second to reach 50 yards, .133 second to reach 100 yards, .205 second to reach 150 yards, and so forth.

Wind Deflection. Projectiles are very susceptible to crosswinds. For that reason the deflections caused by crosswinds of 5, 10, 20, and 30 mph have been calculated for each bullet, and these crosswind deflections are listed in the Ballistics Tables for each muzzle velocity level.

A crosswind tends to drag the projectile along with it, so the deflections can be to the right or left, depending on the direction of the crosswind. For our example bullet, a 5 mph crosswind will carry the bullet sideways 0.66 inch at 100 yards, and 2.82 inches at 200 yards. If the crosswind is 30 mph, these deflections grow to 3.98 inches at 100 yards, and 16.91 inches at 200 yards.

It is evident from the sizes of the wind deflections that strong crosswinds can cause very large misses, if windage corrections are not made. These data may help shooters select loads and improve their "guesstimation" abilities for windy day shooting.

Changing the Zero Range. The bullet trajectory data in the Ballistics Tables of this handbook are computed for a zero range of 100 yards. The bullet path entries in the tables then show you how high or low your bullet will shoot at other ranges out to 600 yards. The 100 yard zero range was used in this handbook because it is a reasonable choice for many shooting purposes. Quite often, though, a shooter would like to use a different zero range, and two questions then arise. The first usually goes something like this:

> "I want to set my zero range at 240 yards, but I need to use a 200 yard target to do it. How high should I shoot at 200 yards to have my gun zeroed at 240 yards?"

The second question is then:

> "After I zero in at 240 yards, where does my gun shoot at other ranges?"

The answers to these questions are given by bullet path data calculated for the new zero range. However, it was not possible in this handbook to include bullet path data in the Ballistics Tables for all possible choices of zero range, because the tables would have been huge! There is a relatively easy method that any shooter can use to calculate bullet path for any cartridge and any load, for any choice of zero range, and for any sight height as well, which is important if he uses a telescope sight instead of iron sights. This method is explained below.

There is a simple and accurate formula for calculation of bullet path height, and it works for both rifles and handguns. To use the formula, you must know drop as a function of range for your bullet and load, and this can be obtained from ballistics tables. Interpolation is necessary if you choose a zero range which is not one of the entries in the table, and the example calculation below will show how to do this. You must also know your sight height. For iron sights 0.75 inch is a good estimate of sight height for most rifles, and 1.50 inches is a reasonable number for rifles with telescope sights.

The calculations can be done by hand, but they are especially easy if you have an electronic calculator. The formula for bullet path height is the following:

$$y_b(R) = y(R) - h_s + \frac{R}{R_z}(h_s - y_z)$$

where $\mathbf{y_b(R)}$ is bullet path height in *inches* at the range **R** (positive values for bullet above the line of sight, and negative values for bullet below the line of sight);

R is the range from the muzzle in *yards* at which bullet path height is to be calculated;

y(R) is the bullet drop (always negative) at the range **R**, measured in *inches*, and taken from the Ballistics Tables);

$\mathbf{h_s}$ is the sight height above the bore centerline in *inches* (always positive);

$\mathbf{R_z}$ is the zero range in *yards*;

$\mathbf{y_z}$ is the drop in *inches* at the zero range $\mathbf{R_z}$ (always negative).

This formula is good for ranges greater than the zero range $\mathbf{R_z}$ as well as for ranges shorter than $\mathbf{R_z}$. The accuracy of the formula is very good for practical shooting purposes. The calculated bullet path will be correct within a small fraction of an inch for rifle bullets to ranges exceeding 1000 yards.

The sketch below will help define the meanings of the terms in the equation above. It is necessary to manipulate negative numbers in the calculations, which might be a little unfamiliar if you're not a whiz at math. The following example calculation shows how these negative and positive numbers combine together. If you follow the method in the example, you can do these calculations even if you aren't a whiz at math!

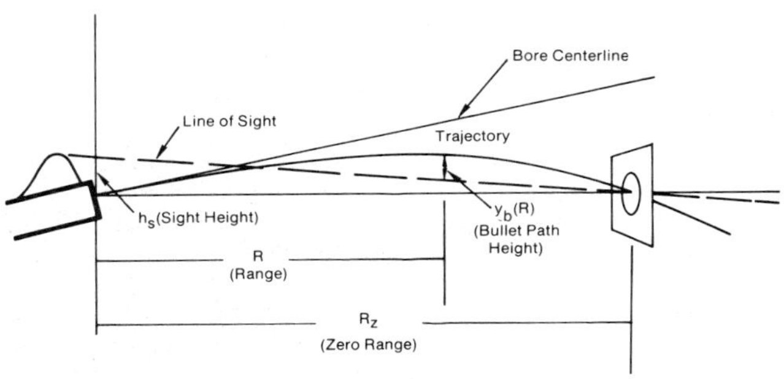

Parameters in Bullet Path Height Calculations

The example we are going to use anticipates a result from the next section of this article, where we describe a procedure for maximizing the point blank range of any cartridge. We suppose there that you will be hunting eastern white tail deer with a rifle chambered for the .308 Winchester cartridge. You intend to use the Lyman paper-patched 301618 bullet weighing 160 grains and loaded to a muzzle velocity of 2400 fps. Also, you use a telescope sight on your rifle. The point blank range procedure tells you that you need to zero your rifle in at 240 yards. However, you only have a 200 yard range available for sighting in, and you need to know where the bullets should strike on the target at 200 yards in order to be zeroed in at 240 yards. You also want to know where your bullet will be at 100 and 300 yards.

Example Calculation

Cartridge:	.308 Winchester with Lyman 301618 paper-patched 160 grain bullet
Muzzle velocity:	2400 fps
Zero range:	240 yards
Sight height:	1.50 inches (telescope sight)
Calculations:	(1) Bullet path height at 200 yards to use for sighting in. (2) Bullet path heights at 100 and 300 yards with rifle sighted in for 240 yards.

Ballistics table for reference: Lyman Cast Bullet Handbook

The first step is to determine the drop $\mathbf{y_z}$ at the zero range $\mathbf{R_z}$ = 240 yards. Turning to the ballistics table for the 301618 bullet at 2400 fps, we notice that trajectory data are listed in 50 yard range steps. So, we have drop figures for 200 and 250 yards, but no figure is listed for 240 yards. This means that we have to interpolate from the data in the table. The easiest way to do this is to use a piece of rectangular graph paper, which can be purchased usually in any store selling school supplies. Let the vertical scale of the graph be for drop and the horizontal scale for range. Plot four points on the graph, the drop at 150 yards (-7.48 inches), at 200 yards (-13.95 inches), at 250 yards (-22.86

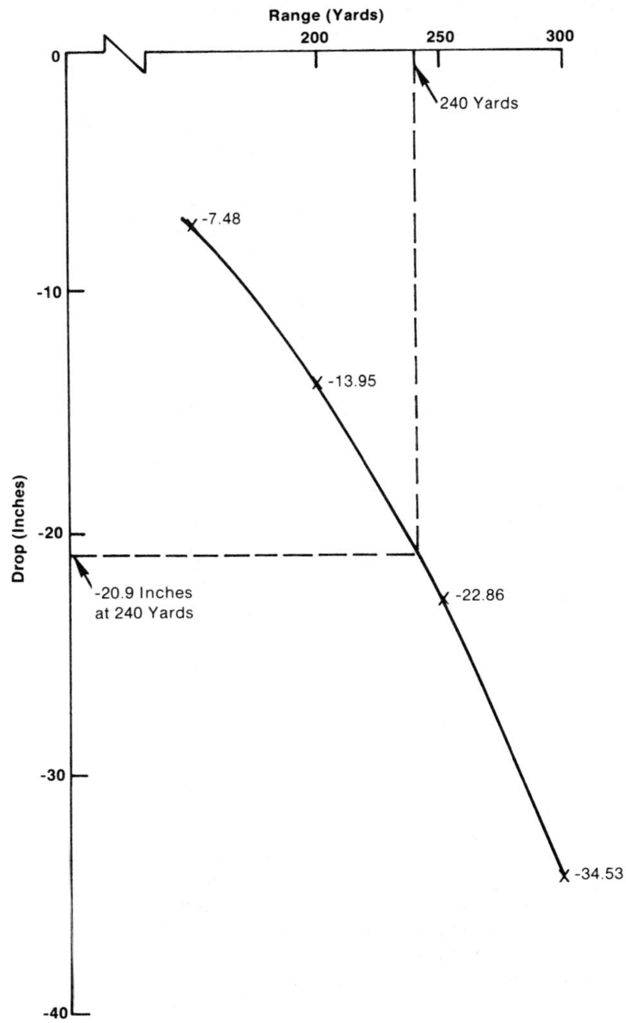

inches), and at 300 yards (-34.53 inches). Always pick four points, like this example, two lying to the left of the desired zero range and two to the right. Draw a smooth curve on the graph paper connecting the four points that you have just located. Then, where the curve crosses the line representing 240 yards you can read the drop from the curve. (Refer to the graph we've drawn as an example.)

If you do this carefully for the example above, you should find that the drop at 240 yards is just about -20.90 inches. This graphical method is accurate enough to give an answer that is correct within 0.1 inch. For our example, then, we have:

$$R_z = 240 \text{ yards}$$

$$y_z = -20.90 \text{ inches}$$

$$h_s = 1.50 \text{ inches}$$

These numbers stay the same for all calculations below.

(1) At R = 200 yards, y(R) = -13.95 inches from the ballistics table. Using the formula:

$$y_b(200) = -13.95 - 1.50 + \frac{200}{240}(1.50 - (-20.90))$$

$$= -15.45 + \frac{200}{240}(+22.40)$$

$$= -15.45 + 1 \times .67$$

$$= +3.22 \text{ inches}$$

Setting your telescope sight to make the bullets print 3.2 inches high at 200 yards, then, will zero the rifle in for 240 yards.

(2) At **R** = 100 yards, y(100) = -3.15 inches from the ballistics table. Using the formula:

$$y_b(100) = -3.15 - 1.50 + \frac{100}{240}(1.50 - (-20.90))$$

$$= -4.65 + \frac{100}{240}(+22.40)$$

$$= -4.65 + 9.33$$

$$= +4.6x \text{ inches (above the line of sight)}$$

At **R** = 300 yards, y(300) = -34.53 inches. From the formula:

$$y_b(300) = -34.53 - 1.50 + \frac{300}{240}(1.50 - (-20.90))$$

$$= -36.03 + \frac{300}{240}(+22.40)$$

$$= -36.03 + 28.00$$

$$= -8.03 \text{ inches (below the line of sight)}$$

Maximizing Point Blank Range. The point blank range concept was explained in an earlier section of this article called "Cast Bullet Performance." As explained there, point blank range can be maximized for any load any size of game animal by choosing a zero range which makes the bullet trajectory rise above the line of sight by an amount just equal to half the vertical dimension of the vital zone for that animal. In this section we'll describe a step-by-step procedure for doing this, and then illustrate it with an example.

The procedure for maximizing the point blank range has five steps. Following these steps will determine the right zero range to use for the size of animal you wish to hunt, and also determine how large the maximum point blank range is. These five steps are as follows.

1. Estimate the vital zone vertical dimension for the game you intend to hunt. The following guidelines may help:

Varmints and small game (squirrels, cotton tails, jackrabbits, woodchucks etc.)	3 to 5 inches
Light game (small deer, javelina, etc.)	6 to 8 inches
Medium game (white tail deer, mule deer, black bear, etc.)	10 inches
Large game (elk, moose, etc.)	15 inches

2. Take half the vital zone dimension and then add a correction for the height of the sights on your rifle. If you use iron sights, this correction is 0.4 inch, and if you have a telescope sight, the correction is 0.8 inch. The number you get after the correction is added to half the vital zone dimension is the midrange height your bullet should have to maximize the point blank range.
3. Turn to the ballistics table for your bullet and muzzle velocity. Look across the Mid-Range values in the table to find the amount you calculated in the preceding step. Usually, you won't find the exact value you want in the table, and you will have to interpolate between a couple of points in the table. The example below will show how to do this. Once you have found the right midrange height, the *range* value for which this midrange height occurs is the *zero range* that you want to use.
4. Zero in your rifle at the range found in the previous step. Usually the zero range that you want to use will not be one listed in the Ballistics Tables, and you will need to use a procedure like the one described in the preceding section.
5. When your rifle is zeroed in, add 40 yards to the zero range you determined in Step 3. This result is the maximum point blank range for your cartridge load.

An an example of you to use this procedure, suppose that you intend to hunt eastern white tail deer with a telescope-sighted .308 Winchester. You decide to use the Lyman 301618 paper-patched 160 grain bullet loaded to a muzzle velocity of 2400 fps, and you want to maximize your point blank range for this load.

Example Calculation

Cartridge:	.308 Winchester with Lyman 301618 paper-patched 160 grain bullet
Muzzle velocity:	2400 fps
Sight height:	1.50 inches (telescope sight)

1. The adult eastern white tail is a medium size game animal, and the vital zone of the animal is about 10 inches in height.
2. Half the vital zone height is 5 inches. The sight height correction for your telescope sight is 0.8 inch. Adding these two figures gives 5.8 inches, which is the midrange height you need to find in the ballistics table for your bullet and muzzle velocity.
3. Turning to the ballistics table for the 301618 bullet at 2400 fps and looking across the Mid-Range row, you'll find that at a range of 200 yards the midrange height is 3.82 inches, and at 250 yards the midrange height is 6.36 inches. Since the midrange height you need is 5.8 inches, it must occur for a range value between 200 and 250 yards, and we need to use interpolation to find the right value. Here again we can use rectangular graph paper to construct a curve from which we can read the answer we are looking for, just like the method used in the last section. This time, the vertical scale of the graph paper is midrange height, and the horizontal scale is range. We select four points, two to the left of where the answer must lie (at 150 and 200 yards) and the

other to the right (250 and 300 yards). We plot the four midrange heights from the ballistics table (2.02 inches at 150 yards, 3.82 at 200, 6.36 at 250, and 9.79 inches at 300 yards). Then we draw a smooth curve through the four points. Looking to see where the curve crosses the line representing 5.8 inches midrange height, we find the crossing at a range of 240 yards. This is the right zero range to use to maximize point blank range for white tails with your load. (See our example graph).

4. You must take your rifle out and zero it in for the 240 yard zero range. This is not a really convenient zero range to use if you must use a public shooting range or club, because these places normally have targets at 100 and 200 yards. The preceding section gave a procedure for figuring out where your rifle should shoot at a range of 200 yards in order to be zeroed in at 240 yards, and the result is 3.2 high. You can always use a procedure like the one in the last section to find a way to zero in your rifle at any range.
5. Now that you have your rifle zeroed in for 240 yards, adding 40 more yards makes your point blank range 280 yards, and it is the maximum that your load will produce for medium game. If you see a white tail buck at any range up to 280 yards, you can center him in your sights and not worry about holding high or low to correct for bullet trajectory. And 280 yards is quite a respectable point blank range!

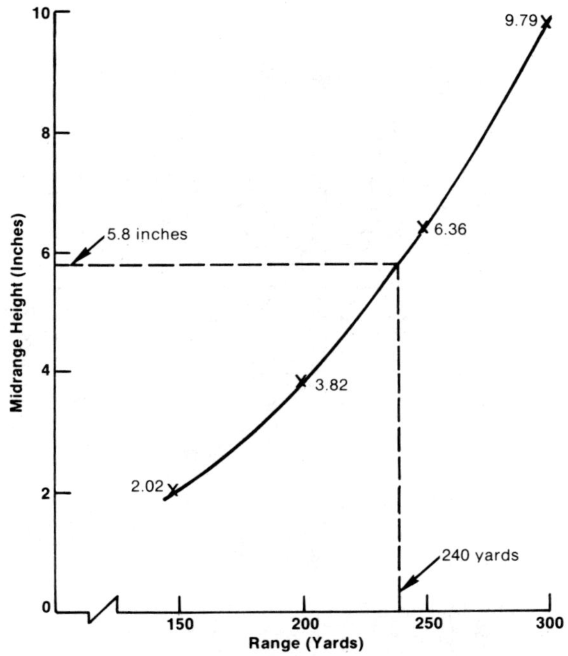

Bullet: Lyman # 225107 37 Gr.
Ballistic Coefficient: .091

RANGE (YDS)	MUZZLE	50	100	150	200	250	300	350	400	450	500	550	600
VELOCITY (FPS)	2700.	2241.	1850.	1531.	1274.	1094.	984.	907.	846.	794.	747.	705.	665.
ENERGY (FT-LB)	599.	412.	281.	193.	133.	98.	79.	68.	59.	52.	46.	41.	36.
DROP (IN)	.00	-.66	-3.07	-8.04	-16.76	-30.86	-52.19	-82.43	-123.16	-175.98	-242.37	-324.29	-423.69
MID-RANGE (IN)	.00	.18	.87	2.41	5.31	10.26	18.06	29.38	44.83	64.97	90.33	121.64	159.66
BULLET PATH (IN)	-.75	.50	.00	-3.05	-9.86	-22.05	-41.47	-69.80	-108.63	-159.54	-224.02	-304.02	-401.51
TIME OF FLIGHT (SEC)	.000000	.061105	.134801	.224125	.331818	.459555	.604687	.763762	.935180	1.118378	1.313365	1.520395	1.739912
WIND DEFLECTION (IN)													
5 MPH	.00	.49	2.08	5.06	9.64	16.00	23.88	32.99	43.18	54.42	66.69	80.02	94.45
10 MPH	.00	.98	4.17	10.11	19.29	31.99	47.76	65.98	86.37	108.83	133.37	160.03	188.89
20 MPH	.00	1.95	8.34	20.23	38.58	63.99	95.52	131.96	172.74	217.67	266.75	320.07	377.78
30 MPH	.00	2.93	12.51	30.34	57.87	95.98	143.27	197.93	259.11	326.50	400.12	480.10	566.67
VELOCITY (FPS)	2600.	2158.	1778.	1474.	1231.	1067.	966.	894.	835.	784.	738.	696.	657.
ENERGY (FT-LB)	555.	383.	260.	178.	124.	94.	77.	66.	57.	50.	45.	40.	36.
DROP (IN)	.00	-.71	-3.31	-8.68	-18.09	-33.25	-56.01	-87.98	-130.78	-185.96	-255.08	-340.11	-443.04
MID-RANGE (IN)	.00	.19	.94	2.61	5.73	11.05	19.33	31.21	47.30	68.14	94.29	126.48	165.52
BULLET PATH (IN)	-.75	.57	.00	-3.33	-10.71	-23.84	-44.56	-74.50	-115.27	-168.42	-235.51	-318.51	-419.41
TIME OF FLIGHT (SEC)	.000000	.063455	.140098	.232874	.344590	.476172	.624322	.785970	.959784	1.145347	1.342734	1.552240	1.774341
WIND DEFLECTION (IN)													
5 MPH	.00	.51	2.17	5.26	10.02	16.52	24.48	33.63	43.85	55.10	67.39	80.75	95.22
10 MPH	.00	1.01	4.35	10.52	20.03	33.04	48.96	67.25	87.69	110.20	134.78	161.50	190.44
20 MPH	.00	2.03	8.70	21.05	40.06	66.07	97.92	134.51	175.38	220.39	269.57	323.00	380.88
30 MPH	.00	3.04	13.05	31.57	60.10	99.11	146.87	201.76	263.07	330.59	404.35	484.51	571.31
VELOCITY (FPS)	2500.	2074.	1709.	1417.	1189.	1042.	949.	881.	824.	774.	729.	688.	650.
ENERGY (FT-LB)	513.	353.	240.	165.	116.	89.	74.	64.	56.	49.	44.	39.	35.
DROP (IN)	.00	-.77	-3.59	-9.40	-19.59	-35.92	-60.22	-94.05	-139.06	-196.74	-268.79	-357.14	-463.84
MID-RANGE (IN)	.00	.21	1.02	2.83	6.21	11.92	20.71	33.18	49.94	71.50	98.47	131.60	171.70
BULLET PATH (IN)	-.75	.65	.00	-3.64	-11.66	-25.83	-47.95	-79.62	-122.46	-177.98	-247.85	-334.03	-438.56
TIME OF FLIGHT (SEC)	.000000	.065984	.145817	.242343	.358327	.493761	.644941	.809229	.985536	1.173579	1.373494	1.585613	1.810449
WIND DEFLECTION (IN)													
5 MPH	.00	.53	2.27	5.49	10.41	17.05	25.07	34.25	44.49	55.75	68.07	81.45	95.96
10 MPH	.00	1.05	4.54	10.97	20.83	34.10	50.15	68.50	88.97	111.51	136.13	162.91	191.92
20 MPH	.00	2.11	9.09	21.94	41.65	68.20	100.30	137.01	177.95	223.02	272.27	325.82	383.84
30 MPH	.00	3.16	13.63	32.92	62.48	102.31	150.45	205.51	266.92	334.53	408.40	488.72	575.76
VELOCITY (FPS)	2400.	1992.	1644.	1363.	1152.	1020.	934.	868.	813.	764.	720.	680.	642.
ENERGY (FT-LB)	473.	326.	222.	153.	109.	85.	72.	62.	54.	48.	43.	38.	34.
DROP (IN)	.00	-.84	-3.88	-10.18	-21.19	-38.74	-64.60	-100.34	-147.58	-207.79	-282.78	-374.49	-485.01
MID-RANGE (IN)	.00	.23	1.11	3.06	6.72	12.83	22.12	35.18	52.60	74.86	102.65	136.71	177.90
BULLET PATH (IN)	-.75	.73	.00	-3.98	-12.68	-27.91	-51.45	-84.88	-129.80	-187.70	-260.37	-349.77	-457.96
TIME OF FLIGHT (SEC)	.000000	.068607	.151716	.252104	.372326	.511384	.665455	.832313	1.011078	1.201586	1.404018	1.618749	1.846323
WIND DEFLECTION (IN)													
5 MPH	.00	.54	2.35	5.69	10.76	17.50	25.56	34.74	44.97	56.24	68.55	81.95	96.48
10 MPH	.00	1.07	4.70	11.37	21.53	35.00	51.12	69.49	89.95	112.48	137.11	163.90	192.95
20 MPH	.00	2.15	9.40	22.74	43.06	70.01	102.24	138.97	179.90	224.96	274.21	327.80	385.91
30 MPH	.00	3.22	14.11	34.11	64.59	105.01	153.36	208.46	269.85	337.44	411.32	491.70	578.86

Bullet: Lyman # 225107 37 Gr. 
Ballistic Coefficient: .091
(Cont'd.)

RANGE (YDS)	MUZZLE	50	100	150	200	250	300	350	400	450	500	550	600
VELOCITY (FPS)	2300.	1903.	1573.	1307.	1114.	997.	917.	854.	801.	754.	710.	671.	633.
ENERGY (FT-LB)	435.	297.	203.	140.	102.	82.	69.	60.	53.	47.	41.	37.	33.
DROP (IN)	.00	-.91	-4.24	-11.13	-23.14	-42.12	-69.78	-107.70	-157.50	-220.61	-298.97	-394.52	-509.40
MID-RANGE (IN)	.00	.25	1.21	3.35	7.33	13.90	23.76	37.47	55.61	78.66	107.37	142.49	184.92
BULLET PATH (IN)	-.75	.84	.00	-4.39	-13.90	-30.38	-55.54	-90.97	-138.27	-198.88	-274.75	-367.80	-480.18
TIME OF FLIGHT (SEC)	.000000	.071722	.158647	.263514	.388425	.531312	.688509	.858201	1.039710	1.232985	1.438255	1.655937	1.886611
WIND DEFLECTION (IN)													
5 MPH	.00	.57	2.48	5.97	11.22	18.06	26.15	35.35	45.58	56.85	69.18	82.59	97.15
10 MPH	.00	1.14	4.97	11.94	22.45	36.12	52.31	70.70	91.16	113.70	138.35	165.18	194.30
20 MPH	.00	2.29	9.93	23.89	44.90	72.24	104.62	141.39	182.33	227.40	276.70	330.37	388.61
30 MPH	.00	3.43	14.90	35.83	67.35	108.36	156.92	212.09	273.49	341.10	415.05	495.55	582.91
VELOCITY (FPS)	2200.	1813.	1503.	1252.	1080.	975.	901.	841.	789.	743.	701.	661.	625.
ENERGY (FT-LB)	398.	270.	186.	129.	96.	78.	67.	58.	51.	45.	40.	36.	32.
DROP (IN)	.00	-1.00	-4.66	-12.21	-25.32	-45.83	-75.41	-115.64	-168.11	-234.32	-316.24	-415.84	-535.31
MID-RANGE (IN)	.00	.27	1.33	3.67	8.00	15.07	25.50	39.87	58.74	82.62	112.29	148.53	192.25
BULLET PATH (IN)	-.75	.95	.00	-4.85	-15.26	-33.07	-59.94	-97.47	-147.24	-210.75	-289.96	-386.86	-503.63
TIME OF FLIGHT (SEC)	.000000	.075125	.166111	.275744	.405363	.551972	.712300	.884879	1.069221	1.265366	1.473590	1.694352	1.928268
WIND DEFLECTION (IN)													
5 MPH	.00	.61	2.62	6.27	11.67	18.57	26.68	35.87	46.09	57.35	69.68	83.10	97.69
10 MPH	.00	1.22	5.24	12.53	23.34	37.15	53.36	71.74	92.18	114.70	139.35	166.21	195.38
20 MPH	.00	2.44	10.47	25.06	46.69	74.29	106.73	143.48	184.37	229.41	278.70	332.41	390.75
30 MPH	.00	3.67	15.71	37.59	70.03	111.44	160.09	215.22	276.55	344.11	418.06	498.62	586.13
VELOCITY (FPS)	2100.	1730.	1434.	1201.	1050.	955.	884.	827.	777.	732.	690.	652.	616.
ENERGY (FT-LB)	362.	246.	169.	119.	90.	75.	64.	56.	50.	44.	39.	35.	31.
DROP (IN)	.00	-1.10	-5.12	-13.42	-27.75	-49.92	-81.53	-124.21	-179.48	-248.98	-334.66	-438.56	-562.91
MID-RANGE (IN)	.00	.30	1.46	4.04	8.75	16.32	27.34	42.39	62.00	86.75	117.42	154.84	199.92
BULLET PATH (IN)	-.75	1.09	.00	-5.36	-16.75	-35.98	-64.66	-104.40	-156.74	-223.30	-306.04	-407.00	-528.41
TIME OF FLIGHT (SEC)	.000000	.078828	.174193	.288872	.423150	.573417	.736904	.912449	1.099733	1.298871	1.510186	1.734180	1.971504
WIND DEFLECTION (IN)													
5 MPH	.00	.65	2.76	6.56	12.09	19.03	27.13	36.30	46.49	57.73	70.04	83.46	98.06
10 MPH	.00	1.30	5.52	13.13	24.19	38.06	54.27	72.59	92.98	115.46	140.08	166.93	196.13
20 MPH	.00	2.60	11.03	26.25	48.38	76.13	108.53	145.18	185.96	230.92	280.16	333.86	392.26
30 MPH	.00	3.91	16.55	39.38	72.57	114.19	162.80	217.77	278.94	346.38	420.24	500.79	588.38
VELOCITY (FPS)	2000.	1650.	1368.	1155.	1022.	935.	869.	814.	765.	721.	680.	643.	607.
ENERGY (FT-LB)	329.	224.	154.	110.	86.	72.	62.	54.	48.	43.	38.	34.	30.
DROP (IN)	.00	-1.21	-5.65	-14.77	-30.39	-54.28	-88.01	-133.22	-191.37	-264.24	-353.80	-462.11	-591.48
MID-RANGE (IN)	.00	.33	1.61	4.44	9.54	17.62	29.23	44.95	65.29	90.92	122.63	161.25	207.73
BULLET PATH (IN)	-.75	1.24	.00	-5.92	-18.33	-39.02	-69.55	-111.56	-166.50	-236.18	-322.54	-427.64	-553.82
TIME OF FLIGHT (SEC)	.000000	.082784	.182771	.302561	.441258	.595036	.761632	.940144	1.130397	1.332568	1.547026	1.774314	2.015117
WIND DEFLECTION (IN)													
5 MPH	.00	.68	2.88	6.83	12.43	19.36	27.42	36.53	46.67	57.87	70.14	83.54	98.13
10 MPH	.00	1.37	5.77	13.65	24.86	38.73	54.85	73.07	93.35	115.73	140.28	167.08	196.26
20 MPH	.00	2.74	11.54	27.30	49.72	77.45	109.69	146.13	186.70	231.46	280.55	334.16	392.52
30 MPH	.00	4.11	17.30	40.95	74.58	116.18	164.54	219.20	280.05	347.20	420.83	501.24	588.78

Bullet: Lyman # 225107 37 Gr.
Ballistic Coefficient: .091

(Cont'd.)

RANGE (YDS)	MUZZLE	50	100	150	200	250	300	350	400	450	500	550	600
VELOCITY (FPS)	1900.	1571.	1305.	1113.	996.	916.	854.	801.	753.	710.	670.	633.	598.
ENERGY (FT-LB)	297.	203.	140.	102.	82.	69.	60.	53.	47.	41.	37.	33.	29.
DROP (IN)	.00	-1.34	-6.25	-16.30	-33.33	-59.05	-95.04	-142.93	-204.13	-280.61	-374.28	-487.30	-622.02
MID-RANGE (IN)	.00	.36	1.78	4.88	10.41	19.00	31.22	47.64	68.74	95.30	128.10	167.99	215.97
BULLET PATH (IN)	-.75	1.41	.00	-6.54	-20.07	-42.29	-74.78	-119.17	-176.87	-249.84	-340.02	-449.53	-580.75
TIME OF FLIGHT (SEC)	.000000	.087047	.192060	.317121	.460127	.617423	.787203	.968798	1.162160	1.367517	1.585290	1.816058	2.060544
WIND DEFLECTION (IN)													
5 MPH	.00	.71	3.01	7.06	12.70	19.60	27.59	36.62	46.69	57.82	70.03	83.39	97.96
10 MPH	.00	1.43	6.01	14.13	25.40	39.19	55.18	73.25	93.38	115.63	140.06	166.78	195.92
20 MPH	.00	2.85	12.03	28.26	50.81	78.39	110.36	146.49	186.76	231.26	280.13	333.57	391.84
30 MPH	.00	4.28	18.04	42.39	76.21	117.58	165.54	219.74	280.15	346.89	420.19	500.35	587.76
VELOCITY (FPS)	1800.	1493.	1245.	1076.	972.	898.	839.	787.	741.	699.	660.	623.	589.
ENERGY (FT-LB)	266.	183.	127.	95.	78.	66.	58.	51.	45.	40.	36.	32.	28.
DROP (IN)	.00	-1.49	-6.93	-18.00	-36.53	-64.18	-102.56	-153.20	-217.66	-297.91	-395.91	-513.86	-654.22
MID-RANGE (IN)	.00	.40	1.97	5.38	11.34	20.45	33.28	50.38	72.30	99.83	133.77	175.00	224.55
BULLET PATH (IN)	-.75	1.60	.00	-7.23	-21.92	-45.74	-80.27	-127.07	-187.70	-264.10	-358.27	-472.38	-608.90
TIME OF FLIGHT (SEC)	.000000	.091611	.201974	.332286	.479440	.640231	.813242	.998009	1.194585	1.403252	1.624476	1.858878	2.107214
WIND DEFLECTION (IN)													
5 MPH	.00	.73	3.11	7.24	12.86	19.67	27.57	36.49	46.46	57.49	69.62	82.91	97.43
10 MPH	.00	1.46	6.21	14.48	25.71	39.35	55.13	72.98	92.91	114.97	139.24	165.83	194.87
20 MPH	.00	2.91	12.43	28.96	51.43	78.69	110.26	145.97	185.83	229.94	278.48	331.66	389.74
30 MPH	.00	4.37	18.64	43.45	77.14	118.04	165.39	218.95	278.74	344.92	417.72	497.49	584.61
VELOCITY (FPS)	1700.	1409.	1184.	1039.	947.	879.	822.	773.	728.	687.	648.	613.	579.
ENERGY (FT-LB)	237.	163.	115.	89.	74.	63.	56.	49.	44.	39.	35.	31.	28.
DROP (IN)	.00	-1.68	-7.78	-20.08	-40.39	-70.29	-111.42	-165.25	-233.51	-318.13	-421.16	-544.87	-691.80
MID-RANGE (IN)	.00	.46	2.21	5.97	12.42	22.11	35.63	53.50	76.36	105.02	140.28	183.08	234.48
BULLET PATH (IN)	-.75	1.84	.00	-8.04	-24.08	-49.72	-86.58	-136.15	-200.13	-280.49	-379.26	-498.70	-641.36
TIME OF FLIGHT (SEC)	.000000	.097067	.213654	.349615	.501210	.665864	.842516	1.030903	1.231167	1.443646	1.668857	1.907465	2.160265
WIND DEFLECTION (IN)													
5 MPH	.00	.78	3.27	7.47	13.05	19.77	27.55	36.37	46.23	57.16	69.21	82.45	96.93
10 MPH	.00	1.55	6.54	14.94	26.10	39.54	55.11	72.73	92.45	114.32	138.42	164.89	193.85
20 MPH	.00	3.11	13.09	29.89	52.19	79.09	110.21	145.47	184.90	228.63	276.85	329.78	387.71
30 MPH	.00	4.66	19.63	44.83	78.29	118.63	165.32	218.20	277.35	342.95	415.27	494.67	581.56
VELOCITY (FPS)	1600.	1328.	1128.	1005.	923.	859.	805.	758.	714.	674.	637.	601.	568.
ENERGY (FT-LB)	210.	145.	105.	83.	70.	61.	53.	47.	42.	37.	33.	30.	27.
DROP (IN)	.00	-1.90	-8.77	-22.44	-44.64	-76.96	-121.02	-178.25	-250.55	-339.86	-448.28	-578.17	-732.16
MID-RANGE (IN)	.00	.52	2.48	6.62	13.55	23.84	38.07	56.75	80.64	110.49	147.17	191.66	245.06
BULLET PATH (IN)	-.75	2.11	.00	-8.91	-26.35	-53.92	-93.21	-145.69	-213.23	-297.77	-401.43	-526.56	-675.80
TIME OF FLIGHT (SEC)	.000000	.103122	.226236	.367673	.523674	.692274	.872720	1.064919	1.269083	1.485611	1.715071	1.958170	2.215741
WIND DEFLECTION (IN)													
5 MPH	.00	.82	3.41	7.61	13.08	19.67	27.30	35.96	45.68	56.48	68.43	81.57	95.99
10 MPH	.00	1.65	6.82	15.21	26.17	39.34	54.60	71.93	91.36	112.97	136.85	163.14	191.97
20 MPH	.00	3.30	13.63	30.42	52.33	78.68	109.20	143.85	182.72	225.94	273.70	326.28	383.94
30 MPH	.00	4.95	20.45	45.63	78.50	118.02	163.80	215.78	274.08	338.90	410.56	489.41	575.91

Bullet: Lyman # 225438 41 Gr.
Ballistic Coefficient: .094

RANGE (YDS)	MUZZLE	50	100	150	200	250	300	350	400	450	500	550	600
VELOCITY (FPS)	2700.	2259.	1894.	1596.	1353.	1164.	1040.	957.	894.	841.	795.	753.	715.
ENERGY (FT-LB)	664.	464.	326.	232.	167.	123.	98.	83.	73.	64.	58.	52.	47.
DROP (IN)	.00	-.66	-3.03	-7.84	-16.09	-29.12	-48.55	-75.99	-112.82	-160.50	-220.42	-293.82	-382.49
MID-RANGE (IN)	.00	.18	.86	2.33	5.01	9.49	16.44	26.54	40.32	58.34	81.09	108.98	142.69
BULLET PATH (IN)	-.75	.48	.00	-2.92	-9.28	-20.42	-37.96	-63.51	-98.45	-144.23	-202.27	-273.78	-360.56
TIME OF FLIGHT (SEC)	.000000	.060900	.133439	.219985	.322240	.442170	.579050	.729758	.892190	1.065378	1.248972	1.442998	1.647645
WIND DEFLECTION (IN)													
5 MPH	.00	.47	1.96	4.69	8.80	14.47	21.62	30.00	39.40	49.75	61.02	73.21	86.33
10 MPH	.00	.94	3.93	9.38	17.60	28.93	43.25	59.99	78.80	99.51	122.04	146.41	172.65
20 MPH	.00	1.88	7.86	18.77	35.21	57.87	86.49	119.99	157.61	199.01	244.08	292.82	345.30
30 MPH	.00	2.82	11.79	28.15	52.81	86.80	129.74	179.98	236.41	298.52	366.12	439.24	517.96
VELOCITY (FPS)	2600.	2181.	1823.	1543.	1309.	1134.	1020.	943.	882.	831.	786.	745.	708.
ENERGY (FT-LB)	615.	433.	303.	217.	156.	117.	95.	81.	71.	63.	56.	51.	46.
DROP (IN)	.00	-.71	-3.26	-8.43	-17.29	-31.24	-51.93	-80.93	-119.60	-169.41	-231.69	-307.83	-399.57
MID-RANGE (IN)	.00	.19	.92	2.50	5.38	10.16	17.54	28.15	42.52	61.18	84.61	113.27	147.85
BULLET PATH (IN)	-.75	.55	.00	-3.16	-10.01	-21.96	-40.65	-67.64	-104.31	-152.11	-212.38	-286.51	-376.25
TIME OF FLIGHT (SEC)	.000000	.063178	.138419	.228005	.333753	.457321	.597274	.750494	.915172	1.090494	1.276205	1.472376	1.679223
WIND DEFLECTION (IN)													
5 MPH	.00	.48	2.03	4.83	9.06	14.86	22.10	30.50	39.92	50.27	61.54	73.72	86.85
10 MPH	.00	.97	4.05	9.67	18.13	29.72	44.20	61.01	79.84	100.54	123.07	147.45	173.70
20 MPH	.00	1.93	8.11	19.33	36.25	59.44	88.39	122.02	159.68	201.08	246.15	294.89	347.39
30 MPH	.00	2.90	12.16	29.00	54.38	89.16	132.59	183.03	239.52	301.63	369.22	442.34	521.09
VELOCITY (FPS)	2500.	2099.	1755.	1486.	1264.	1104.	1001.	928.	870.	821.	777.	737.	700.
ENERGY (FT-LB)	569.	401.	280.	201.	146.	111.	91.	78.	69.	61.	55.	49.	45.
DROP (IN)	.00	-.77	-3.52	-9.10	-18.66	-33.67	-55.78	-86.51	-127.24	-179.42	-244.36	-323.52	-418.67
MID-RANGE (IN)	.00	.21	.99	2.70	5.81	10.94	18.79	29.95	44.96	64.33	88.51	118.03	153.56
BULLET PATH (IN)	-.75	.62	.00	-3.44	-10.86	-23.74	-43.71	-72.31	-110.90	-160.94	-223.75	-300.78	-393.79
TIME OF FLIGHT (SEC)	.000000	.065631	.143901	.236868	.346526	.473999	.617148	.773034	.940132	1.117777	1.305808	1.504335	1.713605
WIND DEFLECTION (IN)													
5 MPH	.00	.50	2.10	5.00	9.37	15.31	22.63	31.07	40.49	50.84	62.11	74.30	87.44
10 MPH	.00	.99	4.21	10.01	18.75	30.62	45.26	62.13	80.98	101.69	124.22	148.60	174.87
20 MPH	.00	1.98	8.41	20.02	37.50	61.25	90.52	124.27	161.97	203.38	248.44	297.21	349.75
30 MPH	.00	2.97	12.62	30.03	56.25	91.87	135.77	186.40	242.95	305.07	372.67	445.81	524.62
VELOCITY (FPS)	2400.	2021.	1694.	1435.	1225.	1078.	984.	915.	859.	811.	768.	729.	692.
ENERGY (FT-LB)	524.	372.	261.	187.	137.	106.	88.	76.	67.	60.	54.	48.	44.
DROP (IN)	.00	-.83	-3.80	-9.81	-20.08	-36.16	-59.67	-92.10	-134.85	-189.33	-256.83	-338.95	-437.41
MID-RANGE (IN)	.00	.22	1.07	2.91	6.24	11.73	20.03	31.73	47.34	67.38	92.26	122.59	159.03
BULLET PATH (IN)	-.75	.70	.00	-3.74	-11.74	-25.54	-46.79	-76.94	-117.41	-169.63	-234.85	-314.70	-410.89
TIME OF FLIGHT (SEC)	.000000	.068111	.149403	.245716	.359165	.490251	.636302	.794662	.964037	1.143890	1.334138	1.534924	1.746521
WIND DEFLECTION (IN)													
5 MPH	.00	.49	2.15	5.12	9.61	15.64	22.99	31.43	40.84	51.16	62.40	74.57	87.69
10 MPH	.00	.99	4.29	10.25	19.21	31.28	45.99	62.86	81.67	102.32	124.81	149.15	175.39
20 MPH	.00	1.98	8.59	20.49	38.43	62.57	91.98	125.72	163.34	204.65	249.62	298.29	350.78
30 MPH	.00	2.96	12.88	30.74	57.64	93.85	137.97	188.58	245.01	306.97	374.43	447.44	526.16

Bullet: Lyman # 225438 41 Gr.

Ballistic Coefficient: .094
(Cont'd.)

RANGE (YDS)	MUZZLE	50	100	150	200	250	300	350	400	450	500	550	600
VELOCITY (FPS)	2300.	1931.	1625.	1377.	1182.	1051.	965.	900.	846.	800.	758.	719.	683.
ENERGY (FT-LB)	482.	339.	240.	173.	127.	100.	85.	74.	65.	58.	52.	47.	42.
DROP (IN)	.00	-.90	-4.15	-10.71	-21.90	-39.30	-64.54	-99.04	-144.21	-201.48	-272.07	-357.75	-460.20
MID-RANGE (IN)	.00	.24	1.17	3.18	6.80	12.71	21.56	33.89	50.21	71.02	96.73	128.02	165.56
BULLET PATH (IN)	-.75	.80	.00	-4.12	-12.85	-27.81	-50.60	-82.64	-125.37	-180.19	-248.32	-331.55	-431.56
TIME OF FLIGHT (SEC)	.000000	.071194	.156154	.256580	.374554	.509731	.659063	.820279	.992320	1.174783	1.367663	1.571137	1.785512
WIND DEFLECTION (IN)													
5 MPH	.00	.53	2.26	5.36	10.00	16.16	23.56	32.01	41.41	51.73	62.96	75.13	88.26
10 MPH	.00	1.05	4.53	10.72	20.01	32.32	47.13	64.02	82.82	103.46	125.93	150.26	176.51
20 MPH	.00	2.10	9.05	21.45	40.02	64.64	84.25	128.04	165.64	206.91	251.85	300.52	353.02
30 MPH	.00	3.16	13.58	32.17	60.03	96.96	141.38	192.06	248.47	310.37	377.78	450.78	529.53
VELOCITY (FPS)	2200.	1841.	1556.	1320.	1141.	1025.	946.	885.	834.	788.	747.	710.	674.
ENERGY (FT-LB)	441.	308.	220.	159.	119.	96.	81.	71.	63.	57.	51.	46.	41.
DROP (IN)	.00	-.99	-4.55	-11.73	-23.93	-42.78	-69.86	-106.55	-154.30	-214.48	-288.40	-377.84	-484.52
MID-RANGE (IN)	.00	.27	1.28	3.48	7.42	13.79	23.20	36.17	53.22	74.82	101.41	133.72	172.40
BULLET PATH (IN)	-.75	.91	.00	-4.54	-14.09	-30.29	-54.73	-88.77	-133.86	-191.39	-262.66	-349.46	-453.50
TIME OF FLIGHT (SEC)	.000000	.074561	.163375	.268228	.390877	.530063	.682652	.846762	1.021541	1.206712	1.402335	1.608618	1.825902
WIND DEFLECTION (IN)													
5 MPH	.00	.56	2.38	5.60	10.40	16.65	24.07	32.52	41.90	52.19	63.41	75.56	88.68
10 MPH	.00	1.12	4.75	11.21	20.79	33.29	48.15	65.03	83.79	104.38	126.81	151.12	177.36
20 MPH	.00	2.25	9.51	22.42	41.59	66.58	96.29	130.06	167.58	208.76	253.62	302.23	354.72
30 MPH	.00	3.37	14.26	33.62	62.38	99.87	144.44	195.09	251.37	313.14	380.43	453.35	532.08
VELOCITY (FPS)	2100.	1756.	1487.	1265.	1104.	1001.	928.	870.	821.	777.	737.	700.	665.
ENERGY (FT-LB)	401.	281.	201.	146.	111.	91.	78.	69.	61.	55.	49.	45.	40.
DROP (IN)	.00	-1.09	-5.00	-12.88	-26.21	-46.64	-75.68	-114.72	-165.21	-228.45	-305.91	-399.36	-510.54
MID-RANGE (IN)	.00	.29	1.41	3.81	8.11	14.96	24.96	38.60	56.39	78.79	106.32	139.68	179.59
BULLET PATH (IN)	-.75	1.04	.00	-5.01	-15.47	-33.02	-59.20	-95.37	-142.98	-203.34	-277.94	-368.51	-476.81
TIME OF FLIGHT (SEC)	.000000	.078237	.171165	.280776	.408205	.551316	.707171	.874239	1.051855	1.239856	1.438352	1.647589	1.867939
WIND DEFLECTION (IN)													
5 MPH	.00	.60	2.49	5.85	10.78	17.09	24.52	32.93	42.28	52.54	63.72	75.84	88.95
10 MPH	.00	1.20	4.98	11.70	21.56	34.17	49.03	65.87	84.55	105.07	127.44	151.69	177.90
20 MPH	.00	2.40	9.96	23.40	43.12	68.35	98.07	131.73	169.11	210.14	254.87	303.38	355.80
30 MPH	.00	3.59	14.95	35.11	64.67	102.52	147.10	197.60	253.66	315.22	382.31	455.07	533.70
VELOCITY (FPS)	2000.	1679.	1422.	1215.	1072.	979.	911.	856.	809.	766.	726.	690.	656.
ENERGY (FT-LB)	364.	257.	184.	134.	105.	87.	76.	67.	60.	53.	48.	43.	39.
DROP (IN)	.00	-1.20	-5.50	-14.14	-28.68	-50.74	-81.80	-123.25	-176.52	-242.86	-323.94	-421.45	-537.20
MID-RANGE (IN)	.00	.32	1.55	4.18	8.84	16.18	26.76	41.05	59.58	82.76	111.23	145.66	186.80
BULLET PATH (IN)	-.75	1.18	.00	-5.51	-16.93	-35.86	-63.79	-102.12	-152.26	-215.48	-293.43	-387.81	-500.44
TIME OF FLIGHT (SEC)	.000000	.082127	.179338	.293789	.425801	.572592	.731590	.901557	1.081987	1.272815	1.474191	1.686397	1.909837
WIND DEFLECTION (IN)													
5 MPH	.00	.63	2.58	6.05	11.07	17.39	24.78	33.14	42.41	52.61	63.73	75.80	88.87
10 MPH	.00	1.25	5.16	12.11	22.14	34.78	49.56	66.27	84.83	105.22	127.46	151.61	177.73
20 MPH	.00	2.51	10.33	24.21	44.28	69.55	99.12	132.55	169.66	210.43	254.92	303.21	355.46
30 MPH	.00	3.76	15.49	36.32	66.42	104.33	148.68	198.82	254.49	315.65	382.37	454.82	533.19

Bullet: Lyman # 225438 41 Gr.

Ballistic Coefficient: .094

(Cont'd.)

RANGE (YDS)	MUZZLE	50	100	150	200	250	300	350	400	450	500	550	600
VELOCITY (FPS)	1900.	1601.	1357.	1167.	1042.	958.	895.	842.	796.	754.	716.	680.	647.
ENERGY (FT-LB)	329.	233.	168.	124.	99.	84.	73.	65.	58.	52.	47.	42.	38.
DROP (IN)	.00	-1.33	-6.07	-15.57	-31.43	-55.26	-88.47	-132.48	-188.72	-258.39	-343.31	-445.16	-565.80
MID-RANGE (IN)	.00	.36	1.71	4.59	9.65	17.51	28.67	43.64	62.93	86.94	116.39	151.98	194.43
BULLET PATH (IN)	-.75	1.33	.00	-6.09	-18.54	-38.97	-68.76	-109.37	-162.19	-228.46	-309.97	-408.41	-525.64
TIME OF FLIGHT (SEC)	.000000	.086255	.188162	.307723	.444284	.594732	.756933	.929900	1.113276	1.307079	1.511496	1.726846	1.953564
WIND DEFLECTION (IN)													
5 MPH	.00	.64	2.66	6.24	11.31	17.60	24.93	33.20	42.39	52.50	63.54	75.54	88.55
10 MPH	.00	1.29	5.33	12.48	22.62	35.20	49.85	66.40	84.78	104.99	127.08	151.08	177.09
20 MPH	.00	2.57	10.65	24.95	45.23	70.40	99.70	132.80	169.56	209.99	254.15	302.17	354.18
30 MPH	.00	3.86	15.98	37.43	67.85	105.60	149.56	199.20	254.34	314.98	381.23	453.25	531.27
VELOCITY (FPS)	1800.	1525.	1295.	1124.	1014.	938.	878.	828.	783.	743.	705.	670.	637.
ENERGY (FT-LB)	295.	212.	153.	115.	94.	80.	70.	62.	56.	50.	45.	41.	37.
DROP (IN)	.00	-1.46	-6.70	-17.14	-34.43	-60.12	-95.59	-142.28	-201.55	-274.77	-363.70	-470.10	-595.85
MID-RANGE (IN)	.00	.39	1.89	5.04	10.52	18.89	30.65	46.31	66.34	91.24	121.73	158.51	202.33
BULLET PATH (IN)	-.75	1.51	.00	-6.72	-20.28	-42.25	-73.99	-116.96	-172.50	-241.99	-327.20	-429.87	-551.90
TIME OF FLIGHT (SEC)	.000000	.090600	.197522	.322282	.463219	.617252	.782663	.958684	1.145090	1.341966	1.549534	1.768151	1.998282
WIND DEFLECTION (IN)													
5 MPH	.00	.64	2.72	6.36	11.43	17.65	24.87	33.03	42.10	52.09	63.03	74.93	87.85
10 MPH	.00	1.28	5.43	12.72	22.86	35.30	49.75	66.06	84.20	104.19	126.05	149.86	175.70
20 MPH	.00	2.56	10.86	25.44	45.72	70.61	99.50	132.12	168.41	208.37	252.10	299.72	351.40
30 MPH	.00	3.84	16.29	38.16	68.58	105.91	149.25	198.19	252.61	312.56	378.15	449.58	527.09
VELOCITY (FPS)	1700.	1439.	1228.	1080.	985.	916.	860.	812.	769.	729.	693.	658.	626.
ENERGY (FT-LB)	263.	189.	137.	106.	88.	76.	67.	60.	54.	48.	44.	39.	36.
DROP (IN)	.00	-1.65	-7.53	-19.18	-38.24	-66.19	-104.42	-154.36	-217.28	-294.80	-388.62	-500.54	-632.52
MID-RANGE (IN)	.00	.45	2.12	5.62	11.59	20.56	33.03	49.50	70.40	96.37	128.12	166.35	211.84
BULLET PATH (IN)	-.75	1.74	.00	-7.51	-22.44	-46.24	-80.34	-126.14	-184.92	-258.30	-347.98	-455.77	-583.61
TIME OF FLIGHT (SEC)	.000000	.095996	.209089	.339842	.485626	.643755	.812914	.992555	1.182588	1.383153	1.594521	1.817087	2.051351
WIND DEFLECTION (IN)													
5 MPH	.00	.68	2.87	6.61	11.68	17.83	24.95	32.99	41.95	51.84	62.67	74.49	87.34
10 MPH	.00	1.37	5.74	13.22	23.35	35.65	49.90	65.98	83.90	103.67	125.34	148.98	174.68
20 MPH	.00	2.73	11.48	26.45	46.71	71.31	99.79	131.97	167.80	207.34	250.68	297.97	349.37
30 MPH	.00	4.10	17.22	39.67	70.06	106.96	149.69	197.95	251.70	311.01	376.02	446.95	524.05
VELOCITY (FPS)	1600.	1356.	1167.	1041.	958.	894.	842.	796.	754.	716.	680.	646.	615.
ENERGY (FT-LB)	233.	167.	124.	99.	84.	73.	65.	58.	52.	47.	42.	38.	34.
DROP (IN)	.00	-1.87	-8.50	-21.50	-42.48	-72.85	-114.03	-167.43	-234.29	-316.39	-415.44	-533.28	-671.94
MID-RANGE (IN)	.00	.51	2.38	6.26	12.74	22.32	35.51	52.81	74.66	101.79	134.87	174.64	221.94
BULLET PATH (IN)	-.75	2.01	.00	-8.38	-24.74	-50.48	-87.04	-135.81	-198.05	-275.53	-369.95	-483.17	-617.21
TIME OF FLIGHT (SEC)	.000000	.101997	.221654	.358297	.508811	.671069	.844090	1.027517	1.221371	1.425840	1.641241	1.868012	2.106687
WIND DEFLECTION (IN)													
5 MPH	.00	.73	3.01	6.78	11.78	17.80	24.78	32.67	41.48	51.22	61.93	73.64	86.39
10 MPH	.00	1.45	6.01	13.56	23.55	35.61	49.56	65.34	82.96	102.45	123.86	147.27	172.78
20 MPH	.00	2.90	12.02	27.12	47.10	71.22	99.12	130.69	165.92	204.90	247.72	294.54	345.55
30 MPH	.00	4.35	18.03	40.68	70.65	106.82	148.68	196.03	248.88	307.34	371.58	441.81	518.33

Bullet: Lyman # 225415 45 Gr.
Ballistic Coefficient: .116

RANGE (YDS)	MUZZLE	50	100	150	200	250	300	350	400	450	500	550	600
VELOCITY (FPS)	2700.	2336.	2048.	1784.	1582.	1402.	1250.	1130.	1044.	980.	929.	886.	848.
ENERGY (FT-LB)	728.	545.	419.	318.	250.	196.	156.	128.	109.	96.	86.	78.	72.
DROP (IN)	.00	-.64	-2.87	-7.18	-14.19	-24.71	-39.66	-60.19	-87.51	-122.77	-167.02	-221.23	-286.48
MID-RANGE (IN)	.00	.17	.79	2.06	4.23	7.63	12.65	19.80	29.56	42.43	58.80	79.05	103.58
BULLET PATH (IN)	-.75	.42	.00	-2.50	-7.71	-16.41	-29.56	-48.28	-73.79	-107.24	-149.68	-202.09	-265.52
TIME OF FLIGHT (SEC)	.000000	.059849	.128434	.206969	.296296	.397089	.510570	.637073	.775473	.923952	1.081276	1.246736	1.419938
WIND DEFLECTION (IN)													
5 MPH	.00	.38	1.52	3.55	6.52	10.50	15.60	21.84	29.13	37.31	46.26	55.93	66.29
10 MPH	.00	.76	3.05	7.09	13.04	21.00	31.19	43.68	58.26	74.62	92.53	111.87	132.58
20 MPH	.00	1.51	6.10	14.19	26.07	42.00	62.39	87.36	116.52	149.23	185.05	223.74	265.15
30 MPH	.00	2.27	9.15	21.28	39.11	63.00	93.58	131.04	174.78	223.85	277.58	335.61	397.73
VELOCITY (FPS)	2600.	2260.	1977.	1729.	1532.	1359.	1215.	1104.	1026.	966.	917.	876.	839.
ENERGY (FT-LB)	675.	510.	391.	299.	235.	185.	147.	122.	105.	93.	84.	77.	70.
DROP (IN)	.00	-.69	-3.08	-7.70	-15.22	-26.47	-42.44	-64.30	-93.26	-130.46	-176.88	-233.56	-301.54
MID-RANGE (IN)	.00	.18	.85	2.21	4.53	8.16	13.52	21.10	31.41	44.91	61.97	82.99	108.33
BULLET PATH (IN)	-.75	.48	.00	-2.70	-8.31	-17.64	-31.69	-51.64	-78.69	-113.96	-158.47	-213.24	-279.30
TIME OF FLIGHT (SEC)	.000000	.062085	.133048	.214385	.306579	.410603	.527523	.657312	.798517	.949408	1.108919	1.276437	1.451634
WIND DEFLECTION (IN)													
5 MPH	.00	.39	1.55	3.64	6.67	10.75	15.96	22.30	29.65	37.86	46.82	56.48	66.82
10 MPH	.00	.77	3.11	7.27	13.34	21.50	31.92	44.61	59.31	75.71	93.63	112.96	133.64
20 MPH	.00	1.55	6.22	14.54	26.69	42.99	63.84	89.22	118.62	151.42	187.26	225.92	267.28
30 MPH	.00	2.32	9.33	21.81	40.03	64.49	95.76	133.83	177.92	227.13	280.89	338.88	400.92
VELOCITY (FPS)	2500.	2181.	1905.	1674.	1483.	1317.	1181.	1080.	1008.	952.	905.	865.	829.
ENERGY (FT-LB)	624.	475.	362.	280.	220.	173.	139.	117.	102.	91.	82.	75.	69.
DROP (IN)	.00	-.74	-3.32	-8.29	-16.36	-28.42	-45.50	-68.80	-99.52	-138.76	-187.47	-246.75	-317.61
MID-RANGE (IN)	.00	.20	.92	2.37	4.86	8.75	14.46	22.51	33.40	47.55	65.32	87.12	113.30
BULLET PATH (IN)	-.75	.54	.00	-2.94	-8.98	-19.00	-34.05	-55.32	-84.00	-121.21	-167.89	-225.13	-293.95
TIME OF FLIGHT (SEC)	.000000	.064420	.138017	.222299	.317556	.424995	.545461	.678528	.822496	.975800	1.137523	1.307141	1.484385
WIND DEFLECTION (IN)													
5 MPH	.00	.39	1.59	3.72	6.82	11.00	16.32	22.75	30.14	38.35	47.30	56.95	67.27
10 MPH	.00	.78	3.17	7.44	13.65	22.00	32.64	45.50	60.28	76.70	94.60	113.90	134.53
20 MPH	.00	1.56	6.34	14.89	27.30	44.00	65.28	91.00	120.56	153.40	189.21	227.79	269.06
30 MPH	.00	2.33	9.51	22.33	40.95	66.00	97.92	136.50	180.84	230.10	283.81	341.69	403.60
VELOCITY (FPS)	2400.	2107.	1837.	1623.	1438.	1279.	1152.	1060.	993.	939.	895.	856.	821.
ENERGY (FT-LB)	575.	444.	337.	263.	207.	164.	133.	112.	98.	88.	80.	73.	67.
DROP (IN)	.00	-.80	-3.56	-8.89	-17.52	-30.39	-48.58	-73.30	-105.73	-146.93	-197.88	-259.64	-333.25
MID-RANGE.(IN)	.00	.21	.98	2.55	5.20	9.34	15.40	23.91	35.34	50.10	68.55	91.08	118.05
BULLET PATH (IN)	-.75	.61	.00	-3.17	-9.66	-20.37	-36.41	-58.97	-89.25	-128.30	-177.09	-236.70	-308.16
TIME OF FLIGHT (SEC)	.000000	.066703	.142964	.230019	.328260	.438977	.562758	.698789	.845254	1.000772	1.164545	1.336123	1.515291
WIND DEFLECTION (IN)													
5 MPH	.00	.37	1.58	3.74	6.89	11.13	16.52	22.99	30.38	38.57	47.48	57.08	67.35
10 MPH	.00	.74	3.16	7.48	13.77	22.26	33.05	45.99	60.76	77.14	94.96	114.16	134.69
20 MPH	.00	1.48	6.32	14.97	27.55	44.52	66.09	91.97	121.53	154.27	189.92	228.32	269.38
30 MPH	.00	2.22	9.49	22.45	41.32	66.78	99.14	137.96	182.29	231.41	284.88	342.47	404.07

Bullet: Lyman # 225415 45 Gr.

Ballistic Coefficient: .116

(Cont'd.)

RANGE (YDS)	MUZZLE	50	100	150	200	250	300	350	400	450	500	550	600
VELOCITY (FPS)	2300.	2014.	1758.	1559.	1382.	1233.	1118.	1035.	973.	924.	881.	844.	810.
ENERGY (FT-LB)	528.	405.	309.	243.	191.	152.	125.	107.	95.	85.	78.	71.	65.
DROP (IN)	.00	-.87	-3.89	-9.71	-19.12	-33.11	-52.83	-79.47	-114.20	-158.02	-211.95	-277.04	-354.31
MID-RANGE (IN)	.00	.23	1.07	2.78	5.67	10.15	16.71	25.83	37.97	53.53	72.87	96.34	124.33
BULLET PATH (IN)	-.75	.70	.00	-3.50	-10.60	-22.26	-39.66	-63.99	-96.39	-137.90	-189.51	-252.27	-327.23
TIME OF FLIGHT (SEC)	.000000	.069693	.149546	.240200	.342493	.457578	.585627	.725354	.874968	1.033318	1.199740	1.373873	1.555563
WIND DEFLECTION (IN)													
5 MPH	.00	.39	1.68	3.92	7.18	11.57	17.10	23.66	31.08	39.28	48.19	57.77	68.02
10 MPH	.00	.79	3.36	7.84	14.37	23.14	34.20	47.31	62.17	78.56	96.37	115.54	136.04
20 MPH	.00	1.58	6.73	15.68	28.73	46.28	68.40	94.63	124.34	157.12	192.74	231.08	272.08
30 MPH	.00	2.36	10.09	23.52	43.10	69.43	102.60	141.94	186.50	235.68	289.11	346.62	408.12
VELOCITY (FPS)	2200.	1922.	1688.	1495.	1328.	1190.	1086.	1012.	955.	908.	868.	832.	799.
ENERGY (FT-LB)	484.	369.	285.	223.	176.	141.	118.	102.	91.	82.	75.	69.	64.
DROP (IN)	.00	-.95	-4.26	-10.62	-20.90	-36.12	-57.48	-86.18	-123.33	-169.89	-226.93	-295.49	-376.58
MID-RANGE (IN)	.00	.26	1.18	3.04	6.19	11.05	18.12	27.88	40.77	57.13	77.34	101.78	130.81
BULLET PATH (IN)	-.75	.80	.00	-3.85	-11.63	-24.34	-43.20	-69.39	-104.03	-148.08	-202.62	-268.67	-347.26
TIME OF FLIGHT (SEC)	.000000	.072959	.156522	.250996	.357565	.477136	.609387	.752667	.905368	1.066534	1.235620	1.412345	1.596610
WIND DEFLECTION (IN)													
5 MPH	.00	.42	1.77	4.09	7.47	11.99	17.63	24.23	31.67	39.85	48.73	58.29	68.50
10 MPH	.00	.84	3.55	8.18	14.93	23.98	35.25	48.47	63.34	79.71	97.47	116.57	137.00
20 MPH	.00	1.68	7.10	16.35	29.86	47.95	70.50	96.94	126.69	159.42	194.94	233.15	274.01
30 MPH	.00	2.52	10.64	24.53	44.79	71.93	105.76	145.41	190.03	239.13	292.41	349.72	411.01
VELOCITY (FPS)	2100.	1830.	1617.	1433.	1275.	1149.	1057.	991.	938.	893.	855.	820.	787.
ENERGY (FT-LB)	441.	334.	261.	205.	162.	132.	112.	98.	88.	80.	73.	67.	62.
DROP (IN)	.00	-1.05	-4.68	-11.65	-22.88	-39.48	-62.63	-93.53	-133.24	-182.72	-243.06	-315.27	-400.26
MID-RANGE (IN)	.00	.28	1.29	3.33	6.76	12.04	19.66	30.09	43.74	60.93	82.05	107.48	137.49
BULLET PATH (IN)	-.75	.92	.00	-4.26	-12.77	-26.64	-47.09	-75.27	-112.26	-159.03	-216.65	-286.14	-368.41
TIME OF FLIGHT (SEC)	.000000	.076536	.163918	.262532	.373656	.497841	.634226	.780991	.936776	1.100796	1.272612	1.452012	1.638953
WIND DEFLECTION (IN)													
5 MPH	.00	.45	1.85	4.25	7.74	12.38	18.10	24.73	32.15	40.30	49.13	58.63	68.80
10 MPH	.00	.90	3.71	8.49	15.48	24.76	36.20	49.45	64.30	80.60	98.27	117.27	137.60
20 MPH	.00	1.80	7.41	16.98	30.96	49.53	72.39	98.91	128.60	161.19	196.53	234.54	275.20
30 MPH	.00	2.70	11.12	25.47	46.43	74.29	108.59	148.36	192.90	241.79	294.80	351.81	412.80
VELOCITY (FPS)	2000.	1747.	1549.	1373.	1226.	1113.	1032.	971.	921.	879.	842.	808.	777.
ENERGY (FT-LB)	400.	305.	240.	188.	150.	124.	106.	94.	85.	77.	71.	65.	60.
DROP (IN)	.00	-1.16	-5.15	-12.79	-25.05	-43.11	-68.16	-101.34	-143.67	-196.16	-259.86	-335.81	-424.83
MID-RANGE (IN)	.00	.31	1.42	3.64	7.38	13.10	21.29	32.40	46.78	64.80	86.82	113.22	144.26
BULLET PATH (IN)	-.75	1.04	.00	-4.69	-14.01	-29.11	-51.21	-81.44	-120.82	-170.36	-231.12	-304.11	-390.19
TIME OF FLIGHT (SEC)	.000000	.080423	.171643	.274574	.390337	.519033	.659310	.809397	.968175	1.135000	1.309523	1.491598	1.681228
WIND DEFLECTION (IN)													
5 MPH	.00	.48	1.90	4.36	7.95	12.67	18.42	25.03	32.40	40.48	49.24	58.66	68.75
10 MPH	.00	.95	3.81	8.72	15.90	25.35	36.84	50.05	64.80	80.96	98.48	117.32	137.50
20 MPH	.00	1.91	7.62	17.45	31.80	50.70	73.68	100.11	129.60	161.92	196.95	234.64	274.99
30 MPH	.00	2.86	11.43	26.17	47.70	76.05	110.52	150.16	194.40	242.88	295.43	351.96	412.49

Bullet: Lyman # 225415 45 Gr. 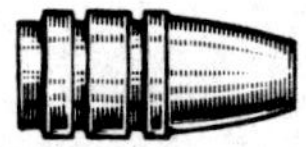
Ballistic Coefficient: .116
(Cont'd.)

RANGE (YDS)	MUZZLE	50	100	150	200	250	300	350	400	450	500	550	600
VELOCITY (FPS)	1900.	1672.	1481.	1316.	1180.	1080.	1007.	951.	905.	865.	829.	796.	766.
ENERGY (FT-LB)	361.	279.	219.	173.	139.	116.	101.	90.	82.	75.	69.	63.	59.
DROP (IN)	.00	-1.28	-5.67	-14.05	-27.46	-47.12	-74.19	-109.80	-154.90	-210.56	-277.81	-357.68	-450.90
MID-RANGE (IN)	.00	.34	1.56	3.99	8.07	14.26	23.05	34.85	49.98	68.84	91.79	119.19	151.26
BULLET PATH (IN)	-.75	1.18	.00	-5.17	-15.38	-31.83	-55.69	-88.10	-129.98	-182.44	-246.48	-323.14	-413.15
TIME OF FLIGHT (SEC)	.000000	.084429	.179802	.287368	.407965	.541150	.685215	.838605	1.000405	1.170094	1.347407	1.532259	1.724688
WIND DEFLECTION (IN)													
5 MPH	.00	.48	1.93	4.45	8.11	12.88	18.61	25.17	32.46	40.44	49.10	58.42	68.40
10 MPH	.00	.96	3.86	8.89	16.22	25.77	37.23	50.33	64.91	80.88	98.20	116.84	136.81
20 MPH	.00	1.93	7.71	17.79	32.45	51.54	74.46	100.66	129.83	161.77	196.39	233.67	273.62
30 MPH	.00	2.89	11.57	26.68	48.67	77.31	111.69	150.99	194.74	242.65	294.59	350.51	410.42
VELOCITY (FPS)	1800.	1597.	1415.	1260.	1138.	1050.	985.	933.	889.	851.	816.	784.	754.
ENERGY (FT-LB)	324.	255.	200.	159.	129.	110.	97.	87.	79.	72.	67.	61.	57.
DROP (IN)	.00	-1.40	-6.24	-15.43	-30.11	-51.49	-80.72	-118.87	-166.90	-225.88	-296.86	-380.71	-478.47
MID-RANGE (IN)	.00	.38	1.71	4.38	8.82	15.52	24.93	37.42	53.33	73.05	96.94	125.30	158.51
BULLET PATH (IN)	-.75	1.34	.00	-5.70	-16.89	-34.78	-60.52	-95.17	-139.71	-195.20	-262.68	-343.04	-437.31
TIME OF FLIGHT (SEC)	.000000	.088515	.188386	.300874	.426405	.563964	.711724	.868400	1.033251	1.205864	1.386050	1.573778	1.769113
WIND DEFLECTION (IN)													
5 MPH	.00	.46	1.91	4.48	8.19	12.96	18.63	25.09	32.26	40.12	48.64	57.83	67.68
10 MPH	.00	.91	3.82	8.95	16.38	25.92	37.26	50.17	64.52	80.23	97.28	115.65	135.36
20 MPH	.00	1.82	7.65	17.91	32.76	51.85	74.53	100.34	129.04	160.46	194.56	231.30	270.73
30 MPH	.00	2.74	11.47	26.86	49.14	77.77	111.79	150.52	193.56	240.70	291.83	346.96	406.09
VELOCITY (FPS)	1700.	1506.	1337.	1197.	1092.	1016.	958.	911.	870.	834.	801.	770.	741.
ENERGY (FT-LB)	289.	227.	179.	143.	119.	103.	92.	83.	76.	69.	64.	59.	55.
DROP (IN)	.00	-1.58	-7.02	-17.33	-33.71	-57.35	-89.38	-130.76	-182.57	-245.81	-321.53	-410.45	-514.01
MID-RANGE (IN)	.00	.42	1.93	4.91	9.84	17.19	27.36	40.70	57.57	78.36	103.42	132.97	167.63
BULLET PATH (IN)	-.75	1.56	.00	-6.43	-18.93	-38.69	-66.83	-104.34	-152.26	-211.62	-283.46	-368.49	-468.17
TIME OF FLIGHT (SEC)	.000000	.093780	.199577	.318349	.449863	.592522	.744677	.905341	1.073948	1.250204	1.434000	1.625363	1.824377
WIND DEFLECTION (IN)													
5 MPH	.00	.49	2.03	4.72	8.53	13.32	18.94	25.32	32.39	40.14	48.54	57.62	67.37
10 MPH	.00	.98	4.07	9.44	17.06	26.64	37.89	50.63	64.78	80.27	97.09	115.24	134.74
20 MPH	.00	1.95	8.13	18.88	34.12	53.27	75.77	101.27	129.56	160.54	194.18	230.48	269.47
30 MPH	.00	2.93	12.20	28.32	51.17	79.91	113.66	151.90	194.34	240.81	291.27	345.72	404.21
VELOCITY (FPS)	1600.	1418.	1262.	1140.	1051.	986.	934.	890.	851.	817.	785.	755.	727.
ENERGY (FT-LB)	256.	201.	159.	130.	110.	97.	87.	79.	72.	67.	62.	57.	53.
DROP (IN)	.00	-1.79	-7.93	-19.54	-37.82	-63.94	-98.95	-143.82	-199.64	-267.42	-348.06	-442.59	-552.35
MID-RANGE (IN)	.00	.48	2.17	5.51	10.98	18.99	29.93	44.14	62.00	83.88	110.09	141.02	177.23
BULLET PATH (IN)	-.75	1.80	.00	-7.27	-21.20	-42.98	-73.65	-114.18	-165.66	-229.10	-305.40	-395.58	-501.01
TIME OF FLIGHT (SEC)	.000000	.099659	.211917	.337223	.474585	.622177	.778701	.943409	1.115883	1.295931	1.483521	1.678714	1.881618
WIND DEFLECTION (IN)													
5 MPH	.00	.52	2.15	4.93	8.76	13.50	19.03	25.27	32.20	39.79	48.05	56.98	66.58
10 MPH	.00	1.04	4.30	9.85	17.53	27.00	38.05	50.54	64.40	79.58	96.10	113.95	133.16
20 MPH	.00	2.08	8.59	19.70	35.05	54.01	76.10	101.08	128.79	159.17	192.20	227.91	266.33
30 MPH	.00	3.12	12.89	29.55	52.58	81.01	114.15	151.62	193.19	238.75	288.30	341.86	399.49

Bullet: Lyman # 225462 54 Gr.
Ballistic Coefficient: .107

RANGE (YDS)	MUZZLE	50	100	150	200	250	300	350	400	450	500	550	600
VELOCITY (FPS)	2700.	2313.	2008.	1737.	1520.	1332.	1180.	1071.	995.	936.	888.	846.	808.
ENERGY (FT-LB)	874.	641.	483.	362.	277.	213.	167.	138.	119.	105.	94.	86.	78.
DROP (IN)	.00	-.65	-2.92	-7.35	-14.66	-25.77	-41.79	-64.05	-93.85	-132.39	-180.74	-240.10	-311.59
MID-RANGE (IN)	.00	.17	.81	2.13	4.42	8.06	13.55	21.43	32.26	46.52	64.60	86.97	114.00
BULLET PATH (IN)	-.75	.44	.00	-2.60	-8.08	-17.36	-31.55	-51.97	-79.94	-116.64	-163.17	-220.69	-290.34
TIME OF FLIGHT (SEC)	.000000	.060205	.129825	.210336	.302720	.408272	.528158	.661928	.807522	.963175	1.127911	1.201220	1.482886
WIND DEFLECTION (IN)													
5 MPH	.00	.41	1.65	3.84	7.08	11.48	17.14	24.03	31.95	40.76	50.37	60.73	71.83
10 MPH	.00	.82	3.29	7.69	14.17	22.97	34.29	48.05	63.90	81.52	100.73	121.46	143.65
20 MPH	.00	1.64	6.59	15.37	28.34	45.93	68.58	96.11	127.80	163.04	201.47	242.92	287.31
30 MPH	.00	2.46	9.88	23.06	42.50	68.90	102.87	144.16	191.70	244.56	302.20	364.38	430.96
VELOCITY (FPS)	2600.	2239.	1939.	1684.	1474.	1293.	1151.	1051.	980.	924.	877.	837.	800.
ENERGY (FT-LB)	810.	601.	451.	340.	260.	201.	159.	132.	115.	102.	92.	84.	77.
DROP (IN)	.00	-.70	-3.13	-7.88	-15.70	-27.56	-44.62	-68.23	-99.67	-140.09	-190.61	-252.39	-326.56
MID-RANGE (IN)	.00	.19	.87	2.28	4.72	8.60	14.43	22.76	34.13	48.99	67.75	90.84	118.66
BULLET PATH (IN)	-.75	.49	.00	-2.81	-8.69	-18.61	-33.73	-55.40	-84.90	-123.39	-171.96	-231.80	-304.03
TIME OF FLIGHT (SEC)	.000000	.062415	.134417	.217662	.312940	.421740	.544948	.681653	.829691	.987508	1.154256	1.329505	1.513094
WIND DEFLECTION (IN)													
5 MPH	.00	.42	1.67	3.92	7.23	11.73	17.49	24.45	32.40	41.21	50.81	61.15	72.23
10 MPH	.00	.83	3.35	7.85	14.46	23.46	34.99	48.89	64.79	82.42	101.61	122.30	144.46
20 MPH	.00	1.66	6.70	15.69	28.92	46.91	69.98	97.79	129.59	164.83	203.22	244.60	288.92
30 MPH	.00	2.49	10.05	23.54	43.39	70.37	104.96	146.68	194.38	247.25	304.83	366.90	433.38
VELOCITY (FPS)	2500.	2160.	1867.	1628.	1424.	1253.	1122.	1031.	964.	911.	866.	827.	791.
ENERGY (FT-LB)	749.	560.	418.	318.	243.	188.	151.	127.	111.	100.	90.	82.	75.
DROP (IN)	.00	-.75	-3.37	-8.48	-16.89	-29.61	-47.88	-73.03	-106.32	-148.87	-201.82	-266.32	-343.46
MID-RANGE (IN)	.00	.20	.93	2.45	5.08	9.24	15.46	24.30	36.27	51.80	71.31	95.20	123.85
BULLET PATH (IN)	-.75	.56	.00	-3.05	-9.40	-20.06	-36.27	-59.36	-90.60	-131.09	-181.98	-244.42	-319.49
TIME OF FLIGHT (SEC)	.000000	.064752	.139447	.225697	.324298	.436767	.563606	.703423	.854082	1.014250	1.183208	1.360602	1.546329
WIND DEFLECTION (IN)													
5 MPH	.00	.42	1.71	4.02	7.42	12.04	17.92	24.94	32.92	41.73	51.32	61.65	72.72
10 MPH	.00	.84	3.42	8.04	14.84	24.07	35.83	49.88	65.84	83.47	102.64	123.31	145.43
20 MPH	.00	1.67	6.85	16.09	29.67	48.14	71.67	99.76	131.68	166.94	205.29	246.61	290.87
30 MPH	.00	2.51	10.27	24.13	44.51	72.21	107.50	149.65	197.52	250.40	307.93	369.92	436.30
VELOCITY (FPS)	2400.	2088.	1801.	1576.	1379.	1217.	1096.	1013.	951.	900.	856.	818.	782.
ENERGY (FT-LB)	691.	523.	389.	298.	228.	178.	144.	123.	108.	97.	88.	80.	73.
DROP (IN)	.00	-.80	-3.61	-9.08	-18.08	-31.67	-51.14	-77.80	-112.90	-157.51	-212.81	-279.93	-359.91
MID-RANGE (IN)	.00	.22	1.00	2.62	5.43	9.87	16.48	25.82	38.37	54.53	74.73	99.38	128.82
BULLET PATH (IN)	-.75	.63	.00	-3.29	-10.11	-21.52	-38.81	-63.29	-96.21	-138.64	-191.76	-256.71	-334.50
TIME OF FLIGHT (SEC)	.000000	.067012	.144377	.233508	.335361	.451347	.581540	.724167	.877232	1.039588	1.210620	1.390042	1.577801
WIND DEFLECTION (IN)													
5 MPH	.00	.40	1.71	4.05	7.51	12.22	18.18	25.23	33.20	41.98	51.53	61.82	72.85
10 MPH	.00	.79	3.41	8.10	15.02	24.44	36.35	50.45	66.39	83.97	103.07	123.65	145.69
20 MPH	.00	1.59	6.82	16.19	30.05	48.87	72.70	100.91	132.79	167.94	206.14	247.29	291.39
30 MPH	.00	2.38	10.23	24.29	45.07	73.31	109.05	151.36	199.18	251.90	309.21	370.94	437.08

Bullet: Lyman # 225462 54 Gr.
Ballistic Coefficient: .107

(Cont'd.)

RANGE (YDS)	MUZZLE	50	100	150	200	250	300	350	400	450	500	550	600
VELOCITY (FPS)	2300.	1996.	1728.	1512.	1325.	1175.	1068.	992.	934.	886.	844.	807.	772.
ENERGY (FT-LB)	634.	477.	358.	274.	211.	166.	137.	118.	105.	94.	85.	78.	71.
DROP (IN)	.00	-.88	-3.94	-9.92	-19.73	-34.51	-55.57	-84.22	-121.66	-168.95	-227.29	-297.81	-381.42
MID-RANGE (IN)	.00	.24	1.10	2.86	5. 2	10.74	17.86	27.83	41.09	58.06	79.14	104.73	135.14
BULLET PATH (IN)	-.75	.72	.00	-3.63	-11.09	-23.52	-42.24	-68.54	-103.63	-148.58	-204.57	-272.74	-354.01
TIME OF FLIGHT (SEC)	.000000	.070022	.151002	.243864	.349956	.470399	.604669	.750677	.906694	1.071768	1.245400	1.427385	1.617722
WIND DEFLECTION (IN)													
5 MPH	.00	.42	1.81	4.24	7.84	12.70	18.78	25.8	33.88	42.66	52.20	62.48	73.4
10 MPH	.00	.85	3.62	8.49	15.68	25.40	37.55	51.77	67.75	85.33	104.41	124.96	146.98
20 MPH	.00	1.69	7.24	16.97	31.36	50.80	75.10	103.54	135.50	170.65	208.82	249.92	293.96
30 MPH	.00	2.54	10.86	25.46	47.04	76.20	112.66	155.31	203.26	255.98	313.22	374.88	440.94
VELOCITY (FPS)	2200.	1904.	1657.	1449.	1273.	1137.	1041.	972.	918.	872.	832.	795.	762.
ENERGY (FT-LB)	580.	434.	329.	252.	194.	155.	130.	113.	101.	91.	83.	76.	70.
DROP (IN)	.00	-.96	-4.32	-10.86	-21.57	-37.65	-60.44	-91.22	-131.11	-181.24	-242.77	-316.83	-404.32
MID-RANGE (IN)	.00	.26	1.20	3.13	6.46	11.69	19.36	29.98	43.98	61.76	83.73	110.29	141.72
BULLET PATH (IN)	-.75	.83	.00	-4.00	-12.17	-25.72	-45.97	-74.21	-111.56	-159.16	-218.16	-289.68	-374.63
TIME OF FLIGHT (SEC)	.000000	.073311	.158018	.254895	.365471	.490452	.628687	.778008	.936971	1.104794	1.281083	1.465707	1.658705
WIND DEFLECTION (IN)													
5 MPH	.00	.45	1.91	4.43	8.16	13.16	19.32	26.46	34.45	43.22	52.74	62.98	73.97
10 MPH	.00	.90	3.81	8.86	16.32	26.32	38.65	52.93	68.91	86.44	105.47	125.96	147.93
20 MPH	.00	1.81	7.62	17.72	32.65	52.64	77.30	105.86	137.81	172.89	210.94	251.93	295.86
30 MPH	.00	2.71	11.43	26.58	48.97	78.96	115.95	158.79	206.72	259.33	316.41	377.89	443.80
VELOCITY (FPS)	2100.	1812.	1585.	1388.	1224.	1101.	1016.	953.	902.	858.	819.	784.	751.
ENERGY (FT-LB)	529.	394.	301.	231.	180.	145.	124.	109.	98.	88.	80.	74.	68.
DROP (IN)	.00	-1.06	-4.75	-11.92	-23.64	-41.16	-65.82	-98.86	-141.37	-194.51	-259.42	-337.13	-428.78
MID-RANGE (IN)	.00	.28	1.32	3.44	7.07	12.74	20.98	32.29	47.05	65.67	88.55	116.05	148.57
BULLET PATH (IN)	-.75	.94	.00	-4.42	-13.38	-28.16	-50.07	-80.36	-120.11	-170.51	-232.67	-307.63	-396.53
TIME OF FLIGHT (SEC)	.000000	.076912	.165498	.266726	.382043	.511608	.653710	.806322	.968264	1.138900	1.317935	1.505305	1.701080
WIND DEFLECTION (IN)													
5 MPH	.00	.48	1.99	4.61	8.48	13.59	19.81	26.96	34.92	43.65	53.12	63.32	74.27
10 MPH	.00	.97	3.98	9.23	16.95	27.19	39.62	53.91	69.84	87.30	106.24	126.65	148.53
20 MPH	.00	1.93	7.97	18.46	33.91	54.37	79.25	107.83	139.69	174.61	212.48	253.30	297.07
30 MPH	.00	2.90	11.95	27.69	50.86	81.56	118.87	161.74	209.53	261.91	318.73	379.94	445.60
VELOCITY (FPS)	2000.	1732.	1515.	1328.	1178.	1069.	993.	935.	887.	845.	807.	773.	741.
ENERGY (FT-LB)	480.	360.	275.	212.	166.	137.	118.	105.	94.	86.	78.	72.	66.
DROP (IN)	.00	-1.17	-5.23	-13.11	-25.93	-45.01	-71.67	-107.08	-152.34	-208.62	-277.05	-358.57	-454.55
MID-RANGE (IN)	.00	.31	1.45	3.77	7.74	13.88	22.73	34.72	50.24	69.70	93.51	121.97	155.61
BULLET PATH (IN)	-.75	1.07	.00	-4.89	-14.72	-30.82	-54.48	-86.91	-129.18	-182.47	-247.91	-326.44	-419.43
TIME OF FLIGHT (SEC)	.000000	.080797	.173443	.279292	.399484	.533528	.679349	.835199	1.000118	1.173601	1.355438	1.545624	1.744254
WIND DEFLECTION (IN)													
5 MPH	.00	.51	2.06	4.78	8.75	13.95	20.18	27.30	35.21	43.88	53.28	63.41	74.29
10 MPH	.00	1.02	4.13	9.56	17.51	27.90	40.37	54.60	70.42	87.75	106.56	126.83	148.59
20 MPH	.00	2.04	8.25	19.11	35.02	55.80	80.73	109.19	140.84	175.51	213.11	252.66	297.18
30 MPH	.00	3.06	12.38	28.67	52.53	83.70	121.10	163.79	211.26	263.26	319.67	380.49	445.77

Bullet: Lyman # 225462 54 Gr.

Ballistic Coefficient: .107

(Cont'd.)

RANGE (YDS)	MUZZLE	50	100	150	200	250	300	350	400	450	500	550	600
VELOCITY (FPS)	1900.	1653.	1447.	1271.	1135.	1040.	971.	917.	871.	831.	795.	761.	730.
ENERGY (FT-LB)	433.	328.	251.	194.	154.	130.	113.	101.	91.	83.	76.	69.	64.
DROP (IN)	.00	-1.29	-5.77	-14.43	-28.49	-49.27	-78.05	-115.97	-164.15	-223.74	-295.87	-381.45	-482.00
MID-RANGE (IN)	.00	.35	1.60	4.14	8.48	15.13	24.60	37.28	53.59	73.91	98.67	128.14	162.95
BULLET PATH (IN)	-.75	1.22	.00	-5.40	-16.20	-33.72	-59.25	-93.91	-138.82	-195.15	-264.03	-346.35	-443.64
TIME OF FLIGHT (SEC)	.000000	.084869	.181934	.292717	.417902	.556312	.705780	.864874	1.032818	1.209225	1.393966	1.587082	1.788688
WIND DEFLECTION (IN)													
5 MPH	.00	.52	2.12	4.92	8.99	14.22	20.42	27.48	35.31	43.89	53.20	63.24	74.04
10 MPH	.00	1.04	4.23	9.83	17.97	28.44	40.85	54.95	70.62	87.77	106.39	126.48	148.07
20 MPH	.00	2.08	8.46	19.67	35.94	56.87	81.70	109.91	141.24	175.54	212.78	252.97	296.14
30 MPH	.00	3.13	12.69	29.50	53.92	85.31	122.55	164.86	211.85	263.31	319.17	379.45	444.22
VELOCITY (FPS)	1800.	1576.	1380.	1218.	1097.	1013.	951.	900.	857.	818.	782.	750.	719.
ENERGY (FT-LB)	388.	298.	228.	178.	144.	123.	108.	97.	88.	80.	73.	67.	62.
DROP (IN)	.00	-1.42	-6.36	-15.89	-31.29	-53.88	-84.91	-125.44	-176.65	-239.69	-315.57	-405.44	-510.74
MID-RANGE (IN)	.00	.38	1.76	4.56	9.28	16.46	26.56	39.94	57.04	78.24	103.90	134.45	170.46
BULLET PATH (IN)	-.75	1.39	.00	-5.97	-17.82	-36.85	-64.32	-101.29	-148.95	-208.43	-280.76	-367.07	-468.81
TIME OF FLIGHT (SEC)	.000000	.089090	.190893	.306826	.436969	.579552	.732579	.894898	1.065893	1.245278	1.432998	1.629128	1.833805
WIND DEFLECTION (IN)													
5 MPH	.00	.51	2.13	5.00	9.12	14.33	20.47	27.42	35.13	43.58	52.77	62.70	73.37
10 MPH	.00	1.01	4.26	10.00	18.24	28.67	40.93	54.84	70.26	87.17	105.54	125.39	146.75
20 MPH	.00	2.03	8.53	20.00	36.48	57.34	81.87	109.67	140.53	174.34	211.08	250.79	293.50
30 MPH	.00	3.04	12.79	30.00	54.72	86.00	122.80	164.51	210.79	261.51	316.62	376.18	440.25
VELOCITY (FPS)	1700.	1487.	1305.	1160.	1057.	984.	928.	880.	839.	802.	768.	736.	706.
ENERGY (FT-LB)	346.	265.	204.	161.	134.	116.	103.	93.	84.	77.	71.	65.	60.
DROP (IN)	.00	-1.60	-7.16	-17.84	-34.96	-59.83	-93.62	-137.41	-192.38	-259.66	-340.14	-435.31	-546.49
MID-RANGE (IN)	.00	.43	1.98	5.10	10.33	18.15	28.97	43.20	61.23	83.48	110.24	142.10	179.63
BULLET PATH (IN)	-.75	1.61	.00	-6.73	-19.90	-40.82	-70.65	-110.49	-161.51	-224.84	-301.37	-392.58	-499.81
TIME OF FLIGHT (SEC)	.000000	.094392	.202203	.324409	.460238	.607542	.764703	.930838	1.105492	1.288488	1.479842	1.679656	1.888100
WIND DEFLECTION (IN)													
5 MPH	.00	.54	2.26	5.25	9.44	14.64	20.71	27.56	35.17	43.50	52.58	62.40	72.98
10 MPH	.00	1.08	4.53	10.51	18.88	29.28	41.41	55.12	70.33	87.01	105.16	124.80	145.95
20 MPH	.00	2.17	9.06	21.02	37.77	58.56	82.82	110.24	140.66	174.02	210.32	249.59	291.91
30 MPH	.00	3.25	13.59	31.52	56.65	87.84	124.23	165.37	210.99	261.03	315.47	374.39	437.86
VELOCITY (FPS)	1600.	1400.	1234.	1108.	1021.	957.	905.	861.	822.	786.	753.	722.	693.
ENERGY (FT-LB)	307.	235.	182.	147.	125.	110.	98.	89.	81.	74.	68.	63.	58.
DROP (IN)	.00	-1.82	-8.09	-20.08	-39.13	-66.47	-103.20	-150.49	-209.47	-281.17	-366.72	-467.54	-585.01
MID-RANGE (IN)	.00	.49	2.23	5.72	11.47	19.95	31.52	46.60	65.61	88.90	116.89	150.18	189.30
BULLET PATH (IN)	-.75	1.85	.00	-7.58	-22.20	-45.12	-77.44	-120.30	-174.86	-242.14	-323.27	-419.68	-532.72
TIME OF FLIGHT (SEC)	.000000	.100306	.214630	.343252	.484566	.636499	.797821	.967867	1.146322	1.333108	1.528291	1.732002	1.944443
WIND DEFLECTION (IN)													
5 MPH	.00	.58	2.39	5.46	9.64	14.76	20.71	27.42	34.88	43.06	51.99	61.67	72.11
10 MPH	.00	1.15	4.77	10.91	19.28	29.52	41.42	54.84	69.75	86.13	103.98	123.33	144.22
20 MPH	.00	2.31	9.55	21.82	38.57	59.05	82.83	109.69	139.51	172.25	207.96	246.66	288.44
30 MPH	.00	3.46	14.32	32.74	57.85	88.57	124.25	164.53	209.26	258.38	311.94	370.00	432.67

Bullet: Lyman # 245496 83 Gr.
Ballistic Coefficient: .202

RANGE (YDS)	MUZZLE	50	100	150	200	250	300	350	400	450	500	550	600
VELOCITY (FPS)	1900.	1722.	1560.	1420.	1296.	1191.	1112.	1050.	1001.	960.	925.	893.	864.
ENERGY (FT-LB)	665.	547.	449.	372.	309.	261.	228.	203.	185.	170.	158.	147.	138.
DROP (IN)	.00	-1.23	-5.39	-13.13	-25.20	-42.46	-65.83	-96.21	-134.44	-181.32	-237.51	-303.81	-380.97
MID-RANGE (IN)	.00	.33	1.46	3.66	7.21	12.47	19.78	29.53	42.02	57.54	76.30	98.58	124.65
BULLET PATH (IN)	-.75	1.09	.00	-4.67	-13.67	-27.86	-48.16	-75.47	-110.62	-154.43	-207.56	-270.78	-344.87
TIME OF FLIGHT (SEC)	.000000	.082934	.174514	.275335	.386010	.506924	.637433	.776402	.922812	1.075932	1.235276	1.400517	1.571447
WIND DEFLECTION (IN)													
5 MPH	.00	.35	1.46	3.39	6.18	9.87	14.41	19.69	25.63	32.16	39.23	46.82	54.92
10 MPH	.00	.70	2.93	6.77	12.36	19.75	28.82	39.38	51.26	64.31	78.46	93.65	109.84
20 MPH	.00	1.40	5.85	13.55	24.72	39.49	57.64	78.77	102.51	128.62	156.92	187.30	219.68
30 MPH	.00	2.10	8.78	20.32	37.08	59.24	86.46	118.15	153.77	192.93	235.38	280.95	329.51
VELOCITY (FPS)	1800.	1629.	1480.	1348.	1234.	1145.	1076.	1022.	977.	940.	906.	877.	849.
ENERGY (FT-LB)	597.	489.	404.	335.	281.	241.	213.	192.	176.	163.	151.	142.	133.
DROP (IN)	.00	-1.38	-6.02	-14.66	-28.08	-47.22	-72.97	-106.21	-147.75	-198.34	-258.63	-329.48	-411.60
MID-RANGE (IN)	.00	.37	1.63	4.08	8.02	13.82	21.83	32.39	45.79	62.29	82.09	105.53	132.83
BULLET PATH (IN)	-.75	1.26	.00	-5.26	-15.30	-31.05	-53.42	-83.28	-121.43	-168.64	-225.55	-293.01	-371.75
TIME OF FLIGHT (SEC)	.000000	.087607	.184284	.290522	.406899	.533324	.668690	.811922	.962150	1.118793	1.281459	1.449895	1.623941
WIND DEFLECTION (IN)													
5 MPH	.00	.38	1.55	3.57	6.47	10.27	14.84	20.12	26.00	32.45	39.44	46.92	54.91
10 MPH	.00	.75	3.10	7.13	12.95	20.53	29.69	40.23	52.01	64.91	78.87	93.85	109.81
20 MPH	.00	1.50	6.20	14.26	25.90	41.06	59.38	80.46	104.01	129.81	157.74	187.70	219.63
30 MPH	.00	2.26	9.30	21.40	38.84	61.60	89.07	120.69	156.02	194.72	236.61	281.54	329.44
VELOCITY (FPS)	1700.	1541.	1402.	1280.	1179.	1103.	1043.	995.	955.	920.	889.	861.	835.
ENERGY (FT-LB)	533.	437.	362.	302.	256.	224.	201.	183.	168.	156.	146.	136.	128.
DROP (IN)	.00	-1.54	-6.76	-16.41	-31.38	-52.59	-80.91	-117.20	-162.23	-216.67	-281.31	-356.92	-444.25
MID-RANGE (IN)	.00	.41	1.84	4.56	8.93	15.31	24.04	35.43	49.73	67.18	88.07	112.66	141.21
BULLET PATH (IN)	-.75	1.46	.00	-5.90	-17.11	-34.56	-59.13	-91.66	-132.94	-183.61	-244.50	-316.36	-399.93
TIME OF FLIGHT (SEC)	.000000	.092778	.194881	.306919	.429187	.560884	.700900	.848239	1.002211	1.162354	1.328360	1.500032	1.677260
WIND DEFLECTION (IN)													
5 MPH	.00	.40	1.62	3.71	6.71	10.53	15.09	20.29	26.08	32.40	39.25	46.59	54.42
10 MPH	.00	.80	3.24	7.43	13.42	21.07	30.18	40.58	52.15	64.81	78.50	93.18	108.84
20 MPH	.00	1.60	6.48	14.86	26.84	42.14	60.36	81.17	104.31	129.62	156.99	186.36	217.69
30 MPH	.00	2.40	9.72	22.29	40.26	63.21	90.55	121.75	156.46	194.43	235.49	279.55	326.53
VELOCITY (FPS)	1600.	1456.	1327.	1216.	1131.	1065.	1013.	970.	933.	901.	872.	845.	820.
ENERGY (FT-LB)	472.	390.	324.	273.	236.	209.	189.	173.	161.	150.	140.	131.	124.
DROP (IN)	.00	-1.74	-7.61	-18.42	-35.11	-58.57	-89.67	-129.20	-177.89	-236.44	-305.67	-386.31	-479.13
MID-RANGE (IN)	.00	.47	2.07	5.10	9.94	16.95	26.41	38.62	53.83	72.28	94.26	120.04	149.90
BULLET PATH (IN)	-.75	1.69	.00	-6.63	-19.14	-38.42	-65.33	-100.68	-145.19	-199.56	-264.61	-341.07	-429.70
TIME OF FLIGHT (SEC)	.000000	.098329	.206335	.324536	.452626	.589455	.733973	.885365	1.043091	1.206786	1.376214	1.551232	1.731772
WIND DEFLECTION (IN)													
5 MPH	.00	.40	1.66	3.81	6.83	10.62	15.09	20.16	25.79	31.95	38.61	45.76	53.40
10 MPH	.00	.81	3.31	7.62	13.66	21.24	30.18	40.32	51.58	63.89	77.21	91.52	106.79
20 MPH	.00	1.61	6.63	15.24	27.32	42.49	60.36	80.65	103.17	127.79	154.43	183.03	213.58
30 MPH	.00	2.42	9.94	22.86	40.99	63.73	90.54	120.97	154.75	191.68	231.64	274.55	320.38

Bullet: Lyman # 245496 83 Gr.

Ballistic Coefficient: .202

(Cont'd.)

RANGE (YDS)	MUZZLE	50	100	150	200	250	300	350	400	450	500	550	600
VELOCITY (FPS)	1500.	1366.	1249.	1156.	1084.	1029.	983.	945.	911.	881.	853.	827.	803.
ENERGY (FT-LB)	415.	344.	288.	246.	217.	195.	178.	164.	153.	143.	134.	126.	119.
DROP (IN)	.00	-2.01	-8.68	-20.93	-39.67	-65.78	-100.07	-143.32	-196.17	-259.45	-333.91	-420.30	-519.39
MID-RANGE (IN)	.00	.53	2.33	5.76	11.16	18.84	29.11	42.23	58.41	77.98	101.18	128.30	159.62
BULLET PATH (IN)	-.75	1.95	.00	-7.54	-21.57	-42.96	-72.55	-111.08	-159.21	-217.79	-287.53	-369.21	-463.58
TIME OF FLIGHT (SEC)	.000000	.104851	.219783	.344849	.479017	.621197	.770474	.926233	1.088061	1.255685	1.428935	1.607720	1.792013
WIND DEFLECTION (IN)													
5 MPH	.00	.43	1.74	3.95	6.95	10.67	15.00	19.91	25.35	31.30	37.75	44.68	52.10
10 MPH	.00	.85	3.48	7.89	13.91	21.33	30.00	39.82	50.70	62.60	75.49	89.36	104.19
20 MPH	.00	1.71	6.96	15.79	27.81	42.66	60.01	79.63	101.40	125.20	150.99	178.72	208.39
30 MPH	.00	2.56	10.45	23.68	41.72	63.99	90.01	119.45	152.10	187.80	226.48	268.08	312.58
VELOCITY (FPS)	1400.	1278.	1178.	1101.	1042.	994.	954.	919.	888.	860.	834.	810.	786.
ENERGY (FT-LB)	361.	301.	256.	224.	200.	182.	168.	156.	145.	136.	128.	121.	114.
DROP (IN)	.00	-2.31	-9.94	-23.83	-44.86	-73.86	-111.63	-158.80	-216.20	-284.58	-364.69	-457.28	-562.87
MID-RANGE (IN)	.00	.61	2.67	6.52	12.49	20.88	31.98	46.00	63.24	83.98	108.48	137.04	169.81
BULLET PATH (IN)	-.75	2.29	.00	-8.54	-24.22	-47.88	-80.30	-122.13	-174.18	-237.21	-311.97	-399.22	-499.47
TIME OF FLIGHT (SEC)	.000000	.112213	.234656	.366514	.506669	.654131	.808214	.968460	1.134561	1.306325	1.483640	1.666464	1.854808
WIND DEFLECTION (IN)													
5 MPH	.00	.45	1.79	3.97	6.87	10.42	14.55	19.22	24.41	30.10	36.27	42.93	50.08
10 MPH	.00	.89	3.59	7.94	13.75	20.84	29.10	38.45	48.83	60.20	72.55	85.87	100.16
20 MPH	.00	1.78	7.17	15.87	27.49	41.68	58.21	76.90	97.65	120.40	145.10	171.74	200.32
30 MPH	.00	2.68	10.76	23.81	41.24	62.52	87.31	115.35	146.48	180.60	217.65	257.61	300.48
VELOCITY (FPS)	1300.	1194.	1114.	1052.	1003.	961.	926.	894.	865.	839.	814.	791.	769.
ENERGY (FT-LB)	311.	263.	229.	204.	185.	170.	158.	147.	138.	130.	122.	115.	109.
DROP (IN)	.00	-2.68	-11.44	-27.17	-50.73	-82.90	-124.37	-175.89	-238.26	-312.20	-398.49	-497.72	-610.58
MID-RANGE (IN)	.00	.71	3.04	7.34	13.92	23.07	35.01	50.02	68.40	90.41	116.35	146.43	180.93
BULLET PATH (IN)	-.75	2.67	.00	-9.64	-27.10	-53.17	-88.54	-133.97	-190.25	-258.10	-338.29	-431.42	-538.19
TIME OF FLIGHT (SEC)	.000000	.120523	.250690	.389355	.535491	.688357	.847458	1.012463	1.183159	1.359420	1.541192	1.728478	1.921313
WIND DEFLECTION (IN)													
5 MPH	.00	.45	1.75	3.80	6.51	9.81	13.65	18.02	22.89	28.24	34.09	40.41	47.23
10 MPH	.00	.90	3.51	7.60	13.02	19.61	27.31	36.04	45.77	56.49	68.17	80.83	94.46
20 MPH	.00	1.81	7.01	15.21	26.03	39.22	54.61	72.08	91.55	112.98	136.35	161.66	188.92
30 MPH	.00	2.71	10.52	22.81	39.05	58.84	81.92	108.12	137.32	169.47	204.52	242.48	283.38
VELOCITY (FPS)	1200.	1119.	1056.	1006.	964.	928.	896.	867.	840.	815.	792.	770.	749.
ENERGY (FT-LB)	265.	231.	205.	186.	171.	159.	148.	138.	130.	123.	116.	109.	103.
DROP (IN)	.00	-3.12	-13.16	-30.97	-57.34	-92.98	-138.62	-195.05	-263.01	-343.27	-436.46	-543.18	-664.56
MID-RANGE (IN)	.00	.81	3.46	8.26	15.51	25.45	38.34	54.49	74.16	97.66	125.25	157.09	193.66
BULLET PATH (IN)	-.75	3.08	.00	-10.85	-30.27	-58.95	-97.64	-147.12	-208.12	-281.42	-367.66	-467.43	-581.85
TIME OF FLIGHT (SEC)	.000000	.129597	.267757	.413446	.565901	.724617	.889252	1.059587	1.235492	1.416909	1.603836	1.796307	1.994372
WIND DEFLECTION (IN)													
5 MPH	.00	.40	1.56	3.38	5.80	8.77	12.25	16.24	20.72	25.69	31.14	37.08	43.50
10 MPH	.00	.81	3.13	6.77	11.60	17.53	24.51	32.49	41.45	51.38	62.28	74.15	87.01
20 MPH	.00	1.62	6.25	13.53	23.20	35.07	49.02	64.97	82.89	102.75	124.55	148.30	174.02
30 MPH	.00	2.43	9.38	20.30	34.80	52.60	73.52	97.46	124.34	154.13	186.83	222.45	261.03

Bullet: Lyman # 245497 90 Gr.
Ballistic Coefficient: .230

RANGE (YDS)	MUZZLE	50	100	150	200	250	300	350	400	450	500	550	600
VELOCITY (FPS)	1800.	1649.	1513.	1391.	1283.	1190.	1119.	1063.	1017.	978.	944.	914.	887.
ENERGY (FT-LB)	647.	544.	458.	387.	329.	283.	250.	226.	207.	191.	178.	167.	157.
DROP (IN)	.00	-1.36	-5.90	-14.27	-27.13	-45.28	-69.56	-100.75	-139.61	-186.86	-243.16	-309.05	-385.31
MID-RANGE (IN)	.00	.36	1.59	3.94	7.66	13.08	20.51	30.29	42.68	57.94	76.29	97.91	123.09
BULLET PATH (IN)	-.75	1.22	.00	-5.04	-14.58	-29.41	-50.36	-78.22	-113.76	-157.68	-210.65	-273.21	-346.15
TIME OF FLIGHT (SEC)	.000000	.087068	.182073	.285489	.397854	.519406	.649478	.787138	.931558	1.082100	1.238320	1.399905	1.566640
WIND DEFLECTION (IN)													
5 MPH	.00	.33	1.36	3.12	5.68	9.04	13.15	17.93	23.31	29.22	35.64	42.52	49.86
10 MPH	.00	.66	2.71	6.25	11.36	18.08	26.31	35.87	46.62	58.45	71.28	85.05	99.73
20 MPH	.00	1.31	5.42	12.49	22.71	36.16	52.62	71.74	93.24	116.90	142.56	170.10	199.46
30 MPH	.00	1.97	8.13	18.74	34.07	54.25	78.92	107.61	139.86	175.35	213.83	255.15	299.19
VELOCITY (FPS)	1700.	1558.	1431.	1318.	1219.	1142.	1081.	1031.	991.	955.	924.	896.	870.
ENERGY (FT-LB)	577.	485.	409.	347.	297.	261.	233.	213.	196.	182.	171.	160.	151.
DROP (IN)	.00	-1.52	-6.64	-16.01	-30.39	-50.62	-77.50	-111.79	-154.23	-205.50	-266.16	-336.94	-418.52
MID-RANGE (IN)	.00	.41	1.80	4.42	8.56	14.56	22.74	33.37	46.72	63.02	82.46	105.29	131.76
BULLET PATH (IN)	-.75	1.42	.00	-5.68	-16.37	-32.90	-56.08	-86.68	-125.43	-173.00	-229.97	-297.06	-374.94
TIME OF FLIGHT (SEC)	.000000	.092227	.192713	.301982	.420411	.547710	.682879	.825078	.973591	1.127915	1.287698	1.452695	1.622740
WIND DEFLECTION (IN)													
5 MPH	.00	.35	1.43	3.28	5.94	9.37	13.51	18.25	23.56	29.37	35.67	42.43	49.62
10 MPH	.00	.70	2.86	6.56	11.87	18.75	27.01	36.51	47.12	58.75	71.34	84.85	99.25
20 MPH	.00	1.41	5.72	13.12	23.75	37.50	54.02	73.02	94.23	117.50	142.68	169.70	198.50
30 MPH	.00	2.11	8.58	19.68	35.62	56.25	81.03	109.52	141.35	176.24	214.02	254.55	297.75
VELOCITY (FPS)	1600.	1470.	1352.	1249.	1164.	1098.	1046.	1003.	966.	933.	905.	878.	854.
ENERGY (FT-LB)	512.	432.	365.	312.	271.	241.	219.	201.	186.	174.	163.	154.	146.
DROP (IN)	.00	-1.72	-7.50	-18.02	-34.13	-56.64	-86.33	-123.94	-170.17	-225.63	-290.95	-366.88	-454.04
MID-RANGE (IN)	.00	.46	2.03	4.95	9.57	16.21	25.15	36.64	50.95	68.28	88.84	112.90	140.69
BULLET PATH (IN)	-.75	1.65	.00	-6.40	-18.39	-36.78	-62.34	-95.83	-137.94	-189.27	-250.48	-322.28	-405.32
TIME OF FLIGHT (SEC)	.000000	.097839	.204282	.319806	.444424	.577210	.717293	.863888	1.016432	1.174532	1.337912	1.506384	1.679825
WIND DEFLECTION (IN)													
5 MPH	.00	.36	1.48	3.39	6.11	9.54	13.62	18.27	23.45	29.11	35.24	41.81	48.82
10 MPH	.00	.72	2.95	6.79	12.22	19.09	27.24	36.54	46.89	58.22	70.47	83.62	97.65
20 MPH	.00	1.44	5.91	13.57	24.44	38.18	54.49	73.09	93.78	116.44	140.94	167.25	195.30
30 MPH	.00	2.16	8.86	20.36	36.66	57.27	81.73	109.63	140.68	174.65	211.42	250.87	292.95
VELOCITY (FPS)	1500.	1379.	1272.	1182.	1113.	1057.	1012.	974.	941.	911.	884.	860.	836.
ENERGY (FT-LB)	450.	380.	323.	279.	247.	223.	205.	190.	177.	166.	156.	148.	140.
DROP (IN)	.00	-1.99	-8.55	-20.49	-38.64	-63.79	-96.68	-138.03	-188.48	-248.60	-319.17	-400.82	-494.22
MID-RANGE (IN)	.00	.53	2.28	5.60	10.77	18.10	27.85	40.27	55.60	74.01	95.80	121.19	150.43
BULLET PATH (IN)	-.75	1.91	.00	-7.30	-20.80	-41.29	-69.54	-106.23	-152.04	-207.51	-273.43	-350.43	-439.18
TIME OF FLIGHT (SEC)	.000000	.104335	.217664	.340163	.471068	.609473	.754558	.905709	1.062498	1.224623	1.391879	1.564127	1.741287
WIND DEFLECTION (IN)													
5 MPH	.00	.38	1.55	3.53	6.25	9.63	13.60	18.10	23.10	28.57	34.49	40.84	47.63
10 MPH	.00	.76	3.11	7.07	12.51	19.27	27.20	36.20	46.20	57.13	68.97	81.69	95.27
20 MPH	.00	1.53	6.22	14.14	25.02	38.53	54.40	72.41	92.40	114.27	137.94	163.37	190.53
30 MPH	.00	2.29	9.33	21.21	37.52	57.80	81.61	108.61	138.60	171.40	206.91	245.06	285.80

Bullet: Lyman # 245497 90 Gr.
Ballistic Coefficient: .230
(Cont'd.)

RANGE (YDS)	MUZZLE	50	100	150	200	250	300	350	400	450	500	550	600
VELOCITY (FPS)	1400.	1290.	1196.	1124.	1067.	1020.	981.	947.	916.	889.	864.	840.	818.
ENERGY (FT-LB)	392.	333.	286.	253.	227.	208.	192.	179.	168.	158.	149.	141.	134.
DROP (IN)	.00	-2.28	-9.80	-23.38	-43.81	-71.86	-108.24	-153.62	-208.55	-273.79	-350.00	-437.83	-537.97
MID-RANGE (IN)	.00	.60	2.62	6.36	12.11	20.15	30.74	44.12	60.46	80.06	103.14	129.95	160.74
BULLET PATH (IN)	-.75	2.24	.00	-8.30	-23.46	-46.23	-77.34	-117.44	-167.09	-227.06	-297.99	-380.55	-475.41
TIME OF FLIGHT (SEC)	.000000	.111679	.232547	.362019	.499138	.643072	.793166	.948963	1.110143	1.276483	1.447835	1.624108	1.805261
WIND DEFLECTION (IN)													
5 MPH	.00	.40	1.61	3.57	6.21	9.45	13.23	17.51	22.26	27.47	33.12	39.21	45.72
10 MPH	.00	.80	3.21	7.14	12.42	18.89	26.45	35.02	44.53	54.95	66.25	78.41	91.44
20 MPH	.00	1.60	6.43	14.29	24.84	37.79	52.91	70.03	89.06	109.89	132.49	156.83	182.88
30 MPH	.00	2.40	9.64	21.43	37.26	56.68	79.36	105.05	133.58	164.84	198.74	235.24	274.32
VELOCITY (FPS)	1300.	1204.	1130.	1071.	1024.	984.	950.	919.	891.	866.	843.	820.	799.
ENERGY (FT-LB)	338.	290.	255.	229.	209.	193.	180.	169.	159.	150.	142.	134.	128.
DROP (IN)	.00	-2.66	-11.30	-26.72	-49.68	-80.92	-121.10	-170.75	-230.65	-301.46	-383.83	-478.43	-585.95
MID-RANGE (IN)	.00	.70	2.99	7.18	13.54	22.35	33.83	48.16	65.64	86.51	111.00	139.37	171.88
BULLET PATH (IN)	-.75	2.62	.00	-9.40	-26.34	-51.55	-85.70	-129.33	-183.21	-248.00	-324.34	-412.92	-514.42
TIME OF FLIGHT (SEC)	.000000	.119992	.248694	.385119	.528434	.677960	.833226	.993899	1.159749	1.330621	1.506420	1.687100	1.872654
WIND DEFLECTION (IN)													
5 MPH	.00	.41	1.58	3.43	5.89	8.89	12.40	16.39	20.83	25.71	31.03	36.77	42.95
10 MPH	.00	.81	3.15	6.86	11.77	17.78	24.80	32.77	41.65	51.42	62.05	73.54	85.89
20 MPH	.00	1.62	6.31	13.72	23.55	35.57	49.60	65.54	83.31	102.84	124.11	147.09	171.79
30 MPH	.00	2.43	9.46	20.57	35.32	53.35	74.40	98.32	124.96	154.26	186.16	220.63	257.68
VELOCITY (FPS)	1200.	1127.	1069.	1022.	982.	948.	918.	890.	865.	841.	819.	798.	779.
ENERGY (FT-LB)	288.	254.	228.	209.	193.	180.	168.	158.	149.	141.	134.	127.	121.
DROP (IN)	.00	-3.10	-13.01	-30.51	-56.30	-91.07	-135.34	-189.90	-255.39	-332.47	-421.81	-524.10	-639.52
MID-RANGE (IN)	.00	.80	3.41	8.10	15.14	24.77	37.16	52.63	71.40	93.71	119.83	150.09	184.42
BULLET PATH (IN)	-.75	3.03	.00	-10.61	-29.52	-57.41	-94.80	-142.47	-201.08	-271.28	-353.74	-449.15	-557.69
TIME OF FLIGHT (SEC)	.000000	.129080	.265846	.409463	.559265	.714786	.875701	1.041783	1.212880	1.388899	1.569796	1.755567	1.946237
WIND DEFLECTION (IN)													
5 MPH	.00	.36	1.39	3.03	5.22	7.90	11.06	14.68	18.73	23.22	28.14	33.49	39.27
10 MPH	.00	.72	2.79	6.07	10.43	15.80	22.12	29.35	37.47	46.45	56.28	66.98	78.54
20 MPH	.00	1.44	5.58	12.13	20.86	31.60	44.25	58.71	74.93	92.89	112.57	133.96	157.08
30 MPH	.00	2.15	8.37	18.20	31.29	47.41	66.37	88.06	112.40	139.34	168.85	200.94	235.61
VELOCITY (FPS)	1100.	1047.	1004.	967.	934.	905.	879.	854.	831.	810.	789.	770.	751.
ENERGY (FT-LB)	242.	219.	201.	187.	174.	164.	154.	146.	138.	131.	125.	118.	113.
DROP (IN)	.00	-3.64	-15.18	-35.32	-64.67	-103.86	-153.63	-214.61	-287.48	-372.91	-471.34	-583.43	-710.16
MID-RANGE (IN)	.00	.94	3.95	9.28	17.16	27.80	41.50	58.48	79.07	103.48	131.81	164.33	201.45
BULLET PATH (IN)	-.75	3.57	.00	-12.18	-33.56	-64.79	-106.60	-159.62	-224.52	-301.98	-392.45	-496.58	-615.34
TIME OF FLIGHT (SEC)	.000000	.139876	.286279	.438641	.596563	.759767	.928061	1.101320	1.279474	1.462498	1.650401	1.843215	2.040977
WIND DEFLECTION (IN)													
5 MPH	.00	.31	1.19	2.60	4.50	6.86	9.67	12.92	16.59	20.70	25.24	30.20	35.61
10 MPH	.00	.62	2.39	5.20	9.00	13.72	19.34	25.83	33.19	41.40	50.47	60.41	71.21
20 MPH	.00	1.24	4.77	10.40	17.99	27.44	38.68	51.66	66.38	82.80	100.94	120.81	142.42
30 MPH	.00	1.85	7.16	15.60	26.99	41.16	58.02	77.50	99.56	124.20	151.41	181.22	213.64

Bullet: Lyman # 245498 95 Gr.
Ballistic Coefficient: .210

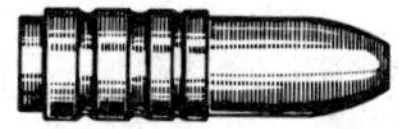

RANGE (YDS)	MUZZLE	50	100	150	200	250	300	350	400	450	500	550	600
VELOCITY (FPS)	1800.	1636.	1492.	1364.	1253.	1163.	1094.	1039.	995.	957.	924.	894.	867.
ENERGY (FT-LB)	683.	564.	469.	393.	331.	285.	252.	228.	209.	193.	180.	169.	159.
DROP (IN)	.00	-1.37	-5.98	-14.53	-27.74	-46.51	-71.70	-104.12	-144.57	-193.78	-252.36	-321.06	-400.62
MID-RANGE (IN)	.00	.37	1.62	4.04	7.89	13.54	21.32	31.56	44.54	60.51	79.67	102.28	128.61
BULLET PATH (IN)	-.75	1.25	.00	-5.19	-15.04	-30.44	-52.26	-81.32	-118.41	-164.26	-219.47	-284.81	-361.00
TIME OF FLIGHT (SEC)	.000000	.087438	.183551	.288744	.403579	.528110	.661298	.802155	.949824	1.103711	1.263407	1.428632	1.599203
WIND DEFLECTION (IN)													
5 MPH	.00	.36	1.49	3.41	6.18	9.81	14.19	19.26	24.92	31.13	37.85	45.05	52.73
10 MPH	.00	.72	2.97	6.82	12.36	19.61	28.39	38.51	49.84	62.25	75.69	90.11	105.46
20 MPH	.00	1.44	5.94	13.64	24.73	39.23	56.78	77.03	99.67	124.51	151.39	180.21	210.92
30 MPH	.00	2.17	8.91	20.46	37.09	58.84	85.17	115.54	149.51	186.76	227.08	270.32	316.38
VELOCITY (FPS)	1700.	1547.	1414.	1295.	1195.	1119.	1059.	1011.	971.	936.	905.	878.	852.
ENERGY (Ft-LB)	610.	505.	421.	354.	301.	264.	237.	216.	199.	185.	173.	162.	153.
DROP (IN)	.00	-1.53	-6.72	-16.26	-31.00	-51.82	-79.55	-114.99	-158.91	-211.98	-274.84	-348.27	-432.97
MID-RANGE (IN)	.00	.41	1.82	4.51	8.78	15.01	23.51	34.57	48.45	65.39	85.60	109.37	136.94
BULLET PATH (IN)	-.75	1.45	.00	-5.81	-16.82	-33.89	-57.89	-89.60	-129.79	-179.12	-238.25	-307.94	-388.90
TIME OF FLIGHT (SEC)	.000000	.092597	.194086	.305023	.425752	.555619	.693531	.838566	.990029	1.147444	1.310488	1.478940	1.652663
WIND DEFLECTION (IN)													
5 MPH	.00	.38	1.55	3.55	6.41	10.07	14.44	19.44	25.00	31.09	37.68	44.73	52.26
10 MPH	.00	.77	3.10	7.10	12.81	20.14	28.89	38.88	50.01	62.19	75.35	89.47	104.52
20 MPH	.00	1.54	6.20	14.19	25.63	40.28	57.77	77.76	100.02	124.37	150.70	178.94	209.03
30 MPH	.00	2.30	9.30	21.29	38.44	60.43	86.66	116.65	150.03	186.56	226.06	268.41	313.55
VELOCITY (FPS)	1600.	1462.	1338.	1230.	1145.	1080.	1028.	985.	949.	917.	888.	861.	837.
ENERGY (FT-LB)	540.	451.	377.	319.	277.	246.	223.	205.	190.	177.	166.	156.	148.
DROP (IN)	.00	-1.73	-7.56	-18.25	-34.69	-57.74	-88.21	-126.87	-174.47	-231.55	-298.94	-377.34	-467.45
MID-RANGE (IN)	.00	.47	2.05	5.03	9.78	16.63	25.86	37.74	52.54	70.44	91.73	116.67	145.51
BULLET PATH (IN)	-.75	1.67	.00	-6.53	-18.82	-37.71	-64.02	-98.53	-141.97	-194.90	-258.13	-332.37	-418.32
TIME OF FLIGHT (SEC)	.000000	.098120	.205457	.322516	.449119	.584132	.726599	.875721	1.030953	1.191920	1.358366	1.530126	1.707102
WIND DEFLECTION (IN)													
5 MPH	.00	.38	1.58	3.63	6.52	10.15	14.44	19.31	24.72	30.64	37.04	43.90	51.22
10 MPH	.00	.77	3.16	7.26	13.04	20.31	28.88	38.63	49.45	61.28	74.07	87.80	102.45
20 MPH	.00	1.54	6.32	14.53	26.09	40.61	57.76	77.25	98.90	122.56	148.14	175.60	204.90
30 MPH	.00	2.31	9.48	21.79	39.13	60.92	86.64	115.88	148.34	183.83	222.22	263.41	307.35
VELOCITY (FPS)	1500.	1372.	1259.	1167.	1097.	1042.	997.	959.	926.	896.	869.	843.	820.
ENERGY (FT-LB)	475.	397.	334.	287.	254.	229.	210.	194.	181.	169.	159.	150.	142.
DROP (IN)	.00	-2.00	-8.62	-20.74	-39.23	-64.90	-98.56	-140.93	-192.64	-254.42	-327.00	-411.09	-507.37
MID-RANGE (IN)	.00	.53	2.31	5.69	10.99	18.51	28.54	41.33	57.09	76.09	98.60	124.86	155.13
BULLET PATH (IN)	-.75	1.93	.00	-7.44	-21.23	-42.23	-71.19	-108.88	-155.90	-213.00	-280.89	-360.29	-451.89
TIME OF FLIGHT (SEC)	.000000	.104630	.218878	.342847	.475536	.615952	.763222	.916739	1.076083	1.240970	1.411209	1.586686	1.767344
WIND DEFLECTION (IN)													
5 MPH	.00	.41	1.66	3.77	6.65	10.20	14.36	19.07	24.30	30.01	36.19	42.83	49.93
10 MPH	.00	.81	3.32	7.54	13.29	20.41	28.73	38.15	48.59	60.01	72.37	85.66	99.85
20 MPH	.00	1.63	6.65	15.08	26.59	40.81	57.45	76.29	97.18	120.02	144.75	171.31	199.70
30 MPH	.00	2.44	9.97	22.62	39.88	61.22	86.18	114.44	145.77	180.03	217.12	256.97	299.56

Bullet: Lyman # 245498 95 Gr.

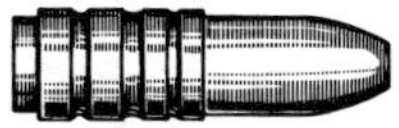

Ballistic Coefficient: .210

(Cont'd.)

RANGE (YDS)	MUZZLE	50	100	150	200	250	300	350	400	450	500	550	600
VELOCITY (FPS)	1400.	1283.	1186.	1112.	1054.	1007.	967.	933.	902.	875.	849.	825.	803.
ENERGY (FT-LB)	413.	347.	297.	261.	234.	214.	197.	184.	172.	161.	152.	144.	136.
DROP (IN)	.00	-2.30	-9.88	-23.64	-44.39	-72.95	-110.06	-156.36	-212.56	-279.39	-357.56	-447.77	-550.73
MID-RANGE (IN)	.00	.60	2.64	6.45	12.31	20.54	31.39	45.09	61.89	82.05	105.83	133.50	165.31
BULLET PATH (IN)	-.75	2.27	.00	-8.44	-23.88	-47.12	-78.91	-119.90	-170.78	-232.29	-305.15	-390.05	-487.69
TIME OF FLIGHT (SEC)	.000000	.111985	.233756	.364520	.503229	.648971	.801080	.959099	1.122716	1.291722	1.465986	1.645438	1.830061
WIND DEFLECTION (IN)													
5 MPH	.00	.43	1.71	3.79	6.57	9.97	13.92	18.40	23.37	28.81	34.72	41.08	47.90
10 MPH	.00	.85	3.43	7.58	13.14	19.93	27.85	36.80	46.74	57.63	69.44	82.17	95.80
20 MPH	.00	1.70	6.85	15.17	26.28	39.87	55.69	73.60	93.48	115.26	128.88	164.34	191.61
30 MPH	.00	2.56	10.28	22.75	39.42	59.80	83.54	110.40	140.22	172.89	208.33	246.51	287.41
VELOCITY (FPS)	1300.	1198.	1122.	1062.	1013.	973.	938.	907.	879.	853.	829.	806.	784.
ENERGY (FT-LB)	356.	303.	266.	238.	217.	200.	185.	173.	163.	153.	145.	137.	130.
DROP (IN)	.00	-2.67	-11.38	-26.97	-50.24	-81.95	-122.77	-173.36	-234.48	-306.84	-391.13	-488.05	-598.07
MID-RANGE (IN)	.00	.71	3.02	7.27	13.74	22.71	34.42	49.09	67.00	88.43	113.62	142.83	176.26
BULLET PATH (IN)	-.75	2.64	.00	-9.52	-26.73	-52.38	-87.14	-131.67	-186.72	-253.02	-331.24	-422.10	-526.05
TIME OF FLIGHT (SEC)	.000000	.120297	.249788	.387362	.532094	.683276	.840426	1.003213	1.171413	1.344885	1.523551	1.707387	1.896410
WIND DEFLECTION (IN)													
5 MPH	.00	.43	1.67	3.63	6.21	9.36	13.03	17.21	21.85	26.97	32.53	38.56	45.04
10 MPH	.00	.86	3.35	7.25	12.42	18.72	26.07	34.41	43.71	53.93	65.07	77.12	90.08
20 MPH	.00	1.73	6.69	14.51	24.84	37.44	52.14	68.82	87.41	107.86	130.14	154.23	180.15
30 MPH	.00	2.59	10.04	21.76	37.25	56.15	78.21	103.23	131.12	161.79	195.20	231.35	270.23
VELOCITY (FPS)	1200.	1123.	1063.	1014.	973.	938.	907.	879.	853.	829.	806.	785.	764.
ENERGY (FT-LB)	304.	266.	238.	217.	200.	186.	174.	163.	154.	145.	137.	130.	123.
DROP (IN)	.00	-3.11	-13.08	-30.72	-56.79	-91.96	-136.88	-192.31	-258.97	-337.54	-428.73	-533.00	-651.21
MID-RANGE (IN)	.00	.81	3.43	8.18	15.31	25.08	37.71	53.49	72.69	95.56	122.41	153.35	188.73
BULLET PATH (IN)	-.75	3.06	.00	-10.72	-29.87	-58.12	-96.13	-144.64	-204.38	-276.04	-360.31	-457.66	-568.96
TIME OF FLIGHT (SEC)	.000000	.129324	.266748	.411345	.562404	.719437	.882111	1.050200	1.223563	1.402119	1.585843	1.774752	1.968882
WIND DEFLECTION (IN)													
5 MPH	.00	.38	1.47	3.20	5.49	8.31	11.63	15.42	19.67	24.39	29.55	35.18	41.26
10 MPH	.00	.76	2.95	6.40	10.98	16.62	23.25	30.84	39.35	48.77	59.11	70.36	82.52
20 MPH	.00	1.52	5.90	12.79	21.97	33.24	46.50	61.67	78.69	97.55	118.22	140.71	165.05
30 MPH	.00	2.28	8.84	19.19	32.95	49.86	69.75	92.51	118.04	146.32	177.33	211.07	247.57
VELOCITY (FPS)	1100.	1044.	999.	960.	927.	897.	869.	844.	821.	798.	777.	757.	738.
ENERGY (FT-LB)	255.	230.	210.	195.	181.	170.	159.	150.	142.	134.	127.	121.	115.
DROP (IN)	.00	-3.65	-15.25	-35.52	-65.12	-104.74	-155.13	-216.98	-291.00	-377.90	-477.98	-592.43	-722.03
MID-RANGE (IN)	.00	.94	3.97	9.35	17.31	28.12	42.04	59.36	80.38	105.32	134.25	167.67	205.89
BULLET PATH (IN)	-.75	3.60	.00	-12.28	-33.88	-65.50	-107.89	-161.75	-227.77	-306.67	-398.76	-505.20	-626.80
TIME OF FLIGHT (SEC)	.000000	.140082	.287050	.440283	.599356	.763976	.933951	1.109162	1.289550	1.475105	1.665855	1.861836	2.063105
WIND DEFLECTION (IN)													
5 MPH	.00	.33	1.26	2.74	4.74	7.23	10.19	13.61	17.48	21.81	26.60	31.84	37.55
10 MPH	.00	.65	2.52	5.49	9.49	14.46	20.38	27.21	34.96	43.62	53.19	63.68	75.11
20 MPH	.00	1.31	5.04	10.98	18.97	28.92	40.75	54.43	69.92	87.24	106.38	127.37	150.21
30 MPH	.00	1.96	7.56	16.47	28.46	43.38	61.13	81.64	104.88	130.86	159.57	191.05	225.32

Bullet: Lyman # 257420 68 Gr.
Ballistic Coefficient: .129

RANGE (YDS)	MUZZLE	50	100	150	200	250	300	350	400	450	500	550	600
VELOCITY (FPS)	2400.	2099.	1849.	1646.	1470.	1316.	1189.	1091.	1021.	965.	920.	881.	846.
ENERGY (FT-LB)	870.	665.	516.	409.	326.	262.	213.	180.	157.	141.	128.	117.	108.
DROP (IN)	.00	-.80	-3.58	-8.89	-17.41	-29.98	-47.58	-71.33	-102.35	-141.68	-190.24	-249.01	-318.95
MID-RANGE (IN)	.00	.22	.99	2.53	5.13	9.11	14.90	22.97	33.77	47.71	65.14	86.42	111.89
BULLET PATH (IN)	-.75	.61	.00	-3.15	-9.50	-19.91	-35.35	-56.94	-85.79	-122.96	-169.36	-225.96	-293.74
TIME OF FLIGHT (SEC)	.000000	.067002	.143162	.229332	.325808	.433740	.553847	.685796	.828144	.979430	1.138749	1.305564	1.479572
WIND DEFLECTION (IN)													
5 MPH	.00	.40	1.60	3.68	6.67	10.67	15.74	21.85	28.88	36.69	45.21	54.39	64.20
10 MPH	.00	.79	3.20	7.36	13.34	21.34	31.48	43.70	57.75	73.38	90.42	108.78	128.40
20 MPH	.00	1.58	6.39	14.72	26.68	42.68	62.95	87.40	115.51	146.76	180.84	217.56	256.81
30 MPH	.00	2.38	9.59	22.09	40.03	64.01	94.43	131.10	173.26	220.14	271.26	326.34	385.21
VELOCITY (FPS)	2300.	2021.	1780.	1591.	1421.	1275.	1156.	1068.	1003.	951.	907.	870.	836.
ENERGY (FT-LB)	799.	616.	478.	382.	305.	245.	202.	172.	152.	136.	124.	114.	105.
DROP (IN)	.00	-.87	-3.87	-9.60	-18.78	-32.29	-51.15	-76.51	-109.46	-151.03	-202.09	-263.66	-336.69
MID-RANGE (IN)	.00	.23	1.07	2.73	5.52	9.79	15.97	24.55	35.95	50.58	68.76	90.85	117.19
BULLET PATH (IN)	-.75	.69	.00	-3.42	-10.28	-21.47	-38.03	-61.07	-91.71	-130.97	-179.71	-238.97	-309.69
TIME OF FLIGHT (SEC)	.000000	.069753	.148921	.238103	.337927	.449499	.573279	.708552	.853726	1.007513	1.169141	1.338152	1.514299
WIND DEFLECTION (IN)													
5 MPH	.00	.40	1.63	3.74	6.78	10.86	16.01	22.18	29.21	37.01	45.49	54.63	64.39
10 MPH	.00	.80	3.25	7.47	13.56	21.72	32.03	44.36	58.43	74.02	90.99	109.25	128.78
20 MPH	.00	1.60	6.51	14.94	27.12	43.44	64.05	88.71	116.86	148.04	181.97	218.51	257.56
30 MPH	.00	2.39	9.76	22.41	40.69	65.16	96.08	133.07	175.29	222.05	272.96	327.76	386.33
VELOCITY (FPS)	2200.	1943.	1719.	1536.	1373.	1235.	1126.	1046.	985.	937.	895.	859.	826.
ENERGY (FT-LB)	731.	570.	446.	356.	285.	230.	191.	165.	147.	132.	121.	111.	103.
DROP (IN)	.00	-.94	-4.19	-10.37	-20.26	-34.78	-55.01	-82.07	-117.06	-160.94	-214.63	-279.13	-355.38
MID-RANGE (IN)	.00	.25	1.15	2.94	5.94	10.52	17.13	26.24	38.27	53.59	72.54	95.45	122.68
BULLET PATH (IN)	-.75	.78	.00	-3.71	-11.13	-23.18	-40.94	-65.53	-98.05	-139.46	-190.68	-252.70	-326.48
TIME OF FLIGHT (SEC)	.000000	.072559	.154849	.247198	.350567	.465927	.593408	.731930	.879885	1.036170	1.200126	1.371369	1.549705
WIND DEFLECTION (IN)													
5 MPH	.00	.39	1.63	3.75	6.85	11.00	16.22	22.41	29.43	37.18	45.61	54.68	64.37
10 MPH	.00	.77	3.25	7.51	13.70	22.00	32.44	44.82	58.86	74.37	91.22	109.36	128.75
20 MPH	.00	1.54	6.51	15.01	27.40	44.01	64.88	89.64	117.72	148.73	182.44	218.72	257.50
30 MPH	.00	2.31	9.76	22.52	41.10	66.01	97.32	134.46	176.58	223.10	273.67	328.08	386.24
VELOCITY (FPS)	2100.	1850.	1647.	1471.	1317.	1189.	1092.	1021.	966.	920.	881.	846.	814.
ENERGY (FT-LB)	666.	517.	410.	327.	262.	214.	180.	157.	141.	128.	117.	108.	100.
DROP (IN)	.00	-1.04	-4.61	-11.38	-22.20	-38.05	-60.03	-89.28	-126.83	-173.62	-230.59	-298.74	-379.02
MID-RANGE (IN)	.00	.28	1.27	3.22	6.50	11.48	18.64	28.41	41.21	57.38	77.25	101.17	129.48
BULLET PATH (IN)	-.75	.89	.00	-4.10	-12.24	-25.41	-44.71	-71.28	-106.16	-150.26	-204.56	-270.03	-347.63
TIME OF FLIGHT (SEC)	.000000	.076109	.162220	.258628	.366486	.486517	.618396	.760684	.911915	1.071185	1.237952	1.411912	1.592929
WIND DEFLECTION (IN)													
5 MPH	.00	.41	1.70	3.90	7.11	11.38	16.70	22.94	29.96	37.69	46.08	55.11	64.75
10 MPH	.00	.82	3.41	7.80	14.22	22.77	33.41	45.88	59.93	75.39	92.17	110.21	129.50
20 MPH	.00	1.65	6.82	15.61	28.43	45.54	66.82	91.76	119.85	150.77	184.33	220.42	259.00
30 MPH	.00	2.47	10.22	23.41	42.65	68.31	100.23	137.64	179.78	226.16	276.50	330.63	388.50

Bullet: Lyman # 257420 68 Gr.
Ballistic Coefficient: .129
(Cont'd.)

RANGE (YDS)	MUZZLE	50	100	150	200	250	300	350	400	450	500	550	600
VELOCITY (FPS)	2000.	1763.	1576.	1408.	1263.	1147.	1061.	998.	947.	904.	867.	833.	802.
ENERGY (FT-LB)	604.	469.	375.	299.	241.	199.	170.	150.	135.	123.	113.	105.	97.
DROP (IN)	.00	-1.15	-5.07	-12.52	-24.37	-41.68	-65.58	-97.16	-137.44	-187.29	-247.74	-319.73	-404.24
MID-RANGE (IN)	.00	.31	1.39	3.54	7.11	12.54	20.27	30.75	44.35	61.37	82.19	107.13	136.54
BULLET PATH (IN)	-.75	1.02	.00	-4.53	-13.48	-27.88	-48.86	-77.53	-114.91	-161.85	-219.38	-288.46	-370.06
TIME OF FLIGHT (SEC)	.000000	.079998	.170041	.270832	.383445	.508253	.644439	.790390	.944871	1.107143	1.276768	1.453516	1.637302
WIND DEFLECTION (IN)													
5 MPH	.00	.44	1.76	4.03	7.34	11.73	17.11	23.35	30.35	38.03	46.36	55.31	64.88
10 MPH	.00	.88	3.53	8.07	14.69	23.45	34.22	46.71	60.70	76.06	92.71	110.62	129.77
20 MPH	.00	1.76	7.05	16.13	29.37	46.91	68.44	93.42	121.39	152.11	185.42	221.24	259.53
30 MPH	.00	2.64	10.58	24.20	44.06	70.36	102.66	140.13	182.09	228.17	278.13	331.86	389.30
VELOCITY (FPS)	1900.	1686.	1506.	1347.	1214.	1110.	1034.	976.	929.	889.	853.	820.	790.
ENERGY (FT-LB)	545.	429.	343.	274.	222.	186.	161.	144.	130.	119.	110.	102.	94.
DROP (IN)	.00	-1.26	-5.58	-13.76	-26.74	-45.63	-71.54	-105.55	-148.62	-201.65	-265.66	-341.60	-430.33
MID-RANGE (IN)	.00	.34	1.53	3.88	7.79	13.68	22.01	33.19	47.57	65.45	87.20	113.16	143.62
BULLET PATH (IN)	-.75	1.15	-.00	-5.02	-14.83	-30.55	-53.30	-84.14	-124.04	-173.91	-234.75	-307.52	-393.09
TIME OF FLIGHT (SEC)	.000000	.084032	.178190	.283570	.401039	.530530	.670780	.820231	.977872	1.143100	1.315569	1.495112	1.681692
WIND DEFLECTION (IN)													
5 MPH	.00	.45	1.79	4.11	7.50	11.95	17.34	23.55	30.47	38.07	46.30	55.15	64.62
10 MPH	.00	.89	3.57	8.22	15.00	23.90	34.69	47.10	60.95	76.13	92.59	110.30	129.24
20 MPH	.00	1.79	7.14	16.45	30.01	47.80	69.38	94.19	121.90	152.27	185.19	220.60	258.48
30 MPH	.00	2.68	10.72	24.67	45.01	71.70	104.07	141.29	182.84	228.40	277.78	330.89	387.72
VELOCITY (FPS)	1800.	1610.	1438.	1289.	1167.	1076.	1009.	956.	912.	873.	839.	808.	778.
ENERGY (FT-LB)	489.	391.	312.	251.	206.	175.	154.	138.	125.	115.	106.	98.	91.
DROP (IN)	.00	-1.39	-6.15	-15.14	-29.38	-49.99	-78.08	-114.68	-160.69	-217.10	-284.88	-364.99	-458.16
MID-RANGE (IN)	.00	.37	1.68	4.27	8.54	14.93	23.90	35.81	50.97	69.74	92.45	119.46	151.00
BULLET PATH (IN)	-.75	1.31	.00	-5.54	-16.32	-33.48	-58.12	-91.28	-133.83	-186.80	-251.12	-327.78	-417.50
TIME OF FLIGHT (SEC)	.000000	.088154	.186818	.297136	.419663	.553813	.698030	.850964	1.011801	1.180056	1.355464	1.537915	1.727419
WIND DEFLECTION (IN)													
5 MPH	.00	.42	1.77	4.15	7.60	12.07	17.43	23.55	30.37	37.84	45.95	54.67	64.01
10 MPH	.00	.85	3.55	8.30	15.19	24.14	34.85	47.10	60.74	75.69	91.89	109.34	128.03
20 MPH	.00	1.70	7.09	16.59	30.39	48.28	69.71	94.21	121.49	151.38	183.79	218.68	256.05
30 MPH	.00	2.55	10.64	24.89	45.58	72.41	104.56	141.31	182.23	227.07	275.68	328.02	384.08
VELOCITY (FPS)	1700.	1519.	1358.	1222.	1116.	1039.	980.	932.	891.	855.	823.	792.	764.
ENERGY (FT-LB)	436.	348.	278.	226.	188.	163.	145.	131.	120.	110.	102.	95.	88.
DROP (IN)	.00	-1.56	-6.92	-17.01	-32.92	-55.78	-86.66	-126.53	-176.30	-236.98	-309.51	-394.79	-493.64
MID-RANGE (IN)	.00	.42	1.90	4.79	9.54	16.58	26.32	39.10	55.23	75.08	98.97	127.24	160.16
BULLET PATH (IN)	-.75	1.52	.00	-6.25	-18.32	-37.35	-64.39	-100.42	-146.36	-203.20	-271.89	-353.34	-448.35
TIME OF FLIGHT (SEC)	.000000	.093395	.197928	.314512	.443165	.582695	.731520	.888591	1.053283	1.225233	1.404263	1.590329	1.783477
WIND DEFLECTION (IN)													
5 MPH	.00	.45	1.89	4.38	7.94	12.45	17.79	23.84	30.57	37.94	45.93	54.54	63.77
10 MPH	.00	.91	3.78	8.77	15.88	24.91	35.57	47.69	61.14	75.88	91.86	109.07	127.54
20 MPH	.00	1.82	7.55	17.53	31.76	49.81	71.14	95.37	122.28	151.75	183.71	218.15	255.08
30 MPH	.00	2.72	11.33	26.30	47.64	74.72	106.71	143.06	183.43	227.63	275.57	327.22	382.62

Bullet: Lyman # 257312 88 Gr.
Ballistic Coefficient: .208

RANGE (YDS)	MUZZLE	50	100	150	200	250	300	350	400	450	500	550	600
VELOCITY (FPS)	1900.	1736.	1593.	1461.	1350.	1258.	1179.	1113.	1060.	1017.	980.	947.	918.
ENERGY (FT-LB)	705.	588.	496.	417.	356.	309.	272.	242.	220.	202.	187.	175.	165.
DROP (IN)	.00	-1.22	-5.33	-12.86	-24.48	-40.86	-62.75	-90.89	-126.02	-168.85	-220.07	-280.33	-350.10
MID-RANGE (IN)	.00	.33	1.44	3.54	6.91	11.81	18.52	27.33	38.53	52.39	69.18	89.09	112.30
BULLET PATH (IN)	-.75	1.06	.00	-4.49	-13.08	-26.42	-45.27	-70.37	-102.46	-142.26	-190.44	-247.66	-314.40
TIME OF FLIGHT (SEC)	.000000	.082729	.172962	.271305	.378310	.493494	.616754	.747802	.886001	1.030601	1.181029	1.336881	1.497866
WIND DEFLECTION (IN)													
5 MPH	.00	.33	1.33	3.03	5.50	8.69	12.59	17.17	22.39	28.17	34.46	41.22	48.44
10 MPH	.00	.67	2.65	6.07	11.00	17.38	25.18	34.35	44.78	56.33	68.91	82.45	96.89
20 MPH	.00	1.33	5.30	12.13	22.01	34.76	50.36	68.70	89.56	112.67	137.83	164.90	193.78
30 MPH	.00	2.00	7.96	18.20	33.01	52.14	75.54	103.05	134.33	169.00	206.74	247.35	290.66
VELOCITY (FPS)	1800.	1653.	1517.	1393.	1296.	1212.	1140.	1082.	1035.	995.	961.	930.	903.
ENERGY (FT-LB)	633.	534.	449.	379.	328.	287.	254.	229.	209.	193.	180.	169.	159.
DROP (IN)	.00	-1.35	-5.88	-14.22	-27.01	-44.99	-68.90	-99.47	-137.45	-183.52	-238.35	-302.48	-376.58
MID-RANGE (IN)	.00	.36	1.59	3.92	7.62	12.96	20.23	29.72	41.71	56.46	74.19	95.08	119.39
BULLET PATH (IN)	-.75	1.21	.00	-5.02	-14.50	-29.16	-49.75	-77.01	-111.67	-154.42	-205.94	-266.76	-337.54
TIME OF FLIGHT (SEC)	.000000	.086970	.181725	.284994	.396690	.516470	.644209	.779410	.921323	1.069287	1.222833	1.381625	1.545426
WIND DEFLECTION (IN)													
5 MPH	.00	.32	1.33	3.08	5.58	8.78	12.69	17.25	22.41	28.10	34.28	40.92	48.00
10 MPH	.00	.64	2.65	6.16	11.15	17.57	25.38	34.51	44.82	56.19	68.55	81.83	96.00
20 MPH	.00	1.28	5.30	12.32	22.30	35.13	50.76	69.02	89.64	112.39	137.10	163.67	191.99
30 MPH	.00	1.92	7.95	18.48	33.45	52.70	76.14	103.53	134.46	168.58	205.66	245.50	287.99
VELOCITY (FPS)	1700.	1560.	1431.	1327.	1238.	1162.	1100.	1049.	1007.	971.	940.	912.	886.
ENERGY (FT-LB)	565.	475.	400.	344.	299.	264.	236.	215.	198.	184.	173.	162.	153.
DROP (IN)	.00	-1.52	-6.63	-15.98	-30.26	-50.24	-76.63	-110.20	-151.63	-201.61	-260.73	-329.57	-408.82
MID-RANGE (IN)	.00	.41	1.79	4.41	8.50	14.39	22.34	32.64	45.56	61.31	80.13	102.20	127.78
BULLET PATH (IN)	-.75	1.42	.00	-5.66	-16.26	-32.55	-55.26	-85.13	-122.88	-169.17	-224.61	-289.75	-365.32
TIME OF FLIGHT (SEC)	.000000	.092131	.192549	.301574	.418693	.543855	.676680	.816471	.962506	1.114259	1.271356	1.433530	1.600594
WIND DEFLECTION (IN)													
5 MPH	.00	.34	1.41	3.24	5.79	9.04	12.96	17.50	22.58	28.17	34.23	40.74	47.68
10 MPH	.00	.69	2.83	6.49	11.57	18.07	25.92	34.99	45.17	56.34	68.46	81.48	95.35
20 MPH	.00	1.37	5.66	12.98	23.14	36.14	51.84	69.99	90.33	112.69	136.93	162.96	190.70
30 MPH	.00	2.06	8.49	19.47	34.72	54.21	77.76	104.98	135.50	169.03	205.39	244.43	286.05
VELOCITY (FPS)	1600.	1468.	1355.	1262.	1183.	1116.	1063.	1019.	981.	949.	920.	893.	869.
ENERGY (FT-LB)	500.	421.	359.	311.	273.	243.	221.	203.	188.	176.	165.	156.	148.
DROP (IN)	.00	-1.72	-7.51	-18.01	-33.99	-56.17	-85.30	-122.11	-167.27	-221.43	-285.07	-358.96	-443.71
MID-RANGE (IN)	.00	.47	2.03	4.94	9.49	15.96	24.64	35.78	49.64	66.46	86.36	109.66	136.55
BULLEP PATH (IN)	-.75	1.65	.00	-6.37	-18.22	-36.27	-61.28	-93.95	-134.98	-185.01	-244.53	-314.29	-394.81
TIME OF FLIGHT (SEC)	.000000	.097910	.204478	.319242	.442087	.572743	.710589	.854869	1.005001	1.160573	1.321290	1.486943	1.657391
WIND DEFLECTION (IN)													
5 MPH	.00	.37	1.49	3.34	5.90	9.15	13.03	17.48	22.44	27.88	33.77	40.10	46.85
10 MPH	.00	.73	2.99	6.69	11.81	18.30	26.06	34.96	44.88	55.76	67.55	80.20	93.70
20 MPH	.00	1.46	5.98	13.37	23.61	36.61	52.13	69.91	89.76	111.52	135.09	160.40	187.40
30 MPH	.00	2.20	8.96	20.06	35.42	54.91	78.19	104.87	134.64	167.28	202.64	240.61	281.10

Bullet: Lyman # 257312 88 Gr.
Ballistic Coefficient: .208
(Cont'd.)

RANGE (YDS)	MUZZLE	50	100	150	200	250	300	350	400	450	500	550	600
VELOCITY (FPS)	1500.	1380.	1285.	1202.	1132.	1075.	1029.	990.	957.	927.	900.	875.	852.
ENERGY (FT-LB)	440.	372.	322.	282.	250.	226.	207.	192.	179.	168.	158.	150.	142.
DROP (IN)	.00	-1.99	-8.53	-20.34	-38.18	-62.78	-94.87	-135.14	-184.25	-242.72	-311.25	-390.47	-481.04
MID-RANGE (IN)	.00	.53	2.27	5.52	10.56	17.66	27.09	39.11	53.95	71.78	92.85	117.40	145.65
BULLET PATH (IN)	-.75	1.90	.00	-7.18	-20.37	-40.33	-67.78	-103.42	-147.89	-201.72	-265.61	-340.19	-426.12
TIME OF FLIGHT (SEC)	.000000	.104392	.217105	.337906	.466623	.602714	.745421	.894108	1.048326	1.207755	1.372165	1.541400	1.715357
WIND DEFLECTION (IN)													
5 MPH	.00	.39	1.51	3.34	5.86	9.04	12.80	17.08	21.85	27.08	32.75	38.84	45.35
10 MPH	.00	.77	3.01	6.67	11.73	18.08	25.59	34.16	43.71	54.16	65.50	77.69	90.70
20 MPH	.00	1.55	6.02	13.34	23.45	36.16	51.19	68.33	87.41	108.33	131.00	155.37	181.41
30 MPH	.00	2.32	9.03	20.01	35.18	54.23	76.78	102.49	131.12	162.49	196.50	233.06	272.11
VELOCITY (FPS)	1400.	1303.	1217.	1145.	1085.	1038.	997.	963.	932.	905.	880.	856.	835.
ENERGY (FT-LB)	383.	332.	289.	256.	230.	210.	194.	181.	170.	160.	151.	143.	136.
DROP (IN)	.00	-2.26	-9.65	-22.92	-42.80	-70.03	-105.31	-149.30	-202.57	-265.73	-339.47	-424.42	-521.21
MID-RANGE (IN)	.00	.59	2.56	6.19	11.75	19.51	29.74	42.67	58.48	77.41	99.70	125.59	155.29
BULLET PATH (IN)	-.75	2.19	.00	-8.07	-22.75	-44.78	-74.86	-113.65	-161.71	-219.68	-288.22	-367.96	-459.56
TIME OF FLIGHT (SEC)	.000000	.111126	.230330	.357514	.492204	.633660	.781203	.934355	1.092770	1.256206	1.424491	1.597514	1.775209
WIND DEFLECTION (IN)													
5 MPH	.00	.35	1.41	3.18	5.60	8.62	12.17	16.22	20.74	25.69	31.07	36.87	43.08
10 MPH	.00	.70	2.82	6.35	11.20	17.24	24.35	32.45	41.47	51.38	62.14	73.73	86.15
20 MPH	.00	1.40	5.65	12.70	22.40	34.48	48.70	64.89	82.94	102.76	124.28	147.47	172.30
30 MPH	.00	2.10	8.47	19.05	33.60	51.72	73.05	97.34	124.41	154.13	186.42	221.20	258.45
VELOCITY (FPS)	1300.	1215.	1143.	1084.	1036.	996.	962.	932.	904.	879.	856.	834.	813.
ENERGY (FT-LB)	330.	288.	255.	230.	210.	194.	181.	170.	160.	151.	143.	136.	129.
DROP (IN)	.00	-2.63	-11.16	-26.32	-48.86	-79.47	-118.80	-167.42	-225.96	-295.10	-375.46	-467.68	-572.39
MID-RANGE (IN)	.00	.69	2.95	7.05	13.27	21.87	33.08	47.09	64.12	84.44	108.26	135.84	167.39
BULLET PATH (IN)	-.75	2.58	.00	-9.21	-25.79	-50.44	-83.82	-126.49	-179.08	-242.26	-316.66	-402.93	-501.68
TIME OF FLIGHT (SEC)	.000000	.119438	.246847	.381744	.523382	.671091	.824395	.982955	1.146527	1.314943	1.488094	1.665913	1.848375
WIND DEFLECTION (IN)													
5 MPH	.00	.36	1.41	3.13	5.44	8.29	11.62	15.42	19.66	24.33	29.41	34.91	40.81
10 MPH	.00	.71	2.83	6.26	10.88	16.57	23.25	30.85	39.33	48.66	58.83	69.82	81.62
20 MPH	.00	1.43	5.66	12.53	21.77	33.15	46.49	61.69	78.65	97.32	117.66	139.63	163.24
30 MPH	.00	2.14	8.49	18.79	32.65	49.72	69.74	92.54	117.98	145.98	176.48	209.45	244.86
VELOCITY (FPS)	1200.	1130.	1074.	1028.	989.	956.	926.	899.	874.	851.	830.	809.	790.
ENERGY (FT-LB)	281.	250.	225.	206.	191.	178.	168.	158.	149.	142.	135.	128.	122.
DROP (IN)	.00	-3.09	-12.96	-30.34	-55.91	-90.34	-134.14	-188.01	-252.59	-328.51	-416.42	-516.94	-630.51
MID-RANGE (IN)	.00	.80	3.39	8.04	15.00	24.50	36.73	51.95	70.38	92.28	117.87	147.43	181.11
BULLET PATH (IN)	-.75	3.01	.00	-10.52	-29.24	-56.82	-93.76	-140.78	-198.49	-267.57	-348.62	-442.28	-548.99
TIME OF FLIGHT (SEC)	.000000	.128891	.265138	.407982	.556792	.711122	.870652	1.035155	1.204477	1.378515	1.557212	1.740549	1.928539
WIND DEFLECTION (IN)													
5 MPH	.00	.34	1.33	2.90	5.00	7.58	10.62	14.09	17.99	22.31	27.03	32.17	37.71
10 MPH	.00	.68	2.66	5.80	10.00	15.16	21.23	28.19	35.99	44.62	54.07	64.34	75.42
20 MPH	.00	1.37	5.33	11.61	19.99	30.31	42.47	56.37	71.98	89.24	108.14	128.67	150.85
30 MPH	.00	2.05	7.99	17.41	29.99	45.47	63.70	84.56	107.96	133.86	162.21	193.01	226.27

Bullet: Lyman # 257464 90 Gr.
Ballistic Coefficient: .204

RANGE (YDS)	MUZZLE	50	100	150	200	250	300	350	400	450	500	550	600
VELOCITY (FPS)	1900.	1728.	1574.	1434.	1315.	1214.	1132.	1067.	1015.	973.	936.	904.	875.
ENERGY (FT-LB)	721.	597.	495.	411.	345.	295.	256.	227.	206.	189.	175.	163.	153.
DROP (IN)	.00	-1.23	-5.36	-13.01	-24.91	-41.84	-64.67	-94.26	-131.46	-177.07	-231.80	-296.33	-371.46
MID-RANGE (IN)	.00	.33	1.45	3.61	7.09	12.22	19.32	28.74	40.82	55.85	74.06	95.70	121.07
BULLET PATH (IN)	-.75	1.08	.00	-4.60	-13.44	-27.31	-47.08	-73.62	-107.76	-150.32	-201.99	-263.46	-335.54
TIME OF FLIGHT (SEC)	.000000	.082849	.173811	.273700	.383087	.501894	.629987	.766673	.910941	1.061996	1.219304	1.382504	1.551358
WIND DEFLECTION (IN)													
5 MPH	.00	.34	1.40	3.24	5.92	9.43	13.75	18.84	24.58	30.93	37.83	45.24	53.15
10 MPH	.00	.69	2.80	6.49	11.84	18.86	27.51	37.67	49.17	61.86	75.65	90.48	106.30
20 MPH	.00	1.37	5.60	12.97	23.69	37.72	55.02	75.34	98.34	123.72	151.30	180.96	212.60
30 MPH	.00	2.06	8.40	19.46	35.53	56.58	82.53	113.01	147.50	185.58	226.95	271.44	318.91
VELOCITY (FPS)	1800.	1641.	1494.	1364.	1257.	1166.	1093.	1037.	991.	952.	918.	887.	860.
ENERGY (FT-LB)	647.	538.	446.	372.	316.	272.	239.	215.	196.	181.	168.	157.	148.
DROP (IN)	.00	-1.36	-5.95	-14.47	-27.65	-46.35	-71.43	-103.76	-144.15	-193.37	-252.04	-320.99	-400.94
MID-RANGE (IN)	.00	.37	1.61	4.02	7.87	13.49	21.24	31.45	44.42	60.43	79.67	102.46	129.04
BULLET PATH (IN)	-.75	1.24	.00	-5.17	-14.99	-30.34	-52.07	-81.04	-118.08	-163.95	-219.27	-284.87	-361.47
TIME OF FLIGHT (SEC)	.000000	.087301	.183140	.288334	.402980	.527034	.660052	.801097	.949239	1.103843	1.264481	1.430864	1.602810
WIND DEFLECTION (IN)													
5 MPH	.00	.35	1.45	3.37	6.13	9.71	14.08	19.16	24.87	31.14	37.94	45.25	53.05
10 MPH	.00	.70	2.90	6.75	12.26	19.42	28.17	38.33	49.73	62.28	75.88	90.50	106.09
20 MPH	.00	1.40	5.80	13.49	24.52	38.85	56.34	76.65	99.47	124.55	151.76	181.00	212.19
30 MPH	.00	2.09	8.70	20.24	36.77	58.27	84.51	114.98	149.20	186.83	227.65	271.50	318.28
VELOCITY (FPS)	1700.	1548.	1410.	1296.	1198.	1119.	1057.	1007.	966.	930.	899.	870.	844.
ENERGY (FT-LB)	577.	479.	397.	335.	287.	250.	223.	203.	186.	173.	161.	151.	142.
DROP (IN)	.00	-1.53	-6.71	-16.26	-31.00	-51.79	-79.49	-114.96	-158.96	-212.19	-275.38	-349.30	-434.68
MID-RANGE (IN)	.00	.41	1.82	4.51	8.79	15.00	23.49	34.56	48.49	65.52	85.90	109.93	137.84
BULLET PATH (IN)	-.75	1.45	.00	-5.82	-16.82	-33.88	-57.86	-89.59	-129.87	-179.37	-238.82	-309.01	-390.66
TIME OF FLIGHT (SEC)	.000000	.092485	.194047	.305111	.425596	.555285	.693392	.838916	.991114	1.149489	1.313705	1.483543	1.658870
WIND DEFLECTION (IN)													
5 MPH	.00	.37	1.55	3.56	6.39	10.04	14.43	19.47	25.10	31.27	37.96	45.14	52.80
10 MPH	.00	.75	3.09	7.11	12.79	20.08	28.86	38.94	50.20	62.55	75.92	90.28	105.61
20 MPH	.00	1.50	6.19	14.22	25.57	40.17	57.72	77.89	100.40	125.09	151.84	180.56	211.22
30 MPH	.00	2.24	9.28	21.33	38.36	60.25	86.58	116.83	150.60	187.64	227.75	270.84	316.82
VELOCITY (FPS)	1600.	1457.	1334.	1231.	1145.	1077.	1023.	980.	942.	909.	880.	853.	827.
ENERGY (FT-LB)	512.	424.	356.	303.	262.	232.	209.	192.	177.	165.	155.	145.	137.
DROP (IN)	.00	-1.74	-7.60	-18.34	-34.83	-57.94	-88.52	-127.37	-175.23	-232.74	-300.73	-379.93	-471.08
MID-RANGE (IN)	.00	.47	2.06	5.06	9.82	16.67	25.92	37.87	52.78	70.86	92.42	117.72	147.02
BULLET PATH (IN)	-.75	1.68	.00	-6.57	-18.89	-37.82	-64.22	-98.90	-142.58	-195.92	-259.74	-334.76	-421.73
TIME OF FLIGHT (SEC)	.000000	.098286	.206035	.323209	.449730	.584999	.728019	.877940	1.034191	1.196384	1.364265	1.537672	1.716520
WIND DEFLECTION (IN)													
5 MPH	.00	.40	1.63	3.69	6.58	10.23	14.57	19.51	25.01	31.03	37.56	44.57	52.05
10 MPH	.00	.80	3.26	7.38	13.15	20.46	29.13	39.02	50.02	62.06	75.11	89.13	104.11
20 MPH	.00	1.60	6.52	14.77	26.30	40.92	58.26	78.04	100.04	124.13	150.22	178.26	208.22
30 MPH	.00	2.39	9.79	22.15	39.46	61.38	87.39	117.05	150.05	186.19	225.33	267.39	312.32

Bullet: Lyman # 257464 90 Gr.

Ballistic Coefficient: .204
(Cont'd.)

RANGE (YDS)	MUZZLE	50	100	150	200	250	300	350	400	450	500	550	600
VELOCITY (FPS)	1500.	1369.	1261.	1169.	1096.	1039.	992.	953.	919.	889.	861.	835.	811.
ENERGY (FT-LB)	450.	375.	318.	273.	240.	216.	197.	182.	169.	158.	148.	139.	131.
DROP (IN)	.00	-2.01	-8.64	-20.76	-39.22	-64.89	-98.59	-141.09	-193.01	-255.17	-328.30	-413.16	-510.47
MID-RANGE (IN)	.00	.53	2.31	5.68	10.97	18.49	28.54	41.40	57.28	76.48	99.26	125.89	156.65
BULLET PATH (IN)	-.75	1.94	.00	-7.42	-21.19	-42.17	-71.17	-108.97	-156.20	-213.67	-282.11	-362.27	-454.88
TIME OF FLIGHT (SEC)	.000000	.104784	.219027	.342681	.475330	.616049	.763897	.918228	1.078606	1.244738	1.416437	1.593595	1.776170
WIND DEFLECTION (IN)													
5 MPH	.00	.42	1.67	3.76	6.63	10.21	14.42	19.20	24.52	30.34	36.65	43.44	50.70
10 MPH	.00	.84	3.35	7.51	13.26	20.42	28.85	38.41	49.03	60.67	73.29	86.87	101.41
20 MPH	.00	1.68	6.70	15.02	26.52	40.85	57.69	76.82	98.07	121.35	146.59	173.75	202.81
30 MPH	.00	2.53	10.05	22.54	39.77	61.27	86.54	115.22	147.10	182.02	219.88	260.62	304.22
VELOCITY (FPS)	1400.	1288.	1191.	1113.	1052.	1004.	963.	928.	896.	868.	841.	817.	794.
ENERGY (FT-LB)	392.	331.	284.	248.	221.	201.	185.	172.	160.	150.	141.	133.	126.
DROP (IN)	.00	-2.29	-9.83	-23.50	-44.15	-72.62	-109.69	-156.03	-212.39	-279.54	-358.21	-449.14	-552.95
MID-RANGE (IN)	.00	.60	2.63	6.40	12.24	20.46	31.35	45.12	62.05	82.41	106.48	134.54	166.78
BULLET PATH (IN)	-.75	2.25	.00	-8.37	-23.73	-46.91	-78.69	-119.74	-170.81	-232.66	-306.04	-391.68	-490.20
TIME OF FLIGHT (SEC)	.000000	.111797	.233017	.363397	.502116	.648179	.800870	.959703	1.124355	1.294614	1.470352	1.651511	1.838089
WIND DEFLECTION (IN)													
5 MPH	.00	.41	1.65	3.69	6.47	9.90	13.91	18.45	23.51	29.07	35.11	41.62	48.61
10 MPH	.00	.82	3.30	7.39	12.94	19.79	27.81	36.91	47.03	58.14	70.21	83.24	97.22
20 MPH	.00	1.64	6.59	14.77	25.89	39.59	55.62	73.82	94.06	116.28	140.42	166.47	194.44
30 MPH	.00	2.46	9.89	22.16	38.83	59.38	83.43	110.72	141.09	174.41	210.63	249.71	291.65
VELOCITY (FPS)	1300.	1202.	1122.	1059.	1009.	967.	931.	900.	871.	844.	820.	797.	774.
ENERGY (FT-LB)	338.	289.	251.	224.	203.	187.	173.	162.	152.	142.	134.	127.	120.
DROP (IN)	.00	-2.66	-11.34	-26.91	-50.20	-82.00	-123.00	-173.92	-235.54	-308.59	-393.82	-491.93	-603.34
MID-RANGE (IN)	.00	.70	3.01	7.26	13.76	22.79	34.59	49.42	67.57	89.31	114.91	144.64	178.66
BULLET PATH (IN)	-.75	2.64	.00	-9.52	-26.76	-52.52	-87.47	-132.35	-187.92	-254.92	-334.10	-426.16	-531.53
TIME OF FLIGHT (SEC)	.000000	.120110	.249441	.387229	.532467	.684401	.842526	1.006499	1.176098	1.351187	1.531700	1.717627	1.908999
WIND DEFLECTION (IN)													
5 MPH	.00	.42	1.64	3.61	6.24	9.46	13.22	17.50	22.27	27.52	33.25	39.46	46.15
10 MPH	.00	.83	3.29	7.23	12.48	18.92	26.44	34.99	44.53	55.04	66.50	78.92	92.29
20 MPH	.00	1.66	6.57	14.46	24.97	37.83	52.88	69.98	89.06	110.08	133.00	157.84	184.58
30 MPH	.00	2.49	9.86	21.69	37.45	56.75	79.32	104.97	133.60	165.12	199.51	236.75	276.87
VELOCITY (FPS)	1200.	1120.	1058.	1008.	966.	931.	899.	870.	844.	819.	796.	774.	753.
ENERGY (FT-LB)	288.	251.	224.	203.	187.	173.	161.	151.	142.	134.	127.	120.	113.
DROP (IN)	.00	-3.12	-13.14	-30.90	-57.19	-92.70	-138.14	-194.29	-261.89	-341.68	-434.37	-540.37	-660.86
MID-RANGE (IN)	.00	.81	3.45	8.24	15.46	25.35	38.17	54.21	73.76	97.07	124.49	156.06	192.29
BULLET PATH (IN)	-.75	3.08	.00	-10.82	-30.16	-58.73	-97.22	-146.43	-207.08	-279.93	-365.67	-464.73	-578.28
TIME OF FLIGHT (SEC)	.000000	.129521	.267476	.412862	.564930	.723178	.887269	1.056979	1.232176	1.412794	1.598826	1.790301	1.987264
WIND DEFLECTION (IN)													
5 MPH	.00	.40	1.54	3.33	5.71	8.64	12.08	16.01	20.43	25.33	30.70	36.55	42.88
10 MPH	.00	.80	3.08	6.66	11.43	17.28	24.16	32.03	40.86	50.65	61.39	73.09	85.76
20 MPH	.00	1.59	6.15	13.33	22.86	34.56	48.32	64.06	81.73	101.30	122.79	146.19	171.52
30 MPH	.00	2.39	9.23	19.99	34.28	51.84	72.48	96.08	122.59	151.96	184.18	219.28	257.28

Bullet: Lyman # 257418 105 Gr.
Ballistic Coefficient: .221

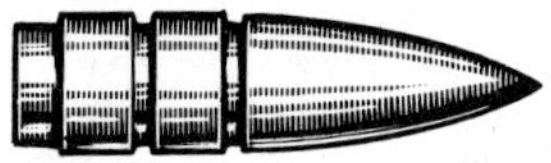

RANGE (YDS)	MUZZLE	50	100	150	200	250	300	350	400	450	500	550	600
VELOCITY (FPS)	2000.	1831.	1687.	1557.	1437.	1338.	1254.	1182.	1120.	1069.	1028.	992.	961.
ENERGY (FT-LB)	932.	782.	663.	565.	481.	417.	367.	325.	292.	267.	246.	229.	215.
DROP (IN)	.00	-1.10	-4.78	-11.50	-21.82	-36.37	-55.75	-80.69	-111.84	-149.90	-195.53	-249.37	-312.03
MID-RANGE (IN)	.00	.30	1.29	3.16	6.13	10.46	16.38	24.18	34.10	46.42	61.40	79.27	100.26
BULLET PATH (IN)	-.75	.91	.00	-3.96	-11.52	-23.30	-39.92	-62.09	-90.48	-125.77	-168.64	-219.72	-279.61
TIME OF FLIGHT (SEC)	.000000	.078384	.163831	.256404	.356723	.465090	.580938	.704234	.834733	.971909	1.115109	1.263808	1.417623
WIND DEFLECTION (IN)													
5 MPH	.00	.30	1.22	2.76	4.99	7.93	11.52	15.77	20.66	26.13	32.13	38.62	45.55
10 MPH	.00	.60	2.43	5.53	9.98	15.86	23.04	31.55	41.31	52.26	64.26	77.23	91.10
20 MPH	.00	1.19	4.87	11.05	19.97	31.71	46.09	63.09	82.63	104.51	128.52	154.46	182.20
30 MPH	.00	1.79	7.30	16.58	29.95	47.57	69.13	94.64	123.94	156.77	192.78	231.69	273.31
VELOCITY (FPS)	1900.	1744.	1610.	1485.	1376.	1288.	1211.	1144.	1089.	1044.	1006.	973.	944.
ENERGY (FT-LB)	842.	709.	604.	514.	441.	387.	342.	305.	277.	254.	236.	221.	208.
DROP (IN)	.00	-1.22	-5.28	-12.69	-24.08	-40.04	-61.25	-88.40	-122.16	-163.23	-212.25	-269.83	-336.55
MID-RANGE (IN)	.00	.33	1.42	3.48	6.76	11.49	17.93	26.34	37.00	50.17	66.08	84.98	107.03
BULLET PATH (IN)	-.75	1.05	.00	-4.40	-12.77	-25.72	-43.91	-68.04	-98.80	-136.85	-182.85	-237.42	-301.13
TIME OF FLIGHT (SEC)	.000000	.082527	.172080	.269114	.374203	.486942	.607145	.734691	.869161	1.009923	1.156381	1.308100	1.464763
WIND DEFLECTION (IN)													
5 MPH	.00	.31	1.25	2.84	5.14	8.11	11.74	16.02	20.91	26.35	32.29	38.69	45.53
10 MPH	.00	.63	2.50	5.68	10.28	16.23	23.49	32.04	41.81	52.69	64.58	77.38	91.06
20 MPH	.00	1.26	4.99	11.36	20.56	32.46	46.98	64.08	83.63	105.39	129.15	154.77	182.12
30 MPH	.00	1.89	7.49	17.04	30.84	48.68	70.47	96.13	125.44	158.08	193.73	232.15	273.18
VELOCITY (FPS)	1800.	1663.	1534.	1416.	1322.	1241.	1170.	1110.	1061.	1021.	986.	955.	928.
ENERGY (FT-LB)	755.	644.	549.	468.	408.	359.	319.	287.	263.	243.	227.	213.	201.
DROP (IN)	.00	-1.34	-5.82	-14.02	-26.54	-44.02	-67.18	-96.67	-133.20	-177.41	-229.94	-291.40	-362.23
MID-RANGE (IN)	.00	.36	1.57	3.85	7.45	12.60	19.57	28.64	40.06	54.09	70.95	90.85	113.93
BULLET PATH (IN)	-.75	1.19	-.00	-4.91	-14.15	-28.35	-48.22	-74.43	-107.67	-148.60	-197.85	-256.02	-323.56
TIME OF FLIGHT (SEC)	.000000	.086718	.180650	.282439	.392172	.509350	.633955	.765688	.903981	1.048192	1.197822	1.352513	1.512000
WIND DEFLECTION (IN)													
5 MPH	.00	.30	1.23	2.85	5.18	8.16	11.79	16.05	20.88	26.24	32.08	38.35	45.06
10 MPH	.00	.60	2.46	5.71	10.36	16.31	23.58	32.09	41.77	52.48	64.15	76.71	90.11
20 MPH	.00	1.19	4.92	11.42	20.71	32.62	47.15	64.19	83.53	104.96	128.30	153.42	180.22
30 MPH	.00	1.79	7.38	17.13	31.07	48.94	70.73	96.28	125.30	157.45	192.45	230.13	270.34
VELOCITY (FPS)	1700.	1569.	1448.	1346.	1262.	1188.	1125.	1074.	1031.	995.	963.	935.	909.
ENERGY (FT-LB)	674.	574.	489.	423.	371.	329.	295.	269.	248.	231.	216.	204.	193.
DROP (IN)	.00	-1.50	-6.55	-15.76	-29.75	-49.22	-74.84	-107.30	-147.28	-195.40	-252.28	-318.43	-394.40
MID-RANGE (IN)	.00	.41	1.77	4.34	8.32	14.01	21.66	31.53	43.89	58.95	76.92	98.00	122.36
BULLET PATH (IN)	-.75	1.40	.00	-5.56	-15.90	-31.71	-53.68	-82.50	-118.82	-163.29	-216.53	-279.02	-351.35
TIME OF FLIGHT (SEC)	.000000	.091862	.191409	.299031	.414164	.536753	.666581	.803146	.945792	1.093978	1.247311	1.405505	1.568346
WIND DEFLECTION (IN)													
5 MPH	.00	.32	1.31	3.02	5.39	8.41	12.07	16.32	21.11	26.39	32.12	38.27	44.84
10 MPH	.00	.64	2.63	6.04	10.78	16.82	24.14	32.65	42.22	52.78	64.23	76.55	89.68
20 MPH	.00	1.28	5.26	12.08	21.55	33.64	48.28	65.30	84.45	105.55	128.47	153.09	179.35
30 MPH	.00	1.91	7.89	18.12	32.33	50.46	72.43	97.94	126.67	158.33	192.70	229.64	269.03

Bullet: Lyman # 257418 105 Gr.
Ballistic Coefficient: .221
(Cont'd.)

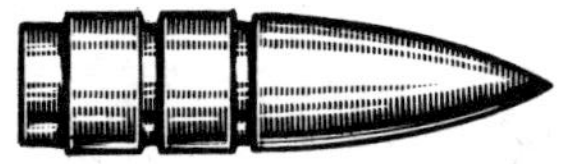

RANGE (YDS)	MUZZLE	50	100	150	200	250	300	350	400	450	500	550	600
VELOCITY (FPS)	1600.	1476.	1368.	1281.	1205.	1140.	1085.	1041.	1003.	971.	942.	915.	891.
ENERGY (FT-LB)	597.	508.	436.	383.	338.	303.	275.	253.	235.	220.	207.	195.	185.
DROP (IN)	.00	-1.70	-7.43	-17.78	-33.43	-55.08	-83.40	-119.08	-162.75	-215.04	-276.49	-347.59	-429.05
MID-RANGE (IN)	.00	.46	2.01	4.86	9.29	15.55	23.92	34.64	47.94	64.05	83.16	105.43	131.12
BULLET PATH (IN)	-.75	1.64	.00	-6.26	-17.82	-35.38	-59.61	-91.19	-130.78	-178.97	-236.33	-303.35	-380.72
TIME OF FLIGHT (SEC)	.000000	.097624	.203326	.316656	.437452	.565565	.700558	.841789	.988674	1.140790	1.297827	1.459557	1.625816
WIND DEFLECTION (IN)													
5 MPH	.00	.34	1.39	3.12	5.50	8.52	12.15	16.33	21.00	26.14	31.71	37.69	44.07
10 MPH	.00	.68	2.79	6.23	10.99	17.04	24.30	32.65	42.01	52.28	63.42	75.38	88.14
20 MPH	.00	1.36	5.57	12.46	21.98	34.08	48.60	65.31	84.01	104.56	126.84	150.76	176.29
30 MPH	.00	2.05	8.36	18.69	32.97	51.12	72.89	97.96	126.02	156.84	190.25	226.15	264.43
VELOCITY (FPS)	1500.	1387.	1298.	1219.	1151.	1095.	1049.	1010.	977.	947.	920.	896.	873.
ENERGY (FT-LB)	524.	448.	393.	346.	309.	280.	257.	238.	222.	209.	197.	187.	178.
DROP (IN)	.00	-1.98	-8.46	-20.10	-37.61	-61.65	-92.91	-132.05	-179.68	-236.40	-302.63	-379.09	-466.36
MID-RANGE (IN)	.00	.52	2.25	5.43	10.34	17.24	26.36	37.95	52.24	69.43	89.67	113.21	140.27
BULLET PATH (IN)	-.75	1.87	.00	-7.04	-19.94	-39.38	-66.04	-100.57	-143.60	-195.71	-257.34	-329.19	-411.86
TIME OF FLIGHT (SEC)	.000000	.104102	.215969	.335295	.461992	.595676	.735728	.881535	1.032645	1.188729	1.349544	1.514916	1.684721
WIND DEFLECTION (IN)													
5 MPH	.00	.36	1.41	3.11	5.46	8.42	11.94	15.98	20.47	25.41	30.76	36.51	42.66
10 MPH	.00	.72	2.81	6.21	10.91	16.84	23.89	31.95	40.95	50.82	61.52	73.03	85.31
20 MPH	.00	1.44	5.62	12.42	21.82	33.68	47.78	63.90	81.89	101.63	123.04	146.05	170.62
30 MPH	.00	2.17	8.43	18.64	32.73	50.52	71.66	95.85	122.84	152.45	184.56	219.08	255.93
VELOCITY (FPS)	1400.	1310.	1230.	1160.	1102.	1055.	1015.	981.	951.	924.	899.	876.	855.
ENERGY (FT-LB)	457.	400.	352.	314.	283.	259.	240.	224.	211.	199.	188.	179.	170.
DROP (IN)	.00	-2.25	-9.56	-22.64	-42.16	-68.81	-103.24	-146.07	-197.92	-259.20	-330.61	-412.76	-506.20
MID-RANGE (IN)	.00	.59	2.53	6.10	11.52	19.06	28.98	41.48	56.79	75.04	96.50	121.37	149.86
BULLET PATH (IN)	-.75	2.16	.00	-7.93	-22.29	-43.78	-73.06	-110.73	-157.42	-213.55	-279.81	-356.80	-445.08
TIME OF FLIGHT (SEC)	.000000	.110813	.229074	.354736	.487456	.626639	.771654	.922028	1.077415	1.237563	1.402288	1.571460	1.744992
WIND DEFLECTION (IN)													
5 MPH	.00	.32	1.30	2.93	5.18	8.00	11.33	15.14	19.38	24.05	29.12	34.57	40.42
10 MPH	.00	.65	2.60	5.86	10.36	16.00	22.67	30.28	38.77	48.10	58.23	69.15	80.83
20 MPH	.00	1.29	5.21	11.72	20.73	32.01	45.34	60.55	77.54	96.19	116.46	138.30	161.67
30 MPH	.00	1.94	7.81	17.59	31.09	48.01	68.00	90.83	116.30	144.29	174.69	207.45	242.50
VELOCITY (FPS)	1300.	1221.	1153.	1096.	1050.	1011.	977.	948.	921.	896.	874.	853.	833.
ENERGY (FT-LB)	394.	348.	310.	280.	257.	238.	223.	209.	198.	187.	178.	169.	162.
DROP (IN)	.00	-2.61	-11.07	-26.05	-48.23	-78.26	-116.78	-164.37	-221.44	-288.74	-366.82	-456.27	-557.64
MID-RANGE (IN)	.00	.69	2.92	6.95	13.04	21.43	32.34	45.99	62.49	82.14	105.15	131.71	162.04
BULLET PATH (IN)	-.75	2.55	.00	-9.07	-25.34	-49.46	-82.07	-123.74	-174.91	-236.29	-308.47	-392.00	-487.47
TIME OF FLIGHT (SEC)	.000000	.119124	.245625	.379126	.519011	.664664	.815629	.971574	1.132253	1.297491	1.467162	1.641183	1.819506
WIND DEFLECTION (IN)													
5 MPH	.00	.33	1.31	2.90	5.06	7.72	10.85	14.42	18.41	22.79	27.57	32.73	38.27
10 MPH	.00	.66	2.61	5.80	10.12	15.44	21.70	28.84	36.82	45.59	55.14	65.46	76.54
20 MPH	.00	1.32	5.23	11.61	20.23	30.88	43.41	57.69	73.63	91.18	110.29	130.93	153.08
30 MPH	.00	1.97	7.84	17.41	30.35	46.33	65.11	86.53	110.45	136.77	165.43	196.39	229.62

Bullet: Lyman # 257325 112 Gr.
Ballistic Coefficient: .235

RANGE (YDS)	MUZZLE	50	100	150	200	250	300	350	400	450	500	550	600
VELOCITY (FPS)	2000.	1841.	1702.	1576.	1459.	1352.	1257.	1175.	1108.	1054.	1010.	973.	940.
ENERGY (FT-LB)	995.	843.	720.	618.	529.	454.	393.	343.	305.	276.	254.	235.	220.
DROP (IN)	.00	-1.09	-4.74	-11.37	-21.52	-35.77	-54.78	-79.31	-110.12	-147.99	-193.66	-247.83	-311.13
MID-RANGE (IN)	.00	.29	1.28	3.12	6.02	10.24	16.02	23.69	33.55	45.90	61.06	79.28	100.79
BULLET PATH (IN)	-.75	.90	.00	-3.89	-11.29	-22.81	-39.07	-60.86	-88.93	-124.06	-166.98	-218.41	-278.96
TIME OF FLIGHT (SEC)	.000000	.078175	.162995	.254597	.353559	.460425	.575585	.699123	.830718	.969662	1.115162	1.266633	1.423664
WIND DEFLECTION (IN)													
5 MPH	.00	.28	1.14	2.60	4.71	7.52	11.05	15.32	20.30	25.93	32.13	38.86	46.08
10 MPH	.00	.56	2.29	5.21	9.43	15.03	22.10	30.65	40.61	51.86	64.27	77.73	92.16
20 MPH	.00	1.12	4.57	10.42	18.85	30.07	44.21	61.29	81.21	103.72	128.54	155.45	184.33
30 MPH	.00	1.68	6.86	15.63	28.28	45.10	66.31	91.94	121.82	155.58	192.81	233.18	276.49
VELOCITY (FPS)	1900.	1751.	1622.	1501.	1390.	1291.	1204.	1131.	1073.	1026.	986.	952.	921.
ENERGY (FT-LB)	898.	763.	654.	560.	481.	414.	360.	318.	286.	262.	242.	225.	211.
DROP (IN)	.00	-1.21	-5.24	-12.57	-23.79	-39.50	-60.44	-87.39	-121.10	-162.34	-211.82	-270.22	-338.07
MID-RANGE (IN)	.00	.32	1.41	3.44	6.65	11.29	17.65	26.04	36.76	50.13	66.41	85.85	108.59
BULLET PATH (IN)	-.75	1.04	.00	-4.33	-12.56	-25.28	-43.22	-67.17	-97.89	-136.13	-182.62	-238.02	-302.88
TIME OF FLIGHT (SEC)	.000000	.082316	.171324	.267469	.371328	.483360	.603785	.732436	.868733	1.011864	1.161160	1.316152	1.476511
WIND DEFLECTION (IN)													
5 MPH	.00	.30	1.18	2.70	4.89	7.80	11.45	15.82	20.87	26.52	32.71	39.40	46.56
10 MPH	.00	.59	2.36	5.39	9.77	15.60	22.90	31.65	41.74	53.04	65.42	78.80	93.13
20 MPH	.00	1.19	4.73	10.78	19.55	31.20	45.80	63.29	83.48	106.07	130.83	157.60	186.26
30 MPH	.00	1.78	7.09	16.17	29.32	46.79	68.69	94.94	125.22	159.11	196.25	236.40	279.39
VELOCITY (FPS)	1800.	1668.	1544.	1429.	1325.	1234.	1156.	1092.	1042.	1000.	963.	932.	903.
ENERGY (FT-LB)	806.	692.	593.	508.	437.	378.	332.	297.	270.	248.	231.	216.	203.
DROP (IN)	.00	-1.34	-5.79	-13.92	-26.31	-43.65	-66.72	-96.28	-133.09	-177.90	-231.38	-294.12	-366.78
MID-RANGE (IN)	.00	.36	1.56	3.81	7.36	12.47	19.44	28.60	40.24	54.64	72.04	92.64	116.67
BULLET PATH (IN)	-.75	1.18	.00	-4.85	-13.97	-28.04	-47.84	-74.12	-107.67	-149.20	-199.41	-258.88	-328.27
TIME OF FLIGHT (SEC)	.000000	.086585	.180086	.281101	.390144	.507538	.633268	.766896	.907651	1.054784	1.207764	1.366216	1.529876
WIND DEFLECTION (IN)													
5 MPH	.00	.29	1.18	2.74	5.00	8.00	11.73	16.15	21.21	26.82	32.95	39.56	46.63
10 MPH	.00	.57	2.36	5.47	10.00	15.99	23.46	32.31	42.41	53.64	65.90	79.12	93.26
20 MPH	.00	1.14	4.72	10.95	20.00	31.99	46.91	64.61	84.83	107.28	131.80	158.24	186.52
30 MPH	.00	1.72	7.09	16.42	30.00	47.98	70.37	96.92	127.24	160.93	197.70	237.36	279.77
VELOCITY (FPS)	1700.	1574.	1457.	1350.	1255.	1174.	1107.	1053.	1009.	972.	939.	910.	884.
ENERGY (FT-LB)	719.	616.	528.	453.	392.	343.	305.	276.	253.	235.	219.	206.	194.
DROP (IN)	.00	-1.50	-6.52	-15.66	-29.56	-49.00	-74.73	-107.54	-148.15	-197.28	-255.54	-323.52	-401.97
MID-RANGE (IN)	.00	.40	1.76	4.30	8.26	13.96	21.71	31.80	44.52	60.10	78.77	100.73	126.25
BULLET PATH (IN)	-.75	1.38	.00	-5.51	-15.77	-31.58	-53.68	-82.85	-119.84	-165.33	-219.96	-284.31	-359.12
TIME OF FLIGHT (SEC)	.000000	.091720	.190810	.297811	.413111	.536787	.668509	.807565	.953165	1.104725	1.261839	1.424217	1.591662
WIND DEFLECTION (IN)													
5 MPH	.00	.31	1.26	2.91	5.29	8.41	12.24	16.71	21.76	27.33	33.39	39.92	46.89
10 MPH	.00	.61	2.52	5.83	10.59	16.83	24.48	33.43	43.52	54.67	66.79	79.84	93.78
20 MPH	.00	1.23	5.05	11.65	21.18	33.65	48.96	66.85	87.04	109.33	133.58	159.68	187.56
30 MPH	.00	1.84	7.57	17.48	31.77	50.48	73.44	100.28	130.57	164.00	200.37	239.52	281.34

Bullet: Lyman # 257325 112 Gr.

Ballistic Coefficient: .235

(Cont'd.)

RANGE (YDS)	MUZZLE	50	100	150	200	250	300	350	400	450	500	550	600
VELOCITY (FPS)	1600.	1481.	1372.	1274.	1190.	1120.	1064.	1018.	979.	946.	916.	889.	864.
ENERGY (FT-LB)	637.	545.	468.	404.	352.	312.	281.	258.	239.	222.	209.	197.	186.
DROP (IN)	.00	-1.70	-7.40	-17.71	-33.39	-55.21	-83.93	-120.31	-165.05	-218.82	-282.14	-355.79	-440.40
MID-RANGE (IN)	.00	.46	2.00	4.84	9.30	15.67	24.25	35.32	49.13	65.92	85.86	109.24	136.27
BULLET PATH (IN)	-.75	1.63	.00	-6.24	-17.84	-35.58	-60.23	-92.53	-133.20	-182.89	-242.14	-311.71	-392.25
TIME OF FLIGHT (SEC)	.000000	.097473	.202751	.316264	.438170	.568227	.705792	.850053	1.000381	1.156337	1.317611	1.483986	1.655318
WIND DEFLECTION (IN)													
5 MPH	.00	.33	1.34	3.08	5.56	8.75	12.61	17.05	22.03	27.51	33.45	39.84	46.67
10 MPH	.00	.66	2.68	6.16	11.12	17.51	25.22	34.11	44.07	55.02	66.90	79.68	93.34
20 MPH	.00	1.31	5.37	12.32	22.24	35.02	50.44	68.22	88.13	110.03	133.80	159.36	186.67
30 MPH	.00	1.97	8.05	18.49	33.35	52.52	75.66	102.33	132.20	165.05	200.70	239.04	280.01
VELOCITY (FPS)	1500.	1389.	1290.	1203.	1131.	1072.	1025.	985.	951.	921.	893.	868.	845.
ENERGY (FT-LB)	559.	480.	414.	360.	318.	286.	261.	241.	225.	211.	198.	187.	177.
DROP (IN)	.00	-1.98	-8.45	-20.17	-37.89	-62.39	-94.43	-134.71	-183.92	-242.58	-311.45	-391.18	-482.42
MID-RANGE (IN)	.00	.52	2.25	5.47	10.49	17.59	27.05	39.13	54.07	72.05	93.34	118.17	146.78
BULLET PATH (IN)	-.75	1.88	.00	-7.11	-20.23	-40.13	-67.57	-103.25	-147.86	-201.92	-266.19	-341.32	-427.95
TIME OF FLIGHT (SEC)	.000000	.103948	.216074	.336593	.465332	.601707	.744905	.894258	1.049302	1.209706	1.375241	1.545753	1.721143
WIND DEFLECTION (IN)													
5 MPH	.00	.35	1.41	3.22	5.75	8.95	12.75	17.09	21.94	27.25	33.02	39.23	45.86
10 MPH	.00	.69	2.83	6.44	11.50	17.90	25.50	34.19	43.88	54.51	66.04	78.45	91.72
20 MPH	.00	1.39	5.66	12.88	23.00	35.80	51.01	68.38	87.75	109.02	132.08	156.90	183.44
30 MPH	.00	2.08	8.49	19.32	34.50	53.70	76.51	102.57	131.63	163.52	198.13	235.36	275.16
VELOCITY (FPS)	1400.	1299.	1211.	1137.	1078.	1029.	989.	954.	924.	896.	871.	847.	825.
ENERGY (FT-LB)	487.	420.	365.	322.	289.	263.	243.	227.	212.	200.	188.	178.	169.
DROP (IN)	.00	-2.27	-9.69	-23.05	-43.11	-70.62	-106.32	-150.87	-204.82	-268.90	-343.77	-430.08	-528.50
MID-RANGE (IN)	.00	.60	2.58	6.24	11.86	19.73	30.11	43.24	59.30	78.56	101.26	127.64	157.94
BULLET PATH (IN)	-.75	2.20	.00	-8.14	-22.97	-45.27	-75.74	-115.07	-163.80	-222.66	-292.31	-373.40	-466.60
TIME OF FLIGHT (SEC)	.000000	.111279	.230944	.358865	.494499	.637036	.785784	.940262	1.100129	1.265145	1.435149	1.610038	1.789760
WIND DEFLECTION (IN)													
5 MPH	.00	.36	1.47	3.29	5.80	8.92	12.58	16.74	21.38	26.48	32.01	37.97	44.36
10 MPH	.00	.73	2.93	6.59	11.60	17.83	25.16	33.49	42.77	52.95	64.01	75.94	88.71
20 MPH	.00	1.46	5.86	13.18	23.21	35.67	50.31	66.97	85.53	105.90	128.03	151.88	177.42
30 MPH	.00	2.18	8.80	19.77	34.81	53.50	75.47	100.46	128.30	158.85	192.04	227.81	266.14
VELOCITY (FPS)	1300.	1212.	1138.	1078.	1030.	989.	955.	924.	896.	871.	847.	825.	804.
ENERGY (FT-LB)	420.	365.	322.	289.	264.	243.	227.	212.	200.	189.	178.	169.	161.
DROP (IN)	.00	-2.64	-11.20	-26.46	-49.17	-80.05	-119.78	-168.91	-228.16	-298.19	-379.66	-473.22	-579.56
MID-RANGE (IN)	.00	.70	2.96	7.10	13.38	22.08	33.43	47.62	64.91	85.55	109.78	137.85	170.00
BULLET PATH (IN)	-.75	2.59	.00	-9.28	-26.01	-50.92	-84.68	-127.83	-181.10	-245.16	-320.65	-408.24	-508.60
TIME OF FLIGHT (SEC)	.000000	.119591	.247442	.383011	.525489	.674183	.828609	.988424	1.153390	1.323344	1.498183	1.677852	1.862340
WIND DEFLECTION (IN)													
5 MPH	.00	.37	1.47	3.24	5.63	8.56	11.99	15.90	20.27	25.07	30.30	35.96	42.04
10 MPH	.00	.74	2.93	6.49	11.26	17.12	23.99	31.81	40.54	50.14	60.60	71.92	84.08
20 MPH	.00	1.48	5.87	12.97	22.51	34.24	47.98	63.62	81.07	100.28	121.21	143.83	168.16
30 MPH	.00	2.22	8.80	19.46	33.77	51.35	71.97	95.43	121.61	150.42	181.81	215.75	252.24

Bullet: Lyman # 280468 122 Gr.
Ballistic Coefficient: .193

RANGE (YDS)	MUZZLE	50	100	150	200	250	300	350	400	450	500	550	600
VELOCITY (FPS)	2500.	2281.	2089.	1917.	1756.	1604.	1479.	1367.	1267.	1181.	1110.	1054.	1009.
ENERGY (FT-LB)	1693.	1409.	1181.	996.	835.	697.	593.	506.	435.	378.	334.	301.	276.
DROP (IN)	.00	-.71	-3.08	-7.44	-14.16	-23.71	-36.60	-53.51	-75.06	-102.04	-135.25	-175.49	-223.52
MID-RANGE (IN)	.00	.19	.83	2.05	4.00	6.86	10.86	16.27	23.36	32.47	43.92	58.05	75.16
BULLET PATH (IN)	-.75	.46	.00	-2.45	-7.26	-14.88	-25.86	-40.85	-60.49	-85.56	-116.85	-155.17	-201.29
TIME OF FLIGHT (SEC)	.000000	.062823	.131666	.206627	.288392	.377790	.475225	.580752	.694822	.817575	.948709	1.087488	1.233063
WIND DEFLECTION (IN)													
5 MPH	.00	.25	1.03	2.34	4.26	6.85	10.14	14.15	18.90	24.43	30.69	37.62	45.15
10 MPH	.00	.50	2.05	4.69	8.52	13.69	20.28	28.29	37.81	48.85	61.37	75.24	90.30
20 MPH	.00	.99	4.11	9.37	17.03	27.38	40.56	56.58	75.62	97.71	122.75	150.48	180.60
30 MPH	.00	1.49	6.16	14.06	25.55	41.07	60.84	84.88	113.43	146.56	184.12	225.71	270.90
VELOCITY (FPS)	2400.	2186.	2010.	1843.	1686.	1546.	1428.	1320.	1227.	1148.	1084.	1033.	991.
ENERGY (FT-LB)	1560.	1294.	1094.	920.	770.	648.	552.	472.	407.	357.	318.	289.	266.
DROP (IN)	.00	-.77	-3.34	-8.07	-15.35	-25.68	-39.65	-57.91	-81.15	-110.18	-145.80	-188.79	-239.91
MID-RANGE (IN)	.00	.21	.90	2.22	4.33	7.43	11.76	17.60	25.23	35.00	47.22	62.23	80.30
BULLET PATH (IN)	-.75	.53	.00	-2.68	-7.91	-16.20	-28.13	-44.34	-65.53	-92.52	-126.10	-167.04	-216.11
TIME OF FLIGHT (SEC)	.000000	.065495	.137066	.215012	.300131	.393183	.494170	.603474	.721433	.847999	.982661	1.124571	1.272966
WIND DEFLECTION (IN)													
5 MPH	.00	.26	1.06	2.42	4.41	7.10	10.49	14.61	19.49	25.12	31.47	38.46	46.02
10 MPH	.00	.53	2.12	4.84	8.82	14.20	20.97	29.21	38.97	50.25	62.95	76.92	92.04
20 MPH	.00	1.05	4.25	9.68	17.65	28.40	41.95	58.42	77.94	100.50	125.90	153.85	184.08
30 MPH	.00	1.58	6.37	14.53	26.47	42.60	62.92	87.63	116.92	150.74	188.84	230.77	276.13
VELOCITY (FPS)	2300.	2105.	1933.	1770.	1618.	1491.	1377.	1276.	1188.	1116.	1059.	1013.	974.
ENERGY (FT-LB)	1433.	1200.	1012.	849.	709.	602.	514.	441.	382.	338.	304.	278.	257.
DROP (IN)	.00	-.83	-3.63	-8.74	-16.63	-27.81	-42.95	-62.67	-87.74	-118.97	-157.14	-203.05	-257.40
MID-RANGE (IN)	.00	.22	.98	2.41	4.69	8.04	12.73	19.03	27.24	37.71	50.76	66.67	85.75
BULLET PATH (IN)	-.75	.60	.00	-2.93	-8.63	-17.63	-30.58	-48.10	-70.99	-100.03	-136.02	-179.73	-231.90
TIME OF FLIGHT (SEC)	.000000	.068289	.142658	.223756	.312409	.409106	.513854	.627113	.749060	.879435	1.017537	1.162505	1.313691
WIND DEFLECTION (IN)													
5 MPH	.00	.27	1.08	2.47	4.54	7.31	10.78	15.01	20.00	25.74	32.15	39.17	46.74
10 MPH	.00	.54	2.15	4.95	9.07	14.61	21.57	30.02	40.01	51.48	64.30	78.34	93.47
20 MPH	.00	1.08	4.30	9.89	18.14	29.22	43.14	60.05	80.02	102.95	128.61	156.68	186.94
30 MPH	.00	1.62	6.45	14.84	27.21	43.83	64.71	90.07	120.03	154.43	192.91	235.02	280.41
VELOCITY (FPS)	2200.	2024.	1856.	1698.	1556.	1437.	1328.	1233.	1153.	1088.	1037.	994.	957.
ENERGY (FT-LB)	1311.	1109.	933.	781.	656.	559.	478.	412.	360.	321.	291.	268.	248.
DROP (IN)	.00	-.90	-3.93	-9.48	-18.03	-30.18	-46.56	-67.86	-94.90	-128.45	-169.32	-218.25	-275.96
MID-RANGE (IN)	.00	.24	1.06	2.61	5.09	8.73	13.80	20.59	29.42	40.62	54.48	71.33	91.42
BULLET PATH (IN)	-.75	.69	.00	-3.21	-9.42	-19.23	-33.27	-52.23	-76.93	-108.15	-146.68	-193.27	-248.64
TIME OF FLIGHT (SEC)	.000000	.071096	.148506	.233024	.325440	.425797	.534433	.651710	.777613	.911668	1.053039	1.200947	1.354844
WIND DEFLECTION (IN)													
5 MPH	.00	.26	1.07	2.51	4.64	7.47	11.03	15.35	20.43	26.23	32.67	39.68	47.23
10 MPH	.00	.51	2.14	5.01	9.28	14.94	22.06	30.70	40.86	52.45	65.33	79.37	94.45
20 MPH	.00	1.03	4.27	10.02	18.55	29.88	44.12	61.40	81.72	104.91	130.67	158.73	188.91
30 MPH	.00	1.54	6.41	15.04	27.83	44.82	66.18	92.10	122.58	157.36	196.00	238.10	283.36

Bullet: Lyman # 280468 122 Gr.

Ballistic Coefficient: .193

(Cont'd.)

RANGE (YDS)	MUZZLE	50	100	150	200	250	300	350	400	450	500	550	600
VELOCITY (FPS)	2100.	1928.	1766.	1614.	1487.	1374.	1273.	1186.	1114.	1058.	1011.	973.	939.
ENERGY (FT-LB)	1194.	1007.	845.	705.	599.	511.	439.	381.	336.	303.	277.	256.	239.
DROP (IN)	.00	-.99	-4.32	-10.44	-19.87	-33.27	-51.28	-74.66	-104.22	-140.76	-185.05	-237.81	-299.69
MID-RANGE (IN)	.00	.27	1.17	2.88	5.61	9.64	15.20	22.63	32.24	44.34	59.25	77.25	98.57
BULLET PATH (IN)	-.75	.79	.00	-3.58	-10.47	-21.34	-36.80	-57.65	-84.68	-118.68	-160.43	-210.65	-270.00
TIME OF FLIGHT (SEC)	.000000	.074544	.155838	.244710	.341648	.446655	.560183	.682398	.813026	.951352	1.096520	1.247889	1.405015
WIND DEFLECTION (IN)													
5 MPH	.00	.27	1.14	2.68	4.92	7.88	11.58	16.05	21.26	27.15	33.64	40.67	48.21
10 MPH	.00	.55	2.28	5.35	9.84	15.75	23.16	32.10	42.52	54.30	67.27	81.34	96.43
20 MPH	.00	1.10	4.57	10.71	19.69	31.51	46.33	64.20	85.04	108.59	134.55	162.69	192.85
30 MPH	.00	1.64	6.85	16.06	29.53	47.26	69.49	96.31	127.56	162.89	201.82	244.03	289.28
VELOCITY (FPS)	2000.	1834.	1677.	1538.	1420.	1314.	1221.	1143.	1080.	1030.	988.	953.	921.
ENERGY (FT-LB)	1083.	911.	762.	641.	546.	468.	404.	354.	316.	287.	265.	246.	230.
DROP (IN)	.00	-1.10	-4.78	-11.54	-21.99	-36.76	-56.57	-82.22	-114.51	-154.23	-202.11	-258.87	-325.07
MID-RANGE (IN)	.00	.30	1.29	3.18	6.22	10.64	16.74	24.85	35.26	48.32	64.29	83.45	105.96
BULLET PATH (IN)	-.75	.92	.00	-4.00	-11.68	-23.69	-40.73	-63.62	-93.14	-130.09	-175.21	-229.20	-292.63
TIME OF FLIGHT (SEC)	.000000	.078337	.163896	.257435	.358951	.468814	.587342	.714456	.849617	.991969	1.140764	1.295485	1.455771
WIND DEFLECTION (IN)													
5 MPH	.00	.29	1.22	2.85	5.19	8.26	12.09	16.67	21.97	27.89	34.39	41.40	48.91
10 MPH	.00	.59	2.45	5.71	10.38	16.51	24.17	33.34	43.93	55.79	68.77	82.81	97.82
20 MPH	.00	1.17	4.89	11.42	20.75	33.02	48.34	66.69	87.87	111.57	137.55	165.61	195.63
30 MPH	.00	1.76	7.34	17.13	31.13	49.53	72.52	100.03	131.80	167.36	206.32	248.42	293.45
VELOCITY (FPS)	1900.	1739.	1590.	1468.	1356.	1257.	1173.	1104.	1049.	1005.	967.	934.	904.
ENERGY (FT-LB)	978.	819.	685.	583.	498.	428.	373.	330.	298.	273.	253.	236.	221.
DROP (IN)	.00	-1.22	-5.31	-12.82	-24.39	-40.69	-62.50	-90.62	-125.84	-168.93	-220.61	-281.50	-352.25
MID-RANGE (IN)	.00	.33	1.44	3.54	6.89	11.75	18.43	27.25	38.53	52.56	69.61	89.89	113.62
BULLET PATH (IN)	-.75	1.06	.00	-4.48	-13.03	-26.29	-45.07	-70.16	-102.35	-142.41	-191.06	-248.92	-316.64
TIME OF FLIGHT (SEC)	.000000	.082527	.172775	.270995	.377358	.492293	.615903	.747835	.887324	1.033532	1.185857	1.343880	1.507311
WIND DEFLECTION (IN)													
5 MPH	.00	.31	1.31	3.01	5.42	8.58	12.52	17.18	22.51	28.42	34.88	41.84	49.27
10 MPH	.00	.63	2.62	6.01	10.84	17.17	25.03	34.36	45.01	56.85	69.76	83.68	98.55
20 MPH	.00	1.26	5.24	12.02	21.67	34.34	50.06	68.71	90.02	113.70	139.53	167.36	197.10
30 MPH	.00	1.89	7.86	18.03	32.51	51.51	75.09	103.07	135.03	170.55	209.29	251.04	295.65
VELOCITY (FPS)	1800.	1646.	1513.	1397.	1294.	1204.	1129.	1069.	1021.	981.	946.	915.	887.
ENERGY (FT-LB)	878.	733.	620.	529.	453.	392.	345.	309.	282.	260.	242.	227.	213.
DROP (IN)	.00	-1.36	-5.92	-14.29	-27.13	-45.16	-69.20	-100.05	-138.47	-185.21	-240.95	-306.26	-381.92
MID-RANGE (IN)	.00	.37	1.60	3.94	7.65	12.99	20.31	29.91	42.11	57.16	75.31	96.74	121.75
BULLET PATH (IN)	-.75	1.22	.00	-5.04	-14.54	-29.25	-49.95	-77.46	-112.55	-155.96	-208.36	-270.34	-342.67
TIME OF FLIGHT (SEC)	.000000	.087172	.182371	.285561	.397187	.517498	.646322	.783030	.926754	1.076799	1.232683	1.394074	1.560741
WIND DEFLECTION (IN)													
5 MPH	.00	.34	1.38	3.13	5.62	8.87	12.88	17.57	22.89	28.76	35.14	42.01	49.35
10 MPH	.00	.68	2.76	6.26	11.24	17.75	25.75	35.15	45.78	57.52	70.29	84.02	98.69
20 MPH	.00	1.35	5.53	12.52	22.48	35.49	51.51	70.29	91.55	115.03	140.57	168.05	197.38
30 MPH	.00	2.03	8.29	18.78	33.71	53.24	77.26	105.44	137.33	172.55	210.86	252.07	296.07

Bullet: Lyman # 280468 122 Gr.
Ballistic Coefficient: .193

(Cont'd.)

RANGE (YDS)	MUZZLE	50	100	150	200	250	300	350	400	450	500	550	600
VELOCITY (FPS)	1700.	1558.	1438.	1330.	1235.	1154.	1089.	1037.	995.	958.	926.	897.	871.
ENERGY (FT-LB)	783.	658.	560.	479.	413.	361.	321.	291.	268.	249.	232.	218.	205.
DROP (IN)	.00	-1.52	-6.63	-15.96	-30.19	-50.15	-76.62	-110.38	-152.20	-202.78	-262.70	-332.70	-413.47
MID-RANGE (IN)	.00	.41	1.79	4.40	8.47	14.35	22.35	32.78	45.90	61.98	81.20	103.83	130.12
BULLET PATH (IN)	-.75	1.42	.00	-5.64	-16.18	-32.45	-55.22	-85.30	-123.42	-170.31	-226.55	-292.85	-369.94
TIME OF FLIGHT (SEC)	.000000	.092294	.192518	.301012	.418143	.543904	.677827	.819079	.966876	1.120669	1.280075	1.444831	1.614758
WIND DEFLECTION (IN)													
5 MPH	.00	.36	1.41	3.19	5.74	9.04	13.06	17.73	22.97	28.74	35.00	41.73	48.92
10 MPH	.00	.71	2.82	6.39	11.48	18.08	26.12	35.45	45.93	57.47	70.00	83.47	97.84
20 MPH	.00	1.43	5.65	12.78	22.95	36.16	52.24	70.90	91.87	114.95	140.00	166.93	195.69
30 MPH	.00	2.14	8.47	19.17	34.43	54.24	78.36	106.36	137.80	172.42	210.00	250.40	293.53
VELOCITY (FPS)	1600.	1477.	1365.	1265.	1179.	1109.	1053.	1008.	969.	936.	906.	879.	854.
ENERGY (FT-LB)	693.	591.	504.	433.	377.	333.	300.	275.	255.	237.	222.	209.	198.
DROP (IN)	.00	-1.70	-7.43	-17.82	-33.66	-55.75	-84.87	-121.81	-167.28	-221.91	-286.35	-361.35	-447.57
MID-RANGE (IN)	.00	.46	2.01	4.88	9.40	15.88	24.62	35.89	49.98	67.08	87.43	111.30	138.91
BULLET PATH (IN)	-.75	1.64	.00	-6.30	-18.05	-36.04	-61.08	-93.92	-135.30	-185.85	-246.20	-317.11	-399.24
TIME OF FLIGHT (SEC)	.000000	.097608	.203316	.317572	.440510	.571814	.710742	.856446	1.008304	1.165887	1.328895	1.497125	1.670445
WIND DEFLECTION (IN)													
5 MPH	.00	.34	1.39	3.20	5.76	9.07	13.05	17.62	22.73	28.35	34.44	41.00	48.00
10 MPH	.00	.68	2.78	6.39	11.53	18.14	26.09	35.23	45.46	56.70	68.89	81.99	96.00
20 MPH	.00	1.36	5.57	12.79	23.06	36.28	52.18	70.47	90.92	113.39	137.77	163.99	192.00
30 MPH	.00	2.04	8.35	19.18	34.59	54.42	78.27	105.70	136.38	170.09	206.66	245.98	288.00
VELOCITY (FPS)	1500.	1385.	1283.	1195.	1121.	1063.	1016.	976.	942.	912.	884.	859.	835.
ENERGY (FT-LB)	609.	520.	446.	387.	341.	306.	280.	258.	240.	225.	212.	200.	189.
DROP (!N)	.00	-1.98	-8.49	-20.29	-38.18	-62.95	-95.39	-136.23	-186.12	-245.68	-315.65	-396.72	-489.54
MID-RANGE (IN)	.00	.52	2.26	5.52	10.60	17.81	27.41	39.69	54.89	73.18	94.87	120.19	149.37
BULLET PATH (IN)	-.75	1.89	.00	-7.18	-20.45	-40.61	-68.43	-104.65	-149.92	-204.86	-270.21	-346.66	-434.86
TIME OF FLIGHT (SEC)	.000000	.104091	.216662	.337921	.467644	.605159	.749599	.900296	1.056790	1.218757	1.385979	1.558311	1.735669
WIND DEFLECTION (IN)													
5 MPH	.00	.36	1.47	3.34	5.95	9.25	13.16	17.63	22.60	28.05	33.97	40.33	47.14
10 MPH	.00	.72	2.93	6.67	11.91	18.51	26.33	35.25	45.19	56.10	67.93	80.66	94.28
20 MPH	.00	1.44	5.86	13.35	23.81	37.02	52.66	70.50	90.39	112.20	135.86	161.33	188.56
30 MPH	.00	2.16	8.80	20.02	35.72	55.52	78.99	105.76	135.58	168.30	203.80	241.99	282.83
VELOCITY (FPS)	1400.	1296.	1205.	1130.	1070.	1022.	981.	946.	916.	888.	862.	838.	816.
ENERGY (FT-LB)	531.	455.	394.	346.	310.	283.	261.	243.	227.	213.	201.	190.	180.
DROP (IN)	.00	-2.28	-9.73	-23.18	-43.41	-71.20	-107.29	-152.38	-207.00	-271.96	-347.93	-435.56	-535.55
MID-RANGE (IN)	.00	.60	2.59	6.29	11.97	19.94	30.46	43.79	60.09	79.67	102.77	129.62	160.49
BULLET PATH (IN)	-.75	2.22	.00	-8.20	-23.19	-45.74	-76.58	-116.43	-165.82	-225.53	-296.26	-378.64	-473.40
TIME OF FLIGHT (SEC)	.000000	.111427	.231537	.360168	.496700	.640265	.790161	.945904	1.107157	1.273687	1.445341	1.622029	1.803709
WIND DEFLECTION (IN)													
5 MPH	.00	.38	1.52	3.41	6.00	9.20	12.96	17.24	22.00	27.23	32.90	39.02	45.58
10 MPH	.00	.75	3.04	6.82	11.99	18.40	25.93	34.48	44.00	54.45	65.81	78.05	91.17
20 MPH	.00	1.51	6.07	13.64	23.98	36.80	51.85	68.96	88.00	108.91	131.62	156.10	182.33
30 MPH	.00	2.26	9.11	20.45	35.97	55.20	77.78	103.44	132.01	163.36	197.43	234.15	273.50

Bullet: Lyman # 280473 124 Gr.
Ballistic Coefficient: .275

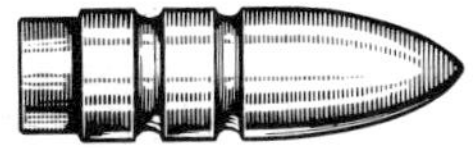

RANGE (YDS)	MUZZLE	50	100	150	200	250	300	350	400	450	500	550	600
VELOCITY (FPS)	2000.	1864.	1737.	1620.	1510.	1409.	1321.	1243.	1175.	1117.	1069.	1029.	994.
ENERGY (FT-LB)	1101.	956.	831.	723.	628.	546.	480.	425.	380.	343.	315.	291.	272.
DROP (IN)	.00	-1.08	-4.65	-11.09	-20.84	-34.43	-52.39	-75.33	-103.92	-138.80	-180.63	-230.04	-287.64
MID-RANGE (IN)	.00	.29	1.25	3.02	5.77	9.73	15.11	22.14	31.13	42.28	55.89	72.19	91.43
BULLET PATH (IN)	-.75	.87	.00	-3.74	-10.79	-21.69	-36.95	-57.19	-83.09	-115.27	-154.40	-201.11	-256.01
TIME OF FLIGHT (SEC)	.000000	.077698	.161112	.250537	.346445	.449310	.559341	.676475	.800708	.931780	1.069193	1.212349	1.360768
WIND DEFLECTION (IN)													
5 MPH	.00	.24	.98	2.25	4.09	6.54	9.62	13.33	17.66	22.60	28.09	34.09	40.55
10 MPH	.00	.47	1.96	4.49	8.17	13.08	19.24	26.66	35.32	45.19	56.18	68.17	81.10
20 MPH	.00	.95	3.91	8.99	16.35	26.16	38.49	53.32	70.65	90.39	112.36	136.35	162.19
30 MPH	.00	1.42	5.87	13.48	24.52	39.24	57.73	79.98	105.97	135.58	168.53	204.52	243.29
VELOCITY (FPS)	1900.	1769.	1651.	1539.	1435.	1343.	1263.	1192.	1131.	1081.	1039.	1003.	972.
ENERGY (FT-LB)	994.	862.	750.	652.	567.	497.	439.	391.	352.	322.	297.	277.	260.
DROP (IN)	.00	-1.19	-5.15	-12.29	-23.12	-38.18	-58.05	-83.40	-114.86	-153.09	-198.74	-252.43	-314.73
MID-RANGE (IN)	.00	.32	1.38	3.34	6.41	10.80	16.73	24.50	34.31	46.45	61.19	78.73	99.31
BULLET PATH (IN)	-.75	1.01	.00	-4.19	-12.07	-24.18	-41.10	-63.49	-92.00	-127.28	-169.98	-220.72	-280.07
TIME OF FLIGHT (SEC)	.000000	.081838	.169617	.263744	.364708	.472846	.588084	.710445	.839739	.975526	1.117208	1.264266	1.416318
WIND DEFLECTION (IN)													
5 MPH	.00	.25	1.03	2.37	4.30	6.87	10.07	13.89	18.32	23.32	28.84	34.83	41.27
10 MPH	.00	.51	2.06	4.73	8.61	13.75	20.13	27.78	36.64	46.64	57.68	69.67	82.54
20 MPH	.00	1.02	4.13	9.47	17.22	27.49	40.27	55.55	73.27	93.28	115.36	139.34	165.07
30 MPH	.00	1.53	6.19	14.20	25.83	41.24	60.40	83.33	109.91	139.92	173.04	209.01	247.61
VELOCITY (FPS)	1800.	1680.	1566.	1460.	1365.	1282.	1209.	1145.	1093.	1049.	1012.	979.	951.
ENERGY (FT-LB)	892.	777.	675.	587.	513.	453.	402.	361.	329.	303.	282.	264.	249.
DROP (IN)	.00	-1.32	-5.72	-13.67	-25.72	-42.42	-64.43	-92.39	-126.95	-168.78	-218.49	-276.67	-343.89
MID-RANGE (IN)	.00	.36	1.54	3.72	7.14	12.00	18.54	27.04	37.76	50.94	66.82	85.62	107.52
BULLET PATH (IN)	-.75	1.16	.00	-4.71	-13.52	-26.99	-45.76	-70.48	-101.81	-140.40	-186.87	-241.81	-305.80
TIME OF FLIGHT (SEC)	.000000	.086273	.178767	.277979	.384320	.497759	.618329	.745904	.880103	1.020350	1.166090	1.316908	1.472500
WIND DEFLECTION (IN)													
5 MPH	.00	.26	1.06	2.46	4.49	7.14	10.41	14.31	18.78	23.79	29.28	35.22	41.58
10 MPH	.00	.52	2.13	4.92	8.97	14.27	20.83	28.61	37.56	47.58	58.57	70.44	83.16
20 MPH	.00	1.03	4.26	9.85	17.95	28.54	41.65	57.22	75.13	95.16	117.13	140.89	166.32
30 MPH	.00	1.55	6.39	14.77	26.92	42.82	62.48	85.84	112.69	142.74	175.70	211.33	249.48
VELOCITY (FPS)	1700.	1585.	1478.	1380.	1296.	1221.	1156.	1101.	1056.	1018.	985.	955.	929.
ENERGY (FT-LB)	796.	692.	601.	524.	462.	410.	368.	334.	307.	285.	267.	251.	238.
DROP (IN)	.00	-1.49	-6.43	-15.39	-28.90	-47.59	-72.13	-103.16	-141.34	-187.30	-241.62	-304.90	-377.55
MID-RANGE (IN)	.00	.40	1.73	4.20	8.02	13.42	20.67	30.04	41.77	56.10	73.22	93.36	116.65
BULLET PATH (IN)	-.75	1.35	.00	-5.37	-15.29	-30.39	-51.34	-78.78	-113.37	-155.73	-206.47	-266.15	-335.22
TIME OF FLIGHT (SEC)	.000000	.091387	.189408	.294512	.406731	.526075	.652464	.785560	.924808	1.069633	1.219600	1.374385	1.533750
WIND DEFLECTION (IN)													
5 MPH	.00	.28	1.14	2.62	4.73	7.47	10.83	14.78	19.27	24.25	29.68	35.53	41.79
10 MPH	.00	.55	2.28	5.25	9.47	14.94	21.66	29.55	38.53	48.49	59.36	71.07	83.59
20 MPH	.00	1.11	4.55	10.49	18.93	29.88	43.31	59.11	77.06	96.98	118.71	142.14	167.17
30 MPH	.00	1.66	6.83	15.74	28.40	44.83	64.97	88.66	115.59	145.47	178.07	213.20	250.76

Bullet: Lyman # 280473 124 Gr.
Ballistic Coefficient: .275
(Cont'd.)

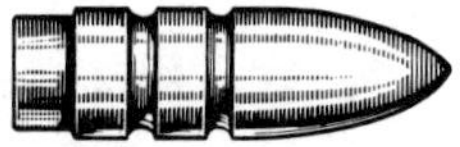

RANGE (YDS)	MUZZLE	50	100	150	200	250	300	350	400	450	500	550	600
VELOCITY (FPS)	1600.	1492.	1392.	1306.	1230.	1163.	1107.	1061.	1022.	989.	959.	932.	908.
ENERGY (FT-LB)	705.	412.	533.	470.	416.	373.	338.	310.	288.	269.	253.	239.	227.
DROP (IN)	.00	-1.68	-7.31	-17.41	-32.60	-53.57	-80.94	-115.37	-157.50	-207.93	-267.23	-335.86	-414.40
MID-RANGE (IN)	.00	.45	1.98	4.73	8.99	15.02	23.06	33.36	46.15	61.64	80.05	101.52	126.26
BULLET PATH (IN)	-.75	1.60	.00	-6.07	-17.23	-34.17	-57.51	-87.91	-126.01	-172.41	-227.68	-292.28	-366.79
TIME OF FLIGHT (SEC)	.000000	.097118	.201272	.312577	.430999	.556491	.688747	.827233	.971363	1.120682	1.274856	1.433636	1.596839
WIND DEFLECTION (IN)													
5 MPH	.00	.30	1.21	2.76	4.93	7.72	11.11	15.05	19.48	24.37	29.69	35.41	41.52
10 MPH	.00	.59	2.42	5.51	9.86	15.44	22.22	30.09	38.96	48.74	59.37	70.82	83.04
20 MPH	.00	1.19	4.85	11.03	19.71	30.88	44.44	60.19	77.92	97.48	118.75	141.64	166.09
30 MPH	.00	1.78	7.27	16.54	29.57	46.33	66.66	90.28	116.88	146.22	178.12	212.46	249.13
VELOCITY (FPS)	1500.	1399.	1313.	1236.	1168.	1112.	1065.	1025.	991.	961.	934.	910.	887.
ENERGY (FT-LB)	619.	539.	474.	420.	376.	340.	312.	289.	270.	254.	240.	228.	217.
DROP (IN)	.00	-1.96	-8.35	-19.77	-36.91	-60.41	-90.91	-129.06	-175.45	-230.67	-295.18	-369.55	-454.39
MID-RANGE (IN)	.00	.52	2.21	5.32	10.11	16.82	25.68	36.94	50.81	67.51	87.18	110.03	136.28
BULLET PATH (IN)	-.75	1.84	.00	-6.88	-19.47	-38.42	-64.37	-97.97	-139.82	-190.49	-250.45	-320.27	-400.57
TIME OF FLIGHT (SEC)	.000000	.103574	.214292	.332119	.457030	.588739	.726729	.870406	1.019304	1.173081	1.331480	1.494315	1.661446
WIND DEFLECTION (IN)													
5 MPH	.00	.31	1.26	2.83	5.02	7.81	11.15	15.00	19.30	24.03	29.17	34.70	40.61
10 MPH	.00	.63	2.52	5.65	10.04	15.62	22.30	29.99	38.60	48.06	58.34	69.40	81.21
20 MPH	.00	1.26	5.03	11.31	20.07	31.24	44.61	59.98	77.20	96.12	116.68	138.80	162.43
30 MPH	.00	1.89	7.55	16.96	30.11	46.85	66.91	89.97	115.79	144.19	175.02	208.20	243.64
VELOCITY (FPS)	1400.	1314.	1236.	1169.	1112.	1065.	1026.	992.	962.	935.	910.	888.	867.
ENERGY (FT-LB)	540.	475.	421.	376.	340.	312.	290.	271.	255.	240.	228.	217.	207.
DROP (IN)	.00	-2.24	-9.51	-22.50	-41.82	-68.15	-102.12	-144.33	-195.36	-255.67	-325.83	-406.46	-498.09
MID-RANGE (IN)	.00	.59	2.52	6.04	11.40	18.82	28.56	40.84	55.86	73.76	94.76	119.09	146.92
BULLET PATH (IN)	-.75	2.14	.00	-7.85	-22.05	-43.25	-72.08	-109.16	-155.06	-210.24	-275.27	-350.77	-437.26
TIME OF FLIGHT (SEC)	.000000	.110646	.228400	.353239	.484881	.622808	.766427	.915271	1.068994	1.227343	1.390127	1.557207	1.728488
WIND DEFLECTION (IN)													
5 MPH	.00	.31	1.24	2.80	4.96	7.66	10.87	14.54	18.64	23.15	28.05	33.32	38.96
10 MPH	.00	.62	2.48	5.60	9.91	15.33	21.75	29.09	37.29	46.30	56.09	66.64	77.93
20 MPH	.00	1.23	4.97	11.20	19.82	30.66	43.50	58.18	74.57	92.60	112.18	133.28	155.86
30 MPH	.00	1.85	7.45	16.80	29.73	45.99	65.24	87.26	111.86	138.89	168.27	199.92	233.78
VELOCITY (FPS)	1300.	1224.	1159.	1104.	1058.	1019.	986.	957.	930.	906.	884.	863.	843.
ENERGY (FT-LB)	465.	413.	370.	335.	308.	286.	268.	252.	238.	226.	215.	205.	196.
DROP (IN)	.00	-2.60	-11.02	-25.90	-47.88	-77.61	-115.68	-162.66	-218.99	-285.29	-362.15	-450.09	-549.70
MID-RANGE (IN)	.00	.69	2.91	6.90	12.92	21.20	31.94	45.37	61.61	80.90	103.46	129.48	159.17
BULLET PATH (IN)	-.75	2.53	.00	-8.99	-25.09	-48.93	-81.12	-122.21	-172.65	-233.07	-304.04	-386.09	-479.82
TIME OF FLIGHT (SEC)	.000000	.118956	.244967	.377707	.516631	.661156	.810838	.965352	1.124453	1.287961	1.455745	1.627716	1.803814
WIND DEFLECTION (IN)													
5 MPH	.00	.31	1.25	2.78	4.85	7.41	10.43	13.87	17.72	21.96	26.57	31.55	36.89
10 MPH	.00	.63	2.50	5.55	9.70	14.82	20.86	27.75	35.44	43.91	53.13	63.09	73.78
20 MPH	.00	1.26	5.00	11.11	19.39	29.65	41.72	55.50	70.88	87.82	106.27	126.19	147.56
30 MPH	.00	1.89	7.50	16.66	29.09	44.47	62.58	83.24	106.33	131.74	159.40	189.28	221.34

Bullet: Lyman # 280412 136 Gr.
Ballistic Coefficient: .245

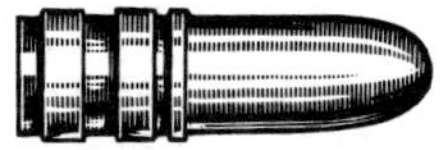

RANGE (YDS)	MUZZLE	50	100	150	200	250	300	350	400	450	500	550	600
VELOCITY (FPS)	2000.	1847.	1706.	1576.	1455.	1348.	1256.	1177.	1111.	1058.	1014.	977.	945.
ENERGY (FT-LB)	1208.	1031.	879.	750.	639.	549.	476.	418.	372.	338.	310.	288.	269.
DROP (IN)	.00	-1.09	-4.71	-11.32	-21.43	-35.67	-54.69	-79.23	-110.05	-147.89	-193.47	-247.49	-310.58
MID-RANGE (IN)	.00	.29	1.27	3.10	6.00	10.22	16.02	23.72	33.59	45.94	61.06	79.21	100.61
BULLET PATH (IN)	-.75	.90	.00	-3.87	-11.25	-22.77	-39.05	-60.86	-88.94	-124.05	-166.91	-218.20	-278.56
TIME OF FLIGHT (SEC)	.000000	.078041	.162568	.254059	.353140	.460329	.575664	.699162	.830517	.969067	1.114051	1.264896	1.421199
WIND DEFLECTION (IN)													
5 MPH	.00	.27	1.11	2.56	4.68	7.51	11.06	15.33	20.29	25.88	32.04	38.71	45.87
10 MPH	.00	.54	2.21	5.11	9.35	15.02	22.12	30.65	40.57	51.76	64.07	77.42	91.73
20 MPH	.00	1.07	4.42	10.23	18.71	30.04	44.23	61.30	81.14	103.51	128.15	154.84	183.46
30 MPH	.00	1.61	6.64	15.34	28.06	45.05	66.35	91.96	121.71	155.27	192.22	232.27	275.19
VELOCITY (FPS)	1900.	1754.	1621.	1497.	1384.	1287.	1203.	1132.	1075.	1028.	989.	955.	925.
ENERGY (FT-LB)	1090.	929.	793.	676.	578.	500.	437.	387.	349.	319.	296.	276.	259.
DROP (IN)	.00	-1.20	-5.23	-12.55	-23.79	-39.57	-60.60	-87.64	-121.44	-162.73	-212.23	-270.58	-338.32
MID-RANGE (IN)	.00	.32	1.41	3.44	6.67	11.34	17.75	26.17	36.92	50.29	66.55	85.92	108.57
BULLET PATH (IN)	-.75	1.03	.00	-4.33	-12.59	-25.38	-43.42	-67.47	-98.28	-136.59	-183.10	-238.46	-303.22
TIME OF FLIGHT (SEC)	.000000	.082197	.171167	.267505	.371814	.484275	.604913	.733562	.869670	1.012464	1.161298	1.315715	1.475392
WIND DEFLECTION (IN)													
5 MPH	.00	.29	1.17	2.70	4.93	7.88	11.55	15.92	20.95	26.57	32.72	39.36	46.47
10 MPH	.00	.57	2.34	5.40	9.86	15.76	23.10	31.84	41.90	53.14	65.44	78.72	92.93
20 MPH	.00	1.14	4.67	10.79	19.72	31.52	46.19	63.69	83.81	106.28	130.88	157.45	185.86
30 MPH	.00	1.72	7.01	16.19	29.58	47.28	69.29	95.53	1D5.71	159.42	196.32	236.17	278.80
VELOCITY (FPS)	1800.	1664.	1536.	1419.	1317.	1229.	1154.	1092.	1043.	1001.	966.	935.	907.
ENERGY (FT-LB)	978.	836.	713.	608.	524.	456.	402.	360.	328.	303.	282.	264.	248.
DROP (IN)	.00	-1.34	-5.81	-14.00	-26.50	-44.01	-67.29	-97.08	-134.13	-179.15	-232.82	-295.72	-368.45
MID-RANGE (IN)	.00	.36	1.57	3.84	7.44	12.60	19.65	28.87	40.57	55.00	72.40	92.98	116.94
BULLET PATH (IN)	-.75	1.19	.00	-4.90	-14.12	-28.35	-48.35	-74.86	-108.63	-150.37	-200.75	-260.37	-329.82
TIME OF FLIGHT (SEC)	.000000	.086691	.180534	.282163	.391960	.509918	.635983	.769726	.910410	1.057321	1.209947	1.367927	1.531002
WIND DEFLECTION (IN)													
5 MPH	.00	.30	1.22	2.83	5.16	8.21	11.97	16.40	21.45	27.04	33.14	39.71	46.73
10 MPH	.00	.59	2.44	5.66	10.32	16.41	23.93	32.81	42.90	54.09	66.28	79.42	93.46
20 MPH	.00	1.18	4.88	11.32	20.64	32.82	47.87	65.61	85.80	108.18	132.57	158.84	186.91
30 MPH	.00	1.77	7.32	16.98	30.96	49.24	71.80	98.42	128.70	162.27	198.85	238.27	280.37
VELOCITY (FPS)	1700.	1570.	1450.	1344.	1252.	1173.	1108.	1055.	1012.	975.	943.	914.	888.
ENERGY (FT-LB)	873.	744.	635.	545.	473.	415.	371.	336.	309.	287.	269.	252.	238.
DROP (IN)	.00	-1.50	-6.55	-15.74	-29.74	-49.32	-75.19	-108.13	-148.84	-198.02	-256.28	-324.18	-402.45
MID-RANGE (IN)	.00	.41	1.77	4.33	8.33	14.07	21.85	31.96	44.67	60.22	78.82	100.67	126.03
BULLET PATH (IN)	-.75	1.39	.00	-5.55	-15.90	-31.83	-54.06	-83.34	-120.41	-165.94	-220.55	-284.80	-359.42
TIME OF FLIGHT (SEC)	.000000	.091833	.191285	.298857	.414576	.538453	.670162	.809028	.954295	1.105399	1.261943	1.423644	1.590303
WIND DEFLECTION (IN)													
5 MPH	.00	.32	1.30	3.01	5.42	8.56	12.39	16.84	21.86	27.39	33.40	39.87	46.77
10 MPH	.00	.63	2.61	6.01	10.85	17.12	24.77	33.68	43.72	54.79	66.81	79.74	93.54
20 MPH	.00	1.27	5.21	12.02	21.70	34.24	49.54	67.37	87.44	109.57	133.62	159.48	187.08
30 MPH	.00	1.90	7.82	18.03	32.54	51.36	74.32	101.05	131.16	164.36	200.42	239.21	280.62

Bullet: Lyman # 280412 136 Gr.
Ballistic Coefficient: .245
(Cont'd.)

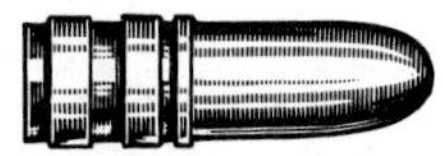

RANGE (YDS)	MUZZLE	50	100	150	200	250	300	350	400	450	500	550	600
VELOCITY (FPS)	1600.	1477.	1367.	1272.	1190.	1122.	1067.	1022.	983.	950.	921.	894.	869.
ENERGY (FT-LB)	763.	659.	564.	489.	428.	380.	344.	315.	292.	273.	256.	241.	228.
DROP (IN)	.00	-1.70	-7.43	-17.79	-33.53	-55.41	-84.17	-120.54	-165.21	-218.86	-281.95	-355.27	-439.43
MID-RANGE (IN)	.00	.46	2.01	4.87	9.34	15.72	24.30	35.33	49.07	65.77	85.57	108.76	135.55
BULLET PATH (IN)	-.75	1.64	.00	-6.27	-17.93	-35.72	-60.39	-92.67	-133.26	-182.81	-241.82	-311.05	-391.12
TIME OF FLIGHT (SEC)	.000000	.097592	.203230	.317018	.438982	.568890	.706138	.849949	.999713	1.154998	1.315498	1.480997	1.651346
WIND DEFLECTION (IN)													
5 MPH	.00	.34	1.38	3.15	5.63	8.81	12.64	17.05	21.97	27.39	33.26	39.58	46.32
10 MPH	.00	.68	2.77	6.30	11.26	17.62	25.28	34.09	43.95	54.78	66.53	79.16	92.64
20 MPH	.00	1.35	5.54	12.59	22.52	35.25	50.56	68.18	87.90	109.56	133.06	158.31	185.27
30 MPH	.00	2.03	8.31	18.89	33.78	52.87	75.84	102.27	·31.85	164.34	199.58	237.47	277.91
VELOCITY (FPS)	1500.	1386.	1290.	1205.	1134.	1076.	1030.	990.	956.	926.	899.	874.	851.
ENERGY (FT-LB)	679.	580.	502.	438.	388.	350.	320.	296.	276.	259.	244.	231.	219.
DROP (IN)	.00	-1.98	-8.47	-20.20	-37.92	-62.37	-94.31	-134.42	-183.37	-241.68	-310.08	-389.19	-479.66
MID-RANGE (IN)	.00	.52	2.26	5.48	10.49	17.56	26.96	38.95	53.77	71.59	92.67	117.23	145.52
BULLET PATH (IN)	-.75	1.88	.00	-7.12	-20.22	-40.07	-67.40	-102.89	-147.24	-200.94	-264.72	-339.22	-425.09
TIME OF FLIGHT (SEC)	.000000	.104073	.216308	.336718	.465148	.601055	.743667	.892332	1.046589	1.206110	1.370664	1.540091	1.714290
WIND DEFLECTION (IN)													
5 MPH	.00	.36	1.44	3.23	5.73	8.89	12.64	16.93	21.70	26.94	32.62	38.73	45.26
10 MPH	.00	.72	2.87	6.46	11.47	17.79	25.29	33.85	43.40	53.88	65.24	77.46	90.51
20 MPH	.00	1.43	5.74	12.92	22.93	35.57	50.57	67.70	86.80	107.75	130.47	154.91	181.03
30 MPH	.00	2.15	8.61	19.39	34.40	53.36	75.86	101.55	130.20	161.63	195.71	232.37	271.54
VELOCITY (FPS)	1400.	1302.	1215.	1143.	1083.	1035.	995.	960.	930.	902.	877.	854.	832.
ENERGY (FT-LB)	592.	512.	446.	394.	354.	324.	299.	279.	261.	246.	232.	220.	209.
DROP (IN)	.00	-2.26	-9.66	-22.95	-42.89	-70.20	-105.60	-149.76	-203.22	-266.65	-340.72	-426.07	-523.34
MID-RANGE (IN)	.00	.59	2.57	6.21	11.78	19.57	29.85	42.84	58.72	77.75	100.16	126.19	156.06
BULLET PATH (IN)	-.75	2.19	.00	-8.09	-22.81	-44.92	-75.12	-114.07	-162.32	-220.55	-289.41	-369.56	-461.62
TIME OF FLIGHT (SEC)	.000000	.111171	.230509	.357909	.492877	.634650	.782548	.936089	1.094931	1.258830	1.427619	1.601188	1.779476
WIND DEFLECTION (IN)													
5 MPH	.00	.35	1.43	3.21	5.66	8.71	12.29	16.38	20.93	25.92	31.34	37.19	43.45
10 MPH	.00	.71	2.86	6.42	11.32	17.41	24.59	32.75	41.85	51.84	62.69	74.38	86.90
20 MPH	.00	1.42	5.71	12.84	22.64	34.83	49.17	65.50	83.70	103.68	125.38	148.76	173.80
30 MPH	.00	2.13	8.57	19.26	33.95	52.24	73.76	98.26	125.55	155.52	188.07	223.14	260.71
VELOCITY (FPS)	1300.	1214.	1141.	1082.	1034.	994.	960.	929.	902.	877.	853.	831.	811.
ENERGY (FT-LB)	510.	445.	393.	354.	323.	299.	278.	261.	246.	232.	220.	209.	198.
DROP (IN)	.00	-2.63	-11.17	-26.36	-48.95	-79.63	-119.09	-167.85	-226.60	-295.99	-376.68	-469.29	-574.48
MID-RANGE (IN)	.00	.69	2.95	7.06	13.30	21.93	33.18	47.24	64.35	84.76	108.70	136.42	168.15
BULLET PATH (IN)	-.75	2.58	.00	-9.23	-25.86	-50.58	-84.08	-126.88	-179.67	-243.10	-317.83	-404.48	-503.71
TIME OF FLIGHT (SEC)	.000000	.119483	.247021	.382115	.524000	.671998	.825632	.984559	1.148540	1.317407	1.491052	1.669412	1.852466
WIND DEFLECTION (IN)													
5 MPH	.00	.36	1.43	3.16	5.50	8.37	11.73	15.56	19.84	24.55	29.67	35.22	41.17
10 MPH	.00	.72	2.86	6.33	10.99	16.73	23.47	31.13	39.68	49.09	59.35	70.43	82.34
20 MPH	.00	1.44	5.72	12.66	21.99	33.47	46.93	62.26	79.36	98.19	118.70	140.86	164.68
30 MPH	.00	2.16	8.58	18.99	32.98	50.20	70.40	93.39	119.04	147.28	178.04	211.30	247.02

Bullet: Lyman # 287448 119 Gr.
Ballistic Coefficient: .165

RANGE (YDS)	MUZZLE	50	100	150	200	250	300	350	400	450	500	550	600
VELOCITY (FPS)	2200.	1969.	1790.	1622.	1489.	1372.	1268.	1179.	1106.	1049.	1003.	964.	930.
ENERGY (FT-LB)	1279.	1025.	846.	695.	586.	497.	424.	367.	323.	291.	266.	246.	229.
DROP (IN)	.00	-.93	-4.09	-9.96	-19.10	-32.21	-49.93	-73.07	-102.46	-138.93	-183.27	-236.22	-298.42
MID-RANGE (IN)	.00	.25	1.11	2.77	5.46	9.45	15.01	22.45	32.13	44.34	59.43	77.67	99.29
BULLET PATH (IN)	-.75	.74	.00	-3.45	-10.18	-20.86	-36.16	-56.88	-83.86	-117.91	-159.82	-210.36	-270.14
TIME OF FLIGHT (SEC)	.000000	.072176	.152083	.240132	.336771	.441768	.555602	.678437	.809939	.949308	1.095643	1.248302	1.406843
WIND DEFLECTION (IN)													
5 MPH	.00	.35	1.38	3.13	5.64	8.88	12.89	17.70	23.27	29.54	36.42	43.85	51.80
10 MPH	.00	.70	2.77	6.26	11.27	17.75	25.79	35.40	46.55	59.08	72.83	87.70	103.60
20 MPH	.00	1.41	5.53	12.53	22.54	35.50	51.57	70.81	93.10	118.16	145.67	175.40	207.21
30 MPH	.00	2.11	8.30	18.79	33.82	53.25	77.36	106.21	139.65	177.23	218.50	263.10	310.81
VELOCITY (FPS)	2100.	1891.	1717.	1562.	1438.	1325.	1228.	1146.	1080.	1029.	986.	949.	917.
ENERGY (FT-LB)	1165.	945.	779.	644.	546.	464.	398.	347.	308.	279.	257.	238.	222.
DROP (IN)	.00	-1.02	-4.46	-10.85	-20.80	-34.97	-54.10	-79.00	-110.51	-149.43	-196.55	-252.60	-318.13
MID-RANGE (IN)	.00	.27	1.21	3.02	5.94	10.23	16.20	24.17	34.45	47.40	63.30	82.43	104.97
BULLET PATH (IN)	-.75	.84	.00	-3.78	-11.12	-22.69	-39.21	-61.51	-90.41	-126.73	-171.24	-224.68	-287.61
TIME OF FLIGHT (SEC)	.000000	.075484	.158736	.250510	.350658	.459383	.577070	.703679	.838660	.981101	1.130202	1.285418	1.446372
WIND DEFLECTION (IN)													
5 MPH	.00	.36	1.40	3.19	5.72	9.00	13.07	17.92	23.52	29.77	36.60	43.97	51.85
10 MPH	.00	.71	2.79	6.38	11.43	17.99	26.14	35.85	47.03	59.53	73.20	87.95	103.70
20 MPH	.00	1.43	5.59	12.75	22.86	35.99	52.27	71.70	94.07	119.06	146.40	175.90	207.41
30 MPH	.00	2.14	8.38	19.13	34.29	53.98	78.41	107.54	141.10	178.59	219.60	263.84	311.11
VELOCITY (FPS)	2000.	1819.	1649.	1510.	1390.	1284.	1192.	1117.	1058.	1010.	970.	936.	905.
ENERGY (FT-LB)	1057.	874.	718.	602.	511.	435.	376.	330.	296.	270.	249.	231.	216.
DROP (IN)	.00	-1.11	-4.84	-11.76	-22.51	-37.77	-58.31	-84.96	-118.55	-159.89	-209.72	-268.71	-337.52
MID-RANGE (IN)	.00	.30	1.31	3.26	6.42	11.02	17.40	25.86	36.76	50.41	67.09	87.03	110.45
BULLET PATH (IN)	-.75	.94	.00	-4.12	-12.08	-24.54	-42.28	-66.14	-96.94	-135.47	-182.51	-238.70	-304.72
TIME OF FLIGHT (SEC)	.000000	.078660	.165296	.260558	.364134	.476482	.597836	.727948	.866081	1.011313	1.162958	1.320547	1.483758
WIND DEFLECTION (IN)													
5 MPH	.00	.32	1.35	3.13	5.64	8.93	13.01	17.86	23.42	29.60	36.34	43.61	51.37
10 MPH	.00	.64	2.69	6.26	11.29	17.86	26.02	35.72	46.83	59.19	72.68	87.22	102.74
20 MPH	.00	1.29	5.38	12.52	22.58	35.72	52.04	71.44	93.66	118.38	145.36	174.43	205.48
30 MPH	.00	1.93	8.08	18.77	33.86	53.58	78.06	107.16	140.49	177.57	218.04	261.65	308.22
VELOCITY (FPS)	1900.	1725.	1567.	1442.	1330.	1231.	1149.	1083.	1030.	987.	950.	918.	889.
ENERGY (FT-LB)	954.	786.	649.	550.	467.	401.	349.	310.	281.	258.	239.	223.	209.
DROP (IN)	.00	-1.23	-5.38	-13.06	-24.94	-41.73	-64.27	-93.39	-129.89	-174.55	-228.10	-291.12	-364.42
MID-RANGE (IN)	.00	.33	1.46	3.63	7.09	12.14	19.08	28.26	40.01	54.61	72.32	93.33	117.94
BULLET PATH (IN)	-.75	1.08	.00	-4.62	-13.43	-27.16	-46.64	-72.69	-106.13	-147.73	-198.22	-258.18	-328.41
TIME OF FLIGHT (SEC)	.000000	.082873	.174247	.274058	.382426	.499747	.626001	.760658	.902812	1.051654	1.206627	1.367351	1.533565
WIND DEFLECTION (IN)													
5 MPH	.00	.35	1.44	3.28	5.86	9.24	13.40	18.31	23.87	30.02	36.71	43.91	51.59
10 MPH	.00	.69	2.88	6.55	11.73	18.48	26.81	36.61	47.74	60.04	73.42	87.81	103.17
20 MPH	.00	1.38	5.76	13.10	23.46	36.96	53.62	73.23	95.47	120.08	146.84	175.62	206.34
30 MPH	.00	2.07	8.63	19.65	35.18	55.45	80.42	109.84	143.21	180.12	220.26	263.43	309.51

Bullet: Lyman # 287448 119 Gr.
Ballistic Coefficient: .165
(Cont'd.)

RANGE (YDS)	MUZZLE	50	100	150	200	250	300	350	400	450	500	550	600
VELOCITY (FPS)	1800.	1632.	1496.	1378.	1273.	1183.	1110.	1052.	1005.	966.	932.	901.	874.
ENERGY (FT-LB)	856.	703.	591.	502.	428.	370.	325.	292.	267.	247.	229.	215.	202.
DROP (IN)	.00	-1.37	-5.99	-14.53	-27.63	-46.12	-70.82	-102.54	-142.10	-190.22	-247.57	-314.84	-392.76
MID-RANGE (IN)	.00	.37	1.62	4.03	7.83	13.33	20.88	30.81	43.41	58.96	77.66	99.77	125.56
BULLET PATH (IN)	-.75	1.25	.00	-5.17	-14.91	-30.03	-51.35	-79.71	-115.89	-160.64	-214.63	-278.52	-353.08
TIME OF FLIGHT (SEC)	.000000	.087542	.183721	.288248	.401591	.523938	.654984	.793946	.939917	1.092239	1.250462	1.414277	1.583478
WIND DEFLECTION (IN)													
5 MPH	.00	.37	1.50	3.37	6.01	9.44	13.64	18.53	24.05	30.12	36.71	43.79	51.35
10 MPH	.00	.74	3.00	6.73	12.01	18.88	27.28	37.07	48.09	60.23	73.41	87.58	102.69
20 MPH	.08	1.48	6.00	13.46	24.03	37.76	54.55	74.14	96.18	120.47	146.83	175.16	205.38
30 MPH	.00	2.22	9.00	20.19	36.04	56.64	81.83	111.20	144.28	180.70	220.24	262.74	308.08
VELOCITY (FPS)	1700.	1548.	1425.	1315.	1218.	1138.	1074.	1024.	982.	946.	914.	885.	859.
ENERGY (FT-LB)	764.	633.	537.	457.	392.	342.	305.	277.	255.	236.	221.	207.	195.
DROP (IN)	.00	-1.54	-6.71	-16.17	-30.66	-51.02	-78.07	-112.63	-155.45	-207.26	-268.64	-340.40	-423.24
MID-RANGE (IN)	.00	.41	1.82	4.46	8.62	14.65	22.87	33.58	47.06	63.57	83.30	106.54	133.55
BULLET PATH (IN)	-.75	1.44	.00	-5.73	-16.50	-33.13	-56.46	-87.28	-126.38	-174.46	-232.11	-300.14	-379.25
TIME OF FLIGHT (SEC)	.000000	.092654	.193669	.303317	.421943	.549458	.685259	.828422	.978178	1.134000	1.295528	1.462516	1.634806
WIND DEFLECTION (IN)													
5 MPH	.00	.39	1.51	3.40	6.07	9.53	13.71	18.55	23.96	29.91	36.36	43.29	50.69
10 MPH	.00	.78	3.03	6.80	12.14	19.06	27.43	37.10	47.92	59.82	72.72	86.58	101.37
20 MPH	.00	1.56	6.05	13.59	24.29	38.12	54.86	74.19	95.85	119.64	145.44	173.16	202.75
30 MPH	.00	2.33	9.08	20.39	36.43	57.17	82.29	111.29	143.77	179.46	218.16	259.74	304.12
VELOCITY (FPS)	1600.	1473.	1357.	1255.	1168.	1098.	1043.	997.	959.	926.	896.	869.	844.
ENERGY (FT-LB)	676.	573.	486.	416.	360.	318.	287.	263.	243.	226.	212.	199.	188.
DROP (IN)	.00	-1.72	-7.47	-17.94	-33.96	-56.34	-85.90	-123.43	-169.67	-225.25	-290.89	-367.33	-455.26
MID-RANGE (IN)	.00	.46	2.02	4.92	9.51	16.10	25.00	36.50	50.87	68.32	89.11	113.50	141.74
BULLET PATH (IN)	-.75	1.64	.00	-6.36	-18.26	-36.53	-61.98	-95.40	-137.53	-188.99	-250.52	-322.85	-406.67
TIME OF FLIGHT (SEC)	.000000	.097754	.203924	.318978	.443018	.575640	.715996	.863215	1.016684	1.175986	1.340832	1.511034	1.686474
WIND DEFLECTION (IN)													
5 MPH	.00	.35	1.45	3.32	5.99	9.41	13.51	18.21	23.47	29.24	35.49	42.22	49.41
10 MPH	.00	.70	2.89	6.64	11.97	18.81	27.02	36.43	46.94	58.47	70.99	84.44	98.82
20 MPH	.00	1.41	5.78	13.28	23.94	37.63	54.03	72.85	93.87	116.95	141.97	168.88	197.64
30 MPH	.00	2.11	8.67	19.92	35.91	56.44	81.05	109.28	140.81	175.42	212.96	253.33	296.46
VELOCITY (FPS)	1500.	1381.	1276.	1186.	1112.	1054.	1007.	967.	933.	902.	875.	849.	825.
ENERGY (FT-LB)	594.	504.	430.	371.	327.	293.	268.	247.	230.	215.	202.	190.	180.
DROP (IN)	.00	-1.99	-8.53	-20.42	-38.49	-63.56	-96.43	-137.86	-188.48	-249.00	-320.15	-402.65	-497.18
MID-RANGE (IN)	.00	.53	2.28	5.57	10.72	18.03	27.80	40.30	55.76	74.40	96.52	122.35	152.16
BULLET PATH (IN)	-.75	1.90	.00	-7.25	-20.68	-41.12	-69.35	-106.14	-152.13	-208.01	-274.52	-352.38	-442.27
TIME OF FLIGHT (SEC)	.000000	.104245	.217293	.339345	.470113	.608827	.754575	.906690	1.064717	1.228342	1.397357	1.571631	1.751094
WIND DEFLECTION (IN)													
5 MPH	.00	.37	1.52	3.46	6.17	9.58	13.60	18.19	23.30	28.89	34.97	41.50	48.50
10 MPH	.00	.75	3.04	6.92	12.34	19.15	27.21	36.38	46.59	57.79	69.93	83.01	96.99
20 MPH	.00	1.49	6.09	13.85	24.68	38.31	54.41	72.75	93.18	115.58	139.87	166.01	193.99
30 MPH	.00	2.24	9.13	20.77	37.02	57.46	81.62	109.13	139.77	173.36	209.80	249.02	290.98

Bullet: Lyman # 287346 135 Gr.
Ballistic Coefficient: .235

RANGE (YDS)	MUZZLE	50	100	150	200	250	300	350	400	450	500	550	600
VELOCITY (FPS)	2000.	1841.	1703.	1577.	1461.	1360.	1276.	1202.	1139.	1087.	1043.	1006.	974.
ENERGY (FT-LB)	1199.	1016.	869.	746.	640.	554.	488.	433.	389.	354.	326.	304.	284.
DROP (IN)	.00	-1.09	-4.74	-11.37	-21.50	-35.73	-54.64	-78.91	-109.18	-146.14	-190.43	-242.69	-303.52
MID-RANGE (IN)	.00	.29	1.28	3.12	6.02	10.22	15.95	23.49	33.09	45.00	59.48	76.78	97.11
BULLET PATH (IN)	-.75	.90	.00	-3.89	-11.28	-22.77	-38.94	-60.46	-87.99	-122.20	-163.76	-213.27	-271.35
TIME OF FLIGHT (SEC)	.000000	.078175	.162982	.254519	.353353	.459938	.573853	.695013	.823268	.958201	1.099203	1.245722	1.397349
WIND DEFLECTION (IN)													
5 MPH	.00	.28	1.14	2.60	4.70	7.47	10.90	14.96	19.65	24.92	30.73	37.02	43.77
10 MPH	.00	.56	2.28	5.20	9.39	14.95	21.80	29.92	39.30	49.84	61.46	74.05	87.53
20 MPH	.00	1.12	4.57	10.39	18.78	29.90	43.60	59.84	78.59	99.69	122.92	148.09	175.07
30 MPH	.00	1.68	6.85	15.59	28.17	44.85	65.39	89.77	117.89	149.53	184.38	222.14	262.60
VELOCITY (FPS)	1900.	1752.	1623.	1503.	1393.	1306.	1229.	1162.	1105.	1059.	1019.	986.	956.
ENERGY (FT-LB)	1082.	920.	790.	677.	582.	512.	453.	404.	366.	336.	311.	291.	274.
DROP (IN)	.00	-1.21	-5.24	-12.56	-23.76	-39.42	-60.16	-86.69	-119.64	-159.68	-207.46	-263.58	-328.63
MID-RANGE (IN)	.00	.32	1.41	3.44	6.64	11.25	17.52	25.70	36.06	48.85	64.31	82.68	104.15
BULLET PATH (IN)	-.75	1.04	.00	-4.33	-12.53	-25.19	-42.95	-66.48	-96.43	-133.48	-178.26	-231.39	-293.44
TIME OF FLIGHT (SEC)	.000000	.082314	.171280	.267323	.371012	.482233	.600688	.726326	.858826	.997632	1.142143	1.291899	1.446564
WIND DEFLECTION (IN)													
5 MPH	.00	.30	1.18	2.68	4.86	7.70	11.18	15.29	20.00	25.27	31.03	37.27	43.93
10 MPH	.00	.59	2.36	5.36	9.72	15.40	22.35	30.57	40.00	50.53	62.07	74.53	87.86
20 MPH	.00	1.18	4.71	10.73	19.44	30.80	44.71	61.14	79.99	101.06	124.14	149.06	175.72
30 MPH	.00	1.78	7.07	16.09	29.16	46.20	67.06	91.71	119.99	151.59	186.21	223.60	263.58
VELOCITY (FPS)	1800.	1669.	1546.	1432.	1337.	1256.	1185.	1124.	1074.	1033.	997.	966.	939.
ENERGY (FT-LB)	971.	835.	716.	614.	536.	473.	421.	379.	346.	320.	298.	280.	264.
DROP (IN)	.00	-1.34	-5.79	-13.90	-26.25	-43.45	-66.19	-95.12	-130.89	-174.17	-225.58	-285.70	-355.05
MID-RANGE (IN)	.00	.36	1.56	3.81	7.34	12.38	19.20	28.05	39.19	52.88	69.34	88.76	111.33
BULLET PATH (IN)	-.75	1.18	.00	-4.84	-13.92	-27.86	-47.33	-72.98	-105.49	-145.50	-193.64	-250.49	-316.57
TIME OF FLIGHT (SEC)	.000000	.086560	.179979	.280847	.389419	.505233	.628279	.758335	.894919	1.037413	1.185302	1.338211	1.495862
WIND DEFLECTION (IN)													
5 MPH	.00	.28	1.17	2.71	4.94	7.79	11.29	15.40	20.09	25.29	30.97	37.10	43.64
10 MPH	.00	.57	2.34	5.43	9.87	15.59	22.58	30.80	40.17	50.58	61.95	74.19	87.27
20 MPH	.00	1.14	4.69	10.86	19.74	31.18	45.15	61.60	80.34	101.17	123.89	148.38	174.54
30 MPH	.00	1.70	7.03	16.29	29.61	46.76	67.73	92.40	120.52	151.75	185.84	222.58	261.81
VELOCITY (FPS)	1700.	1575.	1459.	1358.	1275.	1201.	1138.	1086.	1042.	1006.	974.	945.	919.
ENERGY (FT-LB)	866.	743.	638.	553.	487.	432.	388.	353.	326.	303.	284.	268.	253.
DROP (IN)	.00	-1.50	-6.51	-15.63	-29.45	-48.63	-73.83	-105.72	-144.97	-192.18	-247.98	-312.90	-387.38
MID-RANGE (IN)	.00	.40	1.76	4.29	8.21	13.79	21.28	30.95	43.04	57.77	75.36	96.01	119.86
BULLET PATH (IN)	-.75	1.38	.00	-5.49	-15.67	-31.22	-52.80	-81.06	-116.68	-160.26	-212.42	-273.71	-344.56
TIME OF FLIGHT (SEC)	.000000	.091693	.190696	.297436	.411477	.532764	.661139	.796182	.937282	1.083889	1.235596	1.392105	1.553193
WIND DEFLECTION (IN)													
5 MPH	.00	.30	1.25	2.88	5.15	8.06	11.59	15.71	20.36	25.50	31.09	37.09	43.50
10 MPH	.00	.61	2.50	5.76	10.30	16.12	23.18	31.42	40.73	51.00	62.17	74.19	87.01
20 MPH	.00	1.22	5.01	11.52	20.60	32.24	46.37	62.84	81.45	102.00	124.34	148.37	174.02
30 MPH	.00	1.83	7.51	17.28	30.91	48.36	69.55	94.27	122.18	153.00	186.51	222.56	261.03

Bullet: Lyman # 287346 135 Gr.
Ballistic Coefficient: .235
(Cont'd.)

RANGE (YDS)	MUZZLE	50	100	150	200	250	300	350	400	450	500	550	600
VELOCITY (FPS)	1600.	1482.	1377.	1291.	1216.	1150.	1096.	1051.	1013.	980.	951.	924.	900.
ENERGY (FT-LB)	767.	658.	568.	500.	443.	397.	360.	331.	307.	288.	271.	256.	243.
DROP (IN)	.00	-1.70	-7.39	-17.65	-33.13	-54.50	-82.42	-117.56	-160.54	-211.98	-272.45	-342.37	-422.42
MID-RANGE (IN)	.00	.46	2.00	4.81	9.17	15.34	23.56	34.09	47.15	62.96	81.73	103.59	128.79
BULLET PATH (IN)	-.75	1.62	.00	-6.19	-17.59	-34.90	-58.75	-89.82	-128.73	-176.10	-232.50	-298.35	-374.33
TIME OF FLIGHT (SEC)	.000000	.097444	.202601	.315153	.434949	.561889	.695602	.835502	.981009	1.131689	1.287225	1.447380	1.611981
WIND DEFLECTION (IN)													
5 MPH	.00	.33	1.33	2.98	5.28	8.20	11.71	15.77	20.33	25.34	30.78	36.62	42.85
10 MPH	.00	.65	2.66	5.97	10.55	16.39	23.43	31.55	40.66	50.68	61.55	73.24	85.71
20 MPH	.00	1.30	5.32	11.93	21.10	32.78	46.85	63.10	81.32	101.35	123.10	146.48	171.42
30 MPH	.00	1.95	7.97	17.90	31.65	49.18	70.28	94.65	121.97	152.03	184.65	219.72	257.13
VELOCITY (FPS)	1500.	1391.	1304.	1227.	1160.	1104.	1057.	1018.	985.	955.	928.	904.	882.
ENERGY (FT-LB)	674.	580.	510.	451.	403.	365.	335.	311.	291.	273.	258.	245.	233.
DROP (IN)	.00	-1.97	-8.42	-19.97	-37.31	-61.10	-92.00	-130.65	-177.65	-233.60	-298.92	-374.27	-460.22
MID-RANGE (IN)	.00	.52	2.23	5.39	10.24	17.04	26.03	37.44	51.51	68.44	88.36	111.51	138.11
BULLET PATH (IN)	-.75	1.86	.00	-6.97	-19.73	-38.94	-65.25	-99.32	-141.74	-193.10	-253.85	-324.61	-405.98
TIME OF FLIGHT (SEC)	.000000	.103915	.215318	.333955	.459771	.592435	.731388	.876029	1.025903	1.180677	1.340102	1.503997	1.672230
WIND DEFLECTION (IN)													
5 MPH	.00	.34	1.35	2.99	5.26	8.13	11.56	15.49	19.88	24.70	29.93	35.55	41.56
10 MPH	.00	.69	2.70	5.98	10.52	16.27	23.12	30.98	39.76	49.40	59.86	71.10	83.11
20 MPH	.00	1.38	5.39	11.95	21.04	32.54	46.25	61.96	79.52	98.80	119.72	142.21	166.23
30 MPH	.00	2.07	8.09	17.93	31.56	48.81	69.37	92.94	119.28	148.20	179.57	213.31	249.34
VELOCITY (FPS)	1400.	1312.	1234.	1166.	1109.	1062.	1022.	988.	958.	931.	906.	884.	863.
ENERGY (FT-LB)	587.	516.	456.	408.	368.	338.	313.	292.	275.	260.	246.	234.	223.
DROP (IN)	.00	-2.24	-9.53	-22.55	-41.94	-68.38	-102.51	-144.93	-196.24	-256.89	-327.49	-408.64	-500.90
MID-RANGE (IN)	.00	.59	2.52	6.06	11.44	18.91	28.71	41.06	56.18	74.21	95.36	119.88	147.94
BULLET PATH (IN)	-.75	2.15	.00	-7.88	-22.13	-43.43	-72.42	-109.71	-155.88	-211.39	-276.85	-352.86	-439.98
TIME OF FLIGHT (SEC)	.000000	.110704	.228634	.353759	.485776	.624141	.768248	.917625	1.071929	1.230904	1.394364	1.562172	1.734235
WIND DEFLECTION (IN)													
5 MPH	.00	.31	1.26	2.85	5.03	7.78	11.03	14.75	18.90	23.46	28.42	33.76	39.47
10 MPH	.00	.63	2.53	5.69	10.07	15.56	22.07	29.50	37.80	46.92	56.84	67.51	78.94
20 MPH	.00	1.25	5.05	11.38	20.14	31.13	44.14	59.00	75.60	93.85	113.67	135.03	157.88
30 MPH	.00	1.88	7.58	17.07	30.20	46.69	66.21	88.51	113.41	140.77	170.51	202.54	236.82
VELOCITY (FPS)	1300.	1223.	1157.	1101.	1055.	1017.	983.	954.	927.	903.	880.	859.	840.
ENERGY (FT-LB)	507.	448.	401.	363.	334.	310.	290.	273.	258.	244.	232.	221.	211.
DROP (IN)	.00	-2.61	-11.04	-25.95	-48.00	-77.84	-116.06	-163.25	-219.84	-286.48	-363.76	-452.23	-552.45
MID-RANGE (IN)	.00	.69	2.91	6.92	12.96	21.28	32.08	45.59	61.92	81.33	104.04	130.25	160.16
BULLET PATH (IN)	-.75	2.54	.00	-9.02	-25.18	-49.12	-81.45	-122.75	-173.43	-234.18	-305.57	-388.14	-482.47
TIME OF FLIGHT (SEC)	.000000	.119014	.245195	.378200	.517459	.662377	.812507	.967518	1.127169	1.291279	1.459719	1.632402	1.809272
WIND DEFLECTION (IN)													
5 MPH	.00	.32	1.27	2.82	4.92	7.52	10.58	14.06	17.96	22.25	26.92	31.96	37.37
10 MPH	.00	.64	2.54	5.64	9.84	15.04	21.15	28.13	35.92	44.50	53.83	63.92	74.74
20 MPH	.00	1.28	5.08	11.28	19.68	30.08	42.31	56.26	71.84	88.99	107.67	127.84	149.48
30 MPH	.00	1.92	7.62	16.92	29.53	45.12	63.46	84.39	107.76	133.49	161.50	191.75	224.22

Bullet: Lyman # 287405 150 Gr.
Ballistic Coefficient: .245

RANGE (YDS)	MUZZLE	50	100	150	200	250	300	350	400	450	500	550	600
VELOCITY (FPS)	1800.	1658.	1531.	1419.	1316.	1227.	1154.	1095.	1047.	1008.	973.	943.	915.
ENERGY (FT-LB)	1079.	916.	781.	670.	577.	501.	443.	399.	365.	338.	315.	296.	279.
DROP (IN)	.00	-1.35	-5.84	-14.07	-26.63	-44.20	-67.55	-97.42	-134.51	-179.50	-233.03	-295.68	-367.97
MID-RANGE (IN)	.00	.36	1.58	3.87	7.47	12.65	19.71	28.94	40.63	55.02	72.32	92.73	116.43
BULLET PATH (IN)	-.75	1.20	.00	-4.93	-14.19	-28.46	-48.52	-75.09	-108.89	-150.58	-200.81	-260.16	-329.16
TIME OF FLIGHT (SEC)	.000000	.086832	.181056	.282864	.392680	.510801	.637049	.770605	.910775	1.056904	1.208513	1.365256	1.526883
WIND DEFLECTION (IN)													
5 MPH	.00	.31	1.27	2.89	5.22	8.28	12.06	16.48	21.48	27.01	33.02	39.48	46.37
10 MPH	.00	.62	2.53	5.78	10.45	16.57	24.12	32.96	42.96	54.02	66.03	78.95	92.73
20 MPH	.00	1.23	5.06	11.57	20.89	33.14	48.24	65.92	85.93	108.03	132.06	157.90	185.46
30 MPH	.00	1.85	7.60	17.35	31.34	49.70	72.36	98.88	128.89	162.05	198.09	236.86	278.19
VELOCITY (FPS)	1700.	1567.	1451.	1346.	1252.	1174.	1111.	1061.	1019.	983.	951.	923.	898.
ENERGY (FT-LB)	962.	818.	701.	603.	522.	459.	411.	375.	346.	322.	301.	284.	268.
DROP (IN)	.00	-1.51	-6.56	-15.77	-29.76	-49.32	-75.17	-108.03	-148.60	-197.51	-255.40	-322.72	-400.20
MID-RANGE (IN)	.00	.41	1.77	4.33	8.32	14.05	21.82	31.90	44.54	59.96	78.38	99.95	124.93
BULLET PATH (IN)	-.75	1.40	.00	-5.55	-15.88	-31.79	-53.98	-83.19	-120.10	-165.35	-219.59	-283.26	-357.08
TIME OF FLIGHT (SEC)	.000000	.091971	.191465	.298834	.414450	.538341	.669783	.808054	.952460	1.102472	1.257710	1.417897	1.582835
WIND DEFLECTION (IN)													
5 MPH	.00	.33	1.32	3.00	5.41	8.55	12.35	16.76	21.70	27.14	33.03	39.36	46.11
10 MPH	.00	.66	2.64	6.01	10.83	17.10	24.71	33.51	43.40	54.27	66.06	78.73	92.23
20 MPH	.00	1.31	5.28	12.01	21.65	34.20	49.41	67.02	86.80	108.54	132.13	157.45	184.45
30 MPH	.00	1.97	7.92	18.02	32.48	51.30	74.12	100.53	130.19	162.81	198.19	236.18	276.68
VELOCITY (FPS)	1600.	1482.	1373.	1276.	1193.	1127.	1073.	1029.	992.	960.	931.	904.	880.
ENERGY (FT-LB)	853.	731.	628.	543.	474.	423.	384.	353.	328.	307.	288.	272.	258.
DROP (IN)	.00	-1.70	-7.39	-17.69	-33.33	-55.08	-83.64	-119.72	-163.98	-217.04	-279.42	-351.76	-434.69
MID-RANGE (IN)	.00	.46	2.00	4.83	9.27	15.62	24.13	35.07	48.66	65.12	84.64	107.42	133.71
BULLET PATH (IN)	-.75	1.62	.00	-6.22	-17.80	-35.47	-59.96	-91.97	-132.15	-181.15	-239.46	-307.72	-386.59
TIME OF FLIGHT (SEC)	.000000	.097444	.202630	.315986	.437670	.567142	.703628	.846426	.994955	1.148802	1.307663	1.471323	1.639627
WIND DEFLECTION (IN)													
5 MPH	.00	.33	1.33	3.06	5.51	8.66	12.42	16.74	21.56	26.84	32.57	38.73	45.29
10 MPH	.00	.65	2.66	6.11	11.03	17.32	24.84	33.47	43.11	53.69	65.15	77.45	90.57
20 MPH	.00	1.30	5.33	12.23	22.06	34.63	49.68	66.94	86.22	107.38	130.30	154.91	181.15
30 MPH	.00	1.95	7.99	18.34	33.09	51.95	74.52	100.41	129.34	161.07	195.45	232.36	271.72
VELOCITY (FPS)	1500.	1390.	1291.	1205.	1137.	1081.	1036.	998.	965.	935.	908.	884.	861.
ENERGY (FT-LB)	749.	643.	555.	483.	430.	389.	357.	332.	310.	291.	275.	260.	247.
DROP (IN)	.00	-1.97	-8.44	-20.14	-37.81	-62.19	-93.98	-133.84	-182.41	-240.23	-307.88	-386.03	-475.28
MID-RANGE (IN)	.00	.52	2.25	5.46	10.46	17.51	26.85	38.74	53.39	70.99	91.75	115.89	143.66
BULLET PATH (IN)	-.75	1.87	.00	-7.10	-20.18	-39.96	-67.15	-102.41	-146.38	-199.61	-262.66	-336.21	-420.87
TIME OF FLIGHT (SEC)	.000000	.103918	.215950	.336310	.464595	.599997	.741821	.889458	1.042468	1.200535	1.363429	1.530989	1.703103
WIND DEFLECTION (IN)													
5 MPH	.00	.34	1.40	3.20	5.68	8.80	12.48	16.67	21.34	26.45	31.98	37.93	44.27
10 MPH	.00	.69	2.81	6.39	11.37	17.60	24.96	33.34	42.67	52.89	63.96	75.85	88.55
20 MPH	.00	1.38	5.61	12.78	22.74	35.20	49.92	66.69	85.35	105.79	127.93	151.71	177.09
30 MPH	.00	2.07	8.42	19.17	34.11	52.80	74.88	100.03	128.02	158.68	191.89	227.56	265.64

Bullet: Lyman # 287405 150 Gr.
Ballistic Coefficient: .245

(Cont'd.)

RANGE (YDS)	MUZZLE	50	100	150	200	250	300	350	400	450	500	550	600
VELOCITY (FPS)	1400.	1300.	1212.	1143.	1086.	1040.	1001.	967.	938.	911.	886.	863.	842.
ENERGY (FT-LB)	653.	563.	489.	435.	393.	360.	334.	312.	293.	276.	262.	248.	236.
DROP (IN)	.00	-2.27	-9.68	-23.02	-42.98	-70.30	-105.62	-149.59	-202.77	-265.72	-339.10	-423.53	-519.61
MID-RANGE (IN)	.00	.60	2.57	6.23	11.81	19.58	29.79	42.67	58.40	77.18	99.25	124.85	154.18
BULLET PATH (IN)	-.75	2.20	.00	-8.11	-22.87	-44.96	-75.07	-113.82	-161.79	-219.51	-287.68	-366.89	-457.75
TIME OF FLIGHT (SEC)	.000000	.111247	.230817	.358402	.493170	.634425	.781539	.934062	1.091665	1.254114	1.421239	1.592926	1.769104
WIND DEFLECTION (IN)													
5 MPH	.00	.36	1.45	3.25	5.68	8.69	12.20	16.20	20.64	25.50	30.78	36.46	42.54
10 MPH	.00	.72	2.91	6.51	11.37	17.37	24.41	32.39	41.28	51.01	61.57	72.93	85.08
20 MPH	.00	1.44	5.82	13.01	22.74	34.75	48.82	64.79	82.55	102.02	123.13	145.85	170.15
30 MPH	.00	2.17	8.73	19.52	34.11	52.12	73.22	97.18	123.83	153.03	184.70	218.78	255.23
VELOCITY (FPS)	1300.	1212.	1143.	1086.	1040.	1001.	967.	938.	911.	886.	863.	842.	822.
ENERGY (FT-LB)	563.	490.	435.	393.	360.	334.	312.	293.	276.	262.	248.	236.	225.
DROP (IN)	.00	-2.64	-11.19	-26.37	-48.90	-79.44	-118.63	-167.02	-225.17	-293.77	-373.41	-464.69	-568.21
MID-RANGE (IN)	.00	.70	2.96	7.06	13.26	21.82	32.94	46.84	63.69	83.76	107.27	134.43	165.48
BULLET PATH (IN)	-.75	2.58	.00	-9.22	-25.78	-50.35	-83.57	-126.00	-178.18	-240.81	-314.48	-399.79	-497.35
TIME OF FLIGHT (SEC)	.000000	.119560	.247135	.381891	.523138	.670243	.822755	.980347	1.142784	1.309896	1.481569	1.657732	1.838347
WIND DEFLECTION (IN)													
5 MPH	.00	.37	1.44	3.15	5.42	8.21	11.48	15.19	19.33	23.89	28.84	34.19	39.93
10 MPH	.00	.73	2.88	6.29	10.84	16.42	22.96	30.39	38.67	47.77	57.68	68.38	79.86
20 MPH	.00	1.47	5.76	12.58	21.68	32.85	45.92	60.77	77.34	95.54	115.36	136.75	159.71
30 MPH	.00	2.20	8.64	18.87	32.52	49.27	68.88	91.16	116.01	143.32	173.04	205.13	239.57
VELOCITY (FPS)	1200.	1133.	1078.	1033.	995.	963.	933.	907.	882.	860.	839.	819.	800.
ENERGY (FT-LB)	480.	428.	387.	356.	330.	309.	290.	274.	259.	246.	234.	223.	213.
DROP (IN)	.00	-3.08	-12.91	-30.19	-55.58	-89.72	-133.14	-186.42	-250.24	-325.19	-411.88	-510.90	-622.88
MID-RANGE (IN)	.00	.80	3.37	7.99	14.88	24.28	36.38	51.38	69.53	91.08	116.22	145.19	178.30
BULLET PATH (IN)	-.75	3.00	.00	-10.45	-29.01	-56.31	-92.90	-139.35	-196.33	-264.46	-344.31	-436.50	-541.64
TIME OF FLIGHT (SEC)	.000000	.128731	.264539	.406724	.5546..8	.708001	.866351	1.029514	1.197330	1.369691	1.546532	1.727823	1.913561
WIND DEFLECTION (IN)													
5 MPH	.00	.33	1.28	2.79	4.81	7.30	10.24	13.60	17.37	21.53	26.09	31.05	36.39
10 MPH	.00	.66	2.56	5.58	9.63	14.61	20.48	27.19	34.73	43.07	52.19	62.10	72.79
20 MPH	.00	1.31	5.12	11.17	19.25	29.22	40.96	54.39	69.46	86.13	104.38	124.19	145.57
30 MPH	.00	1.97	7.68	16.75	28.88	43.82	61.43	81.58	104.19	129.20	156.57	186.29	218.36
VELOCITY (FPS)	1100.	1051.	1011.	976.	945.	918.	892.	869.	847.	827.	807.	789.	771.
ENERGY (FT-LB)	403.	368.	340.	317.	297.	280.	265.	252.	239.	228.	217.	207.	198.
DROP (IN)	.00	-3.62	-15.08	-35.02	-64.04	-102.64	-151.52	-211.27	-282.50	-365.80	-461.79	-570.73	-693.44
MID-RANGE (IN)	.00	.94	3.92	9.18	16.94	27.37	40.74	57.27	77.20	100.81	128.26	159.61	195.20
BULLET PATH (IN)	-.75	3.54	.00	-12.02	-33.13	-63.81	-104.78	-156.61	-219.92	-295.31	-383.38	-484.41	-599.21
TIME OF FLIGHT (SEC)	.000000	.139579	.285164	.436262	.592514	.753663	.919523	1.089966	1.264910	1.444309	1.628151	1.816449	2.009232
WIND DEFLECTION (IN)													
5 MPH	.00	.28	1.09	2.39	4.14	6.32	8.92	11.92	15.31	19.10	23.28	27.85	32.81
10 MPH	.00	.57	2.19	4.78	8.28	12.64	17.84	23.83	30.62	38.20	46.55	55.70	65.62
20 MPH	.00	1.13	4.38	9.56	16.56	25.29	35.67	47.67	61.25	76.40	93.11	111.39	131.25
30 MPH	.00	1.70	6.57	14.35	24.85	37.93	53.51	71.50	91.87	114.60	139.66	167.09	196.87

Bullet: Lyman # 287308 162 Gr.
Ballistic Coefficient: .325

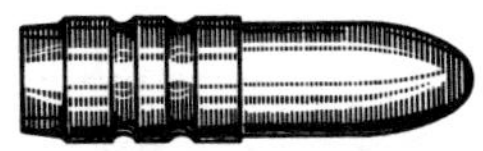

RANGE (YDS)	MUZZLE	50	100	150	200	250	300	350	400	450	500	550	600
VELOCITY (FPS)	1800.	1692.	1590.	1499.	1414.	1335.	1263.	1198.	1145.	1100.	1061.	1028.	998.
ENERGY (FT-LB)	1165.	1030.	909.	808.	719.	641.	573.	517.	472.	435.	405.	380.	359.
DROP (IN)	.00	-1.31	-5.64	-13.39	-25.07	-41.08	-61.96	-88.32	-120.72	-159.73	-205.90	-259.76	-321.84
MID-RANGE (IN)	.00	.36	1.51	3.63	6.89	11.48	17.59	25.44	35.29	47.36	61.87	79.01	98.96
BULLET PATH (IN)	-.75	1.14	.00	-4.56	-13.04	-25.85	-43.54	-66.70	-95.91	-131.72	-174.70	-225.37	-284.25
TIME OF FLIGHT (SEC)	.000000	.085949	.177399	.274573	.377638	.486874	.602473	.724477	.852574	.986303	1.125249	1.268986	1.417182
WIND DEFLECTION (IN)													
5 MPH	.00	.23	.94	2.16	3.90	6.18	9.02	12.42	16.36	20.79	25.69	31.00	36.71
10 MPH	.00	.46	1.89	4.32	7.80	12.36	18.04	24.84	32.72	41.59	51.38	62.01	73.42
20 MPH	.00	.92	3.78	8.65	15.60	24.71	36.07	49.68	65.44	83.18	102.75	124.02	146.85
30 MPH	.00	1.38	5.67	12.97	23.39	37.07	54.11	74.52	98.16	124.77	154.13	186.02	220.27
VELOCITY (FPS)	1700.	1597.	1506.	1420.	1340.	1268.	1203.	1149.	1103.	1064.	1030.	1001.	974.
ENERGY (FT-LB)	1039.	918.	815.	725.	646.	578.	521.	475.	438.	407.	382.	360.	341.
DROP (IN)	.00	-1.47	-6.32	-15.05	-28.10	-45.96	-69.26	-98.56	-134.42	-177.40	-228.04	-286.85	-354.33
MID-RANGE (IN)	.00	.40	1.69	4.08	7.73	12.81	19.58	28.27	39.11	52.31	68.06	86.53	107.90
BULLET PATH (IN)	-.75	1.32	.00	-5.19	-14.70	-29.03	-48.79	-74.55	-106.88	-146.32	-193.43	-248.70	-312.64
TIME OF FLIGHT (SEC)	.000000	.091039	.187775	.290378	.399132	.514240	.635754	.763406	.896717	1.035278	1.178660	1.326522	1.478603
WIND DEFLECTION (IN)													
5 MPH	.00	.25	.99	2.26	4.06	6.43	9.36	12.83	16.79	21.22	26.08	31.32	36.94
10 MPH	.00	.49	1.99	4.52	8.13	12.86	18.72	25.65	33.59	42.44	52.15	62.64	73.88
20 MPH	.00	.99	3.98	9.04	16.26	25.72	37.43	51.31	67.17	84.89	104.30	125.29	147.76
30 MPH	.00	1.48	5.97	13.55	24.39	38.58	56.15	76.96	100.76	127.33	156.45	187.93	221.64
VELOCITY (FPS)	1600.	1508.	1422.	1343.	1270.	1205.	1151.	1104.	1065.	1031.	1001.	975.	951.
ENERGY (FT-LB)	921.	818.	728.	648.	580.	522.	476.	439.	408.	382.	361.	342.	325.
DROP (IN)	.00	-1.65	-7.16	-16.97	-31.59	-51.62	-77.63	-110.19	-149.85	-197.16	-252.62	-316.73	-389.97
MID-RANGE (IN)	.00	.45	1.93	4.59	8.63	14.31	21.85	31.45	43.34	57.69	74.69	94.52	117.36
BULLET PATH (IN)	-.75	1.56	.00	-5.85	-16.51	-32.59	-54.65	-83.25	-118.96	-162.31	-213.81	-273.96	-343.25
TIME OF FLIGHT (SEC)	.000000	.096571	.198999	.307572	.422493	.543822	.671295	.804436	.942841	1.086078	1.233803	1.385754	1.541724
WIND DEFLECTION (IN)													
5 MPH	.00	.25	1.01	2.32	4.18	6.61	9.57	13.04	16.97	21.32	26.07	31.20	36.67
10 MPH	.00	.50	2.02	4.63	8.36	13.21	19.15	26.08	33.94	42.65	52.15	62.39	73.34
20 MPH	.00	.99	4.05	9.27	16.72	26.43	38.30	52.16	67.88	85.30	104.30	124.79	146.69
30 MPH	.00	1.49	6.07	13.90	25.08	39.64	57.44	78.24	101.82	127.95	156.45	187.18	220.03
VELOCITY (FPS)	1500.	1415.	1335.	1263.	1199.	1146.	1100.	1062.	1028.	999.	972.	948.	927.
ENERGY (FT-LB)	809.	720.	641.	574.	517.	472.	435.	405.	380.	359.	340.	324.	309.
DROP (IN)	.00	-1.93	-8.21	-19.34	-35.95	-58.59	-87.83	-124.23	-168.32	-220.61	-281.60	-351.76	-431.37
MID-RANGE (IN)	.00	.52	2.17	5.18	9.77	16.16	24.57	35.19	48.21	63.82	82.21	103.53	127.85
BULLET PATH (IN)	-.75	1.80	.00	-6.66	-18.79	-36.95	-61.71	-93.63	-133.24	-181.06	-237.57	-303.25	-378.38
TIME OF FLIGHT (SEC)	.000000	.102995	.212159	.327683	.449612	.577637	.711296	.850175	.993848	1.141981	1.294317	1.450654	1.610839
WIND DEFLECTION (IN)													
5 MPH	.00	.26	1.07	2.44	4.37	6.83	9.79	13.22	17.06	21.29	25.90	30.86	36.15
10 MPH	.00	.53	2.14	4.87	8.73	13.66	19.59	26.43	34.12	42.59	51.80	61.72	72.31
20 MPH	.00	1.05	4.28	9.74	17.46	27.33	39.18	52.86	68.23	85.18	103.60	123.43	144.62
30 MPH	.00	1.58	6.42	14.62	26.20	40.99	58.76	79.29	102.35	127.77	155.40	185.15	216.92

Bullet: Lyman # 287308 162 Gr.
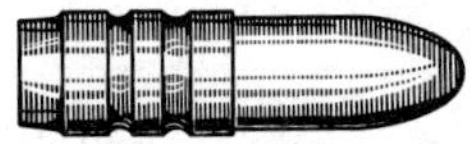

Ballistic Coefficient: .325
(Cont'd.)

RANGE (YDS)	MUZZLE	50	100	150	200	250	300	350	400	450	500	550	600
VELOCITY (FPS)	1400.	1322.	1251.	1189.	1137.	1093.	1055.	1023.	994.	968.	944.	923.	903.
ENERGY (FT-LB)	705.	629.	563.	509.	465.	430.	400.	376.	355.	337.	321.	306.	293.
DROP (IN)	.00	-2.22	-9.41	-22.17	-41.07	-66.67	-99.52	-140.15	-189.08	-246.79	-313.72	-390.22	-476.90
MID-RANGE (IN)	.00	.59	2.48	5.93	11.13	18.28	27.59	39.26	53.47	70.40	90.19	112.95	138.93
BULLET PATH (IN)	-.75	2.11	.00	-7.69	-21.51	-42.03	-69.80	-105.36	-149.21	-201.83	-263.69	-335.11	-416.72
TIME OF FLIGHT (SEC)	.000000	.110287	.226955	.350010	.479061	.613680	.753438	.897923	1.046817	1.199872	1.356897	1.517744	1.682299
WIND DEFLECTION (IN)													
5 MPH	.00	.28	1.11	2.52	4.44	6.86	9.73	13.02	16.69	20.73	25.12	29.85	34.90
10 MPH	.00	.55	2.23	5.03	8.89	13.72	19.46	26.03	33.38	41.46	50.24	59.69	69.80
20 MPH	.00	1.11	4.46	10.06	17.77	27.44	38.92	52.07	66.77	82.93	100.48	119.39	139.60
30 MPH	.00	1.66	6.69	15.09	26.66	41.17	58.39	78.10	100.15	124.39	150.73	179.08	209.40
VELOCITY (FPS)	1300.	1232.	1173.	1123.	1081.	1045.	1013.	986.	961.	938.	917.	897.	878.
ENERGY (FT-LB)	608.	545.	495.	454.	420.	393.	369.	349.	332.	316.	302.	289.	278.
DROP (IN)	.00	-2.58	-10.92	-25.55	-47.06	-75.98	-112.84	-158.13	-212.36	-275.88	-349.18	-432.79	-527.20
MID-RANGE (IN)	.00	.68	2.88	6.78	12.61	20.58	30.87	43.66	59.12	77.38	98.61	123.01	150.76
BULLET PATH (IN)	-.75	2.50	.00	-8.80	-24.47	-47.56	-78.59	-118.05	-166.44	-224.13	-291.59	-369.38	-457.95
TIME OF FLIGHT (SEC)	.000000	.118594	.243513	.374284	.510500	.651720	.797562	.947729	1.101994	1.260179	1.422146	1.587791	1.757035
WIND DEFLECTION (IN)													
5 MPH	.00	.28	1.12	2.48	4.31	6.58	9.26	12.32	15.74	19.51	23.61	28.03	32.77
10 MPH	.00	.56	2.24	4.95	8.62	13.16	18.52	24.65	31.49	39.02	47.22	56.07	65.55
20 MPH	.00	1.13	4.49	9.90	17.23	26.33	37.05	49.29	62.98	78.04	94.44	112.13	131.09
30 MPH	.00	1.69	6.73	14.85	25.85	39.49	55.57	73.94	94.47	117.07	141.66	168.20	196.64
VELOCITY (FPS)	1200.	1147.	1101.	1062.	1028.	999.	973.	949.	927.	906.	887.	869.	852.
ENERGY (FT-LB)	518.	473.	436.	406.	380.	359.	340.	324.	309.	295.	283.	272.	261.
DROP (IN)	.00	-3.04	-12.68	-29.46	-53.92	-86.57	-127.92	-178.43	-238.37	-308.41	-389.00	-480.61	-583.68
MID-RANGE (IN)	.00	.79	3.30	7.73	14.28	23.15	34.50	48.52	65.26	85.01	107.93	134.17	163.92
BULLET PATH (IN)	-.75	2.92	.00	-10.07	-27.82	-53.76	-88.39	-132.18	-185.41	-248.74	-322.62	-407.51	-503.87
TIME OF FLIGHT (SEC)	.000000	.127927	.261492	.400282	.543869	.691918	.844168	1.000420	1.160517	1.324337	1.491790	1.662809	1.837347
WIND DEFLECTION (IN)													
5 MPH	.00	.26	1.01	2.22	3.86	5.89	8.29	11.04	14.13	17.54	21.28	25.33	29.69
10 MPH	.00	.52	2.02	4.45	7.72	11.78	16.57	22.07	28.25	35.08	42.56	50.65	59.37
20 MPH	.00	1.03	4.05	8.90	15.44	23.56	33.15	44.15	56.50	70.17	85.11	101.31	118.75
30 MPH	.00	1.55	6.07	13.35	23.16	35.33	49.72	66.22	84.75	105.25	127.67	151.96	178.12
VELOCITY (FPS)	1100.	1061.	1028.	998.	972.	948.	926.	906.	887.	869.	852.	836.	820.
ENERGY (FT-LB)	435.	405.	380.	358.	340.	323.	308.	295.	283.	271.	261.	251.	242.
DROP (IN)	.00	-3.58	-14.85	-34.32	-62.49	-99.83	-146.62	-203.52	-270.97	-349.45	-439.40	-541.33	-655.72
MID-RANGE (IN)	.00	.93	3.85	8.94	16.40	26.39	38.99	54.47	72.99	94.73	119.86	148.67	181.24
BULLET PATH (IN)	-.75	3.47	.00	-11.67	-32.05	-61.59	-100.58	-149.67	-209.33	-280.01	-362.16	-456.29	-562.88
TIME OF FLIGHT (SEC)	.000000	.138894	.282577	.430713	.583045	.739371	.899538	1.063423	1.230937	1.402012	1.576602	1.754680	1.936233
WIND DEFLECTION (IN)													
5 MPH	.00	.22	.87	1.90	3.31	5.06	7.16	9.58	12.32	15.38	18.74	22.41	26.39
10 MPH	.00	.45	1.73	3.81	6.62	10.13	14.32	19.16	24.64	30.75	37.48	44.82	52.78
20 MPH	.00	.89	3.47	7.61	13.23	20.26	28.64	38.32	49.29	61.51	74.96	89.65	105.55
30 MPH	.00	1.34	5.20	11.42	19.85	30.39	42.96	57.49	73.93	92.26	112.45	134.47	158.33

Bullet: Lyman # 311359 113 Gr.
Ballistic Coefficient: .181

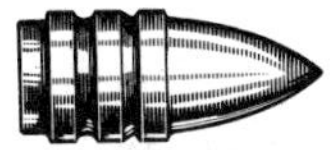

RANGE (YDS)	MUZZLE	50	100	150	200	250	300	350	400	450	500	550	600
VELOCITY (FPS)	2300.	2077.	1868.	1674.	1497.	1341.	1210.	1108.	1034.	976.	930.	890.	854.
ENERGY (FT-LB)	1327.	1082.	876.	703.	562.	451.	367.	308.	268.	239.	217.	199.	183.
DROP (IN)	.00	-.84	-3.70	-9.06	-17.51	-29.87	-47.07	-70.22	-100.42	-138.72	-186.07	-243.39	-311.65
MID-RANGE (IN)	.00	.23	1.01	2.53	5.05	8.90	14.48	22.26	32.70	46.21	63.16	83.89	108.75
BULLET PATH (IN)	-.75	.63	.00	-3.12	-9.35	-19.49	-34.46	-55.38	-83.35	-119.43	-164.55	-219.64	-285.67
TIME OF FLIGHT (SEC)	.000000	.068629	.144785	.229618	.324398	.430336	.548253	.678040	.818438	.967922	1.125510	1.290608	1.462875
WIND DEFLECTION (IN)													
5 MPH	.00	.30	1.26	2.99	5.59	9.17	13.81	19.49	26.11	33.52	41.65	50.44	59.86
10 MPH	.00	.60	2.53	5.98	11.18	18.35	27.62	38.99	52.22	67.05	83.31	100.89	119.73
20 MPH	.00	1.20	5.05	11.96	22.36	36.70	55.25	77.97	104.44	134.10	166.61	201.77	239.45
30 MPH	.00	1.80	7.58	17.93	33.54	55.04	82.87	116.96	156.66	201.15	249.92	302.66	359.18
VELOCITY (FPS)	2200.	1983.	1781.	1594.	1426.	1280.	1161.	1072.	1007.	955.	911.	873.	839.
ENERGY (FT-LB)	1214.	986.	795.	637.	510.	411.	338.	288.	254.	229.	208.	191.	177.
DROP (IN)	.00	-.93	-4.06	-9.94	-19.25	-32.87	-51.80	-75.17	-110.07	-151.51	-202.38	-263.66	-336.30
MID-RANGE (IN)	.00	.25	1.11	2.78	5.57	9.82	15.96	24.47	35.78	50.29	68.32	90.23	116.35
BULLET PATH (IN)	-.75	.73	.00	-3.47	-10.38	-21.59	-38.12	-61.08	-91.57	-130.61	-179.07	-237.95	-308.18
TIME OF FLIGHT (SEC)	.000000	.071823	.151663	.240735	.340312	.451485	.574736	.709419	.853990	1.007149	1.168112	1.336408	1.511780
WIND DEFLECTION (IN)													
5 MPH	.00	.32	1.35	3.18	5.95	9.73	14.58	20.43	27.15	34.63	42.79	51.60	61.04
10 MPH	.00	.64	2.69	6.37	11.89	19.46	29.15	40.86	54.30	69.26	85.59	103.21	122.07
20 MPH	.00	1.28	5.39	12.74	23.79	38.92	58.31	81.72	108.60	138.52	171.18	206.42	244.15
30 MPH	.00	1.92	8.08	19.11	35.68	58.38	87.46	122.57	162.91	207.77	256.76	309.62	366.22
VELOCITY (FPS)	2100.	1889.	1694.	1515.	1356.	1222.	1117.	1040.	982.	934.	894.	858.	825.
ENERGY (FT-LB)	1106.	896.	720.	576.	462.	375.	313.	272.	242.	219.	200.	185.	171.
DROP (IN)	.00	-1.02	-4.47	-10.95	-21.25	-36.28	-57.15	-84.94	-120.74	-165.49	-220.10	-285.56	-362.82
MID-RANGE (IN)	.00	.27	1.22	3.07	6.16	10.85	17.62	26.90	39.12	54.63	73.76	96.86	124.26
BULLET PATH (IN)	-.75	.84	.00	-3.87	-11.56	-23.99	-42.24	-67.42	-100.61	-142.75	-194.75	-257.60	-332.25
TIME OF FLIGHT (SEC)	.000000	.075316	.159181	.252859	.357578	.474223	.602808	.742171	.890757	1.047527	1.211855	1.383378	1.561911
WIND DEFLECTION (IN)													
5 MPH	.00	.34	1.44	3.39	6.32	10.30	15.33	21.31	28.10	35.61	43.79	52.59	62.02
10 MPH	.00	.68	2.87	6.79	12.65	20.61	30.67	42.62	56.20	71.22	87.57	105.19	124.04
20 MPH	.00	1.37	5.75	13.58	25.30	41.21	61.33	85.24	112.40	142.44	175.14	210.38	248.08
30 MPH	.00	2.05	8.62	20.37	37.94	61.82	92.00	127.87	168.61	213.67	262.72	315.57	372.12
VELOCITY (FPS)	2000.	1796.	1608.	1438.	1291.	1170.	1079.	1011.	959.	915.	876.	842.	811.
ENERGY (FT-LB)	1003.	810.	649.	519.	418.	343.	292.	257.	231.	210.	193.	178.	165.
DROP (IN)	.00	-1.12	-4.94	-12.13	-23.54	-40.19	-63.18	-93.61	-132.50	-180.75	-239.34	-309.20	-391.31
MID-RANGE (IN)	.00	.30	1.35	3.41	6.83	12.03	19.46	29.55	42.71	59.24	79.48	103.78	132.48
BULLET PATH (IN)	-.75	.97	.00	-4.34	-12.91	-26.71	-46.86	-74.44	-110.49	-155.90	-211.64	-278.66	-357.92
TIME OF FLIGHT (SEC)	.000000	.079146	.167419	.266097	.376297	.498573	.632378	.776201	.928695	1.089042	1.256751	1.431551	1.613322
WIND DEFLECTION (IN)													
5 MPH	.00	.36	1.53	3.62	6.71	10.87	·6.05	22.11	28.93	36.44	44.59	53.38	62.77
10 MPH	.00	.73	3.07	7.23	13.43	21.75	32.10	44.21	57.85	72.87	89.19	106.75	125.54
20 MPH	.00	1.46	6.13	14.47	26.86	43.50	64.20	88.42	115.70	145.74	178.38	213.51	251.09
30 MPH	.00	2.19	9.20	21.70	40.28	65.25	96.30	132.63	173.55	218.61	267.56	320.26	376.63

Bullet: Lyman # 311359 113 Gr.
Ballistic Coefficient: .181
(Cont'd.)

RANGE (YDS)	MUZZLE	50	100	150	200	250	300	350	400	450	500	550	600
VELOCITY (FPS)	1900.	1704.	1524.	1364.	1229.	1122.	1044.	985.	936.	895.	859.	827.	796.
ENERGY (FT-LB)	906.	728.	583.	467.	379.	316.	273.	243.	220.	201.	185.	171.	159.
DROP (IN)	.00	-1.25	-5.49	-13.50	-26.20	-44.66	-70.01	-103.30	-145.50	-197.51	-260.32	-334.87	-422.10
MID-RANGE (IN)	.00	.33	1.50	3.80	7.61	13.36	21.50	32.45	46.55	64.11	85.49	111.03	141.04
BULLET PATH (IN)	-.75	1.12	.00	-4.90	-14.47	-29.82	-52.04	-82.21	-121.30	-170.18	-229.88	-301.31	-385.42
TIME OF FLIGHT (SEC)	.000000	.083393	.176532	.280654	.396673	.524658	.663505	.811643	.968005	1.131950	1.303104	1.481273	1.666401
WIND DEFLECTION (IN)													
5 MPH	.00	.39	1.64	3.86	7.12	11.43	16.70	22.79	29.61	37.09	45.20	53.93	63.27
10 MPH	.00	.78	3.28	7.71	14.24	22.87	33.41	45.59	59.21	74.17	90.40	107.86	126.55
20 MPH	.00	1.56	6.56	15.42	28.47	45.73	66.82	91.17	118.42	148.34	180.80	215.72	253.10
30 MPH	.00	2.35	9.84	23.13	42.71	68.60	100.23	136.76	177.63	222.51	271.20	323.59	379.65
VELOCITY (FPS)	1800.	1612.	1441.	1293.	1172.	1080.	1012.	959.	915.	877.	843.	811.	782.
ENERGY (FT-LB)	813.	652.	521.	419.	344.	293.	257.	231.	210.	193.	178.	165.	153.
DROP (IN)	.00	-1.39	-6.14	-15.10	-29.27	-49.76	-77.67	-114.04	-159.74	-215.75	-283.03	-362.54	-455.06
MID-RANGE (IN)	.00	.37	1.68	4.26	8.49	14.84	23.74	35.56	50.60	69.22	91.76	118.55	149.86
BULLET PATH (IN)	-.75	1.30	.00	-5.51	-16.24	-33.28	-57.75	-90.67	-132.92	-185.50	-249.33	-325.39	-414.47
TIME OF FLIGHT (SEC)	.000000	.088099	.186579	.296564	.418624	.552233	.695888	.848233	1.008441	1.176018	1.350688	1.532329	1.720940
WIND DEFLECTION (IN)													
5 MPH	.00	.42	1.75	4.10	7.51	11.93	17.24	23.31	30.08	37.49	45.53	54.18	63.44
10 MPH	.00	.84	3.50	8.20	15.01	23.86	34.48	46.62	60.15	74.98	91.05	108.36	126.89
20 MPH	.00	1.68	7.01	16.39	30.02	47.72	68.95	93.24	120.30	149.96	182.11	216.71	253.77
30 MPH	.00	2.52	10.51	24.59	45.03	71.58	103.43	139.87	180.46	224.94	273.16	325.07	380.66
VELOCITY (FPS)	1700.	1521.	1361.	1226.	1120.	1043.	984.	936.	895.	859.	826.	796.	768.
ENERGY (FT-LB)	725.	580.	465.	377.	315.	273.	243.	220.	201.	185.	171.	159.	148.
DROP (IN)	.00	-1.56	-6.91	-16.96	-32.80	-55.54	-86.25	-125.88	-175.34	-235.62	-307.66	-392.39	-490.51
MID-RANGE (IN)	.00	.42	1.89	4.77	9.49	16.49	26.16	38.85	54.87	74.57	98.29	126.36	159.01
BULLET PATH (IN)	-.75	1.52	.00	-6.22	-18.23	-37.14	-64.01	-99.82	-145.44	-201.89	-270.10	-351.00	-445.30
TIME OF FLIGHT (SEC)	.000000	.093336	.197676	.313923	.442127	.581162	.729461	.885968	1.050049	1.221333	1.399628	1.584882	1.777132
WIND DEFLECTION (IN)													
5 MPH	.00	.45	1.87	4.33	7.85	12.32	17.60	23.61	30.29	37.59	45.52	54.06	63.21
10 MPH	.00	.90	3.73	8.66	15.70	24.64	35.21	47.22	60.57	75.19	91.04	108.12	126.42
20 MPH	.00	1.80	7.46	17.32	31.39	49.27	70.42	94.45	121.15	150.38	182.08	216.23	252.84
30 MPH	.00	2.69	11.20	25.99	47.09	73.91	105.63	141.67	181.72	225.57	273.12	324.35	379.27
VELOCITY (FPS)	1600.	1431.	1284.	1165.	1075.	1009.	956.	913.	875.	840.	809.	780.	753.
ENERGY (FT-LB)	642.	514.	414.	340.	290.	255.	229.	209.	192.	177.	164.	153.	142.
DROP (IN)	.00	-1.78	-7.82	-19.15	-36.87	-62.08	-95.80	-138.92	-192.41	-257.23	-334.34	-424.45	-528.72
MID-RANGE (IN)	.00	.48	2.13	5.37	10.61	18.29	28.75	42.32	59.34	80.15	105.09	134.41	168.56
BULLET PATH (IN)	-.75	1.76	.00	-7.04	-20.48	-41.40	-70.84	-109.67	-158.88	-219.42	-292.24	-378.07	-478.06
TIME OF FLIGHT (SEC)	.000000	.099189	.209943	.332774	.467079	.611325	.764193	.924884	1.092919	1.268033	1.450115	1.639169	1.835255
WIND DEFLECTION (IN)													
5 MPH	.00	.48	1.97	4.53	8.10	12.55	17.75	23.64	30.18	37.34	45.11	53.50	62.50
10 MPH	.00	.96	3.95	9.07	16.21	25.09	35.50	47.28	60.35	74.67	90.22	106.99	125.00
20 MPH	.00	1.91	7.90	18.14	32.41	50.19	71.00	94.56	120.71	149.35	180.44	213.99	250.01
30 MPH	.00	2.87	11.85	27.20	48.62	75.28	106.49	141.84	181.06	224.02	270.66	320.98	375.01

Bullet: Lyman # 311441 115 Gr.
Ballistic Coefficient: .143

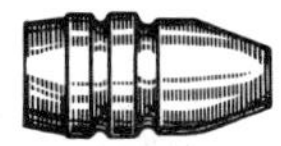

RANGE (YDS)	MUZZLE	50	100	150	200	250	300	350	400	450	500	550	600
VELOCITY (FPS)	2600.	2300.	2030.	1786.	1566.	1383.	1230.	1112.	1029.	967.	917.	874.	836.
ENERGY (FT-LB)	1726.	1351.	1052.	814.	626.	488.	386.	316.	270.	239.	215.	195.	179.
DROP (IN)	.00	-.67	-2.99	-7.43	-14.59	-25.32	-40.61	-61.66	-89.70	-125.92	-171.35	-227.04	-294.07
MID-RANGE (IN)	.00	.18	.82	2.11	4.30	7.75	12.88	20.21	30.26	43.51	60.36	81.20	106.43
BULLET PATH (IN)	-.75	.45	.00	-2.56	-7.85	-16.71	-30.13	-49.30	-75.48	-109.82	-153.38	-207.20	-272.36
TIME OF FLIGHT (SEC)	.000000	.061345	.130831	.209630	.299436	.401474	.516688	.645240	.785718	.936262	1.095710	1.263408	1.439013
WIND DEFLECTION (IN)													
5 MPH	.00	.32	1.36	3.22	6.04	9.95	15.01	21.24	28.53	36.70	45.65	55.33	65.71
10 MPH	.00	.64	2.72	6.43	12.09	19.89	30.01	42.49	57.06	73.40	91.31	110.67	131.42
20 MPH	.00	1.29	5.44	12.87	24.17	39.78	60.03	84.97	114.11	146.79	182.61	221.34	262.84
30 MPH	.00	1.93	8.16	19.30	36.26	59.67	90.04	127.46	171.17	220.19	273.92	332.00	394.26
VELOCITY (FPS)	2500.	2207.	1948.	1710.	1503.	1329.	1187.	1082.	1007.	950.	902.	861.	825.
ENERGY (FT-LB)	1596.	1243.	969.	747.	576.	451.	360.	299.	259.	230.	208.	189.	174.
DROP (IN)	.00	-.73	-3.25	-8.06	-15.84	-27.51	-44.11	-66.88	-97.04	-135.74	-183.96	-242.83	-313.37
MID-RANGE (IN)	.00	.20	.89	2.29	4.68	8.43	14.00	21.91	32.68	46.74	64.47	86.30	112.57
BULLET PATH (IN)	-.75	.52	.00	-2.81	-8.60	-18.27	-32.87	-53.64	-81.81	-118.51	-164.73	-221.59	-290.14
TIME OF FLIGHT (SEC)	.000000	.063870	.136243	.218438	.312162	.418423	.538076	.670764	.814719	.968284	1.130495	1.300815	1.478978
WIND DEFLECTION (IN)													
5 MPH	.00	.34	1.43	3.38	6.35	10.42	15.67	22.07	29.46	37.69	46.68	56.39	66.79
10 MPH	.00	.68	2.86	6.77	12.70	20.84	31.34	44.13	58.91	75.38	93.37	112.78	133.58
20 MPH	.00	1.36	5.72	13.53	25.40	41.68	62.68	88.27	117.82	150.76	186.73	225.57	267.16
30 MPH	.00	2.04	8.58	20.30	38.10	62.53	94.02	132.40	176.73	226.13	280.10	338.35	400.74
VELOCITY (FPS)	2400.	2117.	1866.	1636.	1442.	1278.	1148.	1054.	987.	933.	888.	849.	813.
ENERGY (FT-LB)	1471.	1145.	889.	683.	531.	417.	337.	284.	249.	222.	201.	184.	169.
DROP (IN)	.00	-.79	-3.53	-8.77	-17.25	-29.95	-47.98	-72.59	-104.98	-146.25	-197.40	-259.54	-333.72
MID-RANGE (IN)	.00	.21	.97	2.49	5.10	9.18	15.23	23.75	35.24	50.11	68.75	91.55	118.88
BULLET PATH (IN)	-.75	.60	.00	-3.10	-9.45	-20.00	-35.90	-58.37	-88.62	-127.75	-176.76	-236.76	-308.80
TIME OF FLIGHT (SEC)	.000000	.066592	.142060	.227933	.325723	.436356	.560446	.697091	.844370	1.000877	1.165816	1.338749	1.519483
WIND DEFLECTION (IN)													
5 MPH	.00	.36	1.50	3.56	6.66	10.90	16.32	22.84	30.30	38.58	47.59	57.31	67.71
10 MPH	.00	.72	3.00	7.12	13.33	21.80	32.64	45.69	60.61	77.15	95.18	114.62	135.43
20 MPH	.00	1.44	6.01	14.23	26.65	43.60	65.28	91.38	121.22	154.31	190.37	229.24	270.86
30 MPH	.00	2.16	9.01	21.35	39.98	65.40	97.92	137.06	181.83	231.46	285.55	343.86	406.29
VELOCITY (FPS)	2300.	2030.	1786.	1566.	1383.	1230.	1112.	1029.	967.	917.	874.	836.	802.
ENERGY (FT-LB)	1351.	1052.	814.	626.	488.	386.	316.	270.	239.	215.	195.	179.	164.
DROP (IN)	.00	-.87	-3.84	-9.55	-18.82	-32.66	-52.25	-78.84	-113.61	-157.57	-211.81	-277.38	-355.37
MID-RANGE (IN)	.00	.23	1.06	2.72	5.57	10.02	16.57	25.74	37.98	53.69	73.24	97.03	125.43
BULLET PATH (IN)	-.75	.68	.00	-3.41	-10.39	-21.92	-39.22	-63.52	-95.98	-137.65	-189.59	-252.87	-328.56
TIME OF FLIGHT (SEC)	.000000	.069485	.148281	.238084	.340119	.455329	.583877	.724352	.874891	1.034336	1.202030	1.377630	1.561006
WIND DEFLECTION (IN)													
5 MPH	.00	.38	1.57	3.73	6.97	11.37	16.95	23.57	31.08	39.37	48.39	58.10	68.50
10 MPH	.00	.75	3.14	7.47	13.95	22.75	33.89	47.14	62.15	78.74	96.77	116.20	137.00
20 MPH	.00	1.50	6.28	14.94	27.90	45.49	67.79	94.28	124.31	157.48	193.55	232.40	274.00
30 MPH	.00	2.25	9.42	22.40	41.84	68.24	101.68	141.41	186.46	236.22	290.32	348.61	410.99

Bullet: Lyman # 311441 115 Gr.
Ballistic Coefficient: .143
(Cont'd.)

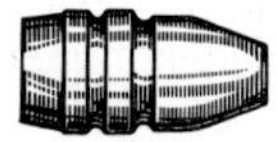

RANGE (YDS)	MUZZLE	50	100	150	200	250	300	350	400	450	500	550	600
VELOCITY (FPS)	2200.	1943.	1706.	1500.	1326.	1185.	1080.	1006.	949.	901.	861.	824.	791.
ENERGY (FT-LB)	1236.	964.	743.	574.	449.	359.	298.	258.	230.	207.	189.	173.	160.
DROP (IN)	.00	-.94	-4.19	-10.44	-20.58	-35.67	-56.95	-85.64	-122.90	-169.68	-227.13	-296.27	-378.14
MID-RANGE (IN)	.00	.25	1.15	2.98	6.09	10.94	18.04	27.86	40.87	57.42	77.89	102.68	132.12
BULLET PATH (IN)	-.75	.78	.00	-3.77	-11.44	-24.06	-42.87	-69.09	-103.87	-148.18	-203.15	-269.83	-349.22
TIME OF FLIGHT (SEC)	.000000	.072559	.154975	.248924	.355400	.475275	.608164	.752288	.906000	1.068346	1.238794	1.417081	1.603140
WIND DEFLECTION (IN)													
5 MPH	.00	.39	1.64	3.91	7.28	11.82	17.52	24.20	31.73	40.01	49.01	58.70	69.08
10 MPH	.00	.77	3.28	7.81	14.55	23.65	35.04	48.40	63.46	80.03	98.03	117.41	138.15
20 MPH	.00	1.54	6.55	15.62	29.10	47.30	70.07	96.81	126.91	160.06	196.06	234.81	276.31
30 MPH	.00	2.31	9.83	23.43	43.65	70.95	105.11	145.21	190.37	240.09	294.08	352.22	414.46
VELOCITY (FPS)	2100.	1850.	1622.	1430.	1268.	1141.	1049.	983.	930.	885.	846.	811.	778.
ENERGY (FT-LB)	1126.	874.	671.	522.	411.	332.	281.	247.	221.	200.	183.	168.	155.
DROP (IN)	.00	-1.04	-4.62	-11.52	-22.69	-39.28	-62.52	-93.64	-133.69	-183.70	-244.77	-317.95	-404.12
MID-RANGE (IN)	.00	.28	1.27	3.29	6.72	12.04	19.75	30.32	44.16	61.62	83.11	108.98	139.54
BULLET PATH (IN)	-.75	.90	.00	-4.21	-12.69	-26.60	-47.16	-75.59	-112.96	-160.28	-218.67	-289.16	-372.65
TIME OF FLIGHT (SEC)	.000000	.076109	.162742	.261363	.372897	.497880	.635301	.783235	.940328	1.105813	1.279273	1.460526	1.649561
WIND DEFLECTION (IN)													
5 MPH	.00	.41	1.75	4.14	7.67	12.38	18.19	24.92	32.46	40.74	49.72	59.38	69.73
10 MPH	.00	.82	3.50	8.29	15.34	24.77	36.38	49.85	64.93	81.48	99.44	118.77	139.47
20 MPH	.00	1.65	7.00	16.57	30.69	49.54	72.77	99.70	129.85	162.96	198.88	237.53	278.93
30 MPH	.00	2.47	10.50	24.86	46.03	74.31	109.15	149.55	194.78	244.44	298.31	356.30	418.40
VELOCITY (FPS)	2000.	1758.	1542.	1363.	1214.	1101.	1021.	961.	912.	869.	832.	798.	766.
ENERGY (FT-LB)	1021.	789.	607.	474.	376.	309.	266.	236.	212.	193.	177.	163.	150.
DROP (IN)	.00	-1.15	-5.11	-12.75	-25.08	-43.32	-68.70	-102.39	-145.40	-198.81	-263.68	-341.09	-431.79
MID-RANGE (IN)	.00	.31	1.41	3.65	7.43	13.25	21.60	32.94	47.62	66.02	88.52	115.50	147.19
BULLET PATH (IN)	-.75	1.03	.00	-4.71	-14.11	-29.42	-51.87	-82.63	-122.71	-173.19	-235.13	-309.61	-397.37
TIME OF FLIGHT (SEC)	.000000	.080013	.171229	.274778	.391596	.521664	.663420	.815065	.975515	1.144160	1.320684	1.504980	1.697082
WIND DEFLECTION (IN)													
5 MPH	.00	.44	1.87	4.38	8.06	12.91	18.78	25.53	33.05	41.29	50.22	59.84	70.14
10 MPH	.00	.88	3.74	8.76	16.12	25.81	37.56	51.05	66.09	82.57	100.44	119.68	140.29
20 MPH	.00	1.76	7.47	17.52	32.24	51.63	75.12	102.10	132.18	165.14	200.88	239.35	280.57
30 MPH	.00	2.65	11.21	26.28	48.36	77.44	112.69	153.15	198.27	247.72	301.32	359.03	420.86
VELOCITY (FPS)	1900.	1667.	1467.	1298.	1164.	1065.	995.	940.	894.	854.	818.	785.	754.
ENERGY (FT-LB)	922.	709.	549.	430.	346.	290.	253.	225.	204.	186.	171.	157.	145.
DROP (IN)	.00	-1.27	-5.68	-14.16	-27.81	-47.86	-75.55	-111.96	-158.12	-215.12	-284.01	-365.74	-461.34
MID-RANGE (IN)	.00	.34	1.57	4.05	8.22	14.58	23.61	35.72	51.25	70.60	94.14	122.19	155.12
BULLET PATH (IN)	-.75	1.19	.00	-5.26	-15.69	-32.53	-57.00	-90.20	-133.14	-186.93	-252.60	-331.12	-423.50
TIME OF FLIGHT (SEC)	.000000	.084321	.180428	.289256	.411533	.546581	.692513	.847821	1.011642	1.183500	1.363172	1.550618	1.745900
WIND DEFLECTION (IN)													
5 MPH	.00	.47	1.98	4.61	8.43	13.36	19.26	25.98	33.45	41.62	50.49	60.03	70.27
10 MPH	.00	.95	3.97	9.22	16.85	26.72	38.51	51.95	66.89	83.24	100.97	120.07	140.54
20 MPH	.00	1.89	7.93	18.45	33.70	53.45	77.03	103.91	133.78	166.49	201.94	240.13	281.08
30 MPH	.00	2.84	11.90	27.67	50.55	80.17	115.54	155.86	200.67	249.73	302.91	360.20	421.62

Bullet: Lyman # 311441 115 Gr.

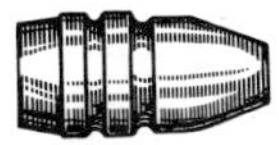

Ballistic Coefficient: .143
(Cont'd.)

RANGE (YDS)	MUZZLE	50	100	150	200	250	300	350	400	450	500	550	600
FELOCITY (FPS)	1800.	1577.	1393.	1238.	1118.	1034.	970.	920.	877.	838.	804.	772.	742.
ENERGY (FT-LB)	827.	635.	496.	391.	319.	273.	240.	216.	196.	180.	165.	152.	141.
DROP (IN)	.00	-1.42	-6.34	-15.77	-30.87	-52.90	-83.03	-122.31	-171.78	-232.53	-305.63	-391.87	-492.59
MID-RANGE (IN)	.00	.38	1.75	4.50	9.09	16.03	25.75	38.62	55.02	75.32	99.91	129.05	163.27
BULLET PATH (IN)	-.75	1.38	.00	-5.87	-17.43	-35.91	-62.50	-98.23	-144.15	-201.36	-270.90	-353.60	-450.78
TIME OF FLIGHT (SEC)	.000000	.089087	.190354	.304743	.432505	.572311	.722276	.881197	1.048397	1.223515	1.406410	1.597099	1.795664
WIND DEFLECTION (IN)													
5 MPH	.00	.51	2.08	4.82	8.73	13.70	19.56	26.21	33.59	41.67	50.43	59.88	70.02
10 MPH	.00	1.01	4.17	9.63	17.45	27.39	39.12	52.42	67.18	83.34	100.86	119.76	140.04
20 MPH	.00	2.03	8.34	19.27	34.91	54.79	78.24	104.85	134.37	166.68	201.72	239.51	280.07
30 MPH	.00	3.04	12.51	28.90	52.36	82.18	117.36	157.27	201.55	250.02	302.58	359.27	420.11
VELOCITY (FPS)	1700.	1494.	1322.	1182.	1078.	1004.	947.	900.	860.	823.	790.	759.	730.
ENERGY (FT-LB)	738.	570.	446.	356.	297.	257.	229.	207.	189.	173.	159.	147.	136.
DROP (IN)	.00	-1.59	-7.11	-17.61	-34.33	-58.50	-91.26	-133.58	-186.59	-251.33	-328.78	-419.89	-526.08
MID-RANGE (IN)	.00	.43	1.96	5.00	10.06	17.60	28.02	41.67	58.96	80.25	105.89	136.21	171.78
BULLET PATH (IN)	-.75	1.59	.00	-6.57	-19.37	-39.61	-68.44	-106.83	-155.91	-216.71	-290.24	-377.42	-479.68
TIME OF FLIGHT (SEC)	.000000	.094273	.201110	.321357	.454583	.598988	.752945	.915515	1.086175	1.264669	1.450933	1.645020	1.847037
WIND DEFLECTION (IN)													
5 MPH	.00	.53	2.17	4.99	8.94	13.89	19.67	26.21	33.47	41.41	50.04	59.35	69.36
10 MPH	.00	1.06	4.34	9.97	17.89	27.77	39.34	52.42	66.93	82.82	100.07	118.70	138.73
20 MPH	.00	2.13	8.67	19.94	35.78	55.55	78.68	104.85	133.86	165.63	200.14	237.40	277.45
30 MPH	.00	3.19	13.01	29.91	53.67	83.32	118.03	157.27	200.79	248.45	300.21	356.10	416.18
VELOCITY (FPS)	1600.	1413.	1254.	1130.	1042.	977.	925.	881.	843.	808.	775.	745.	717.
ENERGY (FT-LB)	654.	510.	401.	326.	277.	244.	218.	198.	181.	167.	153.	142.	131.
DROP (IN)	.00	-1.80	-7.98	-19.70	-38.20	-64.68	-100.19	-145.77	-202.51	-271.47	-353.52	-449.85	-561.82
MID-RANGE (IN)	.00	.49	2.19	5.57	11.12	19.27	30.40	44.86	63.06	85.37	112.08	143.67	180.72
BULLET PATH (IN)	-.75	1.81	.00	-7.36	-21.50	-43.61	-74.76	-115.98	-168.36	-232.95	-310.63	-402.60	-510.21
TIME OF FLIGHT (SEC)	.000000	.099847	.212706	.338988	.477529	.626412	.784356	.950636	1.124861	1.306871	1.496666	1.694324	1.899980
WIND DEFLECTION (IN)													
5 MPH	.00	.54	2.22	5.08	9.02	13.87	19.52	25.91	32.99	40.75	49.21	58.35	68.20
10 MPH	.00	1.07	4.44	10.16	18.05	27.75	39.05	51.81	65.98	81.51	98.41	116.70	136.40
20 MPH	.00	2.15	8.87	20.32	36.09	55.50	78.09	103.62	131.95	163.02	196.83	233.40	272.79
30 MPH	.00	3.22	13.31	30.49	54.14	83.25	117.14	155.44	197.93	244.53	295.24	350.10	409.19
VELOCITY (FPS)	1500.	1327.	1185.	1081.	1006.	949.	902.	861.	824.	791.	759.	730.	703.
ENERGY (FT-LB)	574.	449.	359.	298.	258.	230.	208.	189.	173.	160.	147.	136.	126.
DROP (IN)	.00	-2.06	-9.08	-22.28	-42.89	-72.05	-110.75	-160.10	-221.14	-294.90	-382.26	-484.65	-603.32
MID-RANGE (IN)	.00	.55	2.48	6.27	12.37	21.20	33.10	48.47	67.69	91.12	119.08	152.19	190.91
BULLET PATH (IN)	-.75	2.10	.00	-8.28	-23.98	-48.23	-82.02	-126.45	-182.58	-251.43	-333.87	-431.34	-545.11
TIME OF FLIGHT (SEC)	.000000	.106444	.226284	.359141	.503236	.656919	.8B9236	.989654	1.167909	1.353933	1.547775	1.749543	1.959398
WIND DEFLECTION (IN)													
5 MPH	.00	.57	2.31	5.20	9.08	13.81	19.29	25.49	32.38	39.95	48.20	57.16	66.83
10 MPH	.00	1.13	4.63	10.41	18.17	27.62	38.59	50.98	64.75	79.89	96.41	114.32	133.65
20 MPH	.00	2.27	9.25	20.82	36.34	55.24	77.17	101.96	129.50	159.78	192.82	228.64	267.31
30 MPH	.00	3.40	13.88	31.23	54.51	82.85	115.76	152.94	194.26	239.68	289.23	342.96	400.96

Bullet: Lyman # 311576 120 Gr.
Ballistic Coefficient: .172

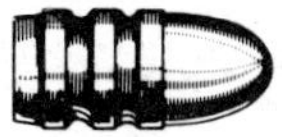

RANGE (YDS)	MUZZLE	50	100	150	200	250	300	350	400	450	500	550	600
VELOCITY (FPS)	2100.	1878.	1687.	1521.	1375.	1259.	1163.	1087.	1029.	981.	941.	907.	875.
ENERGY (FT-LB)	1175.	940.	758.	616.	504.	422.	360.	315.	282.	257.	236.	219.	204.
DROP (IN)	.00	-1.02	-4.51	-11.04	-21.36	-36.27	-56.68	-83.50	-117.65	-159.99	-211.31	-272.32	-343.87
MID-RANGE (IN)	.00	.27	1.23	3.10	6.18	10.78	17.30	26.07	37.46	51.81	69.39	90.43	115.25
BULLET PATH (IN)	-.75	.86	.00	-3.91	-11.60	-23.88	-41.66	-65.86	-97.38	-137.09	-185.79	-244.17	-313.09
TIME OF FLIGHT (SEC)	.000000	.075536	.159975	.253673	.357570	.471669	.595771	.729371	.871400	1.020851	1.177057	1.339578	1.508132
WIND DEFLECTION (IN)													
5 MPH	.00	.36	1.51	3.47	6.32	10.08	14.71	20.18	26.40	33.26	40.72	48.74	57.29
10 MPH	.00	.72	3.01	6.93	12.65	20.16	29.43	40.37	52.79	66.53	81.45	97.48	114.57
20 MPH	.00	1.45	6.03	13.86	25.29	40.31	58.85	80.74	105.59	133.05	162.90	194.96	229.15
30 MPH	.00	2.17	9.04	20.80	37.94	60.47	88.28	121.11	158.38	199.58	244.34	292.44	343.72
VELOCITY (FPS)	2000.	1785.	1611.	1452.	1321.	1214.	1126.	1059.	1006.	963.	925.	892.	862.
ENERGY (FT-LB)	1066.	849.	691.	562.	465.	392.	338.	299.	270.	247.	228.	212.	198.
DROP (IN)	.00	-1.13	-4.97	-12.17	-23.50	-39.82	-62.05	-91.11	-127.89	-173.22	-227.84	-292.53	-368.11
MID-RANGE (IN)	.00	.30	1.36	3.41	6.78	11.81	18.86	28.29	40.44	55.63	74.09	96.12	122.00
BULLET PATH (IN)	-.75	.98	.00	-4.34	-12.81	-26.28	-45.65	-71.84	-105.76	-148.23	-199.99	-261.82	-334.54
TIME OF FLIGHT (SEC)	.000000	.079399	.167872	.265990	.374510	.493098	.621551	.759066	.904515	1.057052	1.216125	1.381370	1.552558
WIND DEFLECTION (IN)													
5 MPH	.00	.39	1.57	3.61	6.56	10.39	15.10	20.60	26.80	33.62	41.02	48.96	57.43
10 MPH	.00	.77	3.15	7.21	13.11	20.79	30.19	41.20	53.59	67.24	82.04	97.92	114.85
20 MPH	.00	1.55	6.29	14.43	26.23	41.57	60.39	82.39	107.19	134.48	164.08	195.84	229.70
30 MPH	.00	2.32	9.44	21.64	39.34	62.36	90.58	123.59	160.78	201.72	246.11	293.76	344.55
VELOCITY (FPS)	1900.	1705.	1537.	1388.	1270.	1172.	1094.	1034.	986.	945.	910.	878.	850.
ENERGY (FT-LB)	962.	774.	630.	513.	430.	366.	319.	285.	259.	238.	221.	206.	192.
DROP (IN)	.00	-1.25	-5.48	-13.42	-25.86	-43.70	-67.85	-99.25	-138.74	-187.15	-245.15	-313.61	-393.31
MID-RANGE (IN)	.00	.33	1.49	3.76	7.44	12.91	20.51	30.60	43.51	59.53	78.88	101.88	128.80
BULLET PATH (IN)	-.75	1.12	.00	-4.82	-14.14	-28.86	-49.90	-78.18	-114.55	-159.84	-214.72	-280.07	-356.65
TIME OF FLIGHT (SEC)	.000000	.083504	.176184	.278978	.392037	.515109	.647765	.788969	.937683	1.093208	1.255085	1.423018	1.596825
WIND DEFLECTION (IN)													
5 MPH	.00	.40	1.61	3.71	6.71	10.59	15.32	20.80	26.94	33.68	40.97	48.80	57.15
10 MPH	.00	.80	3.22	7.42	13.42	21.19	30.64	41.60	53.87	67.35	81.95	97.61	114.30
20 MPH	.00	1.60	6.44	14.83	26.84	42.37	61.28	83.19	107.75	134.70	163.90	195.22	228.61
30 MPH	.00	2.41	9.66	22.25	40.26	63.56	91.91	124.79	161.62	202.06	245.84	292.83	342.91
VELOCITY (FPS)	1800.	1624.	1464.	1330.	1222.	1133.	1064.	1010.	966.	928.	895.	865.	837.
ENERGY (FT-LB)	863.	703.	571.	472.	398.	342.	302.	272.	249.	229.	213.	199.	187.
DROP (IN)	.00	-1.38	-6.06	-14.82	-28.48	-47.98	-74.23	-108.14	-150.53	-202.16	-263.78	-336.23	-420.28
MID-RANGE (IN)	.00	.37	1.65	4.15	8.18	14.11	22.30	33.09	46.79	63.63	83.91	107.91	135.91
BULLET PATH (IN)	-.75	1.27	.00	-5.35	-15.61	-31.71	-54.55	-85.05	-124.04	-172.26	-230.48	-299.53	-380.18
TIME OF FLIGHT (SEC)	.000000	.087744	.185046	.292735	.410510	.538186	.675010	.819852	.971837	1.130394	1.295147	1.465856	1.642386
WIND DEFLECTION (IN)													
5 MPH	.00	.39	1.62	3.76	6.79	10.69	15.40	20.81	26.86	33.47	40.64	48.33	56.53
10 MPH	.00	.78	3.23	7.52	13.58	21.39	30.80	41.63	53.71	66.95	81.28	96.66	113.06
20 MPH	.00	1.55	6.47	15.04	27.17	42.77	61.60	83.25	107.42	133.90	162.56	193.31	226.12
30 MPH	.00	2.33	9.70	22.56	40.75	64.16	92.41	124.88	161.13	200.85	243.84	289.97	339.18

Bullet: Lyman # 311576 120 Gr.
Ballistic Coefficient: .172
(Cont'd.)

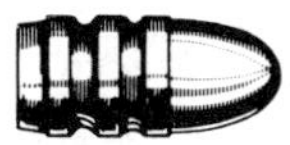

RANGE (YDS)	MUZZLE	50	100	150	200	250	300	350	400	450	500	550	600
VELOCITY (FPS)	1700.	1533.	1384.	1267.	1170.	1092.	1032.	985.	944.	909.	878.	849.	823.
ENERGY (FT-LB)	770.	626.	511.	428.	364.	318.	284.	258.	237.	220.	205.	192.	180.
DROP (IN)	.00	-1.55	-6.82	-16.62	-31.85	-53.42	-82.25	-119.20	-165.09	-220.59	-286.57	-363.81	-453.08
MID-RANGE (IN)	.00	.42	1.86	4.64	9.10	15.60	24.50	36.10	50.70	68.51	89.87	115.04	144.29
BULLET PATH (IN)	-.75	1.49	.00	-6.01	-17.45	-35.23	-60.28	-93.45	-135.55	-187.26	-249.45	-322.91	-408.39
TIME OF FLIGHT (SEC)	.000000	.092958	.196053	.309386	.432731	.565638	.707060	.855967	1.011669	1.173713	1.341805	1.515766	1.695506
WIND DEFLECTION (IN)													
5 MPH	.00	.42	1.72	3.93	7.02	10.95	15.63	20.97	26.91	33.40	40.43	47.98	56.03
10 MPH	.00	.83	3.45	7.86	14.04	21.91	31.27	41.94	53.82	66.81	80.86	95.95	112.06
20 MPH	.00	1.66	6.89	15.73	28.09	43.81	62.53	83.89	107.64	133.62	161.73	191.90	224.11
30 MPH	.00	2.49	10.34	23.59	42.13	65.72	93.80	125.83	161.46	200.43	242.59	287.85	336.17
VELOCITY (FPS)	1600.	1442.	1313.	1207.	1122.	1055.	1003.	960.	923.	890.	861.	833.	808.
ENERGY (FT-LB)	682.	554.	460.	388.	335.	297.	268.	246.	227.	211.	197.	185.	174.
DROP (IN)	.00	-1.76	-7.72	-18.71	-35.67	-59.52	-91.15	-131.38	-180.94	-240.62	-311.25	-393.58	-488.42
MID-RANGE (IN)	.00	.48	2.10	5.19	10.12	17.24	26.86	39.30	54.79	73.63	96.10	122.48	153.06
BULLET PATH (IN)	-.75	1.72	.00	-6.76	-19.49	-39.11	-66.50	-102.50	-147.82	-203.27	-269.66	-347.77	-438.37
TIME OF FLIGHT (SEC)	.000000	.098788	.207970	.327200	.456264	.594318	.740239	.893204	1.052675	1.218298	1.389852	1.567212	1.750331
WIND DEFLECTION (IN)													
5 MPH	.00	.44	1.80	4.04	7.15	11.05	15.64	20.85	26.64	32.96	39.81	47.16	55.03
10 MPH	.00	.89	3.60	8.09	14.30	22.10	31.28	41.70	53.27	65.92	79.61	94.33	110.06
20 MPH	.00	1.77	7.21	16.17	28.61	44.20	62.56	83.41	106.54	131.84	159.23	188.66	220.12
30 MPH	.00	2.66	10.81	24.26	42.91	66.30	93.85	125.11	159.81	197.76	238.84	282.99	330.17
VELOCITY (FPS)	1500.	1358.	1245.	1152.	1078.	1022.	976.	936.	902.	871.	843.	817.	793.
ENERGY (FT-LB)	599.	492.	413.	353.	310.	278.	254.	234.	217.	202.	189.	178.	167.
DROP (IN)	.00	-2.02	-8.75	-21.10	-39.98	-66.31	-100.95	-144.65	-198.15	-262.30	-337.86	-425.63	-526.29
MID-RANGE (IN)	.00	.53	2.35	5.81	11.24	18.99	29.38	42.66	59.09	78.99	102.62	130.27	162.20
BULLET PATH (IN)	-.75	1.98	.00	-7.60	-21.74	-43.32	-73.20	-112.16	-160.91	-220.31	-291.13	-374.14	-470.05
TIME OF FLIGHT (SEC)	.000000	.105277	.220713	.346126	.480917	.623981	.774359	.931419	1.094747	1.264076	1.439248	1.620188	1.806890
WIND DEFLECTION (IN)													
5 MPH	.00	.46	1.82	4.06	7.12	10.91	15.34	20.36	25.94	32.04	38.65	45.78	53.41
10 MPH	.00	.93	3.65	8.12	14.24	21.82	30.69	40.73	51.88	64.08	77.31	91.55	106.81
20 MPH	.00	1.86	7.29	16.24	28.48	43.64	61.37	81.46	103.75	128.15	154.62	183.11	213.63
30 MPH	.00	2.79	10.94	24.35	42.72	65.46	92.06	122.19	155.63	192.23	231.92	274.66	320.44
VELOCITY (FPS)	1400.	1281.	1181.	1101.	1039.	990.	949.	913.	881.	852.	826.	801.	777.
ENERGY (FT-LB)	522.	437.	371.	323.	288.	261.	240.	222.	207.	194.	182.	171.	161.
DROP (IN)	.00	-2.30	-9.91	-23.75	-44.74	-73.76	-111.61	-158.98	-216.72	-285.63	-366.47	-460.05	-566.76
MID-RANGE (IN)	.00	.60	2.65	6.49	12.46	20.88	32.05	46.19	63.62	84.62	109.48	138.50	171.77
BULLET PATH (IN)	-.75	2.28	.00	-8.51	-24.17	-47.85	-80.38	-122.41	-174.82	-238.40	-313.91	-402.16	-503.54
TIME OF FLIGHT (SEC)	.000000	.112097	.234210	.365979	.506406	.654424	.809307	.970576	1.137921	1.311150	1.490163	1.674933	1.865486
WIND DEFLECTION (IN)													
5 MPH	.00	.44	1.75	3.92	6.85	10.45	14.65	19.41	24.71	30.52	36.85	43.68	51.02
10 MPH	.00	.87	3.51	7.84	13.70	20.89	29.30	38.82	49.42	61.05	73.70	87.36	102.04
20 MPH	.00	1.74	7.01	15.68	27.40	41.79	58.59	77.64	98.83	122.10	147.39	174.72	204.08
30 MPH	.00	2.62	10.52	23.52	41.10	62.68	87.89	116.46	148.25	183.14	221.09	262.08	306.12

Bullet: Lyman # 311465 122 Gr.
Ballistic Coefficient: .163

RANGE (YDS)	MUZZLE	50	100	150	200	250	300	350	400	450	500	550	600
VELOCITY (FPS)	2600.	2336.	2092.	1869.	1663.	1491.	1345.	1221.	1123.	1049.	992.	946.	906.
ENERGY (FT-LB)	1831.	1478.	1185.	946.	749.	602.	490.	404.	341.	298.	267.	242.	222.
DROP (IN)	.00	-.66	-2.92	-7.17	-13.91	-23.78	-37.59	-56.20	-80.66	-111.98	-151.18	-199.17	-256.76
MID-RANGE (IN)	.00	.18	.80	2.01	4.03	7.12	11.63	17.91	26.42	37.59	51.81	69.45	90.79
BULLET PATH (IN)	-.75	.42	.00	-2.41	-7.31	-15.34	-27.32	-44.10	-66.72	-96.21	-133.57	-179.72	-235.47
TIME OF FLIGHT (SEC)	.000000	.060873	.128783	.204650	.289743	.385188	.491163	.608316	.736621	.875078	1.022298	1.177301	1.339482
WIND DEFLECTION (IN)													
5 MPH	.00	.28	1.18	2.78	5.19	8.51	12.76	17.99	24.21	31.31	39.19	47.76	56.95
10 MPH	.00	.56	2.36	5.56	10.38	17.02	25.52	35.99	48.41	62.63	78.39	95.51	113.90
20 MPH	.00	1.12	4.72	11.11	20.76	34.05	51.04	71.97	96.83	125.26	156.77	191.03	227.81
30 MPH	.00	1.68	7.07	16.67	31.14	51.07	76.57	107.96	145.24	187.89	235.16	286.54	341.71
VELOCITY (FPS)	2500.	2241.	2007.	1791.	1592.	1435.	1296.	1182.	1093.	1026.	974.	930.	893.
ENERGY (FT-LB)	1693.	1361.	1091.	868.	686.	557.	455.	378.	323.	285.	257.	234.	216.
DROP (IN)	.00	-.72	-3.17	-7.78	-15.10	-25.85	-40.84	-61.01	-87.40	-121.05	-162.93	-213.90	-274.82
MID-RANGE (IN)	.00	.19	.86	2.19	4.38	7.75	12.64	19.44	28.60	40.53	55.62	74.19	96.57
BULLET PATH (IN)	-.75	.49	.00	-2.65	-8.01	-16.80	-29.83	-48.04	-72.48	-104.17	-144.09	-193.09	-252.06
TIME OF FLIGHT (SEC)	.000000	.063370	.134125	.213262	.302155	.401467	.511559	.632931	.765188	.907062	1.057277	1.215016	1.379769
WIND DEFLECTION (IN)													
5 MPH	.00	.30	1.24	2.93	5.47	8.93	13.34	18.74	25.10	32.30	40.24	48.84	58.06
10 MPH	.00	.59	2.49	5.85	10.94	17.86	26.67	37.48	50.19	64.60	80.48	97.68	116.12
20 MPH	.00	1.19	4.97	11.71	21.88	35.72	53.35	74.95	100.39	129.21	160.96	195.37	232.24
30 MPH	.00	1.78	7.46	17.56	32.82	53.57	80.02	112.43	150.58	193.81	241.44	293.05	348.36
VELOCITY (FPS)	2400.	2150.	1924.	1713.	1530.	1379.	1249.	1144.	1065.	1005.	956.	915.	879.
ENERGY (FT-LB)	1560.	1252.	1002.	795.	634.	515.	423.	355.	307.	273.	248.	227.	209.
DROP (IN)	.00	-.78	-3.44	-8.46	-16.43	-28.14	-44.42	-66.29	-94.78	-130.90	-175.61	-229.70	-294.10
MID-RANGE (IN)	.00	.21	.94	2.38	4.77	8.45	13.75	21.11	30.96	43.68	59.66	79.19	102.63
BULLET PATH (IN)	-.75	.56	.00	-2.92	-8.80	-18.41	-32.59	-52.37	-78.75	-112.78	-155.39	-207.38	-269.69
TIME OF FLIGHT (SEC)	.000000	.066069	.139838	.222485	.315341	.418679	.533068	.658705	.794816	.940003	1.093171	1.253639	1.420986
WIND DEFLECTION (IN)													
5 MPH	.00	.31	1.31	3.08	5.75	9.34	13.91	19.47	25.94	33.22	41.20	49.82	59.05
10 MPH	.00	.63	2.61	6.16	11.50	18.69	27.82	38.93	51.89	66.44	82.40	99.64	118.09
20 MPH	.00	1.26	5.22	12.31	23.00	37.37	55.64	77.86	103.78	132.88	164.80	199.28	236.19
30 MPH	.00	1.88	7.83	18.47	34.50	56.06	83.46	116.80	155.66	199.32	247.19	298.92	354.28
VELOCITY (FPS)	2300.	2060.	1839.	1636.	1470.	1327.	1206.	1111.	1040.	985.	940.	901.	867.
ENERGY (FT-LB)	1433.	1149.	916.	725.	585.	477.	394.	334.	293.	263.	239.	220.	203.
DROP (IN)	.00	-.85	-3.75	-9.23	-17.95	-30.71	-48.43	-72.12	-102.83	-141.55	-189.17	-246.52	-314.54
MID-RANGE (IN)	.00	.23	1.02	2.60	5.22	9.22	14.98	22.93	33.47	47.00	63.87	84.38	108.85
BULLET PATH (IN)	-.75	.65	.00	-3.22	-9.69	-20.20	-35.66	-57.10	-85.56	-122.02	-167.40	-222.50	-288.27
TIME OF FLIGHT (SEC)	.000000	.068964	.146035	.232530	.329429	.436939	.555679	.685486	.825248	.973605	1.129643	1.292792	1.462708
WIND DEFLECTION (IN)													
5 MPH	.00	.33	1.37	3.25	6.03	9.76	14.46	20.15	26.71	34.03	42.02	50.64	59.85
10 MPH	.00	.66	2.75	6.49	12.07	19.51	28.93	40.30	53.42	68.05	84.03	101.27	119.70
20 MPH	.00	1.32	5.49	12.98	24.13	39.02	57.86	80.60	106.84	136.10	168.07	202.54	239.39
30 MPH	.00	1.98	8.24	19.47	36.20	58.53	86.79	120.89	160.25	204.15	252.10	303.81	359.09

Bullet: Lyman #. 311465 122 Gr.
Ballistic Coefficient: .163
(Cont'd.)

RANGE (YDS)	MUZZLE	50	100	150	200	250	300	350	400	450	500	550	600
VELOCITY (FPS)	2200.	1971.	1757.	1564.	1410.	1275.	1165.	1080.	1017.	966.	924.	887.	854.
ENERGY (FT-LB)	1311.	1052.	836.	663.	538.	441.	367.	316.	280.	253.	231.	213.	197.
DROP (IN)	.00	-.93	-4.10	-10.09	-19.63	-33.57	-52.86	-78.54	-111.65	-153.15	-203.87	-264.69	-336.54
MID-RANGE (IN)	.00	.25	1.12	2.84	5.72	10.08	16.34	24.91	36.19	50.58	68.36	89.87	115.41
BULLET PATH (IN)	-.75	.75	.00	-3.56	-10.68	-22.19	-39.05	-62.31	-93.00	-132.07	-180.36	-238.76	-308.19
TIME OF FLIGHT (SEC)	.000000	.072047	.152683	.243296	.344356	.456325	.579577	.713550	.856897	1.008419	1.167360	1.333253	1.505819
WIND DEFLECTION (IN)													
5 MPH	.00	.34	1.44	3.41	6.30	10.16	15.00	20.79	27.41	34.74	42.73	51.33	60.51
10 MPH	.00	.68	2.87	6.82	12.61	20.31	30.01	41.58	54.81	69.48	85.46	102.65	121.02
20 MPH	.00	1.36	5.74	13.64	25.21	40.63	60.01	83.17	109.63	138.96	170.91	205.30	242.05
30 MPH	.00	2.04	8.62	20.46	37.82	60.94	90.02	124.75	164.44	208.44	256.37	307.96	363.07
VELOCITY (FPS)	2100.	1877.	1670.	1496.	1350.	1225.	1125.	1051.	994.	947.	907.	872.	841.
ENERGY (FT-LB)	1194.	954.	756.	606.	493.	406.	343.	299.	267.	243.	223.	206.	191.
DROP (IN)	.00	-1.02	-4.52	-11.12	-21.64	-36.93	-58.03	-85.96	-121.74	-166.29	-220.40	-285.02	-361.05
MID-RANGE (IN)	.00	.27	1.24	3.14	6.30	11.08	17.89	27.14	39.23	54.52	73.27	95.82	122.50
BULLET PATH (IN)	-.75	.86	.00	-3.97	-11.85	-24.51	-42.98	-68.28	-101.42	-143.34	-194.81	-256.81	-330.20
TIME OF FLIGHT (SEC)	.000000	.075561	.160298	.255396	.361029	.477825	.605788	.743947	.890905	1.045671	1.207628	1.376397	1.551765
WIND DEFLECTION (IN)													
5 MPH	.00	.36	1.53	3.62	6.63	10.62	15.60	21.47	28.11	35.45	43.41	51.98	61.13
10 MPH	.00	.73	3.07	7.24	13.26	21.24	31.19	42.93	56.23	70.90	86.83	103.96	122.25
20 MPH	.00	1.45	6.14	14.47	26.51	42.48	62.38	85.87	112.46	141.79	173.66	207.92	244.51
30 MPH	.00	2.18	9.21	21.71	39.77	63.72	93.57	128.80	168.68	212.69	260.48	311.88	366.76
VELOCITY (FPS)	2000.	1784.	1586.	1429.	1292.	1178.	1090.	1024.	972.	929.	891.	858.	827.
ENERGY (FT-LB)	1083.	862.	681.	553.	452.	376.	322.	284.	256.	234.	215.	199.	185.
DROP (IN)	.00	-1.13	-5.00	-12.31	-23.90	-40.71	-63.78	-94.14	-132.76	-180.51	-238.23	-306.87	-387.29
MID-RANGE (IN)	.00	.30	1.37	3.47	6.95	12.19	19.58	29.55	42.48	58.67	78.41	102.04	129.86
BULLET PATH (IN)	-.75	.99	.00	-4.44	-13.16	-27.09	-47.29	-74.78	-110.53	-155.40	-210.25	-276.02	-353.56
TIME OF FLIGHT (SEC)	.000000	.079426	.168655	.268344	.378843	.500624	.633255	.775451	.925948	1.083945	1.248942	1.420639	1.598880
WIND DEFLECTION (IN)													
5 MPH	.00	.39	1.64	3.81	6.94	11.05	16.13	22.04	28.68	35.99	43.91	52.42	61.50
10 MPH	.00	.78	3.28	7.63	13.88	22.11	32.25	44.08	57.37	71.97	87.81	104.83	123.00
20 MPH	.00	1.56	6.57	15.26	27.75	44.22	64.51	88.16	114.73	143.95	175.63	209.67	246.01
30 MPH	.00	2.34	9.85	22.89	41.63	66.33	96.76	132.24	172.10	215.92	263.44	314.50	369.01
VELOCITY (FPS)	1900.	1691.	1514.	1365.	1238.	1135.	1058.	999.	952.	911.	876.	844.	815.
ENERGY (FT-LB)	978.	775.	621.	505.	415.	349.	303.	271.	245.	225.	208.	193.	180.
DROP (IN)	.00	-1.26	-5.55	-13.65	-26.43	-44.90	-70.09	-103.01	-144.61	-195.67	-257.15	-329.93	-414.88
MID-RANGE (IN)	.00	.34	1.52	3.85	7.67	13.38	21.39	32.11	45.87	62.95	83.67	108.37	137.35
BULLET PATH (IN)	-.75	1.14	.00	-4.96	-14.59	-29.91	-51.95	-81.73	-120.18	-168.09	-226.42	-296.05	-377.85
TIME OF FLIGHT (SEC)	.000000	.083693	.177651	.282073	.397604	.524350	.661440	.807472	.961399	1.122572	1.290591	1.465224	1.646367
WIND DEFLECTION (IN)													
5 MPH	.00	.42	1.74	3.98	7.20	11.41	16.52	22.43	29.02	36.26	44.10	52.52	61.51
10 MPH	.00	.84	3.48	7.96	14.40	22.81	33.05	44.85	58.05	72.52	88.20	105.04	123.02
20 MPH	.00	1.67	6.95	15.92	28.80	45.62	66.09	89.70	116.10	145.04	176.39	210.07	246.05
30 MPH	.00	2.51	10.43	23.88	43.20	68.44	99.14	134.56	174.15	217.56	264.59	315.11	369.07

Bullet: Lyman # 311465 122 Gr.
Ballistic Coefficient: .163

(Cont'd.)

RANGE (YDS)	MUZZLE	50	100	150	200	250	300	350	400	450	500	550	600
VELOCITY (FPS)	1800.	1600.	1442.	1302.	1186.	1096.	1029.	976.	932.	894.	861.	830.	801.
ENERGY (FT-LB)	878.	693.	563.	459.	381.	325.	287.	258.	235.	217.	201.	187.	174.
DROP (IN)	.00	-1.40	-6.19	-15.18	-29.30	-49.60	-77.10	-112.78	-157.52	-212.16	-277.64	-354.82	-444.60
MID-RANGE (IN)	.00	.38	1.69	4.27	8.46	14.69	23.37	34.86	49.46	67.47	89.22	115.02	145.20
BULLET PATH (IN)	-.75	1.32	.00	-5.52	-16.18	-33.01	-57.04	-89.25	-130.52	-181.70	-243.70	-317.41	-403.73
TIME OF FLIGHT (SEC)	.000000	.088420	.187235	.296792	.417621	.549376	.690818	.840648	.998032	1.162448	1.333583	1.511269	1.695451
WIND DEFLECTION (IN)													
5 MPH	.00	.45	1.81	4.12	7.42	11.68	16.79	22.64	29.16	36.30	44.02	52.33	61.20
10 MPH	.00	.90	3.62	8.24	14.83	23.36	33.58	45.29	58.32	72.59	88.04	104.65	122.40
20 MPH	.00	1.79	7.24	16.47	29.67	46.71	67.17	90.57	116.64	145.18	176.09	209.30	244.80
30 MPH	.00	2.69	10.86	24.71	44.50	70.07	100.75	135.86	174.96	217.77	264.13	313.95	367.20
VELOCITY (FPS)	1700.	1519.	1370.	1242.	1138.	1060.	1001.	953.	913.	877.	845.	816.	788.
ENERGY (FT-LB)	783.	625.	508.	418.	351.	305.	272.	246.	226.	208.	193.	180.	168.
DROP (IN)	.00	-1.57	-6.92	-16.91	-32.56	-54.89	-84.92	-123.59	-171.69	-230.18	-299.93	-381.83	-476.64
MID-RANGE (IN)	.00	.42	1.89	4.74	9.36	16.16	25.54	37.83	53.29	72.26	95.08	122.04	153.40
BULLET PATH (IN)	-.75	1.52	.00	-6.15	-17.96	-36.46	-62.65	-97.48	-141.75	-196.39	-262.31	-340.38	-431.35
TIME OF FLIGHT (SEC)	.000000	.093545	.197583	.312711	.439066	.575811	.721543	.875200	1.036120	1.203895	1.378291	1.559197	1.746604
WIND DEFLECTION (IN)													
5 MPH	.00	.47	1.86	4.22	7.58	11.85	16.91	22.66	29.06	36.06	43.64	51.80	60.52
10 MPH	.00	.93	3.72	8.45	15.16	23.70	33.82	45.33	58.12	72.12	87.29	103.60	121.05
20 MPH	.00	1.87	7.43	16.90	30.32	47.39	67.63	90.66	116.24	144.24	174.57	207.19	242.10
30 MPH	.00	2.80	11.15	25.35	45.47	71.09	101.45	135.99	174.37	216.36	261.86	310.79	363.15
VELOCITY (FPS)	1600.	1442.	1302.	1186.	1096.	1029.	976.	932.	894.	861.	830.	801.	775.
ENERGY (FT-LB)	693.	563.	459.	381.	325.	287.	258.	235.	217.	201.	187.	174.	163.
DROP (IN)	.00	-1.76	-7.73	-18.83	-36.11	-60.59	-93.25	-134.96	-186.58	-249.03	-323.19	-409.94	-509.89
MID-RANGE (IN)	.00	.48	2.10	5.25	10.32	17.72	27.79	40.85	57.18	77.13	101.01	129.14	161.70
BULLET PATH (IN)	-.75	1.73	.00	-6.86	-19.89	-40.13	-68.55	-106.03	-153.40	-211.61	-281.53	-364.05	-459.75
TIME OF FLIGHT (SEC)	.000000	.098815	.208371	.329199	.460952	.602392	.752220	.909600	1.074012	1.245141	1.422822	1.606997	1.797694
WIND DEFLECTION (IN)													
5 MPH	.00	.45	1.84	4.22	7.56	11.76	16.70	22.29	28.51	35.32	42.71	50.67	59.20
10 MPH	.00	.89	3.67	8.44	15.13	23.52	33.39	44.59	57.03	70.64	85.42	101.33	118.39
20 MPH	.00	1.78	7.35	16.88	30.26	47.04	66.78	89.18	114.05	141.29	170.83	202.66	236.79
30 MPH	.00	2.67	11.02	25.32	45.38	70.56	100.17	133.77	171.08	211.93	256.25	303.99	355.18
VELOCITY (FPS)	1500.	1353.	1228.	1128.	1052.	995.	948.	908.	873.	841.	812.	785.	759.
ENERGY (FT-LB)	609.	496.	408.	344.	300.	268.	243.	223.	206.	192.	179.	167.	156.
DROP (IN)	.00	-2.03	-8.81	-21.36	-40.73	-67.92	-103.86	-149.34	-205.30	-272.65	-352.25	-444.83	-551.31
MID-RANGE (IN)	.00	.54	2.38	5.92	11.56	19.66	30.57	44.54	61.92	83.03	108.20	137.67	171.79
BULLET PATH (IN)	-.75	2.00	.00	-7.78	-22.37	-44.78	-75.95	-116.64	-167.83	-230.40	-305.22	-393.03	-494.73
TIME OF FLIGHT (SEC)	.000000	.105361	.221874	.349564	.487482	.634229	.788801	.950574	1.119164	1.294354	1.476047	1.664244	1.858995
WIND DEFLECTION (IN)													
5 MPH	.00	.47	1.92	4.36	7.70	11.81	16.61	22.05	28.09	34.70	41.89	49.65	57.99
10 MPH	.00	.94	3.85	8.72	15.40	23.62	33.23	44.10	56.17	69.41	83.78	99.31	115.98
20 MPH	.00	1.89	7.70	17.45	30.79	47.25	66.46	88.20	112.35	138.81	167.57	198.61	231.97
30 MPH	.00	2.83	11.55	26.17	46.19	70.87	99.69	132.30	168.52	208.22	251.35	297.92	347.95

Bullet: Lyman # 311410 130 Gr.
Ballistic Coefficient: .239

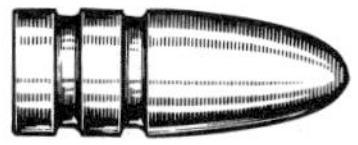

RANGE (YDS)	MUZZLE	50	100	150	200	250	300	350	400	450	500	550	600
VELOCITY (FPS)	1600.	1475.	1361.	1260.	1176.	1109.	1055.	1011.	974.	941.	912.	886.	861.
ENERGY (FT-LB)	739.	628.	535.	458.	399.	355.	321.	295.	274.	256.	240.	226.	214.
DROP (IN)	.00	-1.71	-7.45	-17.88	-33.79	-55.98	-85.22	-122.24	-167.74	-222.34	-286.62	-361.32	-447.07
MID-RANGE (IN)	.00	.46	2.01	4.90	9.44	15.96	24.73	36.02	50.08	67.12	87.33	110.99	138.32
BULLET PATH (IN)	-.75	1.64	.00	-6.32	-18.14	-36.23	-61.36	-94.28	-135.68	-186.18	-246.36	-316.96	-398.61
TIME OF FLIGHT (SEC)	.000000	.097671	.203581	.318184	.441581	.573096	.711916	.857254	1.008529	1.165333	1.327375	1.494456	1.666441
WIND DEFLECTION (IN)													
5 MPH	.00	.35	1.42	3.25	5.86	9.18	13.15	17.69	22.75	28.30	34.31	40.76	47.65
10 MPH	.00	.69	2.83	6.50	11.72	18.36	26.30	35.38	45.50	56.60	68.62	81.52	95.29
20 MPH	.00	1.38	5.66	13.00	23.44	36.73	52.59	70.75	91.00	113.20	137.24	163.05	190.59
30 MPH	.00	2.07	8.49	19.50	35.15	55.09	78.89	106.13	136.50	169.80	205.85	244.57	285.88
VELOCITY (FPS)	1500.	1384.	1280.	1191.	1121.	1065.	1020.	981.	948.	918.	891.	866.	843.
ENERGY (FT-LB)	649.	553.	473.	410.	363.	328.	300.	278.	259.	243.	229.	216.	205.
DROP (IN)	.00	-1.98	-8.50	-20.34	-38.30	-63.14	-95.62	-136.43	-186.24	-245.56	-315.18	-395.71	-487.82
MID-RANGE (IN)	.00	.52	2.27	5.54	10.65	17.87	27.47	39.71	54.82	72.97	94.45	119.47	148.29
BULLET PATH (IN)	-.75	1.89	.00	-7.21	-20.54	-40.76	-68.61	-104.79	-149.98	-204.67	-269.66	-345.57	-433.04
TIME OF FLIGHT (SEC)	.000000	.104158	.216936	.338538	.468438	.605806	.749843	.899920	1.055599	1.216567	1.382609	1.553578	1.729382
WIND DEFLECTION (IN)													
5 MPH	.00	.37	1.49	3.39	6.02	9.31	13.19	17.59	22.49	27.86	33.67	39.91	46.59
10 MPH	.00	.73	2.98	6.78	12.05	18.62	26.37	35.19	44.99	55.72	67.34	79.83	93.17
20 MPH	.00	1.46	5.96	13.57	24.09	37.24	52.74	70.37	89.97	111.43	134.68	159.66	186.34
30 MPH	.00	2.20	8.94	20.35	36.14	55.87	79.12	105.56	134.96	167.15	202.02	239.49	279.51
VELOCITY (FPS)	1400.	1294.	1203.	1131.	1073.	1026.	986.	952.	922.	895.	870.	846.	824.
ENERGY (FT-LB)	566.	483.	418.	369.	332.	304.	281.	262.	245.	231.	218.	207.	196.
DROP (IN)	.00	-2.28	-9.75	-23.23	-43.48	-71.27	-107.28	-152.21	-206.57	-271.11	-346.47	-433.31	-532.29
MID-RANGE (IN)	.00	.60	2.60	6.31	11.99	19.94	30.41	43.64	59.80	79.17	101.97	128.45	158.86
BULLET PATH (IN)	-.75	2.22	.00	-8.23	-23.23	-45.76	-76.53	-116.21	-165.31	-224.60	-294.71	-376.30	-470.02
TIME OF FLIGHT (SEC)	.000000	.111496	.231815	.360567	.496901	.640012	.789241	.944125	1.104337	1.269648	1.439902	1.615001	1.794894
WIND DEFLECTION (IN)													
5 MPH	.00	.38	1.54	3.44	6.01	9.18	12.88	17.08	21.75	26.87	32.43	38.41	44.81
10 MPH	.00	.77	3.09	6.89	12.03	18.36	25.76	34.17	43.51	53.74	64.85	76.81	89.62
20 MPH	.00	1.53	6.17	13.78	24.05	36.71	51.53	68.33	87.01	107.49	129.70	153.62	179.23
30 MPH	.00	2.30	9.26	20.67	36.08	55.07	77.29	102.50	130.52	161.23	194.55	230.43	268.85
VELOCITY (FPS)	1300.	1208.	1134.	1076.	1028.	988.	954.	924.	896.	871.	847.	825.	805.
ENERGY (FT-LB)	488.	421.	371.	334.	305.	282.	263.	246.	232.	219.	207.	197.	187.
DROP (IN)	.00	-2.65	-11.25	-26.59	-49.41	-80.43	-120.33	-169.61	-229.04	-299.26	-380.90	-474.65	-581.15
MID-RANGE (IN)	.00	.70	2.98	7.14	13.45	22.18	33.57	47.78	65.11	85.77	110.02	138.10	170.25
BULLET PATH (IN)	-.75	2.60	.00	-9.34	-26.16	-51.18	-85.07	-128.35	-181.78	-245.99	-321.64	-409.38	-509.88
TIME OF FLIGHT (SEC)	.000000	.119809	.248100	.384016	.526754	.675641	.830205	.990112	1.155127	1.325092	1.499905	1.679513	1.863902
WIND DEFLECTION (IN)													
5 MPH	.00	.39	1.53	3.33	5.74	8.69	12.13	16.05	20.42	25.22	30.45	36.10	42.18
10 MPH	.00	.78	3.05	6.66	11.48	17.37	24.27	32.11	40.84	50.45	60.91	72.21	84.35
20 MPH	.00	1.56	6.10	13.33	22.96	34.75	48.54	64.21	81.68	100.89	121.81	144.42	168.71
30 MPH	.00	2.34	9.15	19.99	34.43	52.12	72.81	96.32	122.52	151.34	182.72	216.63	253.06

Bullet: Lyman # 311440 151 Gr.
Ballistic Coefficient: .134

RANGE (YDS)	MUZZLE	50	100	150	200	250	300	350	400	450	500	550	600
VELOCITY (FPS)	2400.	2105.	1854.	1623.	1439.	1282.	1157.	1064.	997.	944.	899.	860.	826.
ENERGY (FT-LB)	1931.	1486.	1152.	883.	694.	551.	448.	380.	333.	299.	271.	248.	229.
DROP (IN)	.00	-.80	-3.56	-8.86	-17.45	-30.27	-48.38	-72.98	-105.23	-146.18	-196.77	-258.08	-331.10
MID-RANGE (IN)	.00	.21	.98	2.53	5.16	9.28	15.33	23.81	35.16	49.82	68.12	90.45	117.17
BULLET PATH (IN)	-.75	.61	.00	-3.14	-9.58	-20.24	-36.20	-58.65	-88.74	-127.54	-175.98	-235.12	-305.99
TIME OF FLIGHT (SEC)	.000000	.066846	.142784	.229290	.327589	.438155	.561560	.697060	.842907	.997722	1.160701	1.331383	1.509537
WIND DEFLECTION (IN)													
5 MPH	.00	.38	1.57	3.68	6.83	11.06	16.42	22.84	30.18	38.30	47.14	56.66	66.84
10 MPH	.00	.76	3.13	7.36	13.66	22.12	32.83	45.68	60.35	76.60	94.28	113.32	133.68
20 MPH	.00	1.53	6.26	14.71	27.31	44.23	65.67	91.37	120.70	153.20	188.57	226.65	267.36
30 MPH	.00	2.29	9.39	22.07	40.97	66.35	98.50	137.05	181.05	229.80	282.85	339.97	401.04
VELOCITY (FPS)	2300.	2023.	1778.	1560.	1386.	1239.	1124.	1041.	979.	929.	886.	849.	815.
ENERGY (FT-LB)	1773.	1372.	1059.	816.	644.	514.	423.	363.	321.	289.	263.	242.	223.
DROP (IN)	.00	-.87	-3.87	-9.62	-18.96	-32.83	-52.39	-78.80	-113.21	-156.63	-210.04	-274.47	-350.93
MID-RANGE (IN)	.00	.23	1.06	2.74	5.61	10.06	16.57	25.62	37.65	53.06	72.19	95.41	123.08
BULLET PATH (IN)	-.75	.69	.00	-3.44	-10.47	-22.04	-39.29	-63.40	-95.50	-136.61	-187.70	-249.83	-323.98
TIME OF FLIGHT (SEC)	.000000	.069668	.148783	.239021	.341121	.455773	.583189	.722184	.871008	1.028494	1.193967	1.367048	1.547565
WIND DEFLECTION (IN)													
5 MPH	.00	.39	1.61	3.82	7.06	11.41	16.89	23.38	30.74	38.86	47.68	57.17	67.32
10 MPH	.00	.78	3.23	7.63	14.12	22.82	33.77	46.76	61.47	77.71	95.36	114.34	134.63
20 MPH	.00	1.57	6.46	15.27	28.25	45.65	67.54	93.51	122.94	155.42	190.71	228.68	269.26
30 MPH	.00	2.35	9.69	22.90	42.37	68.47	101.31	140.27	184.41	233.13	286.07	343.02	403.90
VELOCITY (FPS)	2200.	1941.	1703.	1501.	1335.	1198.	1094.	1019.	961.	914.	874.	838.	805.
ENERGY (FT-LB)	1623.	1263.	972.	755.	597.	481.	401.	348.	310.	280.	256.	235.	217.
DROP (IN)	.00	-.94	-4.20	-10.46	-20.60	-35.63	-56.72	-85.05	-121.73	-167.70	-224.02	-291.69	-371.71
MID-RANGE (IN)	.00	.25	1.16	2.98	6.10	10.91	17.90	27.54	40.27	56.42	76.38	100.50	129.14
BULLET PATH (IN)	-.75	.78	.00	-3.78	-11.45	-24.00	-42.62	-68.48	-102.68	-146.18	-200.02	-265.21	-342.76
TIME OF FLIGHT (SEC)	.000000	.072590	.155114	.249181	.355256	.474090	.605451	.747804	.899535	1.059668	1.227637	1.403140	1.586056
WIND DEFLECTION (IN)													
5 MPH	.00	.39	1.65	3.93	7.26	11.72	17.28	23.81	31.16	39.25	48.03	57.48	67.57
10 MPH	.00	.78	3.30	7.86	14.53	23.44	34.56	47.61	62.32	78.50	96.06	114.95	135.15
20 MPH	.00	1.55	6.60	15.71	29.05	46.88	69.12	95.23	124.64	157.00	192.13	229.91	270.29
30 MPH	.00	2.33	9.90	23.57	43.58	70.32	103.68	142.84	186.95	235.50	288.19	344.86	405.44
VELOCITY (FPS)	2100.	1849.	1619.	1435.	1279.	1154.	1062.	995.	943.	898.	860.	825.	793.
ENERGY (FT-LB)	1478.	1146.	878.	690.	548.	446.	378.	332.	298.	270.	248.	228.	211.
DROP (IN)	.00	-1.04	-4.63	-11.53	-22.68	-39.16	-62.15	-92.81	-132.21	-181.26	-241.06	-312.59	-396.80
MID-RANGE (IN)	.00	.28	1.28	3.30	6.71	11.97	19.54	29.89	43.42	60.45	81.37	106.52	136.25
BULLET PATH (IN)	-.75	.90	.00	-4.22	-12.68	-26.47	-46.77	-74.74	-111.45	-157.82	-214.92	-283.76	-365.29
TIME OF FLIGHT (SEC)	.000000	.076142	.162889	.261466	.372334	.496041	.631807	.777879	.932894	1.096058	1.266916	1.445243	1.630977
WIND DEFLECTION (IN)													
5 MPH	.00	.41	1.76	4.15	7.62	12.22	17.88	24.45	31.81	39.88	48.63	58.04	68.10
10 MPH	.00	.83	3.53	8.30	15.25	24.45	35.77	48.91	63.62	79.76	97.26	116.08	136.19
20 MPH	.00	1.66	7.05	16.61	30.49	48.89	71.54	97.81	127.24	159.53	194.53	232.15	272.39
30 MPH	.00	2.49	10.58	24.91	45.74	73.34	107.31	146.72	190.85	239.29	291.79	348.23	408.58

Bullet: Lyman # 311440 151 Gr.
Ballistic Coefficient: .134
(Cont'd.)

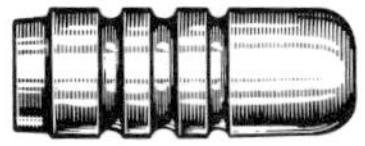

RANGE (YDS)	MUZZLE	50	100	150	200	250	300	350	400	450	500	550	600
VELOCITY (FPS)	2000.	1757.	1542.	1371.	1226.	1114.	1034.	974.	924.	883.	846.	812.	781.
ENERGY (FT-LB)	1341.	1034.	798.	630.	504.	416.	358.	318.	287.	261.	240.	221.	204.
DROP (IN)	.00	-1.15	-5.12	-12.75	-25.03	-43.10	-68.14	-101.28	-143.51	-195.83	-259.27	-334.84	-423.36
MID-RANGE (IN)	.00	.31	1.41	3.65	7.40	13.13	21.32	32.39	46.72	64.64	86.53	112.74	143.54
BULLET PATH (IN)	-.75	1.04	.00	-4.70	-14.05	-29.18	-51.29	-81.49	-120.79	-170.18	-230.68	-303.32	-388.91
TIME OF FLIGHT (SEC)	.000000	.080048	.171364	.274609	.390477	.519056	.659048	.808728	.966989	1.133190	1.306974	1.488186	1.676813
WIND DEFLECTION (IN)													
5 MPH	.00	.44	1.88	4.37	7.96	12.68	18.40	24.97	32.30	40.32	49.01	58.36	68.36
10 MPH	.00	.89	3.76	8.73	15.92	25.35	36.79	49.94	64.59	80.64	98.03	116.72	136.72
20 MPH	.00	1.78	7.52	17.46	31.85	50.71	73.58	99.87	129.18	161.28	196.05	233.44	273.44
30 MPH	.00	2.67	11.28	26.19	47.77	76.06	110.38	149.81	193.77	241.92	294.08	350.16	410.16
VELOCITY (FPS)	1900.	1665.	1472.	1310.	1178.	1079.	1008.	953.	907.	868.	832.	799.	769.
ENERGY (FT-LB)	1210.	930.	726.	575.	465.	391.	341.	305.	276.	252.	232.	214.	198.
DROP (IN)	.00	-1.27	-5.68	-14.13	-27.66	-47.44	-74.66	-110.39	-155.57	-211.26	-278.48	-358.22	-451.21
MID-RANGE (IN)	.00	.34	1.57	4.03	8.15	14.38	23.20	35.01	50.13	68.94	91.80	119.06	150.95
BULLET PATH (IN)	-.75	1.19	.00	-5.24	-15.55	-32.12	-56.11	-88.63	-130.60	-183.08	-247.08	-323.61	-413.37
TIME OF FLIGHT (SEC)	.000000	.084358	.180393	.288537	.409498	.542807	.686802	.839974	1.001431	1.170658	1.347386	1.531522	1.723096
WIND DEFLECTION (IN)													
5 MPH	.00	.48	1.98	4.55	8.25	13.03	18.75	25.29	32.55	40.49	49.10	58.35	68.26
10 MPH	.00	.95	3.96	9.10	16.49	26.06	37.51	50.57	65.09	80.98	98.19	116.71	136.53
20 MPH	.00	1.90	7.92	18.20	32.99	52.12	75.02	101.14	130.19	161.97	196.39	233.41	273.06
30 MPH	.00	2.86	11.88	27.30	49.48	78.18	112.53	151.72	195.28	242.95	294.58	350.12	409.58
VELOCITY (FPS)	1800.	1577.	1401.	1251.	1133.	1047.	984.	933.	890.	852.	818.	787.	757.
ENERGY (FT-LB)	1086.	834.	658.	524.	430.	368.	324.	292.	266.	243.	224.	207.	192.
DROP (IN)	.00	-1.42	-6.34	-15.70	-30.64	-52.33	-81.91	-120.42	-168.81	-228.14	-299.41	-383.51	-481.43
MID-RANGE (IN)	.00	.38	1.75	4.47	8.98	15.76	25.25	37.81	53.76	73.50	97.37	125.69	158.80
BULLET PATH (IN)	-.75	1.37	.00	-5.82	-17.21	-35.36	-61.40	-96.36	-141.21	-196.99	-264.71	-345.27	-439.64
TIME OF FLIGHT (SEC)	.000000	.089125	.190125	.303600	.429874	.567885	.715865	.872588	1.037344	1.209732	1.389564	1.576804	1.771512
WIND DEFLECTION (IN)													
5 MPH	.00	.51	2.06	4.72	8.50	13.31	19.00	25.45	32.62	40.46	48.95	58.09	67.89
10 MPH	.00	1.02	4.13	9.43	16.99	26.61	37.99	50.91	65.24	80.91	97.90	116.18	135.79
20 MPH	.00	2.04	8.26	18.87	33.98	53.23	75.98	101.82	130.48	161.83	195.79	232.37	271.57
30 MPH	.00	3.06	12.39	28.30	50.97	79.84	113.98	152.73	195.72	242.74	293.69	348.55	407.36
VELOCITY (FPS)	1700.	1498.	1333.	1196.	1092.	1018.	961.	914.	873.	837.	804.	774.	745.
ENERGY (FT-LB)	969.	753.	595.	479.	400.	347.	309.	280.	256.	235.	217.	201.	186.
DROP (IN)	.00	-1.59	-7.08	-17.48	-33.95	-57.68	-89.76	-131.17	-182.92	-246.04	-321.53	-410.17	-513.24
MID-RANGE (IN)	.00	.43	1.95	4.95	9.89	17.25	27.41	40.70	57.51	78.18	103.09	132.47	166.85
BULLET PATH (IN)	-.75	1.58	.00	-6.48	-19.04	-38.85	-67.02	-104.50	-152.34	-211.54	-283.12	-367.84	-466.99
TIME OF FLIGHT (SEC)	.000000	.094224	.200475	.319489	.451017	.593509	.745360	.905601	1.073672	1.249271	1.432282	1.622720	1.820663
WIND DEFLECTION (IN)													
5 MPH	.00	.53	2.11	4.82	8.63	13.41	19.00	25.34	32.37	40.05	48.39	57.39	67.04
10 MPH	.00	1.05	4.22	9.64	17.26	26.81	38.01	50.68	64.73	80.11	96.79	114.78	134.08
20 MPH	.00	2.11	8.45	19.28	34.52	53.62	76.01	101.36	129.46	160.21	193.57	229.55	268.17
30 MPH	.00	3.16	12.67	28.93	51.78	80.43	114.02	152.04	194.19	240.32	290.36	344.33	402.25

Bullet: Lyman # 311440 151 Gr.
Ballistic Coefficient: .134
(Cont'd.)

RANGE (YDS)	MUZZLE	50	100	150	200	250	300	350	400	450	500	550	600
VELOCITY (FPS)	1600.	1421.	1267.	1145.	1056.	991.	939.	895.	857.	822.	790.	761.	733.
ENERGY (FT-LB)	858.	677.	538.	440.	374.	329.	295.	268.	246.	227.	209.	194.	180.
DROP (IN)	.00	-1.79	-7.91	-19.45	-37.60	-63.51	-98.23	-142.70	-197.99	-265.10	-344.97	-438.47	-546.98
MID-RANGE (IN)	.00	.48	2.16	5.48	10.89	18.83	29.66	43.72	61.39	83.03	108.97	139.52	175.26
BULLET PATH (IN)	-.75	1.79	.00	-7.21	-21.04	-42.62	-73.01	-113.15	-164.11	-226.89	-302.43	-391.60	-495.78
TIME OF FLIGHT (SEC)	.000000	.099552	.211466	.336210	.472888	.619731	.775434	.939236	1.110705	1.289630	1.475961	1.669749	1.871088
WIND DEFLECTION (IN)													
5 MPH	.00	.51	2.11	4.84	8.61	13.29	18.74	24.90	31.74	39.24	47.38	56.19	65.66
10 MPH	.00	1.02	4.22	9.67	17.23	26.57	37.48	49.81	63.48	78.47	94.77	112.38	131.31
20 MPH	.00	2.04	8.44	19.35	34.46	53.15	74.95	99.61	126.97	156.95	189.54	224.75	262.62
30 MPH	.00	3.06	12.65	29.02	51.69	79.72	112.43	149.42	190.45	235.42	284.31	337.13	393.93
VELOCITY (FPS)	1500.	1334.	1197.	1093.	1018.	961.	914.	874.	837.	804.	774.	745.	718.
ENERGY (FT-LB)	754.	597.	480.	401.	348.	310.	280.	256.	235.	217.	201.	186.	173.
DROP (IN)	.00	-2.05	-9.00	-22.02	-42.27	-70.88	-108.79	-157.05	-216.66	-288.63	-373.74	-473.27	-588.43
MID-RANGE (IN)	.00	.54	2.45	6.17	12.14	20.76	32.38	47.36	66.06	88.85	115.99	148.02	185.42
bULLET PATH (IN)	-.75	2.07	.00	-8.14	-23.52	-47.26	-80.29	-123.67	-178.41	-245.51	-325.74	-420.39	-530.68
TIME OF FLIGHT (SEC)	.000000	.106134	.225028	.356445	.498841	.650608	.810771	.978767	1.154291	1.337227	1.527588	1.725452	1.930939
WIND DEFLECTION (IN)													
5 MPH	.00	.54	2.20	4.97	8.70	13.25	18.55	24.53	31.18	38.48	46.43	55.04	64.32
10 MPH	.00	1.08	4.40	9.93	17.40	26.51	37.10	49.06	62.36	76.95	92.86	110.08	128.65
20 MPH	.00	2.16	8.81	19.87	34.79	53.01	74.19	98.13	124.71	153.90	185.71	220.16	257.29
30 MPH	.00	3.24	13.21	29.80	52.19	79.52	111.29	147.19	187.07	230.86	278.57	330.24	385.94
VELOCITY (FPS)	1400.	1250.	1132.	1047.	983.	933.	890.	852.	818.	786.	757.	729.	703.
ENERGY (FT-LB)	657.	524.	430.	367.	324.	292.	265.	243.	224.	207.	192.	178.	166.
DROP (IN)	.00	-2.35	-10.28	-24.97	-47.56	-79.07	-120.48	-172.82	-237.11	-314.23	-405.17	-511.27	-633.70
MID-RANGE (IN)	.00	.62	2.79	6.94	13.49	22.83	35.27	51.20	70.99	94.97	123.51	157.16	196.37
BULLET PATH (IN)	-.75	2.41	.00	-9.17	-26.24	-52.24	-88.12	-134.95	-193.72	-265.32	-350.75	-451.33	-568.24
TIME OF FLIGHT (SEC)	.000000	.113534	.239865	.377923	.525942	.682698	.847483	1.019896	1.199750	1.387011	1.581736	1.784028	1.994037
WIND DEFLECTION (IN)													
5 MPH	.00	.56	2.25	4.97	8.57	12.93	18.01	23.75	30.15	37.20	44.91	53.28	62.33
10 MPH	.00	1.12	4.50	9.94	17.14	25.87	36.01	47.50	60.30	74.40	89.81	106.56	124.66
20 MPH	.00	2.25	9.00	19.89	34.27	51.74	72.03	95.00	120.60	148.80	179.63	213.12	249.33
30 MPH	.00	3.37	13.51	29.83	51.41	77.61	108.04	142.51	180.90	223.20	269.44	319.68	373.99
VELOCITY (FPS)	1300.	1170.	1074.	1004.	950.	904.	865.	830.	797.	767.	739.	712.	687.
ENERGY (FT-LB)	567.	459.	387.	338.	302.	274.	251.	231.	213.	197.	183.	170.	158.
DROP (IN)	.00	-2.73	-11.80	-28.37	-53.54	-88.21	-133.47	-190.32	-259.76	-342.51	-440.00	-553.40	-683.94
MID-RANGE (IN)	.00	.72	3.17	7.79	14.97	25.03	38.36	55.34	76.34	101.64	131.79	167.24	208.50
BULLET PATH (IN)	-.75	2.79	.00	-10.30	-29.19	-57.58	-96.57	-147.14	-210.31	-286.78	-378.00	-485.12	-609.39
TIME OF FLIGHT (SEC)	.000000	.121830	.255921	.400574	.554324	.716312	.886041	1.063257	1.247876	1.439933	1.639513	1.846750	2.061826
WIND DEFLECTION (IN)													
5 MPH	.00	.57	2.21	4.79	8.17	12.27	17.05	22.49	28.58	35.33	42.74	50.82	59.59
10 MPH	.00	1.13	4.43	9.58	16.33	24.53	34.10	44.98	57.16	70.66	85.48	101.64	119.19
20 MPH	.00	2.27	8.85	19.16	32.66	49.06	68.19	89.96	114.33	141.32	170.95	203.29	238.38
30 MPH	.00	3.40	13.28	28.73	48.99	73.60	102.29	134.94	171.49	211.98	256.43	304.93	357.57

Bullet: Lyman # 311466 151 Gr.
Ballistic Coefficient: .250

RANGE (YDS)	MUZZLE	50	100	150	200	250	300	350	400	450	500	550	600
VELOCITY (FPS)	2300.	2137.	1982.	1833.	1692.	1562.	1444.	1338.	1243.	1163.	1098.	1045.	1002.
ENERGY (FT-LB)	1773.	1532.	1316.	1126.	960.	818.	699.	600.	518.	454.	404.	366.	337.
DROP (IN)	.00	-.82	-3.53	-8.46	-15.96	-26.50	-40.61	-58.91	-82.06	-110.86	-146.06	-188.46	-238.78
MID-RANGE (IN)	.00	.22	.95	2.31	4.45	7.55	11.85	17.59	25.07	34.61	46.53	61.16	78.78
BULLET PATH (IN)	-.75	.57	.00	-2.78	-8.15	-16.55	-28.51	-44.67	-65.68	-92.34	-125.40	-165.66	-213.84
TIME OF FLIGHT (SEC)	.000000	.067653	.140540	.219252	.304436	.396760	.496667	.604631	.721018	.845859	.978755	1.118938	1.265628
WIND DEFLECTION (IN)													
5 MPH	.00	.21	.89	2.08	3.83	6.22	9.27	13.03	17.54	22.78	28.74	35.34	42.51
10 MPH	.00	.43	1.78	4.15	7.67	12.44	18.54	26.07	35.07	45.57	57.48	70.67	85.01
20 MPH	.00	.86	3.56	8.31	15.34	24.88	37.09	52.13	70.15	91.13	114.96	141.34	170.02
30 MPH	.00	1.29	5.34	12.46	23.00	37.32	55.63	78.20	105.22	136.70	172.43	212.02	255.03
VELOCITY (FPS)	2200.	2041.	1890.	1746.	1610.	1489.	1378.	1279.	1193.	1121.	1064.	1018.	979.
ENERGY (FT-LB)	1623.	1397.	1197.	1022.	869.	743.	636.	548.	477.	422.	380.	347.	321.
DROP (IN)	.00	-.89	-3.87	-9.27	-17.52	-29.12	-44.67	-64.80	-90.26	-121.82	-160.28	-206.38	-260.85
MID-RANGE (IN)	.00	.24	1.04	2.54	4.89	8.32	13.06	19.39	27.60	38.04	51.02	66.83	85.75
BULLET PATH (IN)	-.75	.67	.00	-3.09	-9.03	-18.32	-31.56	-49.38	-72.53	-101.78	-137.93	-181.73	-233.89
TIME OF FLIGHT (SEC)	.000000	.070782	.147150	.229733	.319203	.416114	.520887	.633971	.755545	.885381	1.022831	1.167066	1.317441
WIND DEFLECTION (IN)													
5 MPH	.00	.23	.95	2.22	4.09	6.62	9.84	13.79	18.49	23.91	30.01	36.70	43.93
10 MPH	.00	.46	1.90	4.43	8.18	13.24	19.68	27.58	36.98	47.83	60.02	73.40	87.87
20 MPH	.00	.92	3.80	8.87	16.36	26.47	39.35	55.16	73.95	95.65	120.04	146.81	175.74
30 MPH	.00	1.37	5.70	13.30	24.54	39.71	59.03	82.74	110.93	143.48	180.05	220.21	263.61
VELOCITY (FPS)	2100.	1946.	1799.	1660.	1533.	1418.	1314.	1223.	1146.	1084.	1034.	993.	957.
ENERGY (FT-LB)	1478.	1269.	1085.	924.	788.	674.	579.	501.	441.	394.	359.	331.	307.
DROP (IN)	.00	-.98	-4.26	-10.21	-19.31	-32.14	-49.30	-71.49	-99.50	-134.12	-176.10	-226.19	-285.07
MID-RANGE (IN)	.00	.26	1.15	2.80	5.40	9.20	14.44	21.41	30.44	41.84	55.91	72.96	93.24
BULLET PATH (IN)	-.75	.77	.00	-3.45	-10.05	-20.37	-35.03	-54.71	-80.23	-112.34	-151.82	-199.40	-255.78
TIME OF FLIGHT (SEC)	.000000	.074208	.154389	.241202	.335292	.437069	.547011	.665419	.792226	.926931	1.068721	1.216864	1.370847
WIND DEFLECTION (IN)													
5 MPH	.00	.24	1.01	2.37	4.36	7.03	10.42	14.56	19.43	25.00	31.19	37.94	45.21
10 MPH	.00	.49	2.03	4.74	8.73	14.07	20.85	29.11	38.86	50.00	62.38	75.88	90.41
20 MPH	.00	.98	4.06	9.47	17.45	28.13	41.69	58.23	77.72	99.99	124.76	151.76	180.82
30 MPH	.00	1.47	6.09	14.21	26.18	42.20	62.54	87.34	116.58	149.99	187.14	227.65	271.24
VELOCITY (FPS)	2000.	1850.	1709.	1576.	1458.	1350.	1254.	1172.	1105.	1051.	1007.	969.	937.
ENERGY (FT-LB)	1341.	1148.	979.	833.	712.	611.	527.	460.	409.	370.	340.	315.	294.
DROP (IN)	.00	-1.08	-4.70	-11.29	-21.38	-35.60	-54.57	-79.10	-109.95	-147.89	-193.68	-248.03	-311.55
MID-RANGE (IN)	.00	.29	1.27	3.09	5.99	10.20	15.99	23.69	33.59	46.01	61.25	79.57	101.20
BULLET PATH (IN)	-.75	.89	.00	-3.86	-11.23	-22.71	-38.96	-60.77	-88.88	-124.10	-167.17	-218.79	-279.58
TIME OF FLIGHT (SEC)	.000000	.077978	.162348	.253787	.352783	.459778	.575168	.699030	.831014	.970385	1.116341	1.268298	1.425848
WIND DEFLECTION (IN)													
5 MPH	.00	.26	1.09	2.53	4.64	7.46	11.01	15.31	20.33	25.99	32.24	39.01	46.27
10 MPH	.00	.52	2.17	5.07	9.29	14.92	22.03	30.63	40.66	51.99	64.48	78.02	92.55
20 MPH	.00	1.05	4.35	10.13	18.58	29.84	44.06	61.26	81.32	103.98	128.95	156.04	185.10
30 MPH	.00	1.57	6.52	15.20	27.87	44.76	66.09	91.89	121.98	155.96	193.43	234.06	277.65

Bullet: Lyman # 311466 151 Gr.
Ballistic Coefficient: .250
(Cont'd.)

RANGE (YDS)	MUZZLE	50	100	150	200	250	300	350	400	450	500	550	600
VELOCITY (FPS)	1900.	1755.	1619.	1497.	1385.	1285.	1198.	1126.	1068.	1021.	981.	947.	917.
ENERGY (FT-LB)	1210.	1033.	879.	751.	643.	553.	481.	425.	382.	349.	323.	301.	282.
DROP (IN)	.00	-1.20	-5.22	-12.54	-23.79	-39.56	-60.61	-87.72	-121.67	-163.21	-213.07	-271.94	-340.32
MID-RANGE (IN)	.00	.32	1.41	3.44	6.67	11.34	17.76	26.22	37.04	50.54	66.97	86.60	109.55
BULLET PATH (IN)	-.75	1.03	.00	-4.34	-12.60	-25.38	-43.45	-67.57	-98.53	-137.09	-183.96	-239.84	-305.24
TIME OF FLIGHT (SEC)	.000000	.082142	.171123	.267528	.371762	.484284	.605296	.734597	.871565	1.015368	1.165346	1.321039	1.482124
WIND DEFLECTION (IN)													
5 MPH	.00	.28	1.16	2.70	4.93	7.88	11.58	16.01	21.12	26.83	33.08	39.83	47.06
10 MPH	.00	.56	2.33	5.40	9.85	15.76	23.16	32.03	42.24	53.65	66.15	79.66	94.12
20 MPH	.00	1.12	4.66	10.80	19.70	31.52	46.33	64.05	84.48	107.30	132.31	159.32	188.23
30 MPH	.00	1.69	6.98	16.20	29.55	47.28	69.49	96.08	126.71	160.96	198.46	238.98	282.35
VELOCITY (FPS)	1800.	1661.	1534.	1419.	1315.	1224.	1147.	1085.	1035.	993.	958.	926.	898.
ENERGY (FT-LB)	1086.	925.	789.	675.	580.	502.	441.	394.	359.	331.	307.	288.	270.
DROP (IN)	.00	-1.34	-5.83	-14.03	-26.56	-44.12	-67.50	-97.47	-134.80	-180.23	-234.45	-298.02	-371.65
MID-RANGE (IN)	.00	.36	1.57	3.86	7.45	12.64	19.72	29.02	40.84	55.45	73.10	93.97	118.32
BULLET PATH (IN)	-.75	1.20	.00	-4.92	-14.16	-28.43	-48.52	-75.20	-109.24	-151.38	-202.31	-262.59	-332.94
TIME OF FLIGHT (SEC)	.000000	.086760	.180790	.282504	.392378	.510718	.637458	.772100	.913834	1.061924	1.215856	1.375269	1.539911
WIND DEFLECTION (IN)													
5 MPH	.00	.30	1.24	2.86	5.20	8.28	12.10	16.61	21.75	27.45	33.66	40.36	47.51
10 MPH	.00	.60	2.49	5.72	10.39	16.55	24.19	33.22	43.50	54.90	67.32	80.71	95.02
20 MPH	.00	1.21	4.97	11.44	20.78	33.11	48.39	66.45	87.00	109.80	134.65	161.43	190.05
30 MPH	.00	1.81	7.46	17.16	31.18	49.66	72.58	99.67	130.50	164.70	201.97	242.14	285.07
VELOCITY (FPS)	1700.	1569.	1451.	1343.	1248.	1167.	1101.	1048.	1004.	967.	935.	906.	879.
ENERGY (FT-LB)	969.	825.	705.	605.	522.	457.	406.	368.	338.	314.	293.	275.	259.
DROP (IN)	.00	-1.51	-6.55	-15.76	-29.77	-49.38	-75.36	-108.48	-149.49	-199.10	-257.91	-326.55	-405.77
MID-RANGE (IN)	.00	.41	1.77	4.33	8.33	14.09	21.92	32.13	44.98	60.72	79.57	101.75	127.53
BULLET PATH (IN)	-.75	1.40	.00	-5.55	-15.91	-31.87	-54.20	-83.67	-121.03	-166.98	-222.14	-287.13	-362.70
TIME OF FLIGHT (SEC)	.000000	.091901	.191371	.298871	.414780	.539152	.671610	.811401	.957734	1.110038	1.267912	1.431077	1.599342
WIND DEFLECTION (IN)													
5 MPH	.00	.32	1.31	3.01	5.44	8.62	12.51	17.05	22.16	27.80	33.93	40.52	47.57
10 MPH	.00	.65	2.62	6.01	10.88	17.24	25.03	34.10	44.33	55.60	67.86	81.05	95.13
20 MPH	.00	1.29	5.24	12.03	21.77	34.49	50.05	68.20	88.65	111.20	135.72	162.09	190.26
30 MPH	.00	1.94	7.87	18.04	32.65	51.73	75.08	102.30	132.98	166.81	203.58	243.14	285.39
VELOCITY (FPS)	1600.	1480.	1369.	1271.	1186.	1116.	1060.	1015.	976.	943.	913.	886.	861.
ENERGY (FT-LB)	858.	734.	629.	542.	472.	418.	377.	345.	319.	298.	279.	263.	248.
DROP (IN)	.00	-1.70	-7.41	-17.75	-33.48	-55.38	-84.24	-120.79	-165.77	-219.82	-283.51	-357.60	-442.74
MID-RANGE (IN)	.00	.46	2.00	4.85	9.33	15.73	24.37	35.51	49.41	66.30	86.37	109.91	137.13
BULLET PATH (IN)	-.75	1.63	.00	-6.26	-17.91	-35.73	-60.51	-92.98	-133.88	-183.85	-243.46	-313.47	-394.53
TIME OF FLIGHT (SEC)	.000000	.097517	.202935	.316690	.438932	.569397	.707410	.852144	1.002974	1.159463	1.321303	1.488284	1.660265
WIND DEFLECTION (IN)													
5 MPH	.00	.33	1.36	3.12	5.63	8.86	12.75	17.24	22.26	27.78	33.77	40.22	47.10
10 MPH	.00	.66	2.72	6.24	11.25	17.71	25.50	34.48	44.52	55.57	67.55	80.44	94.21
20 MPH	.00	1.33	5.43	12.47	22.50	35.43	51.01	68.95	89.05	111.13	135.10	160.88	188.41
30 MPH	.00	1.99	8.15	18.71	33.76	53.14	76.51	103.43	133.57	166.70	202.65	241.31	282.62

Bullet: Lyman # 301618 160 Gr.
Ballistic Coefficient: .300

RANGE (YDS)	MUZZLE	50	100	150	200	250	300	350	400	450	500	550	600
VELOCITY (FPS)	2700.	2551.	2407.	2272.	2141.	2015.	1893.	1776.	1662.	1555.	1455.	1363.	1279.
ENERGY (FT-LB)	2589.	2312.	2058.	1833.	1628.	1442.	1273.	1120.	982.	859.	752.	660.	581.
DROP (IN)	.00	-.58	-2.49	-5.89	-10.98	-17.95	-27.06	-38.59	-52.88	-70.30	-91.33	-116.49	-146.32
MID-RANGE (IN)	.00	.16	.67	1.59	3.00	4.98	7.64	11.09	15.46	20.94	27.71	36.02	46.09
BULLET PATH (IN)	-.75	.29	.00	-1.79	-5.25	-10.61	-18.10	-28.02	-40.68	-56.48	-75.89	-99.43	-127.65
TIME OF FLIGHT (SEC)	.000000	.057154	.117684	.181839	.249858	.322088	.398906	.480726	.568039	.661338	.761072	.867637	.981335
WIND DEFLECTION (IN)													
5 MPH	.00	.14	.58	1.34	2.43	3.90	5.77	8.08	10.88	14.20	18.09	22.57	27.69
10 MPH	.00	.28	1.16	2.67	4.86	7.80	11.54	16.16	21.75	28.40	36.17	45.15	55.38
20 MPH	.00	.56	2.31	5.34	9.73	15.60	23.08	32.33	43.51	56.79	72.34	90.30	110.76
30 MPH	.00	.84	3.47	8.01	14.59	23.40	34.62	48.49	65.26	85.19	108.51	135.45	166.14
VELOCITY (FPS)	2600.	2454.	2316.	2184.	2056.	1933.	1814.	1699.	1590.	1488.	1392.	1306.	1228.
ENERGY (FT-LB)	2401.	2140.	1905.	1694.	1502.	1327.	1169.	1026.	898.	786.	689.	605.	536.
DROP (IN)	.00	-.62	-2.69	-6.36	-11.86	-19.40	-29.27	-41.77	-57.28	-76.21	-99.12	-126.50	-158.99
MID-RANGE (IN)	.00	.17	.72	1.72	3.24	5.39	8.27	12.02	16.78	22.75	30.15	39.22	50.22
BULLET PATH (IN)	-.75	.34	-.00	-1.96	-5.74	-11.57	-19.71	-30.50	-44.29	-61.50	-82.69	-108.36	-139.13
TIME OF FLIGHT (SEC)	.000000	.059379	.122307	.189002	.259786	.335030	.415134	.500561	.591822	.689372	.793630	.904935	1.023475
WIND DEFLECTION (IN)													
5 MPH	.00	.15	.61	1.40	2.55	4.10	6.07	8.51	11.46	14.97	19.07	23.79	29.14
10 MPH	.00	.30	1.22	2.80	5.11	8.20	12.14	17.02	22.93	29.94	38.14	47.58	58.29
20 MPH	.00	.59	2.44	5.61	10.21	16.39	24.28	34.04	45.86	59.89	76.28	95.15	116.57
30 MPH	.00	.89	3.66	8.41	15.32	24.59	36.42	51.07	68.79	89.83	114.42	142.73	174.86
VELOCITY (FPS)	2500.	2359.	2225.	2096.	1971.	1851.	1735.	1624.	1519.	1422.	1332.	1251.	1181.
ENERGY (FT-LB)	2220.	1976.	1759.	1560.	1380.	1217.	1069.	937.	820.	718.	630.	556.	495.
DROP (IN)	.00	-.68	-2.91	-6.89	-12.85	-21.04	-31.75	-45.35	-62.24	-82.93	-107.92	-137.81	-173.27
MID-RANGE (IN)	.00	.18	.78	1.86	3.52	5.85	8.99	13.07	18.27	24.82	32.93	42.85	54.88
BULLET PATH (IN)	-.75	.40	.00	-2.15	-6.28	-12.64	-21.53	-33.30	-48.36	-67.22	-90.38	-118.44	-152.07
TIME OF FLIGHT (SEC)	.000000	.061784	.127262	.196721	.270519	.349047	.432739	.522108	.617624	.719722	.828772	.945020	1.068498
WIND DEFLECTION (IN)													
5 MPH	.00	.16	.64	1.47	2.69	4.32	6.40	8.99	12.11	15.82	20.13	25.08	30.67
10 MPH	.00	.31	1.28	2.94	5.37	8.63	12.80	17.97	24.22	31.63	40.26	50.16	61.34
20 MPH	.00	.63	2.56	5.89	10.74	17.26	25.60	35.94	48.44	63.26	80.53	100.33	122.67
30 MPH	.00	.94	3.83	8.83	16.11	25.90	38.41	53.91	72.67	94.89	120.79	150.49	184.01
VELOCITY (FPS)	2400.	2265.	2135.	2009.	1887.	1770.	1657.	1550.	1450.	1358.	1275.	1201.	1138.
ENERGY (FT-LB)	2046.	1823.	1619.	1433.	1265.	1113.	975.	854.	747.	655.	577.	513.	460.
DROP (IN)	.00	-.73	-3.15	-7.48	-13.95	-22.86	-34.53	-49.37	-67.83	-90.45	-117.78	-150.47	-189.19
MID-RANGE (IN)	.00	.20	.84	2.02	3.82	6.36	9.79	14.26	19.96	27.15	36.03	46.91	60.06
BULLET PATH (IN)	-.75	.47	.00	-2.37	-6.89	-13.85	-23.57	-36.46	-52.97	-73.64	-99.01	-129.75	-166.52
TIME OF FLIGHT (SEC)	.000000	.064333	.132545	.204984	.282031	.364102	.451698	.545302	.645362	.752271	.866323	.987621	1.116010
WIND DEFLECTION (IN)													
5 MPH	.00	.16	.66	1.54	2.82	4.54	6.75	9.49	12.79	16.70	21.24	26.41	32.21
10 MPH	.00	.32	1.33	3.08	5.64	9.08	13.50	18.97	25.58	33.40	42.47	52.82	64.42
20 MPH	.00	.65	2.66	6.15	11.27	18.16	27.00	37.95	51.17	66.80	84.95	105.64	128.84
30 MPH	.00	.97	3.98	9.23	16.91	27.25	40.50	56.92	76.75	100.20	127.42	158.46	193.25

Bullet: Lyman # 301618 160 Gr.
Ballistic Coefficient: .300
(Cont'd.)

RANGE (YDS)	MUZZLE	50	100	150	200	250	300	350	400	450	500	550	600
VELOCITY (FPS)	2300.	2168.	2041.	1918.	1800.	1686.	1578.	1476.	1382.	1296.	1220.	1154.	1099.
ENERGY (FT-LB)	1879.	1670.	1480.	1307.	1151.	1010.	884.	774.	678.	596.	528.	473.	429.
DROP (IN)	.00	-.80	-3.44	-8.16	-15.24	-25.00	-37.80	-54.10	-74.42	-99.29	-129.35	-165.26	-207.69
MID-RANGE (IN)	.00	.22	.92	2.21	4.18	6.97	10.74	15.66	21.97	29.89	39.68	51.64	66.04
BULLET PATH (IN)	-.75	.55	.00	-2.63	-7.61	-15.27	-25.98	-40.18	-58.41	-81.19	-109.15	-142.97	-183.30
TIME OF FLIGHT (SEC)	.000000	.067168	.138469	.214276	.294993	.381087	.473068	.571392	.676467	.788623	.908021	1.034577	1.167936
WIND DEFLECTION (IN)													
5 MPH	.00	.17	.71	1.64	3.00	4.84	7.20	10.11	13.62	17.75	22.51	27.91	33.91
10 MPH	.00	.34	1.41	3.28	6.01	9.68	14.39	20.22	27.23	35.49	45.03	55.82	67.82
20 MPH	.00	.69	2.83	6.56	12.01	19.36	28.78	40.43	54.46	70.99	90.06	111.65	135.64
30 MPH	.00	1.03	4.24	9.83	18.02	29.04	43.17	60.65	81.70	106.48	135.09	167.47	203.45
VELOCITY (FPS)	2200.	2072.	1948.	1829.	1713.	1603.	1500.	1404.	1316.	1237.	1169.	1111.	1063.
ENERGY (FT-LB)	1719.	1524.	1348.	1188.	1043.	913.	799.	700.	615.	544.	485.	438.	402.
DROP (IN)	.00	-.87	-3.77	-8.95	-16.72	-27.45	-41.55	-59.56	-81.98	-109.41	-142.55	-182.04	-228.55
MID-RANGE (IN)	.00	.24	1.01	2.42	4.59	7.67	11.83	17.29	24.27	33.02	43.83	56.97	72.73
BULLET PATH (IN)	-.75	.63	.00	-2.92	-8.43	-16.90	-28.74	-44.50	-64.66	-89.83	-120.71	-157.94	-202.20
TIME OF FLIGHT (SEC)	.000000	.070264	.144939	.224424	.309169	.399689	.496443	.599856	.710281	.827930	.952781	1.084546	1.222687
WIND DEFLECTION (IN)													
5 MPH	.00	.18	.75	1.75	3.21	5.17	7.69	10.79	14.50	18.86	23.84	29.44	35.60
10 MPH	.00	.37	1.51	3.50	6.41	10.35	15.37	21.57	29.01	37.72	47.69	58.88	71.19
20 MPH	.00	.73	3.02	7.00	12.83	20.69	30.75	43.15	58.02	75.43	95.38	117.76	142.39
30 MPH	.00	1.10	4.53	10.50	19.24	31.04	46.12	64.72	87.03	113.15	143.07	176.64	213.58
VELOCITY (FPS)	2100.	1975.	1855.	1739.	1627.	1522.	1425.	1335.	1254.	1183.	1123.	1073.	1032.
ENERGY (FT-LB)	1566.	1386.	1222.	1074.	941.	823.	721.	633.	558.	497.	448.	409.	378.
DROP (IN)	.00	-.96	-4.14	-9.84	-18.42	-30.26	-45.89	-65.81	-90.60	-120.95	-157.50	-200.92	-251.86
MID-RANGE (IN)	.00	.26	1.11	2.67	5.07	8.47	13.10	19.16	26.88	36.57	48.49	62.89	80.04
BULLET PATH (IN)	-.75	.74	.00	-3.26	-9.38	-18.78	-31.97	-49.44	-71.78	-99.69	-133.79	-174.76	-223.26
TIME OF FLIGHT (SEC)	.000000	.073654	.152026	.235547	.324731	.420047	.521933	.630760	.746780	.870031	1.000283	1.137048	1.279708
WIND DEFLECTION (IN)													
5 MPH	.00	.20	.81	1.87	3.43	5.54	8.22	11.51	15.43	19.99	25.17	30.92	37.19
10 MPH	.00	.39	1.61	3.74	6.87	11.07	16.43	23.01	30.86	39.98	50.34	61.83	74.37
20 MPH	.00	.78	3.23	7.48	13.73	22.14	32.86	46.03	61.72	79.97	100.67	123.67	148.74
30 MPH	.00	1.18	4.84	11.23	20.60	33.21	49.29	69.04	92.59	119.95	151.01	185.50	223.11
VELOCITY (FPS)	2000.	1879.	1762.	1649.	1543.	1444.	1352.	1269.	1197.	1134.	1083.	1040.	1003.
ENERGY (FT-LB)	1421.	1254.	1103.	966.	846.	740.	649.	572.	509.	457.	416.	384.	358.
DROP (IN)	.00	-1.06	-4.57	-10.89	-20.38	-33.55	-50.91	-73.01	-100.52	-134.11	-174.43	-222.15	-277.89
MID-RANGE (IN)	.00	.29	1.23	2.95	5.62	9.42	14.57	21.29	29.88	40.59	53.67	69.40	88.04
BULLET PATH (IN)	-.75	.85	.00	-3.65	-10.48	-20.99	-35.68	-55.12	-79.97	-110.90	-148.56	-193.61	-246.69
TIME OF FLIGHT (SEC)	.000000	.077385	.159823	.247817	.341852	.442371	.549763	.664312	.786104	.914966	1.050457	1.191965	1.338949
WIND DEFLECTION (IN)													
5 MPH	.00	.21	.86	2.01	3.68	5.93	8.78	12.26	16.38	21.12	26.44	32.29	38.63
10 MPH	.00	.42	1.73	4.02	7.37	11.86	17.56	24.52	32.75	42.23	52.88	64.59	77.26
20 MPH	.00	.84	3.46	8.03	14.73	23.71	35.12	49.04	65.51	84.47	105.76	129.17	154.51
30 MPH	.00	1.26	5.19	12.05	22.10	35.57	52.67	73.56	98.26	126.70	158.64	193.76	231.77

Bullet: Lyman # 311291 169 Gr.
Ballistic Coefficient: .202

RANGE (YDS)	MUZZLE	50	100	150	200	250	300	350	400	450	500	550	600
VELOCITY (FPS)	2500.	2290.	2104.	1937.	1779.	1631.	1500.	1384.	1280.	1191.	1118.	1059.	1012.
ENERGY (FT-LB)	2345.	1968.	1661.	1408.	1188.	998.	845.	719.	615.	532.	469.	421.	384.
DROP (IN)	.00	-.70	-3.06	-7.37	-14.00	-23.37	-36.00	-52.52	-73.58	-99.95	-132.45	-171.89	-219.05
MID-RANGE (IN)	.00	.19	.83	2.03	3.94	6.74	10.63	15.89	22.79	31.66	42.85	56.70	73.52
BULLET PATH (IN)	-.75	.45	.00	-2.41	-7.13	-14.60	-25.32	-39.94	-59.09	-83.56	-114.16	-151.69	-196.94
TIME OF FLIGHT (SEC)	.000000	.062691	.131124	.205425	.286230	.374316	.470306	.574438	.687190	.808769	.938927	1.076952	1.221955
WIND DEFLECTION (IN)													
5 MPH	.00	.24	.98	2.24	4.07	6.54	9.71	13.59	18.23	23.65	29.83	36.69	44.17
10 MPH	.00	.47	1.96	4.47	8.14	13.08	19.41	27.18	36.47	47.30	59.65	73.38	88.34
20 MPH	.00	.95	3.92	8.95	16.27	26.16	38.83	54.36	72.93	94.61	119.30	146.77	176.69
30 MPH	.00	1.42	5.87	13.42	24.41	39.24	58.24	81.54	109.40	141.91	178.95	220.15	265.03
VELOCITY (FPS)	2400.	2195.	2024.	1861.	1707.	1566.	1444.	1333.	1236.	1154.	1088.	1036.	993.
ENERGY (FT-LB)	2161.	1808.	1537.	1300.	1094.	921.	782.	667.	573.	500.	444.	403.	370.
DROP (IN)	.00	-.76	-3.32	-7.99	-15.18	-25.33	-39.04	-56.93	-79.71	-108.19	-143.19	-185.49	-235.87
MID-RANGE (IN)	.00	.21	.89	2.20	4.27	7.30	11.52	17.22	24.68	34.25	46.26	61.05	78.90
BULLET PATH (IN)	-.75	.52	.00	-2.64	-7.79	-15.91	-27.58	-43.44	-64.18	-90.63	-123.59	-163.86	-212.21
TIME OF FLIGHT (SEC)	.000000	.065356	.136525	.213816	.297972	.389771	.489541	.597689	.714605	.840292	.974273	1.115690	1.263738
WIND DEFLECTION (IN)													
5 MPH	.00	.25	1.01	2.32	4.22	6.80	10.08	14.10	18.89	24.45	30.74	37.68	45.21
10 MPH	.00	.50	2.03	4.63	8.44	13.60	20.16	28.19	37.77	48.89	61.47	75.36	90.42
20 MPH	.00	1.01	4.06	9.26	16.89	27.20	40.32	56.39	75.54	97.78	122.94	150.72	180.84
30 MPH	.00	1.51	6.09	13.89	25.33	40.80	60.48	84.58	113.31	146.67	184.42	226.08	271.25
VELOCITY (FPS)	2300.	2112.	1945.	1786.	1637.	1507.	1390.	1285.	1195.	1121.	1062.	1014.	975.
ENERGY (FT-LB)	1985.	1674.	1419.	1197.	1006.	852.	725.	620.	536.	471.	423.	386.	356.
DROP (IN)	.00	-.83	-3.60	-8.67	-16.46	-27.48	-42.36	-61.75	-86.40	-117.14	-154.77	-200.09	-253.83
MID-RANGE (IN)	.00	.22	.97	2.38	4.63	7.92	12.51	18.67	26.74	37.04	49.90	65.64	84.55
BULLET PATH (IN)	-.75	.60	.00	-2.89	-8.51	-17.35	-30.05	-47.26	-69.74	-98.30	-133.76	-176.90	-228.47
TIME OF FLIGHT (SEC)	.000000	.068156	.142169	.222652	.310376	.405969	.509673	.621977	.743107	.872841	1.010486	1.155150	1.306154
WIND DEFLECTION (IN)													
5 MPH	.00	.26	1.03	2.38	4.36	7.03	10.42	14.56	19.48	25.16	31.53	38.52	46.07
10 MPH	.00	.52	2.07	4.75	8.71	14.06	20.83	29.12	38.96	50.32	63.06	77.05	92.14
20 MPH	.00	1.03	4.13	9.50	17.43	28.12	41.67	58.24	77.92	100.63	126.13	154.09	184.29
30 MPH	.00	1.55	6.20	14.26	26.14	42.18	62.50	87.36	116.88	150.95	189.19	231.14	276.43
VELOCITY (FPS)	2200.	2029.	1866.	1712.	1570.	1448.	1337.	1239.	1157.	1090.	1037.	994.	957.
ENERGY (FT-LB)	1816.	1544.	1306.	1099.	925.	786.	670.	576.	502.	446.	404.	371.	344.
DROP (IN)	.00	-.90	-3.91	-9.42	-17.88	-29.88	-46.04	-67.06	-93.77	-126.95	-167.43	-215.96	-273.26
MID-RANGE (IN)	.00	.24	1.06	2.59	5.03	8.61	13.60	20.28	29.00	40.08	53.84	70.58	90.59
BULLET PATH (IN)	-.75	.68	.00	-3.18	-9.31	-18.98	-32.81	-51.50	-75.88	-106.74	-144.88	-191.08	-246.05
TIME OF FLIGHT (SEC)	.000000	.071004	.148108	.232057	.323624	.423142	.531023	.647663	.773082	.906816	1.048013	1.195863	1.349800
WIND DEFLECTION (IN)													
5 MPH	.00	.25	1.03	2.42	4.48	7.24	10.73	14.99	20.03	25.80	32.23	39.24	46.78
10 MPH	.00	.50	2.07	4.84	8.96	14.47	21.46	29.99	40.06	51.60	64.45	78.47	93.56
20 MPH	.00	.99	4.13	9.68	17.92	28.95	42.92	59.98	80.12	103.20	128.90	156.94	187.13
30 MPH	.00	1.49	6.20	14.53	26.87	43.42	64.38	89.97	120.19	154.80	193.35	235.42	280.69

Bullet: Lyman # 311291 169 Gr.
Ballistic Coefficient: .202
(Cont'd.)

RANGE (YDS)	MUZZLE	50	100	150	200	250	300	350	400	450	500	550	600
VELOCITY (FPS)	2100.	1933.	1775.	1627.	1497.	1381.	1278.	1189.	1116.	1058.	1011.	972.	937.
ENERGY (FT-LB)	1655.	1403.	1183.	993.	841.	716.	613.	530.	467.	420.	384.	354.	330.
DROP (IN)	.00	-.99	-4.30	-10.37	-19.71	-32.96	-50.76	-73.90	-103.19	-139.43	-183.41	-235.87	-297.47
MID-RANGE (IN)	.00	.27	1.16	2.85	5.55	9.52	15.01	22.35	31.88	43.90	58.74	76.69	97.97
BULLET PATH (IN)	-.75	.79	.00	-3.54	-10.35	-21.08	-36.35	-56.96	-83.72	-117.44	-158.89	-208.83	-267.90
TIME OF FLIGHT (SEC)	.000000	.074446	.155413	.243680	.339869	.444213	.557188	.678989	.809356	.947566	1.092734	1.244198	1.401502
WIND DEFLECTION (IN)													
5 MPH	.00	.27	1.10	2.59	4.77	7.66	11.32	15.75	20.94	26.81	33.30	40.35	47.90
10 MPH	.00	.53	2.21	5.17	9.53	15.32	22.64	31.50	41.88	53.63	66.61	80.69	95.81
20 MPH	.00	1.06	4.42	10.35	19.06	30.65	45.27	63.00	83.75	107.26	133.21	161.39	191.61
30 MPH	.00	1.59	6.63	15.52	28.59	45.97	67.91	94.51	125.63	160.89	199.82	242.08	287.42
VELOCITY (FPS)	2000.	1838.	1686.	1548.	1427.	1318.	1223.	1144.	1080.	1029.	987.	951.	919.
ENERGY (FT-LB)	1501.	1268.	1067.	899.	764.	652.	561.	491.	437.	397.	365.	339.	317.
DROP (IN)	.00	-1.09	-4.76	-11.47	-21.83	-36.47	-56.11	-81.58	-113.68	-153.20	-200.91	-257.52	-323.59
MID-RANGE (IN)	.00	.29	1.28	3.16	6.16	10.53	16.58	24.64	35.01	48.03	63.98	83.15	105.68
BULLET PATH (IN)	-.75	.91	.00	-3.96	-11.56	-23.45	-40.34	-63.05	-92.39	-129.16	-174.12	-227.97	-291.29
TIME OF FLIGHT (SEC)	.000000	.078232	.163442	.256394	.357373	.466804	.585031	.711990	.847130	.989572	1.138547	1.293527	1.454146
WIND DEFLECTION (IN)													
5 MPH	.00	.28	1.18	2.76	5.05	8.08	11.88	16.46	21.75	27.68	34.19	41.23	48.76
10 MPH	.00	.57	2.37	5.53	10.10	16.16	23.77	32.91	43.49	55.36	68.38	82.46	97.53
20 MPH	.00	1.14	4.73	11.05	20.20	32.32	47.53	65.82	86.99	110.73	136.77	164.92	195.06
30 MPH	.00	1.71	7.10	16.58	30.29	48.47	71.30	98.73	130.48	166.09	205.15	247.38	292.59
VELOCITY (FPS)	1900.	1744.	1598.	1473.	1359.	1259.	1173.	1103.	1048.	1003.	964.	931.	901.
ENERGY (FT-LB)	1354.	1141.	958.	814.	693.	594.	516.	457.	412.	377.	349.	325.	305.
DROP (IN)	.00	-1.22	-5.29	-12.75	-24.25	-40.45	-62.15	-90.16	-125.29	-168.31	-219.96	-280.85	-351.68
MID-RANGE (IN)	.00	.33	1.43	3.51	6.84	11.67	18.33	27.12	38.40	52.44	69.53	89.87	113.69
BULLET PATH (IN)	-.75	1.05	.00	-4.44	-12.93	-26.11	-44.79	-69.79	-101.90	-141.90	-190.53	-248.41	-316.22
TIME OF FLIGHT (SEC)	.000000	.082414	.172292	.270097	.376148	.490899	.614465	.746483	.886161	1.032641	1.185310	1.343746	1.507656
WIND DEFLECTION (IN)													
5 MPH	.00	.31	1.27	2.93	5.31	8.46	12.39	17.06	22.40	28.35	34.83	41.83	49.31
10 MPH	.00	.61	2.53	5.85	10.62	16.92	24.78	34.12	44.81	56.69	69.67	83.66	98.61
20 MPH	.00	1.22	5.07	11.71	21.25	33.85	49.55	68.24	89.61	113.38	139.33	167.31	197.22
30 MPH	.00	1.83	7.60	17.56	31.87	50.77	74.33	102.35	134.42	170.08	209.00	250.97	295.83
VELOCITY (FPS)	1800.	1650.	1517.	1400.	1294.	1203.	1127.	1067.	1018.	978.	943.	912.	884.
ENERGY (FT-LB)	1216.	1022.	864.	735.	628.	543.	476.	427.	389.	359.	333.	312.	293.
DROP (IN)	.00	-1.36	-5.89	-14.23	-27.01	-44.99	-68.99	-99.81	-138.24	-185.04	-240.88	-306.36	-382.27
MID-RANGE (IN)	.00	.36	1.59	3.92	7.61	12.95	20.26	29.88	42.11	57.22	75.45	96.99	122.14
BULLET PATH (IN)	-.75	1.22	.00	-5.02	-14.48	-29.13	-49.81	-77.31	-112.42	-155.89	-208.41	-270.57	-343.15
TIME OF FLIGHT (SEC)	.000000	.087052	.181943	.284909	.396438	.516791	.645785	.782764	.926836	1.077294	1.233655	1.395585	1.562855
WIND DEFLECTION (IN)													
5 MPH	.00	.33	1.34	3.07	5.55	8.81	12.83	17.55	22.89	28.80	35.23	42.14	49.53
10 MPH	.00	.65	2.69	6.14	11.11	17.62	25.66	35.10	45.79	57.60	70.46	84.29	99.06
20 MPH	.00	1.31	5.38	12.29	22.21	35.24	51.32	70.20	91.58	115.21	140.91	168.58	198.12
30 MPH	.00	1.96	8.07	18.43	33.32	52.87	76.97	105.30	137.37	172.81	211.37	252.87	297.19

Bullet: Lyman # 311291 169 Gr.
Ballistic Coefficient: .202
(Cont'd.)

RANGE (YDS)	MUZZLE	50	100	150	200	250	300	350	400	450	500	550	600
VELOCITY (FPS)	1700.	1559.	1438.	1328.	1231.	1150.	1085.	1033.	991.	954.	922.	893.	867.
ENERGY (FT-LB)	1084.	912.	776.	662.	569.	497.	442.	401.	368.	341.	319.	299.	282.
DROP (IN)	.00	-1.52	-6.62	-15.95	-30.20	-50.20	-76.75	-110.66	-152.68	-203.53	-263.80	-334.23	-415.55
MID-RANGE (IN)	.00	.41	1.79	4.40	8.48	14.38	22.43	32.92	46.14	62.33	81.70	104.51	131.03
BULLET PATH (IN)	-.75	1.41	.00	-5.64	-16.20	-32.52	-55.39	-85.61	-123.94	-171.11	-227.69	-294.44	-372.07
TIME OF FLIGHT (SEC)	.000000	.092202	.192410	.301025	.418417	.544566	.678968	.820754	.969134	1.123559	1.283651	1.449149	1.619880
WIND DEFLECTION (IN)													
5 MPH	.00	.35	1.40	3.20	5.76	9.10	13.16	17.87	23.17	28.99	35.31	42.11	49.37
10 MPH	.00	.70	2.81	6.39	11.52	18.20	26.32	35.75	46.33	57.98	70.63	84.23	98.75
20 MPH	.00	1.40	5.61	12.78	23.05	36.39	52.64	71.49	92.66	115.96	141.26	168.45	197.49
30 MPH	.00	2.09	8.42	19.18	34.57	54.59	78.97	107.24	139.00	173.95	211.89	252.68	296.24
VELOCITY (FPS)	1600.	1475.	1361.	1260.	1174.	1104.	1049.	1003.	965.	932.	902.	875.	850.
ENERGY (FT-LB)	960.	816.	695.	596.	517.	457.	412.	378.	349.	326.	305.	287.	271.
DROP (IN)	.00	-1.71	-7.45	-17.88	-33.79	-56.00	-85.32	-122.51	-168.32	-223.36	-288.33	-363.96	-450.93
MID-RANGE (IN)	.00	.46	2.01	4.90	9.44	15.97	24.78	36.16	50.37	67.62	88.16	112.26	140.14
BULLET PATH (IN)	-.75	1.64	.00	-6.32	-18.14	-36.25	-61.46	-94.56	-136.26	-187.21	-248.07	-319.60	-402.47
TIME OF FLIGHT (SEC)	.000000	.097671	.203581	.318184	.441602	.573483	.713037	.859405	1.011969	1.170303	1.334115	1.503206	1.677451
WIND DEFLECTION (IN)													
5 MPH	.00	.35	1.42	3.25	5.86	9.22	13.25	17.88	23.05	28.74	34.90	41.53	48.62
10 MPH	.00	.69	2.83	6.50	11.72	18.43	26.49	35.76	46.11	57.47	69.80	83.06	97.23
20 MPH	.00	1.38	5.66	13.00	23.44	36.87	52.99	71.51	92.21	114.95	139.61	166.13	194.46
30 MPH	.00	2.07	8.49	19.50	35.17	55.30	79.48	107.27	138.32	172.42	209.41	249.19	291.69
VELOCITY (FPS)	1500.	1384.	1280.	1191.	1117.	1059.	1012.	972.	938.	908.	880.	855.	831.
ENERGY (FT-LB)	844.	718.	615.	532.	468.	421.	384.	355.	330.	309.	291.	274.	259.
DROP (IN)	.00	-1.98	-8.50	-20.34	-38.31	-63.22	-95.85	-136.94	-187.15	-247.12	-317.61	-399.30	-492.87
MID-RANGE (IN)	.00	.52	2.27	5.54	10.65	17.90	27.58	39.96	55.26	73.71	95.59	121.13	150.59
BULLET PATH (IN)	-.75	1.89	.00	-7.21	-20.55	-40.83	-68.84	-105.30	-150.88	-206.23	-272.09	-349.15	-438.09
TIME OF FLIGHT (SEC)	.000000	.104158	.216936	.338541	.468721	.606761	.751774	.903092	1.060256	1.222948	1.390953	1.564132	1.742408
WIND DEFLECTION (IN)													
5 MPH	.00	.37	1.49	3.39	6.05	9.39	13.36	17.87	22.90	28.42	34.40	40.84	47.73
10 MPH	.00	.73	2.98	6.78	12.09	18.79	26.71	35.74	45.81	56.84	68.81	81.69	95.46
20 MPH	.00	1.46	5.96	13.57	24.19	37.58	53.42	71.49	91.61	113.68	137.62	163.37	190.93
30 MPH	.00	2.20	8.94	20.35	36.28	56.37	80.14	107.23	137.42	170.52	206.42	245.06	286.39
VELOCITY (FPS)	1400.	1294.	1203.	1127.	1067.	1018.	978.	943.	912.	884.	858.	834.	812.
ENERGY (FT-LB)	735.	628.	543.	477.	427.	389.	359.	334.	312.	293.	276.	261.	247.
DROP (IN)	.00	-2.28	-9.75	-23.24	-43.55	-71.46	-107.74	-153.06	-208.02	-273.38	-349.86	-438.11	-538.84
MID-RANGE (IN)	.00	.60	2.60	6.31	12.02	20.04	30.63	44.04	60.46	80.19	103.47	130.54	161.68
BULLET PATH (IN)	-.75	2.22	.00	-8.24	-23.29	-45.96	-76.98	-117.05	-166.76	-226.88	-298.10	-381.10	-476.57
TIME OF FLIGHT (SEC)	.000000	.111496	.231815	.360775	.497723	.641763	.792190	.948518	1.110414	1.277647	1.450070	1.627596	1.810192
WIND DEFLECTION (IN)													
5 MPH	.00	.38	1.54	3.46	6.09	9.33	13.14	17.47	22.29	27.58	33.32	39.51	46.15
10 MPH	.00	.77	3.09	6.93	12.17	18.66	26.28	34.94	44.58	55.15	66.64	79.03	92.31
20 MPH	.00	1.53	6.17	13.85	24.34	37.33	52.57	69.88	89.15	110.30	133.28	158.06	184.62
30 MPH	.00	2.30	9.26	20.78	36.51	55.99	78.85	104.82	133.73	165.45	199.92	237.09	276.92

Bullet: Lyman # 31141 170 Gr.
Ballistic Coefficient: .220

RANGE (YDS)	MUZZLE	50	100	150	200	250	300	350	400	450	500	550	600
VELOCITY (FPS)	2500.	2307.	2131.	1976.	1827.	1687.	1559.	1450.	1350.	1261.	1183.	1118.	1065.
ENERGY (FT-LB)	2359.	2009.	1714.	1473.	1260.	1074.	918.	794.	688.	600.	528.	472.	428.
DROP (IN)	.00	-.70	-3.02	-7.25	-13.71	-22.76	-34.87	-50.55	-70.38	-94.98	-125.06	-161.35	-204.57
MID-RANGE (IN)	.00	.19	.82	1.99	3.83	6.50	10.18	15.09	21.48	29.60	39.77	52.27	67.42
BULLET PATH (IN)	-.75	.44	.00	-2.35	-6.92	-14.08	-24.30	-38.09	-56.04	-78.75	-106.95	-141.35	-182.69
TIME OF FLIGHT (SEC)	.000000	.062463	.130196	.203301	.282256	.367710	.460307	.560084	.667321	.782355	.905270	1.035806	1.173377
WIND DEFLECTION (IN)													
5 MPH	.00	.22	.90	2.05	3.72	5.96	8.83	12.33	16.48	21.33	26.86	33.07	39.90
10 MPH	.00	.43	1.79	4.10	7.44	11.92	17.65	24.65	32.97	42.65	53.73	66.14	79.79
20 MPH	.00	.87	3.59	8.20	14.87	23.83	35.31	49.31	65.94	85.31	107.45	132.28	159.59
30 MPH	.00	1.30	5.38	12.30	22.31	35.75	52.96	73.96	98.91	127.96	161.18	198.43	239.38
VELOCITY (FPS)	2400.	2211.	2050.	1898.	1754.	1618.	1502.	1397.	1303.	1219.	1148.	1090.	1042.
ENERGY (FT-LB)	2174.	1846.	1587.	1360.	1161.	988.	851.	737.	641.	561.	497.	448.	410.
DROP (IN)	.00	-.76	-3.28	-7.87	-14.86	-24.68	-37.80	-54.80	-76.26	-102.84	-135.28	-174.30	-220.62
MID-RANGE (IN)	.00	.20	.88	2.15	4.15	7.04	11.03	16.36	23.27	32.03	42.96	56.36	72.51
BULLET PATH (IN)	-.75	.51	.00	-2.57	-7.55	-15.35	-26.45	-41.44	-60.88	-85.45	-115.88	-152.88	-197.18
TIME OF FLIGHT (SEC)	.000000	.065113	.135590	.211623	.293834	.382892	.479204	.582775	.694002	.813109	.940007	1.074260	1.215169
WIND DEFLECTION (IN)													
5 MPH	.00	.23	.93	2.12	3.86	6.19	9.17	12.78	17.07	22.05	27.72	34.03	40.93
10 MPH	.00	.46	1.86	4.25	7.71	12.39	18.34	25.57	34.14	44.11	55.44	68.07	81.87
20 MPH	.00	.92	3.73	8.49	15.43	24.78	36.68	51.14	68.29	88.21	110.88	136.14	163.74
30 MPH	.00	1.38	5.59	12.74	23.14	37.17	55.02	76.71	102.43	132.32	166.32	204.21	245.61
VELOCITY (FPS)	2300.	2125.	1969.	1821.	1681.	1554.	1445.	1346.	1257.	1180.	1115.	1063.	1020.
ENERGY (FT-LB)	1997.	1704.	1464.	1252.	1067.	911.	788.	684.	596.	525.	470.	427.	393.
DROP (IN)	.00	-.82	-3.57	-8.55	-16.14	-26.81	-41.08	-59.53	-82.77	-111.54	-146.54	-188.50	-238.14
MID-RANGE (IN)	.00	.22	.96	2.34	4.50	7.65	11.99	17.77	25.24	34.71	46.46	60.80	77.99
BULLET PATH (IN)	-.75	.58	.00	-2.82	-8.26	-16.76	-28.88	-45.17	-66.25	-92.86	-125.70	-165.51	-212.99
TIME OF FLIGHT (SEC)	.000000	.067941	.141279	.220491	.306232	.399142	.499259	.606854	.722257	.845536	.976415	1.114296	1.258459
WIND DEFLECTION (IN)													
5 MPH	.00	.24	.95	2.19	3.99	6.43	9.50	13.23	17.65	22.75	28.53	34.93	41.87
10 MPH	.00	.48	1.91	4.37	7.98	12.86	19.00	26.46	35.29	45.51	57.07	69.86	83.75
20 MPH	.00	.96	3.82	8.74	15.97	25.72	38.00	52.92	70.58	91.02	114.13	139.71	167.50
30 MPH	.00	1.44	5.73	13.12	23.95	38.57	57.00	79.38	105.87	136.53	171.20	209.57	251.25
VELOCITY (FPS)	2200.	2041.	1890.	1746.	1610.	1496.	1392.	1298.	1215.	1145.	1087.	1040.	1000.
ENERGY (FT-LB)	1827.	1573.	1348.	1151.	979.	845.	732.	636.	557.	495.	446.	408.	378.
DROP (IN)	.00	-.89	-3.87	-9.27	-17.52	-29.11	-44.61	-64.59	-89.74	-120.79	-158.44	-203.44	-256.45
MID-RANGE (IN)	.00	.24	1.04	2.54	4.89	8.31	13.03	19.28	27.35	37.53	50.12	65.38	83.61
BULLET PATH (IN)	-.75	.67	.00	-3.09	-9.03	-18.31	-31.50	-49.18	-72.01	-100.75	-136.10	-178.78	-229.48
TIME OF FLIGHT (SEC)	.000000	.070782	.147150	.229733	.319203	.415891	.519851	.631486	.751003	.878296	1.012913	1.154146	1.301347
WIND DEFLECTION (IN)													
5 MPH	.00	.23	.95	2.22	4.09	6.60	9.75	13.57	18.09	23.29	29.14	35.56	42.52
10 MPH	.00	.46	1.90	4.43	8.18	13.20	19.49	27.14	36.18	46.58	58.27	71.13	85.04
20 MPH	.00	.92	3.80	8.87	16.36	26.39	38.99	54.28	72.35	93.16	116.55	142.26	170.07
30 MPH	.00	1.37	5.70	13.30	24.54	39.59	58.48	81.42	108.53	139.74	174.82	213.39	255.11

Bullet: Lyman # 31141 170 Gr.
Ballistic Coefficient: .220
(Cont'd.)

RANGE (YDS)	MUZZLE	50	100	150	200	250	300	350	400	450	500	550	600
VELOCITY (FPS)	2100.	1946.	1799.	1660.	1537.	1430.	1332.	1245.	1169.	1107.	1056.	1014.	978.
ENERGY (FT-LB)	1664.	1429.	1221.	1040.	892.	772.	669.	585.	516.	462.	421.	388.	361.
DROP (IN)	.00	-.98	-4.26	-10.21	-19.31	-32.10	-49.16	-71.11	-98.70	-132.63	-173.63	-222.40	-279.60
MID-RANGE (IN)	.00	.26	1.15	2.80	5.40	9.18	14.37	21.22	30.04	41.12	54.72	71.14	90.64
BULLET PATH (IN)	-.75	.77	.00	-3.45	-10.05	-20.33	-34.89	-54.34	-79.42	-110.85	-149.35	-195.62	-250.31
TIME OF FLIGHT (SEC)	.000000	.074208	.154389	.241202	.335222	.436429	.545174	.661756	.786196	.918165	1.057025	1.202071	1.352773
WIND DEFLECTION (IN)													
5 MPH	.00	.24	1.01	2.37	4.36	6.98	10.26	14.23	18.90	24.23	30.16	36.64	43.62
10 MPH	.00	.49	2.03	4.74	8.71	13.95	20.52	28.47	37.80	48.45	60.32	73.28	87.23
20 MPH	.00	.98	4.06	9.47	17.43	27.91	41.04	56.94	75.60	96.91	120.64	146.56	174.46
30 MPH	.00	1.47	6.09	14.21	26.14	41.86	61.57	85.41	113.40	145.36	180.97	219.84	261.69
VELOCITY (FPS)	2000.	1850.	1709.	1577.	1467.	1365.	1274.	1194.	1127.	1073.	1028.	990.	958.
ENERGY (FT-LB)	1510.	1292.	1102.	939.	812.	703.	612.	538.	480.	434.	399.	370.	346.
DROP (IN)	.00	-1.08	-4.70	-11.29	-21.37	-35.52	-5].32	-78.50	-108.77	-145.85	-190.47	-243.28	-304.93
MID-RANGE (IN)	.00	.29	1.27	3.09	5.98	10.16	15.87	23.39	33.01	45.03	59.72	77.35	98.15
BULLET PATH (IN)	-.75	.89	.00	-3.86	-11.21	-22.63	-38.71	-60.16	-87.71	-122.07	-163.95	-214.04	-272.97
TIME OF FLIGHT (SEC)	.000000	.077978	.162348	.253784	.352443	.458495	.572304	.694001	.823383	.959907	1.102848	1.251611	1.405769
WIND DEFLECTION (IN)													
5 MPH	.00	.26	1.09	2.53	4.61	7.35	10.76	14.87	19.66	25.07	31.05	37.54	44.51
10 MPH	.00	.52	2.17	5.07	9.23	14.70	21.53	29.74	39.32	50.14	62.10	75.08	89.02
20 MPH	.00	1.05	4.35	10.13	18.46	29.39	43.05	59.49	78.63	100.29	124.20	150.17	178.03
30 MPH	.00	1.57	6.52	15.20	27.69	44.09	64.58	89.23	117.95	150.43	186.30	225.25	267.05
VELOCITY (FPS)	1900.	1755.	1619.	1503.	1399.	1304.	1220.	1149.	1090.	1043.	1003.	968.	938.
ENERGY (FT-LB)	1362.	1163.	990.	853.	738.	642.	562.	498.	449.	410.	379.	354.	332.
DROP (IN)	.00	-1.20	-5.22	-12.54	-23.74	-39.38	-60.13	-86.74	-119.92	-160.39	-208.83	-265.89	-332.15
MID-RANGE (IN)	.00	.32	1.41	3.44	6.65	11.25	17.53	25.75	36.22	49.23	65.04	83.88	105.94
BULLET PATH (IN)	-.75	1.03	.00	-4.33	-12.54	-25.20	-42.97	-66.59	-96.78	-134.26	-179.72	-233.80	-297.07
TIME OF FLIGHT (SEC)	.000000	.082142	.171123	.267352	.370833	.481967	.600979	.727784	.861950	1.002778	1.149609	1.301965	1.459504
WIND DEFLECTION (IN)													
5 MPH	.00	.28	1.16	2.68	4.84	7.68	11.20	15.41	20.27	25.72	31.69	38.15	45.07
10 MPH	.00	.56	2.33	5.37	9.69	15.35	22.40	30.83	40.55	51.44	63.38	76.30	90.14
20 MPH	.00	1.12	4.66	10.74	19.38	30.71	44.81	61.65	81.09	102.87	126.77	152.61	180.27
30 MPH	.00	1.69	6.98	16.11	29.06	46.06	67.21	92.48	121.64	154.31	190.15	228.91	270.41
VELOCITY (FPS)	1800.	1661.	1538.	1431.	1333.	1245.	1170.	1107.	1057.	1015.	979.	947.	919.
ENERGY (FT-LB)	1223.	1041.	893.	773.	670.	585.	517.	463.	421.	389.	362.	339.	319.
DROP (IN)	.00	-1.34	-5.82	-13.99	-26.43	-43.74	-66.70	-95.98	-132.33	-176.45	-228.98	-290.56	-361.67
MID-RANGE (IN)	.00	.36	1.57	3.84	7.39	12.47	19.36	28.35	39.74	53.79	70.75	90.81	114.14
BULLET PATH (IN)	-.75	1.20	.00	-4.88	-14.03	-28.06	-47.72	-73.72	-106.78	-147.61	-196.85	-255.15	-322.97
TIME OF FLIGHT (SEC)	.000000	.086760	.180721	.281864	.390543	.507056	.631428	.763332	.902134	1.047125	1.197774	1.353702	1.514636
WIND DEFLECTION (IN)													
5 MPH	.00	.30	1.24	2.80	5.03	7.95	11.57	15.84	20.72	26.15	32.07	38.46	45.29
10 MPH	.00	.60	2.47	5.61	10.07	15.91	23.13	31.68	41.44	52.29	64.14	76.92	90.58
20 MPH	.00	1.21	4.95	11.22	20.14	31.82	46.26	63.36	82.88	104.59	128.28	153.84	181.15
30 MPH	.00	1.81	7.42	16.82	30.21	47.73	69.39	95.04	124.33	156.88	192.42	230.75	271.73

Bullet: Lyman # 31141 170 Gr.
Ballistic Coefficient: .220

(Cont'd.)

RANGE (YDS)	MUZZLE	50	100	150	200	250	300	350	400	450	500	550	600
VELOCITY (FPS)	1700.	1570.	1460.	1359.	1269.	1190.	1124.	1070.	1026.	988.	956.	927.	900.
ENERGY (FT-LB)	1091.	931.	805.	698.	608.	535.	477.	432.	397.	369.	345.	324.	306.
DROP (IN)	.00	-1.51	-6.53	-15.66	-29.49	-48.73	-74.11	-106.34	-146.14	-194.18	-251.10	-317.39	-393.74
MID-RANGE (IN)	.00	.41	1.76	4.29	8.21	13.82	21.39	31.22	43.58	58.71	76.82	98.09	122.76
BULLET PATH (IN)	-.75	1.38	.00	-5.49	-15.68	-31.28	-53.01	-81.61	-117.77	-162.17	-215.44	-278.09	-350.80
TIME OF FLIGHT (SEC)	.000000	.091890	.190967	.297462	.411731	.533884	.663698	.800613	.943904	1.092985	1.247441	1.406970	1.571358
WIND DEFLECTION (IN)													
5 MPH	.00	.32	1.28	2.88	5.17	8.16	11.82	16.10	20.95	26.30	32.13	38.40	45.10
10 MPH	.00	.64	2.55	5.77	10.35	16.32	23.63	32.20	41.89	52.60	64.26	76.80	90.21
20 MPH	.00	1.29	5.10	11.53	20.69	32.63	47.27	64.40	83.78	105.20	128.51	153.61	180.41
30 MPH	.00	1.93	7.65	17.30	31.04	48.95	70.90	96.61	125.68	157.80	192.77	230.41	270.62
VELOCITY (FPS)	1600.	1488.	1385.	1291.	1209.	1140.	1083.	1036.	997.	964.	934.	907.	882.
ENERGY (FT-LB)	966.	835.	723.	629.	552.	490.	442.	405.	376.	351.	329.	310.	294.
DROP (IN)	.00	-1.69	-7.34	-17.52	-32.93	-54.28	-82.31	-117.73	-161.21	-213.41	-274.86	-346.19	-428.04
MID-RANGE (IN)	.00	.45	1.98	4.77	9.12	15.32	23.63	34.34	47.68	63.89	83.15	105.68	131.71
BULLET PATH (IN)	-.75	1.61	.00	-6.14	-17.50	-34.81	-58.79	-90.16	-129.60	-177.75	-235.16	-302.44	-380.25
TIME OF FLIGHT (SEC)	.000000	.097241	.201786	.314029	.434158	.562039	.697193	.838904	.986538	1.139639	1.297882	1.461031	1.628920
WIND DEFLECTION (IN)													
5 MPH	.00	.31	1.26	2.88	5.21	8.21	11.85	16.07	20.82	26.04	31.71	37.82	44.34
10 MPH	.00	.61	2.51	5.77	10.41	16.42	23.71	32.15	41.63	52.08	63.43	75.64	88.69
20 MPH	.00	1.23	5.03	11.54	20.82	32.84	47.41	64.29	83.26	104.15	126.85	151.28	177.38
30 MPH	.00	1.84	7.54	17.31	31.24	49.26	71.12	96.44	124.89	156.23	190.28	226.92	266.07
VELOCITY (FPS)	1500.	1396.	1301.	1218.	1147.	1089.	1041.	1002.	967.	937.	910.	885.	862.
ENERGY (FT-LB)	849.	735.	639.	560.	496.	447.	409.	379.	353.	331.	312.	296.	280.
DROP (IN)	.00	-1.96	-8.39	-19.95	-37.39	-61.42	-92.76	-132.09	-180.06	-237.24	-304.21	-381.63	-470.14
MID-RANGE (IN)	.00	.52	2.23	5.40	10.31	17.23	26.43	38.15	52.64	70.08	90.68	114.68	142.31
BULLET PATH (IN)	-.75	1.86	.00	-6.99	-19.86	-39.32	-66.09	-100.85	-144.26	-196.87	-259.26	-332.11	-416.05
TIME OF FLIGHT (SEC)	.000000	.103704	.215070	.334316	.461346	.595718	.736728	.883721	1.036225	1.193901	1.356505	1.523864	1.695862
WIND DEFLECTION (IN)													
5 MPH	.00	.33	1.33	3.02	5.40	8.42	12.03	16.17	20.79	25.86	31.37	37.30	43.64
10 MPH	.00	.65	2.65	6.04	10.80	16.85	24.06	32.33	41.58	51.73	62.74	74.60	87.27
20 MPH	.00	1.30	5.30	12.08	21.59	33.69	48.13	64.67	83.15	103.45	125.49	149.20	174.54
30 MPH	.00	1.96	7.96	18.12	32.39	50.54	72.19	97.00	124.73	155.18	188.23	223.80	261.82
VELOCITY (FPS)	1400.	1305.	1221.	1150.	1091.	1043.	1003.	969.	938.	911.	886.	863.	841.
ENERGY (FT-LB)	740.	643.	563.	499.	449.	411.	380.	354.	332.	313.	296.	281.	267.
DROP (IN)	.00	-2.26	-9.62	-22.83	-42.59	-69.64	-104.65	-148.27	-201.07	-263.63	-336.61	-420.66	-516.38
MID-RANGE (IN)	.00	.59	2.55	6.16	11.68	19.37	29.49	42.29	57.94	76.65	98.67	124.23	153.54
BULLET PATH (IN)	-.75	2.18	.00	-8.02	-22.60	-44.46	-74.28	-112.72	-160.33	-217.70	-285.51	-364.37	-454.90
TIME OF FLIGHT (SEC)	.000000	.111025	.229926	.356622	.490685	.631419	.778159	.930427	1.087878	1.250265	1.417412	1.589202	1.765561
WIND DEFLECTION (IN)													
5 MPH	.00	.34	1.38	3.10	5.47	8.42	11.91	15.88	20.30	25.17	30.45	36.14	42.23
10 MPH	.00	.68	2.75	6.19	10.93	16.84	23.81	31.76	40.61	50.33	60.89	72.27	84.45
20 MPH	.00	1.37	5.51	12.39	21.86	33.69	47.63	63.51	81.22	100.66	121.79	144.54	168.91
30 MPH	.00	2.05	8.26	18.58	32.80	50.53	71.44	95.27	121.83	151.00	182.68	216.81	253.36

Bullet: Lyman # 311407 173 Gr.
Ballistic Coefficient: .270

RANGE (YDS)	MUZZLE	50	100	150	200	250	300	350	400	450	500	550	600
VELOCITY (FPS)	2000.	1861.	1736.	1623.	1516.	1417.	1332.	1256.	1189.	1131.	1083.	1043.	1008.
ENERGY (FT-LB)	1536.	1331.	1157.	1011.	883.	771.	681.	606.	543.	492.	450.	417.	390.
DROP (IN)	.00	-1.08	-4.65	-11.10	-20.85	-34.41	-52.29	-75.06	-103.37	-137.83	-179.06	-227.68	-284.27
MID-RANGE (IN)	.00	.29	1.25	3.02	5.77	9.71	15.04	21.99	30.84	41.80	55.12	71.07	89.85
BULLET PATH (IN)	-.75	.87	.00	-3.75	-10.79	-21.65	-36.83	-56.90	-82.51	-114.26	-152.79	-198.70	-252.60
TIME OF FLIGHT (SEC)	.000000	.077750	.161293	.250692	.346347	.448705	.557984	.674036	.796878	.926322	1.061961	1.203251	1.349709
WIND DEFLECTION (IN)													
5 MPH	.00	.24	.99	2.26	4.08	6.49	9.50	13.12	17.33	22.12	27.45	33.29	39.57
10 MPH	.00	.48	1.99	4.52	8.16	12.97	19.01	26.23	34.65	44.23	54.91	66.57	79.15
20 MPH	.00	.97	3.98	9.04	16.31	25.94	38.01	52.46	69.30	88.47	109.81	133.14	158.30
30 MPH	.00	1.45	5.96	13.57	24.47	38.92	57.02	78.69	103.95	132.70	164.72	199.72	237.45
VELOCITY (FPS)	1900.	1768.	1653.	1545.	1444.	1355.	1276.	1207.	1147.	1096.	1053.	1017.	986.
ENERGY (FT-LB)	1387.	1201.	1050.	917.	801.	705.	626.	559.	505.	461.	426.	397.	373.
DROP (IN)	.00	-1.20	-5.15	-12.28	-23.07	-38.04	-57.74	-82.79	-113.81	-151.43	-196.26	-248.91	-309.94
MID-RANGE (IN)	.00	.32	1.38	3.34	6.38	10.73	16.59	24.22	33.83	45.71	60.09	77.19	97.23
BULLET PATH (IN)	-.75	1.01	.00	-4.18	-12.01	-24.04	-40.78	-62.88	-90.95	-125.61	-167.50	-217.20	-275.28
TIME OF FLIGHT (SEC)	.000000	.081888	.169626	.263489	.363935	.471292	.585423	.706357	.833968	.967909	1.107656	1.252694	1.402631
WIND DEFLECTION (IN)													
5 MPH	.00	.26	1.03	2.34	4.24	6.74	9.83	13.53	17.81	22.65	28.00	33.82	40.06
10 MPH	.00	.52	2.06	4.69	8.47	13.47	19.67	27.06	35.62	45.30	56.00	67.63	80.13
20 MPH	.00	1.04	4.13	9.38	16.95	26.95	39.33	54.11	71.24	90.60	112.00	135.26	160.25
30 MPH	.00	1.55	6.19	14.07	25.42	40.42	59.00	81.17	106.86	135.90	168.00	202.90	240.38
VELOCITY (FPS)	1800.	1684.	1574.	1470.	1377.	1297.	1225.	1162.	1108.	1064.	1026.	994.	965.
ENERGY (FT-LB)	1244.	1089.	951.	830.	728.	646.	576.	518.	472.	435.	405.	379.	358.
DROP (IN)	.00	-1.32	-5.70	-13.59	-25.53	-42.05	-63.73	-91.23	-125.16	-166.13	-214.76	-271.63	-337.27
MID-RANGE (IN)	.00	.36	1.53	3.70	7.07	11.86	18.28	26.59	37.05	49.89	65.34	83.62	104.91
BULLET PATH (IN)	-.75	1.15	.00	-4.67	-13.39	-26.68	-45.14	-69.42	-100.12	-137.88	-183.28	-236.92	-299.35
TIME OF FLIGHT (SEC)	.000000	.086173	.178342	.276975	.382475	.494780	.613882	.739718	.871998	1.010235	1.153888	1.302535	1.455858
WIND DEFLECTION (IN)													
5 MPH	.00	.25	1.03	2.37	4.32	6.87	10.02	13.76	18.07	22.90	28.21	33.96	40.12
10 MPH	.00	.50	2.05	4.75	8.65	13.75	20.04	27.52	36.14	45.80	56.42	67.91	80.23
20 MPH	.00	1.00	4.11	9.50	17.30	27.50	40.09	55.05	72.28	91.60	112.84	135.83	160.46
30 MPH	.00	1.50	6.16	14.24	25.95	41.24	60.13	82.57	108.41	137.40	169.25	203.74	240.69
VELOCITY (FPS)	1700.	1589.	1485.	1389.	1307.	1234.	1170.	1115.	1070.	1031.	998.	969.	942.
ENERGY (FT-LB)	1110.	970.	847.	741.	656.	585.	526.	478.	439.	408.	383.	360.	341.
DROP (IN)	.00	-1.48	-6.40	-15.30	-28.70	-47.17	-71.37	-101.91	-139.41	-184.48	-237.71	-299.64	-370.77
MID-RANGE (IN)	.00	.40	1.72	4.17	7.95	13.26	20.38	29.55	41.01	54.99	71.69	91.30	114.01
BULLET PATH (IN)	-.75	1.35	.00	-5.33	-15.14	-30.04	-50.67	-77.63	-111.55	-153.05	-202.70	-261.06	-328.61
TIME OF FLIGHT (SEC)	.000000	.091279	.188956	.293461	.404817	.522962	.647863	.779262	.916693	1.059611	1.207574	1.360252	1.517398
WIND DEFLECTION (IN)													
5 MPH	.00	.27	1.10	2.53	4.57	7.20	10.42	14.22	18.55	23.36	28.62	34.29	40.35
10 MPH	.00	.54	2.20	5.06	9.13	14.39	20.85	28.44	37.10	46.73	57.24	68.58	80.71
20 MPH	.00	1.07	4.39	10.12	18.26	28.79	41.69	56.89	74.21	93.45	114.48	137.16	161.42
30 MPH	.00	1.61	6.59	15.18	27.39	43.18	62.54	85.33	111.31	140.18	171.72	205.74	242.13

Bullet: Lyman # 311407 173 Gr.

Ballistic Coefficient: .270

(Cont'd.)

RANGE (YDS)	MUZZLE	50	100	150	200	250	300	350	400	450	500	550	600
VELOCITY (FPS)	1600.	1495.	1398.	1316.	1241.	1176.	1121.	1074.	1035.	1001.	972.	945.	921.
ENERGY (FT-LB)	983.	858.	751.	665.	592.	532.	482.	443.	411.	385.	363.	343.	326.
DROP (IN)	.00	-1.67	-7.28	-17.32	-32.36	-53.08	-80.07	-113.95	-155.34	-204.83	-262.96	-330.25	-407.12
MID-RANGE (IN)	.00	.45	1.97	4.70	8.90	14.83	22.72	32.80	45.31	60.43	78.40	99.38	123.49
BULLET PATH (IN)	-.75	1.59	.00	-6.02	-17.05	-33.76	-56.73	-86.59	-123.97	-169.44	-223.56	-286.84	-359.68
TIME OF FLIGHT (SEC)	.000000	.097003	.200795	.311439	.428864	.553059	.683789	.820607	.962966	1.110411	1.262600	1.419278	1.580254
WIND DEFLECTION (IN)													
5 MPH	.00	.29	1.17	2.66	4.74	7.42	10.67	14.46	18.74	23.47	28.61	34.15	40.06
10 MPH	.00	.57	2.34	5.31	9.48	14.84	21.35	28.93	37.48	46.93	57.22	68.29	80.12
20 MPH	.00	1.15	4.68	10.63	18.96	29.68	42.69	57.85	74.96	93.86	114.44	136.59	160.25
30 MPH	.00	1.72	7.02	15.94	28.44	44.52	64.04	86.78	112.45	140.80	171.65	204.88	240.37
VELOCITY (FPS)	1500.	1402.	1319.	1245.	1179.	1123.	1076.	1037.	1003.	973.	946.	922.	899.
ENERGY (FT-LB)	864.	755.	668.	595.	534.	484.	445.	413.	386.	364.	344.	326.	311.
DROP (IN)	.00	-1.95	-8.32	-19.66	-36.65	-59.87	-89.97	-127.54	-173.18	-227.45	-290.85	-363.80	-446.94
MID-RANGE (IN)	.00	.52	2.20	5.28	10.01	16.61	25.32	36.36	49.95	66.28	85.55	107.88	133.50
BULLET PATH (IN)	-.75	1.83	.00	-6.81	-19.27	-37.96	-63.52	-96.56	-137.67	-187.40	-246.27	-314.69	-393.30
TIME OF FLIGHT (SEC)	.000000	.103453	.213789	.330902	.454789	.585227	.721777	.863891	1.011108	1.163082	1.319555	1.480332	1.645267
WIND DEFLECTION (IN)													
5 MPH	.00	.30	1.21	2.72	4.82	7.50	10.72	14.42	18.58	23.15	28.12	33.47	39.18
10 MPH	.00	.61	2.43	5.44	9.64	15.00	21.43	28.84	37.15	46.30	56.24	66.94	78.37
20 MPH	.00	1.22	4.85	10.88	19.29	30.00	42.87	57.69	74.31	92.60	112.48	133.88	156.73
30 MPH	.00	1.82	7.28	16.32	28.93	45.00	64.30	86.53	111.46	138.91	168.72	200.82	235.10
VELOCITY (FPS)	1400.	1317.	1243.	1178.	1122.	1075.	1036.	1002.	972.	946.	921.	899.	878.
ENERGY (FT-LB)	753.	667.	594.	533.	483.	444.	412.	386.	363.	343.	326.	310.	296.
DROP (IN)	.00	-2.23	-9.47	-22.35	-41.49	-67.51	-101.03	-142.62	-192.84	-252.22	-321.15	-400.29	-490.15
MID-RANGE (IN)	.00	.59	2.50	5.99	11.28	18.59	28.16	40.21	54.93	72.51	93.06	116.85	144.05
BULLET PATH (IN)	-.75	2.12	.00	-7.78	-21.81	-42.72	-71.13	-107.61	-152.73	-207.00	-270.82	-344.85	-429.61
TIME OF FLIGHT (SEC)	.000000	.110482	.227743	.351775	.482349	.619024	.761251	.908570	1.060640	1.217203	1.378065	1.543082	1.712148
WIND DEFLECTION (IN)													
5 MPH	.00	.29	1.18	2.67	4.73	7.33	10.42	13.95	17.91	22.26	26.98	32.08	37.53
10 MPH	.00	.59	2.37	5.34	9.46	14.66	20.84	27.91	35.82	44.51	53.97	64.15	75.05
20 MPH	.00	1.18	4.74	10.68	18.93	29.33	41.67	55.82	71.63	89.03	107.94	128.31	150.10
30 MPH	.00	1.76	7.11	16.02	28.39	43.99	62.51	83.73	107.45	133.54	161.90	192.46	225.16
VELOCITY (FPS)	1300.	1228.	1164.	1111.	1066.	1028.	995.	966.	940.	916.	894.	874.	854.
ENERGY (FT-LB)	649.	579.	521.	474.	436.	406.	380.	358.	339.	322.	307.	293.	280.
DROP (IN)	.00	-2.59	-10.98	-25.75	-47.55	-76.98	-114.61	-161.00	-216.62	-281.95	-357.59	-444.09	-541.96
MID-RANGE (IN)	.00	.69	2.89	6.85	12.80	20.97	31.55	44.76	60.76	79.70	101.82	127.31	156.37
BULLET PATH (IN)	-.75	2.52	.00	-8.91	-24.85	-48.41	-80.18	-120.71	-170.47	-229.94	-299.72	-380.35	-472.36
TIME OF FLIGHT (SEC)	.000000	.118791	.244323	.376313	.514283	.657687	.806098	.959193	1.116731	1.278531	1.444457	1.614413	1.788331
WIND DEFLECTION (IN)													
5 MPH	.00	.30	1.19	2.65	4.64	7.11	10.01	13.33	17.04	21.13	25.57	30.38	35.53
10 MPH	.00	.60	2.39	5.31	9.28	14.21	20.03	26.66	34.08	42.25	51.15	60.75	71.05
20 MPH	.00	1.20	4.77	10.62	18.57	28.43	40.05	53.33	68.17	84.50	102.30	121.50	142.11
30 MPH	.00	1.80	7.16	15.92	27.85	42.64	60.08	79.99	102.25	126.76	153.44	182.26	213.16

Bullet: Lyman # 311467 178 Gr.
Ballistic Coefficient: .320

RANGE (YDS)	MUZZLE	50	100	150	200	250	300	350	400	450	500	550	600
VELOCITY (FPS)	2500.	2366.	2237.	2108.	1980.	1858.	1741.	1629.	1523.	1424.	1333.	1251.	1179.
ENERGY (FT-LB)	2470.	2213.	1977.	1755.	1550.	1364.	1198.	1049.	917.	801.	702.	618.	550.
DROP (IN)	.00	-.67	-2.89	-6.84	-12.75	-20.87	-31.50	-44.99	-61.74	-82.28	-107.11	-136.83	-172.13
MID-RANGE (IN)	.00	.18	.77	1.85	3.48	5.80	8.90	12.95	18.12	24.62	32.69	42.57	54.57
BULLET PATH (IN)	-.75	.40	.00	-2.13	-6.22	-12.52	-21.32	-32.99	-47.93	-66.65	-89.66	-117.55	-151.04
TIME OF FLIGHT (SEC)	.000000	.061673	.126873	.195935	.269356	.347554	.430962	.520035	.615255	.717143	.826090	.942355	1.065967
WIND DEFLECTION (IN)													
5 MPH	.00	.15	.60	1.40	2.58	4.18	6.24	8.80	11.90	15.59	19.90	24.85	30.45
10 MPH	.00	.29	1.21	2.80	5.17	8.37	12.49	17.61	23.80	31.18	39.79	49.69	60.89
20 MPH	.00	.59	2.42	5.61	10.33	16.74	24.98	35.21	47.61	62.35	79.58	99.39	121.78
30 MPH	.00	.88	3.63	8.41	15.50	25.11	37.47	52.82	71.41	93.53	119.38	149.08	182.67
VELOCITY (FPS)	2400.	2269.	2141.	2012.	1889.	1770.	1657.	1550.	1448.	1355.	1270.	1196.	1133.
ENERGY (FT-LB)	2276.	2035.	1811.	1600.	1410.	1238.	1085.	949.	829.	725.	638.	566.	507.
DROP (IN)	.00	-.73	-3.14	-7.45	-13.90	-22.78	-34.43	-49.23	-67.67	-90.28	-117.42	-150.36	-189.18
MID-RANGE (IN)	.00	.20	.84	2.01	3.81	6.34	9.76	14.23	19.94	27.13	36.03	46.95	60.16
BULLET PATH (IN)	-.75	.47	.00	-2.36	-6.86	-13.79	-23.50	-36.35	-52.85	-73.51	-98.90	-129.69	-166.56
TIME OF FLIGHT (SEC)	.000000	.064274	.132309	.204585	.281533	.363579	.451173	.544785	.644943	.752073	.866472	.988237	1.117187
WIND DEFLECTION (IN)													
5 MPH	.00	.16	.64	1.50	2.77	4.49	6.70	9.44	12.76	16.68	21.25	26.46	32.31
10 MPH	.00	.31	1.29	3.01	5.55	8.99	13.41	18.88	25.51	33.36	42.50	52.93	64.62
20 MPH	.00	.62	2.57	6.01	11.10	17.98	26.81	37.76	51.02	66.73	85.00	105.86	129.25
30 MPH	.00	.94	3.86	9.02	16.65	26.97	40.22	56.65	76.53	100.09	127.50	158.79	193.87
VELOCITY (FPS)	2300.	2172.	2043.	1918.	1798.	1684.	1575.	1472.	1376.	1290.	1213.	1147.	1092.
ENERGY (FT-LB)	2090.	1864.	1649.	1453.	1277.	1120.	980.	856.	749.	657.	581.	520.	472.
DROP (IN)	.00	-.80	-3.43	-8.15	-15.22	-24.97	-37.79	-54.11	-74.48	-99.43	-129.62	-165.73	-208.42
MID-RANGE (IN)	.00	.22	.92	2.20	4.18	6.97	10.75	15.69	22.02	29.98	39.84	51.89	66.42
BULLET PATH (IN)	-.75	.55	.00	-2.62	-7.60	-15.27	-25.99	-40.22	-58.50	-81.36	-109.46	-143.47	-184.08
TIME OF FLIGHT (SEC)	.000000	.067106	.138325	.214118	.294905	.381131	.473257	.571805	.677233	.789874	.909880	1.037142	1.171264
WIND DEFLECTION (IN)													
5 MPH	.00	.17	.69	1.63	3.00	4.84	7.21	10.14	13.68	17.86	22.68	28.14	34.20
10 MPH	.00	.33	1.39	3.25	5.99	9.69	14.42	20.29	27.37	35.71	45.36	56.28	68.40
20 MPH	.00	.66	2.78	6.50	11.98	19.38	28.85	40.58	54.73	71.43	90.71	112.55	136.81
30 MPH	.00	1.00	4.17	9.75	17.97	29.06	43.27	60.87	82.10	107.14	136.07	168.83	205.21
VELOCITY (FPS)	2200.	2070.	1944.	1823.	1707.	1598.	1493.	1396.	1307.	1228.	1160.	1103.	1056.
ENERGY (FT-LB)	1913.	1693.	1493.	1313.	1152.	1009.	881.	770.	675.	596.	532.	481.	441.
DROP (IN)	.00	-.87	-3.77	-8.97	-16.77	-27.55	-41.73	-59.85	-82.43	-110.09	-143.54	-183.43	-230.45
MID-RANGE (IN)	.00	.24	1.01	2.43	4.61	7.71	11.90	17.41	24.45	33.29	44.22	57.53	73.50
BULLET PATH (IN)	-.75	.64	.00	-2.93	-8.47	-16.99	-28.91	-44.77	-65.09	-90.49	-121.68	-159.31	-204.06
TIME OF FLIGHT (SEC)	.000000	.070299	.145088	.224779	.309814	.400649	.497799	.601746	.712847	.831301	.957057	1.089779	1.228887
WIND DEFLECTION (IN)													
5 MPH	.00	.19	.77	1.78	3.26	5.26	7.81	10.95	14.73	19.15	24.22	29.90	36.14
10 MPH	.00	.37	1.54	3.56	6.53	10.51	15.61	21.91	29.46	38.31	48.44	59.80	72.28
20 MPH	.00	.75	3.07	7.12	13.05	21.03	31.23	43.81	58.92	76.62	96.88	119.60	144.57
30 MPH	.00	1.12	4.61	10.68	19.58	31.54	46.84	65.72	88.38	114.93	145.33	179.40	216.85

Bullet: Lyman # 311467 178 Gr.

Ballistic Coefficient: .320
(Cont'd.)

RANGE (YDS)	MUZZLE	50	100	150	200	250	300	350	400	450	500	550	600
VELOCITY (FPS)	2100.	1973.	1851.	1734.	1623.	1517.	1418.	1327.	1246.	1175.	1116.	1066.	1026.
ENERGY (FT-LB)	1743.	1538.	1354.	1188.	1041.	909.	795.	696.	614.	546.	492.	449.	416.
DROP (IN)	.00	-.96	-4.15	-9.87	-18.47	-30.37	-46.07	-66.11	-91.07	-121.66	-158.53	-202.36	-253.81
MID-RANGE (IN)	.00	.26	1.11	2.68	5.09	8.51	13.17	19.27	27.07	36.85	48.90	63.47	80.83
BULLET PATH (IN)	-.75	.74	.00	-3.27	-9.42	-18.87	-32.13	-49.72	-72.23	-100.37	-134.79	-176.17	-225.17
TIME OF FLIGHT (SEC)	.000000	.073692	.152185	.235915	.325338	.420942	.523244	.632625	.749332	.873379	1.004495	1.142150	1.285707
WIND DEFLECTION (IN)													
5 MPH	.00	.20	.82	1.90	3.49	5.61	8.33	11.67	15.66	20.29	25.54	31.37	37.71
10 MPH	.00	.40	1.64	3.81	6.97	11.23	16.66	23.34	31.31	40.57	51.08	62.73	75.43
20 MPH	.00	.80	3.28	7.61	13.95	22.46	33.32	46.68	62.62	81.14	102.15	125.47	150.85
30 MPH	.00	1.20	4.93	11.42	20.92	33.69	49.99	70.03	93.93	121.72	153.23	188.20	226.28
VELOCITY (FPS)	2000.	1877.	1759.	1646.	1539.	1439.	1346.	1263.	1190.	1128.	1076.	1034.	998.
ENERGY (FT-LB)	1581.	1392.	1222.	1071.	936.	818.	716.	630.	559.	502.	458.	422.	393.
DROP (IN)	.00	-1.06	-4.58	-10.91	-20.44	-33.65	-51.09	-73.31	-101.00	-134.84	-175.49	-223.62	-279.86
MID-RANGE (IN)	.00	.29	1.23	2.96	5.64	9.45	14.63	21.41	30.06	40.88	54.09	70.00	88.85
BULLET PATH (IN)	-.75	.86	.00	-3.66	-10.52	-21.07	-35.84	-55.39	-80.42	-111.59	-149.57	-195.04	-248.62
TIME OF FLIGHT (SEC)	.000000	.077425	.159990	.248149	.342376	.443198	.551027	.666144	.788621	.918250	1.054549	1.196881	1.344710
WIND DEFLECTION (IN)													
5 MPH	.00	.21	.88	2.04	3.73	6.00	8.89	12.42	16.60	21.41	26.80	32.73	39.13
10 MPH	.00	.43	1.76	4.07	7.46	12.00	17.78	24.84	33.20	42.81	53.60	65.45	78.27
20 MPH	.00	.85	3.52	8.15	14.92	24.01	35.56	49.68	66.39	85.62	107.20	130.90	156.54
30 MPH	.00	1.28	5.27	12.22	22.37	36.01	53.34	74.52	99.59	128.44	160.80	196.35	234.81
VELOCITY (FPS)	1900.	1781.	1667.	1559.	1457.	1363.	1278.	1203.	1138.	1085.	1041.	1004.	972.
ENERGY (FT-LB)	1427.	1253.	1099.	961.	839.	734.	645.	572.	512.	465.	428.	398.	373.
DROP (IN)	.00	-1.18	-5.09	-12.12	-22.74	-37.48	-56.69	-81.64	-112.40	-149.85	-194.66	-247.46	-308.86
MID-RANGE (IN)	.00	.32	1.37	3.29	6.28	10.55	16.32	23.87	33.46	45.36	59.85	77.16	97.54
BULLET PATH (IN)	-.75	.99	.00	-4.11	-11.81	-23.63	-40.12	-61.95	-89.79	-124.32	-166.21	-216.09	-274.58
TIME OF FLIGHT (SEC)	.000000	.081549	.168604	.261632	.361164	.467634	.581352	.702435	.830733	.965810	1.107039	1.253854	1.405839
WIND DEFLECTION (IN)													
5 MPH	.00	.23	.94	2.18	3.99	6.41	9.47	13.18	17.53	22.46	27.95	33.92	40.35
10 MPH	.00	.46	1.88	4.36	7.99	12.83	18.95	26.37	35.05	44.93	55.89	67.84	80.69
20 MPH	.00	.92	3.77	8.73	15.97	25.66	37.90	52.73	70.10	89.86	111.78	135.67	161.38
30 MPH	.00	1.37	5.65	13.09	23.96	38.49	56.85	79.10	105.15	134.79	167.67	203.51	242.07
VELOCITY (FPS)	1800.	1686.	1577.	1473.	1378.	1291.	1214.	1148.	1093.	1048.	1010.	977.	948.
ENERGY (FT-LB)	1280.	1123.	983.	858.	750.	659.	583.	521.	472.	434.	403.	377.	355.
DROP (IN)	.00	-1.31	-5.68	-13.55	-25.46	-41.94	-63.64	-91.24	-125.42	-166.85	-216.17	-274.00	-340.91
MID-RANGE (IN)	.00	.36	1.53	3.69	7.05	11.82	18.26	26.65	37.25	50.33	66.15	84.91	106.82
BULLET PATH (IN)	-.75	1.15	.00	-4.65	-13.35	-26.61	-45.09	-69.48	-100.44	-138.66	-184.76	-239.37	-303.06
TIME OF FLIGHT (SEC)	.000000	.086125	.178141	.276570	.381874	.494384	.614259	.741391	.875392	1.015651	1.161577	1.312733	1.468798
WIND DEFLECTION (IN)													
5 MPH	.00	.25	1.01	2.34	4.27	6.84	10.05	13.91	18.37	23.38	28.89	34.85	41.25
10 MPH	.00	.49	2.02	4.68	8.54	13.68	20.11	27.82	36.74	46.75	57.77	69.71	82.51
20 MPH	.00	.98	4.04	9.35	17.09	27.36	40.22	55.64	73.47	93.51	115.54	139.42	165.02
30 MPH	.00	1.47	6.06	14.03	25.63	41.03	60.33	83.45	110.21	140.26	173.31	209.12	247.53

Bullet: Lyman # 311467 178 Gr.
Ballistic Coefficient: .320

(Cont'd.)

RANGE (YDS)	MUZZLE	50	100	150	200	250	300	350	400	450	500	550	600
VELOCITY (FPS)	1700.	1591.	1486.	1390.	1302.	1224.	1156.	1100.	1053.	1014.	981.	951.	924.
ENERGY (FT-LB)	1142.	1000.	873.	763.	670.	592.	528.	478.	438.	407.	380.	357.	338.
DROP (IN)	.00	-1.48	-6.38	-15.27	-28.65	-47.16	-71.49	-102.31	-140.30	-186.09	-240.31	-303.54	-376.22
MID-RANGE (IN)	.00	.40	1.71	4.16	7.94	13.28	20.48	29.79	41.50	55.83	72.99	93.21	116.62
BULLET PATH (IN)	-.75	1.34	.00	-5.32	-15.13	-30.08	-50.84	-78.10	-112.51	-154.74	-205.39	-265.06	-334.17
TIME OF FLIGHT (SEC)	.000000	.091228	.188798	.293193	.404758	.523682	.649892	.783036	.922522	1.067741	1.218238	1.373681	1.533823
WIND DEFLECTION (IN)													
5 MPH	.00	.26	1.08	2.51	4.56	7.26	10.60	14.55	19.86	24.08	29.56	35.47	41.80
10 MPH	.00	.53	2.17	5.01	9.12	14.52	21.20	29.11	38.13	48.16	59.12	70.94	83.60
20 MPH	.00	1.05	4.34	10.03	18.24	29.04	42.41	58.22	76.26	96.32	118.23	141.89	167.20
30 MPH	.00	1.58	6.51	15.04	27.36	43.56	63.61	87.33	114.39	144.47	177.35	212.83	250.80
VELOCITY (FPS)	1600.	1495.	1398.	1309.	1230.	1161.	1104.	1057.	1018.	984.	953.	927.	902.
ENERGY (FT-LB)	1012.	883.	772.	677.	598.	533.	482.	441.	409.	382.	359.	339.	321.
DROP (IN)	.00	-1.67	-7.28	-17.33	-32.45	-53.34	-80.65	-115.07	-157.24	-207.79	-267.31	-336.22	-415.21
MID-RANGE (IN)	.00	.45	1.97	4.71	8.94	14.96	23.00	33.32	46.17	61.76	80.32	101.97	126.96
BULLET PATH (IN)	-.75	1.59	.00	-6.03	-17.14	-34.01	-57.31	-87.71	-125.87	-172.40	-227.91	-292.81	-367.78
TIME OF FLIGHT (SEC)	.000000	.097003	.200795	.311733	.430022	.555615	.688181	.827146	.971889	1.121944	1.276967	1.436708	1.600983
WIND DEFLECTION (IN)													
5 MPH	.00	.29	1.17	2.68	4.84	7.64	11.06	15.04	19.53	24.48	29.87	35.68	41.89
10 MPH	.00	.57	2.34	5.37	9.68	15.29	22.12	30.08	39.05	48.96	59.75	71.36	83.77
20 MPH	.00	1.15	4.68	10.73	19.37	30.58	44.24	60.16	78.10	97.92	119.49	142.72	167.55
30 MPH	.00	1.72	7.02	16.10	29.05	45.86	66.36	90.23	117.16	146.89	179.24	214.08	251.32
VELOCITY (FPS)	1500.	1402.	1313.	1233.	1165.	1107.	1059.	1019.	985.	955.	928.	903.	880.
ENERGY (FT-LB)	889.	777.	681.	601.	536.	484.	443.	411.	383.	360.	340.	322.	306.
DROP (IN)	.00	-1.95	-8.33	-19.74	-36.88	-60.42	-91.04	-129.37	-176.05	-231.67	-296.68	-371.71	-457.39
MID-RANGE (IN)	.00	.52	2.21	5.32	10.11	16.85	25.78	37.13	51.14	68.03	87.92	111.06	137.66
BULLET PATH (IN)	-.75	1.83	.00	-6.87	-19.48	-38.48	-64.56	-98.36	-140.50	-191.58	-252.05	-322.54	-403.68
TIME OF FLIGHT (SEC)	.000000	.103453	.214037	.331967	.457210	.589449	.728116	.872588	1.022390	1.177175	1.336687	1.500738	1.669195
WIND DEFLECTION (IN)													
5 MPH	.00	.30	1.24	2.81	5.03	7.87	11.27	15.19	19.57	24.39	29.63	35.26	41.29
10 MPH	.00	.61	2.47	5.63	10.07	15.74	22.55	30.38	39.14	48.78	59.26	70.53	82.58
20 MPH	.00	1.22	4.94	11.25	20.14	31.49	45.10	60.75	78.28	97.57	118.51	141.06	165.16
30 MPH	.00	1.82	7.41	16.88	30.21	47.23	67.65	91.13	117.42	146.35	177.77	211.59	247.74
VELOCITY (FPS)	1400.	1311.	1232.	1163.	1105.	1058.	1018.	984.	954.	927.	902.	880.	858.
ENERGY (FT-LB)	775.	679.	599.	534.	483.	442.	410.	383.	360.	340.	322.	306.	291.
DROP (IN)	.00	-2.25	-9.55	-22.60	-42.06	-68.62	-102.91	-145.56	-197.16	-258.15	-329.20	-410.90	-503.80
MID-RANGE (IN)	.00	.59	2.53	6.08	11.48	18.99	28.85	41.29	56.52	74.66	95.98	120.69	148.99
BULLET PATH (IN)	-.75	2.15	.00	-7.90	-22.22	-43.62	-72.77	-110.27	-156.72	-212.56	-278.46	-355.01	-442.77
TIME OF FLIGHT (SEC)	.000000	.110764	.228875	.354294	.486698	.625513	.770119	.920044	1.074943	1.234563	1.398718	1.567275	1.740143
WIND DEFLECTION (IN)													
5 MPH	.00	.32	1.28	2.89	5.12	7.90	11.20	14.96	19.17	23.78	28.80	34.21	39.99
10 MPH	.00	.64	2.57	5.78	10.23	15.80	22.40	29.93	38.33	47.57	57.60	68.41	79.98
20 MPH	.00	1.27	5.14	11.57	20.46	31.61	44.80	59.86	76.67	95.14	115.21	136.82	159.96
30 MPH	.00	1.91	7.70	17.35	30.69	47.41	67.19	89.78	115.00	142.71	172.81	205.24	239.94

Bullet: Lyman # 311334 187 Gr.
Ballistic Coefficient: .340

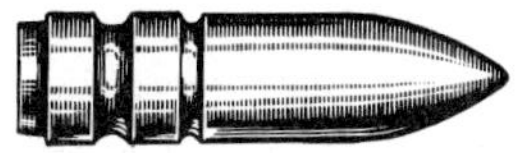

RANGE (YDS)	MUZZLE	50	100	150	200	250	300	350	400	450	500	550	600
VELOCITY (FPS)	2000.	1889.	1782.	1671.	1566.	1467.	1372.	1279.	1197.	1129.	1073.	1028.	989.
ENERGY (FT-LB)	1661.	1482.	1318.	1159.	1017.	893.	782.	679.	595.	529.	478.	438.	406.
DROP (IN)	.00	-1.05	-4.53	-10.75	-20.07	-32.95	-49.90	-71.48	-98.38	-131.35	-171.13	-218.43	-273.94
MID-RANGE (IN)	.00	.29	1.21	2.90	5.50	9.20	14.20	20.74	29.12	39.64	52.61	68.33	87.06
BULLET PATH (IN)	-.75	.84	.00	-3.58	-10.26	-20.49	-34.80	-53.74	-78.00	-108.33	-145.46	-190.12	-242.99
TIME OF FLIGHT (SEC)	.000000	.077168	.158909	.245854	.338617	.437628	.543305	.656605	.777936	.907105	1.043560	1.186551	1.335462
WIND DEFLECTION (IN)													
5 MPH	.00	.19	.78	1.84	3.40	5.51	8.21	11.58	15.66	20.43	25.83	31.82	38.32
10 MPH	.00	.38	1.57	3.67	6.80	11.02	16.42	23.16	31.32	40.85	51.67	63.63	76.64
20 MPH	.00	.76	3.14	7.34	13.59	22.05	32.84	46.32	62.63	81.70	103.33	127.27	153.28
30 MPH	.00	1.14	4.70	11.01	20.39	33.07	49.26	69.49	93.95	122.55	155.00	190.90	229.92
VELOCITY (FPS)	1900.	1793.	1681.	1575.	1476.	1382.	1288.	1205.	1135.	1078.	1032.	993.	959.
ENERGY (FT-LB)	1499.	1335.	1174.	1030.	904.	793.	688.	603.	535.	482.	442.	409.	382.
DROP (IN)	.00	-1.17	-5.04	-11.98	-22.42	-36.89	-55.92	-80.20	-110.48	-147.49	-191.95	-244.55	-305.96
MID-RANGE (IN)	.00	.32	1.35	3.24	6.17	10.34	15.98	23.37	32.82	44.62	59.08	76.48	97.06
BULLET PATH (IN)	-.75	.98	.00	-4.04	-11.59	-23.17	-39.30	-60.69	-88.07	-122.18	-163.75	-213.46	-271.97
TIME OF FLIGHT (SEC)	.000000	.081274	.167676	.259854	.358240	.463233	.575729	.696256	.824656	.960410	1.102774	1.251110	1.404965
WIND DEFLECTION (IN)													
5 MPH	.00	.20	.86	2.03	3.74	6.03	8.98	12.64	16.99	21.99	27.57	33.68	40.27
10 MPH	.00	.41	1.72	4.05	7.47	12.06	17.96	25.28	33.98	43.98	55.14	67.35	80.54
20 MPH	.00	.82	3.44	8.10	14.94	24.11	35.92	50.56	67.96	87.96	110.28	134.71	161.07
30 MPH	.00	1.23	5.16	12.15	22.41	36.17	53.88	75.83	101.94	131.94	165.42	202.06	241.61
VELOCITY (FPS)	1800.	1688.	1582.	1482.	1389.	1293.	1209.	1139.	1081.	1034.	995.	961.	931.
ENERGY (FT-LB)	1345.	1183.	1039.	912.	800.	694.	607.	538.	485.	444.	411.	383.	360.
DROP (IN)	.00	-1.31	-5.66	-13.49	-25.31	-41.66	-63.21	-90.71	-124.90	-166.49	-216.19	-274.65	-342.40
MID-RANGE (IN)	.00	.36	1.52	3.66	6.99	11.72	18.12	26.48	37.14	50.37	66.44	85.59	107.99
BULLET PATH (IN)	-.75	1.15	.00	-4.62	-13.24	-26.38	-44.72	-69.02	-100.00	-138.39	-184.88	-240.13	-304.67
TIME OF FLIGHT (SEC)	.000000	.086061	.177869	.275860	.380431	.492434	.612465	.740390	.875710	1.017684	1.165663	1.319185	1.477919
WIND DEFLECTION (IN)													
5 MPH	.00	.24	.99	2.28	4.14	6.67	9.90	13.82	18.40	23.56	29.25	35.42	42.06
10 MPH	.00	.48	1.97	4.55	8.29	13.34	19.79	27.64	36.79	47.11	58.49	70.84	84.11
20 MPH	.00	.96	3.94	9.10	16.58	26.67	39.59	55.28	73.58	94.22	116.98	141.69	168.23
30 MPH	.00	1.44	5.91	13.65	24.87	40.01	59.38	82.93	110.37	141.34	175.47	212.53	252.34
VELOCITY (FPS)	1700.	1593.	1493.	1399.	1302.	1218.	1146.	1087.	1039.	999.	964.	934.	906.
ENERGY (FT-LB)	1200.	1054.	925.	813.	704.	615.	545.	490.	448.	414.	386.	362.	341.
DROP (IN)	.00	-1.47	-6.36	-15.20	-28.49	-46.91	-71.21	-102.11	-140.35	-186.62	-241.59	-305.80	-379.88
MID-RANGE (IN)	.00	.40	1.71	4.14	7.89	13.20	20.41	29.81	41.69	56.30	73.88	94.62	118.73
BULLET PATH (IN)	-.75	1.33	.00	-5.28	-15.02	-29.89	-50.64	-77.99	-112.67	-155.39	-206.80	-267.46	-337.99
TIME OF FLIGHT (SEC)	.000000	.091159	.188453	.292286	.403463	.522661	.649784	.784369	.925686	1.073064	1.226025	1.384227	1.547425
WIND DEFLECTION (IN)													
5 MPH	.00	.26	1.05	2.43	4.45	7.17	10.59	14.67	19.34	24.55	30.24	36.40	43.00
10 MPH	.00	.51	2.11	4.85	8.89	14.34	21.19	29.34	38.69	49.09	60.49	72.80	85.99
20 MPH	.00	1.03	4.22	9.71	17.78	28.68	42.37	58.69	77.37	98.19	120.97	145.60	171.99
30 MPH	.00	1.54	6.33	14.56	26.68	43.02	63.56	88.03	116.06	147.28	181.46	218.40	257.98

Bullet: Lyman # 311334 187 Gr.

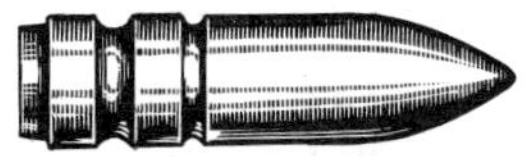

Ballistic Coefficient: .340

(Cont'd.)

RANGE (YDS)	MUZZLE	50	100	150	200	250	300	350	400	450	500	550	600
VELOCITY (FPS)	1600.	1499.	1405.	1308.	1223.	1150.	1090.	1042.	1001.	966.	936.	908.	883.
ENERGY (FT-LB)	1063.	933.	819.	711.	621.	549.	493.	450.	416.	388.	364.	342.	324.
DROP (IN)	.00	-1.66	-7.24	-17.24	-32.31	-53.22	-80.68	-115.42	-158.16	-209.54	-270.15	-340.57	-421.49
MID-RANGE (IN)	.00	.45	1.96	4.68	8.91	14.96	23.10	33.62	46.77	62.78	81.86	104.21	130.07
BULLET PATH (IN)	-.75	1.58	.00	-6.00	-17.08	-33.98	-57.45	-88.20	-126.94	-174.32	-230.93	-297.36	-374.28
TIME OF FLIGHT (SEC)	.000000	.096875	.200263	.310918	.429584	.556193	.690305	.831197	.978189	1.130789	1.288649	1.451517	1.619220
WIND DEFLECTION (IN)													
5 MPH	.00	.28	1.12	2.61	4.80	7.69	11.25	15.40	20.08	25.26	30.90	36.98	43.49
10 MPH	.00	.55	2.25	5.22	9.61	15.39	22.49	30.79	40.16	50.52	61.80	73.97	86.98
20 MPH	.00	1.10	4.49	10.44	19.21	30.78	44.99	61.58	80.32	101.04	123.60	147.93	173.97
30 MPH	.00	1.65	6.74	15.66	28.82	46.17	67.48	92.37	120.48	151.56	185.41	221.90	260.95
VELOCITY (FPS)	1500.	1406.	1310.	1224.	1151.	1091.	1042.	1002.	967.	936.	909.	883.	860.
ENERGY (FT-LB)	934.	821.	712.	622.	550.	494.	451.	417.	388.	364.	343.	324.	307.
DROP (IN)	.00	-1.95	-8.31	-19.73	-36.97	-60.76	-91.83	-130.86	-178.54	-235.43	-302.12	-379.30	-467.60
MID-RANGE (IN)	.00	.52	2.21	5.33	10.18	17.04	26.18	37.85	52.30	69.71	90.30	114.31	141.98
BULLET PATH (IN)	-.75	1.83	.00	-6.90	-19.61	-38.87	-65.41	-99.92	-143.07	-195.43	-257.59	-330.24	-414.02
TIME OF FLIGHT (SEC)	.000000	.103318	.213845	.332377	.458855	.592846	.733630	.880521	1.033028	1.190797	1.353578	1.521195	1.693529
WIND DEFLECTION (IN)													
5 MPH	.00	.29	1.22	2.85	5.18	8.17	11.76	15.89	20.51	25.59	31.11	37.07	43.43
10 MPH	.00	.58	2.44	5.70	10.36	16.34	23.52	31.77	41.01	51.18	62.23	74.13	86.86
20 MPH	.00	1.17	4.87	11.40	20.72	32.68	47.04	63.54	82.03	102.36	124.46	148.26	173.72
30 MPH	.00	1.75	7.31	17.10	31.08	49.02	70.56	95.32	123.04	153.54	186.69	222.39	260.58
VELOCITY (FPS)	1400.	1303.	1218.	1146.	1087.	1039.	999.	965.	934.	907.	882.	858.	836.
ENERGY (FT-LB)	814.	705.	616.	545.	491.	448.	414.	386.	362.	341.	323.	306.	290.
DROP (IN)	.00	-2.26	-9.64	-22.89	-42.74	-69.92	-105.12	-149.00	-202.13	-265.11	-338.63	-423.32	-519.80
MID-RANGE (IN)	.00	.59	2.56	6.19	11.73	19.47	29.67	42.56	58.32	77.19	99.40	125.19	154.78
BULLET PATH (IN)	-.75	2.19	.00	-8.05	-22.70	-44.69	-74.69	-113.38	-161.31	-219.10	-287.42	-366.91	-458.20
TIME OF FLIGHT (SEC)	.000000	.111096	.230212	.357254	.491763	.633009	.780320	.933215	1.091351	1.254483	1.422438	1.595103	1.772409
WIND DEFLECTION (IN)													
5 MPH	.00	.35	1.40	3.15	5.56	8.56	12.10	16.12	20.61	25.54	30.89	36.65	42.83
10 MPH	.00	.70	2.80	6.31	11.12	17.12	24.19	32.25	41.22	51.07	61.78	73.31	85.66
20 MPH	.00	1.39	5.61	12.61	22.24	34.25	48.39	64.49	82.44	102.15	123.56	146.62	171.32
30 MPH	.00	2.09	8.41	18.92	33.37	51.37	72.58	96.74	123.66	153.22	185.33	219.93	256.97
VELOCITY (FPS)	1300.	1215.	1144.	1085.	1038.	998.	963.	933.	906.	881.	857.	836.	815.
ENERGY (FT-LB)	702.	613.	543.	489.	447.	413.	385.	361.	340.	322.	305.	290.	276.
DROP (IN)	.00	-2.63	-11.15	-26.30	-48.80	-79.35	-118.61	-167.13	-225.54	-294.49	-374.64	-466.60	-571.00
MID-RANGE (IN)	.00	.69	2.95	7.04	13.25	21.83	33.01	46.98	63.97	84.22	107.97	135.45	166.89
BULLET PATH (IN)	-.75	2.57	.00	-9.20	-25.75	-50.35	-83.66	-126.23	-178.68	-241.69	-315.89	-401.90	-500.35
TIME OF FLIGHT (SEC)	.000000	.119409	.246733	.381501	.522977	.670496	.823584	.981901	1.145205	1.313327	1.486153	1.663618	1.845692
WIND DEFLECTION (IN)													
5 MPH	.00	.35	1.40	3.11	5.41	8.23	11.55	15.33	19.55	24.19	29.24	34.71	40.57
10 MPH	.00	.71	2.81	6.22	10.81	16.47	23.10	30.66	39.09	48.38	58.49	_9.41	81.15
20 MPH	.00	1.42	5.62	12.44	21.63	32.94	46.21	61.32	78.19	96.75	116.97	138.82	162.30
30 MPH	.00	2.12	8.43	18.66	32.44	49.41	69.31	91.98	117.28	145.13	175.46	208.24	243.45

Bullet: Lyman # 301620 200 Gr.
Ballistic Coefficient: .379

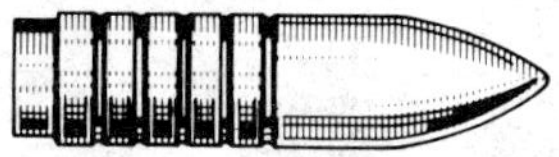

RANGE (YDS)	MUZZLE	50	100	150	200	250	300	350	400	450	500	550	600
VELOCITY (FPS)	2500.	2387.	2277.	2169.	2062.	1958.	1858.	1760.	1665.	1574.	1488.	1407.	1332.
ENERGY (FT-LB)	2775.	2530.	2301.	2089.	1888.	1703.	1532.	1375.	1230.	1100.	983.	879.	788.
DROP (IN)	.00	-.66	-2.83	-6.68	-12.38	-20.11	-30.11	-42.61	-57.92	-76.35	-98.29	-124.21	-154.50
MID-RANGE (IN)	.00	.18	.76	1.80	3.36	5.52	8.37	12.02	16.59	22.20	29.03	37.29	47.14
BULLET PATH (IN)	-.75	.39	.00	-2.06	-5.96	-11.91	-20.11	-30.83	-44.34	-60.98	-81.13	-105.26	-133.76
TIME OF FLIGHT (SEC)	.000000	.061406	.125754	.193252	.264184	.338837	.417488	.500444	.588092	.680776	.778817	.882511	.992105
WIND DEFLECTION (IN)													
5 MPH	.00	.12	.51	1.17	2.13	3.42	5.06	7.08	9.51	12.39	15.74	19.58	23.95
10 MPH	.00	.25	1.01	2.33	4.26	6.84	10.12	14.16	19.02	24.78	31.47	39.16	47.89
20 MPH	.00	.50	2.03	4.66	8.51	13.67	20.24	28.32	38.05	49.55	62.94	78.32	95.78
30 MPH	.00	.74	3.04	7.00	12.77	20.51	30.35	42.47	57.07	74.33	94.42	117.49	143.67
VELOCITY (FPS)	2400.	2289.	2181.	2074.	1970.	1869.	1771.	1676.	1584.	1498.	1416.	1341.	1271.
ENERGY (FT-LB)	2558.	2327.	2113.	1910.	1723.	1551.	1393.	1247.	1114.	996.	891.	798.	718.
DROP (IN)	.00	-.71	-3.08	-7.27	-13.48	-21.92	-32.84	-46.53	-63.30	-83.52	-107.68	-136.16	-169.47
MID-RANGE (IN)	.00	.20	.83	1.96	3.66	6.03	9.15	13.16	18.17	24.36	31.92	41.03	51.89
BULLET PATH (IN)	-.75	.45	.00	-2.28	-6.57	-13.10	-22.11	-33.88	-48.74	-67.05	-89.29	-115.86	-147.26
TIME OF FLIGHT (SEC)	.000000	.063992	.131108	.201630	.275841	.354019	.436460	.523541	.615621	.713022	.816041	.924937	1.039889
WIND DEFLECTION (IN)													
5 MPH	.00	.13	.54	1.24	2.27	3.65	5.41	7.57	10.17	13.25	16.81	20.89	25.51
10 MPH	.00	.26	1.07	2.49	4.55	7.31	10.82	15.14	20.35	26.49	33.62	41.79	51.02
20 MPH	.00	.53	2.15	4.97	9.10	14.61	21.63	30.29	40.70	52.98	67.25	83.58	102.04
30 MPH	.00	.79	3.22	7.46	13.64	21.92	32.45	45.43	61.05	79.48	100.87	125.37	153.06
VELOCITY (FPS)	2300.	2192.	2084.	1980.	1879.	1781.	1685.	1593.	1506.	1424.	1348.	1278.	1215.
ENERGY (FT-LB)	2349.	2134.	1929.	1741.	1567.	1408.	1260.	1127.	1007.	901.	807.	725.	655.
DROP (IN)	.00	-.78	-3.36	-7.94	-14.74	-23.99	-35.97	-51.01	-69.45	-91.78	-118.40	-149.79	-186.52
MID-RANGE (IN)	.00	.21	.90	2.14	4.01	6.60	10.04	14.46	19.99	26.85	35.21	45.27	57.28
BULLET PATH (IN)	-.75	.53	.00	-2.53	-7.26	-14.46	-24.39	-37.37	-53.76	-74.03	-98.59	-127.93	-162.60
TIME OF FLIGHT (SEC)	.000000	.066804	.136979	.210817	.288593	.370601	.457205	.548775	.645633	.748084	.856387	.970733	1.091191
WIND DEFLECTION (IN)													
5 MPH	.00	.14	.58	1.33	2.44	3.92	5.80	8.12	10.90	14.18	17.97	22.29	27.16
10 MPH	.00	.28	1.15	2.67	4.88	7.83	11.60	16.24	21.81	28.36	35.94	44.59	54.31
20 MPH	.00	.56	2.30	5.34	9.76	15.67	23.20	32.47	43.61	56.72	71.88	89.18	108.62
30 MPH	.00	.84	3.46	8.01	14.64	23.50	34.80	48.71	65.42	85.08	107.82	133.76	162.93
VELOCITY (FPS)	2200.	2092.	1987.	1886.	1788.	1692.	1599.	1512.	1430.	1353.	1283.	1219.	1163.
ENERGY (FT-LB)	2149.	1944.	1754.	1579.	1419.	1270.	1136.	1015.	908.	813.	730.	660.	601.
DROP (IN)	.00	-.86	-3.69	-8.72	-16.20	-26.38	-39.60	-56.19	-76.63	-101.33	-130.77	-165.50	-206.10
MID-RANGE (IN)	.00	.24	.99	2.35	4.41	7.27	11.07	15.96	22.12	29.74	39.00	50.16	63.45
BULLET PATH (IN)	-.75	.62	.00	-2.81	-8.06	-16.03	-27.02	-41.39	-59.62	-82.10	-109.31	-141.82	-180.20
TIME OF FLIGHT (SEC)	.000000	.069918	.143480	.220960	.302649	.388912	.480117	.576588	.678631	.786510	.900421	1.020444	1.146503
WIND DEFLECTION (IN)													
5 MPH	.00	.15	.63	1.44	2.63	4.22	6.25	8.74	11.72	15.21	19.24	23.80	28.89
10 MPH	.00	.31	1.25	2.89	5.27	8.45	12.50	17.48	23.44	30.43	38.47	47.60	57.78
20 MPH	.00	.61	2.51	5.78	10.53	16.90	25.00	34.96	46.88	60.85	76.95	95.20	115.57
30 MPH	.00	.92	3.76	8.67	15.80	25.35	37.50	52.44	70.32	91.28	115.42	142.79	173.35

Bullet: Lyman # 301620 200 Gr.
Ballistic Coefficient: .379
(Cont'd.)

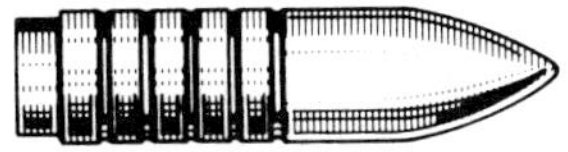

RANGE (YDS)	MUZZLE	50	100	150	200	250	300	350	400	450	500	550	600
VELOCITY (FPS)	2100.	1995.	1893.	1795.	1698.	1606.	1518.	1435.	1358.	1287.	1223.	1167.	1118.
ENERGY (FT-LB)	1958.	1767.	1592.	1431.	1281.	1145.	1023.	915.	819.	736.	665.	604.	555.
DROP (IN)	.00	-.94	-4.06	-9.60	-17.84	-29.08	-43.67	-62.07	-84.71	-112.04	-144.62	-183.03	-227.81
MID-RANGE (IN)	.00	.26	1.09	2.59	4.86	8.02	12.24	17.67	24.52	32.96	43.23	55.58	70.23
BULLET PATH (IN)	-.75	.71	.00	-3.13	-8.96	-17.79	-29.99	-45.98	-66.21	-91.13	-121.31	-157.31	-199.69
TIME OF FLIGHT (SEC)	.000000	.073285	.150466	.231835	.317754	.408591	.504672	.606304	.713754	.827225	.946808	1.072437	1.203885
WIND DEFLECTION (IN)													
5 MPH	.00	.16	.67	1.54	2.82	4.53	6.70	9.35	12.52	16.22	20.46	25.23	30.51
10 MPH	.00	.33	1.34	3.09	5.64	9.05	13.39	18.71	25.05	32.45	40.92	50.46	61.03
20 MPH	.00	.65	2.68	6.18	11.28	18.11	26.79	37.42	50.10	64.90	81.85	100.93	122.05
30 MPH	.00	.98	4.02	9.27	16.92	27.16	40.18	56.13	75.15	97.35	122.77	151.39	183.08
VELOCITY (FPS)	2000.	1898.	1800.	1703.	1610.	1522.	1439.	1362.	1291.	1226.	1169.	1120.	1078.
ENERGY (FT-LB)	1776.	1600.	1438.	1288.	1151.	1029.	920.	824.	740.	668.	607.	557.	516.
DROP (IN)	.00	-1.04	-4.49	-10.62	-19.74	-32.20	-48.44	-68.89	-94.01	-124.36	-160.51	-203.00	-252.41
MID-RANGE (IN)	.00	.29	1.20	2.86	5.38	8.90	13.60	19.66	27.27	36.65	48.06	61.69	77.77
BULLET PATH (IN)	-.75	.83	.00	-3.51	-10.01	-19.85	-33.47	-51.30	-73.80	-101.53	-135.06	-174.93	-221.72
TIME OF FLIGHT (SEC)	.000000	.076988	.158150	.243842	.334436	.430259	.531618	.638784	.751964	.871253	.996596	1.127772	1.264411
WIND DEFLECTION (IN)													
5 MPH	.00	.17	.72	1.66	3.03	4.86	7.18	10.01	13.37	17.27	21.70	26.64	32.07
10 MPH	.00	.35	1.43	3.32	6.06	9.73	14.36	20.03	26.75	34.54	43.40	53.29	64.14
20 MPH	.00	.70	2.87	6.63	12.12	19.45	28.73	40.05	53.49	69.08	86.80	106.58	128.27
30 MPH	.00	1.05	4.30	9.95	18.18	29.18	43.09	60.08	80.24	103.62	130.20	159.86	192.41
VELOCITY (FPS)	1900.	1801.	1705.	1612.	1524.	1441.	1364.	1292.	1228.	1170.	1121.	1079.	1043.
ENERGY (FT-LB)	1603.	1441.	1291.	1154.	1031.	922.	826.	741.	669.	608.	558.	517.	483.
DROP (IN)	.00	-1.16	-4.99	-11.80	-21.95	-35.87	-53.99	-76.77	-104.76	-138.54	-178.66	-225.67	-280.14
MID-RANGE (IN)	.00	.32	1.34	3.18	5.99	9.93	15.19	21.95	30.43	40.87	53.46	68.46	86.08
BULLET PATH (IN)	-.75	.96	.00	-3.94	-11.22	-22.27	-37.52	-57.44	-82.56	-113.47	-150.72	-194.86	-246.46
TIME OF FLIGHT (SEC)	.000000	.081081	.166672	.257158	.352865	.454102	.561140	.674189	.793346	.918560	1.049613	1.186138	1.327688
WIND DEFLECTION (IN)													
5 MPH	.00	.19	.77	1.79	3.26	5.22	7.70	10.70	14.24	18.31	22.89	27.96	33.47
10 MPH	.00	.38	1.54	3.58	6.53	10.45	15.39	21.39	28.47	36.61	45.78	55.92	66.94
20 MPH	.00	.75	3.09	7.15	13.05	20.90	30.78	42.79	56.94	73.23	91.57	111.84	133.87
30 MPH	.00	1.13	4.63	10.73	19.58	31.34	46.18	64.18	85.41	109.84	137.35	167.75	200.81
VELOCITY (FPS)	1800.	1703.	1611.	1523.	1440.	1362.	1291.	1226.	1169.	1120.	1078.	1042.	1011.
ENERGY (FT-LB)	1439.	1288.	1152.	1029.	920.	824.	740.	668.	607.	557.	516.	482.	454.
DROP (IN)	.00	-1.30	-5.58	-13.20	-24.61	-40.22	-60.50	-86.00	-117.31	-154.96	-199.51	-251.53	-311.53
MID-RANGE (IN)	.00	.35	1.49	3.56	6.72	11.16	17.05	24.60	34.05	45.62	59.54	76.02	95.26
BULLET PATH (IN)	-.75	1.12	.00	-4.46	-12.70	-25.14	-42.27	-64.60	-92.74	-127.23	-168.62	-217.47	-274.30
TIME OF FLIGHT (SEC)	.000000	.085675	.176252	.272056	.373396	.480541	.593698	.712965	.838286	.969439	1.106057	1.247692	1.393961
WIND DEFLECTION (IN)													
5 MPH	.00	.21	.84	1.94	3.53	5.62	8.25	11.41	15.10	19.31	24.00	29.13	34.67
10 MPH	.00	.41	1.69	3.88	7.05	11.24	16.49	22.82	30.20	38.62	48.00	58.26	69.34
20 MPH	.00	.82	3.37	7.76	14.10	22.48	32.98	45.63	60.41	77.24	96.00	116.52	138.67
30 MPH	.00	1.24	5.06	11.65	21.15	33.73	49.47	68.45	90.61	115.86	144.00	174.78	208.01

Bullet: Lyman # 311299 200 Gr.
Ballistic Coefficient: .377

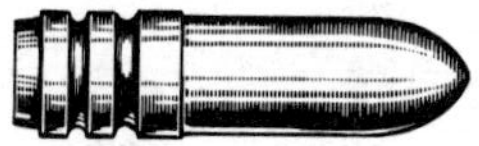

RANGE (YDS)	MUZZLE	50	100	150	200	250	300	350	400	450	500	550	600
VELOCITY (FPS)	2200.	2094.	1991.	1892.	1795.	1697.	1604.	1515.	1432.	1358.	1292.	1232.	1178.
ENERGY (FT-LB)	2149.	1947.	1761.	1589.	1430.	1279.	1142.	1019.	910.	819.	741.	674.	616.
DROP (IN)	.00	-.85	-3.68	-8.70	-16.14	-26.28	-39.43	-55.94	-76.27	-100.85	-130.14	-164.63	-204.88
MID-RANGE (IN)	.00	.24	.99	2.34	4.39	7.24	11.02	15.87	22.00	29.57	38.79	49.85	63.01
BULLET PATH (IN)	-.75	.61	.00	-2.80	-8.03	-15.95	-26.88	-41.18	-59.30	-81.66	-108.73	-141.00	-179.04
TIME OF FLIGHT (SEC)	.000000	.069885	.143342	.220633	.302039	.387987	.478912	.575148	.677005	.784676	.897983	1.016963	1.141574
WIND DEFLECTION (IN)													
5 MPH	.00	.15	.61	1.42	2.58	4.14	6.14	8.61	11.58	15.05	19.02	23.49	28.46
10 MPH	.00	.30	1.23	2.83	5.16	8.29	12.29	17.23	23.15	30.10	38.04	46.99	56.92
20 MPH	.00	.60	2.46	5.66	10.32	16.57	24.58	34.45	46.31	60.21	76.09	93.97	113.83
30 MPH	.00	.90	3.68	8.49	15.48	24.86	36.87	51.68	69.46	90.31	114.13	140.96	170.75
VELOCITY (FPS)	2100.	1997.	1897.	1800.	1703.	1610.	1521.	1437.	1362.	1296.	1235.	1181.	1133.
ENERGY (FT-LB)	1958.	1771.	1598.	1439.	1288.	1150.	1027.	917.	824.	745.	677.	619.	570.
DROP (IN)	.00	-.94	-4.05	-9.58	-17.78	-28.97	-43.51	-61.84	-84.38	-111.60	-143.99	-182.11	-226.45
MID-RANGE (IN)	.00	.26	1.09	2.58	4.84	7.99	12.18	17.59	24.41	32.82	43.02	55.25	69.71
BULLET PATH (IN)	-.75	.71	.00	-3.12	-8.93	-17.72	-29.85	-45.78	-65.92	-90.74	-120.72	-156.44	-198.38
TIME OF FLIGHT (SEC)	.000000	.073249	.150318	.231485	.317133	.407734	.503625	.605119	.712429	.825384	.944009	1.068272	1.198009
WIND DEFLECTION (IN)													
5 MPH	.00	.16	.66	1.51	2.76	4.45	6.60	9.25	12.41	16.06	20.22	24.87	30.00
10 MPH	.00	.32	1.31	3.03	5.53	8.90	13.21	18.50	24.82	32.12	40.43	49.73	59.99
20 MPH	.00	.64	2.63	6.05	11.06	17.81	26.42	37.00	49.63	64.25	80.86	99.46	119.98
30 MPH	.00	.96	3.94	9.08	16.59	26.71	39.63	55.50	74.45	96.37	121.29	149.19	179.98
VELOCITY (FPS)	2000.	1900.	1803.	1706.	1612.	1523.	1439.	1364.	1298.	1237.	1183.	1135.	1093.
ENERGY (FT-LB)	1776.	1603.	1444.	1292.	1154.	1030.	920.	826.	748.	679.	621.	572.	531.
DROP (IN)	.00	-1.04	-4.48	-10.59	-19.68	-32.11	-48.31	-68.72	-93.78	-124.00	-159.94	-202.08	-250.95
MID-RANGE (IN)	.00	.29	1.20	2.85	5.36	8.87	13.56	19.62	27.21	36.54	47.86	61.33	77.16
BULLET PATH (IN)	-.75	.83	.00	-3.50	-9.98	-19.79	-33.38	-51.17	-73.62	-101.22	-134.55	-174.07	-220.33
TIME OF FLIGHT (SEC)	.000000	.076951	.157991	.243504	.333960	.429696	.531028	.638168	.750948	.869397	.993487	1.123058	1.257823
WIND DEFLECTION (IN)													
5 MPH	.00	.17	.70	1.63	2.99	4.81	7.13	9.96	13.28	17.11	21.43	26.23	31.49
10 MPH	.00	.34	1.41	3.26	5.98	9.63	14.26	19.92	26.57	34.21	42.85	52.46	62.98
20 MPH	.00	.69	2.81	6.51	11.95	19.25	28.52	39.84	53.13	68.43	85.71	104.92	125.95
30 MPH	.00	1.03	4.22	9.77	17.93	28.88	42.78	59.75	79.70	102.64	128.56	157.37	188.93
VELOCITY (FPS)	1900.	1803.	1706.	1612.	1523.	1439.	1364.	1298.	1237.	1183.	1135.	1093.	1058.
ENERGY (FT-LB)	1603.	1444.	1292.	1154.	1030.	920.	826.	748.	679.	621.	572.	531.	497.
DROP (IN)	.00	-1.15	-4.98	-11.79	-21.93	-35.84	-53.96	-76.74	-104.67	-138.32	-178.17	-224.76	-278.59
MID-RANGE (IN)	.00	.32	1.34	3.18	5.98	9.93	15.19	21.95	30.41	40.78	53.24	68.03	85.34
BULLET PATH (IN)	-.75	.96	.00	-3.94	-11.21	-22.27	-37.52	-57.43	-82.50	-113.28	-150.27	-193.99	-244.96
TIME OF FLIGHT (SEC)	.000000	.081041	.166553	.257008	.352744	.454075	.561214	.673993	.792442	.916530	1.046099	1.180862	1.320442
WIND DEFLECTION (IN)													
5 MPH	.00	.18	.76	1.77	3.25	5.22	7.70	10.68	14.16	18.13	22.58	27.49	32.83
10 MPH	.00	.37	1.52	3.55	6.50	10.44	15.41	21.36	28.31	36.26	45.17	54.99	65.66
20 MPH	.00	.74	3.05	7.10	13.01	20.89	30.81	42.72	56.62	72.51	90.33	109.98	131.32
30 MPH	.00	1.11	4.57	10.65	19.51	31.33	46.22	64.08	84.94	108.77	135.50	164.97	196.98

Bullet: Lyman # 311299 200 Gr.
Ballistic Coefficient: .377

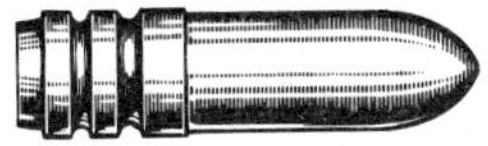

(Cont'd.)

RANGE (YDS)	MUZZLE	50	100	150	200	250	300	350	400	450	500	550	600
VELOCITY (FPS)	1800.	1702.	1609.	1520.	1436.	1361.	1295.	1234.	1180.	1133.	1092.	1056.	1026.
ENERGY (FT-LB)	1439.	1286.	1149.	1025.	916.	823.	744.	677.	619.	570.	529.	496.	467.
DROP (IN)	.00	-1.30	-5.59	-13.22	-24.65	-40.31	-60.64	-86.16	-117.41	-154.88	-199.12	-250.62	-309.89
MID-RANGE (IN)	.00	.35	1.50	3.57	6.74	11.19	17.10	24.65	34.05	45.51	59.25	75.45	94.31
BULLET PATH (IN)	-.75	1.12	.00	-4.47	-12.73	-25.22	-42.38	-64.73	-92.81	-127.12	-168.19	-216.52	-272.63
TIME OF FLIGHT (SEC)	.000000	.085702	.176364	.272319	.373881	.481262	.594288	.712985	.837318	.967120	1.102102	1.241882	1.386099
WIND DEFLECTION (IN)													
5 MPH	.00	.21	.85	1.96	3.57	5.68	8.30	11.41	15.02	19.11	23.65	28.62	33.98
10 MPH	.00	.42	1.71	3.93	7.14	11.37	16.59	22.82	30.03	38.21	47.30	57.24	67.95
20 MPH	.00	.83	3.41	7.86	14.27	22.74	33.19	45.64	60.07	76.43	94.61	114.48	135.91
30 MPH	.00	1.25	5.12	11.78	21.41	34.11	49.78	68.46	90.10	114.64	141.91	171.71	203.86
VELOCITY (FPS)	1700.	1607.	1518.	1434.	1359.	1293.	1233.	1179.	1132.	1091.	1056.	1025.	998.
ENERGY (FT-LB)	1283.	1146.	1023.	913.	821.	743.	675.	617.	569.	528.	495.	466.	442.
DROP (IN)	.00	-1.46	-6.27	-14.89	-27.74	-45.27	-68.01	-96.49	-131.21	-172.70	-221.47	-278.03	-342.84
MID-RANGE (IN)	.00	.40	1.68	4.02	7.60	12.57	19.12	27.49	37.87	50.46	65.46	83.06	103.41
BULLET PATH (IN)	-.75	1.30	.00	-5.11	-14.45	-28.48	-47.70	-72.67	-103.88	-141.86	-187.13	-240.17	-301.47
TIME OF FLIGHT (SEC)	.000000	.090775	.186850	.288538	.396051	.509212	.628045	.752511	.882440	1.017539	1.157426	1.301744	1.450208
WIND DEFLECTION (IN)													
5 MPH	.00	.22	.91	2.10	3.79	5.99	8.68	11.87	15.54	19.66	24.21	29.14	34.44
10 MPH	.00	.45	1.83	4.19	7.59	11.97	17.36	23.74	31.07	39.32	48.41	58.28	68.88
20 MPH	.00	.89	3.65	8.39	15.17	23.95	34.72	47.47	62.15	78.64	96.83	116.57	137.77
30 MPH	.00	1.34	5.48	12.58	22.76	35.92	52.08	71.21	93.22	117.97	145.24	174.85	206.65
VELOCITY (FPS)	1600.	1512.	1428.	1354.	1289.	1229.	1175.	1129.	1088.	1053.	1023.	996.	971.
ENERGY (FT-LB)	1137.	1015.	906.	815.	737.	671.	613.	566.	526.	493.	465.	440.	419.
DROP (IN)	.00	-1.64	-7.13	-16.88	-31.35	-51.05	-76.54	-108.30	-146.87	-192.76	-246.46	-308.46	-379.20
MID-RANGE (IN)	.00	.45	1.92	4.56	8.54	14.08	21.39	30.66	42.09	55.87	72.18	91.18	113.06
BULLET PATH (IN)	-.75	1.55	.00	-5.80	-16.33	-32.09	-53.64	-81.46	-116.09	-158.04	-207.80	-265.85	-332.65
TIME OF FLIGHT (SEC)	.000000	.096466	.198567	.306490	.420055	.539292	.664155	.794463	.929916	1.070128	1.214747	1.363494	1.516154
WIND DEFLECTION (IN)													
5 MPH	.00	.24	.97	2.22	3.96	6.21	8.95	12.16	15.83	19.92	24.40	29.24	34.42
10 MPH	.00	.48	1.95	4.44	7.93	12.42	17.89	24.33	31.67	39.84	48.80	58.48	68.84
20 MPH	.00	.96	3.90	8.88	15.86	24.83	35.78	48.65	63.33	79.69	97.59	116.95	137.69
30 MPH	.00	1.43	5.84	13.33	23.79	37.25	53.67	72.98	95.00	119.53	146.39	175.43	206.53
VELOCITY (FPS)	1500.	1418.	1345.	1280.	1221.	1169.	1123.	1083.	1049.	1019.	992.	968.	946.
ENERGY (FT-LB)	999.	892.	804.	728.	662.	606.	560.	521.	489.	461.	437.	416.	397.
DROP (IN)	.00	-1.92	-8.17	-19.20	-35.55	-57.74	-86.28	-121.69	-164.49	-215.18	-274.21	-342.06	-419.14
MID-RANGE (IN)	.00	.51	2.16	5.12	9.61	15.81	23.94	34.18	46.70	61.70	79.37	99.85	123.29
BULLET PATH (IN)	-.75	1.79	.00	-6.57	-18.46	-36.19	-60.27	-91.23	-129.57	-175.80	-230.37	-293.76	-366.38
TIME OF FLIGHT (SEC)	.000000	.102885	.211579	.325906	.445906	.571515	.702535	.838650	.979472	1.124657	1.273938	1.427105	1.583991
WIND DEFLECTION (IN)													
5 MPH	.00	.25	1.02	2.28	4.04	6.29	9.02	12.20	15.79	19.77	24.11	28.79	33.79
10 MPH	.00	.51	2.04	4.56	8.08	12.59	18.05	24.40	31.59	39.54	48.21	57.57	67.58
20 MPH	.00	1.02	4.08	9.12	16.16	25.17	36.09	48.80	63.17	79.08	96.43	115.14	135.16
30 MPH	.00	1.52	6.11	13.68	24.24	37.76	54.14	73.21	94.76	118.62	144.64	172.71	202.75

Bullet: Lyman # 311299 200 Gr.
Ballistic Coefficient: .377
(Cont'd.)

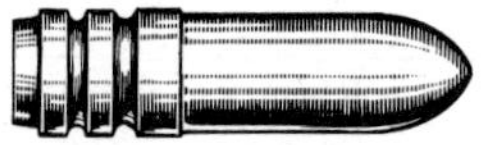

RANGE (YDS)	MUZZLE	50	100	150	200	250	300	350	400	450	500	550	600
VELOCITY (FPS)	1400.	1331.	1267.	1209.	1158.	1114.	1075.	1042.	1013.	987.	963.	942.	922.
ENERGY (FT-LB)	870.	786.	713.	649.	596.	551.	513.	482.	456.	432.	412.	394.	377.
DROP (IN)	.00	-2.20	-9.30	-21.84	-40.33	-65.27	-97.21	-136.65	-184.06	-239.93	-304.71	-378.76	-462.48
MID-RANGE (IN)	.00	.58	2.45	5.81	10.86	17.78	26.77	38.02	51.71	68.01	87.08	109.04	134.03
BULLET PATH (IN)	-.75	2.07	.00	-7.51	-20.97	-40.89	-67.80	-102.21	-144.60	-195.44	-255.19	-324.22	-402.91
TIME OF FLIGHT (SEC)	.000000	.109921	.225478	.346706	.473509	.605664	.742831	.884623	1.030714	1.180849	1.334829	1.492495	1.653724
WIND DEFLECTION (IN)													
5 MPH	.00	.24	.98	2.22	3.95	6.16	8.80	11.85	15.27	19.06	23.18	27.63	32.38
10 MPH	.00	.49	1.97	4.45	7.91	12.31	17.60	23.69	30.55	38.12	46.36	55.25	64.77
20 MPH	.00	.98	3.94	8.90	15.82	24.62	35.19	47.39	61.10	76.23	92.72	110.50	129.54
30 MPH	.00	1.47	5.91	13.35	23.73	36.93	52.79	71.08	91.65	114.35	139.08	165.75	194.31
VELOCITY (FPS)	1300.	1239.	1184.	1136.	1095.	1059.	1028.	1000.	975.	953.	932.	912.	894.
ENERGY (FT-LB)	750.	682.	623.	573.	532.	498.	469.	444.	422.	403.	386.	370.	355.
DROP (IN)	.00	-2.55	-10.80	-25.23	-46.38	-74.75	-110.85	-155.15	-208.12	-270.21	-341.70	-423.12	-514.95
MID-RANGE (IN)	.00	.68	2.85	6.68	12.39	20.16	30.19	42.64	57.68	75.48	96.10	119.76	146.62
BULLET PATH (IN)	-.75	2.47	.00	-8.65	-24.02	-46.62	-76.94	-115.47	-162.66	-218.97	-284.68	-360.33	-446.38
TIME OF FLIGHT (SEC)	.000000	.118225	.242092	.371444	.506001	.645384	.789228	.937242	1.089199	1.244920	1.404265	1.567120	1.733402
WIND DEFLECTION (IN)													
5 MPH	.00	.25	1.00	2.23	3.91	6.02	8.53	11.40	14.62	18.17	22.04	26.21	30.69
10 MPH	.00	.50	1.99	4.45	7.83	12.05	17.06	22.80	29.24	36.34	44.07	52.43	61.39
20 MPH	.00	1.00	3.99	8.90	15.65	24.10	34.12	45.60	58.47	72.67	88.15	104.86	122.77
30 MPH	.00	1.50	5.98	13.35	23.48	36.15	51.17	68.40	87.71	109.01	132.22	157.29	184.16
VELOCITY (FPS)	1200.	1150.	1106.	1069.	1037.	1008.	982.	959.	938.	918.	899.	882.	866.
ENERGY (FT-LB)	639.	587.	544.	507.	477.	451.	429.	408.	390.	374.	359.	345.	333.
DROP (IN)	.00	-3.03	-12.62	-29.28	-53.53	-85.84	-126.68	-176.50	-235.63	-304.54	-383.71	-473.60	-574.66
MID-RANGE (IN)	.00	.79	3.28	7.67	14.14	22.88	34.06	47.82	64.29	83.62	106.02	131.64	160.65
BULLET PATH (IN)	-.75	2.90	.00	-9.98	-27.54	-53.16	-87.32	-130.45	-182.90	-245.12	-317.61	-400.82	-495.19
TIME OF FLIGHT (SEC)	.000000	.127742	.260784	.398769	.541315	.688108	.838905	.993513	1.151781	1.313588	1.478843	1.647474	1.819428
WIND DEFLECTION (IN)													
5 MPH	.00	.24	.95	2.09	3.64	5.55	7.82	10.43	13.36	16.60	20.14	23.98	28.11
10 MPH	.00	.48	1.90	4.18	7.27	11.11	15.65	20.86	26.71	33.19	40.28	47.96	56.22
20 MPH	.00	.97	3.80	8.37	14.54	22.21	31.29	41.72	53.43	66.38	80.55	95.91	112.44
30 MPH	.00	1.45	5.69	12.55	21.81	33.32	46.94	62.57	80.14	99.57	120.83	143.87	168.66
VELOCITY (FPS)	1100.	1064.	1032.	1004.	978.	955.	934.	915.	896.	879.	863.	847.	832.
ENERGY (FT-LB)	537.	502.	473.	447.	425.	405.	388.	372.	357.	343.	331.	319.	308.
DROP (IN)	.00	-3.57	-14.79	-34.15	-62.12	-99.14	-145.53	-201.76	-268.33	-345.69	-434.29	-534.56	-646.94
MID-RANGE (IN)	.00	.93	3.83	8.89	16.27	26.15	38.61	53.84	72.04	93.37	118.00	146.13	177.95
BULLET PATH (IN)	-.75	3.45	.00	-11.59	-31.79	-61.04	-99.65	-148.11	-206.92	-276.51	-357.33	-449.83	-554.44
TIME OF FLIGHT (SEC)	.000000	.138736	.281976	.429419	.580830	.736024	.894853	1.057204	1.222985	1.392130	1.564589	1.740328	1.919327
WIND DEFLECTION (IN)													
5 MPH	.00	.21	.81	1.79	3.11	4.77	6.75	9.03	11.62	14.51	17.68	21.15	24.90
10 MPH	.00	.42	1.63	3.58	6.23	9.54	13.49	18.07	23.25	29.01	35.37	42.30	49.80
20 MPH	.00	.84	3.26	7.16	12.45	19.08	26.99	36.14	46.49	58.03	70.74	84.60	99.60
30 MPH	.00	1.25	4.88	10.73	18.68	28.62	40.48	54.20	69.74	87.04	106.10	126.89	149.40

Bullet: Lyman # 311290 208 Gr.
Ballistic Coefficient: .305

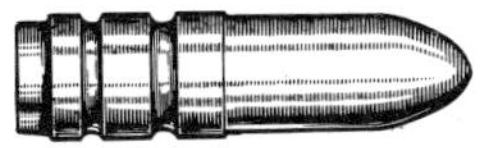

RANGE (YDS)	MUZZLE	50	100	150	200	250	300	350	400	450	500	550	600
VELOCITY (FPS)	2200.	2070.	1944.	1823.	1706.	1595.	1490.	1393.	1297.	1213.	1142.	1083.	1036.
ENERGY (FT-LB)	2235.	1978.	1745.	1534.	1344.	1174.	1025.	896.	777.	679.	602.	542.	496.
DROP (IN)	.00	-.87	-3.77	-8.97	-16.77	-27.56	-41.75	-59.91	-82.55	-110.37	-144.11	-184.51	-232.28
MID-RANGE (IN)	.00	.24	1.01	2.43	4.61	7.71	11.91	17.44	24.51	33.42	44.50	58.05	74.39
BULLET PATH (IN)	-.75	.64	.00	-2.93	-8.47	-17.00	-28.93	-44.83	-65.21	-90.77	-122.25	-160.38	-205.89
TIME OF FLIGHT (SEC)	.000000	.070299	.145088	.224779	.309841	.400803	.498140	.602287	.713960	.833659	.961267	1.096298	1.238017
WIND DEFLECTION (IN)													
5 MPH	.00	.19	.77	1.78	3.27	5.27	7.84	11.00	14.83	19.36	24.59	30.47	36.95
10 MPH	.00	.37	1.54	3.56	6.53	10.54	15.67	22.00	29.66	38.72	49.18	60.95	73.89
20 MPH	.00	.75	3.07	7.12	13.06	21.08	31.35	44.01	59.31	77.45	98.37	121.90	147.78
30 MPH	.00	1.12	4.61	10.68	19.60	31.62	47.02	66.01	88.97	116.17	147.55	182.85	221.67
VELOCITY (FPS)	2100.	1973.	1851.	1733.	1620.	1514.	1415.	1318.	1231.	1157.	1096.	1046.	1005.
ENERGY (FT-LB)	2036.	1798.	1582.	1387.	1212.	1059.	925.	803.	700.	618.	555.	506.	467.
DROP (IN)	.00	-.96	-4.15	-9.87	-18.48	-30.39	-46.12	-66.22	-91.30	-122.15	-159.46	-203.99	-256.44
MID-RANGE (IN)	.00	.26	1.11	2.68	5.09	8.52	13.19	19.32	27.17	37.08	49.34	64.25	82.09
BULLET PATH (IN)	-.75	.74	.00	-3.27	-9.43	-18.89	-32.18	-49.82	-72.46	-100.85	-135.72	-177.80	-227.80
TIME OF FLIGHT (SEC)	.000000	.073692	.152185	.235927	.325449	.421237	.523736	.633548	.751352	.877125	1.010466	1.150674	1.297047
WIND DEFLECTION (IN)													
5 MPH	.00	.20	.82	1.90	3.50	5.64	8.37	11.75	15.83	20.62	26.06	32.12	38.71
10 MPH	.00	.40	1.64	3.81	6.99	11.28	16.75	23.50	31.67	41.23	52.13	64.23	77.42
20 MPH	.00	.80	3.28	7.62	13.99	22.56	33.50	47.01	63.33	82.46	104.26	128.47	154.85
30 MPH	.00	1.20	4.93	11.43	20.98	33.84	50.25	70.51	95.00	123.69	156.38	192.70	232.27
VELOCITY (FPS)	2000.	1877.	1758.	1644.	1536.	1436.	1339.	1249.	1172.	1108.	1056.	1014.	977.
ENERGY (FT-LB)	1847.	1627.	1428.	1248.	1090.	952.	828.	721.	634.	567.	515.	475.	441.
DROP (IN)	.00	-1.06	-4.58	-10.91	-20.45	-33.70	-51.18	-73.50	-101.42	-135.65	-176.94	-225.99	-283.48
MID-RANGE (IN)	.00	.29	1.23	2.96	5.64	9.48	14.68	21.50	30.26	41.25	54.77	71.12	90.56
BULLET PATH (IN)	-.75	.86	.00	-3.66	-10.54	-21.12	-35.93	-55.58	-80.84	-112.40	-151.02	-197.40	-252.23
TIME OF FLIGHT (SEC)	.000000	.077425	.159993	.248226	.342622	.443639	.551784	.667835	.791894	.923639	1.062417	1.207495	1.358326
WIND DEFLECTION (IN)													
5 MPH	.00	.21	.88	2.04	3.75	6.04	8.96	12.57	16.89	21.88	27.49	33.66	40.33
10 MPH	.00	.43	1.76	4.09	7.50	12.08	17.91	25.14	33.77	43.76	54.99	67.32	80.67
20 MPH	.00	.85	3.52	8.18	15.00	24.16	35.83	50.28	67.55	87.52	109.97	134.64	161.33
30 MPH	.00	1.28	5.28	12.26	22.50	36.24	53.74	75.42	101.32	131.28	164.96	201.96	242.00
VELOCITY (FPS)	1900.	1781.	1665.	1556.	1454.	1357.	1265.	1186.	1119.	1065.	1021.	984.	951.
ENERGY (FT-LB)	1667.	1464.	1281.	1118.	977.	851.	740.	649.	579.	524.	482.	447.	418.
DROP (IN)	.00	-1.18	-5.09	-12.14	-22.78	-37.56	-57.06	-82.00	-113.12	-151.15	-196.82	-250.80	-313.74
MID-RANGE (IN)	.00	.32	1.37	3.30	6.30	10.58	16.39	24.03	33.78	45.96	60.85	78.72	99.81
BULLET PATH (IN)	-.75	.99	.00	-4.12	-11.84	-23.70	-40.28	-62.30	-90.49	-125.61	-168.35	-219.41	-279.42
TIME OF FLIGHT (SEC)	.000000	.081550	.168665	.261855	.361581	.468302	.582820	.705363	.835682	.973176	1.117098	1.266863	1.422055
WIND DEFLECTION (IN)													
5 MPH	.00	.23	.95	2.20	4.03	6.47	9.60	13.44	17.96	23.11	28.83	35.06	41.77
10 MPH	.00	.46	1.90	4.40	8.06	12.95	19.21	26.88	35.92	46.23	57.66	70.13	83.54
20 MPH	.00	.92	3.79	8.80	16.12	25.89	38.42	53.76	71.84	92.45	115.32	140.25	167.09
30 MPH	.00	1.37	5.69	13.21	24.18	38.84	57.62	80.64	107.77	138.68	172.99	210.38	250.63

Bullet: Lyman # 311290 208 Gr.
Ballistic Coefficient: .305
(Cont'd.)

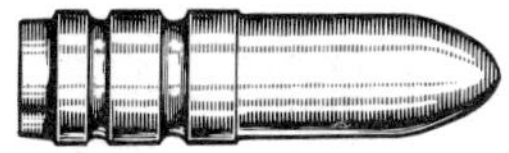

RANGE (YDS)	MUZZLE	50	100	150	200	250	300	350	400	450	500	550	600
VELOCITY (FPS)	1800.	1684.	1574.	1470.	1373.	1279.	1198.	1129.	1073.	1028.	989.	956.	927.
ENERGY (FT-LB)	1496.	1309.	1143.	998.	870.	756.	662.	589.	532.	488.	452.	422.	396.
DROP (IN)	.00	-1.32	-5.70	-13.59	-25.54	-42.09	-63.98	-91.92	-126.66	-168.92	-219.39	-278.72	-347.41
MID-RANGE (IN)	.00	.36	1.53	3.70	7.07	11.88	18.40	26.93	37.79	51.27	67.60	87.04	109.73
BULLET PATH (IN)	-.75	1.15	.00	-4.67	-13.39	-26.73	-45.39	-70.11	-101.63	-140.67	-187.91	-244.01	-309.49
TIME OF FLIGHT (SEC)	.000000	.086173	.178342	.276975	.382516	.495758	.617031	.746144	.882547	1.025489	1.174354	1.328702	1.488219
WIND DEFLECTION (IN)													
5 MPH	.00	.25	1.03	2.37	4.33	6.96	10.30	14.33	19.00	24.24	30.01	36.26	42.96
10 MPH	.00	.50	2.05	4.75	8.66	13.92	20.60	28.65	37.99	48.49	60.02	72.52	85.93
20 MPH	.00	1.00	4.11	9.50	17.31	27.84	41.20	57.31	75.99	96.97	120.04	145.04	171.85
30 MPH	.00	1.50	6.16	14.24	25.97	41.76	61.79	85.96	113.98	145.46	180.06	217.55	257.78
VELOCITY (FPS)	1700.	1589.	1485.	1388.	1292.	1209.	1138.	1081.	1034.	995.	961.	931.	904.
ENERGY (FT-LB)	1335.	1166.	1018.	889.	771.	675.	598.	539.	494.	457.	426.	400.	377.
DROP (IN)	.00	-1.48	-6.40	-15.31	-28.73	-47.36	-71.95	-103.23	-141.93	-188.74	-244.30	-309.17	-383.99
MID-RANGE (IN)	.00	.40	1.72	4.17	7.96	13.35	20.67	30.21	42.24	57.03	74.79	95.72	120.04
BULLET PATH (IN)	-.75	1.35	.00	-5.33	-15.17	-30.24	-51.25	-78.95	-114.08	-157.31	-209.30	-270.59	-341.84
TIME OF FLIGHT (SEC)	.000000	.091279	.188956	.293465	.405536	.525635	.653624	.789003	.931029	1.079054	1.232618	1.391390	1.555136
WIND DEFLECTION (IN)													
5 MPH	.00	.27	1.10	2.53	4.63	7.43	10.93	15.08	19.81	25.07	30.82	37.03	43.68
10 MPH	.00	.54	2.20	5.06	9.26	14.86	21.86	30.16	39.63	50.15	61.65	74.06	87.35
20 MPH	.00	1.07	4.39	10.12	18.51	29.73	43.72	60.32	79.25	100.30	123.29	148.12	174.70
30 MPH	.00	1.61	6.59	15.18	27.77	44.59	65.58	90.48	118.88	150.45	184.94	222.18	262.05
VELOCITY (FPS)	1600.	1495.	1398.	1301.	1216.	1145.	1086.	1038.	998.	964.	933.	906.	881.
ENERGY (FT-LB)	1182.	1032.	902.	782.	683.	605.	544.	498.	460.	429.	402.	379.	358.
DROP (IN)	.00	-1.67	-7.28	-17.34	-32.55	-53.63	-81.34	-116.39	-159.48	-211.27	-272.31	-343.24	-424.71
MID-RANGE (IN)	.00	.45	1.97	4.71	8.99	15.10	23.32	33.94	47.19	63.32	82.52	105.01	131.02
BULLET PATH (IN)	-.75	1.59	.00	-6.05	-17.23	-34.31	-58.00	-89.03	-128.11	-175.88	-232.91	-299.82	-377.28
TIME OF FLIGHT (SEC)	.000000	.097003	.200795	.312081	.431388	.558616	.693297	.834698	.982152	1.135183	1.293450	1.456709	1.624791
WIND DEFLECTION (IN)													
5 MPH	.00	.29	1.17	2.71	4.96	7.91	11.51	15.70	20.43	25.65	31.32	37.44	43.98
10 MPH	.00	.57	2.34	5.43	9.92	15.82	23.02	31.41	40.86	51.29	62.65	74.88	87.96
20 MPH	.00	1.15	4.68	10.85	19.85	31.63	46.04	62.81	81.72	102.58	125.29	149.76	175.93
30 MPH	.00	1.72	7.02	16.28	29.77	47.45	69.06	94.22	122.58	153.88	187.94	224.64	263.89
VELOCITY (FPS)	1500.	1402.	1306.	1220.	1148.	1088.	1040.	1000.	966.	935.	907.	882.	859.
ENERGY (FT-LB)	1039.	908.	787.	688.	609.	547.	500.	462.	430.	404.	380.	359.	341.
DROP (IN)	.00	-1.95	-8.34	-19.82	-37.16	-61.07	-92.29	-131.52	-179.41	-236.53	-303.49	-380.97	-469.60
MID-RANGE (IN)	.00	.52	2.22	5.36	10.24	17.14	26.32	38.04	52.55	70.02	90.68	114.76	142.51
BULLET PATH (IN)	-.75	1.84	.00	-6.94	-19.73	-39.10	-65.77	-100.46	-143.81	-196.38	-258.80	-331.73	-415.82
TIME OF FLIGHT (SEC)	.000000	.103453	.214331	.333226	.460055	.594368	.735440	.882594	1.035344	1.193344	1.356346	1.524177	1.696722
WIND DEFLECTION (IN)													
5 MPH	.00	.30	1.26	2.92	5.28	8.30	11.92	16.07	20.71	25.81	31.36	37.33	43.71
10 MPH	.00	.61	2.52	5.85	10.57	16.61	23.84	32.14	41.42	51.63	62.72	74.66	87.42
20 MPH	.00	1.22	5.04	11.70	21.14	33.22	47.68	64.27	82.84	103.26	125.43	149.31	174.85
30 MPH	.00	1.82	7.57	17.54	31.71	49.83	71.51	96.41	124.26	154.89	188.15	223.97	262.27

Bullet: Lyman # 311290 208 Gr.
Ballistic Coefficient: .305

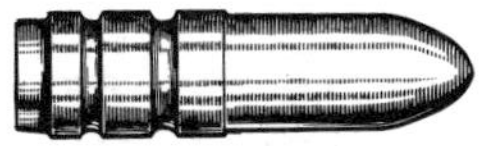

(Cont'd.)

RANGE (YDS)	MUZZLE	50	100	150	200	250	300	350	400	450	500	550	600
VELOCITY (FPS)	1400.	1303.	1218.	1146.	1087.	1039.	999.	965.	934.	907.	882.	858.	836.
ENERGY (FT-LB)	905.	784.	685.	607.	546.	499.	461.	430.	403.	380.	359.	340.	323.
DROP (IN)	.00	-2.26	-9.64	-22.89	-42.74	-69.92	-105.12	-149.00	-202.13	-265.11	-338.63	-423.32	-519.80
MID-RANGE (IN)	.00	.59	2.56	6.19	11.73	19.47	29.67	42.56	58.32	77.19	99.40	125.19	154.78
BULLET PATH (IN)	-.75	2.19	.00	-8.05	-22.70	-44.69	-74.69	-113.38	-161.31	-219.10	-287.42	-366.91	-458.20
TIME OF FLIGHT (SEC)	.000000	.111096	.230212	.357254	.491763	.633009	.780320	.933215	1.091351	1.254483	1.422438	1.595103	1.772409
WIND DEFLECTION (IN)													
5 MPH	.00	.35	1.40	3.15	5.56	8.56	12.10	16.12	20.61	25.54	30.89	36.65	42.83
10 MPH	.00	.70	2.80	6.31	11.12	17.12	24.19	32.25	41.22	51.07	61.78	73.31	85.66
20 MPH	.00	1.39	5.61	12.61	22.24	34.25	48.39	64.49	82.44	102.15	123.56	146.62	171.32
30 MPH	.00	2.09	8.41	18.92	33.37	51.37	72.58	96.74	123.66	153.22	185.33	219.93	256.97
VELOCITY (FPS)	1300.	1215.	1144.	1085.	1038.	998.	963.	933.	906.	881.	857.	836.	815.
ENERGY (FT-LB)	780.	682.	604.	544.	497.	460.	429.	402.	379.	358.	339.	322.	307.
DROP (IN)	.00	-2.63	-11.15	-26.30	-48.80	-79.35	-118.61	-167.13	-225.54	-294.49	-374.64	-466.60	-571.00
MID-RANGE (IN)	.00	.69	2.95	7.04	13.25	21.83	33.01	46.98	63.97	84.22	107.97	135.45	166.89
BULLET PATH (IN)	-.75	2.57	.00	-9.20	-25.75	-50.35	-83.66	-126.23	-178.68	-241.69	-315.89	-401.90	-500.35
TIME OF FLIGHT (SEC)	.000000	.119409	.246733	.381501	.522977	.670496	.823584	.981901	1.145205	1.313327	1.486153	1.663618	1.845692
WIND DEFLECTION (IN)													
5 MPH	.00	.35	1.40	3.11	5.41	8.23	11.55	15.33	19.55	24.19	29.24	34.71	40.57
10 MPH	.00	.71	2.81	6.22	10.81	16.47	23.10	30.66	39.09	48.38	58.49	69.41	81.15
20 MPH	.00	1.42	5.62	12.44	21.63	32.94	46.21	61.32	78.19	96.75	116.97	138.82	162.30
30 MPH	.00	2.12	8.43	18.66	32.44	49.41	69.31	91.98	117.28	145.13	175.46	208.24	243.45
VELOCITY (FPS)	1200.	1131.	1075.	1029.	990.	957.	927.	900.	876.	853.	831.	811.	792.
ENERGY (FT-LB)	665.	591.	533.	489.	453.	423.	397.	374.	354.	336.	319.	304.	289.
DROP (IN)	.00	-3.09	-12.95	-30.31	-55.86	-90.24	-133.57	-187.73	-252.18	-327.94	-415.63	-515.89	-629.15
MID-RANGE (IN)	.00	.80	3.39	8.03	14.98	24.47	36.67	51.85	70.23	92.07	117.58	147.05	180.61
BULLET PATH (IN)	-.75	3.01	.00	-10.51	-29.20	-56.73	-93.61	-140.52	-198.11	-267.02	-347.87	-441.28	-547.68
TIME OF FLIGHT (SEC)	.000000	.128863	.265035	.407766	.556430	.710585	.869912	1.034185	1.203247	1.376996	1.†55373	1.738357	1.925957
WIND DEFLECTION (IN)													
5 MPH	.00	.34	1.32	2.88	4.97	7.53	10.55	14.01	17.89	22.18	26.87	31.98	37.48
10 MPH	.00	.68	2.65	5.77	9.93	15.06	21.10	28.02	35.77	44.35	53.75	63.95	74.97
20 MPH	.00	1.36	5.29	11.53	19.86	30.13	42.21	56.03	71.54	88.70	107.49	127.90	149.94
30 MPH	.00	2.04	7.94	17.30	29.80	45.19	63.31	84.05	107.31	133.05	161.24	191.85	224.91
VELOCITY (FPS)	1100.	1050.	1008.	972.	941.	913.	887.	863.	841.	820.	801.	782.	763.
ENERGY (FT-LB)	559.	509.	469.	437.	409.	385.	363.	344.	327.	311.	296.	282.	269.
DROP (IN)	.00	-3.63	-15.12	-35.13	-64.28	-103.10	-152.32	-212.54	-284.39	-368.50	-465.51	-575.50	-699.69
MID-RANGE (IN)	.00	.94	3.93	9.22	17.02	27.53	41.03	57.73	77.91	101.82	129.65	161.38	197.52
BULLET PATH (IN)	-.75	3.56	.00	-12.08	-33.29	-64.18	-105.47	-157.75	-221.66	-297.84	-386.91	-488.97	-605.23
TIME OF FLIGHT (SEC)	.000000	.139691	.285587	.437165	.594052	.755982	.922766	1.094278	1.270438	1.451208	1.636583	1.826587	2.021247
WIND DEFLECTION (IN)													
5 MPH	.00	.29	1.13	2.47	4.28	6.53	9.20	12.30	15.80	19.71	24.02	28.74	33.87
10 MPH	.00	.59	2.26	4.94	8.55	13.05	18.41	24.59	31.60	39.41	48.04	57.48	67.74
20 MPH	.00	1.17	4.53	9.88	17.11	26.11	36.81	49.19	63.19	78.83	96.08	114.96	135.48
30 MPH	.00	1.76	6.79	14.82	25.66	39.16	55.22	73.78	94.79	118.24	144.12	172.44	203.22

Bullet: Lyman # 311284 210 Gr.
Ballistic Coefficient: .332

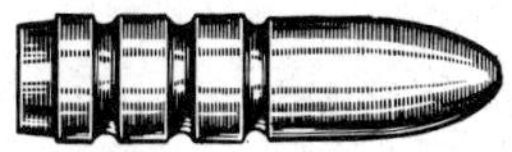

RANGE (YDS)	MUZZLE	50	100	150	200	250	300	350	400	450	500	550	600
VELOCITY (FPS)	2000.	1887.	1778.	1673.	1573.	1480.	1392.	1313.	1241.	1178.	1124.	1078.	1039.
ENERGY (FT-LB)	1865.	1660.	1473.	1305.	1154.	1021.	904.	804.	718.	647.	589.	542.	504.
DROP (IN)	.00	-1.05	-4.54	-10.78	-20.11	-32.94	-49.80	-71.12	-97.48	-129.52	-167.80	-212.96	-265.57
MID-RANGE (IN)	.00	.29	1.22	2.91	5.51	9.17	14.12	20.53	28.63	38.70	50.96	65.64	82.99
BULLET PATH (IN)	-.75	.84	.00	-3.59	-10.27	-20.46	-34.66	-53.34	-77.05	-106.44	-142.08	-184.59	-234.55
TIME OF FLIGHT (SEC)	.000000	.077222	.159134	.246125	.338592	.436919	.541455	.652445	.770018	.894160	1.024652	1.161082	1.302932
WIND DEFLECTION (IN)													
5 MPH	.00	.20	.80	1.86	3.40	5.45	8.05	11.22	14.96	19.29	24.17	29.58	35.46
18 MPH	.00	.39	1.61	3.72	6.79	10.90	16.10	22.43	29.92	38.57	48.34	59.15	70.92
20 MPH	.00	.78	3.22	7.44	13.58	21.80	32.19	44.86	59.85	77.14	96.68	118.30	141.83
30 MPH	.00	1.17	4.82	11.15	20.38	32.69	48.29	67.29	89.77	115.72	145.02	177.45	212.75
VELOCITY (FPS)	1900.	1790.	1685.	1585.	1491.	1402.	1322.	1249.	1185.	1130.	1083.	1044.	1010.
ENERGY (FT-LB)	1683.	1494.	1324.	1171.	1036.	917.	815.	728.	655.	595.	547.	508.	475.
DROP (IN)	.00	-1.17	-5.04	-11.97	-22.35	-36.69	-55.44	-79.15	-108.47	-143.96	-186.25	-235.93	-293.57
MID-RANGE (IN)	.00	.32	1.35	3.24	6.13	10.24	15.75	22.87	31.89	42.98	56.44	72.49	91.35
BULLET PATH (IN)	-.75	.98	.00	-4.03	-11.51	-22.95	-38.80	-59.62	-86.04	-118.63	-158.02	-204.80	-259.55
TIME OF FLIGHT (SEC)	.000000	.081332	.167696	.259488	.357096	.460875	.571085	.687867	.811231	.940981	1.076726	1.217952	1.364203
WIND DEFLECTION (IN)													
5 MPH	.00	.21	.86	1.99	3.63	5.82	8.57	11.90	15.81	20.28	25.28	30.76	36.68
10 MPH	.00	.42	1.72	3.99	7.27	11.64	17.14	23.80	31.62	40.56	50.56	61.52	73.36
20 MPH	.00	.84	3.45	7.97	14.54	23.28	34.29	47.60	63.24	81.12	101.11	123.03	146.73
30 MPH	.00	1.26	5.17	11.96	21.81	34.92	51.43	71.40	94.86	121.68	151.67	184.55	220.09
VELOCITY (FPS)	1800.	1694.	1594.	1499.	1410.	1329.	1255.	1190.	1134.	1087.	1047.	1013.	982.
ENERGY (FT-LB)	1511.	1338.	1184.	1047.	927.	823.	735.	661.	600.	550.	511.	478.	450.
DROP (IN)	.00	-1.31	-5.63	-13.37	-25.02	-41.05	-61.98	-88.45	-121.06	-160.40	-207.08	-261.68	-324.73
MID-RANGE (IN)	.00	.35	1.51	3.62	6.88	11.48	17.62	25.54	35.50	47.74	62.50	79.98	100.39
BULLET PATH (IN)	-.75	1.13	.00	-4.55	-13.01	-25.85	-43.59	-66.87	-96.29	-132.44	-175.93	-227.34	-287.20
TIME OF FLIGHT (SEC)	.000000	.085900	.177195	.274271	.377488	.487118	.603311	.726091	.855284	.990514	1.131269	1.277084	1.427608
WIND DEFLECTION (IN)													
5 MPH	.00	.23	.93	2.14	3.89	6.20	9.09	12.56	16.60	21.17	26.22	31.72	37.63
10 MPH	.00	.45	1.85	4.27	7.77	12.40	18.18	25.13	33.20	42.33	52.44	63.43	75.26
20 MPH	.00	.90	3.71	8.54	15.54	24.80	36.37	50.25	66.39	84.66	104.87	126.87	150.52
30 MPH	.00	1.36	5.56	12.81	23.31	37.20	54.55	75.38	99.59	126.99	157.31	190.30	225.78
VELOCITY (FPS)	1700.	1599.	1504.	1415.	1333.	1259.	1194.	1137.	1089.	1049.	1014.	984.	957.
ENERGY (FT-LB)	1347.	1192.	1054.	933.	828.	739.	664.	603.	553.	513.	480.	451.	427.
DROP (IN)	.00	-1.46	-6.32	-15.06	-28.14	-46.11	-69.58	-99.15	-135.42	-179.00	-230.46	-290.35	-359.20
MID-RANGE (IN)	.00	.40	1.69	4.09	7.75	12.88	19.73	28.55	39.57	53.02	69.12	88.06	110.02
BULLET PATH (IN)	-.75	1.32	.00	-5.21	-14.76	-29.18	-49.12	-75.16	-107.89	-147.94	-195.87	-252.22	-317.53
TIME OF FLIGHT (SEC)	.000000	.090987	.187735	.290605	.399878	.515708	.638129	.766976	.901885	1.042348	1.187893	1.338161	1.492885
WIND DEFLECTION (IN)													
5 MPH	.00	.24	.99	2.28	4.13	6.56	9.57	13.14	17.25	21.84	26.89	32.35	38.20
10 MPH	.00	.48	1.98	4.56	8.26	13.12	19.13	26.28	34.50	43.69	53.77	64.69	76.39
20 MPH	.00	.97	3.96	9.12	16.52	26.23	38.27	52.56	68.99	87.38	107.55	129.39	152.79
30 MPH	.00	1.45	5.95	13.67	24.78	39.35	57.40	78.85	103.49	131.07	161.32	194.08	229.18

Bullet: Lyman # 311284 210 Gr.
Ballistic Coefficient: .332

(Cont'd.)

RANGE (YDS)	MUZZLE	50	100	150	200	250	300	350	400	450	500	550	600
VELOCITY (FPS)	1600.	1505.	1415.	1334.	1260.	1194.	1137.	1089.	1049.	1015.	984.	957.	933.
ENERGY (FT-LB)	1194.	1055.	934.	829.	740.	665.	603.	553.	513.	480.	452.	427.	406.
DROP (IN)	.00	-1.65	-7.19	-17.07	-31.81	-52.07	-78.41	-111.45	-151.79	-200.01	-256.65	-322.25	-397.20
MID-RANGE (IN)	.00	.45	1.94	4.62	8.72	14.48	22.14	31.93	44.08	58.79	76.26	96.68	120.19
BULLET PATH (IN)	-.75	1.57	.00	-5.90	-16.68	-32.96	-55.34	-84.41	-120.78	-165.02	-217.70	-279.32	-350.30
TIME OF FLIGHT (SEC)	.000000	.096692	.199504	.308716	.424482	.546841	.675628	.810480	.950891	1.096386	1.246608	1.401285	1.560213
WIND DEFLECTION (IN)													
5 MPH	.00	.26	1.06	2.42	4.35	6.87	9.96	13.57	17.68	22.23	27.20	32.56	38.30
10 MPH	.00	.52	2.11	4.83	8.71	13.74	19.91	27.14	35.36	44.46	54.40	65.13	76.60
20 MPH	.00	1.04	4.23	9.67	17.42	27.49	39.82	54.29	70.71	88.93	108.81	130.25	153.19
30 MPH	.00	1.55	6.34	14.50	26.13	41.23	59.73	81.43	106.07	133.39	163.21	195.38	229.79
VELOCITY (FPS)	1500.	1411.	1330.	1256.	1191.	1135.	1087.	1047.	1013.	983.	956.	931.	909.
ENERGY (FT-LB)	1049.	928.	825.	736.	661.	600.	551.	511.	478.	450.	426.	404.	385.
DROP (IN)	.00	-1.94	-8.24	-19.44	-36.18	-59.04	-88.63	-125.54	-170.36	-223.63	-285.88	-357.48	-439.04
MID-RANGE (IN)	.00	.52	2.18	5.21	9.85	16.33	24.87	35.69	49.00	65.01	83.90	105.79	130.89
BULLET PATH (IN)	-.75	1.81	.00	-6.71	-18.95	-37.31	-62.41	-94.83	-135.15	-183.93	-241.68	-308.79	-385.85
TIME OF FLIGHT (SEC)	.000000	.103124	.212658	.328751	.451433	.580530	.715668	.856338	1.002072	1.152516	1.307403	1.466531	1.629745
WIND DEFLECTION (IN)													
5 MPH	.00	.27	1.11	2.53	4.53	7.09	10.18	13.76	17.78	22.22	27.05	32.25	37.82
10 MPH	.00	.55	2.23	5.06	9.05	14.17	20.36	27.52	35.56	44.44	54.10	64.51	75.64
20 MPH	.00	1.10	4.46	10.12	18.10	28.35	40.72	55.03	71.13	88.89	108.21	129.02	151.27
30 MPH	.00	1.65	6.68	15.18	27.16	42.52	61.07	82.55	106.69	133.33	162.31	193.53	226.91
VELOCITY (FPS)	1400.	1320.	1247.	1183.	1128.	1082.	1042.	1009.	979.	952.	928.	906.	885.
ENERGY (FT-LB)	914.	812.	725.	653.	593.	545.	507.	474.	447.	423.	402.	383.	366.
DROP (IN)	.00	-2.23	-9.43	-22.26	-41.28	-67.11	-100.35	-141.56	-191.28	-250.06	-318.23	-396.43	-485.21
MID-RANGE (IN)	.00	.59	2.49	5.96	11.21	18.44	27.91	39.82	54.36	71.72	92.00	115.45	142.26
BULLET PATH (IN)	-.75	2.11	.00	-7.74	-21.67	-42.40	-70.55	-106.67	-151.30	-204.99	-268.06	-341.18	-424.86
TIME OF FLIGHT (SEC)	.000000	.110382	.227337	.350870	.480779	.616668	.758020	.904382	1.055415	1.210860	1.370522	1.534252	1.701941
WIND DEFLECTION (IN)													
5 MPH	.00	.29	1.15	2.59	4.59	7.12	10.13	13.59	17.45	21.70	26.32	31.30	36.63
10 MPH	.00	.57	2.30	5.18	9.19	14.25	20.27	27.17	34.90	43.40	52.64	62.60	73.26
20 MPH	.00	1.14	4.59	10.36	18.38	28.50	40.54	54.34	69.79	86.79	105.28	125.20	146.51
30 MPH	.00	1.71	6.89	15.54	27.57	42.74	60.81	81.51	104.69	130.19	157.92	187.80	219.77
VELOCITY (FPS)	1300.	1230.	1168.	1115.	1071.	1033.	1001.	972.	946.	922.	900.	880.	861.
ENERGY (FT-LB)	788.	705.	636.	580.	534.	498.	467.	440.	417.	397.	378.	361.	346.
DROP (IN)	.00	-2.59	-10.95	-25.66	-47.34	-76.58	-113.94	-159.96	-215.15	-279.87	-354.76	-440.35	-537.16
MID-RANGE (IN)	.00	.69	2.89	6.82	12.73	20.82	31.31	44.39	60.23	78.95	100.79	125.96	154.64
BULLET PATH (IN)	-.75	2.51	.00	-8.87	-24.70	-48.09	-79.61	-119.78	-169.12	-227.99	-297.03	-376.78	-467.74
TIME OF FLIGHT (SEC)	.000000	.118690	.243925	.375449	.512821	.655525	.803139	.955348	1.111911	1.272645	1.437415	1.606119	1.778686
WIND DEFLECTION (IN)													
5 MPH	.00	.29	1.16	2.58	4.51	6.92	9.75	12.99	16.62	20.61	24.95	29.65	34.68
10 MPH	.00	.58	2.32	5.16	9.03	13.83	19.51	25.99	33.23	41.22	49.91	59.29	69.36
20 MPH	.00	1.16	4.63	10.31	18.05	27.67	39.01	51.97	66.47	82.43	99.82	118.58	138.71
30 MPH	.00	1.75	6.95	15.47	27.08	41.50	58.52	77.96	99.70	123.65	149.72	177.88	208.07

Bullet: Lyman # 323470 160 Gr.
Ballistic Coefficient: .187

RANGE (YDS)	MUZZLE	50	100	150	200	250	300	350	400	450	500	550	600
VELOCITY (FPS)	2400.	2181.	1989.	1808.	1640.	1494.	1366.	1254.	1161.	1087.	1029.	983.	944.
ENERGY (FT-LB)	2046.	1689.	1406.	1162.	955.	793.	663.	559.	478.	419.	376.	343.	316.
DROP (IN)	.00	-.77	-3.36	-8.15	-15.59	-26.26	-40.85	-60.10	-84.90	-116.14	-154.71	-201.45	-257.16
MID-RANGE (IN)	.00	.21	.91	2.26	4.43	7.67	12.28	18.55	26.86	37.59	51.10	67.70	87.72
BULLET PATH (IN)	-.75	.54	.00	-2.73	-8.12	-16.73	-29.27	-46.47	-69.21	-98.39	-134.90	-179.59	-233.24
TIME OF FLIGHT (SEC)	.000000	.065594	.137618	.216711	.303838	.399812	.504880	.619586	.744068	.877819	1.019804	1.169061	1.324945
WIND DEFLECTION (IN)													
5 MPH	.00	.27	1.11	2.57	4.74	7.68	11.43	16.02	21.48	27.75	34.74	42.38	50.60
10 MPH	.00	.54	2.22	5.14	9.48	15.37	22.86	32.05	42.96	55.50	69.49	84.75	101.19
20 MPH	.00	1.09	4.44	10.28	18.95	30.73	45.72	64.09	85.91	110.99	138.97	169.51	202.38
30 MPH	.00	1.63	6.66	15.42	28.43	46.10	68.58	96.14	128.87	166.49	208.46	254.26	303.57
VELOCITY (FPS)	2300.	2094.	1907.	1731.	1571.	1435.	1314.	1210.	1125.	1060.	1008.	965.	928.
ENERGY (FT-LB)	1879.	1557.	1291.	1065.	876.	732.	613.	520.	450.	399.	361.	331.	306.
DROP (IN)	.00	-.84	-3.66	-8.87	-16.97	-28.60	-44.47	-65.38	-92.23	-125.93	-167.33	-217.27	-276.46
MID-RANGE (IN)	.00	.22	.99	2.46	4.83	8.37	13.37	20.17	29.15	40.67	55.06	72.67	93.76
BULLET PATH (IN)	-.75	.62	.00	-3.00	-8.91	-18.33	-32.00	-50.71	-75.36	-106.84	-146.05	-193.78	-250.76
TIME OF FLIGHT (SEC)	.000000	.068435	.143517	.226090	.317137	.417080	.526372	.645434	.774129	.911672	1.056980	1.209240	1.367915
WIND DEFLECTION (IN)													
5 MPH	.00	.28	1.15	2.68	4.95	8.01	11.89	16.62	22.21	28.57	35.62	43.28	51.51
10 MPH	.00	.57	2.30	5.36	9.90	16.01	23.77	33.25	44.42	57.15	71.25	86.57	103.01
20 MPH	.00	1.13	4.60	10.71	19.81	32.03	47.54	66.50	88.84	114.30	142.49	173.13	206.03
30 MPH	.00	1.70	6.91	16.07	29.71	48.04	71.32	99.75	133.26	171.45	213.74	259.70	309.04
VELOCITY (FPS)	2200.	2008.	1826.	1656.	1507.	1378.	1264.	1169.	1093.	1035.	987.	947.	913.
ENERGY (FT-LB)	1719.	1432.	1184.	974.	807.	675.	568.	485.	425.	380.	346.	319.	296.
DROP (IN)	.00	-.91	-3.98	-9.65	-18.49	-31.18	-48.45	-71.18	-100.25	-136.57	-180.98	-234.29	-297.13
MID-RANGE (IN)	.00	.24	1.08	2.68	5.27	9.13	14.57	21.95	31.63	43.97	59.31	77.95	100.11
BULLET PATH (IN)	-.75	.70	.00	-3.31	-9.78	-20.11	-35.01	-55.37	-82.08	-116.03	-158.08	-209.02	-269.50
TIME OF FLIGHT (SEC)	.000000	.071378	.149733	.236025	.331110	.435241	.548966	.672481	.805344	.946554	1.095115	1.250355	1.411828
WIND DEFLECTION (IN)													
5 MPH	.00	.28	1.18	2.77	5.14	8.30	12.31	17.18	22.87	29.30	36.37	44.03	52.24
10 MPH	.00	.56	2.35	5.54	10.28	16.60	24.62	34.36	45.74	58.59	72.74	88.06	104.48
20 MPH	.00	1.12	4.71	11.08	20.55	33.20	49.24	68.71	91.48	117.19	145.48	176.12	208.96
30 MPH	.00	1.69	7.06	16.62	30.83	49.81	73.85	103.07	137.22	175.78	218.22	264.19	313.45
VELOCITY (FPS)	2100.	1913.	1737.	1575.	1439.	1317.	1213.	1128.	1061.	1009.	966.	929.	896.
ENERGY (FT-LB)	1566.	1300.	1072.	881.	736.	617.	523.	452.	400.	362.	331.	307.	285.
DROP (IN)	.00	-1.00	-4.38	-10.64	-20.40	-34.38	-53.37	-78.27	-110.00	-149.41	-197.33	-254.48	-321.60
MID-RANGE (IN)	.00	.27	1.19	2.95	5.82	10.08	16.05	24.11	34.60	47.89	64.29	84.05	107.43
BULLET PATH (IN)	-.75	.81	.00	-3.69	-10.89	-22.30	-38.73	-61.07	-90.23	-127.07	-172.43	-227.01	-291.57
TIME OF FLIGHT (SEC)	.000000	.074845	.157149	.247902	.347582	.456592	.575367	.703789	.841088	.986181	1.138245	1.296735	1.461292
WIND DEFLECTION (IN)													
5 MPH	.00	.30	1.26	2.96	5.44	8.75	12.92	17.93	23.73	30.21	37.31	44.97	53.17
10 MPH	.00	.60	2.52	5.92	10.89	17.50	25.84	35.87	47.46	60.42	74.62	89.94	106.33
20 MPH	.00	1.20	5.03	11.83	21.78	35.01	51.67	71.73	94.92	120.85	149.23	179.88	212.66
30 MPH	.00	1.80	7.55	17.75	32.67	52.51	77.51	107.60	142.38	181.27	223.85	269.82	318.99

Bullet: Lyman # 323470 160 Gr. 
Ballistic Coefficient: .187
(Cont'd.)

RANGE (YDS)	MUZZLE	50	100	150	200	250	300	350	400	450	500	550	600
VELOCITY (FPS)	2000.	1819.	1649.	1502.	1374.	1261.	1166.	1091.	1033.	986.	946.	911.	880.
ENERGY (FT-LB)	1421.	1175.	966.	802.	670.	564.	483.	423.	379.	345.	318.	295.	275.
DROP (IN)	.00	-1.11	-4.84	-11.76	-22.57	-37.98	-58.88	-86.16	-120.71	-163.40	-215.00	-276.17	-347.77
MID-RANGE (IN)	.00	.30	1.31	3.27	6.44	11.12	17.67	26.44	37.79	52.05	69.52	90.40	115.01
BULLET PATH (IN)	-.75	.94	.00	-4.13	-12.13	-24.75	-42.85	-67.33	-99.10	-138.98	-187.79	-246.16	-314.97
TIME OF FLIGHT (SEC)	.000000	.078660	.165296	.260723	.365198	.479284	.603155	.736345	.877839	1.026655	1.182129	1.343822	1.511446
WIND DEFLECTION (IN)													
5 MPH	.00	.32	1.35	3.14	5.74	9.18	13.48	18.60	24.45	30.95	38.03	45.66	53.81
10 MPH	.00	.64	2.69	6.29	11.47	18.35	26.96	37.20	48.90	61.89	76.05	91.31	107.61
20 MPH	.00	1.29	5.38	12.57	22.95	36.71	53.91	74.39	97.80	123.78	152.11	182.63	215.23
30 MPH	.00	1.93	8.08	18.86	34.42	55.06	80.87	111.59	146.70	185.67	228.16	273.94	322.84
VELOCITY (FPS)	1900.	1725.	1565.	1430.	1310.	1206.	1122.	1057.	1006.	963.	926.	894.	865.
ENERGY (FT-LB)	1282.	1057.	870.	726.	609.	517.	447.	397.	359.	330.	305.	284.	266.
DROP (IN)	.00	-1.23	-5.38	-13.08	-25.05	-42.10	-65.12	-95.03	-132.68	-178.89	-234.38	-299.91	-376.29
MID-RANGE (IN)	.00	.33	1.46	3.64	7.15	12.31	19.48	29.03	41.28	56.57	75.09	97.15	123.02
BULLET PATH (IN)	-.75	1.08	.00	-4.64	-13.55	-27.53	-47.49	-74.33	-108.92	-152.07	-204.49	-266.96	-340.27
TIME OF FLIGHT (SEC)	.000000	.082873	.174258	.274572	.384259	.503724	.632800	.770681	.916285	1.068813	1.227737	1.392706	1.563496
WIND DEFLECTION (IN)													
5 MPH	.00	.35	1.44	3.32	6.03	9.59	14.00	19.19	25.05	31.53	38.57	46.14	54.22
10 MPH	.00	.69	2.88	6.64	12.05	19.18	28.00	38.38	50.11	63.06	77.13	92.27	108.44
20 MPH	.00	1.38	5.76	13.28	24.10	38.36	56.01	76.75	100.22	126.12	154.27	184.55	216.88
30 MPH	.00	2.07	8.64	19.92	36.15	57.55	84.01	115.13	150.32	189.18	231.40	276.82	325.32
VELOCITY (FPS)	1800.	1632.	1488.	1361.	1249.	1157.	1084.	1027.	981.	942.	908.	877.	849.
ENERGY (FT-LB)	1151.	946.	786.	657.	554.	475.	417.	375.	342.	315.	293.	273.	256.
DROP (IN)	.00	-1.37	-6.00	-14.59	-27.87	-46.74	-72.08	-104.80	-145.73	-195.65	-255.23	-325.33	-406.69
MID-RANGE (IN)	.00	.37	1.63	4.06	7.93	13.61	21.45	31.81	45.00	61.28	80.88	104.12	131.26
BULLET PATH (IN)	-.75	1.25	.00	-5.21	-15.12	-30.62	-52.59	-81.93	-119.49	-166.03	-222.24	-288.97	-366.95
TIME OF FLIGHT (SEC)	.000000	.087542	.183929	.289423	.404573	.529490	.663635	.805964	.955527	1.111691	1.274036	1.442286	1.616275
WIND DEFLECTION (IN)													
5 MPH	.00	.37	1.52	3.47	6.27	9.93	14.40	19.59	25.42	31.83	38.78	46.25	54.23
10 MPH	.00	.74	3.04	6.94	12.54	19.86	28.80	39.18	50.84	63.66	77.56	92.51	108.46
20 MPH	.00	1.48	6.08	13.88	25.08	39.71	57.60	78.37	101.68	127.32	155.13	185.02	216.93
30 MPH	.00	2.22	9.11	20.82	37.61	59.57	86.40	117.55	152.52	190.97	232.69	277.53	325.39
VELOCITY (FPS)	1700.	1544.	1411.	1293.	1193.	1112.	1049.	999.	957.	921.	890.	861.	834.
ENERGY (FT-LB)	1027.	847.	708.	594.	505.	439.	391.	355.	326.	302.	281.	263.	247.
DROP (IN)	.00	-1.54	-6.74	-16.31	-31.09	-51.97	-79.87	-115.64	-160.09	-213.90	-277.88	-352.82	-439.47
MID-RANGE (IN)	.00	.41	1.83	4.52	8.81	15.06	23.63	34.83	48.96	66.24	86.97	111.42	139.86
BULLET PATH (IN)	-.75	1.45	.00	-5.83	-16.87	-34.01	-58.16	-90.19	-130.90	-180.96	-241.20	-312.40	-395.30
TIME OF FLIGHT (SEC)	.000000	.092694	.194339	.305441	.426337	.556764	.695837	.842493	.995981	1.155803	1.321630	1.493252	1.670553
WIND DEFLECTION (IN)													
5 MPH	.00	.39	1.57	3.58	6.46	10.17	14.65	19.79	25.53	31.83	38.66	45.99	53.83
10 MPH	.00	.78	3.14	7.17	12.92	20.34	29.29	39.57	51.06	63.66	77.31	91.99	107.66
20 MPH	.00	1.57	6.29	14.34	25.84	40.69	58.58	79.15	102.11	127.31	154.63	183.98	215.33
30 MPH	.00	2.35	9.43	21.51	38.75	61.03	87.87	118.72	153.17	190.97	231.94	275.97	322.99

Bullet: Lyman # 323470 160 Gr.

Ballistic Coefficient: .187
(Cont'd.)

RANGE (YDS)	MUZZLE	50	100	150	200	250	300	350	400	450	500	550	600
VELOCITY (FPS)	1600.	1462.	1338.	1230.	1141.	1072.	1017.	973.	935.	902.	872.	844.	818.
ENERGY (FT-LB)	909.	759.	635.	537.	463.	408.	368.	336.	311.	289.	270.	253.	238.
DROP (IN)	.00	-1.73	-7.56	-18.25	-34.70	-57.80	-88.45	-127.46	-175.58	-233.54	-302.16	-382.19	-474.40
MID-RANGE (IN)	.00	.47	2.05	5.03	9.79	16.66	25.97	38.03	53.09	71.41	93.28	118.96	148.75
BULLET PATH (IN)	-.75	1.67	.00	-6.53	-18.82	-37.77	-64.26	-99.12	-143.08	-196.88	-261.35	-337.22	-425.28
TIME OF FLIGHT (SEC)	.000000	.098120	.205457	.322516	.449289	.585113	.728909	.879795	1.037184	1.200690	1.370060	1.545145	1.725873
WIND DEFLECTION (IN)													
5 MPH	.00	.38	1.58	3.63	6.54	10.24	14.64	19.67	25.27	31.41	38.07	45.22	52.88
10 MPH	.00	.77	3.16	7.26	13.07	20.48	29.29	39.34	50.54	62.82	76.13	90.45	105.75
20 MPH	.00	1.54	6.32	14.53	26.15	40.96	58.58	78.69	101.09	125.64	152.26	180.89	211.51
30 MPH	.00	2.31	9.48	21.79	39.22	61.44	87.86	118.03	151.63	188.46	228.39	271.34	317.26
VELOCITY (FPS)	1500.	1372.	1259.	1164.	1090.	1032.	985.	945.	911.	880.	852.	826.	801.
ENERGY (FT-LB)	799.	668.	563.	482.	422.	378.	345.	317.	295.	275.	258.	242.	228.
DROP (IN)	.00	-2.00	-8.62	-20.75	-39.27	-65.08	-99.03	-141.90	-194.36	-257.27	-331.36	-417.42	-516.23
MID-RANGE (IN)	.00	.53	2.31	5.69	11.01	18.60	28.77	41.80	57.91	77.43	100.60	127.73	159.09
BULLET PATH (IN)	-.75	1.93	.00	-7.44	-21.27	-42.40	-71.66	-109.85	-157.63	-215.84	-285.26	-366.63	-460.75
TIME OF FLIGHT (SEC)	.000000	.104630	.218878	.342907	.476241	.617859	.766783	.922353	1.084135	1.251841	1.425295	1.604404	1.789143
WIND DEFLECTION (IN)													
5 MPH	.00	.41	1.66	3.78	6.71	10.37	14.68	19.57	25.00	30.96	37.43	44.39	51.84
10 MPH	.00	.81	3.32	7.55	13.42	20.74	29.35	39.13	50.01	61.92	74.85	88.78	103.69
20 MPH	.00	1.63	6.65	15.10	26.84	41.49	58.71	78.27	100.02	123.85	149.70	177.55	207.38
30 MPH	.00	2.44	9.97	22.66	40.26	62.23	88.06	117.40	150.02	185.77	224.56	266.33	311.07
VELOCITY (FPS)	1400.	1283.	1185.	1105.	1044.	995.	954.	918.	887.	858.	831.	807.	783.
ENERGY (FT-LB)	696.	585.	498.	434.	387.	352.	323.	300.	279.	261.	246.	231.	218.
DROP (IN)	.00	-2.30	-9.88	-23.66	-44.52	-73.34	-110.90	-157.88	-215.11	-283.37	-363.40	-456.00	-561.65
MID-RANGE (IN)	.00	.60	2.64	6.46	12.38	20.73	31.79	45.79	63.03	83.80	108.37	137.03	169.92
BULLET PATH (IN)	-.75	2.27	.00	-8.46	-24.01	-47.50	-79.75	-121.42	-173.33	-236.27	-310.99	-398.27	-498.60
TIME OF FLIGHT (SEC)	.000000	.111985	.233767	.365023	.504821	.652120	.806195	.966565	1.132915	1.305044	1.482842	1.666270	1.855344
WIND DEFLECTION (IN)													
5 MPH	.00	.43	1.71	3.84	6.71	10.24	14.37	19.06	24.27	29.99	36.20	42.92	50.13
10 MPH	.00	.85	3.43	7.67	13.42	20.49	28.75	38.12	48.54	59.97	72.41	85.83	100.25
20 MPH	.00	1.70	6.86	15.35	26.84	40.97	57.49	76.23	97.07	119.95	144.82	171.67	200.51
30 MPH	.00	2.56	10.29	23.02	40.26	61.46	86.24	114.35	145.61	179.92	217.23	257.50	300.76
VELOCITY (FPS)	1300.	1198.	1116.	1052.	1002.	960.	923.	891.	862.	835.	810.	787.	764.
ENERGY (FT-LB)	600.	510.	442.	393.	356.	327.	303.	282.	264.	248.	233.	220.	207.
DROP (IN)	.00	-2.67	-11.39	-27.07	-50.57	-82.70	-124.15	-175.72	-238.20	-312.33	-398.91	-498.48	-611.90
MID-RANGE (IN)	.00	.71	3.03	7.32	13.89	23.04	35.01	50.07	68.53	90.66	116.76	147.02	181.80
BULLET PATH (IN)	-.75	2.65	.00	-9.60	-27.03	-53.09	-88.47	-133.97	-190.37	-258.43	-338.94	-432.44	-539.79
TIME OF FLIGHT (SEC)	.000000	.120297	.250160	.388732	.534944	.688022	.847456	1.012908	1.184162	1.361097	1.543662	1.731866	1.925750
WIND DEFLECTION (IN)													
5 MPH	.00	.43	1.71	3.75	6.46	9.78	13.65	18.06	22.98	28.39	34.30	40.71	47.62
10 MPH	.00	.86	3.41	7.49	12.92	19.55	27.31	36.12	45.95	56.78	68.61	81.42	95.24
20 MPH	.00	1.73	6.83	14.99	25.84	39.11	54.61	72.24	91.90	113.57	137.22	162.85	190.48
30 MPH	.00	2.59	10.24	22.48	38.76	58.66	81.92	108.35	137.85	170.35	205.82	244.27	285.72

Bullet: Lyman # 323471 214 Gr.
Ballistic Coefficient: .475

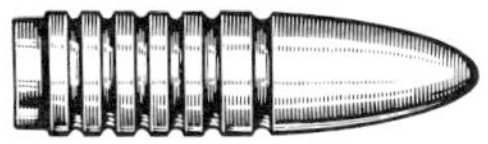

RANGE (YDS)	MUZZLE	50	100	150	200	250	300	350	400	450	500	550	600
VELOCITY (FPS)	2000.	1920.	1843.	1766.	1689.	1615.	1543.	1475.	1410.	1347.	1287.	1232.	1182.
ENERGY (FT-LB)	1900.	1752.	1613.	1482.	1355.	1239.	1132.	1034.	945.	862.	787.	721.	664.
DROP (IN)	.00	-1.01	-4.38	-10.30	-18.99	-30.73	-45.78	-64.49	-87.25	-114.37	-146.28	-183.44	-226.36
MID-RANGE (IN)	.00	.28	1.17	2.76	5.12	8.36	12.60	17.96	24.63	32.72	42.41	53.90	67.40
BULLET PATH (IN)	-.75	.80	.00	-3.35	-9.49	-18.65	-31.15	-47.29	-67.48	-92.04	-121.38	-155.98	-196.33
TIME OF FLIGHT (SEC)	.000000	.076541	.156280	.239411	.326268	.417108	.512142	.611571	.715583	.824431	.938427	1.057631	1.182006
WIND DEFLECTION (IN)													
5 MPH	.00	.14	.55	1.27	2.31	3.71	5.47	7.62	10.17	13.15	16.58	20.47	24.82
10 MPH	.00	.27	1.11	2.54	4.62	7.41	10.94	15.24	20.34	26.30	33.16	40.94	49.63
20 MPH	.00	.54	2.21	5.07	9.25	14.82	21.87	30.47	40.69	52.60	66.33	81.89	99.27
30 MPH	.00	.81	3.32	7.61	13.87	22.23	32.81	45.71	61.03	78.90	99.49	122.83	148.90
VELOCITY (FPS)	1900.	1823.	1746.	1669.	1596.	1525.	1458.	1394.	1331.	1272.	1218.	1170.	1128.
ENERGY (FT-LB)	1715.	1579.	1448.	1324.	1210.	1105.	1010.	923.	841.	769.	705.	651.	604.
DROP (IN)	.00	-1.13	-4.86	-11.44	-21.13	-34.22	-51.08	-72.05	-97.50	-127.85	-163.58	-205.19	-253.13
MID-RANGE (IN)	.00	.31	1.30	3.07	5.70	9.33	14.10	20.15	27.62	36.69	47.57	60.45	75.49
BULLET PATH (IN)	-.75	.93	.00	-3.77	-10.66	-20z94	-34.99	-53.16	-75.80	-103.34	-136.27	-175.07	-220.21
TIME OF FLIGHT (SEC)	.000000	.080601	.164670	.252551	.344471	.440638	.541253	.646494	.756659	.871999	.992547	1.118239	1.248907
WIND DEFLECTION (IN)													
5 MPH	.00	.15	.60	1.38	2.52	4.04	5.95	8.26	11.01	14.21	17.87	21.98	26.54
10 MPH	.00	.29	1.19	2.76	5.05	8.08	11.89	16.52	22.01	28.42	35.74	43.97	53.07
20 MPH	.00	.58	2.38	5.53	10.10	16.16	23.78	33.04	44.03	56.84	71.48	87.94	106.14
30 MPH	.00	.87	3.58	8.29	15.14	24.24	35.68	49.56	66.04	85.26	107.22	131.90	159.21
VELOCITY (FPS)	1800.	1722.	1646.	1574.	1504.	1438.	1374.	1312.	1255.	1203.	1157.	1116.	1080.
ENERGY (FT-LB)	1539.	1408.	1288.	1176.	1075.	982.	897.	818.	748.	688.	635.	591.	554.
DROP (IN)	.00	-1.26	-5.45	-12.83	-23.70	-38.48	-57.46	-81.05	-109.66	-143.84	-184.03	-230.69	-284.32
MID-RANGE (IN)	.00	.35	1.46	3.44	6.41	10.52	15.91	22.72	31.13	41.34	53.54	67.92	84.70
BULLET PATH (IN)	-.75	1.09	.00	-4.28	-12.06	-23.74	-39.62	-60.11	-85.63	-116.71	-153.79	-197.36	-247.89
TIME OF FLIGHT (SEC)	.000000	.085209	.174309	.267513	.365028	.467046	.573756	.685502	.802443	.924582	1.051824	1.183971	1.320738
WIND DEFLECTION (IN)													
5 MPH	.00	.17	.67	1.54	2.79	4.43	6.49	8.99	11.95	15.36	19.23	23.52	28.22
10 MPH	.00	.33	1.35	3.08	5.58	8.87	12.98	17.98	23.90	30.73	38.45	47.05	56.45
20 MPH	.00	.66	2.69	6.16	11.16	17.73	25.96	35.96	47.79	61.45	76.91	94.09	112.90
30 MPH	.00	.99	4.04	9.25	16.73	26.60	38.94	53.95	71.69	92.18	115.36	141.14	169.35
VELOCITY (FPS)	1700.	1625.	1553.	1485.	1420.	1356.	1295.	1239.	1189.	1144.	1105.	1071.	1041.
ENERGY (FT-LB)	1373.	1255.	1146.	1047.	957.	873.	797.	730.	672.	622.	580.	545.	514.
DROP (IN)	.00	-1.42	-6.11	-14.40	-26.70	-43.31	-64.63	-91.13	-123.33	-161.66	-206.61	-258.67	-318.29
MID-RANGE (IN)	.00	.39	1.64	3.87	7.24	11.87	17.91	25.53	34.96	46.39	60.00	75.99	94.52
BULLET PATH (IN)	-.75	1.26	.00	-4.86	-13.73	-26.90	-44.80	-67.87	-96.63	-131.53	-173.06	-221.69	-277.88
TIME OF FLIGHT (SEC)	.000000	.090247	.184657	.283434	.386768	.494879	.608108	.726541	.850156	.978826	1.112327	1.250355	1.392583
WIND DEFLECTION (IN)													
5 MPH	.00	.18	.72	1.65	2.98	4.73	6.93	9.58	12.70	16.25	20.24	24.62	29.37
10 MPH	.00	.35	1.44	3.30	5.95	9.45	13.85	19.17	25.39	32.51	40.48	49.24	58.74
20 MPH	.00	.71	2.88	6.59	11.91	18.90	27.70	38.33	50.78	65.02	80.95	98.48	117.48
30 MPH	.00	1.06	4.32	9.89	17.86	28.35	41.55	57.50	76.18	97.53	121.43	147.72	176.23

Bullet: Lyman # 323471 214 Gr.

Ballistic Coefficient: .475
(Cont'd.)

RANGE (YDS)	MUZZLE	50	100	150	200	250	300	350	400	450	500	550	600
VELOCITY (FPS)	1600.	1529.	1462.	1398.	1334.	1275.	1221.	1173.	1130.	1092.	1060.	1031.	1005.
ENERGY (FT-LB)	1216.	1111.	1015.	928.	846.	773.	709.	653.	606.	567.	534.	505.	480.
DROP (IN)	.00	-1.60	-6.94	-16.39	-30.27	-49.04	-73.16	-103.12	-139.40	-182.46	-232.78	-290.83	-357.04
MID-RANGE (IN)	.00	.44	1.87	4.43	8.19	13.38	20.19	28.81	39.42	52.21	67.36	85.03	105.36
BULLET PATH (IN)	-.75	1.50	.00	-5.60	-15.64	-30.55	-50.83	-76.95	-109.38	-148.59	-195.07	-249.27	-311.64
TIME OF FLIGHT (SEC)	.000000	.095903	.196240	.301192	.411057	.526091	.646332	.771722	.902099	1.037208	1.176726	1.320344	1.467817
WIND DEFLECTION (IN)													
5 MPH	.00	.19	.77	1.75	3.17	5.05	7.38	10.16	13.38	17.02	21.05	25.44	30.17
10 MPH	.00	.38	1.54	3.51	6.35	10.09	14.75	20.32	26.77	34.05	42.10	50.88	60.34
20 MPH	.00	.76	3.08	7.02	12.69	20.18	29.51	40.65	53.54	68.10	84.21	101.76	120.67
30 MPH	.00	1.14	4.61	10.53	19.04	30.28	44.26	60.97	80.31	102.15	126.31	152.64	181.01
VELOCITY (FPS)	1500.	1434.	1370.	1308.	1252.	1200.	1154.	1113.	1078.	1047.	1020.	995.	973.
ENERGY (FT-LB)	1069.	977.	892.	813.	744.	684.	632.	589.	552.	521.	494.	470.	449.
DROP (IN)	.00	-1.89	-8.01	-18.76	-34.57	-55.97	-83.40	-117.34	-158.27	-206.67	-262.98	-327.63	-401.04
MID-RANGE (IN)	.00	.51	2.12	4.98	9.27	15.20	22.94	32.67	44.57	58.79	75.52	94.92	117.13
BULLET PATH (IN)	-.75	1.74	.00	-6.37	-17.80	-34.81	-57.86	-87.42	-123.98	-167.99	-219.92	-280.19	-349.23
TIME OF FLIGHT (SEC)	.000000	.102288	.209285	.321342	.438595	.561042	.688582	.821009	.958035	1.099330	1.244612	1.393658	1.546287
WIND DEFLECTION (IN)													
5 MPH	.00	.20	.82	1.88	3.40	5.37	7.80	10.65	13.91	17.54	21.53	25.84	30.47
10 MPH	.00	.40	1.63	3.76	6.79	10.74	15.59	21.30	27.81	35.08	43.05	51.68	60.95
20 MPH	.00	.81	3.27	7.51	13.59	21.49	31.18	42.60	55.63	70.16	86.10	103.37	121.89
30 MPH	.00	1.21	4.90	11.27	20.38	32.23	46.77	63.89	83.44	105.25	129.16	155.05	182.84
VELOCITY (FPS)	1400.	1336.	1277.	1223.	1174.	1131.	1093.	1061.	1032.	1006.	983.	961.	941.
ENERGY (FT-LB)	931.	848.	775.	711.	655.	608.	568.	534.	506.	481.	459.	439.	421.
DROP (IN)	.00	-2.19	-9.24	-21.63	-39.85	-64.35	-95.64	-134.17	-180.40	-234.78	-297.73	-369.67	-450.91
MID-RANGE (IN)	.00	.58	2.43	5.74	10.69	17.45	26.19	37.10	50.35	66.11	84.51	105.71	129.82
BULLET PATH (IN)	-.75	2.05	.00	-7.40	-20.62	-40.14	-66.43	-99.96	-141.20	-190.59	-248.55	-315.49	-391.75
TIME OF FLIGHT (SEC)	.000000	.109691	.224547	.344611	.469826	.600033	.734982	.874347	1.017821	1.165153	1.316145	1.470638	1.628504
WIND DEFLECTION (IN)													
5 MPH	.00	.22	.90	2.04	3.63	5.66	8.11	10.94	14.14	17.68	21.54	25.70	30.17
10 MPH	.00	.45	1.81	4.08	7.26	11.32	16.21	21.89	28.28	35.35	43.07	51.40	60.33
20 MPH	.00	.90	3.61	8.16	14.52	22.64	32.43	43.77	56.56	70.71	86.14	102.81	120.66
30 MPH	.00	1.35	5.42	12.24	21.78	33.96	48.64	65.66	84.84	106.06	129.21	154.21	180.99
VELOCITY (FPS)	1300.	1244.	1193.	1148.	1108.	1073.	1043.	1016.	992.	969.	949.	930.	912.
ENERGY (FT-LB)	803.	735.	676.	626.	583.	547.	517.	490.	467.	446.	428.	411.	396.
DROP (IN)	.00	-2.53	-10.73	-25.02	-45.89	-73.82	-109.27	-152.70	-204.52	-265.18	-335.06	-414.35	-503.66
MID-RANGE (IN)	.00	.67	2.83	6.61	12.22	19.83	29.62	41.75	56.38	73.65	93.72	116.60	142.56
BULLET PATH (IN)	-.75	2.46	.00	-8.55	-23.68	-45.87	-75.59	-113.28	-159.37	-214.28	-278.43	-351.97	-435.55
TIME OF FLIGHT (SEC)	.000000	.117991	.241168	.369412	.502507	.640151	.782018	.927834	1.077383	1.230491	1.387017	1.546848	1.709895
WIND DEFLECTION (IN)													
5 MPH	.00	.23	.92	2.05	3.61	5.56	7.89	10.57	13.58	16.90	20.52	24.43	28.62
10 MPH	.00	.46	1.83	4.09	7.21	11.13	15.79	21.14	27.16	33.80	41.04	48.86	57.25
20 MPH	.00	.92	3.66	8.19	14.42	22.26	31.58	42.29	54.32	67.59	82.08	97.72	114.50
30 MPH	.00	1.38	5.49	12.28	21.63	33.38	47.37	63.43	81.47	101.39	123.11	146.58	171.75

Bullet: Lyman # 323378 242 Gr.
Ballistic Coefficient: .410

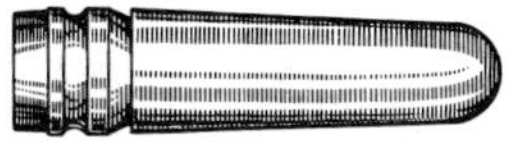

RANGE (YDS)	MUZZLE	50	100	150	200	250	300	350	400	450	500	550	600
VELOCITY (FPS)	1700.	1618.	1541.	1468.	1399.	1335.	1275.	1221.	1172.	1130.	1093.	1061.	1032.
ENERGY (FT-LB)	1553.	1407.	1276.	1158.	1052.	957.	874.	801.	739.	686.	642.	604.	572.
DROP (IN)	.00	-1.43	-6.16	-14.56	-27.04	-43.95	-65.73	-92.86	-125.85	-165.14	-211.22	-264.56	-325.61
MID-RANGE (IN)	.00	.39	1.65	3.92	7.36	12.11	18.30	26.16	35.88	47.65	61.66	78.08	97.08
BULLET PATH (IN)	-.75	1.28	.00	-4.95	-13.97	-27.43	-45.75	-69.43	-98.96	-134.80	-177.43	-227.31	-284.91
TIME OF FLIGHT (SEC)	.000000	.090447	.185459	.285194	.389856	.499623	.614620	.734887	.860343	.990738	1.125812	1.265248	1.408745
WIND DEFLECTION (IN)													
5 MPH	.00	.19	.79	1.80	3.25	5.14	7.50	10.32	13.59	17.30	21.42	25.93	30.79
10 MPH	.00	.39	1.58	3.61	6.50	10.29	15.00	20.63	27.18	34.61	42.85	51.86	61.59
20 MPH	.00	.78	3.16	7.21	12.99	20.57	29.99	41.27	54.37	69.21	85.70	103.72	123.17
30 MPH	.00	1.17	4.75	10.82	19.49	30.86	44.99	61.90	81.55	103.82	128.55	155.58	184.76
VELOCITY (FPS)	1600.	1524.	1452.	1385.	1321.	1263.	1209.	1162.	1121.	1085.	1054.	1026.	1001.
ENERGY (FT-LB)	1375.	1248.	1133.	1030.	938.	856.	786.	726.	675.	633.	597.	566.	538.
DROP (IN)	.00	-1.61	-6.98	-16.50	-30.55	-49.57	-74.07	-104.52	-141.38	-185.14	-236.26	-295.19	-362.36
MID-RANGE (IN)	.00	.44	1.88	4.47	8.30	13.58	20.53	29.32	40.14	53.16	68.56	86.49	107.11
BULLET PATH (IN)	-.75	1.51	.00	-5.66	-15.85	-31.00	-51.64	-78.22	-111.22	-151.11	-198.37	-253.44	-316.74
TIME OF FLIGHT (SEC)	.000000	.096059	.196883	.302679	.413613	.529794	.651239	.777826	.909294	1.045373	1.185742	1.330110	1.478249
WIND DEFLECTION (IN)													
5 MPH	.00	.20	.83	1.89	3.40	5.37	7.81	10.70	14.02	17.74	21.85	26.30	31.09
10 MPH	.00	.41	1.65	3.77	6.80	10.74	15.62	21.40	28.04	35.49	43.69	52.60	62.17
20 MPH	.00	.81	3.30	7.54	13.59	21.49	31.24	42.79	56.07	70.97	87.38	105.20	124.34
30 MPH	.00	1.22	4.95	11.31	20.39	32.23	46.85	64.19	84.11	106.46	131.07	157.80	186.52
VELOCITY (FPS)	1500.	1429.	1363.	1301.	1244.	1193.	1148.	1109.	1074.	1044.	1017.	993.	971.
ENERGY (FT-LB)	1209.	1098.	998.	909.	832.	764.	708.	660.	620.	586.	556.	530.	507.
DROP (IN)	.00	-1.90	-8.05	-18.88	-34.82	-56.43	-84.13	-118.41	-159.73	-208.57	-265.36	-330.53	-404.51
MID-RANGE (IN)	.00	.51	2.13	5.02	9.36	15.36	23.19	33.02	45.04	59.39	76.25	95.78	118.12
BULLET PATH (IN)	-.75	1.75	.00	-6.42	-17.97	-35.18	-58.47	-88.35	-125.28	-169.71	-222.10	-282.88	-352.45
TIME OF FLIGHT (SEC)	.000000	.102454	.209938	.322606	.440536	.563714	.691949	.824982	.962520	1.104243	1.249883	1.399225	1.552098
WIND DEFLECTION (IN)													
5 MPH	.00	.22	.87	1.99	3.57	5.61	8.09	11.00	14.30	17.97	21.99	26.33	30.98
10 MPH	.00	.43	1.75	3.98	7.13	11.21	16.18	22.00	28.60	35.95	43.98	52.66	61.97
20 MPH	.00	.86	3.50	7.96	14.27	22.43	32.37	43.99	57.21	71.89	87.96	105.33	123.94
30 MPH	.00	1.30	5.25	11.94	21.40	33.64	48.55	65.99	85.81	107.84	131.94	157.99	185.91
VELOCITY (FPS)	1400.	1335.	1276.	1221.	1173.	1130.	1093.	1061.	1032.	1007.	983.	962.	943.
ENERGY (FT-LB)	1053.	958.	874.	801.	739.	686.	642.	604.	572.	544.	520.	497.	477.
DROP (IN)	.00	-2.19	-9.24	-21.65	-39.91	-64.46	-95.81	-134.40	-180.70	-235.14	-298.14	-370.11	-451.38
MID-RANGE (IN)	.00	.58	2.43	5.75	10.71	17.49	26.25	37.18	50.44	66.20	84.61	105.80	129.88
BULLET PATH (IN)	-.75	2.06	.00	-7.41	-20.67	-40.23	-66.58	-100.17	-141.48	-190.92	-248.92	-315.90	-392.16
TIME OF FLIGHT (SEC)	.000000	.109721	.224670	.344889	.470296	.600643	.735668	.875056	1.018504	1.165768	1.316655	1.471006	1.628696
WIND DEFLECTION (IN)													
5 MPH	.00	.23	.91	2.06	3.67	5.71	8.17	11.00	14.20	17.73	21.58	25.73	30.18
10 MPH	.00	.45	1.83	4.13	7.34	11.43	16.33	22.01	28.40	35.46	43.16	51.47	60.36
20 MPH	.00	.91	3.66	8.26	14.69	22.85	32.67	44.02	56.80	70.92	86.32	102.94	120.73
30 MPH	.00	1.36	5.48	12.39	22.03	34.28	49.00	66.03	85.20	106.38	129.48	154.41	181.09

Bullet: Lyman # 375248 248 Gr.
Ballistic Coefficient: .290

RANGE (YDS)	MUZZLE	50	100	150	200	250	300	350	400	450	500	550	600
VELOCITY (FPS)	1700.	1586.	1485.	1391.	1306.	1230.	1165.	1111.	1066.	1028.	995.	966.	940.
ENERGY (FT-LB)	1591.	1384.	1214.	1066.	939.	832.	747.	679.	626.	582.	545.	514.	487.
DROP (IN)	.00	-1.49	-6.41	-15.33	-28.72	-47.21	-71.47	-102.12	-139.79	-185.09	-238.59	-300.84	-372.34
MID-RANGE (IN)	.00	.40	1.72	4.18	7.95	13.26	20.41	29.63	41.17	55.26	72.08	91.84	114.70
BULLET PATH (IN)	-.75	1.35	.00	-5.33	-15.14	-30.05	-50.73	-77.80	-111.88	-153.60	-203.51	-262.19	-330.10
TIME OF FLIGHT (SEC)	.000000	.091386	.189166	.293555	.404881	.523324	.648790	.780754	.918701	1.062090	1.210489	1.363578	1.521116
WIND DEFLECTION (IN)													
5 MPH	.00	.28	1.12	2.54	4.57	7.23	10.51	14.35	18.73	23.58	28.88	34.58	40.68
10 MPH	.00	.55	2.23	5.08	9.14	14.46	21.01	28.71	37.46	47.16	57.75	69.17	81.36
20 MPH	.80	1.11	4.47	10.15	18.28	28.92	42.02	57.41	74.91	94.33	115.50	138.33	162.73
30 MPH	.00	1.66	6.70	15.23	27.42	43.37	63.03	86.12	112.37	141.49	173.26	207.50	244.09
VELOCITY (FPS)	1600.	1498.	1404.	1317.	1239.	1173.	1118.	1072.	1033.	1000.	970.	944.	919.
ENERGY (FT-LB)	1409.	1236.	1085.	955.	846.	758.	688.	632.	588.	550.	518.	490.	465.
DROP (IN)	.00	-1.66	-7.25	-17.24	-32.23	-52.90	-79.88	-113.79	-155.24	-204.82	-263.07	-330.52	-407.56
MID-RANGE (IN)	.00	.45	1.96	4.68	8.86	14.80	22.70	32.83	45.39	60.59	78.63	99.69	123.90
BULLET PATH (IN)	-.75	1.59	.00	-5.99	-16.98	-33.66	-56.63	-86.54	-123.99	-169.57	-223.82	-287.27	-360.32
TIME OF FLIGHT (SEC)	.000000	.096896	.200349	.310700	.428156	.552674	.683760	.820905	.963561	1.111281	1.263728	1.420653	1.581866
WIND DEFLECTION (IN)													
5 MPH	.00	.28	1.13	2.59	4.68	7.39	10.67	14.49	18.79	23.54	28.71	34.27	40.20
10 MPH	.00	.55	2.26	5.18	9.36	14.77	21.34	28.98	37.59	47.09	57.42	68.54	80.41
20 MPH	.00	1.11	4.52	10.37	18.71	29.54	42.68	57.96	75.17	94.17	114.83	137.07	160.82
30 MPH	.00	1.66	6.78	15.55	28.07	44.31	64.03	86.94	112.76	141.26	172.25	205.61	241.23
VELOCITY (FPS)	1500.	1405.	1319.	1241.	1174.	1119.	1072.	1034.	1000.	970.	944.	920.	898.
ENERGY (FT-LB)	1239.	1087.	957.	848.	759.	689.	633.	588.	551.	519.	491.	466.	444.
DROP (IN)	.00	-1.95	-8.30	-19.64	-36.65	-59.95	-90.18	-127.94	-173.81	-228.36	-292.09	-365.40	-448.96
MID-RANGE (IN)	.00	.52	2.20	5.28	10.03	16.68	25.45	36.57	50.26	66.70	86.09	108.54	134.30
BULLET PATH (IN)	-.75	1.83	.00	-6.82	-19.31	-38.09	-63.79	-97.03	-138.38	-188.40	-247.61	-316.40	-395.44
TIME OF FLIGHT (SEC)	.000000	.103339	.213572	.330907	.455311	.586295	.723346	.865916	1.013555	1.165925	1.322775	1.483915	1.649203
WIND DEFLECTION (IN)													
5 MPH	.00	.29	1.19	2.72	4.87	7.59	10.85	14.60	18.79	23.40	28.40	33.78	39.53
10 MPH	.00	.59	2.39	5.44	9.73	15.19	21.71	29.20	37.59	46.80	56.81	67.57	79.06
20 MPH	.00	1.18	4.78	10.88	19.47	30.38	43.42	58.40	75.17	93.61	113.62	135.14	158.12
30 MPH	.00	1.76	7.17	16.32	29.20	45.56	65.13	87.60	112.76	140.41	170.43	202.71	237.18
VELOCITY (FPS)	1400.	1314.	1236.	1170.	1116.	1070.	1031.	998.	969.	942.	918.	896.	876.
ENERGY (FT-LB)	1079.	950.	842.	754.	685.	630.	586.	549.	517.	489.	464.	442.	422.
DROP (IN)	.00	-2.24	-9.51	-22.49	-41.80	-68.07	-101.91	-143.89	-194.57	-254.45	-323.96	-403.74	-494.32
MID-RANGE (IN)	.00	.59	2.52	6.04	11.39	18.78	28.46	40.63	55.48	73.20	93.91	117.87	145.26
BULLET PATH (IN)	-.75	2.14	.00	-7.85	-22.03	-43.16	-71.87	-108.72	-154.27	-209.02	-273.40	-348.05	-433.50
TIME OF FLIGHT (SEC)	.000000	.110646	.228400	.353205	.484555	.621941	.764814	.912732	1.065365	1.222464	1.383842	1.549360	1.718916
WIND DEFLECTION (IN)													
5 MPH	.00	.31	1.24	2.80	4.93	7.59	10.73	14.32	18.32	22.72	27.49	32.63	38.12
10 MPH	.00	.62	2.48	5.59	9.85	15.18	21.46	28.64	36.65	45.44	54.98	65.26	76.24
20 MPH	.00	1.23	4.97	11.19	19.71	30.35	42.93	57.28	73.29	90.88	109.97	130.52	152.49
30 MPH	.00	1.85	7.45	16.78	29.56	45.53	64.39	85.92	109.94	136.32	164.95	195.78	228.73

Bullet: Lyman # 375248 248 Gr.
Ballistic Coefficient: .290
(Cont'd.)

RANGE (YDS)	MUZZLE	50	100	150	200	250	300	350	400	450	500	550	600
VELOCITY (FPS)	1300.	1224.	1160.	1107.	1063.	1025.	993.	964.	938.	914.	893.	872.	853.
ENERGY (FT-LB)	930.	825.	741.	675.	622.	579.	543.	512.	485.	460.	439.	419.	401.
DROP (IN)	.00	-2.60	-11.02	-25.87	-47.79	-77.37	-115.20	-161.82	-217.70	-283.32	-359.30	-446.17	-544.45
MID-RANGE (IN)	.00	.69	2.91	6.89	12.88	21.09	31.73	45.00	_1.06	80.08	102.28	127.86	157.03
BULLET PATH (IN)	-.75	2.53	.00	-8.97	-25.00	-48.70	-80.64	-121.38	-171.37	-231.11	-301.20	-382.19	-474.59
TIME OF FLIGHT (SEC)	.000000	.118956	.244908	.377317	.515667	.659420	.808154	.961554	1.119383	1.281463	1.447663	1.617885	1.792067
WIND DEFLECTION (IN)													
5 MPH	.00	.31	1.24	2.74	4.76	7.26	10.19	13.54	17.27	21.38	25.86	30.68	35.86
10 MPH	.00	.63	2.49	5.48	9.53	14.52	20.39	27.08	34.55	42.77	51.71	61.36	71.71
20 MPH	.00	1.26	4.98	10.97	19.05	29.04	40.78	54.16	69.10	85.54	103.42	122.73	143.42
30 MPH	.00	1.89	7.47	16.45	28.58	43.56	61.17	81.24	103.65	128.31	155.14	184.09	215.13
VELOCITY (FPS)	1200.	1141.	1091.	1049.	1013.	982.	955.	930.	906.	885.	865.	846.	829.
ENERGY (FT-LB)	793.	716.	655.	606.	566.	531.	502.	476.	452.	431.	412.	394.	378.
DROP (IN)	.00	-3.06	-12.78	-29.78	-54.66	-87.97	-130.26	-181.93	-243.59	-315.78	-399.05	-493.92	-600.92
MID-RANGE (IN)	.00	.79	3.33	7.84	14.55	23.65	35.35	49.77	67.14	87.67	111.56	139.00	170.20
BULLET PATH (IN)	-.75	2.96	.00	-10.23	-28.34	-54.89	-90.41	-135.32	-190.21	-255.63	-332.14	-420.24	-520.48
TIME OF FLIGHT (SEC)	.000000	.128281	.262839	.403142	.548683	.699084	.854058	1.013391	1.176923	1.344534	1.516141	1.691689	1.871147
WIND DEFLECTION (IN)													
5 MPH	.00	.29	1.13	2.48	4.28	6.52	9.16	12.18	15.57	19.32	23.42	27.87	32.66
10 MPH	.00	.58	2.26	4.95	8.57	13.04	18.31	24.36	31.14	38.64	46.84	55.74	65.32
20 MPH	.00	1.15	4.52	9.91	17.14	26.08	36.63	48.71	62.28	77.28	93.68	111.47	130.64
30 MPH	.00	1.73	6.78	14.86	25.70	39.12	54.94	73.07	93.42	115.91	140.52	167.21	195.97
VELOCITY (FPS)	1100.	1057.	1020.	988.	960.	934.	911.	889.	869.	850.	832.	815.	798.
ENERGY (FT-LB)	666.	615.	573.	538.	507.	481.	457.	435.	416.	398.	381.	365.	351.
DROP (IN)	.00	-3.60	-14.95	-34.63	-63.19	-101.07	-148.79	-206.92	-276.03	-356.63	-449.26	-554.43	-672.72
MID-RANGE (IN)	.00	.93	3.88	9.05	16.64	26.82	39.76	55.69	74.83	97.36	123.56	153.56	187.57
BULLET PATH (IN)	-.75	3.50	.00	-11.83	-32.53	-62.56	-102.43	-152.71	-213.97	-286.72	-371.49	-468.81	-579.25
TIME OF FLIGHT (SEC)	.000000	.139196	.283719	.433168	.587240	.745707	.908399	1.075189	1.245986	1.420730	1.599386	1.781941	1.968400
WIND DEFLECTION (IN)													
5 MPH	.00	.25	.97	2.12	3.68	5.62	7.94	10.62	13.65	17.02	20.75	24.81	29.22
10 MPH	.00	.50	1.93	4.24	7.35	11.24	15.88	21.23	27.29	34.05	41.49	49.62	58.44
20 MPH	.00	1.00	3.87	8.48	14.71	22.49	31.76	42.47	54.59	68.10	82.98	99.24	116.88
30 MPH	.00	1.50	5.80	12.71	22.06	33.73	47.63	63.70	81.88	102.15	124.48	148.86	175.32
VELOCITY (FPS)	1000.	970.	944.	919.	897.	876.	857.	838.	821.	804.	788.	772.	757.
ENERGY (FT-LB)	551.	518.	490.	465.	443.	423.	404.	387.	371.	356.	342.	328.	316.
DROP (IN)	.00	-4.29	-17.75	-40.81	-74.08	-118.11	-173.43	-240.57	-320.05	-412.41	-517.69	-636.93	-770.74
MID-RANGE (IN)	.00	1.12	4.59	10.55	19.29	31.02	45.91	64.15	85.94	111.49	140.74	174.14	211.94
BULLET PATH (IN)	-.75	4.21	.00	-13.80	-37.82	-72.60	-118.67	-176.55	-246.78	-329.89	-425.92	-535.90	-660.46
TIME OF FLIGHT (SEC)	.000000	.152316	.309105	.470177	.635388	.804634	.977843	1.154971	1.335996	1.520916	1.709747	1.902510	2.099230
WIND DEFLECTION (IN)													
5 MPH	.00	.20	.80	1.78	3.11	4.81	6.85	9.24	11.97	15.04	18.46	22.22	26.33
10 MPH	.00	.41	1.60	3.55	6.23	9.62	13.70	18.47	23.94	30.08	36.92	44.44	52.66
20 MPH	.00	.82	3.21	7.10	12.46	19.23	27.40	36.95	47.87	60.16	73.83	88.88	105.33
30 MPH	.00	1.22	4.81	10.65	18.68	28.85	41.10	55.42	71.81	90.24	110.75	133.33	157.99

Bullet: Lyman # 375449 264 Gr.
Ballistic Coefficient: .315

RANGE (YDS)	MUZZLE	50	100	150	200	250	300	350	400	450	500	550	600
VELOCITY (FPS)	1900.	1785.	1674.	1572.	1483.	1400.	1322.	1252.	1190.	1140.	1096.	1059.	1027.
ENERGY (FT-LB)	2116.	1867.	1643.	1449.	1289.	1148.	1025.	919.	831.	761.	704.	657.	618.
DROP (IN)	.00	-1.17	-5.07	-12.05	-22.54	-37.03	-55.94	-79.82	-109.27	-144.84	-187.08	-236.54	-293.73
MID-RANGE (IN)	.00	.32	1.36	3.27	6.20	10.36	15.92	23.08	32.09	43.16	56.51	72.37	90.92
BULLET PATH (IN)	-.75	.99	.00	-4.08	-11.65	-23.23	-39.24	-60.20	-86.74	-119.40	-158.74	-205.28	-259.56
TIME OF FLIGHT (SEC)	.000000	.081464	.168245	.260756	.359000	.463140	.573438	.690063	.813024	.941883	1.076182	1.215518	1.359496
WIND DEFLECTION (IN)													
5 MPH	.00	.22	.91	2.10	3.80	6.02	8.78	12.09	15.97	20.36	25.23	30.54	36.27
10 MPH	.00	.44	1.82	4.21	7.61	12.04	17.56	24.19	31.93	40.72	50.46	61.09	72.53
20 MPH	.00	.89	3.64	8.42	15.21	24.08	35.11	48.38	63.87	81.44	100.92	122.18	145.07
30 MPH	.00	1.33	5.47	12.63	22.82	36.12	52.67	72.56	95.80	122.16	151.38	183.27	217.60
VELOCITY (FPS)	1800.	1689.	1585.	1495.	1411.	1332.	1261.	1198.	1146.	1102.	1064.	1031.	1002.
ENERGY (FT-LB)	1899.	1672.	1472.	1310.	1166.	1041.	932.	841.	770.	711.	663.	623.	588.
DROP (IN)	.00	-1.31	-5.66	-13.44	-25.17	-41.26	-62.23	-88.69	-121.19	-160.30	-206.54	-260.44	-322.51
MID-RANGE (IN)	.00	.36	1.52	3.64	6.93	11.54	17.67	25.55	35.43	47.51	62.01	79.12	99.02
BULLET PATH (IN)	-.75	1.14	.00	-4.58	-13.11	-25.99	-43.76	-67.02	-96.32	-132.22	-175.26	-225.96	-284.82
TIME OF FLIGHT (SEC)	.000000	.086034	.177753	.275213	.378528	.487970	.603724	.725826	.853902	.987466	1.126126	1.269477	1.417195
WIND DEFLECTION (IN)													
5 MPH	.00	.24	.98	2.22	3.98	6.27	9.13	12.54	16.48	20.90	25.77	31.05	36.71
10 MPH	.00	.48	1.95	4.44	7.95	12.55	18.26	25.08	32.95	41.79	51.53	62.09	73.43
20 MPH	.00	.95	3.90	8.88	15.91	25.10	36.51	50.16	65.91	83.59	103.06	124.19	146.85
30 MPH	.00	1.43	5.85	13.31	23.86	37.65	54.77	75.24	98.86	125.38	154.59	186.28	220.28
VELOCITY (FPS)	1700.	1594.	1504.	1419.	1340.	1268.	1204.	1151.	1106.	1067.	1034.	1005.	978.
ENERGY (FT-LB)	1694.	1489.	1325.	1180.	1052.	942.	849.	776.	717.	667.	626.	591.	561.
DROP (IN)	.00	-1.47	-6.34	-15.09	-28.17	-46.07	-69.40	-98.73	-134.59	-177.55	-228.12	-286.80	-354.07
MID-RANGE (IN)	.00	.40	1.70	4.09	7.75	12.83	19.61	28.30	39.13	52.30	67.99	86.38	107.63
BULLET PATH (IN)	-.75	1.32	.00	-5.21	-14.74	-29.10	-48.89	-74.67	-106.99	-146.41	-193.43	-248.57	-312.30
TIME OF FLIGHT (SEC)	.000000	.091130	.188033	.290759	.399589	.514718	.636199	.763723	.896768	1.034951	1.177859	1.325162	1.476604
WIND DEFLECTION (IN)													
5 MPH	.00	.25	1.02	2.29	4.11	6.47	9.40	12.85	16.80	21.19	26.00	31.20	36.76
10 MPH	.00	.51	2.03	4.59	8.21	12.94	18.79	25.71	33.60	42.39	52.01	62.41	73.53
20 MPH	.00	1.02	4.07	9.17	16.42	25.89	37.59	51.42	67.19	84.77	104.02	124.81	147.06
30 MPH	.00	1.53	6.10	13.76	24.63	38.83	56.38	77.13	100.79	127.16	156.03	187.22	220.59
VELOCITY (FPS)	1600.	1509.	1424.	1345.	1272.	1207.	1154.	1108.	1069.	1036.	1006.	980.	956.
ENERGY (FT-LB)	1500.	1335.	1188.	1060.	949.	855.	781.	720.	670.	629.	593.	563.	536.
DROP (IN)	.00	-1.65	-7.15	-16.95	-31.54	-51.52	-77.46	-109.90	-149.40	-196.48	-251.65	-315.38	-388.15
MID-RANGE (IN)	.00	.45	1.93	4.58	8.61	14.27	21.78	31.34	43.16	57.43	74.31	93.98	116.62
BULLET PATH (IN)	-.75	1.55	.00	-5.84	-16.48	-32.51	-54.49	-82.99	-118.53	-161.66	-212.87	-272.65	-341.47
TIME OF FLIGHT (SEC)	.000000	.096546	.198896	.307334	.422062	.543144	.670312	.803025	.940900	1.083525	1.230561	1.381749	1.536886
WIND DEFLECTION (IN)													
5 MPH	.00	.25	1.00	2.30	4.14	6.55	9.49	12.92	16.80	21.10	25.79	30.84	36.25
10 MPH	.00	.49	2.01	4.59	8.28	13.09	18.97	25.83	33.60	42.20	51.58	61.69	72.49
20 MPH	.00	.98	4.01	9.18	16.57	26.19	37.95	51.66	67.20	84.40	103.16	123.38	144.98
30 MPH	.00	1.48	6.02	13.77	24.85	39.28	56.92	77.50	100.80	126.60	154.74	185.06	217.48

Bullet: Lyman # 375449 264 Gr.
Ballistic Coefficient: .315

(Cont'd.)

RANGE (YDS)	MUZZLE	50	100	150	200	250	300	350	400	450	500	550	600
VELOCITY (FPS)	1500.	1415.	1337.	1265.	1201.	1149.	1104.	1066.	1033.	1003.	977.	954.	932.
ENERGY (FT-LB)	1319.	1174.	1047.	938.	846.	774.	714.	666.	625.	590.	560.	533.	509.
DROP (IN)	.00	-1.93	-8.20	-19.32	-35.90	-58.49	-87.64	-123.91	-167.80	-219.83	-280.47	-350.20	-429.30
MID-RANGE (IN)	.00	.52	2.17	5.17	9.75	16.13	24.50	35.06	48.00	63.51	81.75	102.89	127.01
BULLET PATH (IN)	-.75	1.79	.00	-6.65	-18.75	-36.87	-61.55	-93.34	-132.76	-180.31	-236.48	-301.73	-376.36
TIME OF FLIGHT (SEC)	.000000	.102969	.212051	.327438	.449174	.576928	.710187	.848564	.991651	1.139120	1.290716	1.446241	1.605542
WIND DEFLECTION (IN)													
5 MPH	.00	.26	1.06	2.41	4.33	6.77	9.70	13.07	16.87	21.04	25.58	30.47	35.69
10 MPH	.00	.52	2.12	4.83	8.65	13.54	19.39	26.15	33.73	42.09	51.17	60.94	71.38
20 MPH	.00	1.05	4.24	9.66	17.31	27.08	38.79	52.29	67.46	84.17	102.33	121.88	142.75
30 MPH	.00	1.57	6.36	14.49	25.96	40.62	58.18	78.44	101.19	126.26	153.50	182.82	214.13
VELOCITY (FPS)	1400.	1323.	1252.	1191.	1140.	1096.	1059.	1027.	998.	973.	949.	928.	908.
ENERGY (FT-LB)	1149.	1025.	919.	831.	762.	704.	657.	618.	584.	554.	528.	505.	483.
DROP (IN)	.00	-2.22	-9.40	-22.15	-41.01	-66.55	-99.29	-139.76	-188.45	-245.85	-312.42	-388.41	-474.49
MID-RANGE (IN)	.00	.59	2.48	5.92	11.11	18.23	27.50	39.10	53.21	70.01	89.66	112.23	137.96
BULLET PATH (IN)	-.75	2.10	.00	-7.68	-21.47	-41.93	-69.59	-104.99	-148.61	-200.93	-262.43	-333.35	-414.35
TIME OF FLIGHT (SEC)	.000000	.110260	.226845	.349764	.478582	.612839	.752134	.896068	1.044329	1.196676	1.352919	1.512910	1.676534
WIND DEFLECTION (IN)													
5 MPH	.00	.27	1.11	2.49	4.40	6.79	9.62	12.85	16.47	20.45	24.77	29.42	34.39
10 MPH	.00	.55	2.21	4.99	8.80	13.57	19.23	25.71	32.94	40.90	49.54	58.84	68.78
20 MPH	.00	1.10	4.42	9.97	17.60	27.15	38.47	51.42	65.89	81.80	99.08	117.69	137.57
30 MPH	.00	1.65	6.63	14.96	26.41	40.72	57.70	77.12	98.83	122.70	148.63	176.53	206.35
VELOCITY (FPS)	1300.	1232.	1174.	1126.	1084.	1048.	1017.	990.	965.	942.	921.	902.	884.
ENERGY (FT-LB)	991.	890.	808.	743.	689.	644.	607.	574.	546.	520.	498.	477.	458.
DROP (IN)	.00	-2.58	-10.91	-25.52	-46.97	-75.80	-112.51	-157.59	-211.53	-274.70	-347.52	-430.55	-524.23
MID-RANGE (IN)	.00	.68	2.87	6.77	12.58	20.51	30.73	43.44	58.79	76.92	97.96	122.13	149.60
BULLET PATH (IN)	-.75	2.50	.00	-8.78	-24.41	-47.40	-78.29	-117.54	-165.64	-222.99	-289.98	-367.18	-455.03
TIME OF FLIGHT (SEC)	.000000	.118567	.243394	.373941	.509810	.650585	.795891	.945442	1.099014	1.256430	1.417553	1.582278	1.750525
WIND DEFLECTION (IN)													
5 MPH	.00	.28	1.11	2.45	4.25	6.48	9.12	12.12	15.48	19.18	23.21	27.55	32.20
10 MPH	.00	.56	2.22	4.89	8.50	12.96	18.23	24.24	30.96	38.36	46.41	55.10	64.40
20 MPH	.00	1.12	4.44	9.78	16.99	25.93	36.46	48.49	61.93	76.72	92.82	110.19	128.80
30 MPH	.00	1.68	6.67	14.67	25.49	38.89	54.69	72.73	92.89	115.09	139.24	165.29	193.20
VELOCITY (FPS)	1200.	1148.	1103.	1065.	1032.	1003.	977.	953.	931.	911.	892.	875.	858.
ENERGY (FT-LB)	844.	773.	713.	665.	624.	589.	559.	532.	508.	487.	467.	448.	431.
DROP (IN)	.00	-3.04	-12.65	-29.39	-53.76	-86.27	-127.40	-177.63	-237.23	-306.80	-386.81	-477.70	-579.94
MID-RANGE (IN)	.00	.79	3.29	7.70	14.23	23.04	34.32	48.23	64.86	84.43	107.14	133.12	162.57
BULLET PATH (IN)	-.75	2.91	.00	-10.03	-27.70	-53.51	-87.94	-131.46	-184.37	-247.24	-320.54	-404.73	-500.26
TIME OF FLIGHT (SEC)	.000000	.127850	.261198	.399655	.542811	.690340	.841989	.997560	1.156899	1.319886	1.486428	1.656456	1.829922
WIND DEFLECTION (IN)													
5 MPH	.00	.25	.99	2.17	3.77	5.75	8.09	10.79	13.81	17.15	20.81	24.77	29.03
10 MPH	.00	.50	1.97	4.34	7.53	11.50	16.19	21.57	27.61	34.30	41.61	49.54	58.07
20 MPH	.00	1.00	3.94	8.68	15.07	23.00	32.38	43.14	55.23	68.60	83.22	99.07	116.13
30 MPH	.00	1.50	5.91	13.02	22.60	34.50	48.57	64.71	82.84	102.90	124.83	148.61	174.20

Bullet: Lyman # 375167 264 Gr.
Ballistic Coefficient: .315

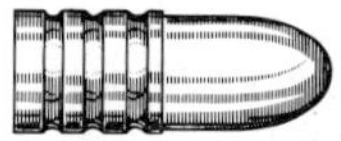

RANGE (YDS)	MUZZLE	50	100	150	200	250	300	350	400	450	500	550	600
VELOCITY (FPS)	1700.	1594.	1486.	1386.	1296.	1216.	1144.	1084.	1036.	996.	961.	931.	903.
ENERGY (FT-LB)	1694.	1489.	1294.	1126.	984.	866.	767.	689.	629.	581.	542.	508.	478.
DROP (IN)	.00	-1.47	-6.37	-15.25	-28.64	-47.22	-71.70	-102.79	-141.26	-187.80	-243.08	-307.66	-382.19
MID-RANGE (IN)	.00	.40	1.72	4.16	7.95	13.33	20.60	30.05	41.99	56.67	74.32	95.15	119.40
BULLET PATH (IN)	-.75	1.34	.00	-5.32	-15.16	-30.17	-51.09	-78.63	-113.54	-156.52	-208.24	-269.25	-340.23
TIME OF FLIGHT (SEC)	.000000	.091130	.188609	.293147	.405114	.524694	.651945	.786765	.928408	1.076191	1.229629	1.388376	1.552188
WIND DEFLECTION (IN)													
5 MPH	.00	.25	1.07	2.50	4.59	7.35	10.78	14.88	19.58	24.82	30.56	36.77	43.42
10 MPH	.00	.51	2.14	5.01	9.18	14.70	21.57	29.76	39.16	49.64	61.12	73.53	86.83
20 MPH	.00	1.02	4.27	10.01	18.36	29.40	43.13	59.53	78.33	99.29	122.24	147.06	173.66
30 MPH	.00	1.53	6.41	15.02	27.55	44.10	64.70	89.29	117.49	148.93	183.36	220.59	260.50
VELOCITY (FPS)	1600.	1492.	1391.	1300.	1220.	1148.	1087.	1038.	998.	963.	932.	904.	879.
ENERGY (FT-LB)	1500.	1304.	1135.	991.	872.	772.	693.	632.	584.	543.	509.	479.	453.
DROP (IN)	.00	-1.68	-7.31	-17.42	-32.68	-53.80	-81.50	-116.53	-159.59	-211.37	-272.41	-343.37	-424.92
MID-RANGE (IN)	.00	.45	1.98	4.74	9.03	15.13	23.33	33.90	47.12	63.22	82.41	104.90	130.96
BULLET PATH (IN)	-.75	1.60	.00	-6.08	-17.31	-34.40	-58.07	-89.07	-128.10	-175.84	-232.86	-299.78	-377.31
TIME OF FLIGHT (SEC)	.000000	.097118	.201273	.312840	.432018	.558853	.693282	.834576	.982041	1.135181	1.293645	1.457185	1.625627
WIND DEFLECTION (IN)													
5 MPH	.00	.30	1.21	2.78	5.02	7.93	11.51	15.69	20.42	25.65	31.34	37.48	44.06
10 MPH	.00	.59	2.42	5.56	10.04	15.86	23.02	31.39	40.84	51.29	62.68	74.96	88.11
20 MPH	.00	1.19	4.85	11.12	20.07	31.72	46.04	62.77	81.68	102.58	125.36	149.93	176.22
30 MPH	.00	1.78	7.27	16.68	30.11	47.57	69.05	94.16	122.52	153.88	188.04	224.89	264.33
VELOCITY (FPS)	1500.	1399.	1307.	1226.	1153.	1091.	1042.	1001.	966.	934.	906.	881.	857.
ENERGY (FT-LB)	1319.	1147.	1002.	881.	779.	698.	636.	587.	546.	512.	482.	455.	431.
DROP (IN)	.00	-1.96	-8.36	-19.84	-37.14	-60.95	-92.04	-131.10	-178.82	-235.77	-302.57	-379.92	-468.45
MID-RANGE (IN)	.00	.52	2.22	5.36	10.21	17.06	26.17	37.83	52.27	69.70	90.34	114.42	142.19
BULLET PATH (IN)	-.75	1.85	.00	-6.93	-19.67	-38.93	-65.46	-99.97	-143.14	-195.53	-257.78	-330.58	-414.55
TIME OF FLIGHT (SEC)	.000000	.103574	.214536	.333101	.459320	.593169	.733946	.880940	1.033641	1.191690	1.354829	1.522882	1.695730
WIND DEFLECTION (IN)													
5 MPH	.00	.31	1.28	2.91	5.22	8.20	11.79	15.92	20.56	25.67	31.22	37.21	43.62
10 MPH	.00	.63	2.56	5.83	10.44	16.40	23.57	31.85	41.12	51.34	62.45	74.43	87.25
20 MPH	.00	1.26	5.12	11.65	20.88	32.80	47.15	63.69	82.24	102.67	124.90	148.85	174.50
30 MPH	.00	1.89	7.67	17.48	31.32	49.19	70.72	95.54	123.36	154.01	187.35	223.28	261.75
VELOCITY (FPS)	1400.	1308.	1227.	1153.	1092.	1042.	1001.	966.	935.	907.	881.	857.	835.
ENERGY (FT-LB)	1149.	1003.	882.	780.	699.	637.	587.	547.	512.	482.	455.	431.	409.
DROP (IN)	.00	-2.25	-9.58	-22.72	-42.36	-69.28	-104.16	-147.70	-200.46	-263.07	-336.21	-420.53	-516.66
MID-RANGE (IN)	.00	.59	2.54	6.12	11.60	19.26	29.37	42.17	57.87	76.67	98.83	124.58	154.17
BULLET PATH (IN)	-.75	2.16	.00	-7.97	-22.45	-44.19	-73.91	-112.28	-159.88	-217.32	-285.30	-364.45	-455.41
TIME OF FLIGHT (SEC)	.000000	.110890	.229382	.355530	.489313	.630031	.776969	.929618	1.087615	1.250705	1.418708	1.591505	1.769029
WIND DEFLECTION (IN)													
5 MPH	.00	.33	1.33	3.00	5.35	8.30	11.80	15.81	20.28	25.20	30.56	36.34	42.53
10 MPH	.00	.66	2.66	6.00	10.69	16.60	23.60	31.61	40.56	50.41	61.12	72.68	85.06
20 MPH	.00	1.32	5.31	12.00	21.38	33.20	47.21	63.23	81.13	100.82	122.24	145.35	170.13
30 MPH	.00	1.98	7.97	18.01	32.07	49.80	70.81	94.84	121.69	151.23	183.36	218.03	255.19

Bullet: Lyman # 375167 264 Gr.
Ballistic Coefficient: .315

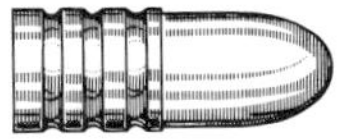

(Cont'd.)

RANGE (YDS)	MUZZLE	50	100	150	200	250	300	350	400	450	500	550	600
VELOCITY (FPS)	1300.	1219.	1147.	1087.	1038.	998.	963.	932.	904.	879.	855.	833.	812.
ENERGY (FT-LB)	991.	872.	771.	692.	632.	583.	543.	509.	479.	453.	429.	407.	387.
DROP (IN)	.00	-2.62	-11.10	-26.16	-48.56	-78.99	-118.14	-166.55	-224.87	-293.79	-373.94	-465.96	-570.50
MID-RANGE (IN)	.00	.69	2.93	7.00	13.18	21.74	32.90	46.87	63.88	84.17	107.98	135.55	167.11
BULLET PATH (IN)	-.75	2.56	.00	-9.14	-25.61	-50.12	-83.34	-125.83	-178.23	-241.23	-315.45	-401.55	-500.16
TIME OF FLIGHT (SEC)	.000000	.119201	.246058	.380505	.521813	.669287	.822432	.980897	1.144434	1.312869	1.486090	1.664031	1.846665
WIND DEFLECTION (IN)													
5 MPH	.00	.34	1.35	3.02	5.30	8.13	11.45	15.24	19.48	24.15	29.24	34.74	40.66
10 MPH	.00	.67	2.69	6.05	10.61	16.26	22.90	30.48	38.96	48.30	58.47	69.48	81.32
20 MPH	.00	1.34	5.38	12.09	21.22	32.51	45.80	60.97	77.92	96.59	116.95	138.97	162.64
30 MPH	.00	2.02	8.07	18.14	31.83	48.77	68.71	91.45	116.88	144.89	175.42	208.45	243.96
VELOCITY (FPS)	1200.	1130.	1073.	1027.	988.	954.	924.	897.	872.	849.	827.	807.	787.
ENERGY (FT-LB)	844.	748.	675.	618.	572.	533.	501.	472.	446.	423.	401.	382.	363.
DROP (IN)	.00	-3.09	-12.97	-30.38	-56.00	-90.51	-134.42	-188.44	-253.22	-329.40	-417.63	-518.54	-632.44
MID-RANGE (IN)	.00	.80	3.39	8.05	15.03	24.56	36.83	52.10	70.61	92.60	118.31	148.02	181.80
BULLET PATH (IN)	-.75	3.02	.00	-10.54	-29.30	-56.95	-94.00	-141.16	-199.08	-268.40	-349.76	-443.81	-550.85
TIME OF FLIGHT (SEC)	.000000	.128933	.265296	.408313	.557344	.711940	.871779	1.036635	1.206352	1.380831	1.560017	1.743895	1.932479
WIND DEFLECTION (IN)													
5 MPH	.00	.35	1.35	2.93	5.05	7.65	10.72	14.22	18.16	22.51	27.28	32.46	38.06
10 MPH	.00	.69	2.69	5.86	10.09	15.30	21.43	28.45	36.32	45.03	54.56	64.93	76.12
20 MPH	.00	1.38	5.38	11.73	20.19	30.60	42.87	56.90	72.64	90.05	109.13	129.85	152.23
30 MPH	.00	2.08	8.08	17.59	30.28	45.90	64.30	85.34	108.95	135.08	163.69	194.78	228.35
VELOCITY (FPS)	1100.	1049.	1007.	970.	939.	910.	884.	860.	838.	817.	797.	778.	759.
ENERGY (FT-LB)	709.	645.	594.	552.	516.	486.	458.	434.	412.	391.	372.	355.	338.
DROP (IN)	.00	-3.63	-15.14	-35.19	-64.41	-103.35	-152.75	-213.21	-285.39	-369.92	-467.42	-578.08	-703.08
MID-RANGE (IN)	.00	.94	3.94	9.24	17.07	27.62	41.18	57.97	78.28	102.35	130.36	162.34	198.79
BULLET PATH (IN)	-.75	3.56	.00	-12.11	-33.39	-64.38	-105.84	-158.35	-222.59	-299.18	-388.73	-491.45	-608.51
TIME OF FLIGHT (SEC)	.000000	.139751	.285810	.437640	.594861	.757201	.924471	1.096544	1.273345	1.454838	1.641025	1.831929	2.027583
WIND DEFLECTION (IN)													
5 MPH	.00	.30	1.15	2.51	4.35	6.63	9.35	12.50	16.05	20.03	24.41	29.21	34.43
10 MPH	.00	.60	2.30	5.02	8.70	13.27	18.71	24.99	32.11	40.05	48.82	58.42	68.85
20 MPH	.00	1.19	4.61	10.05	17.39	26.53	37.41	49.98	64.22	80.10	97.64	116.84	137.71
30 MPH	.00	1.79	6.91	15.07	26.09	39.80	56.12	74.98	96.33	120.15	146.46	175.26	206.56
VELOCITY (FPS)	1000.	965.	934.	906.	880.	856.	834.	813.	793.	774.	756.	739.	722.
ENERGY (FT-LB)	586.	545.	511.	481.	454.	430.	408.	388.	369.	351.	335.	320.	305.
DROP (IN)	.00	-4.33	-17.89	-41.32	-75.32	-120.50	-177.50	-246.95	-329.35	-425.15	-535.40	-660.83	-802.22
MID-RANGE (IN)	.00	1.12	4.62	10.71	19.77	31.92	47.43	66.52	89.36	116.10	147.20	182.93	223.61
BULLET PATH (IN)	-.75	4.24	-.00	-14.11	-38.78	-74.64	-122.32	-182.45	-255.53	-342.01	-442.93	-559.05	-691.11
TIME OF FLIGHT (SEC)	.000000	.152777	.310890	.474082	.642175	.815053	.992646	1.174925	1.361896	1.553586	1.750027	1.951267	2.157366
WIND DEFLECTION (IN)													
5 MPH	.00	.24	.96	2.12	3.71	5.72	8.15	10.99	14.25	17.92	22.00	26.51	31.45
10 MPH	.00	.49	1.92	4.24	7.42	11.45	16.31	21.99	28.49	35.83	44.00	53.02	62.90
20 MPH	.00	.98	3.83	8.48	14.85	22.90	32.61	43.97	56.99	71.66	88.01	106.05	125.79
30 MPH	.00	1.47	5.75	12.72	22.27	34.35	48.92	65.96	85.48	107.49	132.01	159.07	188.69

Bullet: Lyman # 457191 293 Gr.
Ballistic Coefficient: .201

RANGE (YDS)	MUZZLE	50	100	150	200	250	300	350	400	450	500	550	600
VELOCITY (FPS)	2200.	2004.	1819.	1659.	1514.	1383.	1276.	1184.	1109.	1050.	1003.	963.	929.
ENERGY (FT-LB)	3148.	2612.	2152.	1791.	1491.	1245.	1059.	912.	800.	718.	655.	604.	561.
DROP (IN)	.00	-.91	-3.99	-9.69	-18.54	-31.20	-48.41	-70.97	-99.72	-135.52	-179.17	-231.44	-292.97
MID-RANGE (IN)	.00	.25	1.08	2.69	5.28	9.12	14.52	21.80	31.33	43.41	58.38	76.53	98.08
BULLET PATH (IN)	-.75	.71	.00	-3.33	-9.80	-20.10	-34.93	-55.12	-81.51	-114.94	-156.22	-206.11	-265.27
TIME OF FLIGHT (SEC)	.000000	.071444	.150024	.236444	.331117	.434850	.547844	.670029	.801097	.940234	1.086501	1.239223	1.397942
WIND DEFLECTION (IN)													
5 MPH	.00	.29	1.20	2.81	5.14	8.27	12.21	16.96	22.50	28.74	35.61	43.05	51.02
10 MPH	.00	.57	2.40	5.61	10.28	16.53	24.42	33.93	44.99	57.48	71.22	86.10	102.04
20 MPH	.00	1.15	4.81	11.23	20.55	33.07	48.84	67.85	89.99	114.96	142.45	172.21	204.08
30 MPH	.00	1.72	7.21	16.84	30.83	49.60	73.26	101.78	134.98	172.44	213.67	258.31	306.11
VELOCITY (FPS)	2100.	1909.	1736.	1584.	1446.	1327.	1227.	1144.	1078.	1025.	982.	945.	913.
ENERGY (FT-LB)	2869.	2371.	1960.	1633.	1360.	1146.	980.	851.	756.	684.	628.	581.	542.
DROP (IN)	.00	-1.00	-4.39	-10.66	-20.39	-34.30	-53.14	-77.77	-109.03	-147.73	-194.69	-250.61	-316.11
MID-RANGE (IN)	.00	.27	1.19	2.96	5.80	10.02	15.92	23.84	34.12	47.09	63.04	82.25	104.91
BULLET PATH (IN)	-.75	.82	.00	-3.69	-10.86	-22.20	-38.47	-60.53	-89.21	-125.35	-169.73	-223.08	-286.02
TIME OF FLIGHT (SEC)	.000000	.074917	.157414	.247886	.347028	.455457	.573085	.699819	.835077	.977907	1.127494	1.283283	1.444897
WIND DEFLECTION (IN)													
5 MPH	.00	.31	1.28	2.96	5.40	8.65	12.72	17.58	23.20	29.48	36.36	43.79	51.72
10 MPH	.00	.61	2.56	5.91	10.79	17.30	25.43	35.17	46.40	58.97	72.72	87.57	103.44
20 MPH	.00	1.23	5.12	11.83	21.58	34.61	50.87	70.34	92.80	117.94	145.45	175.14	206.89
30 MPH	.00	1.84	7.69	17.74	32.37	51.91	76.30	105.50	139.21	176.91	218.17	262.72	310.33
VELOCITY (FPS)	2000.	1815.	1656.	1511.	1381.	1274.	1182.	1108.	1049.	1002.	963.	928.	897.
ENERGY (FT-LB)	2602.	2143.	1783.	1485.	1241.	1055.	909.	798.	716.	653.	603.	560.	524.
DROP (IN)	.00	-1.11	-4.85	-11.75	-22.50	-37.79	-58.45	-85.33	-119.27	-161.08	-211.52	-271.24	-340.97
MID-RANGE (IN)	.00	.30	1.31	3.26	6.40	11.03	17.47	26.06	37.14	51.01	67.97	88.23	112.03
BULLET PATH (IN)	-.75	.94	.00	-4.11	-12.05	-24.54	-42.41	-66.49	-97.63	-136.64	-184.28	-241.20	-308.13
TIME OF FLIGHT (SEC)	.000000	.078736	.165333	.260202	.364144	.477328	.599703	.730948	.870241	1.016646	1.169495	1.328332	1.492849
WIND DEFLECTION (IN)													
5 MPH	.00	.33	1.35	3.10	5.64	9.00	13.17	18.12	23.78	30.06	36.92	44.29	52.17
10 MPH	.00	.66	2.70	6.20	11.29	18.01	26.35	36.25	47.56	60.13	73.83	88.59	104.34
20 MPH	.00	1.32	5.40	12.39	22.58	36.02	52.70	72.49	95.12	120.26	147.66	177.17	208.68
30 MPH	.00	1.97	8.10	18.59	33.87	54.03	79.04	108.74	142.69	180.39	221.49	265.76	313.02
VELOCITY (FPS)	1900.	1728.	1578.	1440.	1323.	1223.	1141.	1075.	1023.	981.	944.	912.	882.
ENERGY (FT-LB)	2348.	1943.	1619.	1348.	1138.	973.	846.	752.	681.	625.	580.	541.	507.
DROP (IN)	.00	-1.23	-5.36	-13.00	-24.84	-41.66	-64.29	-93.59	-130.38	-175.44	-229.50	-293.18	-367.29
MID-RANGE (IN)	.00	.33	1.45	3.60	7.06	12.13	19.15	28.44	40.34	55.14	73.09	94.40	119.35
BULLET PATH (IN)	-.75	1.08	.00	-4.58	-13.37	-27.12	-46.70	-72.94	-106.67	-148.68	-199.69	-260.31	-331.36
TIME OF FLIGHT (SEC)	.000000	.082883	.173746	.273316	.382161	.500192	.627314	.762923	.906062	1.055928	1.211976	1.373835	1.541254
WIND DEFLECTION (IN)													
5 MPH	.00	.35	1.39	3.21	5.84	9.28	13.52	18.51	24.15	30.40	37.18	44.48	52.26
10 MPH	.00	.69	2.79	6.42	11.68	18.56	27.04	37.01	48.31	60.79	74.36	88.95	104.52
20 MPH	.00	1.39	5.58	12.84	23.36	37.12	54.08	74.02	96.62	121.58	148.72	177.91	209.05
30 MPH	.00	2.08	8.37	19.26	35.04	55.68	81.12	111.03	144.93	182.37	223.08	266.86	313.57

Bullet: Lyman # 457191 293 Gr.

Ballistic Coefficient: .201

(Cont'd.)

RANGE (YDS)	MUZZLE	50	100	150	200	250	300	350	400	450	500	550	600
VELOCITY (FPS)	1800.	1644.	1500.	1372.	1266.	1176.	1103.	1045.	999.	960.	926.	895.	868.
ENERGY (FT-LB)	2108.	1759.	1464.	1225.	1043.	899.	791.	711.	649.	599.	557.	521.	490.
DROP (IN)	.00	-1.36	-5.94	-14.40	-27.47	-45.98	-70.77	-102.69	-142.53	-191.05	-248.90	-316.82	-395.55
MID-RANGE (IN)	.00	.37	1.61	3.99	7.80	13.35	20.99	31.03	43.79	59.53	78.46	100.87	127.00
BULLET PATH (IN)	-.75	1.23	.00	-5.12	-14.85	-30.02	-51.47	-80.04	-116.54	-161.72	-216.22	-280.80	-356.19
TIME OF FLIGHT (SEC)	.000000	.087208	.182743	.287397	.401274	.524335	.656221	.796079	.942989	1.096300	1.255572	1.420504	1.590899
WIND DEFLECTION (IN)													
5 MPH	.00	.34	1.41	3.29	5.98	9.47	13.75	18.72	24.32	30.47	37.16	44.34	52.00
10 MPH	.00	.68	2.83	6.58	11.96	18.95	27.49	37.44	48.63	60.95	74.31	88.68	104.00
20 MPH	.00	1.36	5.66	13.16	23.91	37.90	54.99	74.89	97.27	121.90	148.63	177.35	208.00
30 MPH	.00	2.05	8.49	19.75	35.87	56.85	82.48	112.33	145.90	182.85	222.94	266.03	311.99
VELOCITY (FPS)	1700.	1552.	1416.	1303.	1207.	1128.	1065.	1015.	973.	938.	906.	877.	851.
ENERGY (FT-LB)	1880.	1566.	1304.	1105.	947.	827.	738.	670.	616.	572.	534.	501.	471.
DROP (IN)	.00	-1.53	-6.69	-16.18	-30.80	-51.39	-78.80	-113.85	-157.30	-209.87	-272.21	-345.09	-429.25
MID-RANGE (IN)	.00	.41	1.81	4.49	8.72	14.85	23.22	34.13	47.85	64.64	84.71	108.35	135.82
BULLET PATH (IN)	-.75	1.44	.00	-5.78	-16.68	-33.56	-57.25	-88.57	-128.31	-177.16	-235.78	-304.95	-385.39
TIME OF FLIGHT (SEC)	.000000	.092386	.193627	.304151	.423862	.552593	.689648	.834064	.985090	1.142218	1.305102	1.473512	1.647301
WIND DEFLECTION (IN)													
5 MPH	.00	.37	1.51	3.47	6.24	9.80	14.10	19.04	24.57	30.63	37.20	44.26	51.79
10 MPH	.00	.73	3.02	6.94	12.48	19.61	28.20	38.01	49.14	61.27	74.40	88.51	103.57
20 MPH	.00	1.46	6.04	13.88	24.96	39.22	56.40	76.18	98.28	122.53	148.81	177.03	207.14
30 MPH	.00	2.19	9.06	20.83	37.45	58.83	84.60	114.27	147.42	183.80	223.21	265.54	310.72
VELOCITY (FPS)	1600.	1460.	1339.	1237.	1152.	1084.	1031.	987.	949.	916.	887.	860.	835.
ENERGY (FT-LB)	1665.	1387.	1166.	996.	863.	764.	691.	633.	586.	546.	511.	481.	453.
DROP (IN)	.00	-1.74	-7.57	-18.25	-34.63	-57.54	-87.81	-126.25	-173.59	-230.42	-297.57	-375.75	-465.68
MID-RANGE (IN)	.00	.47	2.05	5.03	9.74	16.51	25.65	37.44	52.16	70.00	91.25	116.17	145.03
BULLET PATH (IN)	-.75	1.68	.00	-6.52	-18.73	-37.48	-63.59	-97.86	-141.04	-193.71	-256.70	-330.72	-416.48
TIME OF FLIGHT (SEC)	.000000	.098181	.205614	.322248	.448016	.582395	.724449	.873331	1.028463	1.189451	1.356030	1.528029	1.705350
WIND DEFLECTION (IN)													
5 MPH	.00	.39	1.59	3.61	6.43	10.00	14.25	19.10	24.50	30.42	36.83	43.72	51.07
10 MPH	.00	.78	3.19	7.22	12.85	20.00	28.50	38.21	49.01	60.84	73.66	87.43	102.14
20 MPH	.00	1.56	6.38	14.43	25.70	40.00	57.01	76.41	98.02	121.69	147.32	174.87	204.28
30 MPH	.00	2.34	9.56	21.65	38.55	60.00	85.51	114.62	147.03	182.53	220.98	262.30	306.42
VELOCITY (FPS)	1500.	1372.	1266.	1176.	1103.	1045.	999.	960.	926.	895.	867.	842.	818.
ENERGY (FT-LB)	1464.	1225.	1042.	899.	791.	711.	649.	599.	557.	521.	489.	461.	435.
DROP (IN)	.00	-2.00	-8.61	-20.66	-39.00	-64.45	-97.84	-139.91	-191.30	-252.77	-325.04	-408.86	-504.92
MID-RANGE (IN)	.00	.53	2.30	5.65	10.89	18.32	28.25	40.96	56.65	75.60	98.07	124.32	154.62
BULLET PATH (IN)	-.75	1.93	.00	-7.37	-21.02	-41.80	-70.50	-107.89	-154.60	-211.39	-278.98	-358.12	-449.50
TIME OF FLIGHT (SEC)	.000000	.104674	.218573	.341655	.473559	.613432	.760354	.913673	1.072951	1.237886	1.408283	1.584023	1.765050
WIND DEFLECTION (IN)													
5 MPH	.00	.41	1.63	3.67	6.47	9.98	14.11	18.80	24.02	29.73	35.93	42.59	49.72
10 MPH	.00	.82	3.27	7.33	12.95	19.96	28.22	37.61	48.04	59.47	71.86	85.19	99.45
20 MPH	.00	1.65	6.54	14.66	25.89	39.93	56.44	75.21	96.08	118.94	143.72	170.38	198.90
30 MPH	.00	2.47	9.81	21.99	38.84	59.89	84.67	112.82	144.12	178.40	215.57	255.56	298.35

Bullet: Lyman # 457191 293 Gr.
Ballistic Coefficient: .201
(Cont'd.)

RANGE (YDS)	MUZZLE	50	100	150	200	250	300	350	400	450	500	550	600
VELOCITY (FPS)	1400.	1290.	1196.	1119.	1058.	1009.	969.	933.	902.	874.	848.	824.	801.
ENERGY (FT-LB)	1275.	1083.	930.	814.	728.	663.	610.	567.	529.	497.	468.	441.	417.
DROP (IN)	.00	-2.28	-9.80	-23.40	-43.91	-72.17	-108.93	-154.87	-210.69	-277.13	-354.95	-444.83	-547.52
MID-RANGE (IN)	.00	.60	2.62	6.37	12.15	20.30	31.07	44.70	61.43	81.54	105.31	132.98	164.83
BULLET PATH (IN)	-.75	2.24	.00	-8.32	-23.56	-46.54	-78.02	-118.69	-169.23	-230.40	-302.94	-387.55	-484.96
TIME OF FLIGHT (SEC)	.000000	.111679	.232548	.362376	.500409	.645690	.797503	.955365	1.118947	1.288029	1.462475	1.642217	1.827237
WIND DEFLECTION (IN)													
5 MPH	.00	.40	1.61	3.60	6.32	9.68	13.61	18.07	23.04	28.49	34.41	40.80	47.65
10 MPH	.00	.80	3.21	7.21	12.64	19.36	27.22	36.14	46.08	56.98	68.82	81.60	95.31
20 MPH	.00	1.60	6.43	14.41	25.29	38.71	54.44	72.29	92.16	113.96	137.65	163.20	190.62
30 MPH	.00	2.40	9.64	21.62	37.93	58.07	81.65	108.43	138.23	170.94	206.47	244.80	285.92
VELOCITY (FPS)	1300.	1204.	1125.	1063.	1014.	972.	937.	905.	876.	850.	826.	803.	781.
ENERGY (FT-LB)	1099.	943.	824.	735.	668.	615.	571.	533.	500.	470.	444.	419.	397.
DROP (IN)	.00	-2.66	-11.31	-26.81	-49.97	-81.56	-122.28	-172.79	-233.87	-306.24	-390.61	-487.72	-597.86
MID-RANGE (IN)	.00	.70	3.00	7.22	13.67	22.63	34.33	49.02	66.97	88.46	113.75	143.11	176.65
BULLET PATH (IN)	-.75	2.63	.00	-9.47	-26.59	-52.16	-86.85	-131.33	-186.37	-252.71	-331.06	-422.13	-526.25
TIME OF FLIGHT (SEC)	.000000	.119992	.248990	.386282	.530904	.682116	.839414	1.002456	1.171014	1.344946	1.524175	1.708682	1.898487
WIND DEFLECTION (IN)													
5 MPH	.00	.41	1.60	3.53	6.10	9.26	12.95	17.14	21.82	26.97	32.59	38.67	45.22
10 MPH	.00	.81	3.21	7.06	12.21	18.51	25.89	34.28	43.64	53.94	65.18	77.34	90.44
20 MPH	.00	1.62	6.41	14.13	24.42	37.03	51.78	68.56	87.27	107.88	130.36	154.69	180.88
30 MPH	.00	2.43	9.62	21.19	36.63	55.54	77.67	102.84	130.91	161.82	195.53	232.03	271.32
VELOCITY (FPS)	1200.	1122.	1060.	1011.	970.	935.	904.	875.	849.	825.	802.	780.	759.
ENERGY (FT-LB)	937.	819.	732.	665.	612.	569.	531.	498.	469.	442.	418.	396.	375.
DROP (IN)	.00	-3.11	-13.11	-30.80	-56.97	-92.28	-137.43	-193.18	-260.25	-339.36	-431.24	-536.19	-655.41
MID-RANGE (IN)	.00	.81	3.44	8.20	15.38	25.20	37.91	53.81	73.16	96.22	123.34	154.52	190.28
BULLET PATH (IN)	-.75	3.07	.00	-10.77	-30.00	-58.38	-96.61	-145.42	-205.57	-277.75	-362.70	-460.72	-573.01
TIME OF FLIGHT (SEC)	.000000	.129411	.267071	.412019	.563527	.721100	.884403	1.053213	1.227389	1.406860	1.591606	1.781651	1.977035
WIND DEFLECTION (IN)													
5 MPH	.00	.39	1.50	3.26	5.59	8.46	11.83	15.68	20.01	24.80	30.06	35.79	41.98
10 MPH	.00	.78	3.00	6.52	11.18	16.91	23.66	31.37	40.02	49.61	60.12	71.57	83.96
20 MPH	.00	1.55	6.01	13.03	22.36	33.83	47.31	62.73	80.04	99.21	120.25	143.14	167.92
30 MPH	.00	2.33	9.01	19.55	33.54	50.74	70.97	94.10	120.06	148.82	180.37	214.71	251.87
VELOCITY (FPS)	1100.	1043.	997.	958.	924.	894.	866.	841.	817.	794.	773.	752.	733.
ENERGY (FT-LB)	787.	708.	647.	597.	556.	520.	488.	460.	434.	410.	389.	368.	349.
DROP (IN)	.00	-3.66	-15.27	-35.60	-65.27	-105.05	-155.66	-217.83	-292.26	-379.57	-480.33	-595.63	-726.26
MID-RANGE (IN)	.00	.95	3.98	9.38	17.37	28.23	42.23	59.68	80.86	105.93	135.12	168.86	207.47
BULLET PATH (IN)	-.75	3.60	.00	-12.32	-33.99	-65.75	-108.35	-162.51	-228.94	-308.23	-400.99	-508.28	-630.90
TIME OF FLIGHT (SEC)	.000000	.140156	.287327	.440873	.600358	.765487	.936066	1.111979	1.293172	1.479644	1.671423	1.868554	2.071096
WIND DEFLECTION (IN)													
5 MPH	.00	.33	1.28	2.80	4.83	7.36	10.37	13.85	17.80	22.21	27.09	32.43	38.26
10 MPH	.00	.67	2.57	5.59	9.66	14.73	20.75	27.71	35.60	44.42	54.17	64.87	76.51
20 MPH	.00	1.34	5.14	11.19	19.33	29.45	41.50	55.42	71.20	88.83	108.34	129.73	153.03
30 MPH	.00	2.00	7.71	16.78	28.99	44.18	62.24	83.12	106.80	133.25	162.51	194.60	229.54

Bullet: Lyman 457122 322 Gr.
Ballistic Coefficient: .274

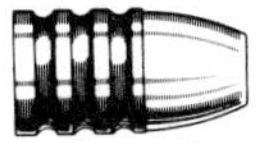

RANGE (YDS)	MUZZLE	50	100	150	200	250	300	350	400	450	500	550	600
VELOCITY (FPS)	2400.	2248.	2101.	1960.	1825.	1697.	1576.	1463.	1360.	1267.	1186.	1118.	1064.
ENERGY (FT-LB)	4118.	3612.	3156.	2747.	2381.	2058.	1775.	1530.	1322.	1148.	1005.	894.	809.
DROP (IN)	.00	-.74	-3.20	-7.62	-14.30	-23.59	-35.89	-51.69	-71.58	-96.17	-126.19	-162.39	-205.52
MID-RANGE (IN)	.00	.20	.86	2.07	3.95	6.65	10.32	15.19	21.49	29.49	39.51	51.86	66.87
BULLET PATH (IN)	-.75	.48	.00	-2.45	-7.16	-14.47	-24.80	-38.62	-56.55	-79.16	-107.21	-141.43	-182.59
TIME OF FLIGHT (SEC)	.000000	.064581	.133602	.207516	.286824	.372070	.463822	.562640	.669027	.783363	.905846	1.036247	1.173943
WIND DEFLECTION (IN)													
5 MPH	.00	.18	.76	1.76	3.24	5.24	7.82	11.01	14.87	19.44	24.71	30.69	37.31
10 MPH	.00	.37	1.51	3.52	6.48	10.48	15.63	22.02	29.75	38.87	49.43	61.38	74.61
20 MPH	.00	.73	3.03	7.05	12.96	20.97	31.27	44.05	59.50	77.74	98.86	122.76	149.23
30 MPH	.00	1.10	4.54	10.57	19.44	31.45	46.90	66.07	89.25	116.62	148.29	184.14	223.84
VELOCITY (FPS)	2300.	2151.	2008.	1871.	1741.	1617.	1501.	1395.	1298.	1212.	1140.	1081.	1034.
ENERGY (FT-LB)	3782.	3309.	2884.	2503.	2166.	1869.	1611.	1390.	1205.	1051.	929.	836.	764.
DROP (IN)	.00	-.81	-3.49	-8.32	-15.64	-25.81	-39.31	-56.69	-78.55	-105.57	-138.51	-178.13	-225.16
MID-RANGE (IN)	.00	.22	.94	2.26	4.33	7.29	11.33	16.70	23.64	32.44	43.44	56.95	73.27
BULLET PATH (IN)	-.75	.56	.00	-2.71	-7.91	-15.96	-27.34	-42.60	-62.34	-87.24	-118.06	-155.56	-200.46
TIME OF FLIGHT (SEC)	.000000	.067433	.139596	.216974	.300097	.389529	.485836	.589548	.701095	.820755	.948467	1.083724	1.225761
WIND DEFLECTION (IN)													
5 MPH	.00	.19	.81	1.88	3.45	5.58	8.32	11.71	15.78	20.57	26.07	32.24	39.00
10 MPH	.00	.39	1.61	3.75	6.90	11.17	16.64	23.41	31.57	41.15	52.15	64.47	77.99
20 MPH	.00	.78	3.22	7.51	13.81	22.33	33.28	46.83	63.13	82.30	104.30	128.95	155.99
30 MPH	.00	1.17	4.84	11.26	20.71	33.50	49.91	70.24	94.70	123.45	156.44	193.42	233.98
VELOCITY (FPS)	2200.	2055.	1916.	1783.	1657.	1538.	1429.	1329.	1239.	1162.	1099.	1048.	1006.
ENERGY (FT-LB)	3460.	3019.	2624.	2273.	1963.	1692.	1459.	1262.	1098.	966.	864.	786.	724.
DROP (IN)	.00	-.88	-3.82	-9.13	-17.16	-28.35	-43.23	-62.39	-86.46	-116.22	-152.40	-195.74	-246.97
MID-RANGE (IN)	.00	.24	1.03	2.48	4.76	8.02	12.49	18.42	26.07	35.78	47.85	62.58	80.25
BULLET PATH (IN)	-.75	.65	.00	-3.02	-8.76	-17.67	-30.26	-47.13	-68.91	-96.39	-130.27	-171.33	-220.27
TIME OF FLIGHT (SEC)	.000000	.070546	.146142	.227303	.314584	.408554	.509761	.618677	.735633	.860717	.993553	1.133429	1.279611
WIND DEFLECTION (IN)													
5 MPH	.00	.21	.86	2.00	3.68	5.95	8.86	12.44	16.74	21.74	27.43	33.74	40.61
10 MPH	.00	.42	1.72	4.01	7.37	11.91	17.72	24.89	33.47	43.49	54.87	67.48	81.21
20 MPH	.00	.83	3.44	8.01	14.73	23.81	35.44	49.77	66.94	86.97	109.73	134.97	162.42
30 MPH	.00	1.25	5.16	12.02	22.10	35.72	53.15	74.66	100.41	130.46	164.60	202.45	243.63
VELOCITY (FPS)	2100.	1959.	1824.	1696.	1575.	1462.	1359.	1266.	1185.	1118.	1063.	1019.	981.
ENERGY (FT-LB)	3153.	2744.	2379.	2056.	1773.	1528.	1320.	1146.	1004.	893.	808.	742.	688.
DROP (IN)	.00	-.97	-4.21	-10.05	-18.91	-31.28	-47.74	-68.90	-95.51	-128.29	-168.02	-215.40	-271.15
MID-RANGE (IN)	.00	.26	1.13	2.74	5.25	8.86	13.82	20.38	28.84	39.53	52.73	68.74	87.83
BULLET PATH (IN)	-.75	.76	.00	-3.36	-9.74	-19.63	-33.62	-52.30	-76.42	-106.73	-143.97	-188.88	-242.14
TIME OF FLIGHT (SEC)	.000000	.073957	.153311	.238610	.330418	.429297	.535749	.650151	.772699	.903161	1.040911	1.185176	1.335360
WIND DEFLECTION (IN)													
5 MPH	.00	.22	.92	2.14	3.93	6.35	9.43	13.21	17.71	22.91	28.74	35.15	42.08
10 MPH	.00	.44	1.84	4.28	7.87	12.70	18.86	26.43	35.42	45.81	57.49	70.31	84.17
20 MPH	.00	.89	3.68	8.56	15.74	25.40	37.73	52.85	70.85	91.63	114.97	140.61	168.33
30 MPH	.00	1.33	5.52	12.84	23.60	38.10	56.59	79.28	106.27	137.44	172.46	210.92	252.50

Bullet: Lyman 457122 322 Gr.
Ballistic Coefficient: .274

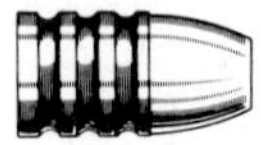

RANGE (YDS)	MUZZLE	50	100	150	200	250	300	350	400	450	500	550	600
VELOCITY (FPS)	2000.	1863.	1733.	1610.	1494.	1388.	1293.	1208.	1136.	1078.	1031.	992.	957.
ENERGY (FT-LB)	2859.	2482.	2147.	1852.	1597.	1378.	1194.	1043.	923.	831.	760.	703.	655.
DROP (IN)	.00	-1.08	-4.65	-11.12	-20.93	-34.67	-52.92	-76.38	-105.80	-141.94	-185.54	-237.29	-297.87
MID-RANGE (IN)	.00	.29	1.25	3.03	5.81	9.84	15.34	22.62	31.98	43.70	58.10	75.46	96.02
BULLET PATH (IN)	-.75	.87	.00	-3.76	-10.87	-21.92	-37.46	-58.22	-84.95	-118.39	-159.28	-208.34	-266.21
TIME OF FLIGHT (SEC)	.000000	.077708	.161195	.251024	.347762	.451935	.563962	.684111	.812291	.947975	1.090391	1.238879	1.392976
WIND DEFLECTION (IN)													
5 MPH	.00	.24	.99	2.29	4.20	6.77	10.03	14.00	18.68	24.02	29.95	36.42	43.38
10 MPH	.00	.48	1.97	4.58	8.41	13.54	20.06	28.00	37.36	48.04	59.91	72.84	86.76
20 MPH	.00	.95	3.94	9.16	16.81	27.08	40.11	56.01	74.73	96.09	119.82	145.69	173.53
30 MPH	.00	1.43	5.91	13.74	25.22	40.62	60.17	84.01	112.09	144.13	179.73	218.53	260.29
VELOCITY (FPS)	1900.	1768.	1643.	1525.	1416.	1318.	1230.	1154.	1093.	1043.	1002.	966.	935.
ENERGY (FT-LB)	2581.	2234.	1929.	1663.	1434.	1241.	1081.	952.	853.	778.	717.	667.	625.
DROP (IN)	.00	-1.19	-5.17	-12.35	-23.30	-38.58	-58.88	-84.94	-117.51	-157.34	-205.13	-261.57	-327.23
MID-RANGE (IN)	.00	.32	1.39	3.37	6.48	10.97	17.09	25.16	35.46	48.31	63.98	82.72	104.74
BULLET PATH (IN)	-.75	1.01	.00	-4.23	-12.22	-24.55	-41.88	-64.99	-94.60	-131.47	-176.31	-229.78	-292.49
TIME OF FLIGHT (SEC)	.000000	.081853	.169892	.264687	.366780	.476629	.594545	.720567	.854271	.994921	1.141801	1.294401	1.452356
WIND DEFLECTION (IN)													
5 MPH	.00	.26	1.06	2.45	4.49	7.21	10.64	14.78	19.60	25.03	31.00	37.49	44.44
10 MPH	.00	.51	2.11	4.90	8.97	14.41	21.27	29.56	39.19	50.05	62.01	74.97	88.88
20 MPH	.00	1.02	4.22	9.80	17.95	28.83	42.54	59.11	78.39	100.11	124.02	149.94	177.76
30 MPH	.00	1.53	6.33	14.70	26.92	43.24	63.81	88.67	117.58	150.16	186.03	224.92	266.63
VELOCITY (FPS)	1800.	1673.	1553.	1442.	1341.	1250.	1172.	1107.	1054.	1011.	975.	942.	914.
ENERGY (FT-LB)	2316.	2001.	1725.	1487.	1286.	1117.	981.	875.	795.	731.	679.	635.	597.
DROP (IN)	.00	-1.33	-5.77	-13.81	-26.06	-43.13	-65.78	-94.76	-130.81	-174.65	-226.97	-288.38	-359.45
MID-RANGE (IN)	.00	.36	1.55	3.77	7.26	12.27	19.08	28.00	39.34	53.38	70.35	90.48	113.94
BULLET PATH (IN)	-.75	1.17	.00	-4.79	-13.78	-27.59	-46.99	-72.71	-105.50	-146.08	-195.14	-253.30	-321.11
TIME OF FLIGHT (SEC)	.000000	.086450	.179516	.279750	.387637	.503531	.627566	.759425	.898426	1.043814	1.195032	1.351684	1.513490
WIND DEFLECTION (IN)													
5 MPH	.00	.27	1.13	2.62	4.78	7.64	11.23	15.50	20.39	25.86	31.83	38.28	45.19
10 MPH	.00	.55	2.26	5.24	9.56	15.29	22.45	30.99	40.79	51.71	63.66	76.56	90.37
20 MPH	.00	1.10	4.52	10.47	19.12	30.58	44.90	61.98	81.58	103.42	127.32	153.13	180.75
30 MPH	.00	1.65	6.78	15.71	28.67	45.86	67.36	92.98	122.37	155.13	190.98	229.69	271.12
VELOCITY (FPS)	1700.	1579.	1466.	1362.	1269.	1188.	1120.	1065.	1020.	982.	949.	920.	893.
ENERGY (FT-LB)	2066.	1782.	1536.	1327.	1151.	1008.	896.	811.	744.	690.	644.	605.	570.
DROP (IN)	.00	-1.50	-6.48	-15.55	-29.28	-48.44	-73.75	-105.97	-145.83	-194.02	-251.21	-317.87	-394.78
MID-RANGE (IN)	.00	.40	1.75	4.26	8.16	13.75	21.33	31.20	43.63	58.88	77.17	98.66	123.64
BULLET PATH (IN)	-.75	1.37	.00	-5.45	-15.57	-31.10	-52.80	-81.40	-117.65	-162.22	-215.79	-278.83	-352.13
TIME OF FLIGHT (SEC)	.000000	.091576	.190204	.296389	.410515	.532788	.662990	.800503	.944553	1.094537	1.250030	1.410728	1.576419
WIND DEFLECTION (IN)													
5 MPH	.00	.29	1.21	2.79	5.07	8.06	11.75	16.09	21.00	26.44	32.36	38.73	45.55
10 MPH	.00	.59	2.42	5.58	10.13	16.12	23.51	32.18	42.01	52.87	64.71	77.46	91.10
20 MPH	.00	1.18	4.83	11.15	20.27	32.25	47.02	64.37	84.01	105.75	129.42	154.93	182.19
30 MPH	.00	1.76	7.25	16.73	30.40	48.37	70.53	96.55	126.02	158.62	194.13	232.39	273.29

Bullet: Lyman 457124 366 Gr.
Ballistic Coefficient: .299

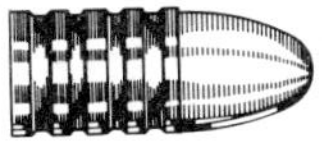

RANGE (YDS)	MUZZLE	50	100	150	200	250	300	350	400	450	500	550	600
VELOCITY (FPS)	2400.	2260.	2125.	1995.	1870.	1750.	1636.	1528.	1428.	1336.	1252.	1178.	1116.
ENERGY (FT-LB)	4680.	4152.	3670.	3234.	2840.	2487.	2174.	1897.	1657.	1450.	1274.	1128.	1012.
DROP (IN)	.00	-.73	-3.16	-7.51	-14.04	-23.05	-34.90	-49.98	-68.82	-91.93	-119.90	-153.45	-193.25
MID-RANGE (IN)	.00	.20	.85	2.03	3.86	6.44	9.93	14.50	20.37	27.75	36.90	48.13	61.73
BULLET PATH (IN)	-.75	.47	.00	-2.39	-6.96	-14.02	-23.90	-37.03	-53.91	-75.06	-101.08	-132.67	-170.51
TIME OF FLIGHT (SEC)	.000000	.064402	.132839	.205687	.283358	.366300	.454986	.549892	.651476	.760130	.876154	.999758	1.130694
WIND DEFLECTION (IN)													
5 MPH	.00	.17	.69	1.60	2.94	4.73	7.04	9.89	13.33	17.39	22.10	27.48	33.50
10 MPH	.00	.33	1.38	3.20	5.87	9.47	14.08	19.78	26.66	34.78	44.20	54.96	67.00
20 MPH	.00	.67	2.76	6.40	11.74	18.94	28.15	39.56	53.32	69.57	88.41	109.91	134.00
30 MPH	.00	1.00	4.14	9.60	17.61	28.41	42.23	59.34	79.98	104.35	132.61	164.87	201.01
VELOCITY (FPS)	2300.	2164.	2032.	1905.	1784.	1668.	1558.	1456.	1361.	1275.	1198.	1133.	1079.
ENERGY (FT-LB)	4298.	3804.	3355.	2949.	2585.	2260.	1973.	1722.	1506.	1322.	1167.	1043.	946.
DROP (IN)	.00	-.80	-3.45	-8.20	-15.35	-25.22	-38.20	-54.78	-75.49	-100.88	-131.63	-168.44	-212.01
MID-RANGE (IN)	.00	.22	.93	2.22	4.22	7.06	10.90	15.94	22.40	30.52	40.60	52.93	67.80
BULLET PATH (IN)	-.75	.55	.00	-2.65	-7.69	-15.46	-26.35	-40.83	-59.43	-82.72	-111.37	-146.08	-187.54
TIME OF FLIGHT (SEC)	.000000	.067242	.138783	.215025	.296406	.383394	.476465	.576089	.682678	.796552	.917983	1.046854	1.182681
WIND DEFLECTION (IN)													
5 MPH	.00	.18	.73	1.70	3.13	5.04	7.49	10.52	14.16	18.44	23.39	28.99	35.21
10 MPH	.00	.36	1.47	3.41	6.25	10.09	14.99	21.04	28.33	36.89	46.78	57.99	70.41
20 MPH	.00	.71	2.94	6.82	12.51	20.17	29.98	42.09	56.65	73.78	93.56	115.97	140.83
30 MPH	.00	1.07	4.41	10.23	18.76	30.26	44.97	63.13	84.98	110.67	140.35	173.96	211.24
VELOCITY (FPS)	2200.	2067.	1939.	1816.	1698.	1587.	1483.	1386.	1298.	1218.	1149.	1092.	1046.
ENERGY (FT-LB)	3933.	3472.	3054.	2679.	2344.	2046.	1786.	1561.	1369.	1205.	1073.	969.	889.
DROP (IN)	.00	-.88	-3.78	-8.99	-16.84	-27.69	-41.99	-60.28	-83.09	-111.07	-144.92	-185.34	-233.03
MID-RANGE (IN)	.00	.24	1.01	2.44	4.64	7.77	12.00	17.58	24.71	33.67	44.77	58.30	74.53
BULLET PATH (IN)	-.75	.64	.00	-2.95	-8.53	-17.11	-29.15	-45.18	-65.72	-91.44	-123.02	-161.18	-206.60
TIME OF FLIGHT (SEC)	.000000	.070343	.145274	.225223	.310651	.402033	.499845	.604517	.716398	.835783	.962688	1.096707	1.237188
WIND DEFLECTION (IN)													
5 MPH	.00	.19	.78	1.82	3.34	5.38	7.99	11.20	15.04	19.55	24.72	30.51	36.87
10 MPH	.00	.38	1.57	3.64	6.67	10.76	15.97	22.40	30.09	39.10	49.43	61.02	73.75
20 MPH	.00	.76	3.14	7.28	13.35	21.52	31.95	44.79	60.17	78.20	98.87	122.04	147.49
30 MPH	.00	1.14	4.70	10.92	20.02	32.27	47.92	67.19	90.26	117.29	148.30	183.06	221.24
VELOCITY (FPS)	2100.	1971.	1846.	1727.	1614.	1508.	1410.	1319.	1237.	1165.	1106.	1057.	1016.
ENERGY (FT-LB)	3583.	3155.	2770.	2424.	2118.	1848.	1614.	1414.	1243.	1103.	993.	907.	839.
DROP (IN)	.00	-.96	-4.16	-9.90	-18.55	-30.52	-46.36	-66.57	-91.77	-122.68	-159.97	-204.35	-256.50
MID-RANGE (IN)	.00	.26	1.12	2.69	5.12	8.57	13.28	19.45	27.34	37.25	49.46	64.25	81.89
BULLET PATH (IN)	-.75	.74	.00	-3.29	-9.48	-19.00	-32.39	-50.14	-72.89	-101.34	-136.18	-178.11	-227.80
TIME OF FLIGHT (SEC)	.000000	.073739	.152384	.236389	.326229	.422381	.525292	.635334	.752804	.877840	1.010119	1.149036	1.293930
WIND DEFLECTION (IN)													
5 MPH	.00	.20	.84	1.95	3.57	5.74	8.51	11.91	15.96	20.68	26.03	31.97	38.44
10 MPH	.00	.41	1.68	3.89	7.13	11.48	17.02	23.82	31.92	41.36	52.07	63.94	76.87
20 MPH	.00	.81	3.35	7.78	14.26	22.96	34.05	47.64	63.84	82.71	104.13	127.89	153.75
30 MPH	.00	1.22	5.03	11.67	21.39	34.45	51.07	71.46	95.77	124.07	156.20	191.83	230.62

Bullet: Lyman 457124 366 Gr.
Ballistic Coefficient: .299

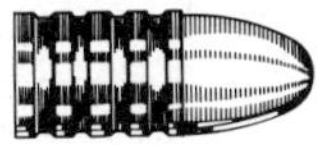

RANGE (YDS)	MUZZLE	50	100	150	200	250	300	350	400	450	500	550	600
VELOCITY (FPS)	2000.	1874.	1754.	1640.	1532.	1432.	1339.	1255.	1181.	1118.	1067.	1025.	989.
ENERGY (FT-LB)	3250.	2855.	2500.	2185.	1907.	1665.	1458.	1280.	1133.	1016.	925.	853.	794.
DROP (IN)	0.0	-1.06	-4.60	-10.95	-20.52	-33.83	-51.39	-73.78	-101.72	-135.89	-176.98	-225.69	-282.64
MID-RANGE (IN)	0.0	0.29	1.23	2.97	5.67	9.52	14.75	21.59	30.34	41.27	54.66	70.80	89.93
BULLET PATH (IN)	-0.75	0.86	0.00	-3.68	-10.58	-21.21	-36.10	-55.82	-81.09	-112.58	-151.01	-197.04	-251.32
TIME OF FLIGHT (SEC)	0.0	0.077475	0.160203	0.248655	0.343310	0.444626	0.552997	0.668724	0.792028	0.922678	1.060123	1.203685	1.352827
WIND DEFLECTION (IN)													
5 MPH	0.0	0.22	0.90	2.08	3.81	6.13	9.06	12.65	16.90	21.80	27.29	33.32	39.85
10 MPH	0.0	0.44	1.80	4.16	7.62	12.25	18.13	25.30	33.80	43.59	54.58	66.65	79.70
20 MPH	0.0	0.87	3.59	8.33	15.25	24.51	36.25	50.59	67.59	87.18	109.16	133.30	159.40
30 MPH	0.0	1.31	5.39	12.49	22.87	36.76	54.38	75.89	101.39	130.77	163.75	199.95	239.09
VELOCITY (FPS)	1900.	1779.	1663.	1554.	1452.	1358.	1272.	1195.	1130.	1077.	1033.	996.	963.
ENERGY (FT-LB)	2933.	2570.	2247.	1962.	1712.	1498.	1315.	1161.	1038.	942.	867.	806.	754.
DROP (IN)	0.0	-1.18	-5.10	-12.16	-22.83	-37.66	-57.19	-82.11	-113.11	-150.90	-196.16	-249.55	-311.68
MID-RANGE (IN)	0.0	0.32	1.37	3.31	6.31	10.61	16.43	24.04	33.72	45.76	60.43	77.98	98.66
BULLET PATH (IN)	-0.75	1.00	0.00	-4.13	-11.88	-23.77	-40.37	-62.37	-90.44	-125.31	-167.64	-218.10	-277.31
TIME OF FLIGHT (SEC)	0.0	0.081603	0.168831	0.262164	0.362066	0.468948	0.583129	0.704872	0.834041	0.970138	1.112485	1.260508	1.413795
WIND DEFLECTION (IN)													
5 MPH	0.0	0.23	0.96	2.23	4.07	6.53	9.63	13.40	17.82	22.85	28.42	34.50	41.05
10 MPH	0.0	0.47	1.92	4.46	8.14	13.06	19.26	26.79	35.63	45.69	56.85	69.01	82.09
20 MPH	0.0	0.93	3.85	8.91	16.29	26.12	38.52	53.59	71.27	91.38	113.70	138.01	164.18
30 MPH	0.0	1.40	5.77	13.37	24.43	39.18	57.79	80.38	106.90	137.07	170.55	207.02	246.27
VELOCITY (FPS)	1800.	1683.	1573.	1469.	1374.	1287.	1208.	1141.	1086.	1040.	1002.	969.	939.
ENERGY (FT-LB)	2633.	2302.	2010.	1755.	1534.	1346.	1186.	1058.	958.	879.	816.	763.	717.
DROP (IN)	0.0	-1.32	-5.70	-13.60	-25.55	-42.11	-63.92	-91.70	-126.13	-167.92	-217.71	-276.14	-343.76
MID-RANGE (IN)	0.0	0.36	1.53	3.70	7.08	11.88	18.36	26.81	37.51	50.74	66.75	85.77	107.96
BULLET PATH (IN)	-0.75	1.16	0.00	-4.67	-13.41	-26.73	-45.32	-69.88	-101.09	-139.65	-186.22	-241.42	-305.82
TIME OF FLIGHT (SEC)	0.0	0.086182	0.178383	0.277072	0.382674	0.495520	0.615898	0.743759	0.878657	1.019926	1.166961	1.319323	1.476694
WIND DEFLECTION (IN)													
5 MPH	0.0	0.25	1.03	2.38	4.34	6.94	10.20	14.12	18.66	23.75	29.36	35.43	41.95
10 MPH	0.0	0.50	2.06	4.76	8.68	13.88	20.40	28.23	37.31	47.51	58.72	70.87	83.90
20 MPH	0.0	1.00	4.12	9.53	17.37	27.76	40.80	56.47	74.62	95.01	117.44	141.73	167.80
30 MPH	0.0	1.50	6.19	14.29	26.05	41.63	61.19	84.70	111.93	142.52	176.16	212.60	251.69
VELOCITY (FPS)	1700.	1589.	1484.	1387.	1299.	1219.	1150.	1093.	1046.	1007.	973.	944.	917.
ENERGY (FT-LB)	2348.	2051.	1790.	1564.	1372.	1207.	1075.	971.	890.	824.	770.	723.	683.
DROP (IN)	0.0	-1.48	-6.40	-15.31	-28.73	-47.30	-71.74	-102.74	-140.98	-187.14	-241.84	-305.67	-379.11
MID-RANGE (IN)	0.0	0.40	1.72	4.18	7.96	13.32	20.56	29.95	41.76	56.25	73.62	94.09	117.81
BULLET PATH (IN)	-0.75	1.35	0.00	-5.33	-15.17	-30.17	-51.03	-78.45	-113.11	-155.69	-206.82	-267.07	-336.94
TIME OF FLIGHT (SEC)	0.0	0.091289	0.189000	0.293566	0.405335	0.524603	0.651394	0.785304	0.925685	1.071908	1.223513	1.380166	1.541625
WIND DEFLECTION (IN)													
5 MPH	0.0	0.27	1.10	2.54	4.61	7.34	10.73	14.75	19.34	24.45	30.02	36.04	42.49
10 MPH	0.0	0.54	2.21	5.08	9.22	14.68	21.47	29.51	38.69	48.89	60.04	72.09	84.97
20 MPH	0.0	1.07	4.41	10.16	18.44	29.37	42.94	59.02	77.37	97.78	120.09	144.17	169.95
30 MPH	0.0	1.61	6.62	15.24	27.66	44.05	64.41	88.52	116.06	146.67	180.13	216.26	254.92

Bullet: Lyman 457124 366 Gr.
Ballistic Coefficient: .299

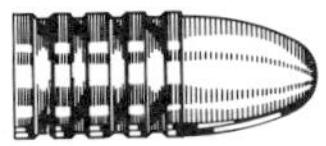

RANGE (YDS)	MUZZLE	50	100	150	200	250	300	350	400	450	500	550	600
VELOCITY (FPS)	1600.	1495.	1397.	1308.	1227.	1157.	1098.	1051.	1011.	977.	947.	919.	895.
ENERGY (FT-LB)	2080.	1815.	1586.	1390.	1223.	1087.	980.	897.	831.	775.	728.	687.	650.
DROP (IN)	.00	-1.67	-7.28	-17.34	-32.47	-53.41	-80.82	-115.40	-157.84	-208.75	-268.74	-338.27	-418.05
MID-RANGE (IN)	.00	.45	1.97	4.71	8.95	14.99	23.07	33.48	46.45	62.20	80.97	102.88	128.21
BULLET PATH (IN)	-.75	1.59	.00	-6.04	-17.16	-34.07	-57.47	-88.04	-126.46	-173.35	-229.33	-294.84	-370.60
TIME OF FLIGHT (SEC)	.000000	.097014	.200841	.311840	.430303	.556310	.689493	.829220	.974846	1.125896	1.282024	1.442978	1.608578
WIND DEFLECTION (IN)													
5 MPH	.00	.29	1.17	2.69	4.87	7.71	11.18	15.22	19.79	24.83	30.32	36.23	42.55
10 MPH	.00	.57	2.35	5.38	9.73	15.41	22.35	30.44	39.57	49.66	60.64	72.46	85.11
20 MPH	.00	1.15	4.70	10.77	19.47	30.82	44.70	60.89	79.15	99.32	121.27	144.93	170.22
30 MPH	.00	1.72	7.04	16.15	29.20	46.23	67.05	91.33	118.72	148.97	181.91	217.39	255.33
VELOCITY (FPS)	1500.	1402.	1312.	1231.	1160.	1101.	1053.	1013.	978.	948.	921.	896.	873.
ENERGY (FT-LB)	1828.	1597.	1400.	1231.	1094.	985.	901.	834.	778.	730.	689.	652.	619.
DROP (IN)	.00	-1.95	-8.33	-19.75	-36.93	-60.55	-91.32	-129.89	-176.92	-233.00	-298.58	-374.37	-460.98
MID-RANGE (IN)	.00	.52	2.21	5.32	10.14	16.91	25.91	37.37	51.53	68.61	88.74	112.19	139.17
BULLET PATH (IN)	-.75	1.84	.00	-6.88	-19.52	-38.60	-64.83	-98.87	-141.35	-192.89	-253.93	-325.18	-407.26
TIME OF FLIGHT (SEC)	.000000	.103465	.214086	.332153	.457773	.590595	.729997	.875327	1.026100	1.181966	1.342669	1.508025	1.677905
WIND DEFLECTION (IN)													
5 MPH	.00	.30	1.24	2.83	5.08	7.97	11.44	15.43	19.90	24.81	30.15	35.91	42.06
10 MPH	.00	.61	2.48	5.66	10.17	15.94	22.88	30.86	39.79	49.63	60.31	71.81	84.11
20 MPH	.00	1.22	4.96	11.32	20.34	31.89	45.76	61.71	79.59	99.25	120.62	143.62	168.22
30 MPH	.00	1.83	7.44	16.98	30.50	47.83	68.64	92.57	119.38	148.88	180.93	215.44	252.33

Bullet: Lyman # 457193 420 Gr.
Ballistic Coefficient: .307

(Cont'd.)

RANGE (YDS)	MUZZLE	50	100	150	200	250	300	350	400	450	500	550	600
VELOCITY (FPS)	1400.	1320.	1247.	1185.	1133.	1090.	1052.	1020.	992.	966.	943.	921.	901.
ENERGY (FT-LB)	1828.	1624.	1451.	1308.	1198.	1107.	1033.	970.	917.	870.	828.	791.	757.
DROP (IN)	.00	-2.23	-9.43	-22.26	-41.26	-67.00	-100.04	-140.91	-190.10	-248.12	-315.37	-392.26	-479.36
MID-RANGE (IN)	.00	.59	2.49	5.96	11.20	18.39	27.76	39.50	53.79	70.81	90.69	113.55	139.64
BULLET PATH (IN)	-.75	2.11	.00	-7.73	-21.64	-42.29	-70.24	-106.02	-150.12	-203.05	-265.21	-337.00	-419.02
TIME OF FLIGHT (SEC)	.000000	.110382	.227337	.350850	.480368	.615422	.755578	.900432	1.049671	1.203055	1.360396	1.521548	1.686399
WIND DEFLECTION (IN)													
5 MPH	.00	.29	1.15	2.59	4.56	7.01	9.92	13.24	16.94	21.01	25.43	30.18	35.26
10 MPH	.00	.57	2.30	5.18	9.12	14.03	19.84	26.48	33.89	42.02	50.86	60.36	70.52
20 MPH	.00	1.14	4.59	10.36	18.23	28.06	39.68	52.95	67.77	84.05	101.72	120.73	141.04
30 MPH	.00	1.71	6.89	15.53	27.35	42.09	59.52	79.43	101.66	126.07	152.57	181.09	211.56
VELOCITY (FPS)	1300.	1230.	1170.	1121.	1079.	1043.	1012.	984.	959.	937.	916.	896.	878.
ENERGY (FT-LB)	1576.	1410.	1277.	1172.	1086.	1015.	955.	904.	858.	818.	782.	749.	718.
DROP (IN)	.00	-2.59	-10.94	-25.62	-47.20	-76.21	-113.19	-158.62	-213.01	-276.71	-350.20	-434.04	-528.68
MID-RANGE (IN)	.00	.69	2.88	6.80	12.66	20.65	30.97	43.80	59.31	77.61	98.89	123.35	151.15
BULLET PATH (IN)	-.75	2.51	.00	-8.83	-24.56	-47.73	-78.86	-118.45	-166.99	-224.84	-292.49	-370.49	-459.28
TIME OF FLIGHT (SEC)	.000000	.118690	.243858	.374879	.511327	.652760	.798800	.949154	1.103596	1.261952	1.424085	1.589891	1.759293
WIND DEFLECTION (IN)													
5 MPH	.00	.29	1.15	2.53	4.38	6.67	9.37	12.45	15.89	19.67	23.78	28.22	32.97
10 MPH	.00	.58	2.30	5.06	8.76	13.35	18.74	24.90	31.77	39.33	47.56	56.44	65.94
20 MPH	.00	1.16	4.61	10.11	17.53	26.69	37.49	49.79	63.54	78.67	95.12	112.87	131.89
30 MPH	.00	1.75	6.91	15.17	26.29	40.04	56.23	74.69	95.31	118.00	142.69	169.31	197.83
VELOCITY (FPS)	1200.	1147.	1101.	1062.	1028.	999.	973.	949.	927.	906.	887.	869.	852.
ENERGY (FT-LB)	1343.	1226.	1130.	1052.	986.	930.	882.	839.	801.	766.	734.	705.	678.
DROP (IN)	.00	-3.04	-12.68	-29.46	-53.92	-86.57	-127.92	-178.43	-238.37	-308.41	-389.00	-480.61	-583.68
MID-RANGE (IN)	.00	.79	3.30	7.73	14.28	23.15	34.50	48.52	65.26	85.01	107.93	134.17	163.92
BULLET PATH (IN)	-.75	2.92	.00	-10.07	-27.82	-53.76	-88.39	-132.18	-185.41	-248.74	-322.62	-407.51	-503.87
TIME OF FLIGHT (SEC)	.000000	.127927	.261492	.400282	.543869	.691918	.844168	1.000420	1.160517	1.324337	1.491790	1.662809	1.837347
WIND DEFLECTION (IN)													
5 MPH	.00	.26	1.01	2.22	3.86	5.89	8.29	11.04	14.13	17.54	21.28	25.33	29.69
10 MPH	.00	.52	2.02	4.45	7.72	11.78	16.57	22.07	28.25	35.08	42.56	50.65	59.37
20 MPH	.00	1.03	4.05	8.90	15.44	23.56	33.15	44.15	56.50	70.17	85.11	101.31	118.75
30 MPH	.00	1.55	6.07	13.35	23.16	35.33	49.72	66.22	84.75	105.25	127.67	151.96	178.12
VELOCITY (FPS)	1100.	1061.	1028.	998.	972.	948.	926.	906.	887.	869.	852.	836.	820.
ENERGY (FT-LB)	1128.	1050.	985.	929.	881.	838.	800.	765.	733.	704.	677.	651.	627.
DROP (IN)	.00	-3.58	-14.85	-34.32	-62.49	-99.83	-146.62	-203.52	-270.97	-349.45	-439.40	-541.33	-655.72
MID-RANGE (IN)	.00	.93	3.85	8.94	16.40	26.39	38.99	54.47	72.99	94.73	119.86	148.67	181.24
BULLET PATH (IN)	-.75	3.47	.00	-11.67	-32.05	-61.59	-100.58	-149.67	-209.33	-280.01	-362.16	-456.29	-562.88
TIME OF FLIGHT (SEC)	.000000	.138894	.282577	.430713	.583045	.739371	.899538	1.063423	1.230937	1.402012	1.576602	1.754680	1.936233
WIND DEFLECTION (IN)													
5 MPH	.00	.22	.87	1.90	3.31	5.06	7.16	9.58	12.32	15.38	18.74	22.41	26.39
10 MPH	.00	.45	1.73	3.81	6.62	10.13	14.32	19.16	24.64	30.75	37.48	44.82	52.78
20 MPH	.00	.89	3.47	7.61	13.23	20.26	28.64	38.32	49.29	61.51	74.96	89.65	105.55
30 MPH	.00	1.34	5.20	11.42	19.85	30.39	42.96	57.49	73.93	92.26	112.45	134.47	158.33

Bullet: Lyman # 457193 420 Gr.
Ballistic Coefficient: .307

RANGE (YDS)	MUZZLE	50	100	150	200	250	300	350	400	450	500	550	600
VELOCITY (FPS)	1800.	1686.	1580.	1487.	1400.	1320.	1247.	1184.	1133.	1089.	1052.	1020.	992.
ENERGY (FT-LB)	3021.	2651.	2327.	2061.	1827.	1624.	1450.	1308.	1197.	1107.	1032.	970.	917.
DROP (IN)	.00	-1.31	-5.67	-13.50	-25.32	-41.55	-62.77	-89.61	-122.63	-162.40	-209.47	-264.37	-327.60
MID-RANGE (IN)	.00	.36	1.52	3.66	6.98	11.65	17.88	25.92	36.00	48.35	63.18	80.68	101.03
BULLET PATH (IN)	-.75	1.15	.00	-4.62	-13.22	-26.24	-44.25	-67.88	-97.68	-134.24	-178.10	-229.78	-289.80
TIME OF FLIGHT (SEC)	.000000	.086106	.178057	.275960	.379975	.490386	.607373	.730920	.860481	.995577	1.135775	1.280672	1.429956
WIND DEFLECTION (IN)													
5 MPH	.00	.24	1.00	2.28	4.10	6.49	9.45	12.99	17.06	21.61	26.61	32.03	37.84
10 MPH	.00	.49	2.00	4.57	8.21	12.97	18.90	25.98	34.11	43.22	53.23	64.06	75.67
20 MPH	.00	.98	4.01	9.14	16.42	25.95	37.80	51.95	68.22	86.44	106.46	128.13	151.34
30 MPH	.00	1.46	6.01	13.71	24.63	38.92	56.69	77.93	102.33	129.66	159.69	192.19	227.02
VELOCITY (FPS)	1700.	1592.	1498.	1410.	1329.	1256.	1191.	1139.	1094.	1057.	1024.	995.	969.
ENERGY (FT-LB)	2695.	2362.	2092.	1854.	1647.	1470.	1323.	1210.	1117.	1041.	977.	923.	875.
DROP (IN)	.00	-1.48	-6.36	-15.17	-28.34	-46.42	-70.05	-99.78	-136.19	-179.84	-231.25	-290.94	-359.40
MID-RANGE (IN)	.00	.40	1.70	4.12	7.81	12.96	19.85	28.71	39.75	53.19	69.20	87.97	109.65
BULLET PATH (IN)	-.75	1.33	.00	-5.26	-14.87	-29.40	-49.47	-75.65	-108.51	-148.60	-196.46	-252.59	-317.49
TIME OF FLIGHT (SEC)	.000000	.091208	.188378	.291619	.401226	.517393	.640146	.768998	.903433	1.043024	1.187358	1.336112	1.489039
WIND DEFLECTION (IN)													
5 MPH	.00	.26	1.05	2.37	4.25	6.71	9.74	13.32	17.38	21.90	26.84	32.17	37.86
10 MPH	.00	.52	2.10	4.74	8.50	13.41	19.49	26.64	34.77	43.81	53.68	64.33	75.72
20 MPH	.00	1.05	4.19	9.47	17.00	26.83	38.98	53.28	69.54	87.62	107.36	128.66	151.44
30 MPH	.00	1.57	6.29	14.21	25.49	40.24	58.47	79.91	104.31	131.42	161.04	193.00	227.15
VELOCITY (FPS)	1600.	1506.	1417.	1336.	1262.	1196.	1143.	1098.	1060.	1026.	997.	971.	947.
ENERGY (FT-LB)	2387.	2114.	1873.	1664.	1484.	1334.	1219.	1124.	1047.	982.	927.	879.	837.
DROP (IN)	.00	-1.65	-7.18	-17.04	-31.75	-51.95	-78.22	-111.11	-151.18	-198.98	-255.01	-319.76	-393.70
MID-RANGE (IN)	.00	.45	1.94	4.61	8.69	14.44	22.07	31.80	43.84	58.36	75.55	95.59	118.63
BULLET PATH (IN)	-.75	1.56	.00	-5.89	-16.64	-32.87	-55.17	-84.09	-120.20	-164.03	-216.09	-276.88	-346.85
TIME OF FLIGHT (SEC)	.000000	.096657	.199355	.308397	.423988	.546172	.674513	.808469	.947620	1.091546	1.239918	1.392479	1.549035
WIND DEFLECTION (IN)													
5 MPH	.00	.26	1.04	2.39	4.31	6.81	9.86	13.40	17.39	21.81	26.61	31.79	37.32
10 MPH	.00	.51	2.09	4.78	8.62	13.63	19.71	26.79	34.78	43.61	53.23	63.58	74.63
20 MPH	.00	1.02	4...7	9.56	17.24	27.25	39.43	53.58	69.56	87.22	106.45	127.15	149.26
30 MPH	.00	1.53	6.26	14.33	25.87	40.88	59.14	80.37	104.34	130.84	159.68	190.73	223.89
VELOCITY (FPS)	1500.	1412.	1331.	1257.	1193.	1140.	1096.	1057.	1025.	995.	969.	946.	924.
ENERGY (FT-LB)	2098.	1859.	1652.	1474.	1326.	1213.	1119.	1043.	979.	924.	876.	834.	796.
DROP (IN)	.00	-1.94	-8.23	-19.42	-36.14	-58.95	-88.42	-125.12	-169.56	-222.27	-283.72	-354.38	-434.57
MID-RANGE (IN)	.00	.52	2.18	5.20	9.84	16.29	24.79	35.51	48.64	64.39	82.92	104.38	128.86
BULLET PATH (IN)	-.75	1.80	.00	-6.70	-18.92	-37.24	-62.23	-94.43	-134.38	-182.60	-239.57	-305.73	-381.43
TIME OF FLIGHT (SEC)	.000000	.103086	.212532	.328535	.451123	.579818	.714103	.853556	.997760	1.146391	1.299198	1.455987	1.616607
WIND DEFLECTION (IN)													
5 MPH	.00	.27	1.10	2.51	4.50	7.02	10.04	13.51	17.40	21.68	26.33	31.33	36.66
10 MPH	.00	.54	2.21	5.02	9.00	14.05	20.08	27.03	34.81	43.36	52.66	62.65	73.32
20 MPH	.00	1.09	4.41	10.04	18.00	28.10	40.16	54.05	69.61	86.73	105.32	125.31	146.65
30 MPH	.00	1.63	6.62	15.07	26.99	42.14	60.25	81.08	104.42	130.09	157.98	187.96	219.97

Bullet: Lyman 457102 426 Gr.
Ballistic Coefficient: .359

RANGE (YDS)	MUZZLE	50	100	150	200	250	300	350	400	450	500	550	600
VELOCITY (FPS)	1900.	1799.	1701.	1608.	1519.	1436.	1358.	1286.	1221.	1164.	1114.	1072.	1037.
ENERGY (FT-LB)	3414.	3059.	2736.	2444.	2182.	1949.	1744.	1565.	1410.	1281.	1174.	1088.	1016.
DROP (IN)	.00	-1.16	-5.00	-11.84	-22.03	-36.02	-54.23	-77.15	-105.33	-139.36	-179.79	-227.20	-282.15
MID-RANGE (IN)	.00	.32	1.34	3.19	6.01	9.98	15.28	22.08	30.63	41.16	53.88	69.03	86.84
BULLET PATH (IN)	-.75	.97	.00	-3.96	-11.27	-22.38	-37.72	-57.77	-83.07	-114.22	-151.78	-196.31	-248.38
TIME OF FLIGHT (SEC)	.000000	.081148	.166915	.257633	.353632	.455224	.562683	.676214	.795954	.921853	1.053667	1.190996	1.333379
WIND DEFLECTION (IN)													
5 MPH	.00	.19	.79	1.83	3.33	5.32	7.83	10.88	14.46	18.60	23.25	28.39	33.97
10 MPH	.00	.39	1.59	3.66	6.66	10.65	15.66	21.75	28.93	37.19	46.50	56.77	67.94
20 MPH	.00	.77	3.18	7.32	13.32	21.29	31.33	43.50	57.86	74.39	93.00	113.55	135.88
30 MPH	.00	1.16	4.76	10.98	19.98	31.94	46.99	65.25	86.79	111.58	139.49	170.32	203.81
VELOCITY (FPS)	1800.	1702.	1609.	1520.	1437.	1359.	1287.	1222.	1165.	1115.	1073.	1037.	1006.
ENERGY (FT-LB)	3064.	2741.	2449.	2186.	1953.	1747.	1567.	1413.	1283.	1176.	1089.	1017.	957.
DROP (IN)	.00	-1.30	-5.58	-13.22	-24.64	-40.29	-60.63	-86.22	-117.66	-155.48	-200.28	-252.61	-313.00
MID-RANGE (IN)	.00	.35	1.50	3.56	6.74	11.19	17.10	24.69	34.19	45.83	59.85	76.46	95.87
BULLET PATH (IN)	-.75	1.12	.00	-4.47	-12.72	-25.20	-42.38	-64.81	-93.07	-127.73	-169.36	-218.52	-275.74
TIME OF FLIGHT (SEC)	.000000	.085695	.176336	.272253	.373759	.481127	.594565	.714207	.840011	.971734	1.108979	1.251283	1.398264
WIND DEFLECTION (IN)													
5 MPH	.00	.21	.85	1.96	3.56	5.67	8.32	11.52	15.25	19.51	24.26	29.45	35.05
10 MPH	.00	.42	1.70	3.92	7.11	11.34	16.64	23.03	30.51	39.03	48.51	58.89	70.09
20 MPH	.00	.83	3.40	7.83	14.23	22.69	33.29	46.07	61.02	78.05	97.03	117.79	140.19
30 MPH	.00	1.25	5.11	11.75	21.34	34.03	49.93	69.10	91.53	117.08	145.54	176.68	210.28
VELOCITY (FPS)	1700.	1607.	1518.	1435.	1357.	1286.	1221.	1163.	1114.	1072.	1036.	1005.	977.
ENERGY (FT-LB)	2733.	2442.	2180.	1947.	1742.	1563.	1409.	1280.	1173.	1087.	1016.	955.	904.
DROP (IN)	.00	-1.46	-6.27	-14.88	-27.73	-45.28	-68.11	-96.78	-131.86	-173.93	-223.54	-281.23	-347.50
MID-RANGE (IN)	.00	.40	1.68	4.02	7.60	12.57	19.17	27.64	38.21	51.08	66.49	84.62	105.64
BULLET PATH (IN)	-.75	1.30	.00	-5.10	-14.44	-28.49	-47.80	-72.97	-104.54	-143.10	-189.20	-243.38	-306.14
TIME OF FLIGHT (SEC)	.000000	.090768	.186820	.288468	.395985	.509576	.629375	.755332	.887200	1.024577	1.167002	1.314093	1.465562
WIND DEFLECTION (IN)													
5 MPH	.00	.22	.91	2.09	3.79	6.02	8.80	12.12	15.96	20.28	25.05	30.23	35.79
10 MPH	.00	.45	1.82	4.18	7.58	12.04	17.59	24.23	31.91	40.56	50.10	60.46	71.59
20 MPH	.00	.89	3.64	8.36	15.15	24.08	35.19	48.47	63.82	81.12	100.20	120.91	143.17
30 MPH	.00	1.34	5.46	12.55	22.73	36.11	52.78	72.70	95.74	121.68	150.29	181.37	214.76
VELOCITY (FPS)	1600.	1512.	1429.	1352.	1281.	1216.	1159.	1110.	1069.	1034.	1003.	975.	951.
ENERGY (FT-LB)	2421.	2162.	1931.	1728.	1551.	1398.	1271.	1166.	1081.	1011.	951.	900.	855.
DROP (IN)	.00	-1.64	-7.13	-16.88	-31.37	-51.18	-76.88	-109.03	-148.20	-194.96	-249.83	-313.33	-385.94
MID-RANGE (IN)	.00	.45	1.92	4.56	8.55	14.15	21.56	31.02	42.73	56.92	73.74	93.41	116.09
BULLET PATH (IN)	-.75	1.55	.00	-5.81	-16.36	-32.23	-53.99	-82.20	-117.43	-160.25	-211.18	-270.74	-339.41
TIME OF FLIGHT (SEC)	.000000	.096458	.198535	.306497	.420547	.540809	.667216	.799507	.937274	1.080055	1.227477	1.379257	1.535176
WIND DEFLECTION (IN)													
5 MPH	.00	.24	.97	2.22	4.01	6.34	9.21	12.61	16.48	20.79	25.52	30.62	36.10
10 MPH	.00	.48	1.94	4.44	8.02	12.68	18.43	25.21	32.96	41.59	51.04	61.25	72.19
20 MPH	.00	.95	3.88	8.89	16.03	25.36	36.86	50.43	65.92	83.18	102.07	122.50	144.38
30 MPH	.00	1.43	5.83	13.33	24.05	38.05	55.29	75.64	98.88	124.77	153.11	183.75	216.57

Bullet: Lyman 457102 426 Gr.
Ballistic Coefficient: .359

RANGE (YDS)	MUZZLE	50	100	150	200	250	300	350	400	450	500	550	600
VELOCITY (FPS)	1500.	1418.	1341.	1271.	1208.	1152.	1104.	1064.	1029.	999.	972.	947.	925.
ENERGY (FT-LB)	2128.	1901.	1702.	1528.	1379.	1255.	1153.	1070.	1002.	944.	893.	849.	809.
DROP (IN)	.00	-1.92	-8.17	-19.24	-35.71	-58.14	-87.10	-123.17	-166.90	-218.81	-279.41	-349.20	-428.45
MID-RANGE (IN)	.00	.51	2.16	5.14	9.68	16.00	24.31	34.83	47.74	63.26	81.56	102.83	127.12
BULLET PATH (IN)	-.75	1.79	.00	-6.61	-18.61	-36.59	-61.09	-92.70	-131.96	-179.41	-235.55	-300.88	-375.67
TIME OF FLIGHT (SEC)	.000000	.102877	.211673	.326576	.447693	.574929	.708001	.846482	.989919	1.137954	1.290311	1.446779	1.607196
WIND DEFLECTION (IN)													
5 MPH	.00	.25	1.03	2.34	4.20	6.59	9.50	12.89	16.71	20.94	25.55	30.52	35.83
10 MPH	.00	.51	2.05	4.68	8.39	13.19	19.01	25.78	33.43	41.88	51.09	61.03	71.67
20 MPH	.00	1.01	4.11	9.35	16.79	26.38	38.02	51.56	66.85	83.76	102.19	122.07	143.33
30 MPH	.00	1.52	6.16	14.03	25.18	39.56	57.02	77.34	100.28	125.64	153.28	183.10	215.00
VELOCITY (FPS)	1400.	1325.	1256.	1194.	1141.	1094.	1056.	1022.	992.	966.	942.	920.	900.
ENERGY (FT-LB)	1854.	1660.	1492.	1349.	1230.	1133.	1054.	988.	932.	883.	839.	801.	765.
DROP (IN)	.00	-2.22	-9.37	-22.07	-40.85	-66.30	-98.97	-139.43	-188.18	-245.74	-312.55	-388.97	-475.66
MID-RANGE (IN)	.00	.58	2.47	5.89	11.05	18.15	27.42	39.05	53.24	70.17	89.98	112.79	138.85
BULLET PATH (IN)	-.75	2.10	.00	-7.63	-21.35	-41.74	-69.35	-104.75	-148.44	-200.94	-262.68	-334.05	-415.67
TIME OF FLIGHT (SEC)	.000000	.110163	.226469	.348989	.477578	.611916	.751555	.896058	1.045089	1.198389	1.355758	1.517043	1.682127
WIND DEFLECTION (IN)													
5 MPH	.00	.27	1.07	2.43	4.31	6.71	9.57	12.85	16.54	20.60	25.02	29.79	34.88
10 MPH	.00	.53	2.14	4.85	8.63	13.41	19.13	25.71	33.08	41.20	50.04	59.57	69.77
20 MPH	.00	1.06	4.29	9.70	17.25	26.82	38.26	51.41	66.16	82.40	100.08	119.14	139.54
30 MPH	.00	1.59	6.43	14.55	25.88	40.23	57.39	77.12	99.24	123.61	150.13	178.71	209.31

Bullet: Lyman 457406 451 Gr.
Ballistic Coefficient: .387

RANGE (YDS)	MUZZLE	50	100	150	200	250	300	350	400	450	500	550	600
VELOCITY (FPS)	2400.	2292.	2186.	2083.	1983.	1887.	1793.	1702.	1616.	1533.	1455.	1381.	1313.
ENERGY (FT-LB)	5767.	5259.	4785.	4346.	3939.	3563.	3218.	2901.	2613.	2353.	2118.	1910.	1725.
DROP (IN)	.00	-.71	-3.07	-7.25	-13.42	-21.79	-32.60	-46.11	-62.61	-82.43	-105.97	-133.65	-165.87
MID-RANGE (IN)	.00	.20	.82	1.95	3.64	5.98	9.06	12.99	17.89	23.90	31.19	39.95	50.33
BULLET PATH (IN)	-.75	.45	.00	-2.26	-6.52	-12.99	-21.89	-33.49	-48.07	-65.98	-87.61	-113.38	-143.69
TIME OF FLIGHT (SEC)	.000000	.063960	.130975	.201260	.275051	.352597	.434166	.520039	.610499	.705833	.806312	.912180	1.023634
WIND DEFLECTION (IN)													
5 MPH	.00	.13	.53	1.21	2.20	3.53	5.21	7.26	9.72	12.61	15.96	19.77	24.08
10 MPH	.00	.26	1.05	2.42	4.41	7.06	10.41	14.53	19.45	25.23	31.91	39.54	48.16
20 MPH	.00	.51	2.10	4.84	8.82	14.11	20.83	29.05	38.90	50.45	63.82	79.09	96.32
30 MPH	.00	.77	3.15	7.27	13.23	21.17	31.24	43.58	58.34	75.68	95.73	118.63	144.48
VELOCITY (FPS)	2300.	2194.	2091.	1991.	1894.	1800.	1709.	1622.	1539.	1460.	1386.	1318.	1254.
ENERGY (FT-LB)	5297.	4821.	4379.	3969.	3591.	3243.	2925.	2635.	2372.	2135.	1925.	1739.	1574.
DROP (IN)	.00	-.78	-3.35	-7.91	-14.65	-23.81	-35.65	-50.46	-68.56	-90.33	-116.22	-146.63	-182.03
MID-RANGE (IN)	.00	.21	.90	2.13	3.98	6.54	9.92	14.24	19.63	26.25	34.30	43.93	55.36
BULLET PATH (IN)	-.75	.52	.00	-2.51	-7.20	-14.32	-24.11	-36.86	-52.91	-72.64	-96.48	-124.83	-158.18
TIME OF FLIGHT (SEC)	.000000	.066771	.136796	.210307	.287553	.368802	.454332	.544427	.639375	.739447	.844891	.955907	1.072645
WIND DEFLECTION (IN)													
5 MPH	.00	.14	.56	1.29	2.35	3.76	5.55	7.74	10.35	13.42	16.96	20.99	25.52
10 MPH	.00	.27	1.12	2.58	4.70	7.52	11.09	15.47	20.70	26.84	33.92	41.98	51.05
20 MPH	.00	.55	2.24	5.16	9.39	15.04	22.19	30.94	41.41	53.68	67.84	83.96	102.09
30 MPH	.00	.82	3.36	7.74	14.09	22.55	33.28	46.41	62.11	80.52	101.75	125.94	153.14
VELOCITY (FPS)	2200.	2097.	1997.	1899.	1805.	1714.	1627.	1544.	1465.	1390.	1321.	1257.	1199.
ENERGY (FT-LB)	4846.	4402.	3991.	3611.	3262.	2942.	2650.	2385.	2148.	1936.	1748.	1583.	1440.
DROP (IN)	.00	-.85	-3.67	-8.66	-16.07	-26.13	-39.14	-55.42	-75.37	-99.40	-127.92	-161.41	-200.43
MID-RANGE (IN)	.00	.23	.99	2.33	4.36	7.18	10.91	15.66	21.62	28.95	37.83	48.46	61.08
BULLET PATH (IN)	-.75	.61	.00	-2.79	-7.98	-15.83	-26.64	-40.71	-58.45	-80.27	-106.58	-137.86	-174.67
TIME OF FLIGHT (SEC)	.000000	.069840	.143153	.220187	.301209	.386496	.476332	.571005	.670789	.775931	.886636	1.003053	1.125277
WIND DEFLECTION (IN)													
5 MPH	.00	.15	.60	1.38	2.51	4.01	5.92	8.25	11.03	14.28	18.02	22.27	27.02
10 MPH	.00	.29	1.19	2.75	5.01	8.02	11.83	16.50	22.06	28.56	36.05	44.54	54.05
20 MPH	.00	.58	2.39	5.51	10.03	16.05	23.67	32.99	44.12	57.13	72.10	89.07	108.10
30 MPH	.00	.88	3.58	8.26	15.04	24.07	35.50	49.49	66.18	85.69	108.14	133.61	162.15
VELOCITY (FPS)	2100.	2000.	1902.	1808.	1717.	1629.	1546.	1467.	1393.	1323.	1259.	1201.	1150.
ENERGY (FT-LB)	4416.	4003.	3623.	3272.	2951.	2658.	2393.	2155.	1942.	1754.	1588.	1444.	1323.
DROP (IN)	.00	-.93	-4.04	-9.54	-17.69	-28.78	-43.14	-61.15	-83.22	-109.78	-141.28	-178.30	-221.35
MID-RANGE (IN)	.00	.26	1.08	2.57	4.81	7.92	12.03	17.31	23.92	32.04	41.86	53.62	67.53
BULLET PATH (IN)	-.75	.71	.00	-3.11	-8.87	-17.57	-29.53	-45.14	-64.82	-88.98	-118.10	-152.72	-193.37
TIME OF FLIGHT (SEC)	.000000	.073201	.150116	.231009	.316159	.405850	.500368	.599988	.704960	.815489	.931727	1.053773	1.181517
WIND DEFLECTION (IN)													
5 MPH	.00	.16	.64	1.47	2.68	4.29	6.32	8.80	11.75	15.19	19.13	23.59	28.54
10 MPH	.00	.31	1.28	2.94	5.36	8.57	12.64	17.60	23.50	30.38	38.27	47.18	57.09
20 MPH	.00	.62	2.56	5.89	10.72	17.14	25.27	35.20	47.00	60.77	76.54	94.36	114.18
30 MPH	.00	.94	3.83	8.83	16.07	25.72	37.91	52.79	70.50	91.15	114.81	141.54	171.27

Bullet: Lyman 457406 451 Gr
Ballistic Coefficient: .387

RANGE (YDS)	MUZZLE	50	100	150	200	250	300	350	400	450	500	550	600
VELOCITY (FPS)	2000.	1903.	1808.	1717.	1630.	1546.	1467.	1393.	1324.	1259.	1201.	1150.	1105.
ENERGY (FT-LB)	4005.	3624.	3274.	2952.	2660.	2394.	2156.	1943.	1754.	1588.	1445.	1324.	1223.
DROP (IN)	.00	-1.03	-4.46	-10.54	-19.57	-31.85	-47.78	-67.78	-92.25	-121.67	-156.61	-197.57	-245.09
MID-RANGE (IN)	.00	.28	1.20	2.84	5.32	8.77	13.34	19.22	26.56	35.56	46.46	59.46	74.77
BULLET PATH (IN)	-.75	.82	.00	-3.47	-9.89	-19.57	-32.89	-50.28	-72.15	-98.97	-131.30	-169.65	-214.56
TIME OF FLIGHT (SEC)	.000000	.076899	.157775	.242907	.332579	.427076	.526676	.631625	.742130	.858345	.980367	1.108087	1.241248
WIND DEFLECTION (IN)													
5 MPH	.00	.17	.68	1.58	2.87	4.58	6.75	9.38	12.51	16.13	20.27	24.91	30.03
10 MPH	.00	.33	1.37	3.15	5.73	9.17	13.49	18.77	25.01	32.27	40.54	49.82	60.06
20 MPH	.00	.67	2.74	6.30	11.47	18.33	26.99	37.53	50.03	64.54	81.09	99.65	120.12
30 MPH	.00	1.00	4.11	9.45	17.20	27.50	40.48	56.30	75.04	96.81	121.63	149.47	180.18
VELOCITY (FPS)	1900.	1806.	1715.	1628.	1544.	1465.	1391.	1322.	1258.	1200.	1149.	1104.	1066.
ENERGY (FT-LB)	3615.	3265.	2944.	2652.	2388.	2150.	1937.	1750.	1584.	1441.	1321.	1221.	1138.
DROP (IN)	.00	-1.15	-4.96	-11.71	-21.74	-35.42	-53.19	-75.44	-102.66	-135.40	-174.18	-219.53	-271.99
MID-RANGE (IN)	.00	.32	1.33	3.15	5.91	9.76	14.88	21.43	29.59	39.61	51.66	65.99	82.80
BULLET PATH (IN)	-.75	.95	.00	-3.90	-11.08	-21.91	-36.82	-56.22	-80.58	-110.47	-146.39	-188.89	-238.49
TIME OF FLIGHT (SEC)	.000000	.080985	.166233	.256029	.350658	.450395	.555489	.666143	.782509	.904680	1.032544	1.165840	1.304191
WIND DEFLECTION (IN)													
5 MPH	.00	.18	.73	1.69	3.07	4.90	7.20	9.99	13.28	17.09	21.39	26.17	31.40
10 MPH	.00	.36	1.47	3.38	6.14	9.80	14.40	19.98	26.56	34.17	42.78	52.35	62.80
20 MPH	.00	.72	2.94	6.75	12.27	19.59	28.80	39.96	53.13	68.34	85.56	104.69	125.60
30 MPH	.00	1.08	4.40	10.13	18.41	29.39	43.19	59.93	79.69	102.51	128.34	157.04	188.40
VELOCITY (FPS)	1800.	1709.	1622.	1539.	1460.	1387.	1318.	1254.	1196.	1146.	1102.	1064.	1032.
ENERGY (FT-LB)	3244.	2925.	2635.	2372.	2136.	1925.	1739.	1574.	1433.	1314.	1215.	1134.	1066.
DROP (IN)	.00	-1.28	-5.53	-13.08	-24.31	-39.65	-59.49	-84.33	-114.74	-151.21	-194.28	-244.50	-302.38
MID-RANGE (IN)	.00	.35	1.48	3.52	6.62	10.95	16.67	23.97	33.06	44.16	57.50	73.28	91.70
BULLET PATH (IN)	-.75	1.11	.00	-4.40	-12.49	-24.69	-41.39	-63.09	-90.35	-123.68	-163.61	-210.69	-265.43
TIME OF FLIGHT (SEC)	.000000	.085521	.175606	.270543	.370604	.476035	.587038	.703762	.826291	.954499	1.088119	1.226768	1.370049
WIND DEFLECTION (IN)													
5 MPH	.00	.19	.79	1.81	3.28	5.22	7.66	10.60	14.05	18.00	22.42	27.29	32.56
10 MPH	.00	.38	1.57	3.62	6.56	10.45	15.32	21.20	28.09	35.99	44.84	54.58	65.13
20 MPH	.00	.77	3.15	7.23	13.12	20.90	30.64	42.39	56.19	71.98	89.68	109.16	130.26
30 MPH	.00	1.15	4.72	10.85	19.68	31.35	45.96	63.59	84.28	107.98	134.53	163.73	195.39
VELOCITY (FPS)	1700.	1613.	1531.	1452.	1379.	1311.	1248.	1191.	1141.	1097.	1060.	1028.	1000.
ENERGY (FT-LB)	2894.	2606.	2346.	2112.	1904.	1720.	1559.	1420.	1303.	1206.	1126.	1059.	1001.
DROP (IN)	.00	-1.44	-6.21	-14.71	-27.36	-44.57	-66.83	-94.71	-128.71	-169.37	-217.23	-272.79	-336.56
MID-RANGE (IN)	.00	.39	1.66	3.97	7.47	12.33	18.70	26.86	36.99	49.32	64.04	81.36	101.45
BULLET PATH (IN)	-.75	1.29	.00	-5.02	-14.19	-27.92	-46.69	-71.09	-101.61	-138.79	-183.16	-235.25	-295.53
TIME OF FLIGHT (SEC)	.000000	.090581	.186042	.286655	.392661	.504256	.621585	.744714	.873501	1.007663	1.146809	1.290551	1.438581
WIND DEFLECTION (IN)													
5 MPH	.00	.21	.84	1.93	3.50	5.55	8.11	11.18	14.75	18.79	23.27	28.16	33.42
10 MPH	.00	.41	1.68	3.86	6.99	11.10	16.22	22.36	29.50	37.58	46.54	56.31	66.84
20 MPH	.00	.83	3.37	7.73	13.98	22.20	32.45	44.73	59.00	75.17	93.09	112.63	133.67
30 MPH	.00	1.24	5.05	11.59	20.97	33.31	48.67	67.09	88.50	112.75	139.63	168.94	200.51

Bullet: Lyman 457406 451 Gr.
Bailistic Coefficient: .387

RANGE (YDS)	MUZZLE	50	100	150	200	250	300	350	400	450	500	550	600
VELOCITY (FPS)	1600.	1518.	1441.	1368.	1301.	1238.	1182.	1133.	1091.	1055.	1024.	996.	971.
ENERGY (FT-LB)	2563.	2307.	2078.	1873.	1694.	1535.	1400.	1286.	1192.	1115.	1049.	993.	944.
DROP (IN)	.00	-1.63	-7.06	-16.70	-30.97	-50.38	-75.49	-106.81	-144.87	-190.22	-243.35	-304.75	-374.90
MID-RANGE (IN)	.00	.45	1.90	4.51	8.42	13.86	21.05	30.18	41.46	55.11	71.29	90.18	111.96
BULLET PATH (IN)	-.75	1.53	.00	-5.73	-16.10	-31.60	-52.80	-80.21	-114.37	-155.80	-205.03	-262.52	-328.77
TIME OF FLIGHT (SEC)	.000000	.096259	.197710	.304589	.417082	.535330	.659368	.789026	.924003	1.063898	1.208333	1.357014	1.509710
WIND DEFLECTION (IN)													
5 MPH	.00	.22	.90	2.05	3.70	5.86	8.52	11.68	15.31	19.37	23.83	28.67	33.85
10 MPH	.00	.44	1.80	4.11	7.41	11.72	17.05	23.37	30.62	38.75	47.67	57.33	67.71
20 MPH	.00	.88	3.59	8.22	14.81	23.44	34.10	46.74	61.25	77.49	95.33	114.67	135.42
30 MPH	.00	1.32	5.39	12.32	22.22	35.15	51.15	70.11	91.87	116.24	143.00	172.00	203.13
VELOCITY (FPS)	1500.	1424.	1352.	1286.	1225.	1170.	1123.	1082.	1048.	1017.	990.	965.	943.
ENERGY (FT-LB)	2253.	2029.	1830.	1655.	1502.	1372.	1263.	1173.	1099.	1036.	981.	933.	890.
DROP (IN)	.00	-1.91	-8.12	-19.07	-35.28	-57.31	-85.67	-120.90	-163.52	-214.04	-272.95	-340.72	-417.73
MID-RANGE (IN)	.00	.51	2.15	5.08	9.52	15.69	23.77	33.97	46.48	61.48	79.17	99.70	123.20
BULLET PATH (IN)	-.75	1.77	.00	-6.51	-18.29	-35.89	-59.81	-90.61	-128.80	-174.89	-229.36	-292.69	-365.27
TIME OF FLIGHT (SEC)	.000000	.102666	.210807	.324595	.444169	.569506	.700406	.836539	.977495	1.122917	1.272526	1.426104	1.583481
WIND DEFLECTION (IN)													
5 MPH	.00	.23	.95	2.16	3.89	6.12	8.84	12.02	15.62	19.62	23.98	28.70	33.75
10 MPH	.00	.47	1.90	4.33	7.77	12.23	17.67	24.03	31.24	39.23	47.96	57.39	67.49
20 MPH	.00	.94	3.80	8.66	15.55	24.47	35.34	48.06	62.48	78.47	95.93	114.79	134.99
30 MPH	.00	1.41	5.71	12.99	23.32	36.70	53.01	72.09	93.72	117.70	143.89	172.18	202.48
VELOCITY (FPS)	1400.	1330.	1265.	1207.	1154.	1109.	1071.	1037.	1008.	982.	958.	936.	916.
ENERGY (FT-LB)	1962.	1772.	1603.	1458.	1334.	1232.	1148.	1077.	1017.	965.	919.	877.	840.
DROP (IN)	.00	-2.21	-9.31	-21.87	-40.40	-65.44	-97.52	-137.16	-184.85	-241.09	-306.32	-380.89	-465.28
MID-RANGE (IN)	.00	.58	2.45	5.82	10.89	17.85	26.89	38.22	52.03	68.47	87.72	109.88	135.12
BULLET PATH (IN)	-.75	2.07	.00	-7.53	-21.03	-41.03	-68.08	-102.69	-145.36	-196.56	-256.77	-326.30	-405.66
TIME OF FLIGHT (SEC)	.000000	.109943	.225580	.347023	.474180	.606806	.744534	.886960	1.033753	1.184655	1.339466	1.498028	1.660216
WIND DEFLECTION (IN)													
5 MPH	.00	.25	.99	2.25	4.01	6.26	8.95	12.05	15.54	19.39	23.59	28.11	32.96
10 MPH	.00	.49	1.99	4.50	8.03	12.51	17.90	24.10	31.08	38.79	47.17	56.22	65.91
20 MPH	.00	.99	3.98	9.01	16.05	25.02	35.79	48.21	62.17	77.57	94.35	112.45	131.82
30 MPH	.00	1.48	5.96	13.51	24.08	37.54	53.69	72.31	93.25	116.36	141.52	168.67	197.74

Bullet: Lyman 457125 464 Gr.
Ballistic Coefficient: .391

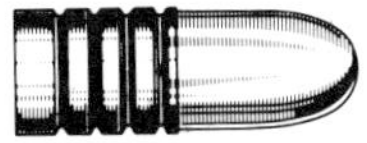

RANGE (YDS)	MUZZLE	50	100	150	200	250	300	350	400	450	500	550	600
VELOCITY (FPS)	2300.	2195.	2093.	1994.	1898.	1805.	1715.	1628.	1546.	1467.	1394.	1325.	1261.
ENERGY (FT-LB)	5449.	4965.	4514.	4096.	3710.	3355.	3028.	2731.	2461.	2218.	2001.	1809.	1638.
DROP (IN)	.00	-.77	-3.35	-7.90	-14.63	-23.77	-35.57	-50.32	-68.33	-90.00	-115.74	-145.94	-181.09
MID-RANGE (IN)	.00	.21	.90	2.12	3.97	6.52	9.89	14.18	19.54	26.12	34.10	43.65	54.98
BULLET PATH (IN)	-.75	.52	.00	-2.50	-7.18	-14.27	-24.03	-36.73	-52.69	-72.31	-96.00	-124.16	-157.26
TIME OF FLIGHT (SEC)	.000000	.066755	.136728	.210146	.287254	.368313	.453594	.543376	.637941	.737558	.842469	.952879	1.068948
WIND DEFLECTION (IN)													
5 MPH	.00	.14	.55	1.28	2.32	3.72	5.48	7.64	10.23	13.25	16.75	20.72	25.20
10 MPH	.00	.27	1.11	2.55	4.64	7.43	10.96	15.29	20.45	26.51	33.49	41.45	50.40
20 MPH	.00	.54	2.22	5.10	9.29	14.86	21.93	30.57	40.90	53.01	66.98	82.89	100.79
30 MPH	.00	.81	3.32	7.65	13.93	22.30	32.89	45.86	61.35	79.52	100.48	124.34	151.19
VELOCITY (FPS)	2200.	2098.	1999.	1902.	1809.	1719.	1632.	1549.	1471.	1397.	1328.	1264.	1205.
ENERGY (FT-LB)	4986.	4534.	4114.	3727.	3370.	3043.	2744.	2473.	2229.	2010.	1817.	1646.	1496.
DROP (IN)	.00	-.85	-3.66	-8.65	-16.04	-26.07	-39.04	-55.26	-75.11	-99.02	-127.37	-160.64	-199.39
MID-RANGE (IN)	.00	.23	.98	2.33	4.35	7.16	10.87	15.60	21.52	28.80	37.61	48.15	60.65
BULLET PATH (IN)	-.75	.61	.00	-2.78	-7.96	-15.79	-26.55	-40.57	-58.20	-79.90	-106.05	-137.11	-173.65
TIME OF FLIGHT (SEC)	.000000	.069823	.143080	.220016	.300890	.385974	.475547	.569889	.669271	.773939	.884096	.999901	1.121507
WIND DEFLECTION (IN)													
5 MPH	.00	.14	.59	1.36	2.48	3.97	5.85	8.15	10.90	14.11	17.80	21.99	26.69
10 MPH	.00	.29	1.18	2.72	4.96	7.93	11.70	16.30	21.79	28.21	35.60	43.98	53.39
20 MPH	.00	.58	2.36	5.45	9.91	15.86	23.39	32.60	43.58	56.43	71.20	87.97	106.77
30 MPH	.00	.87	3.55	8.17	14.87	23.79	35.09	48.90	65.38	84.64	106.80	131.95	160.16
VELOCITY (FPS)	2100.	2001.	1904.	1811.	1720.	1634.	1551.	1472.	1399.	1330.	1265.	1206.	1154.
ENERGY (FT-LB)	4543.	4123.	3735.	3377.	3049.	2750.	2478.	2233.	2015.	1821.	1649.	1499.	1373.
DROP (IN)	.00	-.93	-4.03	-9.52	-17.66	-28.72	-43.03	-60.95	-82.93	-109.35	-140.67	-177.45	-220.21
MID-RANGE (IN)	.00	.26	1.08	2.56	4.80	7.90	11.99	17.23	23.81	31.87	41.61	53.28	67.07
BULLET PATH (IN)	-.75	.71	.00	-3.10	-8.85	-17.52	-29.44	-44.97	-64.56	-88.58	-117.51	-151.91	-192.27
TIME OF FLIGHT (SEC)	.000000	.073183	.150038	.230826	.315818	.405293	.499532	.598805	.703358	.813397	.929083	1.050569	1.177770
WIND DEFLECTION (IN)													
5 MPH	.00	.15	.63	1.46	2.65	4.24	6.24	8.69	11.61	15.01	18.90	23.31	28.22
10 MPH	.00	.31	1.26	2.91	5.30	8.47	12.49	17.39	23.22	30.02	37.80	46.61	56.43
20 MPH	.00	.62	2.53	5.82	10.60	16.95	24.98	34.78	46.44	60.03	75.61	93.23	112.86
30 MPH	.00	.93	3.79	8.73	15.89	25.42	37.47	52.17	69.66	90.05	113.41	139.84	169.29
VELOCITY (FPS)	2000.	1904.	1810.	1720.	1633.	1551.	1472.	1398.	1329.	1265.	1206.	1154.	1109.
ENERGY (FT-LB)	4120.	3732.	3375.	3047.	2748.	2477.	2232.	2013.	1820.	1648.	1498.	1372.	1267.
DROP (IN)	.00	-1.03	-4.46	-10.53	-19.53	-31.78	-47.66	-67.58	-91.94	-121.21	-155.95	-196.67	-243.89
MID-RANGE (IN)	.00	.28	1.20	2.83	5.31	8.75	13.30	19.15	26.44	35.38	46.20	59.10	74.29
BULLET PATH (IN)	-.75	.82	.00	-3.47	-9.87	-19.51	-32.79	-50.11	-71.87	-98.54	-130.67	-168.78	-213.40
TIME OF FLIGHT (SEC)	.000000	.076879	.157692	.242711	.332215	.426484	.525789	.630375	.740449	.856168	.977688	1.104922	1.237625
WIND DEFLECTION (IN)													
5 MPH	.00	.17	.68	1.56	2.83	4.53	6.67	9.27	12.36	15.94	20.04	24.63	29.71
10 MPH	.00	.33	1.35	3.12	5.67	9.06	13.34	18.55	24.72	31.89	40.07	49.27	59.42
20 MPH	.00	.66	2.71	6.23	11.34	18.12	26.68	37.09	49.44	63.77	80.15	98.53	118.84
30 MPH	.00	.99	4.06	9.35	17.01	27.18	40.02	55.64	74.16	95.66	120.22	147.80	178.27

Bullet: Lyman 457125 464 Gr.
Ballistic Coefficient: .391

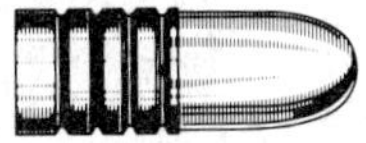

RANGE (YDS)	MUZZLE	50	100	150	200	250	300	350	400	450	500	550	600
VELOCITY (FPS)	1900.	1807.	1717.	1630.	1548.	1469.	1395.	1327.	1263.	1204.	1152.	1107.	1069.
ENERGY (FT-LB)	3719.	3362.	3036.	2737.	2467.	2223.	2006.	1813.	1642.	1493.	1368.	1263.	1177.
DROP (IN)	.00	-1.15	-4.95	-11.69	-21.70	-35.35	-53.06	-75.22	-102.31	-134.89	-173.46	-218.57	-270.74
MID-RANGE (IN)	.00	.32	1.33	3.15	5.90	9.74	14.83	21.34	29.46	39.41	51.38	65.61	82.31
BULLET PATH (IN)	-.75	.95	.00	-3.89	-11.05	-21.85	-36.71	-56.02	-80.26	-109.99	-145.71	-187.97	-237.29
TIME OF FLIGHT (SEC)	.000000	.080964	.166144	.255820	.350271	.449767	.554553	.664832	.780761	.902489	1.029923	1.162815	1.300793
WIND DEFLECTION (IN)													
5 MPH	.00	.18	.73	1.67	3.03	4.84	7.12	9.87	13.13	16.89	21.16	25.91	31.10
10 MPH	.00	.35	1.45	3.34	6.07	9.69	14.23	19.75	26.26	33.79	42.32	51.81	62.20
20 MPH	.00	.71	2.90	6.68	12.14	19.37	28.47	39.49	52.51	67.57	84.64	103.63	124.41
30 MPH	.00	1.06	4.36	10.02	18.21	29.06	42.70	59.24	78.77	101.36	126.96	155.44	186.61
VELOCITY (FPS)	1800.	1710.	1624.	1542.	1464.	1390.	1322.	1258.	1200.	1149.	1104.	1066.	1034.
ENERGY (FT-LB)	3338.	3013.	2717.	2448.	2207.	1991.	1800.	1630.	1483.	1359.	1256.	1172.	1101.
DROP (IN)	.00	-1.28	-5.53	-13.06	-24.26	-39.55	-59.33	-84.08	-114.35	-150.65	-193.53	-243.51	-301.12
MID-RANGE (IN)	.00	.35	1.48	3.51	6.60	10.92	16.61	23.87	32.92	43.96	57.21	72.90	91.23
BULLET PATH (IN)	-.75	1.11	.00	-4.39	-12.45	-24.61	-41.25	-62.86	-89.99	-123.16	-162.90	-209.74	-264.21
TIME OF FLIGHT (SEC)	.000000	.085498	.175511	.270320	.370193	.475372	.586057	.702405	.824554	.952396	1.085670	1.224001	1.366988
WIND DEFLECTION (IN)													
5 MPH	.00	.19	.78	1.79	3.24	5.17	7.57	10.48	13.89	17.81	22.21	27.05	32.29
10 MPH	.00	.38	1.56	3.58	6.49	10.33	15.15	20.96	27.79	35.62	44.41	54.09	64.59
20 MPH	.00	.76	3.11	7.15	12.97	20.66	30.29	41.91	55.58	71.24	88.82	108.18	129.18
30 MPH	.00	1.14	4.67	10.73	19.46	31.00	45.44	62.87	83.36	106.86	133.23	162.27	193.77
VELOCITY (FPS)	1700.	1614.	1532.	1455.	1382.	1314.	1251.	1194.	1143.	1100.	1062.	1030.	1002.
ENERGY (FT-LB)	2977.	2684.	2419.	2180.	1967.	1779.	1612.	1467.	1346.	1245.	1162.	1093.	1033.
DROP (IN)	.00	-1.44	-6.20	-14.69	-27.32	-44.48	-66.66	-94.44	-128.31	-168.82	-216.50	-271.85	-335.38
MID-RANGE (IN)	.00	.39	1.66	3.96	7.45	12.29	18.64	26.76	36.84	49.11	63.77	81.01	101.03
BULLET PATH (IN)	-.75	1.29	.00	-5.01	-14.16	-27.84	-46.55	-70.85	-101.25	-138.28	-182.47	-234.36	-294.41
TIME OF FLIGHT (SEC)	.000000	.090556	.185940	.286418	.392227	.503562	.620585	.743411	.871905	1.005793	1.144687	1.288195	1.436006
WIND DEFLECTION (IN)													
5 MPH	.00	.20	.83	1.91	3.46	5.49	8.02	11.07	14.61	18.63	23.09	27.95	33.19
10 MPH	.00	.41	1.67	3.82	6.91	10.98	16.05	22.13	29.22	37.25	46.17	55.90	66.38
20 MPH	.00	.82	3.33	7.64	13.83	21.96	32.09	44.27	58.44	74.51	92.34	111.80	132.77
30 MPH	.00	1.23	5.00	11.46	20.74	32.94	48.14	66.40	87.66	111.76	138.51	167.70	199.15
VELOCITY (FPS)	1600.	1519.	1442.	1370.	1303.	1241.	1184.	1135.	1093.	1056.	1025.	997.	972.
ENERGY (FT-LB)	2637.	2376.	2142.	1934.	1749.	1585.	1445.	1327.	1230.	1150.	1082.	1024.	973.
DROP (IN)	.00	-1.63	-7.06	-16.68	-30.92	-50.28	-75.32	-106.54	-144.49	-189.70	-242.67	-303.90	-373.86
MID-RANGE (IN)	.00	.45	1.90	4.51	8.41	13.82	20.98	30.08	41.32	54.92	71.06	89.89	111.61
BULLET PATH (IN)	-.75	1.53	.00	-5.72	-16.06	-31.51	-52.65	-79.97	-114.02	-155.32	-204.39	-261.72	-327.77
TIME OF FLIGHT (SEC)	.000000	.096233	.197602	.304340	.416629	.534639	.658441	.787874	.922640	1.062342	1.206597	1.355109	1.507645
WIND DEFLECTION (IN)													
5 MPH	.00	.22	.89	2.03	3.66	5.80	8.44	11.58	15.19	19.24	23.68	28.50	33.67
10 MPH	.00	.44	1.78	4.06	7.33	11.60	16.89	23.17	30.38	38.47	47.36	57.00	67.35
20 MPH	.00	.87	3.56	8.13	14.65	23.19	33.77	46.33	60.77	76.94	94.72	114.00	134.69
30 MPH	.00	1.31	5.33	12.19	21.98	34.79	50.66	69.50	91.15	115.42	142.08	171.00	202.04

Bullet: Lyman 457125 464 Gr.
Ballistic Coefficient: .391

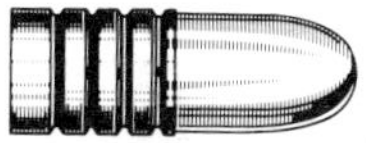

RANGE (YDS)	MUZZLE	50	100	150	200	250	300	350	400	450	500	550	600
VELOCITY (FPS)	1500.	1424.	1353.	1288.	1226.	1172.	1124.	1083.	1048.	1018.	991.	966.	944.
ENERGY (FT-LB)	2318.	2090.	1887.	1708.	1549.	1415.	1302.	1209.	1132.	1067.	1011.	961.	917.
DROP (IN)	0.0	-1.91	-8.11	-19.04	-35.22	-57.21	-85.50	-120.66	-163.19	-213.61	-272.40	-340.03	-416.90
MID-RANGE (IN)	0.0	0.51	2.14	5.07	9.50	15.65	23.71	33.89	46.37	61.35	78.99	99.48	122.95
BULLET PATH (IN)	-0.75	1.77	0.00	-6.50	-18.25	-35.81	-59.67	-90.40	-128.50	-174.49	-228.85	-292.05	-364.49
TIME OF FLIGHT (SEC)	0.0	0.102638	0.210694	0.324337	0.443745	0.568921	0.699667	0.835656	0.976481	1.121780	1.271273	1.424741	1.582013
WIND DEFLECTION (IN)													
5 MPH	0.0	0.23	0.94	2.14	3.85	6.07	8.77	11.94	15.53	19.52	23.87	28.58	33.62
10 MPH	0.0	0.46	1.88	4.28	7.70	12.13	17.54	23.88	31.06	39.03	47.74	57.15	67.23
20 MPH	0.0	0.93	3.76	8.57	15.40	24.26	35.08	47.75	62.12	78.07	95.49	114.31	134.47
30 MPH	0.0	1.39	5.65	12.85	23.10	36.39	52.62	71.63	93.18	117.10	143.23	171.46	201.70
VELOCITY (FPS)	1400.	1331.	1266.	1207.	1155.	1110.	1071.	1038.	1008.	982.	958.	936.	916.
ENERGY (FT-LB)	2019.	1825.	1652.	1502.	1375.	1269.	1182.	1109.	1047.	993.	946.	903.	865.
DROP (IN)	.00	-2.20	-9.30	-21.85	-40.36	-65.36	-97.41	-137.00	-184.64	-240.82	-305.99	-380.48	-464.79
MID-RANGE (IN)	.00	.58	2.45	5.81	10.88	17.83	26.86	38.18	51.97	68.39	87.63	109.76	134.99
BULLET PATH (IN)	-.75	2.07	.00	-7.52	-21.00	-40.98	-68.00	-102.56	-145.18	-196.33	-256.47	-325.93	-405.22
TIME OF FLIGHT (SEC)	.000000	.109914	.225471	.346827	.473900	.606446	.744100	.886458	1.033188	1.184030	1.338784	1.497290	1.659425
WIND DEFLECTION (IN)													
5 MPH	.00	.24	.98	2.24	3.99	6.22	8.91	12.01	15.49	19.34	23.53	28.05	32.89
10 MPH	.00	.49	1.97	4.47	7.98	12.45	17.82	24.02	30.98	38.68	47.05	56.09	65.77
20 MPH	.00	.98	3.94	8.94	15.96	24.90	35.64	48.03	61.97	77.35	94.11	112.19	131.55
30 MPH	.00	1.46	5.91	13.41	23.93	37.35	53.46	72.05	92.95	116.03	141.16	168.28	197.32
VELOCITY (FPS)	1300.	1238.	1182.	1133.	1091.	1055.	1023.	995.	970.	948.	927.	907.	889.
ENERGY (FT-LB)	1741.	1578.	1439.	1322.	1225.	1146.	1079.	1021.	970.	925.	884.	848.	814.
DROP (IN)	.00	-2.56	-10.83	-25.30	-46.53	-75.05	-111.35	-155.93	-209.26	-271.80	-343.81	-425.91	-518.54
MID-RANGE (IN)	.00	.68	2.85	6.70	12.44	20.27	30.37	42.92	58.10	76.05	96.86	120.76	147.92
BULLET PATH (IN)	-.75	2.48	.00	-8.69	-24.13	-46.85	-77.37	-116.16	-163.70	-220.46	-286.67	-362.98	-449.83
TIME OF FLIGHT (SEC)	.000000	.118299	.242386	.372091	.507107	.647036	.791498	.940200	1.092912	1.249453	1.409683	1.573491	1.740793
WIND DEFLECTION (IN)													
5 MPH	.00	.26	1.02	2.28	4.01	6.17	8.73	11.66	14.95	18.57	22.51	26.77	31.34
10 MPH	.00	.51	2.04	4.56	8.02	12.34	17.46	23.32	29.89	37.13	45.03	53.55	62.69
20 MPH	.00	1.03	4.09	9.13	16.04	24.68	34.92	46.64	59.78	74.27	90.05	107.10	125.37
30 MPH	.00	1.54	6.13	13.69	24.06	37.02	52.37	69.96	89.67	111.40	135.08	160.65	188.06

Bullet: Lyman 457132 490 Gr.
Ballistic Coefficient: .384

RANGE (YDS)	MUZZLE	50	100	150	200	250	300	350	400	450	500	550	600
VELOCITY (FPS)	2200.	2096.	1995.	1897.	1802.	1711.	1623.	1539.	1460.	1385.	1316.	1252.	1195.
ENERGY (FT-LB)	5265.	4779.	4330.	3914.	3533.	3183.	2865.	2577.	2318.	2088.	1884.	1706.	1553.
DROP (IN)	.00	-.85	-3.67	-8.67	-16.09	-26.17	-39.22	-55.55	-75.56	-99.69	-128.34	-162.00	-201.23
MID-RANGE (IN)	.00	.23	.99	2.34	4.37	7.20	10.93	15.71	21.69	29.07	38.00	48.69	61.40
BULLET PATH (IN)	-.75	.61	.00	-2.79	-7.99	-15.86	-26.70	-40.82	-58.62	-80.54	-106.98	-138.42	-175.45
TIME OF FLIGHT (SEC)	.000000	.069853	.143208	.220318	.301453	.386895	.476934	.571860	.671952	.777457	.888582	1.005464	1.128153
WIND DEFLECTION (IN)													
5 MPH	.00	.15	.60	1.39	2.53	4.05	5.97	8.32	11.13	14.42	18.20	22.48	27.28
10 MPH	.00	.29	1.20	2.78	5.06	8.09	11.94	16.65	22.26	28.83	36.39	44.96	54.55
20 MPH	.00	.59	2.41	5.55	10.11	16.19	23.88	33.29	44.53	57.66	72.78	89.92	109.11
30 MPH	.00	.88	3.61	8.33	15.17	24.28	35.82	49.94	66.79	86.50	109.17	134.88	163.66
VELOCITY (FPS)	2100.	1999.	1901.	1806.	1714.	1626.	1542.	1463.	1388.	1319.	1255.	1197.	1146.
ENERGY (FT-LB)	4797.	4346.	3930.	3547.	3196.	2877.	2587.	2328.	2096.	1892.	1713.	1559.	1429.
DROP (IN)	.00	-.94	-4.04	-9.55	-17.72	-28.83	-43.23	-61.28	-83.44	-110.10	-141.74	-178.95	-222.21
MID-RANGE (IN)	.00	.26	1.09	2.57	4.82	7.94	12.06	17.36	24.00	32.16	42.04	53.88	67.88
BULLET PATH (IN)	-.75	.71	.00	-3.11	-8.88	-17.60	-29.60	-45.26	-65.02	-89.28	-118.53	-153.34	-194.21
TIME OF FLIGHT (SEC)	.000000	.073215	.150175	.231149	.316419	.406275	.501007	.600894	.706186	.817090	.933749	1.056216	1.184367
WIND DEFLECTION (IN)													
5 MPH	.00	.16	.64	1.48	2.70	4.32	6.37	8.88	11.86	15.33	19.31	23.80	28.80
10 MPH	.00	.31	1.29	2.97	5.40	8.65	12.75	17.76	23.72	30.66	38.63	47.61	57.59
20 MPH	.00	.63	2.58	5.94	10.81	17.29	25.50	35.51	47.43	61.33	77.25	95.22	115.18
30 MPH	.00	.94	3.86	8.90	16.21	25.94	38.25	53.27	71.15	91.99	115.88	142.82	172.77
VELOCITY (FPS)	2000.	1902.	1807.	1715.	1627.	1543.	1464.	1389.	1319.	1255.	1198.	1147.	1103.
ENERGY (FT-LB)	4351.	3934.	3551.	3200.	2880.	2591.	2331.	2099.	1894.	1714.	1560.	1430.	1323.
DROP (IN)	.00	-1.04	-4.47	-10.56	-19.59	-31.90	-47.87	-67.93	-92.49	-122.03	-157.12	-198.26	-246.00
MID-RANGE (IN)	.00	.28	1.20	2.84	5.33	8.79	13.38	19.28	26.65	35.70	46.66	59.73	75.13
BULLET PATH (IN)	-.75	.82	.00	-3.48	-9.91	-19.61	-32.97	-50.42	-72.37	-99.30	-131.78	-170.31	-215.44
TIME OF FLIGHT (SEC)	.000000	.076914	.157839	.243057	.332857	.427529	.527354	.632581	.743417	.860007	.982406	1.110491	1.243995
WIND DEFLECTION (IN)													
5 MPH	.00	.17	.69	1.59	2.89	4.62	6.81	9.47	12.62	16.28	20.45	25.12	30.27
10 MPH	.00	.34	1.38	3.18	5.78	9.25	13.61	18.93	25.24	32.56	40.90	50.25	60.54
20 MPH	.00	.67	2.76	6.36	11.57	18.49	27.23	37.87	50.48	65.12	81.81	100.49	121.09
30 MPH	.00	1.01	4.14	9.53	17.35	27.74	40.84	56.80	75.72	97.68	122.71	150.74	181.63
VELOCITY (FPS)	1900.	1805.	1713.	1626.	1542.	1462.	1388.	1318.	1254.	1197.	1146.	1102.	1064.
ENERGY (FT-LB)	3927.	3544.	3194.	2874.	2586.	2326.	2095.	1890.	1711.	1558.	1428.	1321.	1232.
DROP (IN)	.00	-1.15	-4.96	-11.73	-21.77	-35.48	-53.30	-75.61	-102.92	-135.79	-174.72	-220.26	-272.94
MID-RANGE (IN)	.00	.32	1.33	3.16	5.92	9.79	14.92	21.49	29.69	39.76	51.88	66.28	83.17
BULLET PATH (IN)	-.75	.96	.00	-3.91	-11.10	-21.95	-36.91	-56.37	-80.82	-110.83	-146.91	-189.59	-239.41
TIME OF FLIGHT (SEC)	.000000	.081002	.166301	.256189	.350954	.450876	.556205	.667146	.783842	.906346	1.034532	1.168130	1.306760
WIND DEFLECTION (IN)													
5 MPH	.00	.18	.74	1.70	3.09	4.94	7.26	10.08	13.40	17.23	21.57	26.37	31.63
10 MPH	.00	.36	1.48	3.41	6.19	9.88	14.52	20.15	26.80	34.46	43.13	52.75	63.25
20 MPH	.00	.72	2.96	6.81	12.38	19.76	29.05	40.31	53.60	68.93	86.26	105.50	126.51
30 MPH	.00	1.08	4.44	10.22	18.57	29.64	43.57	60.46	80.39	103.39	129.39	158.25	189.76

Bullet: Lyman 457132 490 Gr.
Ballistic Coefficient: .384

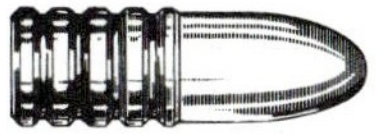

RANGE (YDS)	MUZZLE	50	100	150	200	250	300	350	400	450	500	550	600
VELOCITY (FPS)	1800.	1709.	1621.	1537.	1458.	1384.	1315.	1251.	1194.	1143.	1100.	1062.	1030.
ENERGY (FT-LB)	3525.	3176.	2858.	2571.	2313.	2083.	1880.	1702.	1550.	1422.	1315.	1228.	1154.
DROP (IN)	.00	-1.29	-5.54	-13.09	-24.34	-39.71	-59.61	-84.52	-115.02	-151.62	-194.85	-245.24	-303.32
MID-RANGE (IN)	.00	.35	1.48	3.52	6.63	10.97	16.71	24.04	33.17	44.32	57.71	73.56	92.06
BULLET PATH (IN)	-.75	1.11	.00	-4.41	-12.51	-24.74	-41.49	-63.25	-90.61	-124.06	-164.15	-211.40	-266.34
TIME OF FLIGHT (SEC)	.000000	.085538	.175679	.270714	.370919	.476543	.587789	.704796	.827610	.956093	1.089970	1.228856	1.372357
WIND DEFLECTION (IN)													
5 MPH	.00	.19	.79	1.82	3.31	5.27	7.73	10.69	14.16	18.14	22.58	27.47	32.77
10 MPH	.00	.39	1.59	3.65	6.62	10.54	15.45	21.38	28.33	36.27	45.17	54.95	65.53
20 MPH	.00	.78	3.17	7.29	13.23	21.08	30.90	42.75	56.65	72.54	90.34	109.89	131.07
30 MPH	.00	1.16	4.76	10.94	19.85	31.61	46.35	64.13	84.98	108.82	135.50	164.84	196.60
VELOCITY (FPS)	1700.	1613.	1529.	1451.	1377.	1308.	1245.	1189.	1139.	1096.	1059.	1027.	999.
ENERGY (FT-LB)	3144.	2829.	2545.	2289.	2062.	1862.	1687.	1537.	1411.	1306.	1220.	1148.	1086.
DROP (IN)	.00	-1.44	-6.22	-14.73	-27.40	-44.64	-66.95	-94.91	-129.00	-169.79	-217.78	-273.51	-337.46
MID-RANGE (IN)	.00	.39	1.67	3.97	7.48	12.35	18.74	26.93	37.10	49.47	64.25	81.63	101.78
BULLET PATH (IN)	-.75	1.29	.00	-5.03	-14.21	-27.97	-46.80	-71.27	-101.88	-139.18	-183.69	-235.94	-296.40
TIME OF FLIGHT (SEC)	.000000	.090599	.186120	.286836	.392993	.504787	.622353	.745718	.874732	1.009107	1.148449	1.292373	1.440573
WIND DEFLECTION (IN)													
5 MPH	.00	.21	.85	1.95	3.52	5.60	8.18	11.27	14.86	18.92	23.42	28.32	33.59
10 MPH	.00	.42	1.70	3.89	7.05	11.20	16.36	22.54	29.72	37.84	46.83	56.63	67.19
20 MPH	.00	.83	3.40	7.79	14.10	22.39	32.72	45.08	59.44	75.68	93.67	113.27	134.38
30 MPH	.00	1.25	5.09	11.68	21.15	33.59	49.07	67.62	89.15	113.51	140.50	169.90	201.56
VELOCITY (FPS)	1600.	1517.	1439.	1366.	1299.	1236.	1181.	1132.	1090.	1054.	1023.	995.	970.
ENERGY (FT-LB)	2785.	2505.	2254.	2031.	1834.	1662.	1516.	1394.	1292.	1209.	1138.	1077.	1024.
DROP (IN)	.00	-1.63	-7.07	-16.72	-31.01	-50.46	-75.63	-107.01	-145.17	-190.62	-243.87	-305.41	-375.72
MID-RANGE (IN)	.00	.45	1.90	4.52	8.44	13.89	21.09	30.25	41.57	55.25	71.47	90.41	112.23
BULLET PATH (IN)	-.75	1.53	.00	-5.74	-16.12	-31.66	-52.92	-80.39	-114.64	-156.18	-205.52	-263.15	-329.55
TIME OF FLIGHT (SEC)	.000000	.096279	.197793	.304780	.417428	.535865	.660088	.789924	.925066	1.065114	1.209691	1.358505	1.511328
WIND DEFLECTION (IN)													
5 MPH	.00	.22	.91	2.07	3.73	5.91	8.59	11.76	15.41	19.48	23.95	28.80	34.00
10 MPH	.00	.45	1.81	4.14	7.47	11.81	17.18	23.53	30.81	38.96	47.91	57.60	67.99
20 MPH	.00	.89	3.62	8.28	14.93	23.62	34.35	47.05	61.62	77.92	95.81	115.19	135.99
30 MPH	.00	1.34	5.43	12.42	22.40	35.44	51.53	70.58	92.43	116.88	143.72	172.79	203.98
VELOCITY (FPS)	1500.	1423.	1351.	1284.	1224.	1169.	1122.	1082.	1047.	1016.	989.	965.	943.
ENERGY (FT-LB)	2448.	2203.	1985.	1795.	1629.	1488.	1370.	1273.	1192.	1124.	1065.	1013.	966.
DROP (IN)	.00	-1.91	-8.12	-19.08	-35.32	-57.38	-85.79	-121.07	-163.77	-214.37	-273.37	-341.23	-418.35
MID-RANGE (IN)	.00	.51	2.15	5.09	9.54	15.71	23.81	34.04	46.56	61.59	79.30	99.86	123.39
BULLET PATH (IN)	-.75	1.77	.00	-6.52	-18.32	-35.95	-59.92	-90.77	-129.02	-175.19	-229.75	-293.18	-365.86
TIME OF FLIGHT (SEC)	.000000	.102687	.210894	.324792	.444492	.569953	.700970	.837211	.978268	1.123783	1.273480	1.427142	1.584600
WIND DEFLECTION (IN)													
5 MPH	.00	.24	.96	2.18	3.92	6.16	8.89	12.07	15.69	19.69	24.07	28.79	33.84
10 MPH	.00	.47	1.92	4.36	7.83	12.31	17.77	24.15	31.38	39.39	48.13	57.58	67.69
20 MPH	.00	.95	3.83	8.73	15.66	24.62	35.54	48.30	62.75	78.77	96.27	115.15	135.38
30 MPH	.00	1.42	5.75	13.09	23.49	36.94	53.31	72.45	94.13	118.16	144.40	172.73	203.07

Bullet: Lyman 457132 490 Gr.
Ballistic Coefficient: .384

RANGE (YDS)	MUZZLE	50	100	150	200	250	300	350	400	450	500	550	600
VELOCITY (FPS)	1400.	1330.	1265.	1206.	1154.	1109.	1070.	1037.	1008.	981.	958.	936.	916.
ENERGY (FT-LB)	2132.	1923.	1740.	1582.	1448.	1337.	1246.	1170.	1104.	1048.	998.	953.	912.
DROP (IN)	.00	-2.21	-9.32	-21.89	-40.44	-65.50	-97.61	-137.28	-185.02	-241.30	-306.58	-381.20	-465.65
MID-RANGE (IN)	.00	.58	2.45	5.83	10.90	17.86	26.91	38.26	52.07	68.53	87.79	109.96	135.22
BULLET PATH (IN)	-.75	2.08	.00	-7.54	-21.05	-41.08	-68.15	-102.79	-145.50	-196.74	-256.99	-326.58	-406.00
TIME OF FLIGHT (SEC)	.000000	.109965	.225663	.347173	.474393	.607079	.744863	.887340	1.034181	1.185129	1.339984	1.498587	1.660816
WIND DEFLECTION (IN)													
5 MPH	.00	.25	1.00	2.27	4.03	6.28	8.98	12.09	15.58	19.43	23.63	28.16	33.01
10 MPH	.00	.50	2.00	4.53	8.06	12.56	17.95	24.17	31.16	38.87	47.27	56.32	66.02
20 MPH	.00	.99	4.00	9.06	16.13	25.12	35.91	48.34	62.32	77.74	94.53	112.65	132.04
30 MPH	.00	1.49	6.01	13.59	24.19	37.68	53.86	72.52	93.48	116.61	141.80	168.97	198.05
VELOCITY (FPS)	1300.	1238.	1182.	1133.	1091.	1055.	1023.	995.	970.	948.	927.	907.	889.
ENERGY (FT-LB)	1838.	1666.	1519.	1396.	1294.	1210.	1139.	1078.	1024.	977.	934.	895.	859.
DROP (IN)	.00	-2.56	-10.83	-25.30	-46.53	-75.05	-111.35	-155.93	-209.26	-271.80	-343.81	-425.91	-518.54
MID-RANGE (IN)	.00	.68	2.85	6.70	12.44	20.27	30.37	42.92	58.10	76.05	96.86	120.76	147.92
BULLET PATH (IN)	-.75	2.48	.00	-8.69	-24.13	-46.85	-77.37	-116.16	-163.70	-220.46	-286.67	-362.98	-449.83
TIME OF FLIGHT (SEC)	.000000	.118299	.242386	.372091	.507107	.647036	.791498	.940200	1.092912	1.249453	1.409683	1.573491	1.740793
WIND DEFLECTION (IN)													
5 MPH	.00	.26	1.02	2.28	4.01	6.17	8.73	11.66	14.95	18.57	22.51	26.77	31.34
10 MPH	.00	.51	2.04	4.56	8.02	12.34	17.46	23.32	29.89	37.13	45.03	53.55	62.69
20 MPH	.00	1.03	4.09	9.13	16.04	24.68	34.92	46.64	59.78	74.27	90.05	107.10	125.37
30 MPH	.00	1.54	6.13	13.69	24.06	37.02	52.37	69.96	89.67	111.40	135.08	160.65	188.06
VELOCITY (FPS)	1200.	1149.	1104.	1066.	1033.	1005.	979.	955.	933.	913.	895.	877.	860.
ENERGY (FT-LB)	1566.	1435.	1327.	1237.	1162.	1098.	1042.	992.	948.	908.	871.	837.	805.
DROP (IN)	.00	-3.04	-12.64	-29.35	-53.68	-86.12	-127.16	-177.24	-236.69	-306.03	-385.75	-476.30	-578.13
MID-RANGE (IN)	.00	.79	3.28	7.69	14.20	22.99	34.23	48.09	64.67	84.16	106.76	132.61	161.91
BULLET PATH (IN)	-.75	2.91	.00	-10.01	-27.64	-53.39	-87.73	-131.12	-183.87	-246.51	-319.54	-403.39	-498.53
TIME OF FLIGHT (SEC)	.000000	.127813	.261057	.399352	.542300	.689578	.840937	.996180	1.155154	1.317738	1.483841	1.653392	1.826341
WIND DEFLECTION (IN)													
5 MPH	.00	.25	.97	2.14	3.72	5.68	8.00	10.66	13.65	16.96	20.58	24.50	28.72
10 MPH	.00	.50	1.95	4.29	7.44	11.37	16.00	21.33	27.31	33.92	41.16	49.00	57.44
20 MPH	.00	.99	3.89	8.57	14.89	22.73	32.01	42.66	54.61	67.84	82.31	97.99	114.87
30 MPH	.00	1.49	5.84	12.86	22.33	34.10	48.01	63.98	81.92	101.77	123.47	146.99	172.31

Bullet: Lyman # 515141 422 Gr.
Ballistic Coefficient: .250

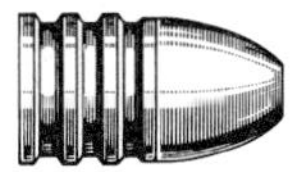

RANGE (YDS)	MUZZLE	50	100	150	200	250	300	350	400	450	500	550	600
VELOCITY (FPS)	1800.	1661.	1531.	1412.	1305.	1212.	1128.	1062.	1009.	966.	929.	896.	867.
ENERGY (FT-LB)	3035.	2585.	2197.	1869.	1596.	1376.	1192.	1056.	954.	875.	809.	753.	704.
DROP (IN)	.00	-1.34	-5.83	-14.05	-26.64	-44.33	-67.96	-98.41	-136.54	-183.17	-239.03	-304.86	-381.49
MID-RANGE (IN)	.00	.36	1.57	3.86	7.49	12.73	19.93	29.45	41.63	56.80	75.19	97.08	122.78
BULLET PATH (IN)	-.75	1.20	.00	-4.93	-14.23	-28.63	-48.98	-76.13	-110.97	-154.32	-206.89	-269.43	-342.77
TIME OF FLIGHT (SEC)	.000000	.086760	.180828	.282863	.393411	.512772	.641159	.778418	.923476	1.075507	1.233967	1.398493	1.568854
WIND DEFLECTION (IN)													
5 MPH	.00	.30	1.25	2.89	5.29	8.46	12.42	17.17	22.60	28.64	35.26	42.40	50.06
10 MPH	.00	.60	2.49	5.78	10.57	16.91	24.84	34.33	45.20	57.29	70.51	84.80	100.12
20 MPH	.00	1.21	4.98	11.57	21.15	33.83	49.69	68.67	90.40	114.58	141.02	169.60	200.24
30 MPH	.00	1.81	7.48	17.35	31.72	50.74	74.53	103.00	135.60	171.87	211.53	254.40	300.35
VELOCITY (FPS)	1700.	1568.	1445.	1334.	1237.	1151.	1079.	1023.	978.	939.	905.	875.	847.
ENERGY (FT-LB)	2708.	2302.	1957.	1668.	1434.	1240.	1091.	981.	896.	827.	768.	717.	673.
DROP (IN)	.00	-1.51	-6.56	-15.80	-29.92	-49.75	-76.12	-109.93	-152.02	-203.14	-263.99	-335.42	-418.16
MID-RANGE (IN)	.00	.41	1.77	4.35	8.40	14.25	22.25	32.76	46.08	62.50	82.25	105.65	132.96
BULLET PATH (IN)	-.75	1.40	.00	-5.58	-16.05	-32.21	-54.93	-85.08	-123.51	-170.98	-228.18	-295.94	-375.03
TIME OF FLIGHT (SEC)	.000000	.091906	.191598	.299663	.416492	.542235	.677042	.819949	.970032	1.126679	1.289480	1.458172	1.632593
WIND DEFLECTION (IN)													
5 MPH	.00	.32	1.33	3.08	5.59	8.89	12.99	17.80	23.25	29.27	35.83	42.91	50.49
10 MPH	.00	.65	2.66	6.15	11.18	17.79	25.98	35.61	46.49	58.53	71.65	85.81	100.98
20 MPH	.00	1.29	5.32	12.30	22.37	35.57	51.97	71.21	92.98	117.06	143.31	171.63	201.97
30 MPH	.00	1.94	7.99	18.46	33.55	53.36	77.95	106.82	139.47	175.59	214.96	257.44	302.95
VELOCITY (FPS)	1600.	1475.	1361.	1260.	1172.	1095.	1036.	989.	948.	914.	882.	854.	828.
ENERGY (FT-LB)	2398.	2038.	1736.	1488.	1286.	1124.	1006.	915.	843.	782.	729.	683.	642.
DROP (IN)	.00	-1.71	-7.45	-17.88	-33.79	-56.04	-85.51	-123.05	-169.45	-225.37	-291.67	-369.11	-458.44
MID-RANGE (IN)	.00	.46	2.01	4.90	9.45	15.99	24.88	36.43	50.93	68.62	89.79	114.73	143.71
BULLET PATH (IN)	-.75	1.64	.00	-6.32	-18.14	-36.29	-61.65	-95.09	-137.40	-189.22	-251.41	-324.75	-409.98
TIME OF FLIGHT (SEC)	.000000	.097671	.203581	.318184	.441645	.574236	.715206	.863551	1.018589	1.179869	1.347095	1.520081	1.698728
WIND DEFLECTION (IN)													
5 MPH	.00	.35	1.42	3.25	5.86	9.28	13.44	18.24	23.64	29.58	36.04	43.02	50.49
10 MPH	.00	.69	2.83	6.50	11.73	18.57	26.88	36.48	47.27	59.16	72.09	86.03	100.98
20 MPH	.00	1.38	5.66	13.00	23.46	37.13	53.75	72.97	94.54	118.31	144.18	172.07	201.95
30 MPH	.00	2.07	8.49	19.50	35.19	55.70	80.63	109.45	141.82	177.47	216.27	258.10	302.93
VELOCITY (FPS)	1500.	1384.	1280.	1190.	1109.	1047.	998.	956.	920.	889.	860.	833.	808.
ENERGY (FT-LB)	2108.	1794.	1535.	1326.	1153.	1027.	932.	857.	794.	740.	692.	650.	612.
DROP (IN)	.00	-1.98	-8.50	-20.34	-38.34	-63.37	-96.30	-137.93	-188.95	-250.17	-322.37	-406.30	-502.75
MID-RANGE (IN)	.00	.52	2.27	5.54	10.66	17.98	27.80	40.45	56.14	75.17	97.82	124.36	155.08
BULLET PATH (IN)	-.75	1.89	.00	-7.21	-20.58	-40.98	-69.28	-106.29	-152.69	-209.28	-276.85	-356.15	-447.98
TIME OF FLIGHT (SEC)	.000000	.104158	.216936	.338546	.469274	.608609	.755498	.909198	1.069218	1.235233	1.407038	1.584518	1.767629
WIND DEFLECTION (IN)													
5 MPH	.00	.37	1.49	3.39	6.10	9.56	13.68	18.41	23.69	29.50	35.82	42.64	49.95
10 MPH	.00	.73	2.98	6.78	12.19	19.12	27.37	36.82	47.38	59.00	71.64	85.28	99.90
20 MPH	.00	1.46	5.96	13.57	24.38	38.23	54.74	73.64	94.76	118.00	143.28	170.55	199.81
30 MPH	.00	2.20	8.94	20.35	36.58	57.35	82.10	110.46	142.15	177.00	214.92	255.83	299.71

Bullet: Lyman # 515141 422 Gr.
Ballistic Coefficient: .250
(Cont'd.)

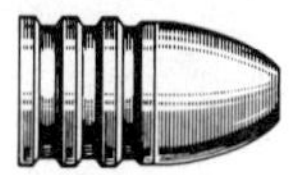

RANGE (YDS)	MUZZLE	50	100	150	200	250	300	350	400	450	500	550	600
VELOCITY (FPS)	1400.	1294.	1203.	1120.	1055.	1004.	962.	925.	893.	864.	837.	812.	788.
ENERGY (FT-LB)	1836.	1569.	1355.	1175.	1044.	945.	867.	802.	747.	699.	656.	617.	582.
DROP (IN)	.00	-2.28	-9.75	-23.26	-43.66	-71.85	-108.62	-154.68	-210.81	-277.82	-356.43	-447.46	-551.49
MID-RANGE (IN)	.00	.60	2.60	6.32	12.08	20.23	31.05	44.80	61.74	82.18	106.37	134.62	167.12
BULLET PATH (IN)	-.75	2.22	.00	-8.25	-23.41	-46.34	-77.87	-118.68	-169.56	-231.31	-304.67	-390.45	-489.23
TIME OF FLIGHT (SEC)	.000000	.111496	.231815	.361181	.499315	.645141	.797867	.956974	1.122114	1.293067	1.469706	1.651978	1.839889
WIND DEFLECTION (IN)													
5 MPH	.00	.38	1.54	3.50	6.23	9.63	13.64	18.21	23.32	28.93	35.05	41.66	48.77
10 MPH	.00	.77	3.09	7.00	12.45	19.26	27.28	36.43	46.63	57.87	70.10	83.32	97.53
20 MPH	.00	1.53	6.17	13.99	24.90	38.52	54.56	72.85	93.27	115.73	140.19	166.64	195.07
30 MPH	.00	2.30	9.26	20.99	37.35	57.78	81.85	109.28	139.90	173.60	210.29	249.96	292.60
VELOCITY (FPS)	1300.	1208.	1124.	1059.	1007.	964.	927.	895.	865.	838.	813.	789.	767.
ENERGY (FT-LB)	1583.	1366.	1184.	1050.	950.	871.	806.	750.	701.	658.	619.	583.	551.
DROP (IN)	.00	-2.65	-11.28	-26.76	-49.96	-81.72	-122.72	-173.74	-235.59	-309.00	-394.78	-493.52	-605.96
MID-RANGE (IN)	.00	.70	2.99	7.22	13.71	22.75	34.60	49.53	67.83	89.78	115.68	145.73	180.26
BULLET PATH (IN)	-.75	2.62	.00	-9.47	-26.66	-52.40	-87.39	-132.40	-188.23	-255.64	-335.40	-428.13	-534.55
TIME OF FLIGHT (SEC)	.000000	.119809	.248643	.386301	.531705	.684047	.842792	1.007586	1.178202	1.354507	1.536446	1.724021	1.917271
WIND DEFLECTION (IN)													
5 MPH	.00	.39	1.57	3.53	6.17	9.43	13.24	17.59	22.45	27.81	33.67	40.02	46.87
10 MPH	.00	.78	3.15	7.07	12.35	18.85	26.49	35.18	44.90	55.62	67.34	80.04	93.75
20 MPH	.00	1.56	6.29	14.13	24.70	37.71	52.97	70.36	89.80	111.25	134.68	160.09	187.49
30 MPH	.00	2.34	9.44	21.20	37.05	56.56	79.46	105.54	134.71	166.87	202.01	240.13	281.24
VELOCITY (FPS)	1200.	1117.	1053.	1003.	960.	924.	892.	863.	836.	811.	787.	764.	743.
ENERGY (FT-LB)	1349.	1170.	1039.	942.	864.	800.	745.	697.	654.	616.	580.	548.	517.
DROP (IN)	.00	-3.13	-13.19	-31.05	-57.54	-93.34	-139.25	-196.04	-264.47	-345.33	-439.17	-546.84	-669.37
MID-RANGE (IN)	.00	.81	3.47	8.29	15.58	25.58	38.57	54.85	74.69	98.43	126.25	158.44	195.44
BULLET PATH (IN)	-.75	3.09	.00	-10.90	-30.42	-59.24	-98.18	-148.00	-209.47	-283.36	-370.23	-470.93	-586.50
TIME OF FLIGHT (SEC)	.000000	.129696	.268121	.414201	.567158	.726478	.891818	1.062963	1.239788	1.422243	1.610335	1.804103	2.003604
WIND DEFLECTION (IN)													
5 MPH	.00	.41	1.59	3.45	5.91	8.93	12.48	16.54	21.10	26.16	31.71	37.76	44.32
10 MPH	.00	.83	3.19	6.90	11.82	17.86	24.96	33.08	42.20	52.31	63.42	75.52	88.63
20 MPH	.00	1.65	6.38	13.80	23.64	35.72	49.92	66.16	84.41	104.63	126.84	151.04	177.27
30 MPH	.00	2.48	9.57	20.70	35.46	53.58	74.88	99.24	126.61	156.94	190.26	226.57	265.90
VELOCITY (FPS)	1100.	1040.	992.	951.	916.	884.	856.	829.	805.	781.	759.	738.	718.
ENERGY (FT-LB)	1134.	1013.	921.	847.	786.	733.	686.	644.	606.	572.	540.	510.	482.
DROP (IN)	.00	-3.67	-15.35	-35.84	-65.80	-106.07	-157.42	-220.60	-296.40	-385.31	-488.31	-406.38	-740.42
MID-RANGE (IN)	.00	.95	4.00	9.46	17.55	28.60	42.87	60.72	82.40	108.05	138.08	172.87	212.79
BULLET PATH (IN)	-.75	3.63	.00	-12.44	-34.35	-66.58	-109.88	-165.01	-232.77	-313.63	-408.58	-518.60	-644.60
TIME OF FLIGHT (SEC)	.000000	.140398	.288228	.442787	.603609	.770387	.942930	1.121132	1.304958	1.494428	1.689584	1.890486	2.097219
WIND DEFLECTION (IN)													
5 MPH	.00	.36	1.36	2.97	5.12	7.79	10.98	14.66	18.84	23.51	28.68	34.36	40.56
10 MPH	.00	.71	2.73	5.93	10.24	15.59	21.96	29.32	37.67	47.02	57.37	68.73	81.11
20 MPH	.00	1.42	5.46	11.86	20.47	31.18	43.91	58.64	75.35	94.04	114.73	137.45	162.22
30 MPH	.00	2.13	8.18	17.79	30.71	46.76	65.87	87.96	113.02	141.06	172.10	206.18	243.33